CONCISE ENCYCLOPEDIA OF

MATERIALS ECONOMICS, POLICY & MANAGEMENT

ADVANCES IN MATERIALS SCIENCE AND ENGINEERING

This is a series of Pergamon scientific reference works, each volume providing comprehensive, self-contained and up-to-date coverage of a selected area in the field of materials science and engineering. The series has been developed primarily from the highly acclaimed *Encyclopedia of Materials Science and Engineering*, published in 1986. Other titles in the series are listed below.

BROOK (ed.)
Concise Encyclopedia of Advanced Ceramic Materials

CAHN & LIFSHIN (eds.)
Concise Encyclopedia of Materials Characterization

CARR & HERZ (eds.)
Concise Encyclopedia of Mineral Resources

CORISH (ed.)
Concise Encyclopedia of Polymer Processing & Applications

EVETTS (ed.)
Concise Encyclopedia of Magnetic & Superconducting Materials

KELLY (ed.)
Concise Encyclopedia of Composite Materials

MAHAJAN & KIMERLING (eds.)
Concise Encyclopedia of Semiconducting Materials & Related Technologies

MOAVENZADEH (ed.)
Concise Encyclopedia of Building & Construction Materials

SCHNIEWIND (ed.)
Concise Encyclopedia of Wood & Wood-Based Materials

WILLIAMS (ed.)
Concise Encyclopedia of Medical & Dental Materials

NOTICE TO READERS

Dear Reader
If your library is not already a standing order/continuation order customer to the series **Advances in Materials Science and Engineering**, may we recommend that you place a standing/continuation order to receive immediately upon publication all new volumes. Should you find that these volumes no longer serve your needs, your order can be cancelled at any time without notice.

CONCISE ENCYCLOPEDIA OF

MATERIALS ECONOMICS, POLICY & MANAGEMENT

Editor

MICHAEL B BEVER†

MIT, Cambridge, MA, USA

Series Executive Editor

ROBERT W CAHN FRS

University of Cambridge, UK

Series Senior Advisory Editor

MICHAEL B BEVER†

MIT, Cambridge, MA, USA

PERGAMON PRESS

OXFORD • NEW YORK • SEOUL • TOKYO

UK	Pergamon Press Ltd, Headington Hill Hall, Oxford OX3 0BW, England
USA	Pergamon Press, Inc., 660 White Plains Road, Tarrytown, New York 10591-5153, USA
KOREA	Pergamon Press Korea, KPO Box 315, Seoul 110-603, Korea
JAPAN	Pergamon Press Japan, Tsunashima Building Annex, 3-20-12 Yushima, Bunkyo-ku, Tokyo 113, Japan

---

First edition 1993

Library of Congress Cataloging in Publication Data
Concise encyclopedia of materials economics, policy & management / editor, Michael B. Bever ; executive editor, Robert W. Cahn ; senior advisory editor, Michael B. Bever. — 1st ed.
p. cm. — (Advances in materials science and engineering)
Includes index.
1. Materials—Encyclopedias. 2. Materials—Economic aspects—Encyclopedias. 3. Materials—Government Policy—Encyclopedias. I. Bever, Michael B. (Michael Berliner) II. Series.
TA402.C6593 1993
620.1'1'03—dc20 91–37830

British Library Cataloguing in Publication Data
A catalogue record for this book is available from the British Library.

ISBN 0–08–037056–X

The paper used in this publication meets the minimum requirements of the American National Standard for Information Sciences—Permanence of Paper for Printed Library Materials, ANSI Z39.48–1984.

Printed and bound in Great Britain by Butler & Tanner Ltd, Frome, Somerset

CONTENTS

HONORARY EDITORIAL ADVISORY BOARD

MICHAEL BERLINER BEVER

Professor Michael Bever, Editor-in-Chief of the *Encyclopedia of Materials Science and Engineering* and Senior Advisory Editor of the Supplementary Volumes and the Concise Encyclopedias, died in Cambridge, Massachusetts, on 17 July, 1992.

He was born in Berlin in 1911 and received a degree of Doctor of Jurisprudence from the University of Heidelberg in 1934, the year in which he emigrated to the USA. In 1937 he was awarded a master's degree in business administration by Harvard University. He then studied metallurgy at the Massachusetts Institute of Technology and received his doctorate in 1944. Soon afterwards, he was appointed to the faculty of the Department of Metallurgy (now the department of Materials Science and Engineering) at MIT, where his profound knowledge and diverse interests, ranging from physical metallurgy to materials economics and management, sustained his research and teaching for nearly half a century.

Professor Bever was one of a small group of people who discussed the concept of an encyclopedia of materials in the 1950s, but the project did not get under way until 1974 when Pergamon Press decided to commit the necessary resources and to appoint Michael Bever as Editor-in-Chief. He devoted the remaining years of his life to the Encyclopedia. It is a fitting memorial to his scholarship and to the central role which he played in the evolution of materials science and engineering.

Walter S Owen
Chairman, Honorary Editorial Advisory Board

Robert W Cahn
Executive Editor

Peter T Shepherd
Publishing Director, Pergamon Press

FOREWORD

In the time since its publication, the *Encyclopedia of Materials Science and Engineering* has been accepted throughout the world as the standard reference about all aspects of materials. This is a well-deserved tribute to the scholarship and dedication of the late Editor-in-Chief, Professor Michael Bever, to the Subject Editors and to the numerous contributors.

During its preparation, it soon became clear that change in some areas is so rapid that publication would have to be a continuing activity if the Encyclopedia were to retain its position as an authoritative and up-to-date systematic compilation of our knowledge and understanding of materials in all their diversity and complexity. Thus, the need for some form of supplementary publication was recognized at the outset. The Publisher has met this challenge most handsomely: both a continuing series of Supplementary Volumes to the main work and a number of smaller encyclopedias, each covering a selected area of materials science and engineering, are in the process of publication.

Professor Robert Cahn, the Executive Editor, was previously the editor of an important subject area of the main work and many other people associated with the Encyclopedia have contributed or will contribute to its Supplementary Volumes and derived Concise Encyclopedias. Thus, continuity of style and respect for the high standards set by the *Encyclopedia of Materials Science and Engineering* are assured. They have been joined by some new editors and contributors with knowledge and experience of important subject areas of particular interest at the present time. Thus, the Advisory Board is confident that the new publications will significantly add to the understanding of emerging topics wherever they may appear in the vast tapestry of knowledge about materials.

The appearance of the Supplementary Volumes and the series *Advances in Materials Science and Engineering* is an event which will be welcomed by scientists and engineers throughout the world. We are sure that it will add still more luster to a most important enterprise.

Walter S Owen
Chairman
Honorary Editorial Advisory Board

EXECUTIVE EDITOR'S PREFACE

As the publication of the *Encyclopedia of Materials Science and Engineering* approached, Pergamon Press resolved to build upon the immense volume of work which had gone into its creation by embarking on a follow-up project. This project ha two components. The first was the creation of a series of Supplementary Volumes to the Encyclopedia itself. The second component of the new project was the creation of a series of Concise Encyclopedias on individual subject areas included in the Main Encyclopedia to be called *Advances in Materials Science and Engineering*.

These Concise Encyclopedias are intended, as their name implies, to be compact and relatively inexpensive volumes (typically 400–600 pages in length) based on the relevant articles in the Encyclopedia (revised where need be) together with some newly commissioned articles, including appropriate ones from the Supplementary Volumes. Some Concise Encyclopedias offer combined treatments of two subject fields which were the responsibility of separate Subject Editors during the preparation of the parent Encyclopedia (e.g., dental and medical materials).

Eleven Concise Encyclopedias have been published. These and their editors are listed below.

Concise Encyclopedia of Advanced Ceramic Materials	Prof. Richard J Brook
Concise Encyclopedia of Building & Construction Materials	Prof. Fred Moavenzadeh
Concise Encyclopedia of Composite Materials	Prof. Anthony Kelly CBE, FRS
Concise Encyclopedia of Magnetic & Superconducting Materials	Dr Jan Evetts
Concise Encyclopedia of Materials Characterization	Prof. Robert W Cahn FRS & Dr Eric Lifshin
Concise Encyclopedia of Materials Economics, Policy & Management	Prof. Michael B Bever†
Concise Encyclopedia of Medical & Dental Materials	Prof. David Williams
Concise Encyclopedia of Mineral Resources	Dr Donald D Carr & Prof. Norman Herz
Concise Encyclopedia of Polymer Processing & Applications	Mr Patrick J Corish
Concise Encyclopedia of Semiconducting Materials & Related Technologies	Prof. Subhash Mahajan & Prof. Lionel C Kimerling
Concise Encyclopedia of Wood & Wood-Based Materials	Prof. Arno P Schniewind

Many of the new or substantially revised articles in the Concise Encyclopedias have been published in one of the three Supplementary Volumes, which are designed to be used in conjunction with the Main Encyclopedia. The Concise Encyclopedias themselves, however, are "free-standing" and are designed to be used without reference to the parent Encyclopedia.

The Executive Editor was personally responsible for the selection of topics and authors of articles for the Supplementary Volumes. In this task, he has had the benefit of the advice of the Senior Advisory Editor and of other members of the Honorary Editorial Advisory Board. The Executive Editor was responsible for appointing the Editors of the various Concise Encyclopedias and for supervising the progress of these volumes.

The entire Advances in Materials Science and Engineering project will be completed at the end of 1992.

Robert W Cahn FRS
Executive Editor

EDITOR'S PREFACE

The *Concise Encyclopedia of Materials Economics, Policy and Management* consists of the relevant articles originally published in 1986 in the *Encyclopedia of Materials Science and Engineering*, most of which have been updated and revised, and new articles written especially for this Concise Encyclopedia. I hope that the volume will be of interest to special groups of users, including those with no background in materials science and engineering, such as economists, management specialists and journalists.

I would like to thank the authors of the new and revised articles for their efforts and cooperation during the editorial process. Several other people deserve to be acknowledged. Professor Robert W. Cahn, the Executive Editor, and Walter S. Owen, the Chairman of the Honorary Editorial Advisory Board, provided enthusiastic encouragement. Dr Colin J. Drayton, Dr Helen D. McPherson, Mr Peter J. Frank and Mr Michael A. Mabe of Pergamon Press guided the project through it various stages. Dr Robert W. Crandall gave invaluable feedback, particularly as coauthor of the Introduction. Finally, I would like to express special gratitude to Ms Aina M. Allen for her constant patient help in every aspect of this project, especially its organization and management.

Michael B Bever†
Editor

GUIDE TO USE OF THE ENCYCLOPEDIA

This Concise Encyclopedia is a comprehensive reference work covering all aspects of materials economics, policy and management. Information is presented in a series of alphabetically arranged articles which deal concisely with individual topics in a self-contained manner. This guide outlines the main features and organization of the Encyclopedia, and is intended to help the reader to locate the maximum amount of information on a given topic.

Accessibility of material is of vital importance in a reference work of this kind and article titles have therefore been selected not only on the basis of article content but also with the most probable needs of the reader in mind. An alphabetical list of all the articles contained in this Encyclopedia is to be found on p. xvii and a systematic outline of the contents appears on p. xx. This outline forms part of the Introduction, which incorporates detailed analysis of the topics covered by the various articles.

Articles are linked by an extensive cross-referencing system. Cross-references to other articles in the Encyclopedia are of two types: in text and end of text. Those in the body of the text are designed to refer the reader to articles that present in greater detail material on the specific topic under discussion at that point. They generally take one of the following forms:

> ...which is fully described in the article *Economic Geography of Materials*.

> ...concentrating on the theoretical concepts (see *Prices of Materials: Theory*).

The cross-references listed at the end of an article serve to identify broad background reading and to direct the reader to articles that cover different aspects of the same topic.

The nature of an encyclopedia demands a higher degree of uniformity in terminology and notation than many other scientific works. The widespread use of the SI system of units has determined that such units be used in this Encyclopedia. It has been recognized, however, that in some fields Imperial units are more generally used. Where this is the case, Imperial units are given with their SI equivalent quantity and unit following in parentheses. Where possible, the symbols defined in *Quantities, Units, and Symbols*, published by the Royal Society of London, have been used.

All articles in the Encyclopedia include a bibliography giving sources of further information. Each bibliography consists of general items for further reading and/or references which cover specific aspects of the text. Where appropriate, authors are cited in the text using a name/date system as follows:

> ...as was recently reported (Smith 1988).

> Jones (1984) describes...

The contributors' names and the organizations to which they are affiliated appear at the end of all articles. All contributors can be found in the alphabetical List of Contributors, along with their full postal addresses and the titles of the articles of which they are authors or coauthors.

The article *An Introduction to Materials Economics, Policy and Management* provides an overview of the subject area and discusses in brief te issues covered in detail by the articles in the body of the work.

The most important information source for locating a particular topic in the Encyclopedia is the multilevel Subject Index, which has been made as complete and fully self-consistent as possible.

ALPHABETICAL LIST OF ARTICLES

AN INTRODUCTION TO MATERIALS ECONOMICS, POLICY AND MANAGEMENT

by Michael B. Bever† and Robert W. Crandall

Virtually all activities that sustain mankind rely to some degree on materials. The materials used in agriculture, the harnessing of energy, manufacturing, health services, transportation and construction account for a substantial share of the world's economic activity. As a result, materials economics, materials management and materials policy have developed into important fields of inquiry. Each could properly be the subject of a separate treatise; however, they lend themselves to a concise, unified treatment in an encyclopedia. Hence, this *Concise Encyclopedia of Materials Economics, Policy and Management*. (A systematic outline of the contents of this encyclopedia is presented in Table 1).

1. General Materials Economics and Materials Policy

Materials, like all resources, are limited in supply. As a result, some mechanism is required to allocate materials among competing uses. The principal alternatives are economic markets or government directives. This part of the article reviews the economics of materials markets and the use of government to influence these markets. Such a review requires an analysis of materials use, substitution, supply and materials availability, the structure of the supplying industry, the geographical characteristics of materials markets, and the prospects for exhaustion of materials supplies in future decades.

The analysis of materials markets begins with the very nature of these goods. Typically, they are homogeneous, fungible products. They can be produced, classified and graded in such a manner as to make brand identification relatively meaningless. Moreover, materials are rarely final products, but rather intermediate products required for the production of other goods and services. Lastly, most materials are derived in some fashion from natural resources—they are extracted, grown or synthesized from products of the land. It is this derivation from natural resources that creates the concern over the sufficiency of supply or the availability of materials, particularly during national defense emergencies, and over environmental externalities.

1.1 Materials Supply

Many materials policy issues involve the current or future supply of materials. There is always a concern that the prospective demand for materials may outrun supply in future years or generations. As a result, assuring an adequate future supply of materials is often an important policy concern.

The supply of most materials depends in part upon the availability of mineral deposits. The mineral resource base is therefore an important first element in analyzing the supply of these materials. Nonfuel mineral resources are limited—in finite supply—but this supply limitation depends upon current knowledge (see *Mineral Resources: Definitions, Uses, Classifications and Future Availability*; *Resource Appraisal*). Over decades or centuries, the size of the identified resource base of any mineral may increase substantially despite man's depletion of existing deposits.

The resource base of any mineral does not equate with economically feasible supply. Reserves of a nonfuel mineral are that subset of the resource base that may be extracted economically with known technology at current prices. An even smaller subset is the quantity of "proved" or "demonstrated" reserves—that is, those whose existence is known with some certainty. In addition to these demonstrated reserves are inferred reserves, reflecting a geological assessment of economically recoverable deposits based upon less certain evidence (see *Mineral Resources: Definitions, Uses, Classifications and Future Availability*).

Renewable resources may appear at first sight to be in unlimited supply, but on closer examination this is not the case. In the long run, agricultural and forestry crops are limited by the available land. In the short run, they are vulnerable to weather, disease and parasites. There has been concern about future supplies of certain types of timber and in certain parts of the world (see *Timber, Trade, Prices and Consumption in the USA*; *Wood: Future Availability*). The supplies of natural fibers and any materials derived from biomass are also limited, in principle.

Materials are produced from nonfuel mineral deposits or renewable resources by a variety of different industries. Often, firms in these industries are vertically integrated, that is, they combine the extraction, smelting, refining or other processing of a nonfuel mineral or the harvesting and processing of a renewable resource such as wood (see *Structure of Materials Industries*). As the technology of extraction or processing changes, these industries change. For instance, direct casting of thin sheets may lead to increased integration in the copper and aluminum industries. Similarly, revolutionary changes in electric furnace technology, direct reduction of iron ore, and thin slab casting may lead to a much less concentrated iron and steel industry.

Materials supply is often augmented by the reuse or recycling of obsolete products containing the material or manufacturing trimmings, turnings and rejected products. Recycling is particularly important in the metals industries, and is likely to increase in relative importance if economic growth proceeds more slowly than in the first three decades after World War II (see *Structure of Metals Industries*). Recycling is also important for other materials, such as paper (see *Recycling of Paper Products*) and even contributes to the supply of building materials (see *Recycling of Demolition Wastes*). In recent years, the slow growth in the demand for steel has seen scrap-based

Table 1
Systematic outline of the encyclopedia

1. General

Data Sources for Materials Economics, Policy and Management
Economic Geography of Materials
Intensity of Use of Materials
Materials in the Industrial System

2. Fundamental Topics

Cartels and Producer Associations
Economic Theory of Mineral Resources and Markets
Economics of Materials, Energy and Growth
Markets for Materials: Principles and Public Policy Implications of Competition

3. Materials Supply

Availability of Materials
Mineral Resources: Definitions, Uses, Classifications and Future Availability
Resource Appraisal
Wood: Future Availability

4. Materials Industries

Economics of the Semiconductor Industry
Natural Fibers in the World Economy
Structure of Materials Industries
Structure of Metals Industries
Timber Use: History
Wood-Based Industries in the USA: Economic Structure in a Worldwide Context

5. Markets

Markets for Materials
Markets for Materials: Modelling
Trade in Materials, International
Trading of Materials on Commodity Exchanges

6. Prices

Prices of Industrial Minerals: History
Prices of Materials: Stabilization
Prices of Materials: Theory
Prices of Metals: History and Current Problems
Timber: Trade, Prices and Consumption in the USA

7. Process Analysis and Materials Flow

By-Products and Coproducts
Energy Considerations in Metals Production
Input–Output Analysis
Input–Output Analysis: Applications
Life-Cycle Analysis of Energy Consumption and Carbon Dioxide Emissions
Process Analysis
Value Added in Processing

8. Materials Use

Economic Effects of Corrosion and Other Degradative Processes
Economic Effects of Fracture
Factor Dominance in Industry
Materials and Innovation
Productivity and Materials
Productivity and Technological Improvements: Managerial Evaluations
Substitution of Materials: Economics
Substitution of Materials: Technology

9. Materials Conservation

Conservation of Materials: Economics
Conservation of Materials: Technology
Reclaiming of Elastomers
Recycling of Demolition Wastes
Recycling of Materials: Economics
Recycling of Metals
Recycling of Paper Products
Recycling of Plastics
Resource Recovery: Economic and Institutional Aspects

10. Environmental Problems

Materials and the Environment
Paper and Paperboard: Environmental Considerations
Residuals Management

11. Policy and Planning

Critical and Strategic Materials
Policy: Australia
Policy: Canada
Policy: Europe
Policy: General
Policy: Japan
Policy: USA
Stockpiling
Stockpiling: Economics
Stockpiling in the USA
Taxation: Effect on Materials Policy

12. Materials Management

Cost Estimating
Life-Cycle Costing
Materials Handling: Automation, Unit Materials
Materials Handling: Bulk Materials
Materials Handling: Unit Materials
Materials Requirements Planning
Multiattribute Utility Analysis
Product Liability and Materials: Europe
Product Liability and Materials: General and USA
Risk Management
Value Analysis

electric furnace production increase its share of total steel output.

Expanding supply in most materials industries requires large lead times because of the necessary investments in mining or processing. As a result, supply tends to respond slowly to sharp increases in demand—such as the 1973–1974 boom—or to sharp downturns in demand—such as in the 1930s or early 1980s. This sluggishness in the response of supply, reflecting large fixed costs of production, is one reason why materials prices tend to be extremely volatile.

1.2 The Use of Materials: Demand and Substitution

Materials are generally used as inputs into the production of more finished economic products. Most have a number of substitutes, given sufficient time to implement the necessary technology and to construct the productive capacity. Moreover, final products themselves have substitutes. The development of these substitutes in response to rising prices of a material is a natural reaction to increasing scarcity (see *Substitution of Materials: Economics*). Rising materials prices induce a search for new technologies and result in the conservation of the remaining supply of the materials (see *Substitution of Materials: Technology*).

As societies develop, they tend to use relatively smaller quantities of materials per unit of final output or consumption. Moreover, they tend to use different types of materials, such as plastics for appliances or silicon for electronic components. With rising wealth, consumers do not increase their purchases of durable goods proportionately; instead, they tend to spend their incomes increasingly on services, such as travel, recreation and amusement. The tendency of mature economies to consume less materials per unit of gross national product is well documented by Malenbaum (see *Intensity of Use of Materials*). This declining "intensity of materials use" is important in forecasting the demand for materials.

The extent of materials use in each sector of the economy is often modelled by input–output (I–O) analysis (see *Input–Output Analysis*). Most investigations by I–O analysis assume fixed coefficients in production, but these coefficients can be changed to allow for technological change and in particular for substitution. Such analyses are useful for tracing the effects of major changes in materials usage through the economic system. Carter (see *Materials in the Industrial System*) provides a detailed analysis of the use of materials in the industrial sector of the economy, relying heavily upon I–O techniques. She demonstrates that the changes in materials intensity derive in large part from different growth rates of materials-using industries.

1.3 Materials Markets: Structure

Materials markets have a number of distinctive features. Since most materials are used as inputs to further production, demand is likely to vary substantially over the business cycle (see *Markets for Materials*). This "derived demand" is likely to be less price sensitive than the demand for most final goods. Varying prices and the large investment and inventory requirements create the need for organized futures markets in most major materials (see *Trading of Materials on Commodity Exchanges*). Firms that must hold substantial raw-material or in-process inventories often wish to hedge the risks of price fluctuations in these materials.

The degree of competition in materials markets depends upon the minimum efficient scale of production, the size of the market and the cost of entry. As Labys (see *Trade in Materials, International*), Gordon (see *Markets for Materials: Principles and Public Policy Implications of Competition*) and Crandall (see *Structure of Metals Industries*) explain, materials markets have become more international in scope in recent years. As a result, few materials firms enjoy sufficiently large shares of their markets to have monopoly power over prices.

In a few cases—such as uranium—cartels of competing producers have been formed. However, as Pindyck (see *Cartels and Producer Associations*) explains, these cartels are often unstable and are susceptible to collapse owing to independent actions by their members. While the OPEC cartel succeeded in raising crude oil prices for a decade, it crumbled in the mid-1980s. The cartel of bauxite producers has never had much success in raising the price of bauxite because of the number of potential competitors in the market. Nor have copper producers been able to cartelize successfully despite their limited numbers, in part because of an ample supply of scrap.

It is always difficult to delineate a market with precision because of the constantly developing substitution possibilities among materials. Rising relative prices of steel led users to substitute aluminum, plastics and even concrete for steel. High real copper prices in the 1970s encouraged substitution of plastics and aluminum for copper. Because of the various substitution possibilities, the boundaries of a market are to a large extent determined by judgement, not by precise measurement (see *Markets for Materials: Principles and Public Policy Implications of Competition*).

1.4 Materials Markets in Action: Prices

The interaction of supply and demand generates a market-clearing price. This price in a materials market is likely to be quite volatile over time because of the inelasticity of supply and the variability of demand over the business cycle. Nevertheless, economic theory suggests that the secular trend of materials prices should be upward. As Newcomb (see *Economic Theory of Mineral Resources and Markets*) explains, the rate at which materials prices rise over time should be the interest rate, assuming unchanging technology and information concerning deposits of the natural resources. Prices may be volatile, but their general trend should be upward in the absence of technological changes or new discoveries of reserves. As Vogely (see *Prices of Materials: Theory*) demonstrates, however, new discoveries have generally been sufficient to depress the real price of most materials over the long run (see also Lovering 1969)

In fact, materials prices have not trended upward over the past century. Detailed research by Potter and Christy (1962), by Manthy (1978) and by Landsberg et al. (1963) demonstrates that most nonrenewable resources have shown a downward real price trend during this period. The reasons for this are obvious: technological change in extraction and processing as well as new resource discoveries. Were these technological breakthroughs to subside and new discoveries of resources to become less frequent, natural resource prices might begin a long upward trend as economic theory predicts.

Slade (see *Prices of Metals: History and Current Problems*) shows that a quadratic function appears to fit the time series data for many metals prices rather well. She finds that the general downward trend through the last decades of the nineteenth century and the early decades of the twentieth century had slowly given way to a rising trend. The 1970s and 1980s, however, were years of extremely volatile metals prices. This volatility obscured any general trend.

The prices of most "renewable" resources have shown a tendency to rise over several decades. For instance, the price of the average forest product trebled in real terms between 1875 and 1975 (Manthy 1978) (see *Timber: Trade, Prices and Consumption in the USA*). The prices of most timber products rose more rapidly than the general price level in the three decades after World War II. The principal exception was the pulpwood used in the manufacture of paper.

If nonfuel minerals prices begin to rise at rates exceeding the inflation rate for a considerable period of time, there will be renewed concern over minerals exhaustion, leading governments to begin to reassess their materials policies.

1.5 Materials Policies: Sufficiency of Supply

For generations, man has worried about depletion and the exhaustion of minerals deposits. Generally, these concerns follow periods of extended economic boom or a period of war. Soon after World War II, the Paley Commission reported on the US government's options for assuring sufficient supplies of materials (US President's Materials Policy Commission 1952). Interest waned, however, until the early 1970s when a commodity boom and the OPEC cartel once again created general concern about the future supply of materials (National Commission on Materials Policy 1973).

In the USA, federal government minerals policies involve the leasing of mineral rights on federal lands, the implementation of environmental policy and the use of tax policy to stimulate exploration and production of minerals (see *Policy: USA*; *Taxation: Effect on Materials Policy*). With little formal government ownership of minerals industries, the US government has little direct role in determining the rate at which mineral deposits are exploited. However, through its leasing policies on federal lands—particularly for coal and timber—the government can have a substantial effect upon some materials industries.

Western European countries are typically more dependent upon imports for their minerals than is the USA. In some countries, such as France, the government has become quite concerned about the growing role of imports. As a result, France has placed increased emphasis upon recycling and conservation (see *Policy: Europe*). Most European countries provide some protection for their minerals industries, particularly their coal industries. But in the UK and Germany, there is very little else in the nature of a formal policy toward materials supply or self-sufficiency.

Japan, which has very few mineral deposits, relies heavily upon imported raw materials. This is especially true for energy resources and metal ores. As a result, the Japanese government encourages Japanese firms to invest in foreign materials industries and participates in programs to assure the security of materials flows from foreign sources (see *Policy: Japan*). Low tariffs on imported raw materials and encouragement of domestic stockpiling are among the more important of these policies.

Unlike Japan, Canada is a large net exporter of materials. Its mining sector accounts for about 6% of total employment and 5% of its gross national product. During the period of rapid development of its mineral industries, foreign investment played a major role. In recent decades, Canada has pursued a policy of attempting to reduce foreign control and ownership of its minerals industries, but in the 1980s, renewed concern for optimal development of the country's minerals industries led to a more relaxed view of foreign investment and a heightened concern for competing in world export markets (see *Policy: Canada*).

Australian materials policy considers resource development and raw materials production from the viewpoint of the competing needs of the country's domestic manufacturing industries and materials exports. Foreign investments and resource ownership are attracting particular attention. Conservation and, in the case of uranium, international political considerations are also concerns. The two major political parties generally favor different policies on most of the issues. Changes of political parties in power have resulted in a materials policy characterized by several reversals in recent decades (see *Policy: Australia*).

The 1980s brought a changed world economic climate from that of the 1970s. Concern over sufficiency of minerals supplies diminished with declining economic growth. With slower growth in the developed countries and major debt problems in the developing countries, concern shifted from that of rationing scarce supplies to exploiting and developing resource deposits to allow mineral-rich countries to grow. This is especially true in many developing countries that are heavily dependent upon sales of basic fuel and nonfuel minerals.

1.6 Materials Policy: Price Stabilization and Stockpiling

The 1970s was a period of extreme price volatility for many materials and particularly for fuels. This led to an increased interest in government policies to mitigate such price volatility or at least to dampen the oscillations. In addition, governments became concerned about maintaining sufficient supplies of critical or strategic materials (see *Critical and Strategic Materials*).

Behrman (see *Prices of Materials: Stabilization*) demonstrates that there can be substantial gains in economic efficiency if governments are able to stabilize materials prices. Among the mechanisms for achieving such stabilization are buffer stocks, direct price controls and the policies to stabilize production levels.

Public policies towards critical and strategic materials have a long history. In the USA, energy reserves were established before World War I because of security considerations (see *Stockpiling in the USA*). More recently the USA has attempted to maintain stockpiles of a number of strategic materials such as titanium, nickel, tungsten, cobalt and rubber. This policy became important during and immediately after the Korean War, but subsided in the 1970s and 1980s. In the mid-1970s, after the sharp rise in commodity prices in 1971–1974, the US Congress established the Commission on Supplies and Shortages to examine public policies for reducing price fluctuations in future business cycles.

Many other countries also maintain stockpiles of strategic and critical materials. France and the UK have modest public stockpiles, while Japan, Sweden and Switzerland pursue government-supported private stockpiling policies (see *Stockpiling*).

One of the most difficult problems in designing a stockpiling policy is identifying the truly critical and strategic materials. Most definitions of critical materials focus on the lack of substitutes for particular uses that are central to maintaining a national defense capability. However, as Clark and Reddy (see *Critical and Strategic Materials*) explain, there are substitutes for most materials, particularly with sufficient lead time. Identifying the speed at which substitution may take place is therefore very important.

Public stockpiles maintained for the purpose of limiting price volatility are even more difficult to manage. Private firms maintain large inventories of most materials to smooth out production and to ensure against price fluctuations (see *Risk Management*). Public stockpiles of materials may simply reduce private inventories and provide little additional protection against supply disruptions or other sources of price volatility.

Another problem with public stockpiles that are designed to reduce price volatility involves the timing of purchases for and sales from the stockpile. Optimal decision rules for reducing price volatility are complex; moreover, political pressures often prevent public authorities from using stockpiles optimally. Sales or purchases designed to smooth out prices have obvious effects on private sellers and buyers and inventories of any material.

2. Topics in Materials Economics

There are a number of specialized topics in materials economics that do not fit neatly into a general analytical framework of supply–demand–prices. These include process analysis, modelling, productivity and innovation. This section surveys some of the relevant concepts and ideas and serves as an introduction to the articles involving each of them.

2.1 Process Analysis: Value Added and Energy Assessment

Materials-producing and materials-using industries differ from most other economic sectors—such as agriculture, transportation, communication and finance—in that they employ physical processes in which one or more materials flow through a progressive set of linked processing stages leading from raw materials to final product. The processes are investigated and assessed by process analysis. Process analysis lends itself to systematic quantification of the important variables in a variety of decisions. In this way, it contributes to allocation and investment decisions as well as to process design and control at levels ranging from unit operations and single plants to the level of an entire economy (see *Process Analysis*).

A manufacturing plant typically produces more than one product. Similarly, many materials-processing plants produce more than one material. In this context, the distinction between by-products and coproducts is relevant (see *By-Products and Coproducts*).

The physical progress through the various operations depicted by a flowchart is accompanied by a cumulative buildup of value embodied in the material or workpiece. This buildup is shown by Pick for aluminum and steel (see *Value Added in Processing*). The value added across all industries can be aggregated into their contribution to the gross national product.

A buildup of the cumulative energy expended in the production and conversion of materials is also possible. A complete energy assessment, however, includes the energy used in conjunction with the use of the material and its disposal after discard. Recycling of the material often results in an energy saving compared with the use of primary material. Such an analysis is a life-cycle energy evaluation (see *Life-Cycle Analysis of Energy Consumption and Carbon Dioxide Emissions*).

2.2 Modelling

Since World War II, there has been a considerable increase in the interest in developing models of individual materials markets, the materials sector of the economy and the entire economy. In part, this has been facilitated by the remarkable improvement in quantitative techniques and the equally rapid improvement in computers. As the cost of developing large models has declined, business and government demand for them has increased. Businesses often develop their own models in order to make production, inventory and investment decisions. Governments use models to forecast prices and supplies so as to aid in stabilization policies, sectoral policies or national defense policies.

Clark and Busch (see *Markets for Materials: Modelling*) identify three types of models: econometric models, engineering models and input–output models. In principle, these models are substitutes for one another, but they are often used for different purposes. Econometric models are generally used by firms or governments to predict prices, output and investment. Engineering models are more often used to analyze the response of individual production units to changes in materials availability, prices or product demands. Input–output models are generally more aggregated than engineering models and are used to trace the impact of structural changes—such as supply shocks or shifts in demand (see *Input–Output Analysis*; *Input–Output Analysis: Applications*).

In the early 1970s, a new form of dynamic model was developed, using techniques drawn from electrical engineering. This type of model involves complex feedback loops that are thought to characterize a variety of social systems from individual markets to the world economy. These models are often developed through a series of simulation exercises on high-speed computers in order to place values on the important parameters. Probably the most publicized of these exercises was the report to the Club of Rome by Meadows et al. (1972) that related scarcity of materials, congestion and pollution to the deceleration of world economic growth.

2.3 Productivity and Materials

Productivity is a measure of the ratio of output to inputs of labor, capital or materials. The most common measure of productivity is output per manhour, but this measure does not capture the full effects of changes in each of the productive factors. Total factor productivity is the ratio of output to a weighted average of all inputs used to produce it.

In most manufacturing industries, including the materials industries, labor productivity has grown more rapidly than total factor productivity because capital has been substituted for labor. Gold argues, however, that decision makers within a firm are likely to be unable to measure total factor productivity because of the difficulty of estimating the rate of change of all productive factors (see *Productivity and Technological Improvements: Managerial Evaluations*). As long as capital is adding rapidly to productive capacity, labor productivity is likely to grow rapidly, but once capital investments are made simply to maintain the capacity of existing plants, productivity growth slows.

Productivity is important in materials industries in several ways. Productivity growth in these industries themselves tends to reduce the real costs of materials over time (see *Productivity and Materials*). Improvements in technology have led to increases in the productivity of materials used as inputs by materials-producing industries and other industries. For example, continuous casting and improved rolling techniques have increased the ratio of finished to crude steel in the steel industry.

Materials usage has declined in many processing and downstream manufacturing industries owing to improvements in technology and production practices (see *Productivity and Materials*). Improved materials can also generate increases in productivity of other inputs in materials-producing industries and in downstream-using industries (see *Materials and Innovation*).

2.4 Materials and Innovation

Innovation increases man's ability to produce a variety of products. It may reduce the cost of producing a current product or yield an entirely new product that is useful in itself. This new product may provide a superior substitute for an existing material, such as the substitution of plastic for lead in plumbing, or it may facilitate the introduction of entirely new final products, such as memory devices used in small computers.

Through the first quarter of the twentieth century, materials innovation largely concentrated on metals. Beginning in the 1930s and throughout the next several decades, research on synthetic polymers was greatly accelerated. This led to a large number of new products. The development of synthetic rubber because of the shortage of natural rubber during World War II was an especially important achievement of polymer research. Since World War II, research on nonsilicate inorganic oxides has led to the development of a new generation of ceramics. Continued refinements in the technology of producing and using semiconductor-grade silicon have been crucial to the rapid progress of the electronics industry (see *Economics of the Semiconductor Industry*; *Materials and Innovation*).

Innovation is driven by the desire to become more competitive through lower costs or improved products. Since the resulting advantage tends to be lost in time as competitors either adopt the innovation or invent new products and processes of their own, the stimulus for innovation persists (see *Markets for Materials: Principles and Public Policy Implications of Competition*).

3. Materials Management

Materials-related management functions include the physical movement of materials, the planning of required supplies and inventories, the estimation of materials costs, value analysis and the disposal of production wastes, such as metallic processing scrap. These functions are the subjects of specialized articles in this encyclopedia. The broader context is provided by articles on economic geography, production of by-products and coproducts, and recycling technology of various materials as well as articles covered in Sects. 1 and 2 of this Introduction.

3.1 Materials Transportation and Materials Handling

The physical inputs and outputs of materials-producing and materials-converting industries must be available in the required amounts at definite times and in definite locations. Fundamentally, the locations involved in the movements of materials, both the origins and destinations, are determined by factors of economic geography (see *Economic Geography of Materials*). The movement of materials, especially over short distances and within plants, is the object of materials handling. Kelly and Laurila (see

Materials Handling: Bulk Materials) describe the handling of bulk materials with special reference to handling systems and materials characteristics and also cover aspects of the transportation of materials such as the various transportation modes available. Bode (see *Materials Handling: Unit Materials*) describes the handling of unit materials with emphasis on moving and storage systems and ancillary operations (see also *Materials Handling: Automation, Unit Materials*).

3.2 Materials Requirements Planning and Materials Risk Management

Materials requirements planning is a technique for managing inventories and controlling the replenishment of materials, components and subassemblies; it is a module in the total manufacturing control system (also referred to as manufacturing resource policy) (see *Materials Requirements Planning*). Materials requirements planning depends in part upon the ability to process large amounts of information. For this reason, the computational speed and memory capacity of computers have been important in developing materials requirements planning in the period since 1960.

Materials requirements planning requires an accurate bill of materials and a master schedule for the final products. A combination of four steps generates planned orders for the purchases of materials and parts. As one way to provide protection for uncertainty, a minimum inventory quantity called a safety stock is specified; this is considered always to be required when calculating net requirements. It ensures materials availability under conditions of uncertainty. An alternative method employs a safety time provision. Each of these methods has its own advantages—the safety time method results in lower average inventory investment when the primary uncertainty is in timing and the safety stock method is preferable when the primary uncertainty is quantity.

Materials risk management is concerned with the strategies by which a firm can try to cope with material scarcities, especially scarcities caused by supply interference of an international nature. It is thus the private-sector equivalent of stockpiling by government and was referred to in this context earlier in this article (Sect. 1.6). Materials risk management is based on vigorous assessments of the firms' needs of critical materials; various companies have developed detailed analytical procedures towards this end (see *Risk Management*).

The "just-in-time" method of managing the supply primarily of components to an assembly-type manufacturing operation is related to materials requirements planning. It has received much attention in the US automobile industry's attempt to become competitive with Japan (see *Materials Handling: Automation, Unit Materials*).

3.3 Cost Estimation and Value Analysis

Materials costs are one item in the estimation of production and manufacturing operations. The mineral content of ores can be an important factor in projecting their processing costs; as Seldon points out (see *Cost Estimating*), this can be used in the parametric method of cost prediction. In the industrial engineering method, materials costs are assessed as an independent item. This should favor the use of this method for projects in which materials costs are a major consideration.

In life-cycle costing, materials can be a decisive factor. As a typical example, a more expensive material, such as stainless steel, may result in lower life-cycle costs because it requires less maintenance and repair than competitive materials. More expensive materials may be more economic because of the increased durability of products made from them (see *Life-Cycle Costing*).

In an analogy to life-cycle costing, a life-cycle energy assessment may favor a lighter, more energy-intensive material because of energy savings during use. This is well illustrated by metals used in transportation applications. The energy saving can be translated into a life-cycle cost saving.

Value analysis measures the absolute value of a material characteristic, or other input feature, in terms of the benefit derived from it and compares it with the material's cost. This value may relate to costs of production or to advantages in the market place. In the latter case, the value may be better quality leading to larger sales; the cost is the cost of producing a higher-quality product (see *Value Analysis*).

3.4 Product Liability

In recent years in the USA, manufacturers have become more concerned about their exposure to product liability suits. As courts move increasingly toward a standard of strict liability and accept weaker evidence as proof of damage to plaintiffs, firms have been forced to reduce their exposure to suits by increasing product quality. In part, this is manifested through an increased demand for high-quality materials that will not expose consumers or workers to risk regardless of the nature of their use. Piehler (see *Product Liability and Materials: General and USA*) points out that these changes have taken place in part because of the contingency-fee system employed by lawyers and the increasing concern over risks with long latency periods.

Product liability suits have increased in number, and the magnitude of the settlements has increased even more rapidly. Large class-action suits such as those involving asbestos have caused manufacturers to become much more risk averse in offering new products (see also *Product Liability and Materials: Europe*).

4. Resource Recovery, Conservation and Recycling

Since the 1970s, concern has increased over the effects of man's use of the environment. This concern has been voiced in various ways. First, the notion that private markets do not provide sufficient incentive to conserve natural resources is pervasive among environmentalists.

Second, there is widespread concern that in the production and use of materials, the air, water and ground may become increasingly polluted. Third, many worry that we are rapidly exceeding our capacity to dispose of waste materials.

4.1 Materials and the Environment

Materials production and consumption can cause environmental degradation unless government establishes the appropriate regulatory rules of property rights. Many production processes—such as mining, extraction, refining or smelting—pollute the air, water or land. Wastes or "residuals" are discarded from both production and consumption. Discards by consumers include a wide range of materials from packaging materials and newsprint in municipal solid waste to the materials in discarded automobiles and discarded capital goods. This waste generation creates a variety of problems for public policy that require technical solutions (see *Residuals Management*).

The immediate or ultimate repository of residuals is the environment. Its absorptive capacity depends on the nature of the discarded material, but it is always limited. The environment is a public good that is usually outside the market pricing system and is therefore subject to excessive use unless it is protected or appropriately rationed by government (see *Materials and the Environment*). Protection can be brought about by regulations that reduce the immediate impact or by the appropriate use of property rights (see Coase 1960).

In recent years, economists have advocated pollution taxes or marketable pollution permits to reduce environmental pollution (see Montgomery 1972, Crandall 1983). Long-range policies reducing residuals generation and encouraging materials recovery can also contribute to environmental protection. Any policy to control environmental pollution has major impacts upon producers and consumers. Any attempt to change regimes from no control to regulation or to a system of private property rights can be disruptive to private economic agents (see *Residuals Management*).

4.2 Recycling and Resource Recovery

Recycling is the recovery of discarded materials for reuse. Complete recovery is generally not possible. It is not technically or economically feasible to recycle most materials used in dispersive and dissipative applications (see *Materials and the Environment*). As an exception, small amounts of dispersed materials can be co-recovered in the recycling of associated major metals; an example is the recovery of solder in the recycling of automobile radiators (see Alward et al. 1976).

By its very nature, recycling contributes to the conservation of materials (see *Conservation of Materials: Technology*). If recycling is conducted efficiently, it generally also yields savings of equipment, pollution abatement expenditures and energy. The use of ferrous scrap in steelmaking in place of hot metal produced in a blast furnace is a case in which all three savings are realized. In addition to its contribution to conservation, the recycling of materials produced and discarded in large volumes also achieves waste disposal; this is especially true of ferrous scrap and paper and paperboard. An extreme case of waste disposal vs materials conservation is resource recovery from municipal solid waste. Often, by-products are captured that would harm the environment if discharged as residuals (Bever 1978).

The recycling industries should be viewed like other materials-producing industries. Their raw materials are of a special kind: residuals from production and residuals from consumption. The quality of these raw materials depends on their physical form and chemical composition. Both the absolute price of the material produced and the price of scrap relative to primary raw materials determine the extent of recycling. This explains the highly developed state of nonferrous metals recycling. Details of the recycling of various materials, including the technology involved, are to be found in the articles on the recycling of metals, polymers, paper, rubber and demolition wastes (see also *Recycling of Materials: Economics*).

In contrast to the recycling of metals and other individual materials, resource recovery from municipal solid waste is driven by the necessity for waste disposal (see *Resource Recovery: Economic and Institutional Aspects*). Institutional barriers and declining materials prices have slowed the development of resource recovery. The course of its development was also affected by the rise of energy prices in the 1970s which made incineration with energy recovery viable; in this way, incineration contributes to the economics of resource recovery but also competes with it for raw materials.

Bibliography

Alward J R, Clark J P, Bever M B 1976 The dissipative uses of lead. *Proc. Council of Economics*. American Institute of Mining, Metallurgical and Petroleum Engineers, New York, pp. 17–26

Ausubel J H, Sladovich H (eds.) 1990 *Technology and Environment*. National Academy Press, Washington, DC

Bever M B 1978 Systems aspects of recycling. *Conserv. Recycling* 2: 1–17

Brooks H 1972 Materials in a steady state world, ASM–TMS Distinguished Lecture. *Metall. Trans.* 3: 759–68

Carter A P 1970 *Structural Change in the American Economy*. Harvard University Press, Cambridge, MA

Coase R H 1960 The problem of social cost. *J. Law Econ.* 3: 1–44

Crandall R W 1983 *Controlling Industrial Pollution*. Brookings Institution, Washington, DC

Fellowship of Engineering 1983 *Modern Materials in Manufacturing Industry*, a study carried out by a working party of fellows with support from the Department of Industry. Fellowship of Engineering, London

Fischman L L 1980 *World Mineral Trends and US Supply Problems*. Johns Hopkins University Press, Baltimore, MD

Herfindahl O C, Kneese A V 1974 *Economic Theory of Natural Resources*. Merrill, Columbus, OH

Landsberg H C, Fischman L L, Fisher J A 1963 *Resources in*

America's Future. Johns Hopkins University Press, Baltimore, MD

Leontief W 1986 *Input–Output Economics*, 2nd edn. Oxford University Press, New York

Lovering T S 1969 Mineral resources from the land. In: Cloud P (ed.) 1969 *Resources and Man*. Freeman, San Francisco, CA, pp. 109–34

Manthy R S 1978 *Natural Resource Commodities: A Century of Statistics*. Johns Hopkins University Press, Baltimore, MD

McMains H, Wilcox L (ed.) 1978 *Alternatives for Growth: The Engineering and Economics of Natural Resources Development*. National Bureau of Economic Research (Ballinger), Cambridge, MA

Meadows D, Meadows D, Randers J, Behrens W 1972 *The Limits to Growth*. Universe, New York

Montgomery W D 1972 Markets in licences and efficient pollution control programs. *J. Econ. Theory* 5: 395–418

MRS Bulletin 17(4) 1992 Issue on materials and the environment

National Academy of Sciences–National Academy of Engineering 1975 *National Materials Policy*, proceedings of a joint meeting, 25–26 October 1973. National Academy of Sciences, Washington, DC

National Commission on Materials Policy 1973 *Materials Needs and the Environment Today and Tomorrow*. US Government Printing Office, Washington, DC

National Commission on Supplies and Shortages 1976 *Government and the Nation's Resources*. US Government Printing Office, Washington, DC

Potter N, Christy F T 1962 *Trends in Natural Resource Commodities*. Johns Hopkins University Press, Baltimore, MD

Smith V K (ed.) 1979 *Scarcity and Growth Reconsidered*. Johns Hopkins University Press, Baltimore, MD

Study Committee on Environmental Aspects of a National Materials Policy 1973 *Man, Materials and Environment*, a report to the National Commission on Materials Policy. National Academy of Sciences–National Academy of Engineering, Washington, DC

US President's Materials Policy Commission 1952 *Resources for Freedom: A Report to the President*. US Government Printing Office, Washington, DC

Vogely W A (ed.) 1975 *Mineral Materials Modeling: A State-of-the-Art Review*. Resources for the Future, Washington, DC

Vogely W A (ed.) 1985 *Economics of the Mineral Industries*, 4th edn. American Institute of Mining, Metallurgical and Petroleum Engineers, New York

A

Availability of Materials

Industrialized countries depend on materials produced from minerals and other primary products to maintain their economies and sustain their living standards. As a result, concern over the future availability of materials, particularly metals, arises in these countries from time to time.

Availability implies the absence of shortages, but what constitutes a shortage is open to different interpretations. To some, particularly economists, a shortage is an excess of demand over supply for a commodity at its prevailing price. With this definition, a shortage can be eliminated simply by allowing the price to rise. At some point, the higher price will stimulate supply and reduce demand sufficiently to bring the two into balance.

For many purposes, this definition is too narrow. A severalfold increase in the price of cobalt in response to an interruption of supplies from Zaire might successfully constrict demand to the available supply, yet still cause serious disruption, at least in certain sectors of the economy. For this reason, the term shortage is used here to encompass those situations where material prices rise sharply in the short run or persistently in real terms over the long run, as well as those situations where demand exceeds supply at the prevailing price.

Materials availability, which prevails when none of these conditions exists, may be threatened in four ways: mineral resources may be depleted or exhausted; investment in the mines and processing facilities required to extract ore and convert it into useful materials may be insufficient; cyclical booms or military buildups may cause unanticipated surges in demand; and interruptions in supply or constraints on trade may arise as a result of embargoes, wars, cartels or other factors.

1. Resource Depletion

Two views of resource depletion are commonly encountered. The first, the physical view, points out that the earth is finite and contains only so much copper, mercury, zinc and other mineral materials. As a result, the supply of such commodities is a fixed stock. Demand, on the other hand, is a flow variable that continues year in and year out. This means that sooner or later demand must consume the fixed amount of supply, at which time the world physically runs out of the commodity.

While logically appealing, this view of depletion suffers from two serious shortcomings. First, with a few trivial exceptions, such as the fissioning of uranium, elemental minerals are not destroyed or transmuted when used. Consumption may diffuse or degrade the material, making recycling difficult and expensive, but it does not diminish the size of the fixed stock. Second, and even more important, long before all of a mineral material found in the earth's crust were consumed, the cost of mining and processing would climb sharply and stifle demand.

For these reasons, the economic view of depletion is generally favored by specialists. This second view of depletion does not see the world literally running out of mineral materials, but rather expects their real costs to rise as society is forced over time to rely on deposits that are of lower grade, more remote and more difficult to process. Eventually, higher prices cause consumers to curtail or abandon their use of materials.

While both views of depletion raise the possibility of severe material shortages, the latter are not inevitable under the economic view of depletion. Other factors, primarily technological change, can offset the cost-increasing effects of depletion.

Since forecasting the future course of technological change and its impact on material costs is fraught with difficulties, it is not surprising to find a wide range of opinion regarding the seriousness of depletion. At one end of the spectrum are the optimists or cornucopians, who maintain that real material costs are not likely to rise substantially in the foreseeable future. They cite Barnett and Morse (1963) and other studies that show the real costs of producing mineral materials have actually fallen, and fallen substantially, during the twentieth century. In addition, they point out that technological change is not totally random and fortuitous. If depletion drives material costs up, this will increase the economic incentives to develop new technologies that offset such cost increases.

At the other extreme are the pessimists, often called Malthusians after the eighteenth century British economist and clergyman who feared the limited supply of land would keep most of humanity in a perpetual state of poverty. The pessimists claim the past is not a reliable guide to the future. In particular, they fear more stringent environmental regulation, rising energy costs and considerably higher levels of demand will in the future overwhelm the cost-reducing effects of new technology.

On the basis of identified deposits and other information, depletion does not appear to pose a serious threat to materials availability for the next 25 to 50 years. Thus, the debate between the cornucopians and the Malthusians focuses for the most part on the more distant future: the latter half of the twenty-first century and beyond. Which eventually will prove closer to the truth will depend largely on the race between the cost-increasing effects of depletion and the cost-reducing effects of new technology. Since the information needed to determine which of these two opposing forces will ultimately dominate is lacking, the controversy over the long-term threat of depletion is not likely to be settled in the near future.

2. *Investment*

No matter how rich the world's endowment of mineral deposits, material shortages may arise if investment in mines and processing facilities is insufficient. Such shortages can, of course, be remedied by building new mines and plants, but this takes time. The expansion of existing facilities often takes two to four years, the construction of entirely new projects five years or longer. In the meantime, shortages can be severe.

Concern over the adequacy of investment may arise for several reasons. Uncertainty associated with government policies is often cited as a deterrent to investment. Expropriation or sharp increases in taxation are feared in some countries, while changes in environmental regulations, energy policy and the availability of public lands for mineral development make planning difficult in others.

While such uncertainties have altered the geographic distribution of investment, apparently away from certain developing countries and towards the USA, Canada and Australia, the effect on the overall level of investment is less clear. In theory, a rise in uncertainty should not deter investors as long as they can realize an increase in their risk premium. While this may require some reduction in overall investment, the amount is presumably small. In practice, however, investors may react more cautiously, preferring to defer investment, at least for a time, in the hope that the uncertainty will decline.

Price controls may also impede investment. For reasons discussed in the next section, the mineral industries are highly cyclical. Profits are high when the economy is strong, and low or negative when the economy is weak. Price controls imposed when the economy is booming and inflationary pressures are greatest prevent firms from earning abnormally large profits during such periods. However, such profits are needed to allow cyclical industries to earn a rate of return over the entire business cycle sufficient to attract capital in the financial markets. For this reason, the use of price controls for macroeconomic stabilization, a practice more common in the 1960s and 1970s than in recent years, falls heavily on the mineral industries, and tends to hamper their ability to compete with other industries for capital.

The cyclical nature of the mineral industries coupled with myopia may impede access to capital during economic recessions. When demand is down, profits low and stock prices depressed, firms may find it difficult to raise investment capital from retained earnings, new equity issues or borrowing, even though they may need to invest at such times to ensure adequate capacity in subsequent years. In addition, the firms themselves may allow current market conditions to color their expectations about the future, and thus voluntarily reduce investment during such periods.

Insufficient investment by restricting capacity eventually produces higher prices and profits, which in turn provide the incentives needed to correct the problem. However, it may be several years before the resulting new capacity comes on stream and shortages are eliminated.

3. *Surges in Demand*

Unanticipated surges in material demand can occur for a number of reasons. In practice, though, most are the consequence either of military buildups or of swings in the business cycle.

Alloy steels, copper, titanium and a number of other materials are heavily used in the production of military aircraft, ships, tanks, ammunition and other armaments. In modern times, the outbreak of hostilities has produced a sharp jump in anticipated material requirements. The resulting shortages have forced governments to impose price controls and to allocate available supplies to priority uses. While such shortages are typically intense, they rarely persist for more than several years, in part because hostilities often cease within a few years and in part because new capacity is eventually brought on stream.

Cyclical instability, which has long plagued material markets, is caused by three basic characteristics of material supply and demand. First, as output approaches the available capacity, short-run supply becomes unresponsive, or inelastic in the terms of the economist, to changes in price. New capacity can be built in the long run, a period of several years or more, but not in the short run. As a result, further increases in price do not add much to supply.

Second, the demand for most materials is similarly unresponsive or inelastic in the short run to changes in price. Materials are intermediate goods whose demand is derived from the demand for final goods and services. Since the cost of a material usually constitutes only a small portion of the selling price of final products, a change in the price of the former normally has little effect on the price or demand of the latter. It is true that a rise in a material's price may encourage manufacturers to substitute other less expensive materials. However, such changes are rarely made in response to short-run and perhaps temporary price increases, for they often require the ordering of new equipment, the retraining of personnel and the redesign of production processes.

Third, the demand for materials tends to be highly responsive in the short run to fluctuations in the business cycle. Most materials are heavily used in the construction, capital equipment, automobile and other transportation, and consumer durable sectors of the economy. As is well known, these sectors tend to fluctuate with, and to a much greater extent than, the overall economy. Their cyclical swings are transmitted back to the demand for materials.

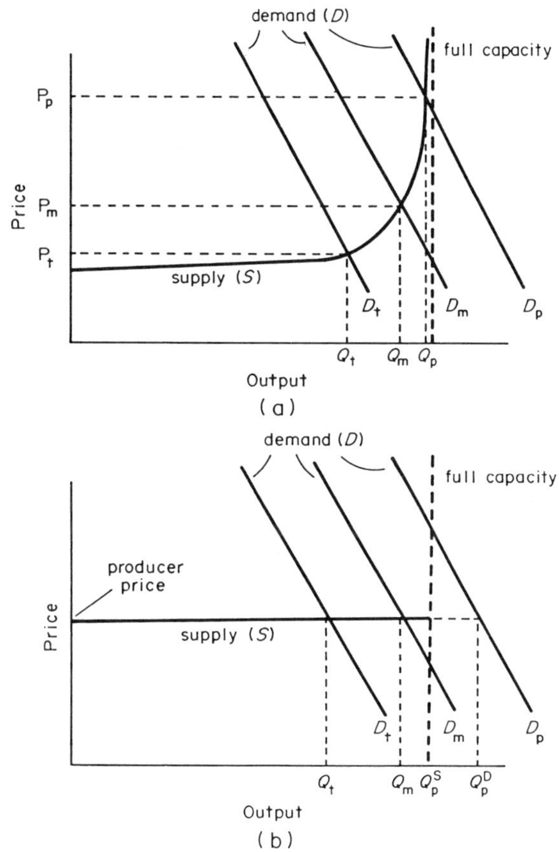

Figure 1
Nature of mineral supply and demand in (a) a competitive market, and (b) a producer market, over the business cycle. The subscripts t, m and p refer to the trough, median and peak of the business cycle, respectively (after Tilton and Vogely 1981)

Cyclical instability resulting from these three characteristics of supply and demand manifests itself in somewhat different ways depending on the type of market in which a commodity is sold. In competitive markets, such as the London Metal Exchange (LME) copper market, price is free to vary as often and as widely as needed to keep short-run supply and demand in balance. As a result, cyclical instability in such markets produces frequent and sizeable fluctuations in price. Output also varies. Since output is down when prices are low and up when prices are high, competitive markets exhibit large fluctuations in revenues and profits.

The nature of cyclical instability in competitive markets is illustrated in Fig. 1a. The demand and supply curves, as capacity is approached, are drawn with steep slopes to reflect their short-run inelasticity with respect to price. The three demand curves illustrate the shifting of demand over the business cycle. At the peak of the cycle, demand is strong as indicated by the curve D_p. The market price P_p is high, and output Q_p approaches capacity. Industry revenues, given by the product of the market price and industry output, are high, as are profits. At the midpoint of the business cycle, the curve D_m reflects demand. Prices, output, revenues and profits are all more modest. At the trough of the cycle, demand portrayed by the curve D_t is depressed. A significant portion of available capacity is idle, price and revenues are low and profits are small or negative.

While competitive markets are important for a number of materials, many materials are sold in markets where the major firms quote a producer price. Actual or transaction prices may deviate from the quoted price due to discounts and other factors, and the quoted price itself may at times change in response to short-run market conditions. However, prices are not allowed to fluctuate to the extent necessary to keep short-run supply and demand in continual balance. For this reason, prices are less volatile in this type of market.

This has led some to conclude that cyclical instability is a problem only for materials sold in competitive markets. However, the same three characteristics of supply and demand that cause instability in competitive markets are shared by materials sold in producer markets. Instability simply manifests itself in different ways.

This is illustrated in Fig. 1b, which shows the consequences of cyclical instability in producer markets. Again, three demand curves are pictured, reflecting the sizeable shift in demand over the business cycle. The slopes of these curves are also steep, since short-run demand is not greatly affected by changes in price. It is the supply curve that is different. It is drawn as a horizontal line at the producer price that terminates once full capacity is reached. This is somewhat of a simplification since, as pointed out earlier, short-run market conditions may encourage some change in the producer price. They may also cause actual prices to deviate from the quoted producer price.

The figure shows that, though price instability is eliminated or considerably reduced in a producer market, output still varies greatly. When the economy is booming, production is at or near industry capacity. When the economy is depressed, much capacity stands idle. These fluctuations in output cause sizeable variations in industry revenues and profits. In addition, at the peak of the business cycle, demand Q_p^D exceeds supply Q_p^S at the producer price, and physical shortages occur. Producers cope with such shortages by drawing down their inventories, lengthening their delivery times and rationing their customers. Consumers may have to do without the material they need, regardless of the price they are willing to pay.

Thus, it is at the peak of the business cycle that cyclical instability poses the greatest threat to materials availability. The resulting shortages are revealed by sharply rising prices in competitive markets and by an excess of demand over supply in producer markets. Though often intense while they last, such shortages tend to be temporary. Within a few years new capacity can be built, and normally the business cycle declines from its peak.

Cyclical instability and the shortages it creates, however, may be growing more severe over time, as the business cycles of the major industrialized countries become more synchronized. Such synchronization limits the opportunities for countries experiencing an economic boom to supplement their traditional sources of material supply by importing from producers who normally serve other geographic regions.

4. Interruptions in Supply and Constraints on Trade

Strikes, accidents, natural disasters and many other factors may interrupt the supply or constrain the trade of materials. This section focuses on those factors of most concern—cartels, embargoes, wars and civil disruptions.

A cartel is an agreement among producers to divide markets, restrict supply, maintain minimum prices or collude in some other way to raise prices above their competitive level. The objective is to earn greater profits, not to cut off supplies to consumers, though supplies must be curtailed to some extent to keep prices artificially high.

Cartels were quite common before World War II in certain industries, such as copper. Generally they lasted for only a few years, before collapsing for one reason or another. In the postwar era, the incidence of cartels declined, in part as a result of antitrust policy in the USA and other countries. The major exception is the Organization of Petroleum Exporting Countries (OPEC), which during the 1970s successfully raised oil prices severalfold.

The success of OPEC raised the fear that similar cartels would arise in copper, bauxite and other material industries. Research on this threat has identified several necessary conditions for a successful cartel: (a) production and exports of the material must be concentrated in a few major producing countries; (b) demand for the material must not fall greatly as its price increases; (c) supply from countries outside the cartel must not expand appreciably as the price of the material increases; and (d) the internal cohesion among cartel members must be strong enough to overcome the temptation to cheat or abandon the cartel.

The first condition is satisfied for many materials. Over half of the world's mine output of iron ore comes from four countries; the same is true of bauxite, copper and nickel. The degree of concentration is even higher for chromium, cobalt, platinum and molybdenum.

The last three conditions vary, not only among materials but over time. Material demand, as noted earlier, is generally unresponsive to price changes in the short run. However, in the long run, few if any materials are immune to the threat of material substitution in at least some of their end uses. As a result, the adverse consequences for demand grow the longer the cartel maintains artificially high prices.

Similarly, high prices are likely to encourage new sources of supply outside the cartel. However, since it takes several years to develop new mines and to construct new processing facilities, such sources will not come on stream in the short run.

Finally, the history of cartels suggests that internal cohesion is often maintained for several years, but eventually breaks down. As material substitution and the entry of new producers take place, disputes arise over how the dwindling market for cartel members should be shared. The temptation to reduce prices secretly or to cheat in other ways increases.

This review of the conditions necessary for a successful cartel suggests that a cartel attempt might succeed for a few years, but is eventually likely to collapse. This conclusion is consistent with the history of most cartels, and implies that material shortages caused by cartels are not likely to last more than a few years.

Since cartels produce only short-term benefits, while encouraging the substitution of alternative materials and the entry of new competitors over the longer run, most of the research on this topic concludes that material producers are not likely to follow OPEC's example. This view, however, assumes that producers are aware of the adverse long-run consequences and are always willing to sacrifice the short-run benefits that cartels may offer to avoid them.

Embargoes, in contrast to cartels, are imposed to interrupt or restrict material trade, not to raise price and earn greater profits. They are primarily motivated by political rather than economic considerations. In some instances, the objective is to punish other countries, or to pressure them into modifying their behavior. The United Nations embargo against chrome exports from Rhodesia during the 1970s is an example. In other instances, embargoes are the result of internal political pressure from domestic consumers who want exports restricted to ensure their access to adequate supplies at low prices. The USA has, on occasions, curtailed exports of metal scrap and soybeans for this reason.

Embargoes tend to be most disruptive in the short run. With time, countries cut off from traditional sources of supply can develop new sources and switch to alternative materials. For this reason, embargoes are rarely effective over the longer term in forcing political concessions from other countries.

Wars and civil disruptions can also interrupt material production and trade, as the invasions in 1977 and 1978 of the cobalt-rich Shaba province of Zaire so clearly illustrate. Normally, the resulting shortages are not deliberately provoked, as is the case for cartels and embargoes, but instead are an unintended by-product of other objectives and activities. There are, though, exceptions. Where mining and material processing are vital to the economic well-being of a country, an adversary may sabotage such activities. In any case, since wars and civil disruptions seldom last longer than a few years, the shortages they produce are not of long duration, even if facilities are damaged or destroyed. Even when hostilities continue over extended periods, consuming countries can, within a few years, turn to new sources of supply and alternative materials.

This examination of the threats of material availability points out several important differences between the first threat—resource depletion—and the last three—inadequate investment, demand surges and interruptions in supply and constraints on trade. The latter induce shortages that tend to arise quickly and unexpectedly. The disruption they cause can be severe. Such shortages are, however, temporary, rarely lasting more than several years. The cessation of the war, economic boom or other underlying cause often results in their termination. When this is not the case, they are eliminated by the development of new sources of supply and the substitution of more available alternative materials. Consumers, if they wish, can protect themselves against the worst effects of these shortages by maintaining sufficient stockpiles.

Shortages arising from resource depletion are very different in nature. They arrive gradually with ample warning in the form of a persistent and long-run climb in real material prices. Once in effect, they are likely to be permanent, or at least of long duration. Fortunately, such shortages appear less likely to occur over the foreseeable future than shortages arising from other threats. Over the very long term, the outlook is less clear. The principal weapon available to society for fending off the threat of shortages from resource depletion is new technology.

See also: Cartels and Producer Associations; Materials Requirements Planning; Prices of Materials: Theory

Bibliography

Barnett H J, Morse C 1963 *Scarcity and Growth.* Johns Hopkins University Press, Baltimore, MD, pp. 1–288

Gordon R B, Koopmans T C, Nordhaus W D, Skinner B J 1987 *Toward a New Iron Age? Quantitative Modeling of Resource Exhaustion.* Harvard University Press, Cambridge, MA, pp. 1–173

Simon J L 1981 *The Ultimate Resource.* Princeton University Press, Princeton, NJ, pp. 1–415

Smith V K (ed.) 1979 *Scarcity and Growth Reconsidered.* Johns Hopkins University Press, Baltimore, MD, pp. 1–298

Tilton J E 1977 *The Future of Nonfuel Minerals.* Brookings Institution, Washington, DC, pp. 1–113

Tilton J E 1991 The changing view of resource availability. In: Hutchinson R W, Grauch R I (eds.) 1991 *Historical Perspectives of Genetic Concepts and Case Histories of Famous Discoveries*, Economic Geology Monograph, Vol. 8. Economic Geology Publishing, New Haven, CT, pp. 133–8

Tilton J E, Vogely W A (eds.) 1981 Market instability in the metal industries. *Mater. Soc.* 5: 243–354

J. E. Tilton
[Colorado School of Mines,
Golden, Colorado, USA]

B

By-Products and Coproducts

A by-product is defined as a material produced secondarily or incidentally to the main product. By-product removal often results in a higher quality or more acceptable primary product. In addition, further processing of the by-product is usually justified by increased revenues from sales that exceed the cost of additional processing.

There is really no satisfactory distinction between by-products and coproducts. The term by-product is usually reserved by accountants to designate a type of joint product. Joint products are produced simultaneously from a single production system. While by-products are joint products with a small total market value relative to the primary product, coproducts usually refer to joint products with a market value large enough to require consideration of the economics of both products.

There are a number of waste materials sometimes referred to as by-products. It is desirable to distinguish between these waste products and those residues treated to recover a marketable product. Some waste not only has no economic value, but may have definite social costs attached to its existence. Examples are burning culm (coal processing) banks in the US coal-mining areas, tailings (ore processing waste) close to ore processing plants, residues from cattle feedlots, sewage sludge, residues from fish and animal processing, fly ash from electric power plants, and waste acids, gases and residues from chemical plants. These wastes may generate dust, pollute water and air, create potential hazards with respect to their containment or retard development of land for other purposes.

The classification of waste, by-products and coproducts is dynamic. Material classified as waste today may be classified as a by-product tomorrow because of changes in price or new technology. In some cases governmental policy such as environmental regulation may require capture of waste materials and, once the wastes are captured, the economics may be changed so that the incremental cost results in a commercial by-product.

By-products and coproducts can be identified in all of the major industries. The US Congress, in 1978, directed the Department of Energy to establish materials recovery targets for metals and metals products, paper and allied products, rubber and textile-mill-products industries. Using information gathered from each industry, recovered material targets were set. In this article examples are taken from the primary metals system, the processing of agricultural products, coal and coal-fired utilities, chemicals and petrochemicals, and waste management.

1. By-Products and Coproducts of Primary Metals Production

There are 24 specific metals produced in significant amounts as by-products of the major metals production systems. From the copper system come selenium, tellurium, arsenic, gold, silver, platinum, palladium, molybdenum, rhenium and the potential for production of uranium and aluminum. From the lead and zinc systems come antimony, bismuth, cadmium, arsenic, silver, gold, germanium and indium. The processing of aluminum ores yields gallium and the production of nuclear-grade zirconium metal yields hafnium. Rubidium is derived from a lithium-producing system. Recycled scrap metal mixed with primary material also yields by-product materials. The "burning" of uranium in nuclear reactors yields man-made by-product materials including plutonium.

Typically there is a wide gap between the cost and selling price of a by-product metal. Often the incremental cost to produce the by-product is small and prices are set by other sources of the material. The by-product producer will market his output at a time and place that yields the highest return. Generally, small volumes of production are involved and thus the price must be high before the total revenue becomes worth the producer's time to bother with by-product production at all. In the case of some metals, by-product supply is a small portion of total supply and the price of the metal is determined by other factors.

Some or all of the by-product costs may be forced on the producer in order to avoid air or water pollution. Dust and fumes have long been separated from blast furnace flue gases at smelters in order to prevent air pollution. Additional processing of the dust and fumes collected may lead to recovery of by-product metals.

By-products help to offset the cost of production of the main process. For example, in the electrolytic refining of copper, a process removing impurities from the acid electrolyte produces nickel sulfate. Marketing of this by-product yields revenues which can be applied towards the costs of the production of copper. Processing of the sulfide ores of copper often yields by-product sulfuric acid, a product with wide industrial application. Blast-furnace slags and, to a lesser extent, steelmaking slags are used as construction materials and soil conditioners. These examples illustrate that the uses of by-products can both recover economic values and aid in the disposal of the wastes of production processes.

Production systems are usually described in terms of the products of major economic importance, such as the copper, lead and zinc production systems. The major source of molybdenum is the mining of molybdenum ores, but there are certain copper ores from which a molybdenum concentrate can be separated by the metallurgical technique of selective flotation. Thus, the copper production system exhibits a split-off point in the milling stage that produces molybdenum concentrates as indicated in Fig. 1.

In the case of the molybdenum by-product, a well-defined split-off point is identified where the copper sulfides and the molybdenum sulfides are split in rougher flotation. Beyond this point further processing of the molybdenum concentrates produces a product acceptable for sale and competitive with molybdenum from primary sources.

To calculate the incremental cost of by-product molybdenum concentrates, it is necessary to calculate costs from the split-off point to the point of conversion to a marketable product. In the example, it is assumed that the molybdenum concentrate itself can be valued. Its value is based on the ultimate revenue derived and the costs incurred from additional processing. Recently, low prices of molybdenum have led producers to discontinue its recovery. As a result the molybdenum is lost to the tailings waste or to the copper concentrate.

In other cases the by-product split-off point occurs during smelting or refining of the major metals. By-product silver, gold, selenium and tellurium are derived from the cleaning of copper anode slimes and sludges from the electrolytic refining tanks in a copper refinery. Further processing of such material yields the by-product metals.

A model of silver by-product production is shown in Fig. 2. The hypothetical system assumes as its raw material 10 tons of leached slimes per week from electrolytic refining. The assumed mud-slimes contents are 7500 oz silver and 150 oz gold per ton of slimes, 10% selenium and 2% tellurium. The slimes are first washed, dried, mixed, packaged and transported from an electrolytic copper refinery to the facility for recovering precious metals.

When the mud-slimes are received at the refinery, they are loaded into a 15 ton cupel furnace and fluxes are added. The cupel furnace is fired with oil or gas over a period of 5–8 days in order to remove impurities. Selenium and tellurium are recovered in gases, dust and slag which may be processed to recover these by-products. The product of the cupel furnace is cast into doré, a gold–silver alloy, then moved to Balbach–Thum electrolytic cells for separation of the silver and gold. Recovery of the silver in the cupel furnace is estimated at 95% of that in the anode mud-slimes.

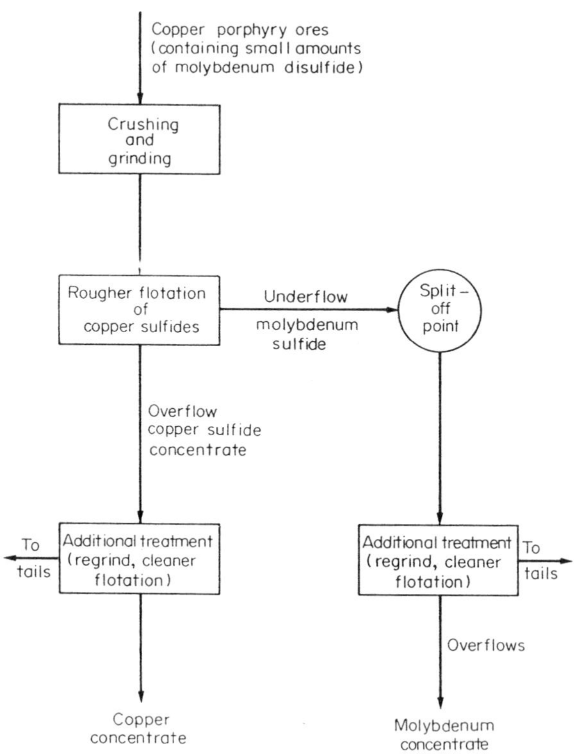

Figure 1
Production of molybdenum concentrates from copper porphyry ores

2. By-Product Materials and Energy from Agricultural Processing

Agriculture is the source of a myriad of wastes, many of which have proven to be commercially attractive by-products.

The meat industry produces protein by-products from the entrails of animals. Fish and animal by-products are sometimes used for the feeding of domestic animals.

Anaerobic digestion of livestock waste has been used to generate methane as an energy source: the residues are fermented and methane is recovered. Research indicates that the fermented liquid effluent and the dewatered effluent are acceptable as feed ingredients providing supplemental protein for cattle feed.

Agricultural crop residues have been investigated as a source of fuel gas, feed and fertilizer. Crop residues, such as wheat straw, corn stover and rice straw, are available in sufficient quantity to provide meaningful supplies of ethanol and fuel gas. Grain dust, a difficult

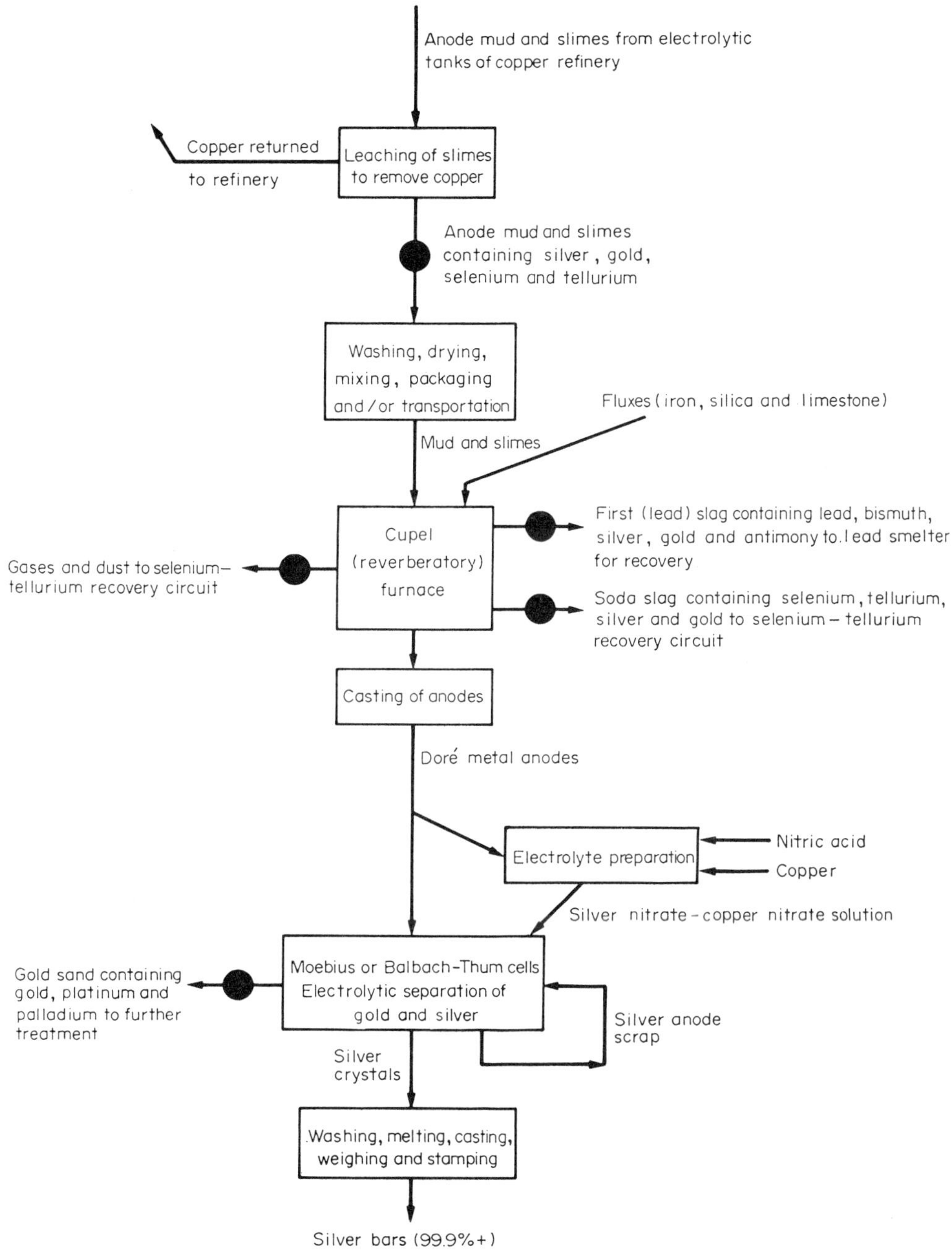

Figure 2
Silver by-product production; ●, split-off point for by-product

and expensive material to handle, has a potential as a feed ingredient. Wood wastes have been investigated as a source of carbohydrate–protein feed additive while bagasse, a residue of the sugar industry, has been used in the production of particleboard.

The brewing industry is also a source of by-product animal feed. Dried distillers' grains have been used to replace corn in the feed for cattle with no significant differences in weight gains, feed efficiency or general health.

3. By-Products of Coal and Coal-Fired Power Plants

The production and utilization of coal yield important by-products and potential by-products. The fly ash from bituminous coals contains about 20 wt% alumina and 16 wt% iron oxide. Recovery of these products would provide an alternative source of aluminum and iron as well as other metals such as germanium, titanium, manganese, uranium and thorium. Power plant fly ash has been considered as a resource for metals and construction materials. Precipitates from fly ash are calcined to produce metallurgical-grade alumina while the residues appear promising as a raw material for the production of Portland cement, concrete building blocks, ceramics and wallboard.

Some coal contains appreciable amounts of mineral materials which must be removed from the coal prior to its utilization as a source of energy. Coal refuse in combination with power plant fly ash has been used to form subbase material for roadway construction.

Coal transported in slurry pipelines can be chemically processed during transport and barite as well as titanium, aluminum, germanium, indium, beryllium and zirconium metals can potentially be recovered in this way.

4. By-Products from the Chemical and Petrochemical Processing Industry and Natural Gas Production

By-products and waste products of the petrochemical industry are used in the production of building materials, including asphalt and other bituminous materials from petroleum refining. Gases such as fluorine, and acids such as formic acid, are also captured to yield commercial by-products. Examples of by-product utilization are recovery of high-protein solids for cattle feed and bituminous materials for construction. The natural gas industry is a source of helium, a scarce and important industrial material.

One of the major problems facing the development of a commercial-scale oil shale industry is the disposal of large amounts of waste shale after processing. By-product uses include the manufacture of construction materials and the utilization of the spent shale as a grout to stabilize underground mining areas.

5. By-Products of the Waste Disposal Industries

Sewage sludges and municipal waste are also sources of by-product production. Gamma-ray irradiation of the sewage sludge can eliminate bacterial, fungal and parasitic pathogens to permit the safe use of the materials. Sewage sludges have been used to generate heat and combustible gas, primarily methane. Digested sludge is concentrated and disposed of on strip mined land during reverberation.

Concern for efficiency in the use of conventional energy sources has led to the development of alternative energy sources. A major cost in cement manufacture is that of fuel. Recently, a cement manufacturer merged with a second company that receives and processes waste fuels for the kilns. With a minimum of processing, waste materials become an economically viable by-product that serves to reduce the costs of both cement manufacture and waste disposal.

Resource recovery from solid waste has been attracting increased attention as a way of alleviating the disposal problem. Recent efforts have been concentrated on the recovery of energy by incineration, but ferrous metals and aluminum in municipal solid waste are products.

See also: Materials and the Environment; Resource Recovery: Economic and Institutional Aspects

Bibliography

Anderson G R 1980 *Potential Products from North Dakota Lignite Fly Ash.* US Department of Energy, Washington, DC

Benn F W, Foot D G Jr 1986 *Recovery of Fluorite and By-Products from the Fish Creek Deposit, Eureka County, Nevada.* US Bureau of Mines, Department of the Interior, Washington, DC

Booz-Allen and Hamilton, Incorporated 1979 *Industrial Recovered Materials Utilization Targets for the Textile-Mill-Products Industry.* US Department of Energy, Washington, DC

Brown P, Evemy P, Ferrero G L 1988 *Seminar on Recovery of Energy from Municipal and Industrial Waste Through Combustion.* Elsevier, Amsterdam

Campbell J A, Russell S H 1986 Resource recovery: technology. In: Bever M B (ed.) 1986 *Encyclopedia of Materials Science and Engineering,* Pergamon Press, Oxford, pp. 4231–8

Dean K C, Froisland L J, Shirts M B 1986 *Utilization and Stabilization of Mineral Wastes,* Bulletin 688. US Bureau of Mines, Department of the Interior, Washington, DC

Filstrup P 1980 Processes and equipment for protein by-products in the meat industry. *Appl. Prot. Chem.* 80: 181–222

Gushue J, Fischer S, Walker J 1979 *Feasibility of Alternatives for Surface Utilization of Coal Wastes,* Final Technical Report. US Department of Energy, Washington, DC

Hansen R W 1983 *Methane Generation from Livestock Wastes.* Colorado State University Extension Service, Fort Collins, CO

Klass D L, Emert G (eds.) 1981 *Fuels from Biomass and Wastes.* Ann Arbor Science Publishers, Ann Arbor, MI

Kreis R D 1979 *Recovery of By-Products from Animal Wastes: A Literature Review.* Robert S Kerr Environmental Research Laboratory, Ada, OK

McDowell W J, Seeley F G 1978 *Recovery of Aluminum and Other Metal Values from Fly Ash.* US Department of Energy, Washington, DC

National Research Council 1981 *Feeding Value of Ethanol Production By-Products.* Committee on Animal Nutrition,

Board of Agriculture and Renewable Resources, Commission on Natural Resources, National Research Council, Washington, DC
Petrick A 1973a *The Economics of By-Product Metals, 1. Copper System*, Information Circular 8569. US Bureau of Mines, Department of the Interior, Washington, DC
Petrick A 1973b *The Economics of By-Product Metals, 2. Lead, Zinc, Uranium, Rare Earths, Iron, Aluminum, Titanium and Lithium Systems*, Information Circular 8570. US Bureau of Mines, Department of the Interior, Washington, DC
Resource Planning Associates 1977 *The Potential for Cogeneration Development in Six Major Industries by 1989*. Resource Planning Associates, Cambridge, MA, for the US Department of Energy, Contract No. CR-04-60172-00
Schwarz S C, Brunner C R 1983 Energy and resource recovery from waste. *Energy Technol. Rev.* 0270–9155; 86
US Bureau of Mines 1986 *Mineral Facts and Problems 1985 Edition*, Bulletin 675. US Bureau of Mines, Department of the Interior, Washington, DC
US General Accounting Office 1981 *The Nations Unused Wood Offers Vast Potential Energy and Product Benefits*. Report to the Congress by the Controller General of the United States, US General Accounting Office, Washington, DC
Vanderborgh N E 1980 *Mineral Recovery from US Coals*. US Department of Energy, Washington, DC
Waddams A L 1969 *Chemicals from Petroleum: An Introductory Survey*, 2nd edn. Chemical Publishing, New York
Wendland R T 1969 *Petrochemicals: The New World of Synthetics*. Doubleday, Garden City, NY
Yates J J, Chaudhry R K, Dewey J A 1982 *An Inventory of Used and By-Product Hydrocarbon Streams*. Center for Environmental Research Information, US Environmental Protection Agency, Washington, DC

A. Petrick
[Petrick Associates, Evergreen, Colorado, USA]

C

Cartels and Producer Associations

Cartels and producer associations are an important factor in international commodity markets. The existence of international cartels is certainly not a new phenomenon. With the appearance of OPEC, however, it is a phenomenon that has received much more public attention.

A historical survey of the experience of some international commodity cartels and producer associations (Eckbo 1975), shows that at one time or another there has been an attempt to cartelize the market for most of the major internationally traded commodities. The large majority of these attempts at cartelization, however, were failures—the cartel either dissolved after a short period of time, or in some cases the cartel remained in force officially, but had little or no real impact on price and member revenues. Of the fifty-one formal cartel organizations documented by Eckbo, only nineteen could be considered successful in the sense of being able to maintain a price significantly higher than what it would have been in the absence of agreements. But even the successful cartels were limited in their durability; the average lifetime of the formal agreements was about five years, and only five of the nineteen cartels lasted ten years or longer.

The late 1970s saw a growing concern that the prospects for successful cartelization had suddenly become greater, and that in the future, world commodity markets would become increasingly dominated by cartels. Much of this concern, of course, was the result of the spectacular success of OPEC in quadrupling world oil prices in 1973–74, and the International Bauxite Association's (IBA) success in tripling the price of bauxite over the same period. (The further increases in oil prices that occurred in 1979–80 also resulted from cartel activity, but this was facilitated by the Iranian revolution and the Iran–Iraq war.) The concern about the cartelization of international markets continues today.

Warranted or not, the concern about cartels casts a shadow over predictions, policy prescriptions and proposals for the international management of commodity markets. Buffer stocks and other instruments for price stabilization, for example, can easily become the vehicles for cartelization and the establishment and maintenance of the monopoly price. And for some, cartelization, or more specifically, an implicit or explicit transfer of monopoly and monopsony power from developed to developing countries, is warranted as a means of redistributing income from rich to poor countries.

Given the historical record of international cartels, is there any reason to expect new attempts at cartelization and the formation of producer associations to succeed where similar attempts in the past have failed? Has the structure of world commodity markets—or the environment surrounding them—changed in such a way as to better facilitate the formation and success of cartels, so that we are likely to witness a proliferation of international cartels that will succeed in raising the prices of a large number of key commodities?

There are no simple answers to these questions. While the interest in cartelization on the part of some less developed countries may indeed be greater, there appears to be no clear change in the structure of commodity markets that would facilitate their cartelization. It is not the case, for example, that the environment has shifted to one in which supplies of raw material commodities are shrinking as demand keeps growing, thereby encouraging cartelization.

In the past some cartels succeeded, while others failed for reasons specific to each market and to each cartel configuration. As an example from the 1970s, IBA succeeded while CIPEC, the copper cartel, did not because of some major differences in the structures of the bauxite and copper markets. Going further back in history, the world mercury market was successfully cartelized from 1928 to 1972 (Spain and Italy were the only major producers) and the world sulfur market was successfully cartelized from 1833 until the 1960s; on the other hand, an international lead cartel formed in 1964 failed to have any significant impact on price (MacKie-Mason and Pindyck 1986). Evaluating the prospects for successful cartelization in the future will have to be done on a market-by-market basis, in each case determining whether a proposed cartel configuration actually has the potential to raise and maintain a monopoly price.

1. Requisites for Successful Cartelization

There are really two issues that determine the potential for cartel success. The first has to do with problems of cartel organization and stability. The success of a cartel depends on its members coming to an agreement on price and production levels, and on the division of revenues. Also required, is a means of detecting and deterring cheating on the part of cartel members. These organizational problems have been the focus of most attempts to evaluate the future prospects for cartelization. Some have argued that existing and potential cartels can solve these problems, and this has often been used as a major argument supporting forecasts of a proliferation of cartels (Bergsten 1974, Mikdashi 1974).

But there is also theoretical and empirical evidence that these problems can often be very difficult to solve (Griffin 1985, MacKie-Mason and Pindyck 1986).

A cartel may indeed solve its organizational problems, but this does not guarantee its success. The existence of monopoly profits depends not only on market concentration, but also on the elasticity of demand. A cartel may account for 100% of the production of a commodity, but if the demand curve facing the cartel is highly elastic, there will be little room to raise price, and little prospect for obtaining significant monopoly profits.

The potential for monopoly profits, assuming that the organizational problems can be solved, is the second issue involved in cartel success. In fact, it is probably the more important issue; if monopoly and competitive prices differ only slightly, no organizational strategy can result in cartel success. On the other hand, the ability to solve the organizational problems may depend very much on the potential for monopoly profits. There are costs involved in the organization and maintenance of a cartel—costs associated with the determination of price and the rationalization of output and revenues, political costs that may result when output is reduced or revenues are used to finance stockpiles, and costs associated with the risk of being undercut by other cartel members that might try to make large short-term profits outside the cartel boundaries. Cartel members must be willing to bear these costs if the cartel is to succeed, and doing so will be much easier when the potential monopoly profits are greater.

To evaluate the prospects for an increase in the cartelization of world commodity markets, the potential gains that cartelization could bring must be evaluated. Calculating these gains, however, may not be straightforward; a simple comparison of the monopoly and competitive prices in static equilibrium is likely to be misleading, since much of the gain from cartelization may depend on the dynamics of the market.

2. Measuring the Potential Gains from Cartelization

The structure of most commodity markets is inherently dynamic. In many markets, for example, the demand for the commodity will respond to price only with long lags, as will the supply from countries that produce competitively. A cartel itself may face long lags in increasing or decreasing its production capacity. A major increase in production by a coffee cartel, for example, could occur only after the 4–5 years needed for new trees to reach maturity, and decreases in production might involve even longer lags, since governments might be able, through incentives, to restrict more tree planting, but might not be able to destroy existing trees. And for many mineral resources, depletion problems may affect critically intertemporal pricing and production decisions.

In theory, the depletability of a resource can limit any potential monopoly power that a cartel might otherwise have. The reason is that when the total volume of a resource is limited, a monopolist has limited flexibility in setting production levels over time (Pindyck 1987). But even for resources such as crude oil, depletability has been quite limited (Fisher 1987), and calculations indicate that it has not resulted in substantial reductions in potential monopoly power (Pindyck 1987).

Because of the dynamic structure of commodity markets, the comparison of short-run or long-run static monopoly and competitive profits is simply not a useful way of measuring the potential gains from cartelization. Large short-run profits might be available to a cartel which has little potential for long-run gains, while resource exhaustion, through its tendency to bring competitive and monopoly prices closer together, might eliminate what a static analysis would indicate are large potential gains. The gains from cartelization can be better measured by assessing the equity value of the production activity under different market structures—that is, by comparing the present discounted value of the flow of current and future profits under cartelization (beginning at the moment the cartel forms) with that under a competitive market, assuming in both cases that production decisions are made optimally.

These comparisons can in fact be made by constructing fairly simple models of the commodity markets in question, and determining the cartel price trajectories and the sum of discounted profits using an appropriate optimal control algorithm. This has been done for three commodities—oil, bauxite and copper—measuring the potential gain as the ratio of the sum of discounted profits under cartelization to that under competition (Pindyck 1977, 1978). Similar calculations have also been done for the mercury and sulfur markets (MacKie-Mason and Pindyck 1986). Such models provide robust estimates of the gains and their distribution over time, and also yield surprisingly accurate ex post forecasts of cartel pricing behavior. The results indicate that the gains from cartelization are considerable for oil and bauxite, but negligible for copper. CIPEC's inability to enjoy large monopoly profits arises not only from its smaller market share (35%), but also from the dynamic response of secondary competitive supply (from scrap), where, because of the stock effect, the short-run elasticity is much larger than the long-run one.

Stylized models such as these are also applicable to the analysis of other commodity markets. Such a model would begin with a dynamic function relating total demand for the commodity to prices and income. This function should capture the important time lags and, for a commodity such as bauxite where demand is extremely inelastic up to a limit price at which point it becomes extremely elastic, may be highly nonlinear.

Next, a function for competitive (noncartel) supply would describe the dynamic dependence of that variable on prices. For an exhaustible resource, the supply function should include the effects of depletion. For commodities such as copper and other metals, it might consist of two or more equations that would explain both primary and secondary supplies. Cartel demand is then total demand minus competitive supply. The model is closed with a cartel cost function, which might depend on reserve levels if the commodity were a depletable resource, or for a commodity such as coffee or timber where the planting of trees is involved, it might depend on current capacity, which in turn would be a function of lagged prices and past capacity.

The cartel price trajectory and sum of discounted profits could then be found as the optimal control solution for such a model. The determination of the price trajectory and sum of discounted profits in the competitive case would depend on the characteristics of the commodity in question. For an exhaustible resource, the price is such that rents grow at the rate of interest, and exhaustion occurs just as the demand for the commodity becomes zero. For a produced good such as coffee, the cartel members act competitively and produce at the point where price is equal to marginal cost. In this case, market clearing yields the price at each point in time.

3. *Cartel Organization and Stability*

Given that the potential gains are present, a cartel can succeed if solutions can be found to the problems of organization and stability. The question, then, is whether a group of producers (with a combined market share that could yield sufficient potential monopoly gains) can agree on an optimal aggregate production level (thereby determining the cartel's contract surface), agree on a division of output and profits (i.e., select a point on the cartel contract surface—a sharing rule), and find a means to detect and deter cheating. As Osborne (1976) has shown, the most critical of these problems are the determination of the contract surface and the detection of cheating; once solutions to these problems have been found, solutions to the sharing problem and the deterrence problem follow readily.

What is important here is that for many real or imagined cartels there are no inherent obstacles to solving the problems of finding the contract surface and detecting cheating. As a result, for many commodities there may be several cartel configurations that could be reasonably stable, so that the real problem, once again, may be the existence of monopoly gains.

There may, of course, be hurdles that must be overcome in solving the organizational problems. As Fog (1976) points out, determination of the contract surface may be complicated by differences in opinion over matters such as the characteristics of demand and competitive supply, and the existence of substitutes. More important, cartel members may operate under different constraints (e.g., reserve levels in the case of mineral resources) and/or have different time preferences. However, as shown by Hnyilicza and Pindyck (1976) for a model of OPEC, these differences become simple differences in objectives (e.g., different discount rates in the calculation of the equity value of the production activity), and solutions exist for the reconciliation of these differing objectives, just as they do for the sharing problem in general.

As for the detection of cheating, this need not be a very serious problem, particularly if the cartel produces a commodity that is fairly homogeneous (e.g., oil), or if sales can to some degree be centralized. Much of the success of the iodine cartel, the longest lived in Eckbo's survey (1878–1939), can be attributed to the fact that all sales were made through a central cartel office in London. Of course, consuming countries might take actions to encourage cheating by making its detection more difficult. For example, by adopting import ticket plans such as the one proposed by Adelman (1976) for crude oil. Under this plan, tickets which give the holders the right to import oil are sold in an anonymous auction, and can be freely transferred or resold. The system would encourage cheating by permitting OPEC countries to establish brokers who would bid for and purchase tickets in the USA, thereby discounting the price of their oil in exchange for assured sales.

Unfortunately, rather than adopting measures to exacerbate cartels' organizational and stability problems, consuming countries have been considering policies for commodity market intervention that would aid cartels in solving these problems. Any measure that has the effect of centralizing sales or disseminating information on transaction prices works to help solve the detection problem—which is why buffer stocks, marketing arrangements and national or international purchasing agencies could work so much to the disadvantage of the consuming countries.

4. *Prospects for Commodity Cartelization*

In the future, the success of international commodity cartels will depend more on the existence of potential monopoly gains than on the ability to solve the organizational and stability problems. As explained, there is no inherent reason why the organizational problems cannot be solved, and the consuming countries, through various policies of market intervention, may even assist in the solution. If the potential monopoly gains do exist, this will provide more motivation to solve the other problems.

For many commodities, however, the potential monopoly gains simply do not exist. This is why CIPEC is for practical purposes a cartel on paper only. And with the exceptions of bauxite, uranium and perhaps mercury, cartels in most other mineral markets would probably have the same problems that CIPEC has had. Bauxite is unusual in that its world demand is extremely inelastic (up to a limit price), even in the long run (Pindyck 1977). Another resource market where cartelization appears to have been successful is uranium; this resulted in congressional hearings in 1976 and antitrust suits against foreign and some US domestic uranium producers during the late 1970s and 1980s. Like bauxite, uranium demand is extremely inelastic.

Cartels in markets for such minerals as iron ore, manganese ore, lead, tin, zinc and nickel would face considerable problems. First, in the long run, the demands for these minerals are relatively price elastic. Second, competitive secondary supply (from scrap) is quite elastic, particularly in the short run. Finally, the income elasticities of demand are below one for many of these metals. As for nonmineral commodities, such as coffee, cocoa, timber and pulp, it is difficult to assess just how large the potential gains might be without actually constructing a market model such as that described earlier. On the one hand, there are considerable lags involved in increasing noncartel production capacity, so that a cartel of, say, the major Latin American coffee producers might be able to make significant short-run profits. On the other hand, the demands for most of these commodities respond much more quickly to price than is the case for many of the mineral resources, so that successful cartelization is unlikely.

The measurement of the potential gains from cartelization requires careful analysis on a market-by-market basis. The indications are, however, that there are not many markets where these gains are likely to be large. Undoubtedly more cartels, producer associations and related organizations will form, but not many of them should be expected to have a significant impact on commodity markets.

Acknowledgement

This is a revised version of the article by the present author "The Cartelization of World Commodity Markets," *American Economic Review*, 1979, Vol. 69(2), pp. 154–8

See also: Economic Theory of Mineral Resources and Markets; Markets for Materials: Principles and Public Policy Implications of Competition; Prices of Materials: Theory

Bibliography

Adelman M A 1976 Oil import quota options. *Challenge* 18: 17–22

Bergsten C F 1974 The threat is real. *Foreign Policy* 14: 84–90

Diaz-Alejandro C F 1978 International markets for LDC's—the old and the new. *Am. Econ. Rev., Papers Proc.* 68: 264–9

Eckbo P L 1975 *OPEC and the Experience of Some Non-Petroleum International Cartels*, MIT Energy Laboratory Working Paper. Massachusetts Institute of Technology, Cambridge, MA

Fisher W L 1987 Can the US oil and gas resource base support sustained production? *Science* 236: 1601–4

Fog B 1976 How are cartel prices determined? *J. Ind. Econ.* 5: 16–23

Griffin J M 1985 OPEC behavior: a test of alternative hypotheses. *Am. Econ. Rev.* 80: 954–63

Hnyilicza E, Pindyck R S 1976 Pricing policies for a two-part exhaustible resource cartel: the case of OPEC. *Eur. Econ. Rev.* 8: 139–54

MacKie-Mason J K, Pindyck R S 1986 Cartel theory and cartel experience in international minerals markets. *Energy Markets and Regulation: What Have We Learned?* Essays in Honor of M. A. Adelman. MIT Press, Cambridge, MA

Mikdashi Z 1974 Collusion could work. *Foreign Policy* 14: 57–68

Osborne D K 1976 Cartel problems. *Am. Econ. Rev.* 66: 835–44

Pindyck R S 1977 Cartel pricing and the structure of the world bauxite market. *Bell. J. Econ.* 8: 343–60.

Pindyck R S 1978 Gains to producers from the cartelization of exhaustible resources. *Rev. Econ. Stat.* 60: 238–51

Pindyck R S 1987 On monopoly power in extractive resource markets. *J. Environ. Econ. Manage.* 14: 128–42

Varon B, Takeuchi K 1974 Developing countries and non-fuel minerals. *Foreign Affairs* 52: 497–510

R. S. Pindyck
[Massachusetts Institute of Technology,
Cambridge, Massachusetts, USA]

Conservation of Materials: Economics

Historically, conservation has implied different things for different people. The Conservation Movement of the late nineteenth and early twentieth centuries called for the adoption of a set of values other than those implied by consumer sovereignty for judging what constituted the appropriate uses of natural resources. To the economist, by contrast, conservation has come to mean a concern over whether natural resources play a unique role in production and consumption activities. Uniqueness, in this case, is defined by judging whether it would be possible to increase, or indeed even to maintain, levels of economic well-being without a minimum supply of some or all of these natural resources. This economic interpretation of the rationale for resource conservation has not been easy to address. Full knowledge of all the earth's stocks of natural resources, including a detailed specification of their attributes, would not by itself be sufficient information to permit a judgement about the essentiality of the resources involved. Thus, to provide an economic interpretation of the problems posed by natural resource conservation requires consideration of the geological availability of those resources, the ways the resources are used to respond to society's needs and the relative importance of those needs requiring natural resources in comparison to other needs. This article summarizes the considerations involved in addressing each of these issues.

1. Conservationist Perspective on Materials Usage

The early conservationists maintained, as a basic assumption, that natural resources were important, not solely because of their role in production and consumption activities, but rather because nature's ecological balance had intrinsic value (Barnett and Morse 1963 pp. 77–8). Resource conservation was also warranted for the conservationists because natural resources were scarce and man's activities wasted them. Barnett and Morse (1963 pp. 80–1) identified three rules as the conservationist's proposed means for realizing an efficient, or nonwasteful, use of natural resources:

(a) the regenerative capacity of renewable resources should not be physically damaged,

(b) renewable resources should be used in place of nonrenewable resources, where physically possible, and

(c) plentiful nonrenewable resources should be used before the less plentiful where physically possible.

The conservationists extended their calls for maximizing the physical yield and minimizing the physical destruction of natural resources to the evaluation of the appropriate end-uses of natural resources as well. These end-uses, if "unwise," contribute to the scarcity of virgin materials. Thus, conservationists called for avoidance of any waste, as well as the full recovery and reuse of extractive materials.

One would not expect that these goals would be achieved through private transactions, motivated by self-interest, and taking place within ideal markets. Indeed, the conservation movement sought an alternative to this perspective (Barnett and Morse 1963):

> "The Conservationists were seriously concerned with ethical revisions of *laissez faire* and economic atomism—with values involving a holistic conception of man and nature, obligation to future generations, so-called scientific management based on physical values and rejection of the individualistic values of the classicists. They were not concerned with the professional economists' maximizing of the present value of the stream of real income over time, based on consumer sovereignty—which professional economists frequently term conservation economics."

This distinction is important because it has affected subsequent economic analyses of resource scarcity and the definition of materials policy. That is, it has very probably promoted the belief that natural resource policy can be based on judgements as to the overall availability of natural resources. Barnett and Morse, for example, called for the use of scarcity indexes to signal the importance of growing limitations on the available stocks of natural resources. Thus, while their approach recognized the dilemma posed at the outset (i.e., knowledge of resource stocks alone does not convey information as to the importance of those resources), it tended to focus attention on the development of these indexes at an aggregate level, for example, all minerals, agricultural products and forestry products.

Page (1977) continued this focus by proposing that the conservationist criterion can be interpreted as calling for the maintenance of constant (or declining) real prices for a composite of virgin materials. He argued that this objective can be considered as assuring a sustainable yield of virgin materials resources. Clearly, this objective will be different from a present-value criterion that would call for maximization of the present value of each material resource individually.

Thus, in discussing materials conservation, it is important to distinguish between conservation as the basis for defining a criterion with which to judge the desirability of alternative allocations of natural resources versus conservation as one of several goals that might be articulated as part of the definition of an overall materials policy. In this latter role one might utilize the conservationists' three rules for avoiding waste as the basis for defining the conservation goals in the larger policy framework. However, this would not imply that conservation was the primary decision criterion in the allocation of natural resources. Equally important, this choice would not suggest that materials policy could be based on broad indexes of resource availability.

2. Measuring Natural Resource Scarcity

There is no reason to believe that private market allocations of natural resources will guarantee that conservationist objectives will be realized. However, it is also unreasonable to conclude that the results of these allocation decisions will be completely at odds with resource conservation. Accordingly, it is desirable to consider whether it is possible to evaluate resource availability and, if so, to judge what results are available from past efforts at this task. The first comprehensive effort to evaluate the adequacy of the US natural resource base was undertaken by Barnett and Morse (1963). Their study used scarcity indexes to signal the growing importance of any natural resource in meeting society's needs. Their primary scarcity index was called the real unit costs of extractive output. It was defined as the real quantity of homogeneous inputs required to produce one unit of a constant quality material (or extractive) output. Using index numbers for labor L, capital K and extractive output Q for each period t relative to a predefined base period B, the Barnett–Morse real-unit-cost index UC_t is given as:

$$UC_t = \left[w_B\left(\frac{L_t}{Q_t}\right) + r_B\left(\frac{K_t}{Q_t}\right)\right] \Bigg/ \left[w_B\left(\frac{L_B}{Q_B}\right) + r_B\left(\frac{K_B}{Q_B}\right)\right] \qquad (1)$$

where w_B and r_B are the base levels of the input prices for labor and capital, respectively (for the derivation see Smith 1980).

Their examination of the trends in this index (from 1870 to 1957) for the aggregate extractive sector and for three subsectors (i.e., minerals, agriculture and forestry) indicated that, with the exception of the forestry subsector, there was no evidence of increasing resource scarcity. This conclusion was consistently upheld under a wide variety of modifications to the ways the data were used in the index. Their analysis has had a profound effect on the methods used to consider questions associated with natural resource scarcity and on the judgements made as to long-term resource availability.

Recently this work has been subjected to detailed reevaluation (Brown and Field 1978, Fisher 1979, and Smith 1978a), suggesting that there are conceptual and empirical limitations to Barnett and Morse's ideal scarcity index. Moreover, given the complexity of the issues involved in judging resource scarcity, there appears to be a need to consider disaggregate analyses of individual resources using indexes of resource scarcity that can be directly linked to the behavioral decisions of the firms involved in extracting the resources. These indexes should be evaluated with detailed supplementary information including: (a) the geological conditions governing current and prospective future deposits of each resource, (b) the market structure governing the patterns of exchange of the resource and the prospects they provide for strategic behavior on the part of market participants, and (c) an understanding of the process by which economic agents form expectations on the state of geological and market conditions.

These conclusions follow from three interrelated aspects of recent research. Firstly, the appropriate scarcity index is one which summarizes the direct and indirect sacrifices made to obtain one unit of the resource (Fisher 1979). This definition implies that attention be given to what is considered to be the resource under study, that is, the choice will be different if one considers *in situ* resources versus processed natural resource commodities. Recent analyses suggest that both the Barnett–Morse real unit costs and neoclassical unit costs of extractive outputs are inappropriate measures of resource scarcity. The first is intended to measure the time path of the physical inputs required to obtain a unit of the output without regard to changes in the opportunity costs of these inputs (Smith 1980). The second takes account of changes in these opportunity costs, but does not fully meet the needs of a scarcity index. Indeed, neither measure conveys information concerning the cost features of those resources remaining for extraction (Brown and Field 1978, Hanson 1980).

Secondly, the empirical evidence on the movements of scarcity indexes is not clear-cut. Replications of the Barnett–Morse analysis with updated data (Barnett 1979, Johnson et al. 1980) generally agree with the earlier findings. By contrast, evidence from relative prices (Smith 1978a) seems to cast doubt on the conclusiveness of the original findings. Recursive fitting of linear and nonlinear models using relative prices as the scarcity index indicated that the trend coefficient was quite sensitive to sample composition (Slade 1982).

Finally, theoretical models of the extractive firm's behavior suggest that the past conclusions on these firms' behavioral responses are quite sensitive to the modelling of: (a) the geological conditions governing the availability of the extractive resource (Harris and Skinner 1982); (b) the treatment of the inherent uncertainty in the exploration process (Devarajan and Fisher 1982); and (c) the character of the market structure for the firm's output, especially in cases where the most reasonable characterization of the market is one of limited competition (i.e., those market forms falling between perfect competition and monopoly).

Thus, it is not clear that the private market allocation of natural resources has succeeded in maintaining a constant supply price for these resources. Indeed, the recent analyses of the properties of scarcity indexes suggest that a much more detailed level of enquiry is needed if these indexes are to be useful signals of resource availability.

3. Economic Mechanisms for Resource Conservation

Ideally a scarcity index will reflect both the character of the demand for a resource and the incremental costs of obtaining it. Given the difficulties identified in the previous section with developing and using such indexes it is reasonable to consider alternative approaches to judging the importance of particular natural resources and to gauging the prospects for conservation. One such alternative involves investigation of how natural resources individually contribute to production activities (Brown and Field 1978). In economic models these judgements are based on: (a) measures of the prospects for substituting other inputs for nonrenewable resources within existing technologies, and (b) the pattern of nonneutral innovations to these technologies and their implications for the use of nonrenewable resources. Both of these factors can be described in either of two ways. The first of these, an engineering analysis of materials usage, would describe each of these mechanisms for conserving natural resources (i.e., input substitutions and resource-saving innovation) with examples of specific cases. For example, in iron and steel production, one might cite the presence of an electric-arc steelmaking furnace along with a basic oxygen furnace provides the basis for substituting scrap steel for iron ore in composing the set of iron-bearing materials used to produce molten steel (Russell and Vaughan 1976). For resource-saving innovations, the introduction of scrap preheating to steelmaking practices (Vaughan et al. 1976) enhances the feasible substitutions between scrap and iron ore within basic oxygen furnace production technology. Similarly, within certain electrical conductor uses, it is possible to replace copper with aluminum. However, this approach faces a significant limitation: it is unable to summarize adequately the quantitative nature of the conversion rates between the inputs over different levels of input usage, or to gauge the quantitative impact of the resource-saving innovations that would accompany a given change in relative prices of inputs.

The economic approach to summarizing input substitution is based on a neoclassical description of production activities. As such, it maintains that engineering production processes can be approximated adequately with smooth functions relating the inputs to these activities to the output(s). In this framework, the substitution between inputs permitted within a given technology is generally measured by an Allen (1938) partial elasticity of substitution. In heuristic terms, this index describes how the rate of conversion between any two inputs changes with changes in the relative usage of those inputs, while maintaining the output rate at a constant level and allowing other inputs to adjust conformably (for a discussion of the technical details see Mundlak 1968, and for a heuristic review, Smith 1978b). Large positive values for the elasticity imply that the relevant two inputs are substitutes and that each factor can readily be used in place of the other. An elasticity value of zero implies that the two inputs involved must be used in fixed proportion. Negative values suggest complementary relationships between inputs. While this index has the advantage of providing a convenient quantitative index of input conversion relationships, its descriptive ability is contingent on the authenticity of the neoclassical production model as an approximation for the underlying engineering production processes (Marsden et al. 1974). Neoclassical models also offer indexes of the pace of nonneutral technical change (Bingswanger 1974, Jorgenson and Fraumeni 1981). In this case, they generally measure the rate at which the innovation increases the effectiveness of (or augments) the existing inputs.

The available evidence of the quality of neoclassical models as descriptions of the substitution relationships present in the underlying engineering technologies suggests that the performance depends on the level of aggregation of the input measures used and on the degree of disaggregation in technologies under study (Kopp and Smith 1980a, 1980b; Lau 1982). Under ideal conditions a neoclassical cost function, used as a means for describing production activities, can provide a reasonably good description of the input associations implied by the underlying technology. The available experimental evidence indicates that the estimated substitution elasticities generally identify the correct input association (i.e., substitution or complementarity) and tended to be conservative in their appraisal of input relationships when there were conflicts in the underlying true associations at different input prices.

These findings suggest that, in evaluating the available estimated neoclassical elasticities for actual production processes, one should focus on those with disaggregated input descriptions that have been conducted within carefully defined, disaggregate technologies. Unfortunately, the available data for statistical estimation of production models have greatly limited the ability of empirical efforts to respond to these guidelines. Table 1 summarizes a representative sampling of the available estimates to date. The industries are represented by Standard Industrial Classification (SIC) codes. This is a way in which the US Government classifies manufacturing industries, depending on the degree of detail in the definition of the industry. Thus, for example, a two-digit specification is a highly aggregative one. Within the two-digit specification, there would be a number of different categories, and these would be specified using a three-digit code. Within each three-digit code there would be further specification up to a four-digit code. The specific codes used in Table 1 for the SIC-coded industries are defined in Table 2. Based on the criteria discussed earlier, only the Moroney–Trapani (1981a) estimates approach the conditions necessary for being judged as reasonable approximations of the underlying input associations. These results clearly suggest that the role of natural resource inputs varies greatly across production processes. Moreover, when they were used in simulation analyses (Moroney and Trapani 1981b) designed to evaluate the implications of increased resource scarcity, it was found that substitution provided very limited relief from increased resource scarcity. Indeed, Moroney and Trapani (1981b) concluded that

> "Rapid increases in resources prices are translated into substantial commodity inflation, and factor substitution provides little relief. ... From these results, we conjecture that the most promising paths for exhaustible resource conservation are likely to be found in substitution among semifinished materials and changes in final demand."

The available empirical evidence on the prospects for resource-saving technological change is even more limited. Moreover, it is important to recognize clearly the limitations on the interpretation of these studies. The existing empirical results provide a description of

Table 1
Summary of selected Allen elasticities of input substitution[a]

Study	Character of data used in study	Estimated Allen elasticities of substitution σ_{KN}[b]
Humphrey–Moroney (1975)	cross-sectional data for 4-digit SIC-coded industries grouped by 2-digit category in 1963	0.64 (SIC 20) 1.12 (SIC 24)
Berndt–Wood (1975)	time series (1947–1971) aggregate manufacturing	0.56
Moroney–Toevs (1977)	time series for seven 3- and 4-digit SIC-coded industries (1954–1971)	0.370 (SIC 202) 0.602 (SIC 2011) −0.320 (SIC 2041, 2043) 1.292 (SIC 3241) 0.565 (SIC 3275)
Hudson–Jorgenson (1978)	annual interindustry accounts (1947–71)	
	agriculture, nonfuel mining construction	0.61
	manufacturing	0.10
	commercial transport	0.58
	services, trade communications	0.07
Wills (1979)	time series for the primary metals industry (1947–1974)	0.48
Moroney–Trapani (1981b)[c]	time series for seven 4-digit SIC-coded industries (1954–1974)	0.18 (SIC 3351) 1.08 (SIC 3312) 0.47 (SIC 3334) 0.02 (SIC 3691) 0.08 (SIC 3275) −0.38 (SIC 3241)

a Smith (1980) b σ_{KN} is the partial elasticity of substitution between capital K and the natural resource input N c These estimates are different from those reported in Smith (1980). The present results relate to the Moroney–Trapani model without variables to measure nonneutral technical change. The previous summary reported the estimates derived from models with measures of nonneutral technical change

Table 2
Some SIC-coded industries and their primary resource inputs

SIC code	Industry	Resource input
20[a]	food and beverages	meat, grain, dairy and nondairy fluids
2011[a]	meat packing	slaughtered cattle, calves, lambs and hogs
202[a]	dairy products	cream and fluid milk
2041[a]	flour milling	wheat, rye, oats and corn sold to produce foodstuffs
2043[a]	cereals	
24[a]	lumber products	forestry products
3241	hydraulic cement	gypsum (uncalcined)
3275	gypsum products	gypsum (calcined)
3312	blast furnaces and basic steel	crude iron ore and scrap
3334	primary aluminum	bauxite
3351	copper rolling and drawing	refined copper
3691	storage batteries	lead

a Uses renewable natural resources.

the pattern of technical change that has taken place over the period of analysis. It does not permit a judgement as to what innovations would be adopted under alternative patterns of input availability. Moreover, these studies have been conducted at fairly aggregate levels of analysis, both in terms of the input definitions used and the descriptions of the technologies involved. Jorgenson and Fraumeni's (1981) analysis offers the most detailed evidence available. It considered 36 industrial sectors over the period 1957–1974 and indicated that nearly all the industries experienced materials-saving and energy-using technical change. Nonetheless, the available empirical evidence remains too limited to draw conclusive judgements on the prospects that technological innovations might relax specific resource needs on an industry-by-industry basis or a specific material basis.

Thus, while economic analysis identifies, in principle, input substitution and nonneutral technical change as mechanisms for resource conservation, the available empirical evidence (from neoclassical economic models) cannot, as yet, provide detailed judgements on the magnitude of these effects for specific natural resources and production activities.

4. Definition of Natural Resources and Conservation

This discussion has implicitly maintained that natural resources and materials were synonymous. This has been conventional practice. However, it is reasonable to ask whether this definition for resources reflects all the implications of alternative patterns of materials usage for economic well-being; Krutilla (1967) has argued it does not. This limited perspective fails to recognize the contribution made to economic well-being by the amenity services provided by unspoilt natural environments. It also fails to recognize that the ways in which economic processes respond to increased materials needs can affect the quality of environmental resources over time as well. To the extent that both of these types of service contribute to society's economic well-being, it is important to consider them as an integral part of the definition and evaluation of resource-conservation policies. It is within this more general framework that the economic analyses of the importance of natural resources must be conducted. Since the framework is new little can be concluded, at this point, as to its implications for the adequacy of our natural resource base.

See also: Conservation of Materials: Technology; Resource Appraisal; Resource Recovery: Economic and Institutional Aspects; Substitution of Materials: Economics; Substitution of Materials: Technology

Bibliography

Agbeyegbe T D 1989 Interest rates and metal price movements: further evidence. *J. Environ. Econ. Manage.* 16: 184–92

Allen R G D 1938 *Mathematical Analysis for Economists.* Macmillan, London

Barbier, E B *Economics, Natural Resource Scarcity and Development: Conventional and Alternative Views.* Earthscan Publications, London

Barnett H J 1979 Scarcity and growth revisited. In: Smith V K (ed.) 1979 *Scarcity and Growth Reconsidered.* Johns Hopkins University Press, Baltimore, MD, pp. 163–97

Barnett H J, Morse C 1963 *Scarcity and Growth.* Johns Hopkins University Press, Baltimore, MD

Berndt E R, Wood D O 1975 Technology, prices and the derived demand for energy. *Rev. Econ. Stat.* 57: 259–68

Bingswanger H P 1974 The measurement of technical change biases with many factors of production. *Am. Econ. Rev.* 64: 964–76.

Brown G M Jr, Field B C 1978 Implications of alternative measures of natural resource scarcity. *J. Polit. Econ.* 86: 229–43

Devarajan S, Fisher A C 1982 On measures of natural resource scarcity under uncertainty. In: Smith and Krutilla 1982, pp. 327–46

Farrow S 1985 *Testing the efficiency of extraction from a stock resource. J. Polit. Econ.* 93: 452–87

Fisher A C 1979 Measure of natural resource scarcity. In: Smith V K (ed.) 1979 *Scarcity and Growth Reconsidered.* Johns Hopkins University Press, Baltimore, MD, pp. 249–71

Hall D C, Hall J V 1984 Concepts and measures of natural resource scarcity with a summary of recent trends. *J. Environ. Econ. Manage.* 11: 363–79

Halvorsen R, Smith T R 1984 On measuring natural resource scarcity. *J. Polit. Econ.* 92: 954–64

Halvorsen R, Smith T R 1991 A test of the theory of exhaustible resources. *Q. J. Econ.* 106: 123–40

Hanson D A 1980 Increasing extraction costs and resource prices: some further results. *Bell J. Econ.* 11: 335–42

Harris D P, Skinner B J 1982 The assessment of long-term supplies of minerals. In: Smith and Krutilla 1982, pp. 247–326.

Hazilla M, Kopp R J 1984 A factor demand model for strategic nonfuel minerals in the primary metals sectors. *Land Econ.* (November): 328–39

Hazilla M, Kopp R J 1986 Systematic effects of capital service price definition on perceptions of input substitution. *J. Bus. Econ. Stat.* 4: 209–24

Hudson E A, Jorgenson D W 1978 The economic impact of policies to reduce US energy growth. *Resour. Energy* 1: 205–29

Humphrey D B, Moroney J R 1975 Substitution among capital, labor, and natural resource products in American manufacturing. *J. Polit. Econ.* 83: 57–82.

Johnson M H, Bell F W, Bennett J T 1980 Natural resource scarcity: empirical evidence and public policy. *J. Environ. Econ. Manage.* 7: 256–71

Jorgenson D W, Fraumeni B M 1981 Relative prices and technical change. In: Berndt E R, Field B C (eds.) 1981 *Measuring and Modeling Natural Resource Substitution.* MIT Press, Cambridge, MA, pp. 17–47

Kopp R J, Smith V K 1980a Measuring factor substitution with neoclassical models: an experimental evaluation. *Bell J. Econ.* 11: 631–55

Kopp R J, Smith V K 1980b Input substitution, aggregation, and engineering descriptions of production activities. *Econ. Lett.* 5: 289–96.

Krutilla J V 1967 Conservation reconsidered. *Am. Econ. Rev.* 57: 777–86

Lau L J 1982 On the measurement of raw materials inputs. In: Smith and Krutilla 1982, pp. 167–200

Marsden J, Pingry D, Whinston A 1974 Engineering foundations of production functions. *J. Econ. Theory* 9: 124–40

Miller M, Upton C 1985 The Hotelling vacation principle. *J. Polit. Econ.* 93: 1–25

Moroney J R, Toevs A 1977 Factor costs and factor use: an analysis of labor, capital and natural resource inputs. *South. Econ. J.* 44: 222–39

Moroney J R, Trapani J M 1981a Alternative models of substitution and technical change in natural resource intensive industries. In: Berndt E R, Field B C (eds.) 1981 *Measuring and Modeling Natural Resource Substitution.* MIT Press, Cambridge, MA, pp. 48–69

Moroney J R, Trapani J M 1981b Factor demand and substitution in mineral intensive industries. *Bell J. Econ.* 12: 272–84

Mueller J J, Gorin D R 1985 Informative trends in natural resource commodity prices: a comment on Slade. *J. Environ. Econ. Manage.* 12: 89–95

Mundlak Y 1968 Elasticities of substitution and the theory of derived demand. *Rev. Econ. Stud.* 35: 225–36

Page T 1977 *Conservation and Economic Efficiency: An*

Approach to Materials Policy. Johns Hopkins University Press, Baltimore, MD
Pearce D W, Barbier E, Markandya A 1989 *Sustainable Development: Economics and Environment in the Third World*. Elgar, London
Russell C S, Vaughan W J 1976 *Steel Production: Processes, Products and Residuals*. John Hopkins University Press, Baltimore, MD
Slade M E 1982 Trends in natural resource commodity prices: an analysis of the time domain. *J. Environ. Econ. Manage.* 9: 122–37
Smith V K 1978a Measuring natural resource scarcity: theory and practice. *J. Environ. Econ. Manage.* 5: 150–71
Smith V K 1978b Duality principles and measuring the production technology: a heuristic introduction. *Socio-Econ. Plann. Sci.* 12: 161–6
Smith V K 1980 The evaluation of natural resource adequacy: elusive quest or frontier of economic analysis? *Land Econ.* 56: 257–98
Smith V K, Krutilla J V (eds.) 1982 *Explorations in Natural Resource Economics*. Johns Hopkins University Press, Baltimore, MD
Vaughan W J, Russell C S, Cochrane H C 1976 *Government Policies and the Adoption of Innovation in the Integrated Iron and Steel Industry*. Resources for the Future, Washington, DC
Wills J 1979 Technical change in the US primary industry. *J. Econom.* 10: 85–98

V. K. Smith
[North Carolina State University, Raleigh, North Carolina, USA]

Conservation of Materials: Technology

Materials conservation comprises those strategies by means of which present or projected materials shortages are addressed by effectively reducing demand rather than increasing supply. This approach is relied on more heavily as the limits to increasing supply, such as added energy demands and environmental damage, become more evident. The major technical strategies employed to accomplish materials conservation are: (a) effective materials utilization, (b) materials substitution and (c) materials recycling.

Effective materials utilization is the broadest of these strategies since it addresses not only product design, development and deployment, but also manufacturing materials and processes. Materials substitution is probably the most important of these strategies, however, since it directly alters materials usage. Recycling is an obvious strategy but one still hampered by a variety of economic, technical and institutional factors.

1. Total Materials Cycle

Materials conservation and its supporting strategies are best understood in the context of the total materials cycle shown in simplified form in Fig. 1. In this global system, as first discussed in *Materials and Man's Needs* (National Academy of Sciences 1974), natural resources extracted from the earth are refined into basic materials and then further processed into engineering materials. These, in turn, are designed and manufactured, or assembled, into the myriad products required by society. At the end of their useful life, they are either returned to the earth as waste or recycled. Materials, energy and the environment interact strongly at virtually every point in this materials cycle and events or decisions occurring at any given point may affect other points. Furthermore, every operation in the materials cycle requires energy and, at every stage in the cycle, material (and energy) losses are encountered. These losses for common metals are estimated at 50–70% of the amounts that enter the cycle and, in fact, materials conservation can alternatively be defined as the reduction of these losses from the materials cycle. The concept of the total materials cycle provides a useful framework for examining the global materials–energy–environment system though its operation is far more complicated than that shown in Fig. 1. This complexity has been discussed by Bever (1978) for those parts of the materials cycle related to recycling. The economics of materials conservation is treated in the article *Conservation of Materials: Economics*.

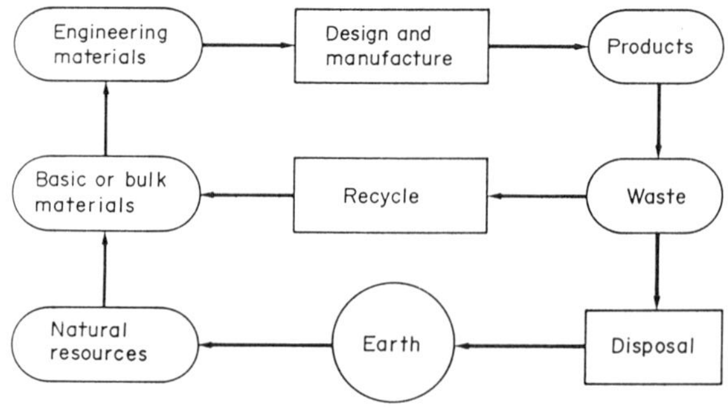

Figure 1
Simplified diagram of the total materials cycle

2. Effective Materials Utilization

Effective materials utilization is broadly recognized as an essential materials conservation strategy. It refers to the development, selection and design into products of materials that most efficiently meet application requirements, that have optimum durability and life, and that are recyclable. It further refers to the obtaining, processing and fabrication of materials so as to consume, waste or disperse the least amount of material (or energy) for equivalent performance. In short, materials (and energy) conservation calls for processes that minimize the input of resources per unit of output. The specific tactics for achieving this fall conveniently under the headings of research and development, design and manufacture.

Research and development are expected to provide the stockpile of technical information needed to implement any of these conservation strategies. Examples that apply to effective materials utilization include the reduction of material losses due to corrosion and wear, the development of new and improved materials and processes and higher manufacturing yields obtained through improved flaw detection and materials characterization techniques.

Design is the preeminent tactic for achieving effective materials utilization since it dictates not only the kinds and amounts of materials to be used in a product, but also the likely manufacturing and assembly methods to make it. These tactics include miniaturization, designing to closer tolerances, using more abundant and renewable materials for scarcer ones, designing for longer life and easier maintenance, facilitating recycling and the use of reclaimed materials, simplifying and standardizing material and product variants and making greater use of composites while avoiding wholly dissipative uses of materials that thwart conservation.

Dissipative uses were of central concern in a NATO (1976) study of eleven potentially scarce metals which estimated that some 30% of the mine production of these metals was dissipated beyond possible recovery. They included such dissipative uses as silver in photographic prints, cobalt in paints, tin in solders and toothpaste and zinc in galvanizing. They further estimated that another 40% of the mine production of these scarcer metals was dissipated by virtue of being technically but not economically recoverable. One example is the platinum-group metals required in catalytic converters for automobiles to control exhaust emissions. This is the largest current use of such metals in the USA (45% in 1988) and only about 12% of these platinum-group metals are recovered from discarded converters due to the high collection costs for recycling. By contrast, over 85% of these metals used as catalysts in chemical plants and petroleum refineries are recycled (Frosch and Gallopoulos 1989). Such combined dissipation, which is estimated to be some two thirds of all metals produced, can only be reduced by the substitution or recycling strategies discussed in the following.

Improved manufacturing processes that minimize materials (and energy) waste are the third and final tactic used in achieving effective materials utilization. Examples of the kinds of manufacturing processes to be preferred where a choice is available would include:

(a) additive over subtractive processes,

(b) solventless over solvent processes,

(c) closed-loop over open-loop processes,

(d) net-shape over material-removal processes, and

(e) feedback-controlled over sampling processes.

In each of these cases, the preferred process results in less materials waste and usually in less energy waste, since the energy content of the material itself is large in comparison with process energy.

More specific information relative to the material and energy losses associated with various metalworking processes has been collated by van den Kroonenberg (1979) among others. Ranking these in terms of decreasing material losses, the order is: machining > stamping > hot forging > cold or warm pressing or diecasting. If the processes are ranked in terms of decreasing energy required per unit weight of chips produced or product formed, the order is: grinding > forging > hot forming > planing > turning > drilling. Such guidelines are helpful, though obviously not determinant, in selecting preferred manufacturing processes.

A necessary corollary of effective materials utilization is that the material remaining is more highly stressed and, therefore, this strategy requires higher performance materials, better processing control and understanding, better reliability information and better materials characterization methods. These requirements can serve, in fact, as impediments in implementing the strategy of effective materials utilization.

3. Materials Substitution

Materials substitution is common and open to many possibilities because it focuses normally on the function(s) a material provides rather than the material itself. It will, however, only occur if the substitute material has properties that permit the required function to be achieved at lower cost under the prevailing circumstances.

Several kinds of substitution may be identified that can contribute to the overall goal of materials conservation. They are listed in the following together with selected examples:

(a) abundant materials for scarcer materials (e.g., the use of aluminum for copper in electrical applications or ceramics for high-temperature alloys in turbine engines),

(b) nonmetals for metals (e.g., plain or reinforced plastics for zinc or aluminum in structural applications, or glass for metals in containers),

(c) renewable materials for nonrenewable materials (e.g., wood for metals in construction),

(d) synthetic materials for natural materials (e.g., hydrothermally grown quartz crystals for electronic use and synthetic diamonds for cutting and polishing), and

(e) composite materials for monolithic materials—these include all those materials where needed surface properties are conferred by coatings or claddings.

Though many substitutions offer the possibility of materials conservation, the time required to bring them into use can be substantial because of the need to examine alternative materials, to develop design and reliability data and to evaluate the substitute in a manufacturing environment. This is especially relevant to back-up technologies that might be needed if supplies of critical or strategic materials are vulnerable to unanticipated interruption. In addition, the proposed substitute may require novel processing or may affect related technologies that interact with the substitute material in its intended use. All of these factors make the substitution process slower than desired (see *Substitution of Materials: Technology*). Furthermore, where certain unique features of a material are involved, substitution may be difficult, if not impossible, to effect. This would include, for instance, those uses that depend on the unique catalytic properties of platinum already cited or the thermal properties of helium. Equally difficult are those cases where the scale of use or cost cannot be met by any other material. For example, no other material could replace steel in its structural uses nor lead in its battery uses.

The altered material flows caused by substitution also create other problems in various parts of the materials cycle, particularly in basic materials supply and in recycling. Thus, a metal obtained solely as a by-product of another will have its own supply reduced if the latter is substituted for in any important use. The changes in recycling practices required as cars become smaller, shift to a greater use of high-strength low-alloy steels, aluminum and plastics and use more platinum-group metals to control exhaust emissions are other examples of the problems brought about by substitution.

4. Recycling

Recycling has long been an integral part of the materials industry and is a strategy which, by its very nature, adds to material supply and modestly alleviates resource depletion. It has the advantage over effective use and substitution strategies that no product or manufacturing changes are needed to implement it, though such changes can be used to enhance recyclability. Recycling can also contribute to energy conservation, since material produced from scrap sources requires much less processing energy than that from virgin sources. For example, the energy required to obtain aluminum from scrap sources can be as low as 5% of that required to obtain it from its principal ore, bauxite.

Scrap recycling occurs at three stages in the material cycle of Fig. 1. The first is at the production stage for the basic material where scrap, known as home, revert or runaround scrap in the case of metals, is returned directly to the basic production process. Having known composition and little contamination, this scrap is the most desirable though it cannot be said to add to materials supply. The second stage at which recycling occurs is in manufacturing, where scrap (new, prompt-industrial or processing scrap) is collected as the next most desirable type. Like home scrap, it does not really add to supply though it is a very important component of recycling. The third and most familiar place where recycling occurs is at the postuser or waste stage. Here, the old or obsolete scrap recovered does add to materials supply. The availability of home and new scrap depends on current production levels, whereas that for old scrap depends on the life cycles of the products from which it is recovered and on the number of products then produced. The recovered new or old scrap is returned to the materials cycle at the basic production stage as shown in Fig. 1 or to a secondary production stage not shown. In fact, the secondary materials industry handles the majority of all new and old scrap, since it specializes in this activity.

In addition to the conventional type of scrap recycling just described, other types should be noted. The first is the closed-loop recycling of such items as paper or aluminum cans back into the same products. The second is the *de facto* recycling that occurs with the rebuilding and reuse of products or the dismantling or cannibalizing of them to provide useful parts.

Recycling can postpone the exhaustion of a nonrenewable resource and a simple expression for estimating this was developed in a NATO study (1976). It shows clearly that for all but very low growth rates, recycling contributes only modestly to resource extension. In the USA only about 20% of the available old metal scrap is recycled, though it has been estimated that this figure could be easily doubled. The level of recycling for an individual material depends on the ease of collection and separation in the waste stream, the value of the scrap material, the difficulty of recovering pure material from the scrap and its degree of dependence on corecovery. For example, 90–95% of all cars in the USA are recycled principally for the scrap iron and steel they contain, but copper, lead, aluminum and other nonmetals can be recovered at the same time. The level of this corecovery will be highly dependent on the current market for old iron and steel scrap.

Plastics recycling presents a special challenge at the postuser stage due to the large number of resins and additives used, the wide array of products and forms in which they are found, especially as composites, and the difficulties they present in collection and separation. There is broad agreement, however, of the need to either recycle or safely dispose of these important materials through better technology and appropriate incentives.

5. A New Conservation Ethic

Materials conservation and its supporting strategies are part of a broader conservation ethic, or industrial ecosystem, in which the consumption of materials (and energy) is optimized and the generation of waste and environmental pollution is minimized. The analogy with biological ecosystems points to a more closed-loop industrial ecosystem in equilibrium with the environment and sustainable over a long period. This, in turn, calls for a new philosophy of industrial design that treats a product as a materials system and seeks to optimize it in terms of the total materials cycle. This optimization requires some governmental incentives and changed attitudes on the part of society to be realized, since present market incentives are insufficient to do so. Finally, this ethic emphasizes that there are certain inherent limits in our ability to continuously expand supplies and that it, therefore, is the most viable means for meeting these future needs of society.

See also: Conservation of Materials: Economics; Recycling of Demolition Wastes; Recycling of Metals; Substitution of Materials: Economics; Substitution of Materials: Technology

Bibliography

Bever M B 1978 System aspects of materials recycling. *Conserv. Recycling* 2: 1–17

Campbell J A, Russell S H 1986 Resource recovery: technology. In: Bever M B (ed.) 1986 *Encyclopedia of Materials Science and Engineering*, Pergamon Press, Oxford, pp. 4231–8

Frosch R A, Gallopoulos N E 1989 Strategies for manufacturing. *Sci. Amer.* 261: 144–52

National Academy of Sciences 1974 *Materials and Man's Needs*. National Academy of Sciences, Washington, DC, pp. 4–6

North Atlantic Treaty Organization 1976 *Rational Use of Potentially Scarce Metals*. NATO Scientific Affairs Division, Brussels

van den Kroonenberg H H 1979 Solutions to future resource problems by design engineering. *Mater. Soc.* 3: 271–9

T. D. Schlabach
[AT&T Bell Laboratories, Murray Hill, New Jersey, USA]

Cost Estimating

Cost estimating or costing is the process of forecasting the monetary cost of performing a specified task to a specified schedule by a particular organization. This process uses three basic methods: analogy to known tasks, parametric estimating and the industrial engineering or bottom-up method.

Estimates are normally generated in response to an invitation to bid, as an evaluation tool for estimates received as a result of an invitation to bid, as a basis for formulating a price for a product or service, or budgeting, and for cost control by providing an estimate-to-completion for ongoing tasks.

Various terms define different applications/methods of costing. Contract cost analysis refers to the evaluation of a bid to estimate the bidder's costs, usually using the industrial engineering method. Price analysis refers to a general review by analogy or parametrics, rather than an industrial engineering estimate of a prospective program cost by a potential customer. Rough order of magnitude estimates provide a simple, quick cost estimate usually based on analogy by a qualified expert. Estimate-to-completion, conducted while a job is performed, provides the costs to complete the job, usually by the industrial engineering method though a rough order of magnitude estimate may be useful in some situations.

To a producer or seller, price consists of his cost plus profit; or it may be said that the producer's (seller's) price is the consumer's (buyer's) cost. Pricing, therefore, is the same process as cost estimating, with the producer's profit added after the cost estimate.

1. Methods

The three methods of estimating all start from past experience with other similar tasks and the cost of performing those tasks.

Analogy deals with these experiences in a gross fashion, attempting to estimate how the new project differs from selected historical projects. This method requires experience and mature judgement in the particular field to evaluate quantitatively the similarities and differences between various projects. Frequently a number of expert opinions are sought and combined to provide a more reliable estimate.

Parametric estimating systematically and mathematically examines past experience and uses correlation methods to develop various formulae which provide estimates of cost based on the characteristics (parameters) of the task under consideration. The depth of an oil well might be the parameter used to estimate the cost of drilling the well or to estimate the cost of maintaining the well. For greater accuracy, this cost per unit of depth may be increased at various depths. Similarly, the cost of processing ores might be estimated on the basis of mineral content weight with modifiers for particular ore bodies.

The industrial engineering or bottom-up method splits each task into subtasks, attempting to find recognizable operations which have been done before and for which the costs are known. The performing organization usually provides such bottom-up estimates, as they are the sum of many smaller estimates from the lower reaches of an organization.

The calculation of cost estimates, particularly the industrial engineering method, uses a number of specialized techniques. Specific techniques develop cost estimates for design work, field operations, manufacturing, maintenance and so on. The standard hours procedure for manufacturing costs draws on a file of common operations, such as "drill hole" or "rivet" or "paint," and assigns an experimentally determined number of man-hours to perform that operation. The effort required to perform a required task, such as "fabricate bracket," would be the sum of the standard hours for each of the required operations.

On account of training, interruptions, skill levels and other factors, production shops or departments usually do not perform to the assigned standard(s). When labor costs relative to time are greater than the assigned standards, an actual-to-standard hour ratio can be calculated. This ratio or its inverse (realization) is a conversion rate which can be used to transform standard hours to those hours expected to be generated, given a lot quantity/unit number and the standard hour value.

Repetitive tasks exhibit an improvement or learning curve, as changes in the process and learning of the operator make possible reductions in the required time to perform the task. This learning curve has been mathematically formulated and shows surprising consistency across a broad range of industries and organizations. When the required labor has been determined, multiplying by the labor rate provides the quantity based labor cost. These may be modified with provisions for expected overtime pay, travel costs and hazard bonuses. Materials costs, services (e.g., computer service) and time-based labor costs (e.g., quality control, tooling maintenance, sustaining engineering, etc.) must be added to obtain the total of direct costs. Improvement or learning curve factors may be applied to the quantity based portion of the direct costs. The suggested calculations develop the recurring costs of the delivered product. In addition, various nonrecurring costs must be estimated and added to the total. These nonrecurring costs include design, test, manufacturing operations planning, tooling, test equipment and any capital equipment which is to be charged to the job. Indirect costs, such as administrative and general costs (overheads), are added either directly or as a multiplicative factor. Other such factors include taxes, insurance and interest costs and fee. The total of all of these is the cost estimate.

Computers can perform many of the repetitive calculations required for a cost estimate. Their memory banks provide convenient repositories for the historical data needed to prepare an estimate. These memory banks can help in the preparation of parametric cost-estimating relations. Cost estimators should systematically collect and store cost-experience data for such purposes. Such a data bank, carefully indexed, facilitates cost estimating and heightens credibility.

Cost estimates derived from the above methods are often modified by evaluations of risk and competition. Sources of risk include risk in the technical feasibility of the project, risk in meeting the proposed schedule or risk in the dependability of the cost estimate. Quantitative methods promise improved objectivity over the more usual intuitive approach to risk compensation. Competition, on the other hand, often mandates a decrease in cost estimates. Rather than change the cost estimate, many managements prefer to adjust the profit (which is to be added to the cost estimate) to reflect the risk and competitive situation. Furthermore, considerations of the investment in capital required will mandate changes in the required profit to obtain a satisfactory return on investment. Management pricing actions usually consider advice by the cost estimators, marketing, finance, contracts and production departments before adjusting the internal cost estimate to a final bid price (Fig. 1).

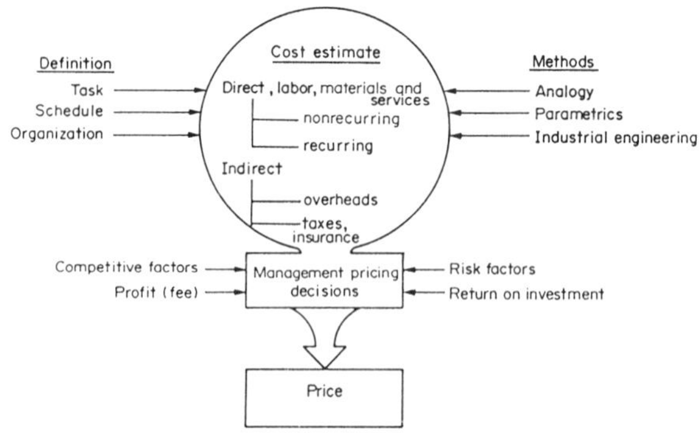

Figure 1
The cost-estimating process and pricing decisions

A price estimate, when accepted by a customer, becomes part of the contract between the parties. The estimate, as submitted to a prospective customer, should contain the specific assumptions and conditions for the performance of the job. The legal aspects of an estimate require careful attention. Records of each estimate must be maintained for possible later challenges, particularly for government contracts.

Most industrial firms retain specialists to prepare cost estimates. Some firms assign these specialists to the functional departments, such as manufacturing, engineering and purchasing. A preferable organizational arrangement uses a centralized estimating department which prepares cost estimates for the entire firm. Such an arrangement tends to enhance the expertise and flexibility of the estimating process and is recommended.

2. *Limitations of Cost Estimating*

A statement of assumptions and conditions usually accompanies a cost estimate in addition to the task statement, schedule and organizational arrangement, and is an integral part of that estimate. These express the limitations on that estimate and may include such factors as:

(a) economic conditions: assumptions of the monetary inflation in the general economy (and any adjustments to be made if actual inflation rates differ from the assumed rate); assumptions on the international currency exchange rates; interest rate levels and the availability of credit; discount rates used;

(b) costs that are excluded from the estimate: customer supplied materials or services; governmental taxes, fees or charges;

(c) the calculation of future changes to the task statement or schedule, accompanied with methods of establishing the price for such changes;

(d) capital equipment and facility costs, their definition and whether these are part of the contract; disposition of such equipment after completion of the contract;

(e) situations that may invalidate the estimate, such as wars, insurrections, strikes and acts of God;

(f) data sources and references (e.g., learning curve, experience basis); and

(g) any changes or additions necessary to clarify the statement of work or schedule for estimating purposes.

These assumptions and conditions constitute the limitations to the estimate as any changes to the assumptions and conditions warrant changes to the estimate. These limitations may be incorporated as part of a resultant contract; they should be precisely worded. Further, most estimators will decline to quantify the effect of violating the assumptions and conditions. They avoid such quantification because the effect of a war or of finding unexpected rock strata in a foundation site cannot be evaluated in advance. After the fact, the impact of changes in the assumptions and conditions must be quantified in order to evaluate the validity of the original estimate.

3. *Accuracy of Cost Estimates*

The difference between the cost estimate and the actual costs experienced measures the accuracy of the cost estimate, excluding any violation of the assumptions and conditions of the estimate or other changes:

$$\text{Estimating error} = \text{estimated cost} - \text{actual cost}$$

Firms organized for profit, as noted earlier, set their price by the cost estimate plus an allowance for profit. Errors in the cost estimate can, therefore, affect the actual realized profit. For example, a 5% error in the cost estimate and, consequently, in the derived price can change an intended 10% profit to either a 5% or a 15% profit depending on the direction of the error. Note that a ±5% estimating error results in a three to one change in profit. Not surprisingly, therefore, entrepreneurs value accurate cost estimates highly.

The motivation for accurate cost estimates has resulted in remarkably dependable estimating procedures. For most routine industrial jobs, cost estimates are dependable to within 2% (though even this accuracy changes profit considerably). Research and development tasks, including design and test jobs, tend to have larger variances, in the 5–10% range. For large, complex, novel jobs, cost estimates tend to be less accurate. A large new pipeline through hostile terrain using innovative technology, for example, might be expected to have significant errors (more than 10%) in the cost estimate. If some of the assumptions and conditions, for example with regard to the organization of the workers, were changed, then even greater variances (as high as ±50%) could be expected in the actual costs from the cost estimate. Unfortunately, the errors usually result in higher actual costs than the estimated costs.

The cost estimating method and the time and effort expended on the cost estimate seem to influence the accuracy of the estimate. Conventional wisdom holds that the industrial engineering method, which requires the most time and effort of the three estimating methods, is the most accurate. With a well-defined and well-understood task, this is generally true. For more novel or advanced programs, parametric methods or analogy can yield equal or greater accuracy. Because of this uncertainty, each estimate should include an estimate of accuracy or risk.

Quantitative risk estimation has been of growing interest in recent years with many proposed methods, none of which have general acceptance as yet. Despite this, the effort expended to estimate risk usually improves the basic estimate, and so makes the risk estimation procedure worthwhile. Cost risk includes:

(a) the possible error in the cost estimating process,

(b) possible errors in the schedule and their impact on costs, and

(c) technical risks and their impact on costs.

A probability distribution can express the possible variations in cost. Decision tree analysis provides a method of deriving such distributions.

Acknowledgement

The publishers wish to thank Mr E I Tinkham for having reviewed and updated this article.

See also: Life-Cycle Costing

Bibliography

Lee Saylor Inc. 1977 *Current Construction Cost*. Lee Saylor, Walnut Creek, CA
Miller I, Freund J E 1965 *Probability and Statistics for Engineers*. Prentice-Hall, Englewood Cliffs, NJ
Ostwald P F 1974 *Cost Estimating for Engineers and Management*. Prentice-Hall, Englewood Cliffs, NJ
RCA Corporation 1975 *Price System Model (Programmed Review of Information for Costing and Evaluation 1975)*. RCA, Cherry Hill, NJ
Richardson Engineering Services 1977 *General Construction Estimating Standards*. Richardson Engineering Services, Solana Beach, CA
Seldon M R 1979 *Life Cycle Costing: A Better Method of Government Procurement*. Westview, Boulder, CO
Trucker S A 1962 *Cost Estimating and Pricing with Machine Hour Rates*. Prentice-Hall, Englewood Cliffs, NJ

The *Journal of Parametrics*, published by the International Society of Parametric Analysis (ISPA), contains numerous articles on costing through parametrics

M. R. Seldon[†]
[General Dynamics Corporation, Pomona, California, USA]

Critical and Strategic Materials

This article originally appeared in Carr and Herz (1989) and is reprinted here for its historical interest, as the international situation has changed considerably since its initial publication.

The terms critical materials and strategic materials are often used jointly or interchangeably. In general, a material is considered critical by a country if future events involving it threaten to inflict serious damage on the nation. A material that is needed for military purposes is considered strategic, although a strategic material need not be critical, nor a critical material strategic. Assessments of whether materials should be considered critical and/or strategic can vary from country to country. Metals like platinum, manganese and chromium are often considered critical in the USA, but in South Africa, a major source of all three, they are not. This article focuses on the US perspective concerning critical and strategic materials, and it will use the terms critical and strategic jointly. It also concentrates on metals and minerals rather than organic materials such as natural rubber and opium.

US legislation concerning stockpiles of critical and strategic materials defines them as those that: (a) would be needed to supply the military, industrial, and essential civilian needs of the USA during a national emergency; and (b) are not found or produced in the USA in sufficient quantities to meet such a need. This definition is fairly precise, but it leaves considerable room for disagreement on practical grounds. Some materials (such as nickel) are defined as critical and strategic, even though all US requirements during a national emergency could be satisfied by imports from a relatively secure source of supply, such as Canada. Although technically a critical and strategic material, nickel is seldom considered a serious problem for the USA. Criteria are needed for measuring criticality and for comparing it across different materials.

1. Criteria for Evaluating Criticality

Many criteria have been proposed for analyzing the degree of criticality of materials. They can generally be grouped into one of three categories: availability of foreign supplies, availability of domestic supplies, and means for reducing domestic consumption.

1.1 Foreign Supplies

Several issues can be raised concerning foreign supplies of a material that might be considered critical or strategic: how likely is a disruption, how severe might a disruption be, and how long might a disruption last. These general issues are usually analyzed by considering a number of factors.

A major concern is whether production is concentrated in a small number of countries. If production is dominated by a small number of suppliers, the possibility exists for cartelization of the market. This risk is considered quite small for most mineral markets, particularly in the long term. Of greater practical concern is the diversification of supplies. Materials-processing operations can experience disruptions caused by labor unrest, fires, floods, earthquakes and local political or economic problems. If a market is supplied by a large number of small producers, one or more is likely to have production problems at any given time, but the variance of total production will be relatively low. If the market is dominated by just a few producers, disruptions will be more unusual events, but the variance of total production will be relatively high. Markets dominated by a small number of supply sources are therefore of relatively high risk.

The location of these supply sources is also of concern. Some, such as Canada, might be considered relatively safe on the grounds of both political orientation and security of supply routes. A country like Australia might be considered secure politically, but the length of the supply route from Australia to the USA raises some concerns. Supplies of materials from many African countries might be considered insecure on both grounds.

The speed with which foreign producers could increase production during a shortage is also of interest. Most mineral projects have long lead times, in the range of five to ten years for new properties. If adequate reserves are available, existing capacity can be expanded much more quickly, sometimes in as little as one to two years.

1.2 Domestic Supplies

Issues concerning domestic supplies are somewhat different. The availability of domestic raw materials and processing capacity are of major importance, but these two factors must be considered separately. In the USA, some materials (e.g., copper) are primarily mined and processed domestically. Others, like aluminum, are primarily processed domestically from foreign raw materials. Still others, such as cobalt, are primarily imported in metallic form. The ability of the domestic industry to expand during an emergency will depend on the availability of raw materials, the availability of domestic processing capacity, and the speed with which both can be expanded. Expanding domestic processing capabilities is of little benefit, however, if additional raw materials would not be available during a crisis. The costs of new or additional production are important from a policy standpoint, because the cost of producing domestically during an emergency must be compared with the costs of other policies, such as stockpiling.

1.3 Domestic Demand

Demand issues include substitution, conservation, and recycling possibilities, the normal business cycle, and effects of new technologies. In some instances, substitute materials that provide equivalent properties may be readily available with little or no increase in costs. However, this is seldom the case. Potential substitutes are typically either more expensive or provide somewhat inferior properties, and they almost always require time to develop and/or implement. These lags can be as long as ten years, depending on the amount of development required. Substitutions can be either direct or functional. Consider, for example, the use of cobalt as a binder in tungsten carbide cutting materials. A direct substitution might involve replacing some of the cobalt with nickel, while a functional substitution might involve replacing a machined part with a forged part.

Conservation possibilities also tend to take time to explore and implement, and usually entail either inferior properties or higher costs. For example, the amount of manganese used in carbon steel might be reduced in several ways. One approach would involve more precise control of the manganese ferroalloy additions to the steel furnace, which would permit the steelmaker to aim for the low end of the acceptable manganese range, rather than the midpoint. This increases the costs of making the steel but does not affect the steel's properties, and could be accomplished rather quickly. Another approach would be to lower the manganese specification for steels in which the manganese is used primarily as a desulfurization rather than alloying agent. This would affect the properties of the steel, but not unacceptably so in many applications. Educating steel consumers to use lower quality steel where acceptable could take some time. A much more expensive, time-consuming approach to conserving manganese might involve recycling steel furnace slags. This approach conserves manganese by changing processing methods, rather than by just using less with the same processing methods. For some materials, the conservation allowed by more material-efficient (but more expensive) processing and fabrication methods can be substantial.

Recycling, if technically and economically feasible, can be an effective way to reduce domestic demand during an emergency. Recycling of some materials, like platinum, is extensive even during normal market conditions. The situation can be very different for other materials, however. For example, prior to 1978 virtually no superalloy scrap, except some of that generated by the superalloy producers themselves, was recycled. Technically, recycling mixed superalloy scrap is at best marginally feasible, and in the past the costs of segregating and storing scrap by alloy type exceeded the value of the scrap. When cobalt prices rose to very high levels in 1978 and 1979, segregating the scrap by type of superalloy became economically feasible, and recycling of scrap generated in fabrication became quite common. Recycling of old jet engine parts made of superalloys, however, was still minimal because of technical and economic problems.

Variations in demand over the business cycle can also cause problems for consumers of materials. For example, the aerospace industry has occasionally experienced large surges in demand. This has led to corresponding increases in the demand for titanium castings and forgings, sometimes outstripping the capacity of the industry. Analysis of issues concerning the demand for critical and strategic materials must cover the possibility that supply emergencies could coincide with upswings in the business cycle.

Another potential problem with demand is the emergence of new technologies that require large amounts of materials. If these technologies emerge suddenly, shortages can develop. Usually, however, firms assure themselves of a steady supply of raw materials before they commit themselves to new products. One example of this is the use of platinum in automobile catalytic converters in the USA. Before the US automobile manufacturers committed themselves to this product, they negotiated contracts with the South African platinum producers to assure themselves of continued supplies, and even promoted capacity expansions by the smaller producers to diversify their supplies somewhat. Catalytic converters are now the largest single use of new platinum in the USA.

2. *Methods for Ranking Criticality*

After all the factors important in analyzing criticality have been analyzed, methods for ranking the relative importance of different materials are needed. A number of approaches have been suggested, all of which have drawbacks.

The simplest approach is to rank materials by import dependence, the percent of normal demand that is met through imports. Although it is relatively easy to obtain data and use import dependence as a measure of criticality, the approach ignores so many important issues that it is generally useful only for rough comparisons across materials. For instance, this method does not account for scrap as a source of supply or changes in governmental stockpiles in estimating import dependence.

Purely judgemental approaches have been used at times to rank materials by their degree of criticality. The US Congress, Office of Technology Assessment (1976) used such an approach in a study of stockpiling policies. The OTA study relied on a consensus of expert opinion to compare materials. Judgemental methods may be acceptable for simply ranking materials by their perceived degree of criticality, but they can be criticized for not providing numerical estimates of the degree to which some materials can be considered more critical than others.

Modified judgemental methods have been proposed to produce numeric indices of criticality. In these studies, such as those by King and Cameron (1974), King (1977) and Hughes (1975), a number of important factors were identified that affect criticality and, for each material, numeric scores (on arbitrary scales) were assigned to each factor. Summing the scores across factors yields a criticality index for each material. This approach is an improvement on the simple judgemental methods, but the arbitrary numeric indices are difficult to estimate and interpret.

One method for measuring the criticality of materials in monetary terms involves using input–output models of a national economy. An input–output model can be used to estimate the change in an aggregate measure of the economy, such as gross national product (GNP), that arises when imports of a material deviate from some normal or base level (Levine and Yabroff 1975).

Multiplying the monetary measure of the costs of a disruption in supplies by the probability that such a disruption will occur yields an estimate of the expected costs of a disruption for a material. This expected disruption cost can be used as a criticality index. This approach requires much more information than the judgemental and relative weighting approaches, but it makes explicit, visible assumptions concerning important issues, and it leads to monetary measures of criticality that can be used directly in performing policy analyses. The input–output approach is probably best suited for studying short-run criticality issues, because input–output models are ill suited for reflecting inventory, supply and demand responses to shortages.

The market model approach to measuring criticality overcomes these objections to the input–output approach, but it requires even more information. A market model for a material can be constructed using engineering and economic information on domestic and foreign supply, demand and inventory behavior. The model can be used to produce forecasts of market conditions during both normal times and a variety of market emergencies. Measures of economic losses during these emergencies can be constructed, usually relying on the concept of economic surplus (consumer surplus and producer surplus). As with the input–output approach, the measures of economic loss are weighted by the disruption probabilities to estimate a monetary index of criticality. The major drawback of this approach is its large information requirements. Detailed market models are needed for each of the materials to be studied.

3. Policy Options

Many policies have been implemented or proposed to deal with perceived problems of critical and strategic materials. Stockpiling has been the most commonly followed practice. Other policies have generally been designed to reduce domestic consumption or increase domestic production.

In most cases stockpiling has proved to be the most cost-effective policy in providing supplies for potential emergencies. The carrying costs for large stockpiles can be quite high, however, even if storage costs are negligible, so other options have been investigated regularly.

The other policies most commonly implemented in the USA have been designed to encourage domestic production. Such policies have included price guarantees for firms considering whether to develop domestic capacity, guaranteed loans, and direct capital subsidies. Federally funded research programs to investigate methods for making use of low-grade domestic resources have been quite common.

Policies directed at reducing consumption or increasing recycling have primarily involved research and development, and information programs. Although never implemented, direct taxes on primary consumption would be one method to encourage recycling and discourage primary consumption.

Some policies, such as import tariffs and import quotas, affect both consumption and production. Their effects are similar: both raise the domestic price of a material, relative to the world price. This encourages recycling and domestic production and discourages domestic consumption. Import tariffs and quotas have often been imposed on industries, but usually for purely protective reasons. The US oil import quota in the 1960s was, however, called a strategic policy.

4. Markets for Chromium, Manganese and Cobalt

The markets for chromium, manganese and cobalt are of concern to policymakers in some developed countries because they are considered to be essential materials for some defense applications and because supplies are highly concentrated in a few potentially unstable or hostile countries. Moreover, most developed countries import more than 95% of their supplies of these materials.

Before discussing these markets, however, it is useful to consider briefly what is meant by the terms requirements and demand for materials.

The term demand refers to the quantity of consumption that exists at varying levels of price, other factors being constant. It is usually expected that the demand for a material is inversely correlated with its price and directly proportional to the level of economic activity in the industrial sectors that use the material. If prices rise, demand is expected to decrease if economic activity and other factors remain unchanged. Requirements refer to the level of consumption of a material by domestic customers that is expected to occur based on technological and market growth assumptions. the concept of requirements thus implicitly assumes that the price elasticity of demand is small enough to be neglected.

Much of the debate concerning critical and strategic materials has focused on projections of the requirements for these materials and not on an analysis of demand. However, it is the demand analysis that is most useful for evaluating future materials consumption. It is desirable to evaluate the demand for these materials this year and in the future if prices rise (e.g., because of a supply disruption).

4.1 Chromium

Chromium production is highly concentrated in a few countries. The most important supplier to the market economies of the world is South Africa, which is of concern to some because of the potential for political disruptions in the future. South Africa accounted for about 32% of total world production in 1987. The second most important supplier to the West in the 1980s was the USSR, which accounted for 29% of total world production in 1987. Other important western producers have been Zimbabwe, the Philippines and Turkey, all of which have experienced severe financial or political difficulties in recent years. This suggests that a moderate disruption in chromium supplies is likely to occur within the next decade or so, and a severe disruption is also possible.

Metallurgical grade chromite ores are found primarily in the USSR, Turkey and Zimbabwe. Although South African production is of a lower grade, most useful for chemical uses, technological developments have made it possible to use these ores in metallurgical applications. In the 1970s, South Africa gradually assumed an increasingly important position in the world market. The chromite ores found in the Philippines are primarily useful for refractory applications.

In the developed world chromium is used for three main applications: production of metallic alloys, chemicals and refractories. In the USA, about two thirds of the chromium consumed in a typical year is in the form of ferrochromium for metallurgical applications. In 1987, 79% of this was used for the production of stainless steels, and 18% was consumed in alloy steels; 3% was used in superalloys. Most of the chromium used by the steel industry is in the form of high-carbon ferrochromium, which cannnot be used in superalloys because of requirements for high-purity materials.

It is often stated that there is no substitute for chromium in the production of stainless steels and superalloys. For instance, it is estimated that only 10% of current superalloy usage, which accounts for 1–2% of total chromium consumption in the USA, could be substituted for, even at very high prices (e.g., 50 times current chromite prices). Moreover, since most of these materials are used in aircraft and industrial gas turbine applications, which require long testing and qualification programs before new alloys can be used, there is a 3–5 year lag time before a significant amount of the substitutions could be accomplished. Moreover, only a limited amount of substitution for stainless steels by other materials would occur, even at higher prices. More importantly, however, there is opportunity for considerable substitution among grades of stainless steels (e.g., ferritic grades for austenitic grades) and for conservation in the production processes. Tables 1–3 summarize the results of a study of the potential for various substitution and conservation options at prices that are various multiples of a base price. The prices are expressed in terms of the form of the material that is actually consumed in one application, and therefore the price elasticity of demand cannot be estimated in terms of equivalent prices. However, it has been shown that a twofold increase in the price of chromite would lead to an increase of ~6% in the cost of producing an austenitic stainless steel. There is considerably more elasticity of demand with respect to price in refractory applications than in stainless steel applications.

4.2 Manganese

Manganese is used as a bulk ferroalloy even more than chromium. In the USA over 90% of the manganese consumed is used by the metallurgical industry, mostly in carbon steels. Alloy steels, particularly the Hadfield steels, use more manganese per ton of steel than do

Table 1
Stainless steel consumption relative to base consumption at different stainless steel price levels

Price relative to base price	Years after price increase 0	1	2	3	4
1.00	1.00	1.00	1.00	1.00	1.00
1.10	0.97	0.96	0.95	0.94	0.93
1.25	0.92	0.88	0.86	0.84	0.83
1.50	0.83	0.76	0.70	0.68	0.66
2.00	0.75	0.66	0.60	0.55	0.51
3.00	0.60	0.51	0.45	0.40	0.37

Table 2
Chromium consumption in alloy steels, relative to base consumption at different ferrochromium price levels

Price relative to base price	Years after price increase								
	0	1	2	3	4	5	6	7	8
1.00	1.00	1.00	1.00	1.00	1.00	1.00	1.00	1.00	1.00
1.67	0.92	0.92	0.92	0.92	0.92	0.92	0.92	0.92	0.92
2.67	0.80	0.79	0.77	0.76	0.74	0.72	0.70	0.69	0.67
4.33	0.75	0.72	0.70	0.67	0.63	0.60	0.55	0.51	0.47
9.33	0.72	0.66	0.60	0.55	0.50	0.45	0.36	0.29	0.23
17.67	0.72	0.65	0.58	0.52	0.47	0.42	0.31	0.23	0.17

carbon steels, but the much higher production levels for carbon steels make them the most important market for manganese. Manganese is an essential input in modern steelmaking, for desulfurization, which is why manganese is of vital concern to all steel producers.

Most manganese is consumed in the form of standard high-carbon ferromanganese, which contains about 78% manganese and 5% carbon. Standard ferromanganese is well suited for use in producing carbon steel, but alloy steels and other specialized metallurgical products often require more refined forms. Medium- and low-carbon ferromanganese directly follow standard ferromanganese in importance, and electrolytic manganese has only a very small market. Manganese ore is used directly by the chemical industry, particularly for use in dry cell batteries.

The world's largest producer of manganese is the USSR, but East–West trade is minimal. The most important producer in the noncommunist world, from the standpoint of both current output and reserves, is South Africa. In 1987 it accounted for 26% of noncommunist supply, and 62% of noncommunist reserves. Brazil, Australia and Gabon are all important, accounting for 54% of noncommunist supply. Other important producers include Mexico and India. The USA currently produces no manganese.

Manganese reserves (1987) are estimated to be of the order of 1.0×10^9 t, with South Africa and the USSR accounting for about 40% and 32% of the total, respectively. There are considerable additional tonnages of land-based manganese deposits that could potentially be recovered at higher prices, but these are also located primarily in South Africa and the USSR.

There is not much concern about short-term problems with manganese supplies because the leading free-market producer—South Africa—accounts for only 13.5% of total production. Moreover, the other major producers (Gabon, Australia, Brazil and India) account for >10% of noncommunist output each. Therefore, a supply disruption in any individual country would not be of great consequence. The major concern is over the potential for long-term problems. Of particular concern is that developed market economy countries may be dependent upon only two major suppliers of manganese (South Africa and the USSR) by the year 2000. The potential situation may not be as troublesome as some fear, however; there are conservation and substitution options available to consumers in developed countries if manganese prices escalate. For instance, it has been estimated that if the price of manganese ore doubles, consumption of manganese by the carbon steel industry in the USA would decline by about 10% in one year and by approximately 24% at the end of five years (Table 4). A tenfold increase in the price of ore would result in a 15% decrease in consumption in the first year and a decline of almost 50% by the end of seven years. Similar decreases in the consumption of manganese in other industrial sectors and in other geographical regions would be expected at higher prices.

Table 3
US chromite consumption in refractories relative to base consumption at different chromite price levels

Price relative to base price	Years after price increase			
	0	1	2	3
1	1.00	1.00	1.00	1.00
2	0.70	0.59	0.48	0.35
5	0.49	0.44	0.36	0.24
10	0.46	0.38	0.29	0.19
25	0.45	0.35	0.24	0.13
50	0.45	0.33	0.22	0.10

4.3 Cobalt

Cobalt is considered by many to be both a strategic and a critical material for the USA; strategic because of its use as a high-temperature material in jet engines and industrial gas turbines, and as a catalyst for petroleum desulfurization; critical, because the USA imports virtually all of its supply of primary cobalt, and a large percentage of these imports comes from countries with potentially unstable political environments.

Table 4
Manganese demand in carbon steel production

Price factor[a]	Time lag (years) 0	2	4	7
USA (Base steel production = 127 630 000 short tons)				
1 ×	810450			
2 ×	750000	725000	700000	619000
5 ×	680000	640000	595000	505000
10 ×	635000	585000	515000	420000
25 ×	615000	500000	440000	330000
50 ×	600000	490000	420000	325000
Japan (Base steel production = 110 135 000 short tons)				
1 ×	610147			
2 ×	585000	574000	558000	544000
5 ×	572000	512000	480000	462000
10 ×	556000	482000	426000	405000
25 ×	550000	465000	410000	263000
50 ×	544866	461465	406398	257715
EEC (Base steel production = 155 058 000 short tons)				
1 ×	1052840			
2 ×	981000	954000	924000	883000
5 ×	929000	852000	815000	766000
10 ×	906000	810000	751000	688000
25 ×	890000	770000	695000	485000
50 ×	882800	765987	688458	479129
Other (Base steel production = 92 734 000 short tons)				
1 ×	634000			
2 ×	612000	594000	578000	532000
5 ×	577000	524000	495000	440000
10 ×	558000	480000	444000	398000
25 ×	540000	450000	415000	330000
50 ×	532293	439559	397829	263365

a Base manganese price = 55 US cents kg^{-1}

Zaire is the dominant producer in the cobalt market, accounting for 50–60% of world production in most years. Cobalt is a coproduct with copper in Zaire, and serious declines in production occurred from 1975 through 1978. In Zambia, cobalt is a by-product of copper and production declined after 1975 as copper prices fell. In Canada, the Philippines, Australia, Indonesia, New Caledonia and Botswana, cobalt is recovered as a by-product of nickel mining and processing operations. Disposals of cobalt from the US strategic stockpile played an important role in the cobalt market in the early 1970s, but sales of cobalt ended in 1976.

Zaire has experienced severe problems in supplying cobalt in recent years. In 1975, the Benguela railroad from Zaire through Angola to the Atlantic was cut off due to the civil war in Angola. In 1977, the Shaba province of Zaire, where cobalt is produced, was invaded. Cobalt production was not affected that year, but the invasion in 1978 did affect production. The damage to the facilities in 1978 was slight, but virtually all of the foreign workers fled, and many facilities were in a state of poor repair. Low copper prices have left Zaire almost bankrupt, unable to improve its domestic transportation network or to maintain the copper and cobalt facilities. The International Monetary Fund and the World Bank have been attempting to remedy Zaire's financial difficulties, but it remains to be seen if their attempts will prove successful. It is possible that Zaire could just manage for the next several years until the copper market recovers and the efforts by the IMF and World Bank bear fruit. Despite attempts by Zaire to improve its economy and mining industry, another cutoff in supplies, although unlikely, could happen at any time.

The three most important uses of cobalt on a tonnage basis in the USA are in superalloys, as an alloying element in permanent magnets and in chemicals. Other important uses include cemented carbides, hard-facing and welding materials and as an alloying element in steels.

A recent study of the world cobalt market found that there is considerable price elasticity in the demand for cobalt, because, as the price rises, consumers shift their preferences to alternative materials and designs. Furthermore, the increase in price encourages conservation of cobalt through increased recycling and improvements in processing techniques. For instance, the study showed that if the price of cobalt were to stabilize at US\$55 kg^{-1} (1978) the demand for cobalt would decline to less than half of its initial value within five years. The short-term response would be less, but there would still be a decrease of approximately 15% in the total demand for cobalt within one year if the price were to increase from US\$22 to US\$55 kg^{-1}. The results of the substitution analysis, which was developed in 1978, have undergone the test of time well, and were used as part of an overall supply–demand framework to predict, with considerable accuracy, world cobalt prices over the period 1978–1982.

The results of the substitution analysis are crucial to an understanding of the economic and national security consequences to the USA of a supply disruption in the world cobalt market. It is conceivable, even likely, that another supply disruption will occur because of the political instability in southern and central Africa. In such an event, it is clear that the price of cobalt will again rise as the result of psychological factors and actual production curtailments. The magnitude of the price increase is uncertain, but it is quite unlikely to increase above US\$110 kg^{-1} for any appreciable period because of the downward pressure from substitution and additional supplies from other nations. The results of the analysis may be interpreted as follows. If the price were to increase by a factor of 2.5, and other parameters (such as industrial activity and prices of substitutes) were to remain unchanged, the total demand for cobalt in the USA would decrease by 15% in a one-year period. A price increase by a factor of five would bring a demand decrease of 20% in the first year. Currently, about 15% of cobalt supplies in the USA come from recycled (secondary) sources. Thus even if supplies were restricted by 50% in the first year of the disruption only up to 20% of demand would have to be met by releases from the government stockpile. The greater the price, the less the demand and the greater the contribution from recycled material.

If the supply restriction were to last for more than a year, demand would have more time to adjust and there would be more substitution. Thus the overall demand for cobalt in the USA would decrease by 50% in years two through seven, with a price increase factor of 2.5, if other factors were to remain unchanged. In the event of larger price increases, substitution would be greater. If the disruption were to continue, there would also be time for additional sources of supply to respond to the higher price.

Indeed, such responses have been shown by both the demand and supply sides of the market to the 1978 disruption. Following the disruption, the producer price was set at US\$55 kg^{-1} and an attempt was made to maintain the price at that level. However, because of substitution and increased supply from other sources outside Zaire and Zambia, large stocks were built up by producers, and the free market price declined to less than US\$22 kg^{-1} in 1982. Consequently, extreme pressure was placed on the price, which gradually decreased to about US\$11 kg^{-1} in early 1983, before rebounding to about US\$25 kg^{-1} in late 1984. The price declined to US\$11.40 kg^{-1} in 1987.

Two conclusions may be drawn from this analysis. First, a disruption in cobalt supplies from Zambia and/or Zaire would not have a measurable effect on the national security of the USA. The increased prices would result in substantial substitution away from cobalt in nonessential uses. However, there would be a time delay, and it might be necessary to release some cobalt from the government stockpile. Second, there would be an economic cost to the USA resulting from a supply disruption because of the loss in consumer surplus due to higher market prices and reduction in quantities consumed. The costs of such a disruption could be substantial but would be short lived because of the downward pressure on the price.

See also: Risk Management; Stockpiling; Trade in Materials, International

Bibliography

Carr D D, Herz N 1989 *Concise Encyclopedia of Mineral Resources.* Pergamon, Oxford

Charles River Associates 1978 *Measuring Materials Criticality for National Policy Analysis*, prepared for the US General Accounting Office. Charles River Associates, Boston, MA

Charles River Associates 1981 *The Effects of Supply Restrictions on the Demand for Manganese, Chromium, Cobalt and Platinum*, report to the US Department of the Interior. Charles River Associates, Boston, MA

Hughes E E 1975 *Strategic Resources and National Security: An Initial Assessment*, prepared for the Defense Advanced Research Projects Agency. Stanford Research Institute, Stanford, CA

King A H 1977 *Materials Vulnerability of the United States—An Update.* US Army War College, Strategic Studies Institute, Carlisle Barracks, PA

King A H, Cameron J R 1974 *Materials and the New Dimensions of Conflict.* US Army War College, Strategic Studies Institute, Carlisle Barracks, PA

Levine M D, Yabroff I W 1975 *Department of Defense Materials Consumption and the Impact of Material and Energy Resource Shortages*, prepared for the Advanced Research Projects Agency. Stanford Research Institute, Stanford, CA

US Bureau of Mines 1988 *Mineral Commodity Summaries 1988.* USBM, Washington, DC

US Congress, Office of Technology Assessment 1976 *An Assessment of Alternative Economic Stockpiling Policies.* US Government Printing Office, Washington, DC

J. P. Clark
[Massachusetts Institute of Technology, Cambridge, Massachusetts, USA]

B. Reddy
[National Economic Research Associates, Cambridge, Massachusetts, USA]

D

Data Sources for Materials Economics, Policy and Management

A manager in a materials enterprise has information needs quite different from those in his likely prior role as a materials technologist and is thus dealing with unfamiliar data sources. In addition to information on materials economics and business matters, he increasingly needs information on regulatory affairs, tariffs and trade, patents, standards and specifications, people and organizations, among other topics. Information gleaned from such sources will facilitate his goal setting, strategic planning, sourcing of materials, and compliance with safety, environmental and trade rules and regulations.

As with any subject area, intelligent and effective use of information resources makes demands on the user to

(a) thoroughly understand the problem under study,

(b) have a clear view of how the information can help,

(c) identify the specific information needed, and

(d) find and select the most appropriate source(s) of information to further his work.

The purpose of this article is to assist in requirement (d). It begins with a review of guides to general sources of technical information on materials and then proceeds to a discussion of topics related to managerial concerns. In a brief article such as this, it is impossible to be exhaustive in the listing of information sources. Rather, an attempt has been made in each area to cite a sufficient number of sources (including the most important ones) to give the reader an idea of what is available and to afford a place of beginning in his search for information. Occasionally it has been found desirable to list a particular source in more than one category when it might otherwise be overlooked.

1. Guides to Technical Information

There are numerous guides and referral centers for technical information, most of which are familiar to any reference librarian. Here the discussion is confined to those particularly relevant to materials. The several guides cited also include reference to other guides of both broader (e.g., all science and technology) and narrower (e.g., corrosion of metals or textile science and engineering) scope.

Ash J E, Hyde E (eds.) 1975 *Chemical Information Systems.* Ellis Horwood, Chichester, UK and Halsted Press, New York

Boodson K 1972 *Non-Ferrous Metals: A Bibliographic Guide.* MacDonald, London

Carr M J, Gavert R B, Moore R L, Westbrook J H 1976 *Critical Surveys of Data Sources: Electrical and Magnetic Properties of Metals*, NBS Special Publication 396-4. National Bureau of Standards, Washington, DC

Carter G C, Bennett L H, Cuthill J R, Kahan D J 1968 *The NBS Alloy Data Center: Function, Bibliographic System, Related Data Centers and Reference Books*, NBS Technical Note 464. National Bureau of Standards, Washington, DC

Chaney J F, Ramdas V (eds.) 1982 *Thermophysical Properties Research Literature Retrieval Guide 1900–1980*, Vols. 1–7. Plenum, New York

Diegle R B, Boyd W K 1976 *Critical Surveys of Data Sources: Corrosion of Metals.* NBS Special Publication 396-3. National Bureau of Standards, Washington, DC

Directory of Engineering Documentation Sources, 3rd edn. 1985 Global Engineering, Santa Ana, CA

Farnfield C 1974 *Guide to Sources of Information in the Textile Industry.* State Mutual, New York

Gavert R B, Moore R L, Westbrook J H 1974 *Critical Surveys of Data Sources: Mechanical Properties of Metals*, NBS Special Publication 396-1. National Bureau of Standards, Washington, DC

Gibson E G, Tapia E W 1965 *Guide to Metallurgical Information*, 2nd edn., SLA Bibliography No. 3. Special Libraries Association, New York

Guide to Materials Engineering Data and Information 1986 American Society for Metals, Materials Park, OH

Hall D M, Achee N Jr, Persons C C 1972 *Guide to Literature on Textile Engineering and Textile Science.* American Society for Engineering Education, Washington, DC

Hyslop M R 1973 *A Brief Guide to Sources of Metals Information.* Information Resources Press, Washington, DC

Johnson D M, Lynch J F 1975 *Critical Surveys of Data Sources: Properties of Ceramics*, NBS Special Publication 396-2. National Bureau of Standards, Washington, DC

Kahan D J, Harris J F, Bennett L H 1983 Compilations of alloy phase diagrams and related data. *Bull. Alloy Phase Diagrams* 3

Kaplan S R (ed.) 1965 *A Guide to Information Sources in Mining, Minerals, and Geosciences.* Wiley, New York

Mailloux E N 1989 Engineering information systems. *Annu. Rev. Inf. Sci. Technol.* 24: 239–66

Milek J T 1951 *Guide to Foreign Sources of Metallurgical Literature.* Richard Rimbach Association, Pittsburgh, PA

Minerals Data Source Directory 1982, Bureau of Mines, Information Circular 8881. US Government Printing Office, Washington, DC

Patten M N (ed.) 1990 *Information Sources in Metallic Materials.* Bowker-Saur, New York

Reynard K W 1992 *UK Materials Information Sources*, 2nd edn. The Design Council, London

Sommar H G 1973 *A Brief Guide to Sources of Fiber and Textile Information.* Information Resources, Arlington, VA

Thornton C H, Harper S, Bowers J E 1983 *A Critical Survey of Available High Temperature Mechanical Property Data for Copper and Copper Alloys*, INCA Monograph 12. International Copper Association, New York

Touloukian Y S, Gerritsen J K, Moore N Y (eds.) 1967

Thermophysical Properties Research Literature Retrieval Guide, Vols. 1–3 (plus 1973, supplement Vols. 1–6). Plenum, New York

Wawrousek H, Westbrook J H, Grattidge W 1986 Data sources of mechanical and physical properties of engineering materials. *Phys. Daten* 30(1) (Fachsinformationszentrum-Karlsruhe)

White D 1970 *How to Find Out in Iron and Steel.* Pergamon, Oxford

Wilcox V L 1970 *Guide to Literature on Metals and Metallurgical Engineering.* American Society for Engineering Education, Washington, DC

Yescombe E R 1968 *Sources of Information on the Rubber, Plastics and Allied Industries.* Pergamon, Oxford

2. *Materials Economics and Business Information*

While this article is directed primarily towards the interests of materials managers, such individuals frequently have need for information outside technology *per se*, in the world of materials business and trade. Answers are sought to such questions as: where can this material be bought? at what price? under what name? in what unit quantities? which are the leading producing companies for this material? what are their addresses? what are the time trends of price and production of various commodities? and so on.

(*a*) *General Guides*

Daniels L M 1985 *Business Information Sources*, revised edition, University of California Press, Berkeley, CA

Larson H M 1964 *Guide to Business History: Materials for the Study of American Business History and Suggestions for Their Use.* Cahners, Newton, MA

Lovett R W 1971 *American Economic and Business History Information Sources.* Gale Research, Detroit, MI

Woy J (ed.) 1988 *Encyclopedia of Business Information Sources*, 7th edn. Gale Research, Detroit, MI

(*b*) *Directories*

Bunsell A R, Kelly A 1985 *Composite Materials: A Directory of European Research.* Butterworth, Guildford, UK

Directory of American Research and Technology, 23rd edn. 1988 Bowker, New York

Directory of Contract Laboratories, 2nd edn. 1988 Dekker, New York

Directory of Steel Foundries 1990 Steel Founders Society of America, Des Plaines, IL

Directory of Testing Laboratories, 1992 American Society for Testing and Materials, Philadelphia, PA.

Dun's Business Rankings 1988 Dun's Marketing, Parsippany, NJ

Dun's Census of American Business 1988 Dun's Marketing, Parsippany, NJ

Dun's Industrial Guide: The Metalworking Directory 1988 Dun's Marketing, Parsippany, NJ

Encyclopedia of American Business History and Biography 1989 Facts on File, New York

European Plastics Directory 1990 Waterlow Directories (in association with RAPRA), London

Ferro-Alloy Directory and Databook, 2nd edn. 1989 Metal Bulletin, New York

Gillemott L F 1970 *Materials Testing Laboratories.* UNIPUB (UNESCO), New York

Harrington J P 1990 *1990–1991 US Foamed Plastics Markets and Directory.* Technomic, Lancaster, PA

International Ceramics Directory. London and Sheffield Publishing, London

International Refractories Handbook and Directory. London and Sheffield Publishing, London

International Trade Names Dictionary 1989 Gale Research, Detroit, MI

Iron and Steel Industry Profiles (ceased publication in 1983) Institute of Metals, London

Kawata S (ed.) (annual) *Japan's Iron and Steel Industry.* Kawata, Tokyo

Lockwood's Directory of the Paper and Allied Trades (annual) Vance, New York

Material Handling Engineering Handbook and Directory (biennial) Compiled by Material Handling Engineering Magazine, Industrial Publishing, Cleveland, OH

Materials Research Centers, 2nd edn. 1986 Gale Research, Detroit, MI

Metal and Mining Industry, Directory of Iron and Steel Works of the United States and Canada, (published every three years) American Iron and Steel Institute, Washington, DC

Metal Finishing, Guidebook and Directory (annual) Metals and Plastics Publications, Hackensack, NJ

Mines Register (biennial) Atlas, New York

Modern Plastics Encyclopedia, A Buyer's Guide to the Plastic Industry (annual) McGraw-Hill, New York

Nelson D 1986 *Mines and Mining Equipment and Service Companies Worldwide, 1986–1987.* Brookfield, VT

Stainless Steel Directory and Handbook. Modern Metals Publishing, London

Textile Bluebook (annual) Davison, Ridgewood, NJ

Thomas' Register of American Manufacturers (annual) Thomas, New York

World Directory: Secondary Lead Plants. International Lead and Zinc Group, London

World Directory: Secondary Zinc Plants. International Lead and Zinc Group, London

(*c*) *Trade Associations*

Koek K et al. (eds.) 1989 (biennial) *Encyclopedia of Associations*, Vols. 1–3. Gale Research, Detroit, MI

Verrel B, Opitz H (1985) *World Guide to Trade Associations*, 3rd edn. Saur, Ridgewood, NJ

Young M L (ed.) 1985 *Scientific and Technical Organizations and Agencies Directory*, Vols. 1, 2. Gale Research, Detroit, MI

Aluminum Association, Inc., 900 19th St NW, Washington, DC 20006. Contact: P Pollack, Tel: (202) 862 5100

American Concrete Institute, PO Box 19150, Detroit, MI 48219

American Die Casting Institute, Des Plaines Ave., Des Plaines, IL 60018

American Iron and Steel Institute, 1133 15th St NW, Washington, DC 20005

American Paper Institute, 260 Madison Avenue, New York, NY 10016

Association of Metal Sprayers, 5 Keats Road, Stratford-upon-Avon, Warwickshire CV37 7JL, UK. Tel: 0789 299661

Association of Plastics Manufacturers in Europe, Box 73, 250 Avenue Louise, 1050 Brussels, Belgium.
British Cement Association, Wexham Spring, Slough, Berkshire SL3 6PL, UK. Tel: 0753 662727
British Investment Casting Trade Association, Bordesley Hall, Alvechurch, Birmingham B48 1QA, UK. Tel: 0527-584770
Cast Steel Technology Associates, Inc., Steel Founders Society of America, 455 State St, Des Plaines, IL 60016. Tel: (312) 299 9160
Copper Development Association, Orchard House, Mutton Lane, Potters Bar, Hertfordshire EN6 3AP, UK. Tel: 0707 50711. Also: Box 1840, Greenwich, CT 06836 (17 national offices, worldwide)
Gold Institute, Suite 1140, 1001 Connecticut Avenue NW, Washington, DC 20036
International Copper Association (ICA), 160 Madison Ave., 16th Floor, New York, NY 10016. Tel: (212) 251 7240
International Lead Zinc Research Organization, Research Triangle Park, NC 27709-2036
Lead Industries Association, 292 Madison Avenue, New York, NY 10017
Material Handling Institute, 1326 Freeport Road, Pittsburgh, PA 15238
Metal Powder Industries Federation, 105 College Road East, Princeton, NJ 08540
Metal Treating Institute, Suite 200, 1311 Executive Center Drive, Tallahassee, FL 32301
Methods and Materials Standards Association, c/o H B Fuller Co., 315 South Hicks Road, Palatine, IL 60067
Mineral Insulation Manufacturers Association, 182 Springfield Ave., Summit, NJ 07901
Refractories Institute, Suite 1517, 301 Fifth Ave., Pittsburgh, PA 15222
Silver Institute, 1001 Connecticut Ave. NW, Washington, DC 20036
Society of the Plastics Industry, 355 Lexington Avenue, New York, NY 10017
Technical Association of the Pulp and Paper Industry, Technology Park/Atlanta, PO Box 105113, Atlanta, GA 30348
Zinc Institute, 292 Madison Ave., New York, NY 10017

(*d*) *On-Line Business Information*

Materials Business File (Jan 1985–) Institute of Materials, London
Plasnews (plastics business information), available from Science and Technology Network (STN) Columbus, OH, Karlsruhe, Germany, and Tokyo
RAPRA Tradenames, available on Orbit

(*e*) *Statistics*

AISI Annual Statistical Report (annual) American Iron and Steel Institute, Washington, DC
Aluminum Statistical Review (annual) Aluminum Association, Washington, DC.
Annual Census of Production (annual, covers UK only) Central Statistical Office, Great George Street, London SW1. Tel: 071 233 3000
Bruchey S (ed.) *Mineral Foreign Trade of the United States in the Twentieth Century: A Study in Mineral Economics.* Ayer, Salem, NH
Business and Trade Statistics Ltd, Lancaster House, More Lane, Esher KT10 8AP, UK. Tel: 0372 463121
Business Statistics (biennial) US Department of Commerce, Government Printing Office, Washington, DC
Chemical and Engineering News (weekly, the first June issue contains the annual *Facts and Figures for the Chemical Industry*). American Chemical Society, Washington, DC
Chemical Statistics Handbook 1991 (quinquennial with annual *Statistical Summary* in intervening years) Manufacturing Chemists Association, Washington, DC
Curry S, Rich S (eds.) 1980 *The Kline Guide to the Chemical Industry*, 4th edn. Kline, Fairfield, NJ
Handbook of Basic Economic Statistics (annual, with monthly supplements) Economic Statistics Bureau, Washington, DC
Huber J E (ed.) 1979 *The Kline Guide to the Paper Industry*, 4th edn. Kline, Fairfield, NJ
Iron Age (weekly, the *Annual Statistical Review* is in the first January issue) Chilton, Radnor, PA
Kawata S (ed.) (annual) *Japan's Iron and Steel Industry.* Kawata, Tokyo
Knoerr A W (ed.) 1988 (quinquennially) *Mineral Facts and Problems.* US Bureau of Mines, Government Printing Office, Washington, DC
Kollonitsch V, Dietsch M (eds.) *The Kline Guide to the Plastics Industry*, 2nd edn. Kline, Fairfield, NJ
Lead and Zinc Statistics (monthly) World Bureau of Metal Statistics, Ware, Hertfordshire SG12 9BA, UK. Tel: 0920 461274
Metal Statistics (annual) American Metal Market, Fairfield, New York
Mineral Information Directorate, US Bureau of Mines, Washington, DC (data center for production and consumption of nonfuel minerals). Contact: D S Colby, Tel: (202) 634 1187
Minerals Handbook 1986–1987 (statistics and analysis of the world's mineral industries) Stockville/Macmillan, New York and London
Minerals Yearbook, Vols. 1–3 (annual) US Bureau of Mines, Government Printing Office, Washington, DC
Misra K C 1986 *Mineral and Energy Resources: Current Status and Future Trends.* University of Tennessee Press, Knoxville, TN
Non-Ferrous Metal Data (annual) American Bureau of Metal Statistics, New Jersey. Tel: (201) 863 6900
Roskill's Metals Databook, 7th edn. 1991 (prices, production and producers of 32 metals worldwide). Metallurgical Society of AIME, Warrendale, PA
Rubber Industry Facts (annual) Rubber Manufacturers Association, New York
Statistical Abstract of the United States (annual) US Bureau of the Census, Government Printing Office, Washington, DC
Statistics of Paper and Paperboard (annual) American Paper Institute, New York
Statistics of World Trade in Steel 1984 1986 United Nations, New York
Textile Organon (monthly, the Jan/Feb issue is the annual statistical summary) Textile Economics Bureau, New York
Tin Statistics (annual) UNCTAD, Geneva
Trends in Production of Lead and Zinc, World Bureau of Metal Statistics, Ware, Hertfordshire SG12 9BA, UK. Tel: 0920 461274
United Nations Statistical Yearbook (annual) United Nations, New York (also publishes monthly *Bulletin of Statistics*)
US Industrial Outlook (annual) US Department of Commerce, Government Printing Office, Washington, DC

Eurostat, Statistical Office of the European Communities, Batiment Jean Monnet, L-2920, Luxembourg. Tel: 352 4301 4567. Eurostat does not collect statistical data itself but obtains them from the EC member states. Selected data from the EC's major trading partners are also used for purposes of comparison. Publications include: *Basic Statistics of the Community*, *Eurostatistics* (monthly) and *Regiostat*, a selection of socioeconomic data for regions of the EC.

(*f*) *Purchasing, Prices and Sourcing of Materials*

American Metal Market (5 per week) Fairchild, New York
Basil D C et al. (eds.) 1977 *Purchasing Information Sources*. Gale Research, Detroit, MI
Brady G S, Clauser H R 1991 *Materials Handbook: An Encyclopedia for Managers, Technical Professionals, etc.* 13th edn. McGraw-Hill, New York
Ceramic Sources (annual) American Ceramic Society, Westerville, OH
Chemical Economics Handbook 1962–1984 Stanford Research Institute, Palo Alto, CA
Frick J (ed.) 1990 *Woldman's Engineering Alloys*, 7th edn. American Society for Metals, Materials Park, OH
Grant A H *Guide to Buying Castings*. State Mutual, New York
Leenders et al. (eds.) 1989 *Purchasing and Materials Management*, 9th edn. Irwin, Homewood, IL
Product Source (annual) R&D Magazine, Cahners, Newton, MA
Roskill's Metals Databook, 7th edn. 1991 (prices, production and producers of 32 metals worldwide) Metallurgical Society of AIME, Warrendale, PA
World Market Costs (irregular, market forecasts) Predicasts, Cleveland, OH

(*g*) *Market Studies and News*

Allen P 1985 *World Market for High Technology Industrial Ceramics*. FIND SVP (an information service, largely computerized), Information Clearing House Inc., New York
Corey A R 1956 *The Development of Markets for New Materials*. Harvard University Press, Cambridge, MA
Frost and Sullivan (market reports and market research) Sullivan House, 4 Grosvenor Gardens, London, SW1D 0DH. Tel: 071 730 3438
Guide to Export Controls (annual) Business Laws, Chesterland, OH
Materials Business File (an on-line version of *Steels Alert*, *Non-Ferrous Metals Alert* and *Polymers/Ceramics/Composites Alert*) Institute of Materials, London. Contact: Mr W G Jackson, Materials Information Manager
Metallurgical Industry Consumers Market 1984 Frost and Sullivan, London
Metra Martech (market reports and market research) Glenthorne House, Hammersmith Grove, London W6 0LG, UK. Tel: 081 563 0666
Predicast Europe (database with broad international coverage of all manufacturing and service industries with information on products, markets, technology, etc.) 8–10 Denman Street, London, W1V 7RF. Tel: 071 494 3817
Robbins P 1982 *Guide to Non-Ferrous Metals and Their Markets*. Nicholls, New York

(*h*) *Investment Data*

Moody's Industrial Manual, Vols. 1–6. Moody's Investor Service, New York
Standard Corporation Records, Vols. 1–6 (looseleaf with bimonthly supplements) Standard and Poor's Corporation, New York
Value Line Ratings and Reports (weekly)

3. *Materials Policy*

Various events since the early 1970s, both nationally and internationally, have intensified attention on materials policy. The issues involved include materials shortages, materials and the environment, materials and energy, materials and defense, international technological competition and the contributions of materials to the quality of life. All of these topics and related areas have been vigorously discussed and analyzed in various conferences and specially commissioned studies, a sampling of which is given below. The primary organizations which have been active in this field in the USA include the National Materials Advisory Board and the Science and Technology Information Board (formerly NDAB) of the National Academy of Sciences, the Engineering Foundation, the Federation of Materials Societies, the Office of Technology Assessment of the Congress, and the National Commission on Supplies and Shortages. Internationally, such organizations as OECD, IAEA, CODATA, NATO–AGARD, UNESCO and the World Federation of Engineering Organizations have been similarly involved.

(*a*) *Conferences*

Federation of Materials Societies 1977 *Materials and National Issues*. Federation of Materials Societies
Federation of Materials Societies 1979 *Materials Aspects of World Energy Needs*. Federation of Materials Societies
National Academy of Sciences/National Academy of Engineering 1975 *National Materials Policy*, Proceedings of a Joint Meeting of the National Academy of Sciences and the National Academy of Engineering, 25–26 October 1973. NAS/NAE, Washington, DC.

The following are from the Henniker Conference Series:

US Library of Congress 1970 *Problems and Issues of a National Materials Policy*, Henniker I. US Government Printing Office, Washington, DC
US Library of Congress 1973 *Resolving Some Selected Issues of a National Materials Policy*, Henniker II. US Government Printing Office, Washington, DC
US Office of Technology Assessment 1975 *Requirements for Fulfilling a National Materials Policy*, Henniker III. US Government Printing Office, Washington, DC
US Office of Technology Assessment 1977 *Engineering Implications of a Chronic Materials Scarcity*, Henniker IV. US Government Printing Office, Washington, DC
US Office of Technology Assessment Congressional Research Service and General Accounting Office 1978 *Building a Consensus on Legislation for a National Materials Policy*, Henniker V. US Government Printing Office, Washington, DC

(*b*) *Reports*

Committee on Science, Engineering and Public Policy 1991 *Finding Common Ground: US Export Controls in a Changed Global Environment*. National Academy Press, Washington, DC
Congressional Research Service 1977 *Materials Policy Handbook*. US Government Printing Office, Washington, DC

National Academy of Sciences, Committee on the Survey of Materials Science and Engineering 1974 *Materials and Man's Needs: Materials Science and Engineering.* National Academy of Sciences, Washington, DC

National Research Council, Commission on Natural Resources, Committee on Mineral Resources and the Environment (COMRATE) 1975 *Mineral Resources and the Environment.* National Academy of Sciences, Washington, DC

US National Commission on Materials Policy 1973 *Materials Needs and the Environment Today and Tomorrow.* US Government Printing Office, Washington, DC

US National Commission on Supplies and Shortages 1978 *Government and the Nation's Resources.* US Government Printing Office, Washington, DC

US Office of Technology Assessment 1976 *An Assessment of Information Systems Capabilities Required to Support US Materials Policy Decisions.* US Government Printing Office, Washington, DC

(c) Books

Altenpohl D G, Daugherty T S 1980 *Materials in World Perspective.* Springer, Berlin

Anderson J C et al. 1981 *Materials Requirements Planning: A Study of Implementation and Practice.* American Production and Inventory Control Society, Falls Church, VA

Beekman-Love G, Neiger L 1978 *Materials Management.* Kluwer, Dordrecht, The Netherlands

Bergsten C F, Cline W R *Trade Policy of the Nineteen Eighties.* Books on Demand, University Microfilms International, Ann Arbor, MI

Compton W D, et al. (eds.) 1988 *The Technological Dimensions of International Competitiveness.* National Academy of Engineering, Washington, DC

Future Metal Strategy 1980 The Metals Society, London

Klass M W et al. 1980 *International Mineral Cartels and Embargoes: Policy Implications for the United States.* Praeger, New York

Materials Science and Engineering for the 1990s: Maintaining Competitiveness in the Age of Materials 1990 National Academy Press, Washington, DC

Rational Use of Potentially Scarce Metals 1976 NATO, Brussels

Szuprowicz B O 1981 *How to Avoid Strategic Materials Shortages (Dealing with Cartels, Embargoes, and Supply Disruption).* Wiley, New York

(d) Journals

Conservation and Recycling 1976–1988 Bever M B, Henstock M E (eds.) Pergamon, Oxford

Materials and Society 1977– Hibbard W R Jr (ed.) Pergamon, Oxford

Resources and Conservation 1975–1987 Alter H (ed.) Elsevier, Amsterdam

Resources, Conservation and Recycling 1988– Alter H, Henstock M E (eds.) Pergamon/Elsevier, Amsterdam

4. Safety and Health

Concerns over worker safety and environmental protection have mounted dramatically in recent years. The resultant legislation as well as moral, ethical and business considerations have placed a new emphasis on access to complete and reliable information on safety and health aspects of materials, not only with respect to workers, but also to all elements of the biosphere. The following is a brief list of representative sources of information in this field, including journals, codes, information centers and databases, as well as print publications.

(a) Journals

Chemical Regulations Reporter (weekly plus a basic reference file) Bureau of National Affairs, Washington, DC

OSHA Safety and Health Reporter (weekly plus a basic reference file) Bureau of National Affairs, Washington, DC

(b) Codes

Boiler and Pressure Vessel Code. American Society of Mechanical Engineers, New York

Code of Federal Regulations

- Title 21: Radiological Health
- Title 29: Occupational Safety and Health
- Title 40: Environmental Protection Administration
- Title 49: Transportation of Hazardous Materials and Pipeline Safety

All available from Superintendent of Documents, US Government Printing Office, Washington, DC

(c) Information Centers and Databases

Advisory Center of Toxicology, National Academy of Sciences, 2101 Constitution Ave., Washington, DC 20418. Tel: (202) 389 6751. Director: R Wands

BDT (Data Bank in Toxicology, Banque de Donnees), INSERM, 78 Rue du General-LeClerc, 94 270 Le Kremlin-Bicetre, Paris, France. Director: Mme R Gerday

Clearinghouse for Occupational Safety and Health Information Center (COSHI), 4676 Columbia Parkway, Cincinnati, OH 45226. Contact: T F Schoenborn, Tel: (513) 684 8326

Hazardous Materials Service, Building 151, Atomic Energy Research Establishment, Harwell, Didcot, UK

International Register of Potentially Toxic Chemicals (IRTC), United Nations Environment Programme, Pavilon du Petit Sacannex, 16 Ave Jean Trembly, 1209 Geneve, Switzerland

NIOSH Registry of Toxic Effects (RTECS). Accessible on-line via Chemical Information Systems, Inc., 7215 York Road, Baltimore, MD 21212. Tel: (301) 321 8440 or (800) CIS-USER

Physical and Chemical Properties of Hazardous Materials, Occupational Health Services, Inc., 515 Madison Ave., New York, NY 10022

Toxic Materials Information Center (TMIC), Oak Ridge National Laboratory, PO Box X, Oak Ridge, TN 37830. Contact: Ms E D Copenhaver, Tel: (615) 483-8611

Toxicology Data Bank, National Library of Medicine, NIH, 9000 Rockville Pike, Bethesda, MD 20014. Contact: H Kissman, Tel: (301) 656 4000

(d) Print Publications

Bretherick L 1985 *Handbook of Reactive Chemical Hazards*, 3rd edn. Butterworth, Boston, MA

Christensen H E, Fairchild E J (eds.) 1986 *Registry of Toxic Effects of Chemical Substances.* US Department of Health and Human Services, Washington, DC

Clayton G D, Clayton F E (eds.) 1985 *Patty's Industrial Hygiene and Toxicology*, 3rd edn., Vols. 1–3. Wiley, New York

Dominguez G S, Bartlett K G 1986 *Hazardous Waste Management*, Vol. 1. CRC Press, Boca Raton, FL

Friberg L, et al. 1980 *Handbook of the Toxicology of Metals.* Elsevier, Amsterdam
Gosselin R E 1984 *Clinical Toxicology of Commercial Products*, 5th edn. Williams and Wilkins, Baltimore, MD
Gross P, Brown D C 1984 *Toxic and Biomedical Effects of Fibers: Asbestos, Talc, Inorganic Fibers, Man-Made Vitreous Fibers and Organic Fibers.* Noyes, Park Ridge, NJ
Haley T J, Berndt W O 1987 *Handbook of Toxicology.* Hemisphere, Washington, DC
Jorgensen S E (ed.) 1979 *Handbook of Environmental Data and Ecological Parameters.* Pergamon, Oxford
Martin W F et al. 1987 *Hazardous Waste Handbook for Health and Safety.* Butterworth, Guildford, UK
Matheson 1971 *Effects of Exposure to Toxic Gases.* Matheson Gas Products, East Rutherford, NJ
Matheson 1971 *Gas Data Book.* Matheson Gas Products, East Rutherford, NJ
National Fire Protection Association 1973 *Hazardous Chemicals Data.* NFPA
National Safety Council 1978 *Industrial Safety Data Sheets.* National Safety Council
Neely, W B, Blau G E 1985 *Environmental Exposure from Chemicals.* CRC Press, Boca Raton, FL
Nielsen J R, Kania C J 1983 *Materials Safety Data Sheet Collection.* General Electric Company, Schenectady, NY
O'Donoghue J L 1985 *Neurotoxicity of Industrial and Commercial Chemicals*, Vols. 1–2. CRC Press, Boca Raton, FL
Sax N R, Feiner B 1984 *Dangerous Properties of Industrial Materials*, 6th edn. Van Nostrand Reinhold, New York
Seiler H G, Sigel H (eds.) 1987 *Handbook on Toxicity of Inorganic Compounds.* Dekker, New York
Sittig M 1985 *Handbook of Toxic and Hazardous Chemicals and Carcinogens*, 2nd edn. Noyes, Park Ridge, NJ
Soderman J V (ed.) 1982 *CRC Handbook of Identified Carcinogens and Non-Carcinogens: Carcinogenicity–Mutagenicity Database*, Vol. 1: Chemical Class File; Vol. 2: Target Organ File. CRC Press, Boca Raton, FL
Toxic and Hazardous Industrial Chemicals Safety Manual 1986 The International Technical Information Institute, Tokyo
Verschueren K 1977 *Handbook of Environmental Data on Organic Chemicals.* Van Nostrand Reinhold, New York
Yoshida T 1987 *Safety of Reactive Chemicals.* Elsevier, Amsterdam

5. Materials Forecasting

The topic of materials forecasting embraces a number of related subtopics including such questions as: to what degree can properties of new or improved materials be predicted? what techniques can be employed for this purpose? how and at what rate will a new material substitute for a traditional material? what impact will new materials have on other technologies and on the quality of life? can the computer be used to assist in the design and synthesis of new materials? The following publications can provide some answers to these questions.

Cetron M J 1969 *Technological Forecasting: A Practical Approach.* Gordon and Breach, London
Doyama M, Suzuki T, Kihara J, Yamamoto R 1991 *Computer Aided Innovation of New Materials.* Elsevier, Amsterdam
Easterling K 1988 *Tomorrow's Materials.* Institute of Materials, London
Fisher J C, Pry R H 1971 A simple substitution model of technological change. *Tech. Forecasting Soc. Change* 3: 75–88
Hondros E D, Kelly A (eds.) 1987 *Technology in the 1990's: The Promise of Advanced Materials.* The Royal Society, London
Jantsch E 1966 *Technological Forecasting in Perspective.* Organisation for Economic Co-operation and Development, Paris
Lenz R C, Lanford H W 1972 The substitution phenomenon. *Bus. Horiz.* February 63–8
Martino J P 1972 *Technological Forecasting for Decision Making.* Elsevier, Amsterdam
Modern Materials in Manufacturing Industry 1983 The Fellowship of Engineering, London, UK
Nichols R W et al. (eds.) 1988 *Advanced Materials by Design.* Office of Technology Assessment (of US Congress), US Government Printing Office, Washington, DC
Westbrook J H 1970 Materials for tomorrow. In: Bromwell A B (ed.) 1970 *Science and Technology in the World of the Future.* Wiley, New York
Wilson W D, Asaro R J, Dutton R W, Sanchez J M, Srolovitz D J (eds.) 1988 *The Impact of Supercomputing Capabilities on US Materials Science and Technology*, NMAB Report 451. National Academy Press, Washington, DC

6. Sources on People

In any technical field it is often necessary to find information on individuals. This may range from a full biographical sketch to the current address, affiliation, and telephone and fax numbers. While many professional societies in the materials field publish directories of their own members, there are scarcely any full biographical sources that focus on the materials community or any of the subgroups within it. A representative listing of useful sources is given below. A comparatively recent development is the preparation, either in print or on-line, of directories of "who knows what," indicating, by area of expertise, people available for consultation.

Abbott D (ed.) 1986 *Biographical Dictionary of Scientists: Engineers and Inventors.* Bedrick, New York
American Men and Women of Science: Physical and Biological Sciences 1989–90 17th edn. Vols. 1–8. Bowker, New York
Asimov I 1982 *Asimov's Biographical Encyclopedia of Science and Technology*, 2nd edn. Doubleday, Garden City, New York
Consulting Services—A Directory (annual) Association of Consulting Chemists and Chemical Engineers, Inc., New York
Davis G (ed.) 1988 *Who's Who in Engineering*, 7th edn. American Association of Engineering Societies, New York
Directory of Experts and Consultants in Science and Engineering, 2nd edn. Research Publications, Woodbridge, CT

Directory of Metallurgical Consultants and Translators 1984 American Society for Materials, Materials Park, OH
Gillespie C C (ed.) 1970–1980 *Dictionary of Scientific Biography*, Vols. 1–16. Scribners, New York
Ingham J N 1983 *Biographical Dictionary of American Business Leaders*, Vols. 1–4. Greenwood, Westport, CT
International Directory of Research and Development Scientists (annual) Institute for Scientific Information, Philadelphia, PA
International Engineering Directory (annual) American Consulting Engineers Council of the United States, Washington, DC
ISI's Who is Publishing in Science (annual) Institute for Scientific Information, Philadelphia, PA
Kennedy J H (ed.) 1985 *Directory of Management Consultants*, 4th edn. Consultants News, Fitzwilliam, NH
McLean J (ed.) 1988 *Consultants and Consulting Organizations Directory*, 9th edn. Gale Research, Detroit, MI
Miles W D (ed.) 1976 *American Chemists and Chemical Engineers*. American Chemical Society, Washington, DC
Roysdon C M, Khatri L A (eds.) 1978 *American Engineers of the Nineteenth Century, A Biographical Record.* Garland, New York
Teltech Resource Network Corporation, 2850 Metro Drive, Bloomington, MN 55425. Tel: (612) 851 7500 (maintains a computerized file of several thousand experts in many technological fields accessible to users in client companies)
Tomlinson D M (ed.) 1989 *European Technical Consultancies.* Longman, Harlow, UK
Unruh B, Cornog M (eds.) 1987 *Directory of Consultants and Contractors.* NFAIS
Unterburger L (ed.) *Who's Who in Technology (US and Canada)*, 6th edn., Vols. 1–2. Gale Research, Detroit, MI
Who Knows and What 1954 Marquis, Chicago, IL
Who's Who in Science and Engineering Marquis, Wilmette, IL
Who's Who in Science in Europe 1989 Longman, Harlow, UK
Who's Who in Technology (International) 1984 Vols. 1–3. UNIPUB, New York
Who's Who in Technology: Who's Who in Mechanical Engineering and Materials 1981 Research Publications, Woodbridge, CT

(*b*) *Publications*

American Society for Testing and Materials *Guide for the Development of Standards Relating to the Proper Use of Recycled Plastics*, ASTM Standard D5033. ASTM, Philadelphia, PA
Diaz L F, Savage G M *Resource Recovery from Municipal Solid Wastes*, Vols. 1–2. CRC Press, Boca Raton, FL
Directory of Engineering Firms Engaged in Resource Recovery and Solid Waste Management 1979 American Consulting Engineers, Washington, DC
Hollander H I (ed.) *Thesaurus on Resource Recovery Terminology*, ASTM STP 832. American Society for Testing and Materials, Philadelphia, PA
Mason B (ed.) 1981 *Directory of Recyclable Wastes.* Kelso, Greenville, MS
Materials: Renewable and Non-Renewable Resources. American Association for the Advancement of Science, Washington, DC
Recyclable Scrap: Places to Locate; An International Directory 1983 Prosperity and Profits
Recycling Commodity Exchange Encyclopedia 1985 Prosperity and Profits
Recycling: Where to Find or Locate Information on Recycling 1984 Prosperity and Profits
US Department of Defense *Waste Disposal Methods for Military Packaging Materials*, Military Handbook 742. US Government Printing Office, Washington, DC
Voight R L 1986 *Waste Management and Resource Recovery Information Database.* International Research and Evaluation, Eagan, MN
Waste Management and Recycling 1983 Business Communications Company, Norwalk, CT
Wolfe J (ed.) 1981 *Waste Management.* Kluwer, Dordrecht, The Netherlands

(*c*) *Journals*

Conservation and Recycling 1976–1988 Bever M B, Henstock M E (eds.) Pergamon, Oxford
Resources and Conservation 1975–1987 Alter H (ed.) Elsevier, Amsterdam
Resources, Conservation and Recycling 1988– Alter H, Henstock M E (eds.) Pergamon/Elsevier, Amsterdam

7. Recycling

Information on materials recycling is important to individuals, to businesses, to communities and even to nations. Our concerns in this area have been driven by the need for resource conservation and reduction of the waste stream, the opportunity for energy generation from waste, and the requirement for regulatory compliance with respect to protection of workers and the environment. Representative associations and publications treating such topics are given in the following.

(*a*) *Associations*

Institute of Scrap Recycling Industries, Inc., 1627 K Street NW, Washington, DC 20006. Tel: (202) 466 4050
National Recycling Coalition, Room 2350, 45 Rockefeller Plaza, New York, NY 10111
National Resource Recovery Association, 1620 Eye Street NW, Washington, DC 20006. Tel: (202) 293 7330

8. Standards and Specifications

The many hundreds of standards-making bodies in the USA comprise branches of state and federal governments, trade associations, professional and technical societies, consumer groups, and institutions in the safety and insurance fields. Specifications may originate with these same organizations or even with individual companies. The principal coordinating body, clearinghouse and information center in the USA is the American National Standards Institute (ANSI). It also represents the USA in international standards affairs dealing with other national standards bodies in the International Organization for Standardization (ISO). This is discussed further in

Westbrook J H 1986 Standards and specifications for materials. In: Bever M B (ed.) 1986 *Encyclopedia of Materials Science and Engineering*, pp. 4579–85

The principal guides to standards and specifications are

Chumas S J 1977 *Directory of US Standardization Activities*, NBS SP 417. National Bureau of Standards, Washington, DC

Struglia E J 1973 *Standards and Specifications—Information Sources*. Gale Research, Detroit, MI

Technical Indexes Limited (providing a database of 100 indexed microfilm or microfiche files to product data, standards and specifications, updated monthly), Willoughby Road, Bracknell RG12 1BR, UK. Tel: 0344 426 311

A special catalog for materials reference standards is

NBS Standard Reference Materials Catalog 1984–1985 Superintendent of Documents, US Government Printing Office, Washington, DC

A very comprehensive collection of standards and specifications of all types is maintained by Information Handling Services, Englewood, Colorado.

A sample of various directories and cross-references to materials standards and specifications is listed below.

British and Foreign Specifications for Steel Castings 1980 Steel Castings Research and Trade Association, Sheffield, UK

Department of Defense Index of Specifications and Standards, Superintendent of Documents, US Government Printing Office, Washington, DC

European Committee for Iron and Steel Standardisation (ECISS), Index of Standards, PSP2 1986 British Steel plc. Head Office Standards, London, SE1 7SN

Frick J P 1990 *Woldman's Engineering Alloys*, 7th edn. American Society of Metals, Materials Park, OH

German Standards (English translation of more than 4750 popular DIN Standards plus 1500 DIN-EN, DIN-IEC and DIN-ISO Standards). Beuth, Postfach 1145, D-1000 Berlin 30, Germany

Handbook of Comparative World Steel Standards 1990 International Technical Information Institute, Tokyo

Hucek H, Wahl M 1990 *Handbook of International Alloy Compositions and Designations*, Vol. 1: Titanium, MCIC-HB-09. Metals and Ceramics Information Center, Battelle Memorial Institute, Columbus, OH

Hufnagel W (ed.) 1991 *Key to Aluminium* (in German), 4th edn. Aluminium AG, Dusseldorf, Germany

Index of Aerospace Materials Specifications 1986 Society of Automotive Engineers, Warrendale, PA

Iron and Steel Specifications, 7th edn. 1989 British Steel plc. Head Office Standards, London, SE1 7SN (steel grades to BS970 together with US, French, German, Japanese and Swedish standards)

Kehler W 1981 *Handbook of International Alloy Compositions and Designations*, Vol. 3: Aluminium. Aluminium Verlag, Heyden, Germany

Metals and Alloys in the Unified Numbering System, 5th edn. 1989 ASTM Publication DS56C. American Society for Testing and Materials, Philadelphia, PA

Potts D L, Gensure J G (eds.) 1988 *International Metallic Materials Cross-Reference*. Genium, Schenectady, NY

Reynard K W (ed.) 1992 *Inventory of Materials Designation Systems*. Versailles Advanced Materials and Standards (a project of TWA-10) National Physical Laboratory, Teddington, UK

Ricci P L, Perry L 1991 *Standards: A Resource and Guide for Identification, Selection and Acquisition*. Stirz, Minneapolis, MN

Ross R B (ed.) 1992 *Metallic Materials Specifications Handbook*, 4th edn. Chapman and Hall, London

Schmitz H 1987 *Stahl-Eisen-Listen/Steel-Iron Lists* 7th edn. Verein Deutscher Eisenhuttenleute (VDEh), Postfach 8209, Sohnsstrasse 65, D-4000 Düsseldorf 1, Germany. Tel: 0211 6707 0 (2500 German and 1000 foreign materials covered, also in computerized database)

Simmons W F, Gunia R B 1972 *Compilation and Index of Trade Names, Specifications and Producers of Stainless Alloys and Superalloys*, ASTM Publication DS45A. American Society for Testing and Materials, Philadelphia, PA

Standards Cross-Reference List, 2nd edn. 1977 MTS Corp., Minneapolis, MN (includes standards issued by an agency but adopted and renumbered or redesignated by another, compiled and cross-referenced to aid in their location and identification)

Unterweiser P M (ed.) 1987 *Worldwide Guide to Equivalent Irons and Steels*, 2nd edn. American Society for Metals, Materials Park, OH

Unterweiser P M 1987 *Worldwide Guide to Equivalent Non-Ferrous Metals and Alloys*, 2nd edn. American Society for Metals, Materials Park, OH

Wegst C W 1982 *Key to Steel* (in German), 16th edn. Verlag Stahlschlüssel, Marbach/Neckar, Germany

World Metal Index, Sheffield City Libraries, Central Library, Surrey Street, Sheffield, Yorkshire S1 1XZ, UK. Tel: 0742 734 714 (a service based on an index of over 70 000 grades of metallic materials, all British and US standards and most European and national standards)

World Steel Standard Specifications 1974 Foreign Technology Division, US Air Force, FTD-NC-23-856-74

9. Patents

Information on what patents exist on a given subject and on where copies may be obtained is often important to the materials manager whose organization may be seeking patent protection for its own work, for the sale or licensing of its patents to others, or for securing license to use technology covered by the patents of others. Listed below is a sample of indices, guides and services offering access to patent literature.

British Patent Office, Sales Branch, Unit 6, Nine Mile Point, Cross Keys, Newport, NP1 7HL, UK (copies of all British specifications and patents)

British Science Library, Science Reference Library, Computer Search Service, 25 Southampton Buildings, Chancery Lane, London WC2A 1AW, UK. Tel: 01 405 8721 (photocopies from collection of over 22 million specifications from all over the world)

Central Patents Index Derwent Publications, London

Derwent Publications Ltd, Patents Documentation Services, Rochdale House, 128 Theobalds Road, London WC1X 8RP, UK (abstracts, indexes and distributes worldwide patent specification information in hardcopy, microform and machine-readable form)

European Patent Office, EDP Department, EPO Data Banks, PO Box 5818, Patentlaan 2, NL-2280, HV Rijswijk ZH, The Netherlands (patent information on microfilm or on-line access via Euronet)

Foltz R D, Penn T A 1989 *Protecting Engineering Ideas and Inventions*. Penn Institute, Cleveland, OH

German Patent Information System, SRZ, Berlin, Lützowstrasse 105, D-1000 Berlin 30, Germany (machine-readable database of all patents registered with the German patent office since 1973, full text since 1984)

Haughton B (ed.) 1972 *Technical Information Sources: A Guide to Patent Specifications, Standards and Report Literature.* Bingley, London

Japan Patent Information Center, Bansui Building, 1-5-16 Toranomon, Minato-ku, Tokyo, Japan (offers an on-line service (PATOLIS) providing patent and trademark information on Japanese, US and INPADOC databases)

Konold W G 1989 *What Every Engineer Should Know About Patents*, 2nd edn. Dekker, New York

North O S 1982 *Mineral Exploration, Mining and Processing Patents, 1980.* Society of Mining Engineers Littleton, CO

Official Gazette of the US Patent Office (weekly) US Patent Office, Washington, DC

Official Journal (Patents) (weekly) UK Patent Office, Orpington, Kent, UK

Overseas Patent Information Center, Kowa Building No. 3, 11-45, 4 Chome, Akasaka Minato-ku, Tokyo, Japan (copies of Japanese patents and published Japanese patent specifications)

Patent Information Clearing House, Building 7, 1500 Partridge Ave., Sunnyvale, CA 94087. Tel: (408) 738 5580 (US patents only)

Patent Information Office, PO Box 309, Rijswijk ZH 2109, The Netherlands (copies of patent publications from 12 countries, access to Dutch patent office library)

Pergamon Infoline, Inc. *Patsearch* (an on-line database of abstracts, bibliographic data and other information for US patents and international patent application files under the Patent Cooperation Treaty (PCT)). Tel: (800) 336 7575

Polysearch Service, 20 Creekwood Drive, West Hill, Ontario, Canada M1E 4L7. Tel: (416) 282 4494 (photocopies of patent specifications from 18 countries)

Rapid Patent Service of Research Publications, Inc., PO Box 2577, Eades Station, Arlington, VA 22202. Tel: (703) 920 5050 (offers file histories on US and foreign patents)

Research Publications, Inc., 12 Lunar Drive, Drawer AB, Woodbridge, CT 06525. Tel: (203) 397 2600 (offers US and international patent documentation from ten countries in microform, monitoring and document retrieval services for paper copies)

Rimmer B M 1980 *Guide to US Patent and Trademark Literature.* Longwood, Wakefield, NH

J. H. Westbrook
[Brookline Technologies, Ballston Spa, New York, USA]

K. W. Reynard
[Wilkinson Consultancy Services, Newdigate, UK]

E

Economic Effects of Corrosion and Other Degradative Processes

Corrosion is defined as the degradative interaction of a material with its environment. Although usually applied to metals, the term may also be applied to other materials, particularly inorganic ones such as stone, brick and concrete. Corrosion is only one of a number of degradative processes that affect materials in use. Others that may be mentioned are wear, fatigue, fracture, ultraviolet degradation, mildew and rot. These degradative processes cause the users of durable goods to incur costs for their maintenance, repair and early replacement. On a national level, these costs have an economic effect in that they represent resources in the form of materials, capital, energy and labor, all of which, in the absence of these degradative effects, could be used for other purposes.

Corrosion, and more specifically metallic corrosion, is the only degradative process for which a number of studies have been carried out in various countries and these costs determined on a national level. Studies of various degrees of thoroughness have been carried out in the UK, the USSR, Germany, Finland, Sweden, India, Australia, Japan and the USA. The most thorough of these studies were carried out in the UK (the famous Hoar report (1971)) and the USA (the National Bureau of Standards report by Bennett et al. (1978)).

1. Definitions

The National Bureau of Standards (NBS) study, carried out in cooperation with the Battelle Memorial Institute, (and the Hoar study before it, although less explicitly) distinguishes between three types of cost: total costs, avoidable costs and presently unavoidable costs. The total costs are all the costs incurred because corrosion occurs; that is, the reference is a hypothetical world of no corrosion. This world serves only as a basis for accounting of costs, and there is no implication that such a world could exist. The avoidable costs are those that are incurred because best economic corrosion-control practices are not always used by everyone in the economy. Best economic practice means a minimum life-cycle cost. It therefore balances initial costs against the lifetime of a durable good and total maintenance cost; it does not mean the use of the most corrosion-resistant material. Hence, the baseline for this accounting is a hypothetical (but theoretically realizable) best-practice world in which everyone uses best economic practices. The difference between the total costs and the avoidable costs are the unavoidable costs. These costs cannot be further reduced with a given state of corrosion-control technology. Their reduction requires an advance in this technology and hence requires further research and development.

It should be noted that the accounting of both total costs and avoidable costs in this way means a comparison of the present world with a corrosion-free world and a best-practice world, respectively. What is measured is the amount by which a country's Gross National Product (GNP) would decrease in going from the present world to either of the two hypothetical worlds with nothing else changing. This decrease in the GNP is just the measure of the resources that would become available for other purposes if corrosion were to cease, if best practice were to become more widely adopted, or if the presently unavoidable costs were to be reduced by more effective corrosion-control technology. In the real world, of course, the GNP would not decrease (and might actually increase) because these resources of capital, materials, energy and labor would in fact be used for other purposes.

2. Elements of the Cost of Corrosion

The NBS study used an input–output model of the economy. The importance of this model for an economic study of this kind is that indirect effects are appropriately accounted for. For example, a reduction of consumer purchases in automobiles brought about because of increased durability through reduction of corrosion has a direct effect on automobile dealers and the automobile industry, and large indirect effects on related industries (e.g., the tire, steel and chemical industries). In addition to accounting for these indirect effects, the model gives a means for determining in which sector the costs are accounted for. To this end, the model divides the economy into two principal parts: the producing part and the consuming part. The producing part consists of all agricultural, industrial and commercial sectors, whereas the consuming part consists of the so-called final demand sectors: personal consumption expenditures (PCE, the private consumer); government expenditures, which in the USA are in turn broken down into federal government expenditures (FGE) and state and local government expenditures (SLGE); and producer fixed capital formation (PFCF), that is, the expenditures of the producing part of the economy for capital equipment.

In the NBS study, the producing sectors are

identified as intermediate output (IO). It should be noted that any costs borne by the producing sectors are eventually passed through to final demand sectors.

The elements of the costs of corrosion which apply to both the total and avoidable costs are as follows:

(a) Capital costs—these are associated primarily with the shortened lifetime (defined as the age at scrapping) of capital equipment in the presence of corrosion or if best practice is not used. They include costs for consumer durable goods, and are accounted for in all the final demand sectors (PCE, FGE, SLGE and PFCF where they are the only costs). Since the lifetime of equipment in service is only poorly known, and the amount that that lifetime would be extended in the absence of corrosion has to be an estimate, a large error exists in their magnitude. In addition to these lifetime costs there are costs associated with any excess capacity or redundant equipment that may be necessary because of corrosion. These latter aspects are accounted for in the producing sectors (IO) and in two of the final demand sectors (FGE, where they are significant for military equipment, and SLGE).

(b) Control costs—these are associated with maintenance and repair (e.g., painting, replacement of corroded parts of structures) when caused by corrosion, and costs of corrosion control (e.g., cathodic protection, water treatment). Because of the ubiquitous nature of this activity, these costs are accounted for in all sectors except PFCF which contains only capital replacement costs.

(c) Design costs—these are associated with special corrosion-resistant materials such as stainless steel or copper in place of carbon steel; allowances for extra-thick sections made because of corrosion; special surface treatments such as shot peening; and special coatings such as galvanizing or plating, or corrosion-resistant paints in their first application. Since they are incurred only at manufacture, they are accounted for solely in the producing sectors (IO).

(d) Associated costs—these include such items as loss of product, research and development, insurance, and the cost of maintaining the inventory of parts and equipment needed for corrosion control. They do not include the cost of the items in the inventory since this is already accounted for under control costs. These associated costs are relatively small and are mainly accounted for in IO, FGE and SLGE.

The NBS study gives results for the US economy in 1975 for each of the sectors, along with an estimated range of uncertainty (Table 1). These results indicate that for the USA about 4% of the GNP is used because of corrosion, and about 0.7% of the GNP of these expenditures could be avoided. The latter figure is, however, known only very imprecisely because it is difficult to provide a quantitative definition of economic best practice that is applicable to all sectors of the economy. Financial considerations, primarily the cost of borrowing money and the associated discount rate, vary from firm to firm, leading to a large uncertainty. Nevertheless, the results are remarkably similar to the results of the Hoar study for the UK when allowance is made for the range of uncertainty in this type of study. As previously mentioned, the results of other, less complete studies are also similar to these results. It appears then that in modern industrial nations total costs of corrosion equivalent to about 3–4% of the GNP, and avoidable costs equivalent to about 0.7% of the GNP, are incurred. Only continued development of more effective corrosion-control technology, and more widespread applications of this technology can reduce these costs.

Table 1
Results of the NBS study in billions of US dollars (after Bennett et al. 1978)

Sector	Estimated total cost	Range of uncertainty[a]	Estimated avoidable cost	Range of uncertainty
PCE	15.8	10.3–21.3	4.9	2.8–15.9
FGE	7.9	6.2–9.6	1.7	0.8–2.5
SLGE	2.4	1.2–3.6	0.9	0.5–1.4
PFCF	24.5	23.5–25.0	−4[b]	−6.0–2.0
IO	19.1	11.5–26.7	6.2	3.0–19.1
Total	69.7	52.7–86.2	9.7	2.1–40.9
% of GNP	4.3		0.65	

a Uncertainty ranges are calculated from the estimated cost and include systematic and random uncertainties b This value is only poorly known

3. Costs of Wear

An important study on the costs of wear was carried out in the UK in 1966 (Jost 1966). Extrapolating these costs to the USA, Jost (1976) estimated that twelve to sixteen billion US dollars could have been saved in 1971 by the application of tribology. This amounts to 1.1–1.5% of the GNP of the USA for that year. From the definition used, this represents costs between avoidable costs and total costs as defined above, but the results indicate that wear costs are probably roughly comparable to corrosion costs.

No studies after the Battelle–National Bureau of Standards report that lead to different conclusions are known to the author. The impact of the Battelle–NBS work is discussed further by Payer and Ugiansky (1986).

See also: Economic Effects of Fracture

Bibliography

Bennett L H, Kruger J, Parker R L, Passaglia E, Reiman C, Ruff A W, Yakowitz H, Berman E B 1978 *Economic Effects of Metallic Corrosion in the United States: A Report to the Congress by the National Bureau of Standards*, NBS Special Publication No. 511-1. National Bureau of Standards, Washington, DC

Hoar T P 1971 *Report of the Committee on Corrosion and Protection: A Survey of Corrosion and Protection in the United Kingdom.* HMSO, London

Jost H P 1966 *Lubrication (Tribology): A Report on the Present Position and Industry's Needs.* HMSO, London

Jost H P 1976 *Economic Impact of Tribology*, NBS Special Publication No. 423. National Bureau of Standards, Washington, DC, pp. 117–39

Leontief W W 1951 Input–output economics. *Sci. Am.* 185: 15–21

Miernyk W H 1965 *The Elements of Input–Output Analysis.* Random House, New York

Payer J H, Ugiansky G M 1986 Impact of the NBS–Battelle cost of corrosion study of the USA. *Proc. Symp. International Approaches to Reducing Corrosion Costs.* National Association of Corrosion Engineers, Houston, TX

Pludek V R 1986 Design for corrosion control. In: Bever M B (ed.) 1986 *Encyclopedia of Materials Science and Engineering*, Pergamon Press, Oxford, pp. 1108–17

E. Passaglia
[Washington, DC, USA]

Economic Effects of Fracture

Along with corrosion, wear, oxidation, ultraviolet degradation, mildew, rot and other microbial processes, fracture (and fatigue) is a phenomenon that causes materials to fail in service. All these failure processes have in common that their occurrence and their prevention, both during design and manufacture and in the use of finished goods, require the use of resources in the form of materials, energy and labor that would not otherwise be necessary and hence could be used for other purposes. These extra resource requirements associated with manufacture, maintenance and repair, and early replacement, may be termed the "economic effects" of these failure processes, and since the economic effects may be accounted for in terms of money, they are usually called the *costs* of these processes. Studies carried out at the National Bureau of Standards (now the National Institute for Standards and Technology) estimated the costs of corrosion (see *Economic Effects of Corrosion and Other Degradative Processes*) and of fracture. The costs of wear were also estimated, but less accurately.

Fracture is a particularly important process since its occurrence can lead to the catastrophic failure of large structures and equipment, sometimes with significant loss of life. Thus, the principal aim of engineering design is the prevention of fracture. As will be seen, this is the main factor that determines the cost of fracture.

As described in the article *Economic Effects of Corrosion and Other Degradative Processes*, the method by which the costs associated with these processes are estimated is to compare the costs in the world (i.e., the United States economy) as it presently exists to worlds in which the process does not exist, or a world in which best economic practice is used. These comparisons lead to what are termed "total costs" and "avoidable costs." The latter are costs that would not be incurred if everyone used best control practice. The comparisons are done by an input–output model of the present world to what the input–output model would be under these other two conditions. This is described in the cited article on corrosion and will not be repeated here. Suffice it to say that the accounting of the costs is done in different sectors of the economy. The sectors are: intermediate output, or manufacturing, where any special or extra materials, treatment, or design are used because of the degradative process; in final demand, that is, costs to the final user associated with extra maintenance, repair or control, and early replacement of equipment; and private fixed capital formation, or the early replacement of industrial capital equipment because of the degradative process. The last is listed here under final demand.

Now, corrosion-free and wear-free worlds are not too difficult to envisage. Special materials and treatment might be used, but by and large equipment would be much the same as it now is. A totally fracture-free world would be quite different. It would lead to the concept of glass automobiles and to other unlikely equipment changes that would produce fanciful results. As a result, the fracture-free world was defined as one in which design was still load or stiffness limited, but one in which knowledge and materials were perfect. As a result, all the safety factors that are presently employed

Table 1
Costs of fracture for the United States economy in 1978. Costs are in billions of US dollars

Sector	Total cost	Avoidable cost
Manufacturing	72.3	21.1
Final demand		
personal consumption	4.6	1.6
government	2.2	0.23
capital equipment	8.4	2.94
Total final demand	15.2	4.8
Total	87.5	25.8
Percentage of GNP	3.9	1.2

because of materials variability and imperfect knowledge of the science of fracture would be 1.0 rather than the presently used values. Moreover, all material properties would be four standard deviations above the present mean; there would be no time-dependent fracture such as fatigue or stress corrosion; and no inspection in service would be required. Fracture would still exist, but only when design specifications were exceeded. These more complex definitions arise from the fact that the principal goal of engineering design is to insure that fracture does not occur.

The results of the study are given in Table 1. They show that, for the United States economy in 1978, the total costs of fracture were US\$87.5 billion, or 3.9% of the gross national product (GNP). The division of these costs is, however, instructive. US\$72.3 billion are incurred in manufacturing (or intermediate output) and are, of course, passed on to the final consumer, and US\$15.2 billion are incurred by the final consumer in the form of maintenance and repair and early equipment replacement (US\$6.8 billion) and early capital replacement in industry (US\$8.4 billion). Thus, the bulk of the costs are incurred during manufacture in the form mainly of extra materials occasioned by the use of large safety factors because of materials variability and imperfect knowledge of fracture science. The avoidable costs are US\$25.8 billion, or 1.2% of the GNP, with approximately the same distribution. These are the costs that could be saved by the use of present best practice. For corrosion, the total costs are 4.2% of the GNP (doubtless a fortuitous similarity to the fracture results), but the distribution (in 1975 US dollars) is US\$24.5 billion in intermediate output, and US\$45.2 billion in final demand. Thus, for fracture as compared to corrosion, the necessity to prevent catastrophic failure shifts the costs to the manufacturing sector.

See also: Economic Effects of Corrosion and Other Degradative Processes

Bibliography

National Bureau of Standards 1983 *The Economic Effects of Fracture in the United States.* NBS Special Publication No. 647-1, 647-2. NBS, Washington, DC

Passaglia E 1985 The economic effects of degradative processes on materials. *American Chemical Society Special Symp. Effects of Acid Rain on Materials.* ACS, Washington, DC

E. Passaglia
[Washington, DC, USA]

Economic Geography of Materials

Economic geography has been defined as the study of the conditions, methods and effects of man's work in its spatial extension. The economic geography of materials is concerned with the following:

(a) the location of production of materials at various stages from raw material to finished product;

(b) the patterns of transportation and trade that link material producing and processing locations with one another and with markets;

(c) the effects of such factors as technical change, the availability of capital and of skill, differing forms of industrial organization, the actions of governments, international agreements and environmental considerations on the location and output of mining and processing activities; and

(d) the influences that material deposits, supplies and processing operations exert on the location, growth and decline of industries, settlements and industrial areas.

This article provides an introduction to the problems and approaches in studying the changing locations of raw materials production, especially of minerals, and of the considerations taken into account in providing explanations of these changes. Some features of the current geography of materials are outlined. Attention is also given to the theory of location of economic activity with reference to the influence of materials on the geography of processing and manufacturing.

1. Approaches to the Subject

The subject may be approached in a number of ways. The mapping of the geographical distributions of material sources and production levels, and of changing patterns of production is a prime task. Some atlases, such as the *Oxford Economic Atlas of the World* (1972), display patterns of raw material sources, production and known reserves at world or regional scales. Distribution maps at larger scales for particular commodities at country or local levels of scale are also produced, though not usually in uniform ways.

Some sources stress the relations between mineral deposits and geological conditions (Dixon 1979). Distribution maps indicate existing sources of production and the location of known reserves. Areas with future potential are also shown on some maps. However, the distributions shown reflect past and present economic conditions and technical knowledge; patterns may change significantly as new scientific knowledge is acquired or as market conditions alter.

A number of books exist that describe at world levels the principal sources of production, often placed in the context of historical reviews of past activity and evolving technology, and of the effects of changing economic, political and technical conditions on patterns of production and trade (Zimmermann 1951, Carlson 1956, Alexandersson and Klevebring 1978, Blunden 1985). Such works may be arranged in terms of individual materials or material groups. Standard commercial geographies (Stamp 1980) provide mainly descriptive accounts of the geography of occurrence and production and are useful for reference. Some studies reflect traditional geographical approaches in placing a strong emphasis on the influence of physical conditions of climate, soil, water and geological conditions on the patterns of production of raw materials. However, most studies in economic geography attempt a more balanced approach considering how, in the light of changing conditions of demand, technology, business organization, capital availability and international agreements, organizations and societies evaluate and exploit the possibilities of the natural world and convert raw materials into usable commodities. Important in this approach are studies of technical change and of how this has led to the revaluation of resources and materials and of market factors and accessibility, the latter leading to the examination of transport developments. Geographical conditions may be seen to exert themselves through spatial variations in the costs of production and in price/cost ratios (Buchanan 1952). A further concern of economic geography is with the effects of the production of materials on the environment. This ranges from the examination of the devastation and dereliction of land, and of pollution, to the creation of new settlements, mining and industrial areas and the quality of the environments offered to the communities that dwell in them.

2. *Concepts of Natural Resources*

The economic geography of materials may be regarded as an important aspect of the wider study of the geography of resources. This itself depends on the accumulation and analysis of knowledge about the earth itself with its extremely diverse conditions of climate, soil, relief, vegetation, geology, mineralization and relative location. Even with perfect knowledge about the nature of the earth, estimates of future resource availability would be subject to degrees of error arising from lack of knowledge of future scientific and technical developments, future movements of demand and price and the effects of various forms of control and arrangement, both national and international, over production and trade.

Natural resources may be classified into biological resources, including the various organisms and levels of animal and plant life of interest or use to man, and physical resources, such as the metallic ores present at or near the surface of the earth. A further division is often made into renewable resources, such as crops or forest products, which are thought to be capable of regeneration into usefulness within a short time scale, and nonrenewable resources, such as coal or crude oil, which are regarded as wasting assets. A somewhat more refined division is into three categories. These are: (a) flow resources, such as forests, which may be depleted, sustained or increased by human action and in which the rate of depletion or increase becomes a very important consideration; (b) continuous resources, such as tidal power or solar energy, which are available in nature independently of human use but which may become available or potentially available for human use; and (c) stock resources, including all the minerals and the carbon and hydrocarbon fuels (Blunden 1985).

Many industrial materials, such as timber, mangrove, bark products, vegetable oils, natural rubber, cotton, fish products and gums, fall within the class of renewable resources. It is not surprising, in view of the traditional interest in economic geography in studies of societal–environmental relationships, that much attention has been given to the methods of production and processing of such resources: emphasis has been given to the impact of the production methods and rates on natural conditions, such as soils, and on the resource base itself, such as tropical forests. Important also are studies that point to ways to improving standards of living, especially in developing countries. Many problems of resource development, which have given rise to adverse consequences for the natural environment and for the resource base, have arisen from the failure to recognize the complexity of the interrelationships both within and between societies and the natural environments. The concept of the "resource system" has been employed both to provide a framework for the further investigation of such problems and also to give a basis for communication to planners and decision makers of the necessity for well-informed analyses and scientific bases for planned actions. Underlying this concept are the general principles of conservation, essentially those of the wise management of resources and an application of systems analysis. In addition to many individual studies, such an approach has been the mainspring of an international programme of studies by the United Nations University, designed to improve management practices. The programme has embraced research studies on the humid tropical forests, on dryland environments, on coastal zones and on the interactions between highland and lowland resource systems. The general field is of importance at a time when economic development is being actively sought. The aim is to maximize the productivity of resource systems with full regard to the maintenance of the natural resource base. Existing resource systems should be thoroughly understood before attempts are made to introduce new technologies (Ruddle and Manshard 1981).

Influential approaches regard resources not so much as physical or biological substances but as expressions of human appraisal or need. For example, E. W. Zimmermann has argued for a functional concept of resources:

> "... the word 'resources' does not refer to a thing or a substance but to a function which a thing or a substance may perform or to an operation in which it may take part."

Resources may thus be regarded as "cultural appraisals" for which the demand changes through time, in the light of new discoveries, or through space, according to transport changes or to alterations in regional or local economic and social needs. The substitution of new materials for old may have vitally important geographical consequences. The chemical revolution of the nineteenth century, for example, helped to create an enhanced value for many coalfield industrial areas and brought into production and trade hitherto unrecognized resources, such as the sodium nitrates of Peru and northern Chile, while destroying the value of earlier vegetable sources of materials, for example, indigo farming in India. The invention of vulcanization by Charles Goodyear in 1839 opened the way to the development of the rubber industry in Amazonia, a region which was itself to decline following the transference of seeds and the successful commercial exploitation of rubber in Malaya.

Such approaches have influenced work by economic geographers in the wide-ranging debates on the adequacy of the earth's resource base in the light of population growth and increasing economic demands. The *Atlas of Earth Resources* (1979), for example, took a balanced view. Other studies focused on individual minerals or on national or regional economies. Issues facing developing countries were also identified, though differences of view emerged on why and how problems of poverty had developed. In these debates the point was frequently made that "what constitutes a resource or a reserve is not fixed over time" (Blunden 1985).

More recently, environmental issues have provided an important focus for discussion especially following the publication in 1987 of the Report of the World Commission on Environment and Development (the Brundtland Commission). There has been a renewed attempt to link economic studies with environmental problems (Pearce and Turner 1990). Adoption of the concept of "sustainable development," involving approaches to economic progress and the production of materials that do not impair the value of the environment for future generations, should result in enhancing the importance of environmental quality in all aspects of decision making on the production of materials.

Finally in this section, mention should be made of a valuable review by Rees (1990) of approaches to the study of resources by economic geographers.

3. Changing Patterns of Mineral Production

The description and explanation of changing patterns of mineral production have been important tasks of economic geography. Distinctions are made between fuel and nonfuel minerals, as also between metallic and nonmetallic minerals. In the latter group, substances such as clay, sand, gravel, gypsum and limestone tend to be quarried and consumed within restricted geographical areas for reasons of bulk, low value in relation to weight and high transport costs in relation to value. Much interesting work has been done, for example, on the availability of sand and gravel resources for building and construction needs at times of great urban expansion and infrastructure development. Studies of the changing production of metallic minerals have often taken a broader geographical scale and have often been placed within a historical perspective. Following the Industrial Revolution of the eighteenth century, the demand for materials grew rapidly and production of metals such as copper, lead, tin, mercury and the precious metals, despite some fluctuations, accelerated in the nineteenth century. The growing use of electricity, for example, was a great stimulus to the demand for copper. By the end of the nineteenth century demand was increasing for newer metals such as aluminum, chromium and nickel. Manganese, tungsten, titanium and uranium are examples of still "newer" metals for which demand has grown strongly in the twentieth century.

In the early stages of the Industrial Revolution, much of the demand was met by supplies found and exploited in the industrializing countries themselves. Nonferrous metals, for example, were mined in Cornwall, from the Parys Mountain in Anglesey, and in Saxony and Sweden. Smelting and refining were carried out in the rising industrial areas such as South Wales and Lancashire. However, as demand increased and high-quality local ores became more expensive to mine, new mining fields were opened in countries overseas. In the case of nonferrous metals, tin production grew in Bolivia and Malaya, copper in Arizona and Utah using lower-grade ores and, after World War I, in Chile, Canada and the copper belt in the Congo and the then

Northern Rhodesia (now Zambia). Goldfields were opened in California, Australia, the South Island of New Zealand, Alaska and the Witwatersrand. The increased popularity of aluminum, particularly after the lowering of production costs following the introduction of the Hall–Héroult process, led to the opening of new sources of bauxite. France had been the leading early source, the Arkansas fields followed and, after World War II, Jamaica, Brazil and Surinam gained eminence. More recently, the Soviet Union, Guinea and Australia have become important producers. Such changes in patterns of extraction may be drawn for other minerals (Schmitz 1979).

A number of models have been presented to generalize stages in the industrial development of a country possessing metallic ores. A five-stage model partly based on that of Hewett (1929) postulates:

(a) a period of the discovery of new deposits, the opening of small mines and increasing output;

(b) the introduction of smelting operations, the exhaustion of earlier mines and the opening of a smaller number of new ones with the development of larger mining companies;

(c) a decline in domestic mine production but, as industrial concentrations develop, an increase in ore imports;

(d) a growth in the import and consumption of metals smelted and refined abroad; and

(e) almost complete dependence on imports of refined metals.

Such a model is based on the experience of strongly industrialized countries; the dates at which different countries have embarked on the sequence vary. Nor, indeed, is it necessary to assume that all countries will experience the complete sequence.

Chisholm (1982) has also provided a schematic model for the geography of mineral prospecting and exploitation, embodying two general cases. In the first case the consuming region, meeting an increasing need to import to satisfy its rising industrial production, prompts exploration and development in countries that have a settled population but are at an early stage of economic development or in countries that have little habitation but are suited to settlement on a substantial scale. He cites the development and European settlement in Australia, South Africa, Canada and Argentina. The early settlement of the South Island of New Zealand prompted by gold mining is another relevant case. The development of tin mining on the Jos Plateau, Nigeria, was linked to colonial expansion. The second case arises from the building of communications across uninhabited or wilderness areas. The development of Sudbury and Cobalt followed the construction of the Canadian Pacific Railway. There has also been a propensity to explore inhospitable or difficult environments: the Sahara Desert, northern Alaska, northern Siberia, the North Sea and northwest Australia are relevant examples.

A historical perspective is also useful in accounting for the differences between countries and areas in patterns of mine, smelter and refinery production. In some areas geographical inertia operates: the tradition of processing and refining of metallic ores remains long after the initial location factors have ceased to operate. The influence exerted by former colonial powers and their companies may continue long after former colonies have achieved independence. But patterns do change, sometimes rapidly, and the interest lies in deciding how far political or economic pressures have been responsible and, of the latter, how much importance can be attached to transport costs for ores, fuels and energy, how much to capital availability, entrepreneurship and technical skill, or to labor availability or to market considerations. What, in short, are the comparative advantages of each possible location?

4. Geographical Concentrations of Mineral Resources and Production

A significant feature of the geography of mineral fuels and nonfuel minerals is the concentration of known resources and production in a relatively small number of countries among whom the developed countries figure strongly. Chisholm (1982) pointed out the importance of the Soviet Union, the USA, Canada, South Africa and Australia. These countries and, to a lesser extent, China and Brazil feature prominently on the map of mineral wealth in *The Economist Atlas* (1989). Rees (1990) also demonstrates the "dominant position of the currently advanced nations in both the production and consumption of minerals." Such countries also appear strongly in estimates of reserves of nonfuel minerals. Their significance may be ascribed only partly to geological conditions (which are, to some extent, a reflection of the state of geological exploration and knowledge); economic conditions, features of industrial organization, political stability and government policies are among the other factors involved. Since the pattern of mineral search has been heavily oriented to the developed countries, remarks Rees, "the present reserve position in many less developed countries may bear very little relationship to the real resource endowment." Other countries assume more importance if some individual nonfuel minerals are examined. About three-quarters of reserves of iron ore are in the Soviet Union, Brazil, Canada, Australia, China and India. For reserves of copper attention turns primarily to the USA, Chile, the Soviet Union, Peru, Canada, Zambia and Zaire. Some three-quarters of bauxite reserves are to be found in Guinea, Australia, Brazil, Jamaica, India, Guyana and Cameroon. Notable omissions from such lists are the EC countries and Japan; increasing dependence on imported supplies may be expected to play a significant role in the economic and foreign policies of these countries (Manners 1981, Blunden 1985). However, the developed nations are themselves the chief mineral producers and the chief sources of their own mineral consumption.

Developing countries have, not surprisingly, been keen to obtain the advantages of processing their own raw materials. Jamaica, for example, produces alumina, though not aluminum. However, established patterns are strong and not easy to break down. Most progress has been made in copper where major ore producers such as Chile, Zaire and Zambia smelt and refine about 60% of their ore (Rees 1990).

5. *Mineral Resources and Development*

Many less-developed countries are heavily dependent on their exports of just one or a small number of minerals. Classic cases include Bolivia's dependence on tin and other nonferrous ores, Chile's reliance on copper ores and concentrates, Zambia's and Zaire's reliance on copper and Mauritania's dependence on iron ore. Dangers of overreliance on a small range of minerals clearly exist. On the other hand, it may be argued that the possession of such resources provides opportunities for development, given that the resulting funds are wisely used and that a high proportion of revenue is retained within the country for beneficial use. How such opportunities may best be employed has been the subject of considerable debate. Pressure of expected demand on reserves will probably prompt more vigorous mineral exploration. However, decisions to invest in production processes will rest on conditions additional to those of mineral availability and quality, including technical knowledge and skills, accessibility and transfer costs, government policy, political stability and ability to compete on cost and quality terms in wider markets.

It is tempting to regard the quality of a mineral deposit as the major factor influencing decisions to exploit a resource and this is sometimes the case. However, quality is relative and may change dramatically as new techniques evolve. A classic case of the effect of the development of the basic process of steelmaking on the value of the phosphoric minette ores of Lorraine and of eastern England may be indicative in this respect. The location of production will depend also on what is known at the time about availability rather than on what may actually be there. A study of the location of mineral extraction in Mexico placed most emphasis on the state of geological exploration and on the installation of transport facilities to local markets. The potential offered by a local market may prompt the development of small-scale fields. Economic considerations may, however, be modified by political considerations such as the desirability of national self-sufficiency or by the policies of actual or potentially involved mining and development companies.

As has been indicated earlier, an increasingly important set of considerations now influencing development decisions arises from environmental concerns and their expression in legislation. Stress need not be laid on the issues involved in uranium mining and processing for environmental controversy may also surround plans to work iron ore, coal, nonferrous metals or even to quarry sand and gravel. Conflicts arise, as in the UK, between applications to explore or develop minerals in National Parks or in areas of outstanding natural beauty and the arguments of the environmentalists.

Environmental questions arise too as decisions to work lower-grade or less-accessible sources are taken. To produce one tonne of refined copper in the USA from certain low-grade deposits, it may be necessary to remove and dispose of some 350 t of overburden and to handle about 150 t in the forms of tailings and gases (Manners 1981). Nor is environmental protection a concern only of the more industrially developed countries, as the work of the United Nations Environmental Programme has shown: such concerns are increasingly becoming worldwide. Note should also be made of the energy implications of transforming minerals into usable products. Broadly, it requires twice as much energy to produce a tonne of titanium than of aluminum; twice as much to produce a tonne of aluminum as of copper from medium-grade ores and six times as much as steel. Manners (1981) has argued that such considerations may give advantages in a world becoming more energy and environmentally conscious to those minerals, metals and processes requiring low energy inputs in conversion. One can expect increasing concern with pollution problems arising from the discharge of sulfur dioxide and other gases into the atmosphere, from the discharge of polluting substances into rivers, inland seas and the oceans, and from the despoliation of land areas.

6. *Materials and the Location of Industry*

What role do material resources and especially mineral deposits play in influencing the location of industry and, consequently, the growth of industrially based cities and conurbations? This question has lain at the center of much work in the field of the location of economic activity.

Early theoretical work by J. H. von Thünen in 1826 established principles that would explain the disposition of agricultural land uses and crops around an isolated urban market, but it was not immediately followed up. In the later nineteenth century the economist Alfred Marshall regarded the location of raw materials as the prime determinant of industrial location. Alfred Weber's theory of the location of industry (1909) marked a considerable advance, first by taking account not only of the costs of procuring localized materials but also of transporting them to the chosen place of production. He distinguished between "pure" materials, which impart their total weight to the product, and "gross" materials, of which only a part of the weight is employed. Through an analysis of the relative costs of transporting materials and products, Weber hypothesized a point of minimum transport costs at which the production process would tend to localize. Secondly, Weber also examined the costs involved in the transformation of materials, including labor costs. He gave attention, thirdly, to the advantages that could be obtained from the agglomeration of industries at particular places. Savings in the costs of production could pull manufacturing away from the point of minimum transport cost.

Weber's ideas were very influential and are still of some value in considering, in the contexts of their times, the locations of industrial areas based on heavy industries relying on large inputs of coal, iron ore, nonferrous metals, limestone, clay and other materials that lose weight or bulk in processing. However, his ideas have been subjected to severe criticism. August Lösch (1954) attempted to remedy Weber's overemphasis on supply-side considerations by incorporating demand factors, especially the size of the market. Hoover (1948) introduced more realistic ideas on transport costs and refined Weber's ideas. Isard (1956) and Greenhut (1956) examined the factors that would determine the optimum location, that is, the point that would combine the lowest costs with the highest attainable revenue and Isard, in particular, linked the field of the location of economic activity with that of regional development, formulating ideas that collectively formed the basis of "regional science." However, ideas derived from Weber still find expression in later writing, for example (Estall and Buchanan 1980):

> "... if the material loses greatly in weight or bulk in manufacture it will cause the industry to be attracted to the point of material production."

Beet sugar, pulp and paper, and pig iron production are industrial activities cited in illustration. Attention is also given to the value per tonne of the raw material, for materials of high value in relation to weight (e.g., wool) can more easily bear the costs of transport, and to the degree of perishability of the material. The location decision will also depend on the extent to which materials are substitutable, for example, one metal for another, or plastic materials for metals.

In an age in which industrial processes have become more complex, adding high value in the course of manufacture, relying on the assembly of a wide variety of manufactured or semimanufactured materials and depending on large inputs of capital and technical skill and on efficiently organized production systems, greater emphasis has been placed on the organizational and locational relationships between enterprises and plants in moving materials through long chains of processing from raw materials to finished product. An important idea in explaining the geographical association of industries has been that of industrial "linkage," the grouping of interrelated industrial concerns in one area. Florence (1948), for example, identified four types of linkage: (a) vertical, where materials are passed through consecutive stages from raw material to finished product, as in the steel or nonferrous metal industries; (b) horizontal, where firms produce individual parts that come together on the assembly line; (c) diagonal, where firms make products or provide specialized processes that are required at various stages in the manufacturing chain; and (d) a further general type where local industrial specializations depend on the availability of materials, capital, services and labor skills available in one area but not elsewhere. The concept of linkage has significantly assisted understanding of the reasons for the growth of highly concentrated industrial areas, such as the Ruhr or the Black Country.

Such ideas as industrial linkage, while still helpful in observing flows of materials through industrial systems and in assisting explanations of industrial locations, have, to some extent, lost their former potency as more flexible transport and communication systems have developed, thereby reducing the need for the close colocation of manufacturing plants. However, the ideas are still helpful in explaining the geographical concentration of industries making advanced products and employing a high degree of innovative skill both in long established trades such as jewelry manufacture and the automobile industry and in more recent science-based industries (Oakey 1981, Pacione 1985).

Summarizing, it may be said that the role of materials supply in location decisions varies according to:

(a) the extent to which materials lose weight or bulk in processing or manufacture;

(b) the number of materials and material sources that are to be employed and their relative importance;

(c) the extent to which the availability of materials is general or localized;

(d) the degree of perishability of the materials;

(e) the value of the material in relation to its weight or bulk;

(f) the possibility of using substitute materials;

(g) the availability of transport and communication facilities and the structure of freight rates and other charges; and

(h) the availability of suitable technology, capital equipment, labor skills and productivity.

However, considerations of local raw material availability and weight loss have, generally speaking, become less important in the light of the striking trend towards economy in raw material and fuel use. The iron and steel industries, formerly dependent on local supplies of coal (coke) and iron ore, provide an example. Technical progress brought dramatic economies in the requirements for coal and coke. Transport costs declined relative to other costs from the mid-nineteenth century until well into the present century. The development of bulk carriers facilitated imports and long-distance transport generally. Beneficiation of ores at or near the mine made available concentrated ores for shipment. Improvements in electricity transmission and the installation of oil and gas pipelines have generally reduced dependence on local fuel and energy sources. In the nineteenth century international trade in iron ore was negligible; by the late 1960s it had risen to more than one-third of world production, much of it long-haul traffic from Venezuela to the USA or from Australia to Japan. Further, new methods of organization and control have, in many countries, reduced the scale of the demand on manpower, and skill has become a more important attribute of the workforce than numbers.

Such developments have reduced the importance of locational explanations based on Weberian material availability and transportation considerations. New directions of work have appeared. One such direction (Manners 1971) has produced more sophisticated examinations of market conditions and of supply costs. A second direction has studied the political and institutional bases of materials production. This takes into account the very different assumptions that lie behind decision making in capitalist societies as compared with those organized on state socialist principles. Structuralist approaches have been adopted in studies of industrial change. Some students of materials production have directed attention to the nature and effects of national government policies; there has also been a growth of interest in the outcomes of policies adopted by regional trade and economic groups, such as the EC, COMECON or LAFTA. Attention has been given to the influence of international commodity agreements. The effects of the policies of OPEC on world fuel markets have been examined. A very important direction of work has been into the process of decision making by large firms, especially by multinational companies (Watts 1980). At least 9500 multinational firms operate information and trading systems across national boundaries and are responsible for major capital investments influencing the location and scale of materials production and processing. International investments by Japanese companies have attracted interest in this connection.

The geographical differences to which reference has been made earlier become still more apparent when the use of materials in manufacturing industry is considered. The major manufacturing industrial systems are concentrated in North America, Europe and Japan. They are, as has been shown, dependent on imports of raw and semiprocessed materials from one another and from elsewhere. It is mainly in these regions that high value is added in the course of manufacture; the materials pass through many stages of refining, processing and assembly, for example in the high-technology industries. In these regions, too, lie the major capital investments and the most advanced skills. Economies of scale are obtained, where appropriate, through the establishment of large plants or the agglomeration of related enterprises or the close interconnections that may now be obtained from the latest forms of electronic communication.

The distribution of manufacturing industry may be seen as the latest stage of development in a process of geographical change and diffusion that began in the English Midlands in the eighteenth century and was taken up, adapted in form, structure and organization in Western Europe, the USA, the Soviet Union and Japan (Hamilton and Linge 1981).

Table 1, compiled from estimates by the United Nations Industrial Development Organization (1988), gives an indication of the relative importance of developed and developing economies in the world distribution of manufacturing industry.

It must be remembered that over this period the global value of industrial production has greatly increased, and probably has more than doubled, so that in absolute terms the gap between developed and developing countries, which seems on a percentage basis to have narrowed, has in actual terms widened. The centrally planned economies have achieved a small increase in share. The developing countries as a whole have achieved only very limited improvement. However, manufacturing growth rates have been high in some countries: Brazil, Mexico and countries in south-east and east Asia such as South Korea, Hong Kong, Singapore and Taiwan are examples. Some of these countries which began their industrialization with relatively unsophisticated processes such as textile and clothing manufacture have developed more sophisticated industries such as electrical equipment, electronics, petrochemicals, mechanical engineering and automobiles.

Table 1
Distribution of world manufacturing industry by economic groupings (value added)

	Developing countries	Centrally planned economies	Developed market economies
1970	10.4	16.0	73.6
1975	12.0	19.5	68.5
1980	13.0	20.2	66.8
1985	12.8	21.2	66.0

Source: United Nations Industrial Development Organization (1988)

Despite the emergence of the newly industrializing countries, the concentration of high-technology materials use in the so-called developed industrial world is very strong. Under current economic and international conditions, the transition to the postulated New International Economic Order, in which developing countries would have a much larger share in the processing and transformation of raw materials, seems likely to be a very difficult objective to achieve.

7. Conclusion

This article attempts to illustrate the variety and complexity of the considerations that must be examined to obtain an understanding of the geography of materials extraction, manufacture and use. In the cases of raw materials obtained from agriculture, fishing, forestry and mineral extraction, the conditions of the natural environment must be fully examined. Variations that arise from changes in technology, especially those affecting demand, changes in relative accessibility, transport costs and the effects of economic conditions and the terms of trade must also be studied. As materials technology becomes more sophisticated, other considerations intrude, such as the nature of industrial organization, the linkages between processing operations and marketing operations, and the availability of high-level technology and capital. In all cases, the influence of the policies exerted by governments, multinational corporations and by international trade agreements must be examined. Studies of the ways in which materials pass through the industrial system from place to place and from plant to plant before finally emerging transformed into finished products can play a useful part in formulating general principles to assist understanding of the complex industrial world and in increasing the efficiency of the industrial system itself.

See also: Economics of Materials, Energy and Growth; Resource Appraisal; Structure of Materials Industries

Bibliography

Alexandersson G, Klevebring B 1978 *World Resources.* de Gruyter, New York

Atlas of Earth Resources 1979 Mitchell Beazley, London

Bliss C, Boserup M (eds.) 1980 Economic Growth and Resources, Vol. 3. Natural Resources, Cambridge University Press, Cambridge

Blunden J 1985 *Mineral Resources and their Management.* Longman, Harlow, UK

Wise M J, Rawstron E M (eds.) 1952 *R O Buchanan and Economic Geography.* Bell, London

Carlson A S 1956 *Economic Geography of Industrial Materials.* Van Nostrand Reinhold, New York

Chisholm M 1982 *Modern World Development: A Geographical Perspective.* Hutchinson, London

Dicken P 1986 *Global Shift.* Harper and Row, London

Dixon C J 1979 *Atlas of Economic Mineral Deposits.* Chapman and Hall, London

Economist Atlas, The 1989 Hutchinson, London

Estall R C, Buchanan R O 1980 *Industrial Activity and Economic Geography*, 4th edn. Hutchinson, London

Florence P S 1948 *Investment, Location and Size of Plant.* Cambridge University Press, Cambridge

Greenhut M L 1956 *Plant Location in Theory and Practice: The Economics of Space.* University of North Carolina Press, Chapel Hill, NC

Hamilton F E I, Linge G J R 1981 *Spatial Analysis, Industry and the Industrial Environment*, International Industrial Systems, Vol. 2. Wiley, New York

Hewett D F 1929 *Cycles in Metal Production*, Technical Publication 183. American Institute of Mining, Metallurgical and Petroleum Engineers, New York

Hoover E M 1948 *The Location of Economic Activity.* McGraw-Hill, New York

Industry and Environment published quarterly by United Nations Environment Program, New York

Isard W 1956 *Location and Space-Economy.* Wiley, New York

Leith C K 1931 *World Minerals and World Politics.* Whittlesey House, New York

Lösch A 1954 *The Economics of Location.* Yale University Press, New Haven, CT

McDivitt J F, Manners G 1974 *Minerals and Men.* Johns Hopkins University Press, Baltimore, MD

Manners G 1971 *The Changing World Market for Iron-Ore 1950–1980.* Johns Hopkins University Press, Baltimore, MD

Manners G 1981 Our habitable planet—a geographical appraisal. 2. Our planet's resources. *Geogr. J.* 147 (3): 1–22

Mikesell R F 1979 *The World Copper Industry.* Johns Hopkins University Press, Baltimore, MD

Mikesell R F, Whitney J W 1987 *The World Mining Industry, Investment, Strategy and Public Policy.* Allen and Unwin, London

Oakey R P 1981 *High Technology Industry and Industrial Location.* Gower, Farnborough, UK

Oxford Economic Atlas of the World 4th edn. 1972 Oxford University Press, Oxford

Pacione M (ed.) 1985 *Progress in Industrial Geography.* Croom Helm, London

Pearce D W, Turner R K 1990 *Economics of Natural Resources and the Environment.* Harvester Wheatsheaf, New York

Rees J A 1990 *Natural Resources: Location, Economics and Policy*, 2nd edn. Routledge, London

Richards T 1983 *Metallic Mineral Mining: An Introductory Analysis.* Causeway Press, Ormskirk, UK

Ruddle K, Manshard W 1981 *Renewable Natural Resources and the Environment.* Tycooly, Dublin and United Nations University, Tokyo

Schmitz D J 1979 *World Non-Ferrous Metal Production and Prices 1700–1976.* Cass, London

Stamp L D 1980 *Chisholm's Handbook of Commercial Geography*, 20th edn. Longman, Harlow, UK

United Nations 1989 *Industry and Development: Global Report 1988/89.* UN, Vienna

United Nations Conference on Trade and Development 1988 *Commodity Year Book.* UN, New York

United Nations Industrial Development Organization 1988 *Handbook of Industrial Statistics.* UN, Vienna

Watts H D 1980 *The Large Industrial Enterprise.* Croom Helm, London

Weber A 1909 *Theory of the Location of Industries.* University of Chicago Press, Chicago, IL

Wolfe J A 1984 *Mineral Resources: A World Review.* Chapman and Hall, London

World Commission on Environment and Development 1987 *Our Common Future.* Oxford University Press, Oxford

Zimmermann E W 1951 *World Resources and Industries.* Harper, New York

M. J. Wise
[London School of Economics and Political Science, London, UK]

Economic Theory of Mineral Resources and Markets

Theoretical economics provides analytical frameworks to aid the understanding of how nonrenewable resources are efficiently discovered, developed and valued over time through market operations. Theories permit empirical tests of hypotheses; for example, interpretations of data on mineral reserves, the rates of depletion, technical change and the trends in prices or production. They also offer current insights into the uncertain future value of resources.

1. Equilibrium and Capital Theory

Because mineral resource information is costly to collect, the efficiency of a given theoretical framework and its correlates, the information's structure and the data, are always important considerations to both industry and public policymakers. The efficiency of an economic information system is defined by its ability to reach equilibrium solutions to problems and by the degree to which a particular theory models well important aspects of the market, that is, with parsimony, special insights and some measure of confidence. How well a theory models reality describes its validity. Measures of confidence depend on how well the important relations of the market, such as supply and demand, are specified and identified, and on how robust the results are under changing conditions. The sensitivity of an application is examined by describing the nature of the equilibrium and its stability under the specified market adjustment mechanism.

A materials industry's market structure is defined by the behavior of its buyers and sellers and varies by industry and over time. For example, the dominance of a few sellers in a national market has characterized particular nonfuels at times. The behavior of a firm in such a market will be different from that of one among many sellers, namely, prices will be higher and levels of output lower than in the more competitive case. Furthermore, both the industry's technologies and information on the relevant economic geology of deposits change. Thus, a single static or dynamic model is not likely to accommodate all the interesting questions addressable to resource information, requiring that the objectives of different mineral economics applications be properly distinguished. The building of an appropriate theoretical model can assist an industry firm in negotiating contract terms, in predicting the movements of spot market prices in the short run, or in critically assessing the longer-run policy decisions of agencies concerned with mineral conservation, regulation or price stability.

Most fundamentally, mineral deposits are a form of capital and their transformation from the asset side of the industry's balance sheet into flows of materials from producers to consumers is a major problem of mineral economics analyses. This article is limited to four principal issues of applied capital theory under perfectly competitive market conditions:

(a) the valuation of specific depletable resources;

(b) the dynamics of capital investment's growth or decline under conditions of risk and uncertainty;

(c) the problems of interregional trade, including transfers of mineral wealth; and

(d) the problem of growth under rent conditions arising from market imperfections or rising costs of adjustment.

This division emphasizes the analysis of mineral and energy resources as nonrenewable capital stocks and flows and hence the inevitability of exhaustion, the uncertainties and high risks of exploration, environmental damage and other economic externalities and the prominence of spatial equilibrium and economic rents. A glossary of convenient terms describing mineral markets is provided in Table 1. In the table and text all variables, and often parameters as well, can be functions of time t but, for convenience, the t is generally omitted in the notation.

2. Mineral Resource Valuation

In the long run, future expected prices $\hat{p}_t$ form the basis for measuring the extent of resources and their potential for transformation to reserves over time. This transformation takes place largely through investment I in exploration and mine development. Reserves are valued by finding the present discounted worth of their expected net income stream $\hat{V}$, given the rate of interest r_t and the risk premium, just as in the case of renewable capital assets. However, the expected price path will rise, due to depletion as deposits are mined to exhaustion. It is optimal for each firm to equate in all time periods the present discounted net return received over its deposit life. Assuming constant deposit quality and zero extraction costs for simplicity, this means that in the time interval $(t, t + \theta)$ the return to conserving the resource is $(p_{t+\theta} - p_t)/p_t$. The optimal rate of depletion will thus equal the return for holding a bond

Table 1
Glossary of terms and symbols

Symbol	Term
a	Vector of supply entry costs in consuming center
b	Capacity and demand constraints vector
c	Adjustment cost
d	Own-price slopes of demand vector
e	Variable extraction costs vector
f	Demand intercept vector
g	Transportation costs
h	Production function, $q = h(z, k)$
i	ith input industry or region
j	jth receiving industry or region
k	Capital stock (balance sheet asset)
l	Output-to-capital ratio, q/k
m	Number of producing industries $(1, \dots, i, \dots, m)$
n	Number of receiving industries $(1, \dots, j, \dots, n)$
q^i	Quantity supplied at time t from the ith source
p^i	Supply price $= f(q^i)$ at time t at the ith source
q^j	Quantity demanded at time t at the jth sink
p^j	Demand price $= f(q^j)$ at time t at the jth sink
$\bar{q}$	Initial quantity vector $(\bar{q}^j, \bar{q}^i)$
$\bar{p}$	Initial price vector $(\bar{p}^j, \bar{p}^i)$
$q^j - \Sigma_i x_{ij}$	Excess demand in j (if positive) Excess demand in i (if negative)
$\dot{p}$, dp/dt	Mechanism of price adjustment
$\hat{p}$	Price expectations $(\hat{p}^j, \hat{p}^i)$
$\hat{q}$	Quantity expectations $(\hat{q}^j, \hat{q}^i)$
r	"The" rate of interest
s	"The" cost of a unit of capital
t	Time
u	Vector of errors in observed price
v	Vector of errors in observed lags
w	Vector of input flow prices
x	Aggregate industry output $= (z, k_0, I)$
y	Vector of individual deposit outputs
z	Vector of interindustry input or regional flows
$E(V)$	Expected value of V; mean value $\hat{V}$
G	Matrix of zeros and vectors routing flows
H	Hamiltonian; $H(\bar{k}_0, \Gamma) = \sup_{w,I} H(z, I)$
I	Gross investment; net $= k = (I - \delta k)$
Q	Matrix of intercepts, slopes and bounds
R	Matrix of constraints
S	Matrix of own-price slopes of supply
V	Present discounted value of an income stream
α	Discount factor
β	System parameter; $\beta(t)$ when a function of time e.g., $\beta(t) = \alpha_{t+1}/\alpha_t$ is a discount rate when $\beta(t)$ a constant < 1, $\alpha(t) = \beta^t$
Γ	System shadow price (Lagrangean)
Φ	Cash flow $= pq - wz - sI + s(1 - \delta)k_{t-1}$, levellized $= pq - wz - sI$, not levellized
θ	Time difference lag

of equivalent risk the same amount of time, $r_t\theta$ (Hotelling 1931). This "pure theory of exhaustion," therefore, produces the basic arbitrage equation forecasting an upward path for mineral prices:

$$\dot{p}/p = r_t \qquad (1)$$

If the interval describes the life of a single mine, the typical grade–tonnage relation will reinforce Hotelling's result, and there will always be an economic rent or "user cost" assignable to the deposit above monopoly or other rents and costs. If the interval describes long-run industry supply, then offsetting factors such as technological change, substitution or capital stock effects can offset or diminish the effects of depletion. However, in the very long run, resource depletion impacts are inevitable and imply difficult choices for society in the determination of levels of conservation.

In the more immediate run, current market prices p_t determine the decision of both the level of producers' output and its allocation over existing capacities. Figure 1 depicts how conventional theories of market equilibrium and growth portray a typical materials market by a system of equations and an adjustment mechanism. These can be used to explain the pattern of observed price variations in specific investigations. At point A($\bar{p}$, $\bar{q}$), an initial equilibrium, excess demand is zero for both short- and long-run supply and demand. Assuming a shift in long-run "permanent" demand from some initial level OA to a higher level O*f*, due, say, to materials-intensive innovation, positive excess demand ($\hat{q} - \bar{q}$) results. Such shortages are removed in the short run by firms increasing outputs from existing supply capacities to the expected point $\hat{\mathrm{B}}$($\hat{p}$, $\hat{q}$).

Emphasis is placed on the mechanism of such adjustments, which typically relate the magnitude of excess demand to the speed with which changes in price occur from A to $\hat{\mathrm{B}}$ and to the initial variable values:

$$\dot{p} = f[(\hat{q}_t^j - \bar{q}_t^i), \bar{p}_t, \bar{q}_t] \tag{2}$$

Here *i* identifies the *i*th commodity source and *j* the *j*th consuming center. In the longer run, capacity changes are indicated by the long-run supply function AC. New capital in the form of exploration or development will increase reserve response to the higher price expectations, forcing a limit to profits at $\hat{p}$ and eventually a decline, following the intermediate or "lagged" supply schedule AB. However, the ultimate long-run equilibrium price and quantity after all such adjustments at C(p', q') are expected to be higher than at A, supporting

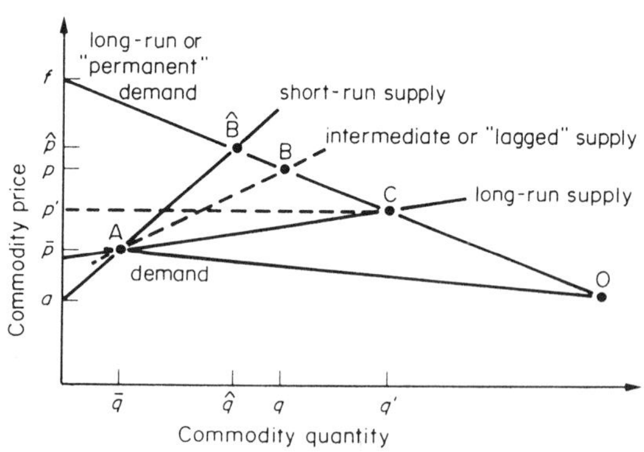

Figure 1
Short- and long-run supply adjustments to a change in "permanent" demand

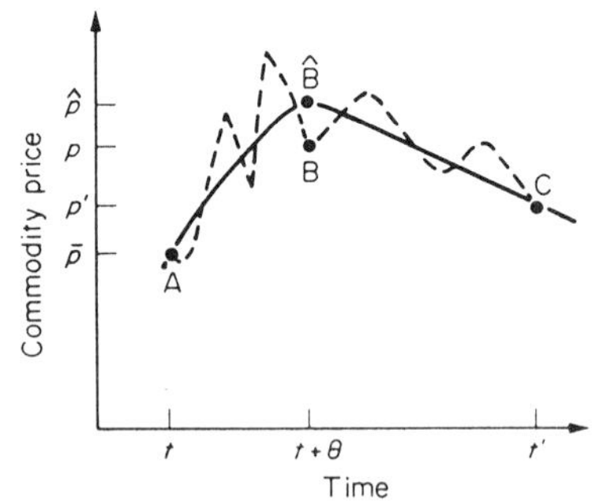

Figure 2
Expected (——) and observed (– – –) price paths under dynamic demand and uncertain conditions

the solid-line pattern of price and output changes over time in Fig. 2 forecast by depletion theory. Such forecasts assume that technological change does not offset depletion entirely by shifting AC downward and to the right.

Price, quantity and other key random variables are forecast as probability distribution functions with finite variance and mean. Let the mean values or expectations $\hat{\mathrm{B}}$($\hat{p}$, $\hat{q}$) be accepted as "certainty equivalents" in Fig. 1 for the received price and market clearing quantities supplied from a specific deposit or project. Then the "wedge" given by $\hat{p}\hat{\mathrm{B}}a = \Phi$ represents net cash flow:

$$\Phi = pq - wz - sI + s(1-\delta)k \tag{3}$$

where Φ represents the area *py* (total revenues) less area $a\hat{\mathrm{B}}\hat{q}$ (operating costs including levellized capital investment, viz. *sI* net of undepreciated asset) approximating surplus available for return on capital. The deposit valuation from this income approach comes by adding up the period discrete cash flows over the expected asset life, that is, the sequence of "wedges" $\Sigma_t\, \Phi(t)$. This sequence is simulated by forecasting shifts in supply and demand, and discounting by the rate β_t, the firm's cost of capital:

$$\beta_t = \alpha_{t+1}/\alpha_t$$

Equation (3) employs the process engineer's and accountant's data on the production function *h* relating output flows *q* to variable input flows *z* and the stock of capital *k*, that is, $q = h(z, k)$, and on the cost function, relating input prices *w*, the price *s* of units of gross capital investment *I*, and depreciation δ to output. dk/dt then gives the growth rate of net assets k_t and $\dot{k} = 0$ in the limit. This is a classical balanced growth assumption when gross investment is limited to the depreciation plus the surplus generated and borrowings restricted by capital/output (debt/equity) ratios (Newcomb 1979):

$$k_t = I_t + (1-\delta)k_{t-1} \tag{4}$$

Assume, following Nerlove (1972), a rising cost of capital to the firm which is proportionate to investment and quadratic, $(c/2)I^2$. Then, in discrete time, given the discount factor $\alpha_t = \beta^t$ constant, and h log–linear quadratic, the present value of its deposit over a life of interval (t, t') is:

$$\hat{V} = \sum_{t}^{t'} \beta^t[ap_tq_t - w_tz_t - s_tI_t - (c/2)I_t^2] \qquad (5)$$

Within this "rational expectations" framework, following Muth (1961), the firm with its use of "certainty equivalents" attempts to maximize the conditional expectation of each project. The engineering estimate of the project's value from the replacement cost approach is, of course, sI. The ratio of expected benefits to costs $\hat{V}/sI$ is the expected yield or marginal efficiency of the investment to maturity. When marginal yield exceeds average yield increased investment is signalled. If the firm chooses to increase projects ranked and undertaken until yield of the marginal project approaches the firm's cost of capital, this strategy in open markets is optimal. If the firm cannot, because of some constraint expected to be maintained, increase its investment, the value of the firm's capital stock will be raised by the market to bring its yield in line with those of similar risk. If markets are protected or restricted, other strategies may be preferred. In any case, the optimal investment limit considers the risk premium and level of risk aversion, which determine the proportion of risky to risk-free assets the firm strives to maintain on its balance sheet.

Economic geologists (see *Mineral Resources: Definitions, Uses, Classifications and Future Availability*), in collaboration with mineral economists (see *Resource Appraisal*), have recently addressed the problems of integrating geological information more explicitly into the modelling of changes in the resource–reserve base of large regions. Following Harris (1982), such are termed "broad" mineral appraisals, because they broaden market theory applications to include the assessment of a region's stock of deposits or endowment and the partitioning of the set of potential and known deposits into economic and uneconomic subsets. The discovery of these reserves is modelled, ordering them by their net present values. Such a stocks schedule is a "potential supply function." The forward prices indicated by such appraisals supplement the results of the conventional economic models described above.

3. Dynamic Equilibrium under Uncertain Conditions

While benefits can be derived from the insights of conventional geostatistical and economic theory applications as described, the equilibrium conditions of the competitive models remain essentially "stationary," and the pretense is required that the futures they forecast are "certain." Even with such simplifications, the mathematical requirements to ensure modelling solutions are quite severe. For instance, production must be "well behaved," in the sense that no interdependencies can exist among agents or activities; externalities, such as pollution, are not easily accommodated; no economies are permitted such as those of scale; no costs of information, transport or transactions are included; and so on. The mineral industries offer abundant evidence that these restrictions do not portray realistically the dynamic and uncertain decisions of actual markets. For instance, investment choices require the undertaking of unknown risks of exploration, based on expectations of prices and profits as far removed as 30 years.

The question of how conventional theory might be rendered more dynamic and econometrically relevant has occupied economists since Jevons (1865) analyzed Britain's coal industry. A step toward realism taken by more recent theorists to reflect uncertainties relies on the observation that the price expectations of suppliers depend on assumptions about past observed prices. This leads to difference equation adjustment mechanisms which produce fluctuating price paths because of lags:

$$p_t = \beta_1 + \beta_2 p_{t-\theta} \qquad (6)$$

The constants β of this equation are reductions, being functions of the slope and intercept parameters of the structural supply and demand equations depicted in Fig. 1. Depending on the β values, distributed lag models produce oscillating price behavior more suggestive of the observed short-term cyclical variations of mineral prices about an expected trend. A case is shown by the dotted path of prices over the interval (t, t') in Fig. 2. In this formulation, the stability conditions are open to empirical inquiry, while explicit tests of the inflexibilities and lags in supply or demand associated with observed market fluctuations can be econometrically attempted.

The excess demand differences can similarly be employed to measure the extent to which intermediate price expectations on the way to longer-run target outputs fail to be met in actuality $(\hat{p} - p)$. This can be the result of inflexibilities in capacity adjustments in capital-intensive segments of a minerals industry. In such a case, the estimation of the distributed lags can be based on the relationships observed between ongoing investment and the previous output prices. When the shifts of supply or demand can be exactly identified, the estimation of the β coefficients in Eqn. (6) permits the representation of actual price oscillations p_t as the addition to expected prices $\hat{p}$ of disturbed departures of observed prices from that expectation, p_t^*, plus an error term. In a simple example by Nerlove (1972)

$$p_t = \hat{p} + p_t^* + u_t \qquad (7)$$

where

$$p_t^* = \beta p_{t-\theta}^* + v_t \tag{8}$$

Under conditions where the error terms u and v are normal and independent with zero means and constant variances, $(\hat{p} + p_t^*)$ may be interpreted as the long-run normal or trend from initial price $A(\bar{p}, \bar{q})$ to target price $C(p', q')$ in Fig. 2. Then, the mean square error forecasts of $(p_{t+\theta} - \hat{p})$, that is, $(\hat{B} - B)$, will be the conditional expectations of the price deviations at $t + \theta$ through time t. Under these simple conditions, and with target prices fixed, well behaved trends and measures of confidence can be stated for the statistical estimates of the distributed lags obtaining for actual materials market price data.

4. Spatial and Temporal Market Equilibrium

Empirically useful valuations of regional materials flows often require a spatial framework for the description of market equilibrium. Following Samuelson (1952), multimarket models offer a mathematical representation of allocative efficiency that is consistent with activity analysis (input–output) approaches as well as the maximization of surplus (rents) in interregional trade. These approaches reduce multicommodity and multimarket equilibria to the joint minimization of total delivered resource costs, and recognize economic rents as an essential part of the allocation problem. Rents can be computed by region, as in the seminal programming study by Henderson (1958).

Programming theory presents an ideal model of linked markets as multiple sources i and sinks j interconnected by transport costs g, in which both aggregated (x) and disaggregated (y) supplies are defined in optimal interregional flow patterns G. In his simple linear program, Henderson fixed demands. Then the matrices of parameters for supply singly (S) and jointly with other constraints (R) are easily identified and the net social payoff is equivalent to economic rent. The surplus is maximized by finding $(y, z) > 0$, such that

$$\min_{y,z} \; e^{\mathrm{T}}y + g^{\mathrm{T}}z$$

$$\text{subject to } Ry + Gz \geqslant b \quad y, z > 0 \tag{9}$$

The equivalent quadratic programming objective function is

$$\min_{y,z} \; a^{\mathrm{T}}x + \tfrac{1}{2}x^{\mathrm{T}}Sx + g^{\mathrm{T}}z$$

$$\text{subject to } Qx + Gz \geqslant b \quad x, z > 0 \tag{10}$$

where T indicates the vector transpose.

As reformulated, Eqn. (10), following Takayama and Judge (1971) and their log–linear version (dynamic program), has been increasingly employed to present an efficient spatial informational system for comparative studies of the reassignment of mineral rents, for example, through taxation or changes in the transportation system. An example is provided by oil markets where such rents, represented by the area $\bar{p}Cp'$ in long-run competitive equilibrium, can be significant. Clearly they are most impressive in the short run, represented by the area $a\hat{B}f$ in Fig. 1.

5. General Theory of Resource Policy and Control

While the theories described have led to valuable insights into decisions under the stylized "linear" and competitive conditions assumed by mineral economists for the study of market and investment behavior in the basic resource industries, they do not always cope well with the presence of temporal and spatial constraints in actual markets, including subsidized technological changes, and environmental and nontariff barriers in international trade, when, as hypothesized in Figs. 1 and 2, these introduce significantly rising costs of adjustment into real-time and real-space analyses. Then, the nonlinearities created can readily confer very large surpluses or rents to some firms and countries engaged in trade so restricted (with corresponding losses to other regions and firms). The modifications required to cope with real-time rising adjustment costs, pioneered early by Eisner and Strotz (1963) and Lucas (1967) among others, offer a new paradigm of resource policy analysis on both broad (macro) and industrial (micro) resource appraisals. Based on the variational calculus, and the maximum principal of the Pontryagin et al. (1962) or Bellman (1957) type, applications have taken two interesting directions, facilitated by the very large advances achieved since 1980 in computational equipment and skills. In simplistic terms of Fig. 1, the balance equation permits short-run equilibria such as $\hat{B}$ and B to be "optimally off" any real-time average long-run cost path such as $\overline{AC}$ because the present value of adding a unit of variable input at any time t on the short-run (operating) margin is exactly equal to the present value of adding one more unit of capital along the long-run (investing) margin.

One direction taken in regional science policy analyses, exemplified by Nagurney and Aronson (1988), applies techniques of the variational calculus to more general dynamic spatial price equilibrium formulations in which the static models of Sect. 4 can accommodate rising costs of transportation, the feedback of inventory changes and the like. These formulations make explicit some of the mechanisms of change left to *ad hoc* assertion in applications of distributed lag theory. The increase in realism promises also to apply effectively to evaluating strategies in marketing and exchange when open market competition is restricted.

An alternative direction, opened to economists interested in the investment capital problems characteristic of the resource industries by Dorfman (1969) and Nerlove (1972) is that of optimal control theory, in which the real-time path of variables in nonlinear differential equation systems can be simulated under different assumptions and constraints. These applications (see, for example, Murata 1982) are closely related to improvements in engineering techniques for solving problems in dynamic nonlinear differential equation systems, and seem appropriate for the more realistic modelling of policy impacts, such as taxes, tariffs and the like. Of equal interest are technical shifts, such as that of biased (capital intensive) technological change capable of increasing productivities and competitiveness along with rising real wages. Technical applications encourage the collaborations of quantitative economics and engineering which have long characterized investigations of material sciences and engineering.

In brief, assuming a production function in which the vector x represents z, k and I, that is, $q = h(z, k, I) = h(x)$, where $dk/dt = I - \delta k$, the problem considered is to maximize the present expected value EV in continuous terms. As before, the t subscript is understood for stock and flow variables, all of which are subject to variation over time:

$$EV = \hat{V} = E\int_t^{\infty} (pq - wz - sI)\alpha_t \, dt \qquad (11)$$

To solve the problem for constant α_t requires forming the present value Hamiltonian

$$\begin{aligned} H(\bar{k}_0, \Gamma) &= \sup_{w,\, I} \quad H(z, I) \\ &= ph(x) - wz - sI + \Gamma(I - \delta k) \qquad (12) \end{aligned}$$

The solution is characterized by the short-run rising-cost curve associated by diminishing returns on the n-dimension internal margin of $h(x)$ and, on the $(n + 1)$th dimension of time, by $\Gamma(t)$ on the rising-cost long-run supply function which is the optimal, really nonlinear path of potential total cost. Optimization via the differentiation of H reflects the balanced growth Pontryagin rule: at every moment t the aggregate cost to achieve, or the value of, an added unit of output on the short-run margin by the variation of current input flows will equal the value of the integral representing the potential value of the capital contribution to target time t' as the industry expands capacity. A necessary condition (transversality) at all times is that the Lagrangean not decline. If it rises at a constant rate over the interval (t, t') the problem may be reduced to a general solution by quadrature methods. If it rises at varying rates, engineering techniques given by numerical control methods may approximate solutions of interest. In this theory of policy, the interesting policy recommendations stem from the equilibrium conditions giving the values of the shadow prices. For evaluations of imperfectly competitive environments, the recommendations stem from the implications of differential game theory (see, for example, Ulph and Ulph 1989).

See also: Markets for Materials: Modelling; Prices of Materials: Theory; Prices of Metals: History and Current Problems; Resource Appraisal

Bibliography

Bellman R 1957 *Dynamic Programming.* Princeton University Press, Princeton, NJ

Dorfman R 1969 An economic interpretation of optimal control theory. *Am. Econ. Rev.* 4: 817–31

Eisner R, Strotz R 1963 Determinants of business investment. *Impacts of Monetary Policy.* Prentice-Hall, Englewood Cliffs, NJ

Harris D 1982 Mineral endowment, geostatistical theory and methods for appraisal. In: Newcomb R (ed.) 1982 *Future Resources: Their Geostatistical Appraisal.* West Virginia University Press, Morgantown, WV, pp. 83–135

Henderson J M 1958 *The Efficiency of the Coal Industry.* Harvard University Press, Cambridge, MA

Hotelling H 1931 The economics of exhaustible resources. *J. Polit. Econ.* 3: 137–75

Jevons W S 1865 *The Coal Question.* MacMillan, London

Lucas R E 1967 Optimal investment and the flexible accelerator. *Int. Econ. Rev.* 8: 78–85

Murata Y 1982 *Optimal Control Methods for Linear Discrete-Time Economic Systems.* Springer, New York

Muth J F 1961 Rational expectations and the theory of price movements. *Econometrica* 29(1): 315–35

Nagurney A, Aronson J 1988 A general dynamic spatial price equilibrium model: formulation, solution, and computational results. *J. Comput. Appl. Math.* 22: 339–57

Nerlove M 1972 Lags in economic behavior. *Econometrica* 40: 221–50

Newcomb R 1967 Measuring technical progress in the resource industries. *Council of Economics Proceedings.* American Institute of Mining, Metallurgical and Petroleum Engineers, New York, pp. 53–68

Newcomb R 1979 Growth and change in the American coal industry. *Growth and Change* 10(1): 111–37

Pontryagin L S, Blotyanskii V G, Gamrelidze R B, Mishchenko E F 1962 *The Mathematical Theory of Optimal Processes.* New York

Samuelson P A 1947 *Foundations of Economic Analysis.* Harvard University Press, Cambridge, MA

Samuelson P A 1952 Spatial price equilibrium and linear programming. *Am. Econ. Rev.* 52: 283–303

Samuelson P A 1978 The canonical classical model of political economy. *J. Econ. Lit.* 16(4): 1415–28

Takayama T, Judge G G 1971 *Spatial and Temporal Price and Allocation Models.* North-Holland, Amsterdam

Ulph A, Ulph D 1989 Gains and losses from cartelization in markets for exhaustible resources. In: Van der Ploeg F, de Zeeuw A J (eds.) 1989 *Dynamic Policy Gains in Economics.* Elsevier, Amsterdam, pp. 157–89

R. Newcomb
[University of Arizona, Tucson, Arizona, USA]

Economics of Materials, Energy and Growth

The principal determinants of the relations observed between a country's materials usage, its energy consumption and the production of final goods demanded are access to natural resource endowments and the structure of its economy. Both assume at any time the country's level of capital investment and stage of industrialization are given, as well as the degree of free trade linking its economy to the rest of the world through imports and exports. Over time, changes occur in technologies and in other dynamic factors, including the concentration of technical superiority in a few globally dominant firms, at times fostered by industrial policies distorting free trade, so that the structure of the world economies will change. For this reason, a historical review would appear to offer the simplest approximation to and best insights into the changing patterns of materials and energy use as a developed economy grows. This article employs the Leontief structure of the postwar US economy and data on corresponding national accounts from 1947 to 1982 for measures of materials and energy input and final output values. Then, noting the break in trends from 1972, it looks at these and future changes and implications, dividing the discussion into three parts. First, the years from World War II through to 1972 represent a period of unparalleled growth in which increasing free trade in materials and fuels helped to establish stable prices and rather consistent patterns of use among these sectors and the rest of the economy. Second, beginning in 1973, these patterns were disturbed as both world trading conditions and the organization of the international markets changed significantly. A surge of cartel activity and an increasing number of restrictions on open market exchange reversed the trends toward trade liberalization and growth that prevailed earlier. New technologies appeared, some of them influenced by the threat of resource scarcities. From 1973 to 1992 the effect of these changes on the competitiveness of the US economy are notable illustrating the more dynamic factors underlying structural change in its economy, inducing important changes in some materials supplies and demands. Third, the period ahead, through the turn of the century and beyond, may make these changes more pronounced. Some of the implications for policies on materials, energy and growth for the future are explored in this final part.

It is important at the outset to establish the data and the model employed. Statistics on individual consuming and producing industries are taken from the national accounts in the *Survey of Current Business* and the *Census of Manufactures*, both its subject series on fuels and electric energy consumed and its industry series on mineral industries. These reports provide data for the model in the following form. They publish aggregate values of gross industrial output (x), the array of input values as interindustry flows (x_{ij}) from producing to consuming sectors, which form the transactions matrix of the economy (X), and the net final output values (y) of all goods produced in the economy during the given year. The value of the vector of final goods demand y, called gross national product (GNP), is equal to the sum of values added in the input industries (v). The input–output framework of the economy (see *Input–Output Analysis*) then permits the construction of input–output relations among the economy's industries (a_{ij}). These are economic analogues to the process engineering estimates of specific materials-use rates, and give the array of technical coefficients, $A = X_{ix}^{-1}$, describing the structure of the economy in terms of three accounting identities:

$$x = (I - A)^{-1}y = Ly \tag{1}$$

$$x = Xi + y \tag{2}$$

$$x = X^{\mathrm{T}}i + v \tag{3}$$

where I and i are the identity matrix and unit vector, respectively; x is the vector of 81 sectors defined at the two-digit level of the US national accounts; y is the vector of final demands, and T indicates transpose. The inverse matrix L is the "Leontief matrix" of the economy, defining its structure, so-called after its originator, Wassily Leontief.

In the first equation the national accounts distinguish the gross output of each materials or energy industry (x_i) required to produce its own product (y_i) and the rest of the final demands (y_j). These demands are net of the interindustry contributions ($-x_{ij}$) given by the inputs from the economy as a whole in Eqn. (2). When the vector of input prices and final goods prices is known, the value of demands y is clearly the same as the value of the aggregate inputs, v, which go to make it up. This identity is shown in Eqn. (3). Data from these identities are summarized in Tables 1–3 from 1958 through to 1982, providing an overview of the materials- and energy-use rates and responses to changes which occurred in market prices and technologies over this 25-year first period. After 1972, a surge of restrictions on open markets left shortages in fuel supply sectors and excess supplies in metals, suggesting an alternative accounting in which the interindustry flows, $\mathbf{A}x$, are defined proportional to the selling industry's output, rather than to the output of the purchasing industry. The resulting output–input coefficients are written $\mathbf{a}_{ij} = x_{ij}/x_i$ to distinguish them from Leontief's input–output coefficients ($\mathbf{a}_{ij} = x_{ij}/x_j$). This modifies Eqn. (1) to become

$$x^{\mathrm{T}} = v^{\mathrm{T}}(I - \mathbf{A})^{-1} = v^{\mathrm{T}}G \tag{1'}$$

where G is the array of output–input coefficients. The importance of the energy sector to the economy is revealed more clearly and analyzed in this fashion in

Table 1
The relationship of minerals and energy sectors to the gross national product (GNP), 1958–1982

Industry classification (BEA two-digit number)	Value added (millions of constant 1982 US dollars)			
	1958[a]	1967[b]	1972[c]	1982[d]
Materials processing machinery and equipment				
Construction and mining (45); materials handling (46); and Metalworking (47) machinery and equipment	11470	23669	23182	26237
Electrical wiring (53); and industrial (55) machinery and equipment	11445	19385	19812	19957
Metal containers (39); heating, plumbing (40); screw machine (41); and other fabricated metal (42) products	26089	46612	49658	48619
	49004	89666	92652	94813
Basic metals manufacturing and mining				
Nonferrous primary metals (38); and mining (6)	10529	18963	18117	11914
Primary iron and steel (37); and mining (5)	25680	38648	38588	21220
	36209	57611	56705	33134
Basic nonmetal manufacturing and mining				
Stone and clay products (36); and mining (9)	14599	19023	22600	16663
Chemicals and selected products (27)	14815	25599	26285	20378
	29414	44622	48885	37041
Energy				
Petroleum refining and related industries (31)	11436	20653	18867	20556
Electric, gas, water and utilities (68)	31423	53093	63203	89983
Coal mining (7)	5084	5528	7972	15056
Crude petroleum and natural gas (8)	21144	25812	29313	124531
	69087	105086	119355	250126
Total energy and minerals sectors	183714	296985	312577	415114
Total final demand	1417857	2384256	2956912	3166245
Energy and minerals fraction of total demand	0.13	0.13	0.11	0.13

a National Economics Division Staff 1965. The transactions table of the 1958 input–output study and revised direct and total requirements data. *Surv. Curr. Bus.* 45(9) b Interindustry Economics Division 1974. The input–output structure of the US economy: 1967. *Surv. Curr. Bus.* 54(2) c Ritz P M, Roberts E P, Young P C 1979 Dollar-value tables for the 1972 input–output study. *Surv. Curr. Bus.* 59(4) d *Surv. Curr. Bus.* 68(4) Annual input–output accounts of the US economy, 1982. All data reflated by Producer Price Index (1982 = 100)

Sect. 2, Tables 4 and 5. In Sect. 3, the implications for the growth of the US and world economies are briefly discussed.

1. Materials and Energy Growth in a Free Trading World: 1945–1972

It took more than a decade after World War II for Europe and Japan to restore their industrial structure, and for the less developed countries to re-establish their primary exports of materials and fuels. The period 1945–1972 saw an unprecedented and continuous expansion in the world in materials trade and use, which took place under relatively open market trade conditions. In the US national accounts defined by Eqns. (1) and (2) at the two-digit level, the domestic mineral and energy sectors comprise 20 of the total of 78 industries. The tables exclude the trade accounts. Materials are defined by the metals numbered 6, 38, 37, 5; the basic nonmetals net of glass and polymers numbered 36, 9, 27; and the materials processing industries numbered 46, 45, 47, 55, 53, 40, 41, 39, 42. Energy is defined by the fossil fuel extractive and refining industries and electricity numbered 31, 68, 7, 8. Twenty-five percent of the structure thus accounts for about 13% of the national product. This share is small and falling (Table 1). Energy or mining alone accounts for less than 4%. For perspective, it is interesting to note that these mineral and energy sectors account for by far the largest part of pollutant emissions, making environmental damage and its control a major industry concern and cost.

The influence of energy in the interindustry flow matrix is broadly spread across the economy. This implies that when international ore and energy markets are functioning freely, the lack of ample energy resource endowment will mean little to economic growth. This

Table 2
Mineral and energy sectors, services and the costs of energy, 1958 (in 1958 US$ million at producer prices for energy sector accounts and major interindustry transactions with Leontief coefficients of direct requirements)

		Energy sector account (inputs)[a]						
		Petroleum refining and related industries (31)	Electric, gas, water and sanitary services (68)	Coal mining (7)	Crude petroleum and natural gas (8)	Total energy inputs	Total gross output of industry	Sum of energy sector technical coefficients
Basic metal[b]	Primary nonferrous metal manufacturing (38)	42 (0.0042)	217 (0.0215)	12 (0.0012)		271	10097	0.0269
	Nonferrous metal ores mining (6)	8 (0.0062)	30 (0.0227)	1 (0.0008)		39	1319	0.0296
	Primary iron and steel manufacturing (37)	147 (0.0076)	466 (0.024)	507 (0.0261)		1120	19391	0.0578
	Iron and ferroalloy ores mining (5)	10 (0.0080)	17 (0.0137)	5 (0.004)		32	1245	0.0257
Basic nonmetals	Stone and clay products (36)	85 (0.0112)	213 (0.028)	63 (0.0083)		361	7609	0.0474
	Stone and clay mining and quarrying (9)	5 (0.0031)	39 (0.024)	3 (0.0018)		47	1624	0.0289
	Glass and glass products (35)	7 (0.0032)	80 (0.0363)	3 (0.0014)		90	2206	0.0408
	Paper and allied products except containers (24)	121 (0.0115)	189 (0.018)	73 (0.0069)		383	10508	0.0364
	Forestry and fishery products (3)	19 (0.013)	c			19	1451	0.013
	Miscellaneous agricultural products (2)	898 (0.0384)	172 (0.0074)	1		1071	23395	0.0458
	Plastics and synthetic materials (28)	56 (0.0131)	41 (0.0096)	23 (0.0054)		120	4272	0.0281
	Chemicals and selected chemical products (27)	658 (0.0543)	288 (0.0238)	61 (0.005)	24 (0.002)	1031	12107	0.0852
	Chemical and fertilizer minerals mining (10)	5 (0.0089)	24 (0.0426)	c	1 (0.0018)	30	563	0.0533

Energy sector	Petroleum refining and related industries (31)	1243 (0.0691)	262 (0.0146)	10 (0.0006)	9291 (0.5163)	10806	17997	0.6004
	Electric, gas, water, sanitary services (68)	245 (0.0121)	3380 (0.1666)	546 (0.0269)	1166 (0.0575)	5337	20289	0.2630
	Coal mining (7)	30 (0.0109)	65 (0.0236)	471 (0.1711)		566	2752	0.2057
	Crude petroleum and natural gas (8)	52 (0.0048)	78 (0.0072)	c	242 (0.0223)	372	10852	0.0343
Services	Federal government enterprises (78)	6 (0.0015)	73 (0.0178)	47 (0.0114)		126	4105	0.0307
	Transportation and warehousing (65)	1518 (0.0445)	147 (0.0043)	27 (0.0008)		1692	34119	0.0496
	State and local government enterprises (79)	40 (0.0084)	376 (0.0786)	79 (0.0165)	20 (0.0042)	515	4784	0.1077
	Hotels, personal and repair services excluding auto (72)	144 (0.0118)	235 (0.0193)			379	12171	0.0311
	Automobile repair and services (75)	27 (0.0034)	151 (0.0191)	10 (0.0013)		188	7913	0.0238
	Medical and educational services and nonprofit organizations (77)	70 (0.0031)	418 (0.0184)	c		488	22703	0.0215
	Wholesale and retail trade (69)	727 (0.0076)	1916 (0.0201)	4	c	2647	95250	0.0278
	Maintenance and repair construction (12)	375 (0.0222)	25 (0.0015)			400 (0.0237)	16875	(0.0474)
Energy accounts	Gross raw sales	6538	8902	1946	10744	28130	345597	
	% of total sales	36	44	71	99			
	Net product (total final demand)	8836	8929	631	−12			

a The two-digit industrial classifications are as given in Bureau of Economic Analysis 1965 *Surv. Curr. Bus.* 45(9) b In parentheses below transactions are given the technical coefficients, that is, the amount of inputs required from each industry at the top of the column to produce one dollar's worth of the output of the industry on the left c Less than US$500 000.

Table 3
Mineral and energy sectors, services and the costs of energy, 1972 (in 1972 US$ million at producer prices for energy sector accounts and major interindustry transactions with Leontief coefficients of direct requirements) and 1982 (last column only, in 1982 US$ million)

		Energy sector account (inputs)[a]						
		Petroleum refining and related industries (31)	Electric, gas, water and sanitary services (68)	Coal mining (7)	Crude petroleum and natural gas (8)	Total energy inputs	1972 total gross output of industry	1982 total gross output of industry
Basic metal[a]	Primary nonferrous metal manufacturing (38)	112 (0.0047)	618 (0.0268)	11 (0.0004)		741	23353	46232
	Nonferrous metal ores mining (6)	25 (0.0113)	78 (0.0341)	c		103	2208	3804
	Primary iron and steel manufacturing (37)	173 (0.0047)	1223 (0.0346)	763 (0.0208)		2159	35380	57704
	Iron and ferroalloy ores mining (5)	13 (0.0101)	82 (0.0667)	4 (0.0036)		99	1295	2478
Basic nonmetals	Stone and clay products (36)	147 (0.0096)	500 (0.0328)	93 (0.0061)		740	15374	31665
	Stone and clay mining and quarrying (9)	86 (0.0037)	106 (0.0372)	2 (0.0001)		194	2911	6795
	Glass and glass products (35)	20 (0.0035)	223 (0.0400)	2 (0.0027)		245	5642	12452
	Paper and allied products except containers (24)	223 (0.0115)	579 (0.0005)	109 (0.0055)		911	19425	58767
	Forestry and fishery products (3)	49 (0.0249)	1 (0.0005)			50	2245	7609
	Miscellaneous agricultural products (2)	877 (0.0250)	248 (0.0073)	1		1126	34163	93205
	Plastics and synthetic materials (28)	80 (0.0084)	208 (0.0064)	43 (0.0044)	10 (0.0010)	341	9468	30934
	Chemicals and selected chemical products (27)	279 (0.0116)	1184 (0.0458)	110 (0.0046)	57 (0.0024)	1630	25869	92724
	Chemical and fertilizer minerals mining (10)	13 (0.0169)	69 (0.0884)	1 (0.0014)		83	492	2232

Energy sector	Petroleum refining and related industries (31)	2332 (0.0742)	638 (0.0208)	35 (0.0011)	14981 (0.4765)	17986	30664	199879
	Electric, gas, water, sanitary services (68)	1370 (0.0235)	11382 (0.195)	2308 (0.0460)	3686 (0.0735)	18746	58383	274017
	Coal mining (7)	78 (0.0143)	110 (0.0214)	656 (0.1205)		844	5439	28321
	Crude petroleum and natural gas (8)	66 (0.0039)	235 (0.013)		596 (0.0335)	897	16730	144971
Services	Federal government enterprises (78)	51 (0.0043)	241 (0.0205)	171 (0.0145)		463	9568	25359
	Transportation and warehousing (65)	2610 (0.0063)	580 (0.0078)		32 (0.0004)	3223	76618	218802
	State and local government enterprises (79)	249 (0.0194)	1463 (0.1114)	138 (0.0108)		1850	2799	10265
	Hotels, personal and repair services excluding auto (72)	325 (0.0106)	902 (0.0056)	8 (0.0003)		1235	30291	73545
	Automobile repair and services (75)	181 (0.0074)	150 (0.0062)			331	24551	63916
	Medical and educational services and nonprofit organizations (77)	429 (0.0051)	1728 (0.0204)	21 (0.0002)		2178	84900	311302
	Wholesale and retail trade (69)	1543 (0.0071)	3280 (0.0152)	c	c	4823	218236	577751
	Maintenance, repair and construction (12)	1137 (0.0312)	65 (0.0018)			1202	36417	109095
Energy accounts	Gross raw sales	12468	25893	4477	19362	62200	772421	
	% of total sales	41	44	82	116[d]			
	Net product (total final demand)	12905	24406	771	−2634			

a The two-digit industrial classifications are as given in Bureau of Economic Analysis 1974 *Surv. Curr. Bus.* 79(2), (4) and 1988 *Surv. Curr. Bus.* 68(4) b In parentheses below transactions are given the technical coefficients, that is, the amount of inputs required from each industry at the top of the column to produce one dollar's worth of the output of the industry on the left c Less than US$500 000 d Due to positive net import balance.

Table 4
Total supply multipliers of the 20 top-ranked US sectors (GNP), 1974[a, b, c]

Sector number		Total supply multipliers	Rank
5	Iron and ferroalloy ores mining	4.0106	1
6	Nonferrous metal ores mining	3.9546	2
67	Radio and TV broadcasting	3.5115	3
10	Chemical and fertilizer mineral mining	3.4993	4
28	Plastic and synthetic materials	3.2469	5
38	Primary nonferrous metal manufacturing	3.2116	6
4	Agriculture, forestry and fishery services	3.1879	7
27	Chemical and chemical products	3.1438	8
7	Coal mining	3.1285	9
37	Primary iron and steel manufacturing	3.1158	10
8	Crude petroleum and natural gas	3.0927	11
9	Stone and clay mining, and quarrying	3.0496	12
24	Paper and allied products, except containers	2.9487	13
3	Forestry and fishery products	2.9433	14
30	Paints and allied products	2.9382	15
16	Broad and narrow fabrics	2.9080	16
21	Wooden containers	2.8502	17
50	Machine shop products	2.8491	18
25	Paperboard containers and boxes	2.8068	19
20	Lumber and wood products, except containers	2.8012	20

a Range of total supply multipliers: the largest is iron and ferroalloy ores mining, value 4.0106; the smallest is medical, educational services and nonprofit organizations, value 1.0922 b Source: Giarratani (1976) c Using the Leontief interindustry matrix for 1967, BEA

Table 5
Purchased fuels and electric energy consumed by major industry groups, 1977[a, b]

SIC code	Major group	Quantity (10^{12} J)	Total cost (million dollars)
	All industries	12929.0	33379.6
20	Food and kindred industries	952.4	2537.8
21	Tobacco products	20.8	63.9
22	Textile mill products	339.2	1136.6
23	Apparel and other textile products	65.6	285.8
24	Lumber and wood products	227.7	726.5
25	Furniture and fixtures	52.7	198.3
26	Paper and allied products	1308.4	2960.4
27	Printing and publishing	92.3	421.7
28	Chemicals and allied products	2978.5	6448.7
29	Petroleum and coal products	1303.4	2445.4
30	Rubber and miscellaneous plastics products	272.3	962.4
31	Leather and leather products	23.1	82.6
32	Stone, clay, and glass products	1251.8	2587.4
33	Primary metal industries	2539.4	7043.6
34	Fabricated metal products	395.3	1348.7
35	Machinery, except electrical	339.6	1268.0
36	Electric and electronic equipment	249.3	986.7
37	Transportation equipment	389.9	1402.6
38	Instruments and related products	78.4	272.5
39	Miscellaneous manufacturing industries	49.1	199.9

a Source: *Census of Manufacturers* 1978, Subject Series, Fuels and Electric Energy Consumed, pp. 4–9 b In Btu 10^9 and US$$10^6$ for two-digit SIC sectors

case is confirmed in the period outside the US by the successful revival of countries without domestic endowments of these resources, such as Japan and much of Europe.

Unlike energy, chemicals and metals processing industries are clustered in the L matrix. To appreciate the importance to the economy of such metals and processing complexes, it is necessary only to observe that together the firms of each group form separable parts of the structure. They are "strategic" because of their forward linkages and the fact that one either finds them in an economy functioning together or not at all. Their important role is confirmed again by the experience of postwar growth in Germany, Japan and the US among developed countries. The underdeveloped nations attempted less successfully to stimulate industrialization by building protected materials sectors.

In any case, the interaction between materials industries and energy at both initial and advanced industrialized stages of an economy can be seen summarily in Tables 2 and 3. These give account values for the midpoint (1958) and the end of the period of growth (1972) as well as the midpoint (1982) of the most recent period of stagnation or decline in current US dollars. Table 1 summarizes the story of sectoral shifts in 1982 constant US dollars. The absolute volume of the flows (X) and the Leontief coefficients (in parentheses, A) are also shown. In 1958 these specific flows in constant 1982 dollars summed to US\$184 billion of inputs required of a US\$1418 billion economy. By comparison, in 1972, US\$318 billion of materials and energy were required for a US\$2384 billion economy. The tables confirm a declining share of US materials and energy in the value product. This is especially notable in principal metals industries such as the nonferrous and ferrous metals manufacturing and steel industries which stagnated after 1963. In part this decline is due to the shift to lighter material technologies, such as aluminum or plastic materials, that is, a change in A. In part the decline is due to larger import components of certain metals in fabricated or end-product forms now shown in the table, for example steel strip and automobile imports. The relative shares of energy in the materials and processing industries remained constant over the period, despite the significant transformation of most of the relevant technologies. Finally, the economy as a whole grew faster than its materials sectors and this helped to produce declining value added shares in the US.

While the energy coefficients were relatively large for a number of the materials industries, only the exceptional industry, such as aluminum, glass and paper, shows a coefficient high enough to warrant its relocation were fuel prices to suddenly rise. This result stands in contrast with the energy sector's own consumption in the refining, petrochemicals and electric power industries, and suggested to economists at the end of the period of cheap energy in 1973 that higher OPEC tariffs and the rising costs of environmental control would not have severe or lasting impacts on growth or levels of demand outside the energy sector itself. As the period closed, the government sought to mitigate impacts with energy regulations. From 1962 to 1972, awareness of the dominant role played by the mineral and energy sectors in pollution grew.

2. *Materials and Energy in a Restricted Trading World: 1973–1982*

From 1972 to 1982, the basic fuels and nonfuels industries achieved dramatic successes in emissions control. However, over the same period the metals manufacturing industries exhibited a dramatic decline in real output, associated with the "hollowing out" of US manufactures in consumer durables and capital goods (Table 1). Basic metals manufacturing, nonferrous and ferrous primary metals with associated mining, and steel declined 42% from US\$57 billion to US\$33 billion in real terms (constant 1982 US dollars). Basic nonmetal manufacturing and mining, along with chemicals, declined 24% from US\$49 billion to US\$37 billion. The value added in materials processing, electrical wiring, construction and industrial machinery and equipment along with fabricated metals products stagnated at US\$95 billion. In the economy as a whole, final demand growth slowed to less than 1% annually reaching US\$3166 billion, and only the rise of energy factor costs masked further predicted declines in the sectoral share by domestic materials and energy producers (last column of Table 3).

Beginning with the 1973 OPEC cartel restrictions on oil exports, a decade of restricted materials and energy supply emerged. Energy prices quadrupled. Further, many nontariff barriers to world materials trade surfaced as negotiations lowered tariff walls. The increased distortions in world markets turned economists to the consideration of the problems of materials and energy interrelations in a regime of regulated trade.

The analysis of such a world of manipulated shocks suggests a model which places emphasis on the production relations less in terms of final demands than in terms of the supply availabilities of inputs, as would be typical of technologies forced to favor energy substitutions or materials conservation. Redefining Eqn. (1) using Eqn. (2) and A, yields in Eqn. (1′) the input value added of $v^{T}G$, where G is a matrix of supply side multipliers relating unit changes in primary inputs to the sellers' gross output. Table 4 applies such an interindustry supply model to energy issues showing that when total supply multipliers are calculated and ranked by size, the extractive energy sectors become very crucial. Coal and crude petroleum and natural gas rank ninth and eleventh in the top 20 of the 82 sectors in the structure. The importance of the materials and energy supply is reinforced by the finding that ten of the top 20 are basic fuels, metals and nonmetals industries. True, this framework also confirms that the materials sectors do not have large impacts on fuel production. Nevertheless, the direct and indirect multiplier effects or "elasticity" of energy impacts on supply are very high, approximately 1.4 using the Bureau of Economic Analysis (BEA) 1974 input–output matrix. In confirmation of this analysis Table 5 shows four of the five largest major industrial users of purchased fuels and energy are chemicals, fuel products, stone, clay, glass and primary metals. In 1977 these absorbed 62 vol.% of the purchased fuels sold to all industries.

Clearly, then, restrictions of the free access to fuels has a very large impact on the materials and processing sectors as well as the use of energy itself, and rises in energy prices such as those which occurred since 1973 can be expected to force greater changes in the competitiveness and technologies of the materials producers than on other industries.

3. Some Implications for the Future: 1992–2000 and Beyond

As of 1992, with a US$4184 billion GNP (in 1982 US dollars), a falling US exchange rate and commercial replacing strategic concerns globally, the US trade balance with its European common market and hemispheric trading partners is improving. However, imports of final goods intensive in materials continue to replace domestic manufactures. Thus industrial policies are gaining attention equal to that accorded controversial environmental issues in the discussions concerning global pollution, competition and economic growth.

Development economists have long noted the various stages of economic growth and their effect on the distribution of sectoral output and final demands. More recent studies of materials and energy demands confirm that materials sectors may experience rising and then declining shares as they mature. The contrasting Japanese and US consumption rates for materials also offer examples of how the dynamic growth of support industries as well as the shift to service sectors over time may affect materials-use rates at different stages of maturity. Exchange rate distortions will affect such trends. However, the disparate rates of growth since the mid-1960s in lighter materials such as plastics or aluminum in contrast to heavier metals such as steel (all relatively intensive in energy use) show clearly that technological changes and time permit no easy generalizations to laws of demand about materials consumption rates. Indeed, differential growth in productive capacities and switches in international trade flows are subject to myriad real and monetary influences. These increase the difficulty of relating trends in materials propensities to consume primarily to stages of growth or the composition of final demands. Economists dispute the basic claim that a country's growth in income per capita can be disassociated from either materials intensive or energy intensive industrial growth. Some conjecture that the ability to support ever increasing service sector values may require continuing comparative advantages and strengths in the materials and processing industries. Notice is given to the fact that the metals, materials and chemical industrial complexes are decomposable in the X and A matrices of most economies.

Thus, a country tends either to have the entire set of closely linked metals industries or to lack them entirely. This implies that in the dynamics of growth there are technological or other externalities and economies defined in these industries at the microlevel which are less discernible at the macrolevel. Such a finding supports concerns over the decline of important materials and energy processing sectors in developed countries such as Germany, France and the USA, and prompts the adoption of industrial policies that restrict free trade.

In the materials sector some countries maintain high subsidies or nontariff barriers to trade, and these restrictions combined with aggressive export policies can warp the growth of domestic materials sectors in other countries more open to free trade. The implications of such asymmetries in trade and restrictions in the materials and energy markets are significant for the economic analysis of commercial strategies in both the private and public policy arenas.

On the energy side, the dynamics of energy substitution and conservation have been debated by economists as avidly as the materials substitution issues. The most recent input–output analyses of world materials use and forecasts suggest that trade restrictions, depletion and technological changes can effect notable differences in observed regional propensities to produce given materials, not to mention major differences in final investment and consumption demands. The range of such predictions based on energy models and different scenarios in the US forecast of materials use is very great. It is important to note that the applied models of the sort most useful to the analysis of materials, energy and economic policy are largely static. This reflects a large degree of ignorance about the dynamics of technical innovation and engineering design and the influence of such changes in the future on the maintenance of national comparative advantages or national economic growth. A simple example from automotive engineering is illustrative. The rise in gasoline prices and concerns over exhaust emissions motivate current research on new automotive engines by the industry, which portends a significant shift toward higher-temperature combustion. This would greatly increase fuel efficiency and reduce some residuals emissions, but would require a shift away from conventional engine materials. The switch would favor the use of ceramic materials with higher heat resistance than currently used metallic materials. The expected result would reduce weight and improve performance while lowering operating costs. The elimination of conventional metals in the course of such system redesign, however, cannot be predicted by comparing static input–output coefficients at the two-digit industry level or by observing the trend in aggregate cost indices employed by economists to measure technical progress. Finally, such switches cannot become industrially feasible without a vital metals, materials and tool manufacturing complex operating within the economy.

See also: Energy Considerations in Metals Production; Intensity of Use of Materials; Life-Cycle Analysis of Energy Consumption and Carbon Dioxide Emissions; Materials in the Industrial System

Bibliography

Ghosh A 1958 Input–output approach to an allocative system. *Econometrica* 25: 58–64

Giarratani F 1976 Application of an interindustry supply model to energy issues. *Environ. Plann.* 8: 447–54

Leontief W 1986 *Input–Output Economics*, 2nd edn. Oxford University Press, New York

R. Newcomb
[University of Arizona, Tucson, Arizona, USA]

Economics of the Semiconductor Industry

Semiconductors, which are the heart of almost all final electronic systems today, are miniaturized electronic circuits imprinted on tiny chips with dimensions of only a few millimeters. While most semiconductor chips are fabricated from silicon, there is an increasing use of other materials, principally gallium arsenide, due to their superior properties in very specialized applications.

The size of the total world market for electronics in 1990 was almost US\$800 × 10^9—nearly four times the size of the industry in 1983—the semiconductor industry comprising roughly US\$50 × 10^9 of this. Table 1 compares the size of the global electronics industry with the size of the semiconductor industry from 1985 to 1988. Semiconductors, on average, account for 7% of the total value of final electronic equipment. There is considerable variation in this share across different types of electronic systems. Among the systems with the largest share of semiconductor content are computer workstations, mainframe computers and supercomputers. Automobilies are experiencing a sharp increase in semiconductor content. As of 1990, for example, approximately 1–2% of the total cost of an automobile was related to the semiconductor content needed for engine controllers, advanced braking systems and sophisticated entertainment packages. This share is projected to increase to over 5% by the late 1990s with some estimates as high as 10% by the early 2000s.

Table 1
World electronic equipment and semiconductor content, 1985–1988 (US\$ × 10^9)

	1985	1986	1987	1988
Electronic	456	529	651	760
Semiconductor demand	27	33	40	55
Pervasiveness[a]	6.0%	6.2%	6.1%	7.2%

Source: DATAQUEST *Semiconductor Industry Insights—1990*
a Pervasiveness equals semiconductor revenue divided by electronics revenue

1. Overview of Semiconductor Technology

Semiconductor technology can be defined according to the process technology used in device manufacture or by dividing it along the lines of major product families according to end application.

Process technology refers to the manufacturing "recipes" used to fabricate or manufacture semiconductors. The particular process technology applied in manufacturing a semiconductor device defines, as well as limits, the particular functions and characteristics of the device. There are literally an infinite variety of potential manufacturing recipes which, in practice, can be grouped into three major categories: (a) complementary metal–oxide–semiconductors (CMOS), a widely favored processing technique because it allows devices to be fabricated that use less power and, hence, emit less heat; (b) an older major process technique designated bipolar, most frequently applied in chips which require faster speeds; and (c) a hybrid manufacturing technique which is a mix of both CMOS and bipolar processing steps called BIMOS. These three general process characteristics are also one way to categorize different types of semiconductor products.

Product family designations for semiconductors can most readily be understood in the context of a family tree (see Fig. 1). Semiconductor device technology has two major divisions: discrete devices and integrated circuits (ICs). Discrete device technology, which is the older branch of semiconductor technology, has its roots in the initial discovery of the transistor at Bell Laboratories in the late 1940s. The demand for discrete devices continues to expand, but at a very slow rate.

Integrated circuits, however, are a fast growing segment. As discussed previously, ICs can be categorized by the process technology used in their manufacture. The next division of ICs is into major product areas: memory, microdevices, logic, linear and

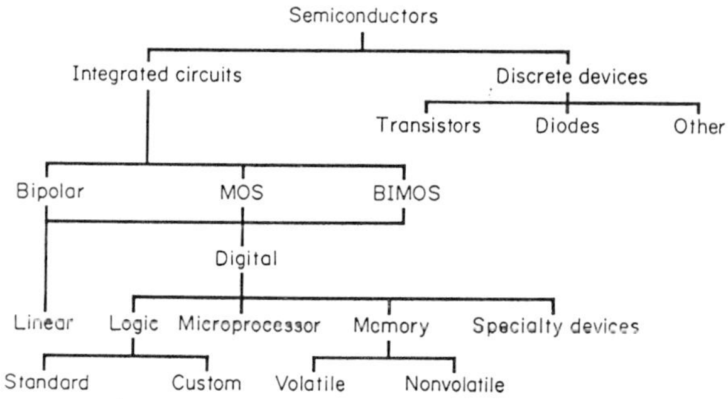

Figure 1
Family tree of semiconductor technology and device types

specialty products. Each of the four specific product types can be further divided into major product families. For example, logic devices can be subdivided into standard and nonstandard or custom devices; memory devices can be subdivided into those types that continue to store data when power is turned off, that is, nonvolatile memory, and those that lose their stored data, or volatile memory. The last product area, specialty devices, accounts for the balance of device types that do not fall directly into one of the other four product categories.

Table 2 shows the division of 1990 worldwide semiconductor consumption among these major product families. By far the largest single product family is MOS memory representing approximately one-third of total semiconductor consumption. The older technology segments of discrete devices and bipolar ICs accounted for only about one-quarter of total 1990 consumption.

1.1 The Semiconductor Manufacturing Process

The manufacturing process for semiconductors is probably the most complex mass production process ever attempted. Furthermore, the technical sophistication of the process is continually increasing. The inherent complexity of the manufacturing process, together with continuous improvement in manufacturing capabilities, generates a particular set of characteristics for the production process.

(*a*) *Process yields.* The success of the semiconductor manufacturing process is measured by yield, which is the percentage of initial product or die that is acceptable at the end of the manufacturing sequence. Because of the complexity of the manufacturing process, yields tend to be well below 100%. For example, if manufacturing extremely complex devices with newly introduced mass production techniques, the overall yield would be around 40%. For simpler products that have been manufactured for a long period, yields could be as high as 90%. This differential provides enormous economic incentives for semiconductor manufacturers to focus on improving yields.

(*b*) *Learning economies.* These are also known as dynamic economies of scale and occur as production experience accumulates. The rule of thumb widely used in the semiconductor industry is that for each doubling of cumulative production volume, the cost of manufacturing falls by 30%. The source of learning economies is closely intertwined with yield improvements. Due to the multiple manufacturing steps and the continuous processing of successive batch lots, there are a multitude of opportunities for semiconductor manufacturers to gradually increase yield.

(*c*) *The manufacturing process.* There are four basic stages in semiconductor manufacturing process: design, fabrication, assembly and testing.

Design requires the detailed specification of exactly what the circuit is expected to do. The design of complex devices can take several years and tens of thousands of man-hours. Due to pressures to reduce the design cycle time, IC design is becoming increasingly automated through the application of computer-aided design (CAD) tools. Between 1980 and 1990 the CAD tool industry grew from almost nothing to a US\$300 × 10^6 industry.

The second step in the production sequence, fabrication, is the most complex and costly of the four. Most semiconductors are fabricated on disks of silicon called wafers. A process known as photolithography transfers the pattern of the circuit to the wafer. Individual circuits are embedded in a chip or die in the wafer with, typically, hundreds of dies being fabricated on each wafer. The processing steps for each wafer are sequential; that is, the wafer will move through several hundred process operations during the entire wafer fabrication sequence. Successive layers on the wafer are used to construct a multilayered circuit. Since each layer requires essentially similar types of process steps, a number of steps in the sequence will be repeated in order to build up the necessary number of layers. The number of layers constructed in process sequence

Table 2
Worldwide semiconductor consumption by major product segments, 1990 (US\$ × 10^9)

Total semiconductor	58
Total IC	49
bipolar IC	4.2
memory	0.5
logic	3.8
MOS digital	35.4
memory	18.5
microprocessor	7.6
logic	9.3
linear	9.3
Total discrete	7.5

Source: DATAQUEST *Semiconductor Industry Insights—1990*

depends on the complexity of the circuit being fabricated. For simpler circuits, ten layers are common, but increasingly there are products requiring over 20 layers.

Once the semiconductor dies have been fabricated onto the surface of the silicon wafer, the wafer is sawed, permitting separation of the individual dies. At this assembly stage the die must be placed in a package both to protect it and to permit electrical connections to be attached.

Testing is the final stage of the manufacturing sequence. While testing is conducted throughout the entire design–fabrication–assembly cycle, the complete, packaged die is still subjected to a final round of electrical, mechanical and visual inspections.

1.2 The Role of Successive Generations of Products

Over time, as process capabilities improve and product designs evolve, successive generations of a product family appear. In the memory product area, successive generations of memory devices are designated by their storage capacity. For logic devices, the designation of generations is more complex. For example, for microprocessor families successive generations are designated by the size of the instruction set that can be processed (e.g., 4 bit, 8 bit).

Product generations signify new plateaus of process technology capability since each new generation requires a higher level of processing performance. For example, one indicator of a manufacturing difficulty is the minimum feature size of a device, that is, the smallest geometry that is being fabricated in the manufacturing sequence. Typically, each successive generation would have a finer minimum feature size specified—the current standard is 0.8 μm for some types of memory devices, with the leading edge prototype technologies calling for a minimum feature size below 0.5 μm. A second feature of a new generation product is the capability of the product itself. In the area of memory products, for example, the storage capacity of the device and the average time required to access data on the device are two measures that improve with each generation.

Table 3 illustrates a number of attributes of successive product generations using the highest volume semiconductor device currently manufactured—the dynamic random access memory (DRAM). The parameter used to designate a new DRAM generation is storage capacity.

1.3 The Economics of Semiconductor Production

The complexity of the semiconductor manufacturing process creates a special set of economic conditions related to the production operation. To start with, the semiconductor device user, for a given level of device performance, selects for incorporation into the final electronic systems those devices which cost the least per unit of functionality. A functional unit or element refers to, for example, the storage unit used to hold information or the processing element used in logic devices to perform logical operations. Lower costs per functional element are achieved over time primarily by two means. First, within a given generation, costs decline as productivity rises with increased manufacturing yield. However, there are practical limits to this type of gain. Once the overall process yield exceeds roughly 70%, there are likely to be only marginal further improvements in productivity and, therefore, only marginal cost reductions.

The second way of achieving significant cost reductions is to put succeeding device generations into production with an even greater density of functional elements, that is, not only including more elements per device, but also increasing the number of elements per unit area (or density). The result of these increases in device functionality and density is the observed learning phenomenon measured by the steady decline in the cost per function.

2. General Industry Characteristics

The remarkable aspect of the semiconductor industry is not so much its absolute size, but rather its continued rapid rate of growth. In 1990, the semiconductor industry, worldwide, was nearly a US\$60 × 10^9 a year

Table 3
Successive generation technical developments: illustration for the dynamic random access memory (DRAM)

DRAM generation	Process development time	First customer qualification	Year of peak production
4 Mb (0.80 μm)	1984–1993	1988/89	1994
16 Mb (0.50 μm)	1986/87–1996	1991	1997
64 Mb (0.35 μm)	1989–1999	1993/94	2000
256 Mb (0.25 μm)	1991/92–2002	1996	2003
1 Gb (0.15 μm)	1994–2005	1998/99	2006

Source: R Kopp 1988 Device architecture and process technology. *SEMI ISS Conf.*

business. As recently as 1980, the worldwide scale of the industry was only US\$14 $\times$ 10^9. Historically, the industry has enjoyed an average annual compound growth rate of around 14%, but the IC portion of the business has expanded at a much higher rate, averaging between 15% and 20% annually. In some explosive expansionary stages, such as in 1984, industry IC revenues have grown by over 40% in a single year.

2.1 Brief History of Industry Development

The semiconductor industry was created with the creation of the first workable transistor at Bell Laboratories in December 1947. Five years later, the transistor became commercially viable. In the mid-1950s, Western Electric, holder of Bell Laboratories' patents on semiconductor technology, under a consent decree, licensed any firm interested in acquiring the basic technology. The results were spectacular. By the early 1960s, there were over 100 firms worldwide engaged in manufacturing semiconductor devices. In 1961, the next major technical breakthrough occurred with the development of the IC. This single development was to revolutionize the entire electronics industry.

Initially, the US military was the principal market for these new forms of semiconductor devices. The military remained the major end-use consumer throughout most of the 1960s. Gradually, as manufacturing costs fell, commercial applications emerged with the computer market becoming the major market by the early 1970s.

European and Japanese semiconductor industries developed differently from the US industry, principally due to the absence of large support from military buyers which characterized the US market up to the late 1960s. Both the European and Japanese industries became oriented towards serving industrial and consumer applications. Since these markets did not require as high a level of performance as devices used in military applications, both industries lagged behind the US industry in advanced developments. The European industry ultimately became dominated by subsidiaries of US producers.

The development of Japan's semiconductor industry was affected by severe government restrictions on foreign entry. By the early 1970s the number of firms in the Japanese semiconductor industry totalled only about 20, of which the six largest firms controlled over 80% of production. This structure changed significantly over the next 15 years as a large number of indigenous and foreign firms entered the Japanese market. By 1990, the Japanese industry had grown to approximately 80 firms, but the ten largest Japanese firms still accounted for about 80% of production.

2.2 Structural Characteristics of The Industry

The semiconductor industry is global in scope. Producers with headquarters in one region have production plants located in other major producing regions. Table 4 shows the distribution of worldwide IC production by the major consuming regions. (By convention, the industry reports production results by base of producer; therefore, in reviewing Table 4, this convention should be kept in mind.) Note the especially rapid growth of Japanese IC production between 1980 and 1989—a tenfold increase. Although part of this increase in the post-1985 period was due to the change in the yen–US dollar exchange rate, this only accounted for about one-third of the growth. IC production in the rest of the world between 1985 and 1989 also showed an explosive rate of growth; most of this was due to the rapid expansion of IC production capacity in Korea.

Table 4 also indicates the distribution of semiconductor production between IC and discrete devices. Production of the latter represented 40% of worldwide production in 1975; this had declined to under 20% by 1989.

Table 5 shows the distribution of semiconductor consumption by the major consuming regions for 1990. Mirroring the fact that Japan is the largest base for IC production in the world, in 1990 it was also the largest consuming region accounting for over 40% of worldwide consumption.

Table 4
Worldwide semiconductor production (captive and merchant) by base of producer and major product segments, 1975–1989 (US\$ $\times$ 10^9)

Producer base	1975	1980	1985	1989
IC production				
USA	2.0	9.1	14.1	24.7
Europe	0.3	0.7	1.5	5.8
Japan	0.4	2.3	7.0	28.0
rest of world	0.1	0.1	0.3	3.3
Total IC	2.8	12.2	22.9	61.8
Total discrete	2.1	4.5	5.1	9.4
Total	4.9	16.7	28.0	71.2

Source: Integrated Circuit Engineering *Status 1990*

Table 5
Distribution of semiconductor consumption by major regions, 1990

Region	Value (US\$ $\times$ 10^9)	Percentage share
North America	18.0	31.2
Japan	22.5	40.8
Europe	10.4	16.7
Rest of world	7.6	11.3
Total	58.5	100.0

Source: DATAQUEST *Semiconductor Industry Insights—1990*

(*a*) *Captive and merchant producers.* These are the two usual categories of semiconductor producers. A captive producer is a semiconductor firm whose production is almost completely consumed internally by the parent organization. By contrast, a merchant producer is one that sells most of its output to outside customers. While it is increasingly rare to find a captive semiconductor producer that does not sell some of its production to outside parties, captive operations typically lack the marketing infrastructure to support a large external sales base. At the other extreme, there are firms that sell a substantial proportion of their output to the parent firm but are still classified as merchant producers. For example, all of the major Japanese IC producers support a large internal demand with between 10% and 20% of their IC production being consumed by related operations.

The USA has the largest base of captive IC production, this representing perhaps as much as one-third of its semiconductor industry. This reflects, in part, the large number of major electronics systems and military producers located in the USA—two end-consumer markets whose competitive status is heavily dependent on advanced semiconductor technology. Firms that support a captive semiconductor manufacturing operation seek to obtain several strategic benefits:

(i) leverage to the associated systems business in developing and producing low-volume, custom circuits;

(ii) minimization of the time to market for new systems;

(iii) support of exotic technologies that do not have a large, general market; and

(iv) a guaranteed supplier of last resort.

Since captive operations exist to support these strategic objectives, they are not normally the most efficient producers.

Table 6 lists the ten largest merchant semiconductor producers in 1989. Japan-based firms account for five of these producers reflecting the major shift in the locus of production activity away from operations based in the USA towards Japan and other Asian locations during the 1980s. Of the ten firms given, only the US firms are stand-alone producers as opposed to semiconductor producers who are part of a larger vertically integrated operation.

There are currently an estimated 200 to 250 firms producing semiconductors in the USA of which one-quarter to one-third (roughly 75 firms) are captive IC manufacturing facilities. Table 7 lists the leading US IC captive producers in 1990. IBM at an estimated US\$4.3 $\times 10^9$ in total IC production ranks as the largest IC producer in the world—captive or merchant.

(*b*) *Onshore and offshore production.* Another way to divide semiconductor production is by the location of various production activities. The four basic semiconductor production steps of design, fabrication, assembly and testing can be segmented into two basic blocks. The first block, consisting of design and fabrication of the circuit on wafers, is called the front end of the production process. The second block consists of assembly and testing and is called the back end of the production process. Whereas front-end production operations are located principally in the USA, Japan, Europe and, to an increasing degree, Korea and Taiwan, back-end operations are located throughout the world. This bifurcation of the production process results from the low cost of shipping silicon wafers and assembled devices to any location in the world combined with significant factor cost differences between the fabrication and the assembly–testing stages of production.

The bifurcation of the production process leads to a large worldwide trade in semiassembled devices. For example, wafers can be fabricated in one location, say the USA, and assembled in another, say Malaysia, with final testing occurring in yet another, say Japan. Final consumption does not necessarily occur in the country where final testing occurs. Thus, the import and

Table 6
The ten largest merchant semiconductor producers, 1989 (US\$ $\times 10^6$)

NEC	4825
Toshiba	4680
Hitachi	3775
Motorola	3200
Texas Instruments	2900
Mitsubishi	2590
Fujitsu	2490
Intel	2400
Matsushita	2050
Philips	1680
Total	30590 (55% of world total)

Source: Integrated Circuit Engineering *Status 1990*

Table 7
Major US captive IC producers, 1990 (US\$ $\times 10^6$)

IBM	4300
GM Hughes	300
Hewlett Packard	300
Rockwell	290
DEC	160
Northern Telecom	125
Commodore	70
Total	5510

Source: Integrated Circuit Engineering *Status 1990*

Table 8
Projection of future semiconductor demand by major product group (US$ × 10^9)

	1990	1991	1992	1993
Discrete/ digital bipolar and analog	20.6	22.1	24.2	25.6
MOS microprocessor	8.7	9.9	12.4	14.2
MOS standard and custom logic	4.1	4.5	5.0	5.4
MOS semicustom	4.1	5.1	6.6	7.7
MOS volatile memory	8.8	10.5	14.0	17.0
MOS nonvolatile memory	3.2	3.6	4.4	5.1
Total	49.5	55.7	66.6	75.0

Source: Semiconductor Industry Association 1990 Annual Forecast Report

export statistics for semiconductor devices can be very misleading measures of the competitive condition of a particular location.

3. Future Expansion of Demand

The short-term outlook for growth across the major semiconductor product families is shown in Table 8. The MOS memory families—both volatile and nonvolatile—will continue to be the largest market segment. However, MOS logic families are expected to show strong growth as well, reflecting the growing demand for custom and semicustom circuits in a wider base of applications.

Between 1990 and 1993, worldwide semiconductor demand is expected to increase by an average of 15% per year (a rate of growth in line with the historic average annual rate of growth of 15–20%). By 1993, the industry will reach total worldwide sales of US$75 × 10^9. In the longer term, as the pervasiveness of semiconductor technology in final systems reaches saturation, the rate of growth of semiconductor consumption will track the overall rate of growth of electronics. The overall consumption of electronics has averaged about 8% long-term growth, implying that there will be a gradual slowing in the rate of expansion of the semiconductor industry during the 1990s. Even so, the semiconductor industry could, by the year 2000, exceed US$150 × 10^9 in worldwide sales.

Bibliography

Finan W, Amundsen C 1986 Modeling US–Japan competition in semiconductors. *J. Policy Model.* 8(3): 305–26

Finan W, Frey J 1989 The Effectiveness of the Japanese Research–Development–Commercialization Cycle: Engineering and Technology Transfer in Japan's Semiconductor Industry. Semiconductor Research Corporation

Integrated Circuit Engineering 1991 *Status 1991*. Integrated Circuit Engineering, Scottsdale, AZ

Kikuchi M 1983 *Japanese Electronics*. Simul International

McWhorter G 1984 *Understanding Digital Electronics*. Texas Instruments, Dallas, TX

US International Trade Commission 1991 *Global Competitiveness of U.S. Advanced-Technology Manufacturing Industries: Semiconductor Manufacturing and Testing Equipment*, USITC Publication 2435. USITC

Wilson R, Ashton P, Egan T 1981 *Innovation, Competition, and Government Policy in the Semiconductor Industry*. Lexington Books, Cambridge, MA

W. F. Finan
[Technecon Analytic Research
Washington, DC, USA]

Energy Considerations in Metals Production

Production of primary metals from ores, through the stages of mining, ore beneficiation, smelting and refining, requires large energy inputs in the form of fuels, electrical energy and energy-intensive reagents and supplies. Some metals (copper, molybdenum, gold) are energy-intensive primarily because the low grade of their ores requires that many tonnes of ore be processed to recover one tonne of metal. Other metals (aluminum, magnesium) are energy-intensive because of inherent difficulties in the processes for reduction to the metallic state. High-grade ore and ease of processing combine to make steel the least energy-intensive metal. Titanium, produced from relatively low-grade ore and involving complex reduction chemistry and energy-intensive reagents, is the most energy-intensive of the tonnage metals (i.e., those having a world production $>50\,000$ t per year).

Rapid escalation of energy prices since 1973 and the recognition that world supplies of oil and natural gas are limited have stimulated the search for new metal production processes that are less energy-intensive and the application of energy-conservation practices to conventional processes for primary metal production. Secondary metals (i.e., those produced from scrap metal) require far less energy for production than the primary metals. Increased production of secondary metal from obsolete metal scrap is certain to play a larger role as the twenty-first century approaches, since it conserves both energy and metallic resources and reduces both litter and the cost of waste disposal (see *Recycling of Metals*).

1. Methodology for Energy Analysis

Energy requirements reported herein are based on the following methodology (Kellogg 1974). The energy for any given process in the production flowsheet is evaluated as the process fuel equivalent (PFE), defined as

$$\mathrm{PFE} = F + E + S - B \tag{1}$$

in which all terms have units of $\mathrm{J\,t^{-1}}$ of product. F is the heating value (gross) of all fuels used, E is the fuel equivalent of electrical energy used (evaluated at the US average fuel rate of 11.1×10^6 J per kWh), S is the fuel equivalent for production of major supplies and reagents consumed by the process (explosives, oxygen, acids, lime, ammonia, reducing agents, for example) and B is the fuel equivalent of useful by-products and surplus steam produced by the process.

The energy requirement for a primary metal in unfabricated form (ingot, sponge, slab, cathodes and so on) is given as a material fuel equivalent (MFE), $\mathrm{J\,t^{-1}}$, evaluated as the sum of all process energies leading to that product, including mining of raw material, ore beneficiation, smelting or other means of metal extraction, refining and those means for pollution abatement normally associated with each step.

Evaluation of electrical energy used at the equivalent fuel rate (fossil or nuclear), as in Eqn. (1), has the effect of charging the inefficiency of electrical energy production against the process or material in question. This is justified for most of the world's industrial economies, where electrical energy is predominantly generated from fuel, since it properly designates electricity as a premium form of energy, obtained at a premium cost in primary fuels. For metals, such as aluminum and magnesium, requiring large amounts of electrical energy for production, almost half of the total MFE represents the energy loss in electrical energy production.

Table 1
Material fuel equivalent for selected primary metals[a]

Metal	Product form	MFE ($10^9\ \mathrm{J\,t^{-1}}$)
Iron	steel slabs	28
Lead	refined ingot	31
Manganese	ferromanganese	58
Zinc	slab (electrolytic)	70
Chromium	ferrochrome (high C)	71
Silicon	ferrosilicon (50%)	89
Sodium	metal	107
Copper	wire bar, refined	128
Cobalt	electrolytic	144
Nickel	electrolytic, from sulfide ore	167
Molybdenum	molybdic oxide (crude)	168
Cadmium	electrolytic	178
Niobium	ferroniobium	220
Tin	refined ingot	221
Aluminum	ingot	284
Tantalum	metal powder	289
Tungsten	metal powder	402
Magnesium	electrolytic ingot	416
Mercury	refined metal	459
Titanium	sponge metal	474
Hafnium	sponge metal	768
Uranium	yellow cake (U_3O_8)	902
Zirconium	sponge metal	1390
Silver	refined bars	1710
Beryllium	cast metal	6100
Gold	refined bars	68400

a For production by technology typical of US industry in the early 1970s (Battelle–Columbus Laboratories 1975a,b, 1976). Estimated accuracy 10–30%. Includes minor amounts of energy for transportation of raw materials to and within the USA

2. *Range of Energy Intensity for Metals*

Table 1 lists estimated values of MFE for a selected list of primary metals and shows a range from $28 \times 10^9\ \mathrm{J\,t^{-1}}$ for steel slabs to $68\,400 \times 10^9\ \mathrm{J\,t^{-1}}$ for refined gold bars. This extreme range—a ratio of >2400: 1—reflects widely different ore grades and widely different processing technology applicable to the various metals, as discussed in the following. The magnitude of the energy requirements in Table 1 can be better appreciated if it is recognized that one barrel (159 liters) of crude petroleum has an energy content of $\sim 6.6 \times 10^9$ J; thus production of one tonne of steel slabs requires about four barrels of oil, whereas one tonne of aluminum requires ~40 barrels of oil and one tonne of refined gold requires >10 000 barrels of oil.

It should also be noted that even the least energy-intensive of the metals (steel, lead, ferromanganese) are more energy-intensive than many nonmetallic mineral-based commodities such as those listed in Table 2. Characteristic of the commodities listed in

Table 2
Material fuel equivalent for some nonmetallic mineral commodities[a]

Commodity	MFE ($10^9\ \mathrm{J\,t^{-1}}$)
Sand and gravel (washed)	0.065
Rock salt	0.20
Sulfuric acid (from Frasch sulfur)	1.2
Calcined gypsum	1.7
Common building brick	4.1
Gaseous oxygen	4.2
Asbestos fiber	8.0
Portland cement	8.8
Quicklime	9.9
Phosphoric acid	13
Glass containers	20
Gaseous chlorine	21

a See footnote, Table 1

Table 2 are raw materials (ores) that are abundant and high-grade, product specifications that lack the stringent purity requirements of many refined metals and processing technology that consists of few stages of low or moderate energy intensity.

3. *Energy for Mining and Primary Processing*

Energy requirements for mining depend strongly on the nature of the ore and its location. Unconsolidated sand and gravel (alluvial ore), lying near the surface, requires the least energy, since it needs only to be excavated by shovel or dredge and processed for heavy mineral separation. Dredging of alluvial ore, including physical concentration of heavy minerals (with values of tin, or titanium and zirconium) on board the dredge, requires a total of only $0.023–0.07 \times 10^9$ J t^{-1} of ore excavated and processed. In contrast, a hard rock ore that is deeply buried, requiring underground mining, crushing, fine grinding and mineral concentration by flotation, may require as much as 1.0×10^9 J t^{-1} of ore. Intermediate between these extreme cases are typical hard rock ores that can be mined from the surface by open pit methods, crushed and ground to about 100 μm particle size and concentrated by flotation; such ores require $0.35–0.45 \times 10^9$ J t^{-1} of ore mined and processed (Kellogg 1977a).

For mining, the most energy-intensive operations are loading of ore, elevation out of the pit or mine and haulage of the ore (Pitt and Wadsworth 1981). Lesser amounts of energy are associated with drilling, blasting, and ventilation and dewatering of underground mines. For primary physical beneficiation of ores, the most energy-intensive operation is fine grinding to liberate the mineral values, which requires $0.11–0.28 \times 10^9$ J t^{-1} (10–25 kWh t^{-1}) of ore, depending on the hardness of the rock and the fineness of grind. Coarse crushing of rock and mineral separation by flotation, magnetic separation or gravity concentration together require only $0.07–0.10 \times 10^9$ J t^{-1} of ore.

The above mentioned energy values for mining and primary processing apply to ores that are mineralized in such a way (discrete grains of valuable mineral in a matrix of gangue minerals) that physical concentration methods (flotation, gravity separation, magnetic separation) will yield a valuable mineral concentrate with acceptable recovery (generally 80–95% of the value in the ore). Where applicable, this physical concentration step usually leads to lower costs and lower energy consumption for metal extraction. There are important ore types, however, where the nature of the mineralization requires that primary processing be of a chemical nature.

Two categories of primary chemical treatment of ores can be distinguished, based both on what the chemical treatment accomplishes and on the energy requirements of the process. If the metal values of the ore are selectively soluble in relatively dilute acid or alkaline solution, leaving the gangue unaffected for the most part, and if the metal values may be removed from solution by simple means (precipitation, ion exchange or solvent extraction), then the energy requirement for mining plus selective leaching, to the stage of a concentrate containing the metal, is usually $\sim 0.35–1.4 \times 10^9$ J t^{-1} of ore. This range of energy is comparable to that for mining of hard rock plus physical beneficiation ($0.35–1.0 \times 10^9$ J t^{-1} of ore). Selective leaching of gold and uranium ores are examples that require energy near the low and high end of the range, respectively.

In contrast, if the ore is of such a type that the entire rock matrix must be attacked by chemical digestion in hot, concentrated aqueous solution, or by smelting to yield molten phases, then the range of energy requirements rises to $4.5–14 \times 10^9$ J t^{-1} of ore for mining plus primary processing. Examples are the production of pure aluminum oxide from bauxite ore by the Bayer process (strong alkali digestion at elevated temperature and pressure), with an energy requirement $\sim 8.5 \times 10^9$ J t^{-1} of ore, and smelting of nickel–laterite ores to produce molten ferronickel and slag, with energy requirements of $8.5–10 \times 10^9$ J t^{-1} of ore. The large energy requirements in these cases stem from the need for fuels or electric energy to heat solutions or furnaces, and from the energy equivalent of major reagents (caustic, ammonia, acids, ferrosilicon and so on).

The energy requirements for mining and primary processing of ores to yield concentrates for further processing can be summarized in the four categories shown in Table 3.

4. *Effect of Ore Grade on Energy Intensity*

Metallic ores range in grade from that of high-grade, direct-shipping iron ores (~65% Fe) to the extreme low grade of typical gold ores (<0.001% Au). Ore grade G (% metal content) and recovery during processing R (% metal recovered) determine the number of tonnes of ore T that must be processed to recover one tonne of metal, through the relation

$$T = 10\,000/(GR) \qquad (2)$$

Table 3
Unit energy for mining and primary processing

Ore type and processing method	Unit energy (10^9 J t^{-1} ore)
Alluvial ore, physical concentration	0.023–0.07
Hard rock ore, physical concentration	0.35–1.0
Hard rock ore, selective leaching	0.35–1.4
Hard rock ore, chemical digestion or smelting	4.5–14.0

For a typical recovery of 90%, Eqn. (2) shows that it requires only 1.7 t of 65% iron ore to produce 1 t of iron, but 111 000 t of 0.001% gold ore to produce 1 t of gold.

If the unit energy U for mining and primary processing of 1 t of ore (from Table 3) is multiplied by T, the result is the energy requirement E for recovery of 1 t of metal through the stages of mining and primary processing only:

$$E = UT = 10\,000U/(GR) \tag{3}$$

The significance of Eqn. (3) is illustrated in Fig. 1, where E is plotted against G for a recovery value of 80% and for three of the unit energy categories of Table 3. The width of the bands represents the typical range of unit energy for any given ore type (Table 3). The upward slope from right to left shows the effect of ore grade on the energy for recovery of 1 t of metal (the "tyranny of ore grade"). About 73×10^9 J are required for mining and concentration to recover 1 t of copper from a hard rock ore of 0.6% grade, but $44\,000 \times 10^9$ J are required for mining and selective leaching to recover 1 t of gold from ore of 0.001% grade (the band for selective leaching of ores is almost identical to that for concentration of hard rock ores, as shown in Table 3).

The three bands of Fig. 1, spaced apart by a little more than a factor of ten in energy intensity, illustrate the very significant effect of ore type on energy requirements (the "tyranny of ore type"). It requires more than ten times the energy to recover 1 t of nickel as ferronickel from laterites by smelting as it does to recover 1 t of nickel as concentrate from a sulfide nickel ore of the same grade. It should be recognized, however, that the energy values of Table 3 and Fig. 1 include only mining and primary ore processing; the energy required to convert nickel sulfide concentrate to saleable metal has not yet been considered.

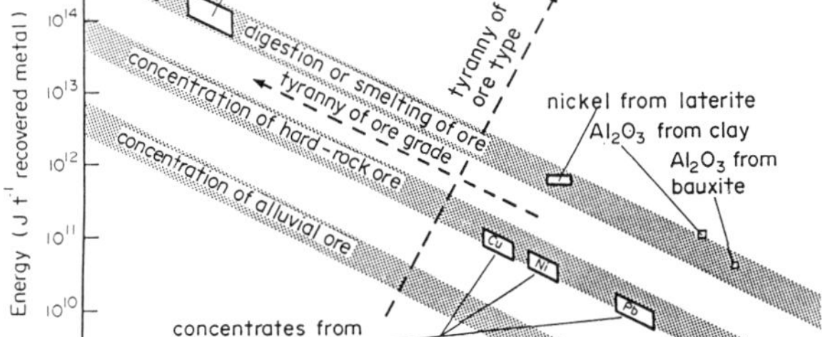

Figure 1
Energy for mining and primary ore processing (Kellogg 1978)

It is characteristic of nonrenewable resources, like metallic ores, that high-grade resources are used first, so that the average grade of ores mined decreases with time. The average grades of iron and copper ores mined in the USA around 1940 were 55% and 1.1%, respectively; today, the values are about 34% for iron ores and 0.6% for copper ores. Figure 1 shows that as the grade of ores decreases further in the future, the energy required for recovery of 1 t of metal will increase because of the larger tonnage for ore that must be mined and processed. The trend may be counteracted, in part, by technological efforts to reduce the unit energies of Table 3. The use of belt conveyors to elevate and haul ore from open pit mines may save energy relative to the truck haulage used today. Improved computer control of grinding circuits may reduce the energy required for fine grinding. These and numerous other energy conservation measures can be expected to reduce the unit energies of Table 3, but a realistic goal of 25% reduction in unit energy will only compensate for a similar reduction in ore grade.

Common rock is known to contain metals like copper, lead and zinc in parts per million and has frequently been suggested as a future source of scarce metals. Taking copper as an example, basaltic and gabbroic rocks containing 100 ppm copper (0.01% Cu) are abundant; if they were to be exploited for copper recovery, the processing would require digestion or smelting of the rock, because the copper is contained not as discrete copper minerals but as ionic replacements in the silicate structure. The top band of Fig. 1 indicates an energy requirement of $\sim 10^{14}\ J\,t^{-1}$ of copper recovered from such a source. This is 1000 times the energy now required to recover copper, and a yearly production of 2 million t of copper to satisfy the present US market would require 2×10^{20} J, which is more than twice the energy used for all purposes in the USA in 1981.

5. Energy for Metal Reduction and Refining

With a few exceptions, the primary processing considered in Sect. 4 produces a metal concentrate or other intermediate that must subsequently be converted to metal by pyrometallurgical processes (iron, copper, lead) or electrowinning (zinc, aluminum, magnesium) and in some cases further refined by pyrometallurgical processes (steel, lead, tin) or electrorefining (copper, silver, gold). The energy requirements for these final processing stages are independent of ore grade, because the raw material input is already a concentrate or intermediate rich in metal (e.g., copper concentrate is typically 20–30% Cu, purified Al_2O_3 for aluminum reduction is 52.9% Al, zinc concentrate is typically 50–60% Zn).

Process energy for reduction and refining of metal concentrates or intermediates to saleable metal is compared in Table 4 with the theoretical minimum free

Table 4
Process energy for metal production from concentrates

Metal (process)	Process energy[a]	Free energy[b]	Ratio free/process
	(10^9 J t^{-1} metal)		
Titanium sponge (Na reduction)	417	19	0.045
Magnesium (electrolytic)	394	23	0.058
Aluminum (Hall process)	229	29	0.13
Sodium (electrolytic)	106	8.1	0.077
Nickel (sulfide smelt)	103	3.6	0.035
Ferrochrome, high C (arc furnace)	64	10	0.16
Ferromanganese (arc furnace)	56	8.5	0.15
Copper (smelt, electrorefine)	56	2.1	0.038
Zinc (electrolytic)	56	4.9	0.087
Steel (blast furnace, BOF)	26	6.6	0.25
Lead (smelt, fire refine)	21	0.93	0.044

a Adapted from data prepared by Battelle–Columbus Laboratories (1975a,b, 1976). Energy for mining and primary processing has been subtracted from total energy to yield the process energy for reduction and refining b Free energy of formation of the stable oxide, from standard sources, converted to unit shown

energy for decomposition of the pure metal oxide to metal. This comparison shows one inherent reason for differences among the process energies for different metals: the theoretical minimum energy for reduction to the metallic state differs from metal to metal; even if carried out under idealized (i.e., reversible) conditions, it will always require more than four times as much energy to reduce Al_2O_3 to aluminum as to reduce Fe_2O_3 to iron. The free energy per tonne of metal, listed in Table 4, is higher, the more stable the metal oxide, the lower the atomic weight and the larger the valence of the stable oxide. Production of aluminum, magnesium and titanium will always be far more energy-intensive than production of lead, copper and nickel because of these intrinsic differences in chemical behavior.

The ratios of free to process energy, shown in Table 4, range from 0.035 for nickel to 0.25 for steel, and these are rough measures of theoretical energy efficiency for the actual production processes. It remains to explain the sevenfold range of energy efficiencies for metal production (nickel, copper, lead and titanium at very low efficiencies and steel at the highest efficiency). A number of practical process considerations interact to determine these efficiencies. Process features leading to lower energy efficiency are as follows.

(a) The need for large amounts of electrical energy, produced at low efficiency from fuels, for electrowinning (aluminum, magnesium, sodium, zinc) or for electric furnace processing (ferrochrome, ferromanganese). Of the metals listed in Table 4, those using the least electrical energy for production are steel, lead and copper.

(b) Necessity for a complex flowsheet, involving numerous process steps, either because of the difficulty of metal reduction (titanium and magnesium production require that chloride intermediate materials be produced from oxide raw materials) or because of the need to separate impurities and by-products by special processes (production of copper, lead and nickel suffer particularly from this difficulty).

(c) Low grade of feed material. Most of the metals in Table 4 are produced from feed materials that contain 50–70% of metal. Nickel and copper are exceptions, where the typical sulfide concentrates are only ~8% Ni and ~25% Cu. More energy is required to process larger amounts of low-grade material to produce 1 t of metal.

(d) Necessity for energy-intensive measures against pollution. Raw materials for production of copper, nickel, zinc and lead all consist of metal sulfides. The sulfur is oxidized to SO_2 during processing, and most industrialized countries now require capture of the process gases and conversion to sulfuric acid. For copper production, the gas handling circuits, including capture of fugitive gas, and conversion to sulfuric acid add as much as 8×10^9 J to the total process energy for production of 1 t of copper. Aluminum production faces similar problems with abatement of pollution from fluorine and fluorides. The iron and steel industry must contain air pollution from coke ovens and steelmaking dusts and water pollution from a variety of sources.

(e) Necessity for small-scale, batch production methods. Titanium is produced today by small-scale, batch processes (in the region of 1 t of metal per batch), in plants having an annual capacity of 5000–10 000 t. The energy efficiency of production on this scale cannot hope to rival that of large-scale, continuous methods, such as the iron blast furnace (2 000 000 t per year for a large modern furnace), or large-scale, batch methods like basic oxygen steelmaking (200 t for a batch cycle of only 40 min). Large-scale operations serve to reduce heat losses per unit of production and to make feasible the recovery of waste heat from process streams.

6. Process Design to Conserve Energy

The process technology leading to the energy values reported in Table 4 are those typical of US industry in the early 1970s (Battelle–Columbus Laboratories 1975a,b, 1976). In large part, these industries were employing technology designed in the early part of the century, at a time when energy conservation and pollution abatement were not important criteria in process design. Improved technology, capable of minimizing costs for pollution abatement and at the same time reducing energy consumption, is now available for many metals and further developments of this kind are to be expected.

In pyrometallurgical operations the most important developments leading to improved energy efficiency have been combinations of the use of commercial purity oxygen (95–99.5% O_2) with reactor designs capable of high-intensity reactions (Kellogg 1977b). Basic oxygen steelmaking, which has largely replaced open hearth steelmaking in the two decades since 1960, is a prime example of this trend in process design. Oxygen is blown at supersonic velocity into a bath of molten pig iron (hot metal) through a lance positioned above the metal bath. Oxidation of carbon, silicon and other impurities in the bath occurs very rapidly due to intimate mixing of metal, slag and gas phases. A batch of 200 t of steel can be produced in 35–40 min by oxygen steelmaking, compared with 8–12 h by the open hearth process. The specific capacity (tonnes of product per day per cubic meter of reactor volume) of a basic oxygen converter is ~65, compared with two for the open hearth process. Not only is the productivity of the process greatly enhanced, but the energy requirement (mainly for production of oxygen from air) is reduced to $\sim 2.6 \times 10^9\ \mathrm{J\,t^{-1}}$ of steel, compared with $5.8 \times 10^9\ \mathrm{J\,t^{-1}}$ for the open hearth (mainly for fuel to maintain the furnace at temperature over many hours). Basic oxygen steelmaking can even be operated with little or no net energy requirement in those modern plants which are designed to capture and use the energy associated with the hot carbon monoxide that issues from the mouth of the vessel.

A comparable example, for smelting sulfide concentrates of copper or nickel, is flash smelting with oxygen-enriched air or undiluted oxygen, one of the new technologies that has rapidly replaced reverberatory smelting since 1970. In the process, finely divided dry concentrate is blown into the furnace with the air or oxygen stream. Oxidation of sulfur and iron in the concentrate occurs within a fraction of a second, yielding molten sulfide matte containing copper or nickel and iron silicate slag. Compared with the conventional reverberatory furnace, the oxygen flash furnace has a specific capacity about four times higher and a process fuel equivalent 35–50% lower (Kellogg and Henderson 1976).

Conventional technology for smelting lead sulfide concentrate (sinter roasting followed by the blast furnace) suffers from severe environmental problems associated with control of toxic fumes and dust. Two new, intensive, oxygen-smelting processes have reached commercial production (1989), and both are capable of greatly improved environmental control as well as reduced energy consumption. The Kivcet process (Perillo and Carminate 1990) employs flash smelting; the QSL process (Lewis 1989) injects oxygen into the molten bath through tuyeres in the bottom of the smelting furnace. These processes, or similar new developments, are likely to rapidly replace the lead blast furnace.

Many design improvements in the Hall process for electrolytic production of aluminum have led to reduction of the energy requirement from 22 $\mathrm{kWh\,kg^{-1}}$ of aluminum (1940 average) to as little as 14 $\mathrm{kWh\,kg^{-1}}$ in the best modern plants (Cochran 1981). Efforts to develop a radically new process for aluminum smelting have, so far, not met with success.

Unlike the ore-limited sectors of metal production (mining and primary processing), the reduction and refining sectors seem to offer considerable opportunities for energy conservation. Only the inertia resulting from the need for very large capital expenditures prevents the wider adoption of modern technology, capable of significantly lower energy requirements coupled with improved pollution abatement.

7. Energy for Secondary Metal Production

Secondary metal produced from scrap has always played a significant role in the overall metal cycle of industrialized countries. For example, in the USA in 1976, 340 000 t of refined copper plus 780 000 t of brass and bronze were produced from secondary sources. Secondary copper and brass in the USA account for almost 40% of the total yearly supply of these metals (Kusik and Kenahan 1978).

For most metals the energy required to process scrap, including energy for collection, sorting, transportation, preparation and processing to saleable secondary metal, as well as the energy equivalent for major consumable supplies and reagents, is only a small fraction of the energy to produce primary metals from ores. Table 5 summarizes the MFE values for secondary production of selected metals and compares these with the MFE for primary metals given in Table 1.

The purity of the scrap processed determines the type

Table 5
Material fuel equivalent for production of secondary metals[a]

Secondary metal	Type of scrap used	MFE (10^9 J t^{-1})	MFE (% of primary)[b]
Aluminum (alloy 380)	various	17.5	6.2
Aluminum (can stock)	scrap cans	10.1	3.6
Copper (refined)	no. 1 scrap	4.4	3.4
Copper (refined)	no. 2 scrap	20.1	16
Coppper (refined)	low-grade	49.3	39
Brass, bronze	new scrap	8.2	6.4
Steel billets	various	9.7	35
Lead (soft)	batteries, etc.	9.4	30
Lead (hard)	batteries, etc.	11.2	36
Nickel alloy	alloy scrap	12.9	7.7
Tin (refined)	tinned steel, new	201	91
Zinc (cast alloy)	new scrap	3.8	5.4
Zinc (slab)	various	22.0	31

a Kusik and Kenahan (1978) b MFE for secondary metal divided by MFE for primary metal (Table 1) times 100

of processing and the energy requirement to produce a given quality of secondary metal. The three qualities of copper scrap listed in Table 5, for example, require the following processing. Low-grade copper scrap (also called copper-bearing scrap) usually averages ~30% copper but may contain material with as little as 5% copper. It consists of skimmings, grindings, ashes, residues, slags, brass, electric motors, switchboards and so on. Processing requires smelting in a cupola furnace to black copper, converting, anode refining, electrolyte refining, melting and casting. No. 2 copper scrap (~96% Cu) consists of burnt wire, and copper contaminated with solder, iron, oil and the like; processing requires only anode refining, electrolytic refining, melting and casting. No. 1 copper scrap consists of clean, uninsulated copper that requires only melting and casting to yield high-quality refined metal. The effect of scrap quality on energy for secondary production resembles the effect of ore grade on energy for primary production.

Of the metals listed in Table 5, only tin recovered from tinned steel scrap requires energy in an amount approaching that for the primary metal. All other secondary metals require far less energy than for the primary metal, as would be expected from the fact that most scrap metals are very high-grade raw materials (>90% metal) and already exist in the metallic state (the reduction of metal oxide to metal, as discussed in Sect. 5, need not be performed). Tin recovered from tinned steel is the exception because the scrap is very low grade (i.e., contains only 0.2–0.3% Sn) and because the treatment requires dissolution of the tin in caustic solution and energy-intensive electrowinning to reconstitute the metallic state. The major economic incentive for detinning steel is to produce steel scrap (detinned) suitable for steel manufacture.

Scrap metal may be conveniently divided into two categories: (a) new (prompt industrial) scrap, consisting of materials generated by the metal fabricating industries (e.g., punchings, trimmings, off-specification products, borings, turnings); and (b) old (obsolete, post-consumer) scrap, consisting of the myriad metal products that have served their useful life and have been discarded by industry or the consumer. New scrap is mostly (>90%) recycled as secondary metal today, because it is generated in large amounts at relatively few locations and is usually well segregated as to metal and alloy content. Collection and sorting of new scrap requires relatively little cost and the quality of the raw material is dependable.

Old scrap is recycled to secondary metal at far less than the rate at which it is generated in the USA. Crude estimates of the rate of obsolescence of metal products suggest that the percentage of old scrap that is recycled ranges from ~13% for aluminum to ~42% for silver (Kellogg 1976). The huge quantities of municipal solid waste disposed to landfills are known to contain large tonnages of ferrous metals, aluminum and other nonferrous metals. Although many practical problems remain to be resolved, it would seem likely that significant tonnages of old scrap, now discarded to landfills, will become available for secondary metal production in the years ahead. Such scrap may be recovered by processing of mixed solid wastes or by collection of segregated wastes from households and commercial enterprises. The economic justification for this will come from increasing cost of energy, increasing scarcity of some primary metals, increasing cost of landfill disposal and the desire to reduce litter.

Old scrap is certain to be of lower grade than most new scrap. It is contaminated by paint, paper, grease, food particles, glass and other metals and alloys that cannot be completely separated. For some secondary metal industries (notably steel and aluminum) new technology will be required to process such scrap into saleable secondary metal. Today the secondary aluminum industry has no means to remove iron, silicon, copper and other contaminants from scrap aluminum, with the result that it can accept only limited amounts of scrap containing such impurities, which is then diluted to product specifications with cleaner scrap or primary aluminum. Likewise, the steel industry must limit its intake of scrap containing copper and tin. New technology for processing to remove deleterious impurities from low-grade, old scrap is likely to be more energy-intensive than current secondary metal recovery, but still far less energy-intensive than primary metal production.

For a more comprehensive treatment of the topics discussed in this article the reader is referred to the monograph *Metal Resources and Energy* (Chapman and Roberts 1983).

See also: Economics of Materials, Energy and Growth

Bibliography

Battelle–Columbus Laboratories 1975a *Energy Use Patterns in Metallurgical and Nonmetallic Mineral Processing. High-Priority Commodities*, PB 245 759. US National Technical Information Service, Springfield, VA

Battelle–Columbus Laboratories 1975b *Energy Use Patterns in Metallurgical and Nonmetallic Mineral Processing. Intermediate Priority Commodities*, PB 246 357. US National Technical Information Service, Springfield, VA

Battelle–Columbus Laboratories 1976 *Energy Use Patterns in Metallurgical and Nonmetallic Mineral Processing. Low-Priority Commodities*, PB 261 150. US National Technical Information Service, Springfield, VA

Chapman P F, Roberts F 1983 *Metal Resources and Energy*, Butterworths Monographs in Materials. Butterworths, London

Cochran C N 1981 Energy balance of aluminum—from production to application. *J. Met.* 33(7): 45–8

Kellogg H H 1974 Energy efficiency in the age of scarcity. *J. Met.* 26(6): 25–9

Kellogg H H 1976 The role of recycling in conservation of metals and energy. *J. Met.* 28(12): 29–32

Kellogg H H 1977a Sizing up the energy requirements for producing primary metals. *Eng. Min. J.* 178(4): 61–5

Kellogg H H 1977b Conservation and metallurgical process design. *Trans. Inst. Min. Metall.* 86: C47–58

Kellogg H H 1978 Toward a materials-conservation ethic. *Metall. Trans. B* 9: 491–500

Kellogg H H, Henderson J M 1976 Energy use in sulfide smelting of copper. In: Yannopoulos J C, Agarwal J C (eds.) 1976 *Extractive Metallurgy of Copper*, Vol. 1, American Institute of Mining, Metallurgical and Petroleum Engineers, New York, pp. 373–415

Kusik C L, Kenahan C B 1978 *Energy Use Patterns for Metal Recycling*, Information Circular 8781. US Bureau of Mines, Washington, DC

Lewis W 1989 Cominco improves lead production. *Eng. Min. J.* 190(9): 35–8

Perillo A, Carminate A 1990 The Kivcet furnace construction at Portovesme. In: Mackey T S, Prengaman R D (eds.) 1990 *Lead–Zinc '90*. Minerals, Metals and Materials Society, Warrendale, PA, pp. 903–17

Pitt C H, Wadsworth M E 1981 Current energy requirements in the copper producing industries. *J. Met.* 33(6): 25–34

H. H. Kellogg
[Columbia University, New York, USA]

F

Factor Dominance in Industry

The production and cost characteristics of industries are heavily influenced by their basic technologies and, in particular, by the comparative roles of labor, materials and capital facilities in determining their production capabilities. Actual industrial processes cover a wide range in respect of such patterns of "factor dominance."

This concept is fundamentally different from the traditional idea of "factor intensity," which relates to cost proportions. Thus, classifying an industry as "capital intensive" implies a high ratio of capital costs to total costs, as in the case of electricity generating plants where such ratios may be 40% or more. But "capital intensive" is also often mistakenly applied to operations requiring large amounts of fixed investment, although these need not result in large ratios of capital costs. For example, this term is frequently applied to the steel industry because of its enormous capital requirements. But these investments yield such huge productive capacities that total depreciation and interest charges for major steel companies in the USA usually average less than 10% of total costs. On the other hand, steel may well be classified as a "capital-dominated" industry because the scale and modernity of its capital facilities and equipment are the overwhelming determinants both of the effectiveness with which materials, energy and labor inputs can be utilized and of the production capacity of the mill.

The "capital-dominated" extreme would include processes in which production workers are chiefly engaged in starting, stopping, loading, unloading, setting controls or otherwise simply facilitating the functioning of the equipment and capital facilities. In this category, materials merely play the passive role of providing the relatively standardized inputs being acted upon by the capital goods. Similarly, the "labor-dominated" extreme would encompass processes in which tools and other capital facilities serve essentially as accessories to wage earners, whose manual efforts and related skills represent the fundamental core of the production process—materials once again serving passively as the standardized inputs being acted on by manual operations. Similarly, a "materials-dominated" extreme may be envisaged, covering operations in which productive capacity is dominated by the resources to which labor and capital facilities are applied—both of the latter merely facilitating attainment of the production capacity determined by materials and related inputs, as in the case of changes in the metallic yield of smelters owing to variations in ore tenor, or in the case of farms differing in soil fertility.

What is the significance for management of the distinction between "factor dominance" and "factor intensity"? Unit costs are determined, of course, both by the volume of various inputs per unit of output and by their respective factor prices. Because factor intensity refers to cost proportions, or the relative magnitude of various unit costs, changes are attributable in large measure, especially in the short run, to market-determined adjustments in factor prices. Factor dominance, however, reflects the relationship among the various unit input requirements as determined by management's choice of production technologies. Thus, factor dominance is an active determinant of factor intensity, and tends to exert an increasing influence on it over the longer run as technological changes are introduced, whereas changes in factor intensity are merely the passive result of internal changes in factor dominance and of external changes in factor prices. Thus, despite material cost proportions exceeding 80%, blast furnaces, petroleum refineries and flour mills are basically capital-dominated operations.

1. Factor Dominance and Productivity Relations

1.1 Single Factor Dominance

The powerful influence of factor dominance on productivity relations is readily illustrated. In operations approximating the capital-dominated extreme (e.g., power plants, petroleum refineries, cement mills and blast furnaces), increases in labor productivity are more likely to reduce man-hour requirements than to increase productive capacity or the productivity of fixed capital; increases in the productivity of fixed capital are likely to leave direct labor requirements per unit of fixed capital input unchanged or reduced, and hence induce parallel increases in physical output per man-hour; and major increases in the productivity of the operations as a whole are thus much more likely to come from—and to be better measured by—adjustments in the productivity of fixed capital than in the productivity of labor. In operations approximating the labor-dominated extreme (e.g., bricklaying, custom tailoring and carpentry), increases in the productivity of fixed capital are more likely to reduce requirements for such equipment than to increase productive capacity or the productivity of direct labor; increases in labor productivity are likely to leave fixed-capital requirements unchanged, and hence induce parallel increases in the productivity of fixed capital; and major increases in the apparent productivity of the operation as a whole are thus much more likely to come from—and to be better measured by—adjustments in the productivity of direct labor than in the productivity of fixed capital.

Moreover, both in labor-dominated and in capital-dominated fabrication processes, materials requirements tend to vary with output levels, thus tending to increase the ratio of materials inputs to labor inputs with an increase in the productivity of labor, and to increase the ratio of materials inputs to effectively utilized fixed-capital inputs with an increase in the productivity of fixed capital. Also in such processes, decreases in the input requirements of materials per unit of output are more likely to reduce the volume of the input requirements of materials than to increase the productivity of other factors, or the productive capacity of the operation as a whole.

Materials-dominated processes are encountered most commonly in the extractive industries. Such dominance tends to be greatest in operations involving the direct extraction of natural resources, including agriculture, fisheries, forestry and mining. In each of these, the output resulting from the application of given levels of labor and capital inputs varies with the quality of the natural resources being worked, namely, the fertility of the land, the abundance of fish in the seas being combed, the volume of timber on the lands being lumbered, and the richness of the ore deposits being mined.

The degree of dominance by material inputs is, however, being reduced even in these industries, primarily as a result of growing reliance on capital goods to increase attainable rates of production. More machinery enables farmers to increase the productive capacity of their operations, not only through more intensive cultivation of existing farmland, but also through farming additional acreage. Fishing vessels with improved fish-locating, catching, handling and refrigeration equipment, as well as larger storage capacity and more powerful engines, can increase their average catch per month. Similar means exist for increasing the production capabilities of mines and lumbering operations through greater investments in facilities and equipment.

Increases in production rates from unchanged geographical boundaries are usually generated by greater capital inputs accompanied by smaller increases, or even significant decreases, in labor inputs. Thus, output per unit of material resources being worked increases, output per man-hour tends to increase either more or less rapidly than the preceding, and output relative to capital investment tends to decrease. The effects on total unit costs obviously depend on the extent to which decreases in unit material and unit labor costs are offset by increases in average capital charges per unit of output. Increases in the production of a farm or mine or timber operation due solely to extending the boundaries of such operations, and applying unchanged rates of labor and capital to similar additional areas, would obviously leave all productivity relationships unchanged. But in the more common case where such boundaries are extended because the production capabilities of new machinery, which increases output and reduces labor requirements, are not fully utilized by applications to past working areas, such extensions may be expected to increase output per man-hour further and also to reduce the initial decrease in capacity relative to fixed investment.

1.2 Jointly Dominated Processes

Within the enormous diversity to be found in industry, the extremes of capital-dominated, labor-dominated and materials-dominated processes may well account for only a limited proportion of actual operations. A large number of these might well be characterized as "jointly dominated," representing the determination of production capabilities either jointly by labor and capital, or jointly by materials and capital. The former may be illustrated by a wide array of manufacturing operations involving significant contributions either by skilled labor utilizing essentially general purpose equipment to make a variety of products, or by operators on assembly lines involving substantial reliance on the speed and quality of labor efforts to determine acceptable output. Jointly dominated materials and capital operations are most commonly exemplified in manufacturing by operations involving the initial refining of natural raw materials, such as ore smelting, petroleum refining and the processing of milk into dairy products.

Such essentially extractive operations in manufacturing are jointly dominated because changes in the quality of the materials input may alter the relationship of attainable output to fixed investment within a significant range. This results from the fact that any given capital facilities limit the flow of materials which can be processed within a given period. Hence, even if such facilities are fully utilized, the marketable output will depend on the metal content or chemical composition or butter-fat content of the raw materials being treated. On the other hand, changes in the technology, processing capabilities and speed of the capital facilities can also alter the marketable output of a plant, while leaving the quality of the materials being processed unchanged. Thus, the actual capacity will be determined jointly by the processing capabilities of the capital facilities and by the quality of the materials being processed, but the relative contributions of each are usually readily determinable.

In smelting operations, for example, an increase in the metal content of the ores being processed by a given plant would tend to increase output relative to man-hours and fixed investment as well as materials volume. Investments in improving the extractive capabilities of the facilities as well as the volume of materials which can be processed would tend to have the following effects in the absence of any change in the quality of the ores being treated: a significant increase in output per man-hour, some increase in output relative to material volume and a probable decrease in capacity relative to fixed investment.

2. Cost Effects of Factor Dominance

The cost effects of changes in the proportioning of inputs under different conditions of factor dominance depend not only on accompanying interactions with input factor prices, but also on any resulting changes in outputs (see *Productivity and Technological Improvements: Managerial Evaluations*). If output per unit of materials volume is increased by decreasing scrap and wastage, resulting savings will be paralleled by lower unit material costs. If such economies are also effected by other users through increasing diffusion of the technologies and practices involved, resulting decreases in demand may engender reductions in the price of such materials, thereby tending to enhance the initial reductions in unit material costs.

On the other hand, if increases in output per unit of materials volume are achieved through the use of better or more highly processed materials or purchased components, resulting benefits in unit material costs may be at least partly offset by the higher prices charged for such improved inputs. In such instances, however, two additional ramifications would be of concern to management. First would be the possibility of partially or wholly offsetting reductions in other processing costs per unit of output. Secondly, consideration would also have to be given to the possibility that the changes in the quality of material inputs make possible improvements in the attractiveness of the final products to customers which permit increases in the prices charged for them or in the quantities which can be sold.

Changes in output per man-hour may also produce a variety of cost effects, depending on accompanying changes in wage rates and in other unit costs. But increases in output per man-hour are more likely to engender at least partially offsetting increases in wage rates, although they are less likely to be attributable to increases in the direct contributions of labor to production per man-hour than to the partial replacement of such contributions by inputs of capital and of better or more highly processed materials. Moreover, product quality improvements in most industrial processes are seldom due to enhanced labor contributions.

The cost effects of changes in the productivity of capital goods (i.e., in productive capacity per dollar of net fixed assets) depend heavily on the capacity utilization rate, of course, as well as on the rate of charges to cover interest and capital recovery. But these cost effects may be affected to an even greater degree by resulting changes in other unit input requirements and in product qualities. Major changes in the productivity of fixed assets are usually the result of advances in technological processes, or in the scale of modernity of facilities and equipment. Such innovations frequently involve at least temporary decreases in the productivity of net fixed assets because the facilities being replaced may have cost less initially and because they have been largely depreciated—as a result of which the increase in new investment is likely to exceed the gain in capacity. Moreover, interest and depreciation rates may also be higher because of inflation and recognition of the need to adjust to faster rates of technological obsolescence in many industries. Hence, the key determinants of the economic desirability of such new capital inputs center around the magnitude of resulting decreases in other unit costs and increases in output and average price levels due to improvements in product qualities and mix. Detailed illustrations of the cost effects of various kinds of productivity adjustments in industries representing differing patterns of factor dominance and under alternative assumptions concerning attendant capacity and output adjustments are available (Gold 1955 pp. 202–32).

3. Competitive Effects of Technological and Productivity Improvements

The economic effects of improvements in a firm's productivity and technology cannot be determined by comparisons with its past performance. Resulting benefits and burdens can only be appraised within a broader framework which encompasses recent and prospective changes in market demand, in factor prices and in the developmental efforts of competitors. Changes in the product composition of demand or in the relative prices of different input factors often affect different producers unevenly; and although early adopters of major technological advances may gain significant cost and market share benefits, late adopters may merely reduce their hitherto growing competitive disadvantages.

Moreover, prospective means of improving productivity and technology may be rooted not only in managerial efforts to improve current performance, but also in anticipation of unfavorable pressures from factor or product markets. In such cases, as illustrated by major jumps in energy prices and by shortages of certain critical materials, achievements may be measured by slower or smaller increases in unit costs rather than by decreases.

See also: Productivity and Materials; Productivity and Technological Improvements: Managerial Evaluations; Value Added in Processing

Bibliography

Baird R N 1977 Production functions, productivity and technological change. In: Gold 1977, pp. 11–41

Dogramaci R (ed.) 1981 *Productivity Analysis: A Range of Perspectives*. Martinus Nijhoff, Boston, MA

Erdilek A 1977 Productivity, technological change and input–output analysis. In: Gold 1977, pp. 43–85
Gold B 1955 *Foundations of Productivity Analysis: Guides to Economic Theory and Managerial Control.* University of Pittsburgh Press, Pittsburgh, PA
Gold B 1971 *Explorations in Managerial Economics: Productivity, Costs, Technology and Growth.* Macmillan, London
Gold B (ed.) 1977 *Research, Technological and Economic Analysis.* Heath-Lexington, Lexington, MA
Gold B 1982 *Productivity, Technology and Capital: Economic Analysis, Managerial Strategies and Governmental Policies.* Heath-Lexington, Lexington, MA
Gold B, Peirce W S, Rosegger G, Perlman M 1984 *Technological Progress and Industrial Leadership.* Heath-Lexington, Lexington, MA
Gold B, Rosegger G, Boylan M G 1980 *Evaluating Technological Innovations: Methods, Expectations and Findings.* Heath-Lexington, Lexington, MA
Guile B, Brooks H (eds.) 1987 *Technology and Global Industry.* National Academy Press, Washington, DC
Gunn P G 1988 *Manufacturing for Competitive Advantage.* Ballinger, New York
Hogan J D, Craig A M (eds.) 1981 *Dimensions of Productivity Research.* American Productivity Center, Houston, TX
Nabseth L, Ray G F (eds.) 1974 *The Diffusion of New Industrial Processes: An International Study.* Cambridge University Press, London
Skinner W 1985 *Manufacturing: The Formidable Competitive Weapon.* Wiley, New York
Tornatzky L G et al. 1983 *The Process of Technological Innovation: Reviewing the Literature.* National Science Foundation, Washington, DC

B. Gold
[Claremont Graduate School, Claremont, California, USA]

I

Input–Output Analysis

Input–output analysis is a method of systematically quantifying the mutual interrelationships among the various sectors of a complex economic system. In practical terms, the economic system to which it is applied may be as large as a nation or even the entire world economy, or as small as the economy of a metropolitan area or even a single enterprise.

In all instances the approach is essentially the same. The structure of each sector's production process is represented by an appropriately defined vector of structural coefficients that describes in quantitative terms the relationship between the inputs it absorbs and the output it produces. The interdependence among the sectors of the given economy is described by a set of linear equations expressing the balances between the total input and the aggregate output of each commodity and service produced and used in the course of one or several periods of time.

The technical structure of the entire system can accordingly be represented concisely by the matrix of technical input–output coefficients of all its sectors. It constitutes at the same time the set of parameters on which the balance equations are based.

1. Input–Output Tables

An input–output table describes the flow of goods and services between all the individual sectors of a national economy over a stated period of time, say, a year. A simplified example of an input–output table depicting a three-sector economy is shown in Table 1. The three sectors are agriculture, whose total annual output amounts to 100 bushels of wheat; manufacture, which produced 50 yards of cloth; and households, which supplied 300 man-years of labor. The nine (3 × 3) entries inside the main body of the table show the intersectoral flows. Of the 100 bushels of farm products turned out by agriculture, 25 bushels were used up within the agricultural sector itself, 20 were delivered to and absorbed, as one of its inputs, by manufacture, and 55 were taken by the household sector. The second and third rows of the table describe in the same way the allocation of outputs of the two other sectors.

The figures entered in each column of the table thus describe the input structure of the corresponding sector. To produce the 100 bushels of its total output, agriculture absorbed 25 bushels of its own products, 14 yards of manufactured goods, and 80 man-years of labor received from the households; for the manufacturing sector to be able to produce the 50 yards of its total output, it had to receive and use up 20 bushels of agricultural—and 6 yards of its own (i.e., of manufactured)—products as well as 180 man-years of labor from households. In their turn, the households have spent the incomes—which they have received for supplying 300 man-years of labor—to pay for the consumption of 55 bushels of agricultural and 30 yards of manufactured commodities and 40 man-years of direct services of labor.

All entries in Table 1 are supposed to represent quantities, or at least physical indices of the quantities, of specific goods or services. A less aggregative, more detailed input–output table describing the same national economy in terms of not three but of 50, 100 or even 1000 different sectors would permit a more specific qualitative identification of all the individual entries. In a larger table, manufacturing would, for example, be represented not by one but by many distinct industrial sectors; its output—and consequently also the inputs of the other sectors—would be described in terms of "yards of cotton cloth," "tons of paper products," or even "yards of percale," "yards of heavy cotton cloth," as well as "tons of newsprint" and "tons of writing paper."

Table 1
Simplified input–output table for a three-sector economy

into / from	Sector 1: Agriculture	Sector 2: Manufacture	Sector 3: Households	Total output
Sector 1: Agriculture	25	20	55	100 bushels of wheat
Sector 2: Manufacture	14	6	30	50 yards of cloth
Sector 3: Households	80	180	40	300 man-years of labor

1.1 Input–Output Tables and National Income Accounts

Although in principle the intersectoral flows, as represented in an input–output table, can be thought of as being measured in physical units, in practice most input–output tables are constructed in value terms. Table 2 represents a translation of Table 1 into value terms on the assumption that the price of agricultural products is US\$2 per bushel, the price of manufactured goods is US\$5 per yard and the price of services supplied by the household sector is US\$1 per man-year. Thus the values of total outputs of the agricultural, the manufacturing and the household sectors are shown in the new translated table as being equal to, respectively, US\$200 ($=100 \times 2$), US\$250 ($=50 \times 5$) and US\$300 ($=300 \times 1$). The last row shows the combined value of all outputs absorbed by each of the three sectors. Such column totals could not have been shown in Table 1 since the physical quantities of different inputs absorbed by each sector cannot be meaningfully added.

The input–output table expressed in value terms can be interpreted as a system of national accounts. The US\$300 showing the value of services rendered by the households over the period of the years obviously represent the annual national income. It equals the sum total of the income payments—shown in row 3—received by the households for services rendered to each sector; it also equals the combined value of goods and services—as shown in column 3—purchased by the households from themselves and from the other sectors. To the extent to which the column entries (showing the input structure of each productive sector) cover the current expenditures but not purchases made on capital account, the latter—being paid out of the net income—should be entered in the households column.

All figures in Table 2—except the column sums shown in the bottom row—can also be interpreted as representing physical quantities of the goods or services to which they refer. This only requires that the physical unit in which the entries in each row are measured be redefined as being equal to that amount of output of that particular sector which can be purchased for US\$1 at prices which prevailed during the interval of time for which the table was constructed.

National input–output tables are now constructed in some 80 countries; a large number of regional and metropolitan input–output tables have also been compiled. The number of sectors that describe the economic system has increased dramatically in recent years. Some of the more detailed tables describe a national economy in terms of 500 or 600 separate sectors.

2. Technical Coefficients

Let the national economy be subdivided into $n + 1$ sectors: n industries, that is, producing sectors, and the $(n + 1)^{\text{th}}$ final demand sector, represented in input–output Tables 1 and 2 by the households. For the purposes of mathematical manipulation, the physical output of sector i is usually represented by x_1, while the symbol x_{ij} stands for the amount of the product of sector i absorbed—as its input—by sector j. The quantity of the product of sector i delivered to the final demand sector $x_{i,n+1}$ is usually identified in short as y_i.

The quantity of the output of sector i absorbed by sector j per unit of its total output j is described by the symbol a_{ij} and is called the input coefficient of the product of sector i into sector j:

$$a_{ij} = x_{ij}/x_j \tag{1}$$

A complete set of the input coefficients of all sectors of a given economy arranged in the form of a rectangular table—corresponding to the input–output table of the same economy—is called the structural matrix of that economy. Table 3 represents the structural matrix of the economy whose flow matrix is shown in Table 1. The flow matrix constitutes the usual—although not necessarily the only possible—source of empirical information on the input structure of the various sectors of an economy. The entries in Table 3 are computed according to Eqn. (1) from figures presented

Table 2
Simplified input–output table expressed in value terms

into / from	Sector 1: Agriculture (US\$)	Sector 2: Manufacture (US\$)	Sector 3: Households (US\$)	Total output (US\$)
Sector 1: Agriculture	50	40	110	200
Sector 2: Manufacture	70	30	150	250
Sector 3: Households	80	180	40	300
Total input (US\$)	200	250	300	

Table 3
Simplified structural coefficient matrix of three-sector economy

from \ into	Sector 1: Agriculture	Sector 2: Manufacture	Sector 3: Households
Sector 1: Agriculture	0.25	0.40	0.108
Sector 2: Manufacture	0.14	0.12	0.100
Sector 3: Households	0.80	3.60	0.133

in Table 1. For example, $a_{11} = 25/100 = 0.25$, and $a_{12} = 20/50 = 0.40$.

In practice the structural matrices are usually computed from input–output tables described in value terms, such as Table 2. In any case, the input coefficients (for analytical purposes described below) must be interpreted as ratios of two quantities measured in physical units. To emphasize this fact, the structural matrix (Table 3) in this example was derived from Table 1, not Table 2.

2.1 Static Input–Output System

The balance between the total output and the combined inputs of the product of each sector, as shown in the example in Tables 1 and 2, can be described by the following set of n equations:

$$\begin{aligned}
(x_1 - x_{11}) \quad - x_{12} - \cdots \quad & - x_{1n} = y_1 \\
- x_{21} + (x_2 - x_{22}) - \cdots \quad & - x_{2n} = y_2 \\
\cdots \quad \cdots \quad \cdots \quad & \cdots \quad \cdots \\
- x_{n1} \quad - x_{n2} - \cdots + (x_n - x_{nn}) & = y_n
\end{aligned} \tag{2}$$

Substitution of Eqn. (1) in Eqns. (2) yields n general equilibrium relationships between the total outputs, $x_1, x_2, \ldots, x_n$, of all producing sectors and the final bill of goods, $y_1, y_2, \ldots, y_n$, absorbed by households, government and other final users:

$$\begin{aligned}
(1 - a_{11})x_1 \quad - a_{12}x_2 - \cdots \quad & - a_{1n}x_n = y_1 \\
- a_{21}x_1 + (1 - a_{22})x_2 - \cdots \quad & - a_{2n}x_n = y_2 \\
\cdots \quad \cdots \quad \cdots \quad & \cdots \quad \cdots \\
- a_{n1}x_1 \quad - a_{n2}x_2 - \cdots + (1 - a_{nn})x_n & = y_n
\end{aligned} \tag{3}$$

If the final demand, $y_1, y_2, \ldots, y_n$, that is, the quantities of all the different kinds of goods absorbed by households and all other sectors whose outputs are not represented by the variables appearing on the left-hand side of Eqns. (3), is assumed to be given, the system can be solved for the n total outputs, $x_1, x_2, \ldots, x_n$.

The general solution of these equilibrium equations for the "unknown" x's in terms of the given y's can be presented in the following form:

$$\begin{aligned}
x_1 &= A_{11}y_1 + A_{12}y_2 + \cdots + A_{1n}y_n \\
x_2 &= A_{21}y_1 + A_{22}y_2 + \cdots + A_{2n}y_n \\
\cdots & \quad \cdots \quad \cdots \quad \cdots \quad \cdots \\
x_n &= A_{n1}y_1 + A_{n2}y_2 + \cdots + A_{nn}y_n
\end{aligned} \tag{4}$$

The constant A_{ij} indicates by how much the output x_i of the ith sector would increase if y_j, that is, the quantity of good j absorbed by households (or any other final users), had been increased by one unit. Such an increase would affect sector i directly (and also indirectly) if $i = j$, but when $i \neq j$ the output x_i is affected only indirectly, since sector i has to provide additional inputs to all other sectors which in their turn—directly or indirectly—must contribute to the increase in the delivery y_j made by sector j to the final users. From the computational point of view, that means that the magnitude of each coefficient A in the "solution" (4) in general depends on all the input coefficients a_{ij} appearing on the left-hand side of the system of equilibrium equations (3).

In mathematical language, the matrix

$$\begin{bmatrix}
A_{11} & A_{12} & \cdots & A_{1n} \\
A_{21} & A_{22} & \cdots & A_{2n} \\
\vdots & \vdots & & \vdots \\
A_{n1} & A_{n2} & \cdots & A_{nn}
\end{bmatrix}$$

of constants appearing on the right-hand side of the "solution," Eqns. (4), is identified as the "inverse" of the matrix

$$\begin{bmatrix}
(1 - a_{11}) & -a_{12} & \cdots & -a_{1n} \\
-a_{21} & (1 - a_{22}) & \cdots & -a_{2n} \\
\vdots & \vdots & & \vdots \\
-a_{n1} & -a_{n2} & \cdots & (1 - a_{nn})
\end{bmatrix}$$

of constants appearing on the left-hand side of Eqns. (3). The computation involved in finding such a solution is called the inversion of the coefficient matrix of these original equations. The "inverse" of the matrix

$$\begin{bmatrix} (1-0.25) & -0.40 \\ -0.14 & (1-0.12) \end{bmatrix}$$

based on Table 3 is

$$\begin{bmatrix} 1.4570 & 0.6623 \\ 0.2318 & 1.2417 \end{bmatrix}$$

Inserted in Eqns. (4) this yields two equations:

$$\left.\begin{aligned} x_1 &= 1.4570\,y_1 + 0.6623\,y_2 \\ x_2 &= 0.2318\,y_1 + 1.2417\,y_2 \end{aligned}\right\} \qquad (5)$$

which permits us to determine what total outputs of agricultural and manufacturing sectors, x_1 and x_2, would correspond to any given combination of the deliveries of their respective products, y_1 and y_2, to the exogenous sector, households. To verify this result by comparing them with the corresponding entries in Table 1, set $y_1 = 55$ and $y_2 = 30$ on the right-hand sides of Eqns. (5) and they will yield $x_1 = 100$ and $x_2 = 50$ on the left.

Only if all elements A_{ij} of the inverted matrix are non-negative will there necessarily exist for any given set of final deliveries, $y_1, y_2, \ldots, y_n$, a combination of positive total outputs, $x_1, x_2, \ldots, x_n$, capable of satisfying it. A sufficient condition for this is that the determinant of the matrix

$$\begin{bmatrix} (1-a_{11}) & -a_{12} & \cdots & -a_{1n} \\ -a_{21} & (1-a_{22}) & \cdots & -a_{2n} \\ \vdots & \vdots & & \vdots \\ -a_{n1} & -a_{n2} & \cdots & (1-a_{nn}) \end{bmatrix}$$

and of all its principal submatrices

$$(1-a_{11}) > 0 \qquad \begin{bmatrix} (1-a_{11}) & -a_{12} \\ -a_{21} & (1-a_{22}) \end{bmatrix} > 0 \ldots$$

$$\begin{bmatrix} (1-a_{11}) & -a_{12} & \cdots & -a_{1n} \\ -a_{21} & (1-a_{22}) & \cdots & -a_{2n} \\ \vdots & \vdots & & \vdots \\ -a_{n1} & -a_{n2} & \cdots & (1-a_{nn}) \end{bmatrix} > 0$$

should be positive. If this so-called Hawkins–Simon condition is satisfied for one arbitrarily numbered sequence of sectors, it is necessarily satisfied for any other sequence too. The material interpretation of that condition is that if an economic system in which each sector functions by absorbing output of other sectors directly and indirectly is to be able not only to sustain itself but also to make positive delivery to final demand, each one of the smaller and smaller subsystems contained in it must necessarily be capable of doing so too. If even one of them cannot pass that test, it is bound to cause a leak that will destroy the sustainability of the entire system.

A simpler sufficient, but not necessary, condition of sustainability of an economy is that the sum of the coefficients of each column of its structural matrix should be ≤ 1, with at least one of the column sums strictly < 1.

In most cases in which the structural matrix of a national economy has been derived from a set of actually observed value flows such as are represented, for example, in Table 2, the condition stated above will be satisfied.

Since in an open input–output system the households are usually treated as a final, that is, an exogenous sector, its total output, x_{n+1}, that is, the total employment, usually does not appear as an unknown variable on the left-hand side of Eqns. (3) and on the right-hand side of its solution, Eqns. (4). After the outputs of the endogenous sectors, $x_1, x_2, \ldots, x_n$, have been determined, the total employment can be computed from the following equation:

$$x_{n+1} = a_{n+1,1}x_1 + a_{n+1,2}x_2 + \cdots + a_{n+1,n}x_n + y_{n+1} \qquad (6)$$

The technical coefficients, $a_{n+1,1}, a_{n+1,2}, \ldots, a_{n+1,n}$, represent the inputs of labor absorbed by various industries (sectors) per unit of their respective output; y_{n+1} is the total amount of labor directly absorbed by households and other exogenous sectors. Such an employment equation constructed for the three-sector system with the structural matrix shown in Table 3 is:

$$x_3 = 0.80\,x_1 + 3.60\,x_2 + y_3 \qquad (7)$$

Households must not necessarily be considered to be part of the exogenous sectors as they are in the example used above. In dealing with problems of income generation in its relation to employment, the quantities of consumer goods and services absorbed by households can be considered to be structurally dependent on the total level of employment in the same way as the quantities of coke and ore absorbed by blast furnaces are considered to be structurally related to the amount of pig iron produced by them. With households shifted to the left-hand side of Eqns. (2, 4), the exogenous final demand appearing on their right-hand side will comprise only such items as governmental purchases and exports and, in any case, additions to or reductions in stocks of goods, that is, real investment or disinvestment.

When all sectors and all purchases are considered to

be endogenous, the input–output system is called closed. A static system cannot be truly closed, since endogenous explanation of investment or disinvestment requires consideration of structural relationships between inputs and outputs which occur in different periods of time (see Sect. 3).

2.2 Exports and Imports

In an input–output table of a country or a region which trades across its borders, exports can be entered as positive and imports as negative components of final demand. If the economy described in Table 1 ceased to be self-sufficient and started to import, say, 20 bushels of wheat and to export 8 yards of cloth—while letting the households consume the same amounts of both products as before—a new balance between all inputs and outputs would be established, as described in Table 4.

The input coefficients of the endogenous sectors, and consequently also the structural matrix of the system and its "inverse," remain the same as before. To form the new column of final demand, we have to add to the quantity of each good absorbed by the households the amount that was exported and subtract the amount that was imported (i.e., imports can be treated as negative exports):

$$y_1 = x_{1n} + e_1, \quad y_2 = x_{2n} + e_2, \ldots, y_n = x_{nn} + e_n \qquad (8)$$

The corresponding sectoral outputs can then be derived (see above) from the general solution (4). For the present numerical example, we can use directly Eqns. (5). The total labor requirement of the economy—300 man-years—remains in this particular case unchanged after it enters foreign trade, because the total direct plus indirect labor content of the 20 bushels of imported wheat happens to be equal to the labor content of the 8 yards of exported cloth.

If the imports of good i—that is, the negative e_i—happen to exceed the final domestic consumption of that good x_{in}, the corresponding "net" final demand y_i will turn out to be negative. As y_i diminishes, the total output of all sectors and in particular the total output x_e must (*ceteris paribus*) diminish. At some point, that output will be reduced to zero, which means that the entire direct and indirect demand for that particular commodity will be covered by imports. The corresponding domestic industry will be automatically eliminated from the endogenous part of the input–output table. The imports of such goods are called noncompeting, particularly when—as in the case of coffee and certain minerals—even a large increase in demand does not call forth their domestic production. The magnitude of total domestic demand for noncompeting imports can be computed in the same way as the total demand for labor can be derived from Eqn. (6).

2.3 Prices in an Open Static Input–Output System

Prices are determined in an open input–output system from a set of equations which states that the price which each productive sector of the economy receives per unit of its output must equal the total outlays incurred in the course of its production. These outlays comprise not only payments for inputs purchased from the same and from the other industries, but also the "value added," which essentially represents payments made to the exogenous sectors:

$$\begin{array}{lllll} (1-a_{11})p_1 & -a_{21}p_2 - \cdots & & -a_{n1}p_n = v_1 \\ -a_{12}p_1 + (1-a_{22})p_2 - \cdots & & & -a_{n2}p_n = v_2 \\ \cdots & \cdots \quad \cdots & & \cdots \quad \cdots \\ -a_{1n}p_1 & -a_{2n}p_2 - \cdots + (1-a_{nn})p_n = v_n \end{array} \qquad (9)$$

Each equation describes the balance between the price received and the payments made by each endogenous sector per unit of its product; p_i represents the payments made by sector i—per unit of its product—to all exogenous (i.e., the final demand) sectors. These usually comprise wages, interest on capital and entrepreneurial revenues credited to

Table 4
Simplified input–output table showing the effects of imports and exports on a three-sector economy

from \ into	Sector 1: Agriculture	Sector 2: Manufacture	Final demand: Sector 3: Households	Final demand: Exports (+) or Imports (−)	Final demand: Total final demand	Total output
Sector 1: Agriculture	19.04	22.12	(55)	(−20)	35	76.16 bushels
Sector 2: Manufacture	10.66	6.64	(30)	(+8)	38	55.30 yards
Sector 3: Households	60.93	199.07	(40)		40	300 man-years

households, taxes paid to the government and other final demand sectors.

In analogy to the solution (4) of output equations (3), the solution of the price equation (9) permits the determination of prices of all products from the given values added (per unit of output) in each sector:

$$\begin{aligned} p_1 &= A_{11}v_1 + A_{21}v_2 + \cdots + A_{n1}v_n \\ p_2 &= A_{12}v_1 + A_{22}v_2 + \cdots + A_{n2}v_n \\ &\cdots \quad \cdots \quad \cdots \quad \cdots \quad \cdots \\ p_n &= A_{1n}v_1 + A_{2n}v_2 + \cdots + A_{nn}v_n \end{aligned} \tag{10}$$

The constant A_{ij} measures the dependence of the price p_i of the product of sector j on the value added, v_i, earned per unit of its output in sector i.

Each row of the a_{ij} coefficients appearing in the output equations (3) makes up the corresponding column of coefficients appearing in price equations (9); the A_{ij} coefficients appearing in each row of the "output solution" equations (4) make up the corresponding coefficient column in the "price solution" equations (10).

Thus, inserting the inverse computed in the example used above in the Eqn. (10) solution of the price equation, we have:

$$\left.\begin{aligned} p_1 &= 1.4570\, v_1 + 0.2318\, v_2 \\ p_2 &= 0.6623\, v_1 + 1.2417\, v_2 \end{aligned}\right\} \tag{11}$$

From Tables 2 and 3 we can see that in our example the values added paid out (i.e., the wages) by agriculture and in manufacture per unit at their respective outputs amounted to US\$0.8 and US\$3.6. According to the two equations above, this yields $p_1 =$ US\$2, $p_2 =$ US\$5, which are the prices of agricultural and manufactured products used in deriving the value figures presented in Table 2 from Table 1, which described the input–output flows only in physical units.

The internal consistency of the price and the quantity relationships within an open input–output system is confirmed by the following identity derived from Eqns. (4, 9):

$$x_1v_1 + x_2v_2 + \cdots + x_nv_n = y_1p_1 + y_2p_2 + \cdots + y_np_n \tag{12}$$

On the left-hand side stands the sum total of value added paid out by the endogenous to the exogenous sectors of the system; on the right-hand side are the combined values (quantities times prices) of their respective products delivered by all endogenous sectors to the final (exogenous) demand. This identity confirms, in other words, the accounting identity between the national income received and the national income spent, as shown in Table 2.

For the purposes of more detailed price analysis, the technical "cooking recipe" for producing, say, one ton of bread not only has to specify the requisite amounts of current inputs such as flour, milk and yeast but also has to list needed pots and pans and other kinds of capital goods required for that purpose. Thus the matrix A of technical flow coefficients has to be supplemented by a corresponding matrix of capital stock coefficients, B:

$$B = \begin{bmatrix} b_{11} & b_{12} & \cdots & b_{1n} \\ b_{21} & b_{22} & \cdots & b_{2n} \\ \vdots & \vdots & & \vdots \\ b_{n1} & b_{2n} & \cdots & b_{nn} \end{bmatrix}$$

A capital coefficient b_{ij} represents the technologically determined stock of the particular kind of goods—machine tools, equipment, industrial buildings, "working inventories"—produced by industry i that industry j has to employ per unit of its output. In other words, each column of matrix B describes the physical capital requirements (per unit of its total output) of a particular industry, in the same way that the corresponding column of matrix A describes its "current inputs" requirements.

The price analysis outlined above can be advanced one step further by splitting each of the "values added," appearing on the right-hand side of each Eqn. (9), into two parts: the returns on capital invested in buildings, machinery and other stocks of goods required for production of the output in question, on the one hand, and wages, on the other. The return on capital can be represented as the value (price times quantity) of all productive stocks (used per unit of its output) in each industry multiplied by the given rate of return.

The relationship between wage rates, the rate of return on capital (i.e., the "price" of capital) and the price of different goods and services takes on the following form:

$$p = (1 - A' - rB')^{-1}W \tag{13}$$

where W is a column vector of wage costs paid by different industries per unit of their respective outputs.

Insertion in Eqn. (13) of the numerical values of the flow coefficient matrix A' as given in Table 3 and of capital coefficient matrix B as given above and "inverting" the bracketed expression on the right-hand side yields an explicit solution of that equation for various values of the rate of return on capital, r. For instance,

if $r = 10\%$	if $r = 20\%$
$p_1 = 1.55\, w_1 + 0.26\, w_2$	$p_1 = 1.65\, w_1 + 0.30\, w_2$
$p_2 = 0.76\, w_1 + 1.34\, w_2$	$p_2 = 0.87\, w_1 + 1.45\, w_2$

The magnitudes of all numerical parameters increase as the value of r is raised from 10% to

20%. That means that with given wage costs, w_1 and w_2, a rise in the rate of return, that is, in the costs of capital, must obviously result in higher prices p_1 and p_2. With higher prices, the purchasing power of money wages (i.e., the real wages) must necessarily fall.

3. Dynamic Input–Output Analysis

The following set of linear difference equations

$$X(t) - AX(t) - B[X(t+1) - X(t)] = Y(t) \quad (14)$$

provides the simplest possible description—in input–output terms—of the relationship between the column vectors of current outputs, $X(t), X(t+1), \ldots$, final consumption, $Y(t)$, and the increase from one time period to the next of necessary investment in productive stocks, $B[X(t+1) - X(t)]$.

Given appropriate initial conditions, the solution of these equations traces the path of even, fully equilibrated economic growth of the economic system described. This growth process can, however, be proven to be inherently unstable. If it is displaced from that path by even a smallest upward or downward shock, the process will become explosive, in the sense that some sectors of the economy will begin to expand very fast, while the growth rate of other sectors is likely to become negative. Thus, total levels of both output and investment in these sectors will eventually tend to fall below zero—which of course is impossible.

While such a process is unstable and explosive—if traced from a given set of initial conditions forward—it proves to be stable and convergent when traced from any given set of initial conditions backward. Thus, given the levels of final deliveries for the year $Y(0)$, the solution of the same equations for the years $t-1, -2, \ldots$, and so on—called their "dynamic inverse"—can be used to compute what the levels of output and investment would have to be in the preceding years in order to enable the economy in question to arrive at the prescribed final state.

A more realistic dynamic input–output model has to account for the fact that the stocks of durable producers' goods held by any given sector are at times kept partly idle and at other times pressed to produce more than their "normal" (i.e., optimal "full capacity" output). This is due to the fact that new investment is undertaken to satisfy not the actual, but rather the anticipated, demand, which inevitably leads at some time to shortfall or excess.

A distinction has to be made at any point (or rather interval) of time between the actual and the anticipated (i.e., originally targeted) full capacity output. When the latter exceeds the former, some equipment remains idle or is operated on a less than "full time" basis. When the opposite is true it means that the existing equipment has been pushed to produce more than its optimal, full capacity output. Only in this case can new investment aimed at an increase of the full capacity output be expected to occur. In the other case, since negative investment cannot take place, the stock of durable capital goods held by the particular sector remains constant and gross investment is limited to normal maintenance.

If $C_i(t)$ represents the potential "full capacity" output of sector i in time period t (i.e., the output that the sector could, but not necessarily does, produce by making the most efficient use of its equipment), vector $B_t C_i(t)$ would represent the stocks of capital goods of which that equipment consists. The subscript t indicates that the composition of these stocks reflects the technology employed by that sector at that time.

An increase in the full capacity output of sector i from time period t to time period $t+1$—let it be $O_i(t)$—cannot be negative since, within a relatively short time interval, stocks of durable structures and equipment cannot easily diminish, although they certainly can become partly idle:

$$O_i(t) = \max[0, C_0(t) - C_i(t-1)] \quad (15)$$

where $C_0(t)$ represents the next years' targeted full capacity output.

Iterative computations tracing the sequence of demand projection, investment and production decision involved for each successive year have to be carried out in the following five consecutive steps.

Step 1. Projection of the future level of demand on the basis of the following (or some other similar) extrapolation formula:

$$C_1^0(t) = \min\left[\frac{X_i(t-1)tX_i(t-2)}{X_i(t-2)tX_i(t-3)}, 1 + \delta_i X_i(t-1)\right] \quad (16)$$

This equation states that the full capacity output targeted in time period $t-1$ for the next time period t depends on the average growth rate of output observed over the previous three years provided it does not exceed the exogenously prescribed upper limit, δ_i.

Step 2. Using Eqn. (15) the meaning of which has been explained above, compute $O(t)$ the change in the full capacity outputs of each sector from time period $t-1$ to time period t.

Step 3. With the magnitudes appearing on the right-hand side of the matrix in Eqn. (16) determined in the first two steps, solve that equation for the unknown vector of sectoral outputs levels $X(t)$:

$$(1 - A)X(t) = B_t O(t+1) + Y(t) \quad (17)$$

$$X(t) = (1 - A)^{-1}[B_t O(t+1) + Y(t)]$$

The same computation procedure can be repeated to trace the development of the given economic system forward to the next and the following periods of time.

Structural changes are described in great detail by appropriate changes in the magnitude of the elements of the technological matrices A_t and B_t.

The somewhat simplistic method of targeting the level of future demand through Eqn. (16) can, of course, be replaced by some other method.

Depending on the amount of unused or overused capacity appearing at successive stages of that dynamic process, the rate of growth of the entire system can be higher or lower. However, the automatic self-correcting trial and error mechanism built into the model prevents the process from becoming completely unbalanced or explosive.

The short-run fluctuation of sectoral outputs can be expected to be erratic and the stocks of capital goods can be expected to increase by irregular steps; however, in the long run the system will not deviate very much from the path of equilibrated (i.e., balanced) expansion.

4. Technological Change

Since the technological structure of each sector of the economy is represented by a column vector of input coefficients and the corresponding column vector of capital coefficients, technological change can be described concisely as a change in the magnitudes of the elements of these vectors. Introduction of new commodities or industries is represented through introduction of new and disappearance of old commodities (or industries) through elimination of old vectors from the structural matrix of the economy in question.

The choice between two (or more) alternative processes that might be available for production of a particular good or service must obviously be based on a comparison of the effects of a hypothetical shift from one technology to another. For instance, a shift from coal-generated to atomic energy would affect the cost of production, prices and the level of output and input of goods directly or indirectly. To determine these effects, several input–output computations have to be carried out, each based on the introduction into the flow (A) and into the capital (B) matrix of the economy in question of coefficient vectors characterizing the alternative technologies available for the industries in question. In the case where the choice of appropriate technology can be based on maximizing or minimizing an explicitly defined function of some variable—such as the aggregate input requirements for labor or specific natural resources, investment requirements or the cost of production of various goods (whose magnitudes can be determined by means of appropriate input–output computations)—it can be formalized and carried out with the help of an appropriate linear programming algorithm. Dr. George Danzig, the inventor of the well-known Simplex method of linear programming, actually developed it first as a means of automating input–output computations involving sequential substitution of alternative column vectors into square input–output coefficient matrices.

5. The Scenario Approach

Practical application of input–output analysis often takes the form of comparisons of the implications—described in terms of complete projected input–output tables—of several alternative scenarios, each based on a different set of assumptions concerning the level and the composition of the final demand, changes in the magnitudes of input coefficients incorporated in various column vectors of the flow and capital coefficient matrices, or a combination of both.

Shortly before the end of World War II, President Roosevelt asked the US Labor Department to assess the probable effect on the American economy of the impending transition from the war to a peacetime footing. A static input–output model was constructed on the basis of a matrix of structural input coefficients derived from the 1939 input–output table of the US economy, the first such table compiled in the USA under government auspices. A comparison of the output and unemployment levels attained in all industries under war conditions with the hypothetical output and employment levels, computed on the assumption that a vector of final demand representing normal civilian consumption would be substituted for the vector of final delivery dominated by military goods, provided a detailed and internally consistent answer to the question raised. To the great surprise of experts who predicted a slump in steel—conventionally considered to be a "war industry"—these input–output computations led to the conclusion that substitution of a normal peacetime vector for the wartime vector of final demand would lead to a sharp rise in output and employment level in the steel sector. Subsequent development demonstrated that that conclusion was indeed correct.

Many, if not most, studies aimed at assessment of future energy demand and the effects of a shift from oil to coal or to atomic power have involved the use of supply and demand "elasticities" derived by means of simple or multiple correlation analyses applied to time series describing past changes in energy input, energy prices and prices of other goods. The contribution of an input–output approach to a consideration of the energy problem has consisted on the other hand in the construction of several alternative scenarios, each involving a different combination of input–output vectors describing the technical structure of various methods of producing and using energy.

Such computations have, for instance, shown that while alcohol distilled from grain does indeed improve the energy balance of Brazil, it would not do so in the USA. Taking into account the amounts of energy absorbed directly and indirectly in operating agricultural machinery and producing chemical fertilizer in the USA, it is found that more than one thermal unit would be used to supply one unit in the form of alcohol. A similar computation based on the Brazilian input–output table yields the opposite result.

In this connection, it has to be pointed out again that such computations necessarily take into account the entire input–output structure of the economy in question, including that of its foreign trade.

The first practical application of the input–output method to systematic study of materials flow was carried out at the end of World War II in the USA by the Western Electric company. Lead was, next to copper, one of the principal materials used at that time to manufacture electric cable. Expecting a rapid rise in demand for its own products, as well as the products of many other industries depending on the supply of lead, the management of that company carried out an input–output projection of production and consumption a few years ahead and came to the conclusion that critical shortages were bound to develop. On the basis of that finding, Western Electric initiated a crash research program aimed at replacing lead with a suitable polymeric material in cable manufacture.

A recent application of the input–output methodology to systematic study of materials has been reported by Leontief et al. (1983). It is based on a modified input–output model of the US economy imbedded into the previously constructed multiregional input–output model of the world economy. The core of the structural matrix consists of the official US input matrix with the five of its 106 sectors which depict production of nonferrous minerals expanded to 36 sectors, describing production, processing and consumption of 26 nonfuel metallic and nonmetallic minerals. Mine output supplemented by product output resulting from other mining operations and reprocessing of scrap is described in great detail in the corresponding matrix of technical coefficients.

The system of 321 equations (containing 328 variables) describing the balance between the total supply including imports and the total use including exports of various goods, among them all nonferrous metals in different forms, as well as the generation and elimination of major pollutants and allocation of labor, is presented in a schematic form below.

The system is described by the following set of equations:

$$p_1^{(f)} = (I - A_c) \cdot q_1^{(f)} + I \cdot q_3^{(f)} \tag{18}$$

$$p_2^{(f)} = -B_c \cdot q_1^{(f)} + (I + C_c) \cdot q_2^{(f)} + G_c \cdot q_4^{(f)} \tag{19}$$

$$p_3^{(f)} = L_c \cdot q_1^{(f)} - I \cdot q_3^{(f)} \tag{20}$$

$$p_4^{(f)} = M_c \cdot q_1^{(f)} + N_c \cdot q_0^{(f)} + H_c \cdot q_4^{(f)} \tag{21}$$

$$p_5^{(f)} = D_c \cdot q_1^{(f)} - I \cdot q_5^{(f)} \tag{22}$$

where the subscript c indicates a matrix of coefficients while superscript f refers to flows. Equation (18) states that the gross domestic output of each commodity plus imports minus intermediate consumption must satisfy final demand. Similarly, Eqn. (19) states that the domestic output of minerals (own industry plus by-product) plus competitive imports minus intermediate consumption must equal final demand for minerals. Equation (20) states that the level of imports for each commodity is equal to a specified fraction of domestic (own industry) output. Equation (21) states the same proposition for noncompetitive minerals (primary and scrap). Equation (22) states that the sum of each industry's value added, labor inputs and emissions "output" equals the respective total for the economy as a whole. As explained previously, noncompetitive imports are goods used to satisfy intermediate or final demand for which there is no corresponding domestic producing sector.

A solution vector has the following form:

$$\boldsymbol{Q}^f = \begin{cases} q_1^{(f)} = \text{a } 106 \times 1 \text{ vector of commodity output levels in time } t \\ q_2^{(f)} = \text{a } 36 \times 1 \text{ vector of mineral and scrap output levels in time } t \\ q_3^{(f)} = \text{a } 106 \times 1 \text{ vector of commodity imports levels in time } t \\ q_4^{(f)} = \text{a } 39 \times 1 \text{ vector of mineral and scrap import levels in time } t \\ q_5^{(f)} = \text{a } 34 \times 1 \text{ vector of value added, labor requirements, energy consumption, pollution emission and new scrap generation levels in time } t \end{cases}$$

and

$$\boldsymbol{P}^f = \begin{cases} p_1^{(f)} = \text{a } 106 \times 1 \text{ vector of final demand components minus imports for 106 commodities valued in dollars for time } t \\ p_2^{(f)} = \text{a } 36 \times 1 \text{ vector of final demand components minus imports for 36 mineral commodities in physical units for time } t \\ p_3^{(f)} = \text{a } 106 \times 1 \text{ null vector} \\ p_4^{(f)} = \text{a } 39 \times 1 \text{ vector with zeros everywhere except in rows 253, 254, 265 and 288, whose elements give final demand minus import levels for noncompetitive imports in time } t \\ p_5^{(f)} = \text{a } 34 \times 1 \text{ null vector} \end{cases}$$

The symbols used in Eqns. (17–21) are defined as follows.

A_c a commodity-by-commodity matrix of input–output coefficients (106 × 106) whose elements $a_{ij}^{(c)}$ give dollar amounts of input i required to produce one dollar's worth of output j (valued in base year prices).

0 a null matrix or vector.

I an identity matrix.

$-B_c$ an input–output coefficient matrix (36 × 106) whose elements $b_{ij}^{(c)}$ give the physical amount of mineral (primary or scrap) input i required to produce one dollar's worth of output j (only minerals which are produced in the USA are included in this submatrix).

C_c a diagonal by-product coefficients matrix (36 × 36) whose elements $c_{ij}^{(c)}$ give the physical amount of each mineral (primary or scrap) produced as a by-product per physical unit of its "own industry" output.

G_c a step-diagonal matrix (36 × 40) whose nonzero elements $g_{ij}^{(c)} = 1$.

L_c a diagonal import coefficient matrix (106 × 106) whose elements $l_{ij}^{(c)}$ give the dollar amount of imports per dollar's worth of j ($i = j$).

M_c a matrix (39 × 106) whose only nonzero elements appear in IEA–USMIN rows 253, 254, 265 and 288; the elements of these four rows, $m_{ij}^{(c)}$, give the physical or dollar amount of noncompetitive import i per dollar's worth of output j ($i = j$).

N_c a step-diagonal matrix (39 × 36) whose nonzero elements $n_{ij}^{(c)}$ give the physical amount of mineral (primary or scrap) i imported per physical unit of mineral j's "own industry" output ($i = j$).

H_c a diagonal matrix (39 × 39) whose nonzero elements $h_{ij}^{(c)} = -1$, except in rows 253, 254 and 288, where $h_{ij}^{(c)} = 1$.

D_c a matrix (34 × 106) whose elements d_{ij} give the amounts in dollars or physical units of value added, labor, energy, pollution emissions and new scrap associated with a dollar's worth of output of commodity j.

Such a system of equations was solved (to check its internal consistency) for the base year 1972 and for the years 1980, 1990, 2000 and 2030. Systematic projections of future changes in all sets of technological coefficients, particularly those reflecting efficient methods of extraction and refining and substitution between different materials, were the most demanding part of that task. The estimates of future changes in the exports and imports (that enter into the system as vectors of exogenously determined variables) were obtained by incorporating the system into the multiregional input–output analysis of the world economy constructed for the United Nations several years earlier.

Alternative projections were computed, based on eleven different scenarios. Each of these represented a different combination of specific assumptions concerning the dependence of the US economy on imports of nonferrous metals and future rates of technological change. Final conclusions were summarized in the form of separate observations on the present and expected future supply and demand for each nonferrous mineral on the domestic US and international markets.

One of the most ambitious applications of the input–output approach was the construction of a multiregional, multisectoral, dynamic input–output model of the entire world economy referred to above. That model was employed in the preparation of long-run projections based on alternative scenarios of prospective developments of the economic relationship between the developed and less developed regions. It also provided the basis for long-run projections of the economic growth (or decline as the case may be) of the various regions under alternative assumptions concerning population growth, technical change—particularly in the field of agriculture and energy production—and in the uncertain supply of various natural resources.

See also: Economics of Materials, Energy and Growth; Input–Output Analysis: Applications; Materials in the Industrial System

Bibliography

Carter A P 1970 *Structural Change in the American Economy.* Harvard University Press, Cambridge, MA

Davis C H, Lofting E M 1979 *Development of Interindustry Transactions Data on the Structure of United States Mining Industries for 1967 and a Comparison of Techniques for Updating Related Input–Output Coefficients.* US Department of the Interior, National Technical Information Service, Springfield, VA

Leontief W 1966 *Input–Output Economics.* Oxford University Press, New York (Expanded version, Oxford University Press 1985)

Leontief W 1970 The dynamic inverse. In: Carter A P, Brody A (eds.) 1970 *Contributions to Input–Output Analysis.* North-Holland, Amsterdam

Leontief W 1977 *The Structure of the American Economy, 1919–39.* International Arts & Sciences Press, White Plains, NY

Leontief W, Carter A, Petri P 1977a *The Future of the World Economy.* Oxford University Press, Oxford

Leontief W, Chenery H B, Clark P G, Duesenberry J S, Ferguson A R, Carter A P, Grosse R N, Holzman M, Isard W, Kistin H 1977b *Studies in the Structure of the American Economy.* International Arts & Sciences Press, White Plains, NY

Leontief W, Koo J, Nasar S, Sohn I 1983 *The Future of Nonfuel Minerals in the US and World Economy.* Lexington Books, Lexington, MA

Economic Systems Research is the journal of the International Input–Output Association and is published by the Carfax Publication Company, UK

W. W. Leontief
[New York University, New York, USA]

Input–Output Analysis: Applications

Input–output models, also known as interindustry models, can be used in many areas of economic analysis including the demand and supply of materials. Major applications are structural analysis, impact analysis, projections of future events and marketing analysis. These applications are useful management and policy analytic tools for decision makers to evaluate the probable consequences of alternative options and actions.

1. Structural Analysis

A major question that is often asked is: How important are the materials industries to the economy? Input–output tables permit such a structural analysis, which may be defined as the use of an economic model to evaluate the interdependency of the economy. This type of analysis generally involves the use of the concepts of forward linkage and backward linkage. In the official US input–output tables, compiled every five years by the US Department of Commerce, industries are categorized according to Standard Industrial Classification (SIC) codes. If SIC industry 3331 (primary copper fabrication) is taken as an example, the commodity output distribution of the most recent table shows that about 60% of total output is sold to other intermediate industries, thus representing the direct forward linkage effect. In contrast, SIC industry 121 (bituminous coal mining) indicates a direct forward linkage of 74%. Because the bituminous coal mining industry has a higher forward linkage than that of copper fabrication, a strike in the bituminous coal industry will have greater first round (short-run) disruption effects on the national economy than a strike in the primary copper industry. However, the copper metal industry shows a direct backward linkage of 92%, the sum of its purchases of intermediate inputs, whereas bituminous coal mining shows only 41%. This means that the development of a new bituminous coal mine is likely to have less influence on the economy in the short run than developing a new copper primary metal plant with the same magnitude of new investment. A complete analysis would involve total forward and backward linkages summarizing the various rounds of spending involved as a result of overall interdependence.

Using the US input–output tables available for every census year since 1958, the examination of the input–output relationships over various intervals would show the trend in the changing economy as a whole, as well as in individual industrial sectors. Shifting patterns of input requirements and output distributions reflect the impact of changing technology, innovation or consumers' taste. It could also indicate shifts due to substitution of raw materials and other inputs, including capital equipment, resulting from price fluctuations. Furthermore, structural changes may be accounted for by changes in labor productivity. Input–output coefficients may also be altered by changes in relative wage levels and relative price changes of inputs, although these phenomena are actually valuation adjustments rather than real structural adjustments.

A formal approach to analyzing structural change is now referred to as "input–output structural decomposition analysis" (SDA), or the examination of the transformation of economic structure by means of sequential comparative static changes in key parameters of an input–output table. Some sophisticated versions of this methodology have been formulated to examine international trade patterns (Skolka 1977), material use (Devine 1988) and energy use (Rose and Chen 1991). The latter formulation is able to yield insights comparable to those obtained from KLEM models estimated using flexible functional forms (e.g., translog production functions), but with much less restrictive data requirements (input–output tables for the initial and final years, as opposed to a time series of similar input data). Applying their SDA methodologies, Rose and Chen (1991) found that the major sources of change in energy use in the intermediate sectors of the US economy between 1972 and 1982 were: changes in the level of final demand (or economic growth); changes in the mix of final demand (shifts away from more energy-intensive goods and services); KLEM substitution (shifts in the relative usages of capital, labor, energy and materials); direct technological change in energy (conservation); and technological change in materials (lower material composition of product). The latter effect is also one of the key factors in explaining major changes in the material intensity of use (see *Intensity of Use of Materials*). Input–output analysis holds great promise for understanding this phenomenon (Auty 1985; Rose and Chen 1990).

Using input–output tables for two or more countries or regions within a country, a comparative analysis of the structure of their economies can be undertaken. Such an analysis will not only show a simple comparison of the complexity of the economies involved but also can be used by the policymakers in the less developed countries or regions to assist in determining an economic development program based on an emulation strategy. Currently, some 80 member nations of the United Nations have published input–output tables.

The statistical description of the interdependency of industry is in itself a valuable contribution to understanding the structure of the economy. The systematic reconciliation of a vast amount of data on output and cost of production not only improves national statistics, such as revealing data inconsistencies and gaps, but also provides a sounder basis for estimating the National Income and Product Accounts. The structural analysis of the National Income and Product Accounts would be greatly facilitated by the application of interindustry analysis, such as the deflation of gross national product by the industrial origin and also the analysis of production changes by sector, based on technical changes.

2. *Impact Analysis*

The interindustry structure of the economy permits the computation of the quantitative impact of any specific changes in output of one industry on all other industries in the economy. Such impacts may be estimated by using the direct backward linkage and interindustry (direct and indirect) backward linkage tools. The impact of direct requirements in relation to the output of the mining industries, or any other materials producing industries, may be ascertained from a direct requirement coefficients table, whereas the impact of interindustry requirements of both the direct and indirect requirements in relation to the output of the mining industries can be quickly determined by looking at a total requirement coefficients (also called the Leontief inverse) table. It is easy to calculate the direct impact, but it is relatively difficult to ascertain the interindustry impact without input–output tables.

Thus, input–output tables can be readily used to assay the probable impact economy wide of changes in technology and innovation, consumers' choice and taste, price substitution and changes, foreign demand and supply, government policies and/or business and industry decisions on investment expenditures. An example is the switching over from using high explosives to using ammonium nitrate for mine blasting operations. Similarly, the interindustry impact of an increase in foreign demand for phosphate rock (Wang and Opyrchal 1981) can be calculated. Another example would be if the government were to impose a tariff or quota on copper imports, which would result in an increase in output of the domestic copper industry. More subtly, the total impact on the sand and gravel (nonmetallic mineral) industry due to the government fiscal policy to maintain economic growth through a tax cut can be calculated.

So far, only the measurement of the total impact upon output resulting from a given change in final demand has been discussed, but business and government decision makers are also interested in measuring the total impact on income and employment. One of the modern economic analytical concepts developed by Keynesian economics is that of the multiplier. Since Keynes dealt in broad aggregates, his income and employment multipliers were also highly aggregated. Aggregate multipliers are useful analytical tools, but they do not indicate the details of how multiplier effects are worked out and at times both government and business decision makers are more interested in the sector details than in the overall impact.

Income effects trace the sectoral household income changes (wages and salaries generated in the row industry) resulting from a change in final demand; this is necessary because industry coefficients react differently to market influences than household coefficients react to changes in demand. Interindustry, or "type I," income multipliers account for the direct and indirect effects on income with respect to the direct effect of a change in final demand in a particular industry. That is, as final demand changes in industry *i*, the income originating in that sector will be affected directly by this change. As all other industries adjust to a change in industry *i*, the income originating in these industries will be affected indirectly by the original change in industry *i*.

The total, or "type II," income multipliers account for the direct and indirect effects on income plus the induced income effects, which arise as consumers adjust their spending to changes in personal income with respect to the direct income effect. Consequently, the total income multiplier will always be larger than interindustry income multipliers.

To generate income multipliers, the "open" input–output model must be "closed" with respect to households. That is, the household sectors must be part of the endogenous mechanism of the model. In this respect, households are analogous to production processes requiring inputs from various sectors, in fixed proportions, in order to produce an output called labor. The household coefficients are established when personal consumption expenditures in the final demand vector are balanced with wages and salaries in the value-added row. These household coefficients serve as the basis from which income multipliers are calculated.

Also, the distribution of the income from public policies affecting materials has become an important consideration. Input–output models can be used for these purposes as well by disaggregating consumption columns and household income rows according to income bracket. Various distributional multipliers can be calculated from this formulation as shortcuts to analysis (Miyazawa 1976). Input–output distributional analysis can be applied to the examination of normative aspects of public policy, such as in the evaluation of their equity implications, or for positive economic analysis, such as predicting the extent of public support for specific policies, including those affecting natural resources and materials produced from them (Rose et al. 1988). Distributional implications according to other major groupings (such as institutions, racial or ethnic groups, rural vs urban) can be analyzed by further extending an input–output table into a social accounting matrix (Pyatt and Round 1985).

Type I and type II multipliers have proved to be widely applicable for general magnitude estimating purposes. Specific impact analysis can be conducted for energy, materials and water requirements, changing supplies and prices of materials, adding or shutting down industries, pollutant output and so on. For energy requirement impact analysis, the following procedure may be used: (a) divide the particular industry energy requirement by the industry gross output to derive the sectoral average energy requirement coefficient; (b) develop the energy requirement coefficient for the whole economy (a principal diagonal matrix of energy coefficients); (c) premultiply the diagonal matrix with the Leontief inverse matrix $(I - A)^{-1}$ to derive the energy requirements interaction matrix; and (d) premultiply the energy interaction matrix with the final demand vector. The result will be the interindustry (direct and indirect) energy requirement by sector for the whole economy. The ratio of interindustry energy requirements divided by the direct energy requirement yields the type I energy multiplier.

To derive the total energy requirement multiplier which embodies the direct, indirect and income-induced requirements, the energy requirements diagonal matrix is premultiplied with the Leontief inverse of the "closed" input–output model. The total energy requirement interaction matrix can be premultiplied with the final demand vector to derive the total energy requirement by industry sectors. The total energy requirement is then divided by the original direct energy requirement to derive the type II energy requirement multipliers. This and other approaches are discussed by Miller and Blair (1985). Similarly, using a diagonal matrix of a specific material or natural resource requirement coefficients, the interindustry specific material or natural resource requirement and total specific material requirement—the type I and type II specific material multipliers, respectively—can be derived.

Also, by dividing the industry-by-industry output of pollutants of solid, liquid and gaseous waste with column totals of gross pollutant outputs a principal diagonal matrix of average pollutant coefficients would be obtained. Following a similar procedure, the interindustry and total pollutant output, as well as type I and type II pollutant output multipliers, can be calculated.

Similar procedures can be used to develop interindustry (direct plus indirect) labor requirements and total labor requirements that embody direct plus indirect and income-induced labor requirements, or type I and type II labor or employment multipliers, respectively.

By the use of supplementary data for capital investment requirements and a summary of available capacity pertaining to both construction and equipment costs, including the material inputs consumed for the new plant and equipment construction and lead time, capital coefficients for every sector can be calculated and can be used to indicate the possible needs for and impact of additional plant and equipment. With the capital coefficients and the trend in production technology projections, the static input–output models can be transformed into dynamic models (see *Input–Output Analysis*). Linear programming techniques can be adapted to the model to find the implications of various constraints and ranges.

Industries are not located uniformly throughout the USA. For example, production of domestic copper ores is concentrated in the Rocky Mountain states. Production of oil and gas is concentrated in the Gulf Coast states, the Rocky Mountain region and in California. With additional information on the geographic distribution of industry, regional input–output commodity flows (including trade and income accounts) can be used for developing models to obtain regional impacts. Multiregional and interregional input–output models can analyze the probable impact of changes in final demands, including investment in one region on the industrial outputs of another region(s). Thus, interindustry analysis can be used to shed light on the regional output and on income and employment implications of many national and regional programs. The various impacts of adding new industries to a region or the nation can readily be calculated by expanding the existing input–output model with relevant input data for new industries.

A new industry, such as a synthetic fuel industry, can be added to the input–output model with new industry input columns and new industry output distribution row coefficients. An old industry can be eliminated in a similar fashion and input–output applications can be conducted on the revised models.

Another area of impact analysis is the examination of repercussions of materials shortages due to trade embargoes, strikes, regulatory prohibitions, mineral depletion or natural disasters. A variant of the standard input–output model, referred to as the *supply-driven* or *allocation* model (Ghosh 1958), has been utilized effectively for this purpose (e.g., Davis and Cherniak 1987). Allocation coefficients are defined as the ratio of each intersectoral flow in a given row of an input–output transactions table to its row sum (total sales). Thus, the model assumes stability in trading patterns, but is still able to maintain relatively stable production relations in most empirical applications (Chen and Rose 1991). With the use of the allocation model, analysts are able to estimate the direct, indirect and induced effects of a supply shortage. Typically, the results yield less severe cutbacks than with the use of the more standard input–output version. For example, using the latter, a 50% decrease in copper would lead to a 50% decrease in the output of all sectors using copper in the short run, as well as further multiplier reductions of output throughout the economy. However, the supply-driven version incorporates small implicit substitutions that mute the impact significantly. Moreover, the model can be modified to include more explicit substitution and planning strategies as part of a more long-run planning approach (Oosterhaven 1989).

Also, supply-driven output multipliers can be calculated by simply summing the elements of each row of the inverse of the allocation coefficients matrix. These multipliers can be used as shortcuts for the estimation of impacts of shortages. In addition, they can be used to determine the economywide impacts of positive stimuli to the economy, as in the case of new mineral discoveries or material patents. Finally, it should be noted that the allocation coefficients are essentially forward linkages, and that the allocation model is considered the best basis for the calculation of total forward linkage effects (Augustinovics 1970).

Note that there is an important asymmetry between backward and forward linkages. The stimulus to production from the former is highly certain because of the need for material inputs to deliver any bill of final goods (demand brings forth the necessary supply in an input–output model, assuming productive capacity is in place). However, the mere existence of new basic input will not necessarily lead to its upstream utilization (supply does not create its own demand). That is, the discovery of copper in a region will not necessarily lead to its extraction, smelting and utilization. Economic conditions must be favorable for the full extent of forward linkages to be realized.

The US national input–output tables are usually seven or more years late in completion. Because of the long delay and prohibitive cost of collecting survey data, the US Bureau of Mines Quantitative Economics Project funded research in the 1970s and developed a series of interactive computer programs and methodologies to update national input–output tables by using the RAS nonsurvey method to update the 1972 national input–output tables to 1977 (Lofting and Schink 1981). They were able to project regional technologies, value added and final demands to construct regional input–output tables with detailed mineral sectors for 1977 (Lofting and Davis 1979, 1980). Applying US Bureau of Mines developed interactive computer program methodologies and available secondary data, Lofting and his associates simulated 1982 and 1986 regional final demands, trade flows and other relevant input–output data at the 460-sector level for all 50 states and 3150 counties in the USA and also constructed many regional input–output analyses.

Since 1979, the US Forest Service, under the direction of G. S. Alward, has developed the popular IMPLAN, impact analysis for planning system (Alward and Palmer 1983). With Lofting's 1982 data and a series of interactive computer programs, the IMPLAN system can produce detailed regional nonsurvey interindustry models and multipliers for any county or county group in the US at a modest cost and in a short time. IMPLAN-based PC tools have been developed to assess economic impacts of earthquakes, land-use planning, regional trade flows, and so on. The current IMPLAN database year is 1982, but the 1985 year database will be available in late 1990. IMPLAN-generated input–output tables and multipliers have been compared with similar tables and multipliers derived from survey-based state models and were found to be reasonably accurate.

IMPLAN has been used literally thousands of times since its inception. Numerous federal, state and local government agencies, universities and research institutes and private concerns are current IMPLAN users, conducting economic impact assessments; preparedness planning for natural disasters; natural resource and material policy option analysis; economic strategic planning; environmental impact studies; land-use option assessments; business incubators and venture financing; social service planning; industrial marketing surveys; government expenditure and capital investment analyses; and many other extended applications. The IMPLAN Development and Applications Group was established by the US Forest Service at the Department of Agricultural and Applied Economics at the University of Minnesota to provide IMPLAN data, software and help services for non-Forest Service users.

3. *Projections of Future Events*

It has been shown how the level and structure of production and its various requirements corresponding to a given final demand schedule (the autonomous sector) may be derived from an input–output matrix under the assumption of a constant input–output ratio. The basic tool used for this type of application is the variation of both the level and composition of the final demand schedule in which the outputs of more than one industry are changed simultaneously.

The principal distinction between impact analysis and projections is that impact analysis generally involves only the change of one component of the final demand or the direct requirements placed on one or two industries.

A given or assumed change in the level and composition of the autonomous final demand schedule will project a new transactions table that would be consistent with the input–output matrix. This is also known as consistent forecasting. There is no assurance that consistent forecasting will turn out to be correct unless the structural input–output relations of the economy do not alter significantly over the projection period or unless proper allowance can be made for expected changes in the structural relations of the economy. The potentialities of input–output analysis for forecasting and programming or planning purposes are suggested by this production character of the input–output techniques (e.g., Almon et al. 1974).

These projections are used only to compute the various production requirements necessary to sustain a certain level of final demand, but cannot be used in analyzing the size and character of final demand. Autonomous final demand must either be given as a goal or "plan" or else estimated by some techniques quite different from input–output analysis. The composition of final demand is determined independently, that is, outside of the input–output model.

Despite some of its limitations, input–output analysis is probably one of the best available techniques for tracing the interindustry and total implications in terms of production levels and input requirements of an economic plan or investment program for a country or region. These derived data in economic programming may be of utmost importance for judging the reasonableness, for testing the feasibility and for trying to locate potential bottlenecks. Analysis may indicate, for example, that a program is too ambitious and not feasible because the limited supply of some scarce resources such as skilled labor, foreign exchange, specific materials or specialized capital equipment is insufficient to support the given or projected bill of final demand. The fast growing or more desirable industries can also be identified by this analysis.

A given investment program of a country or region, such as the construction of a multipurpose hydroelectric power dam or of a new integrated iron and steel manufacturing complex, can be analyzed and evaluated by the use of input–output techniques based on the probable total requirements of domestic production, imports and other scarce resources. In these cases, the direct requirements for investment in new or additional production capacity must be determined beforehand and must be added in with the additional final demand that will be generated from the higher level of investment activity to arrive at the new or desired bill of final demand.

Since the 1960s, the Division of Economic Growth Bureau of Labor Statistics (BLS) of the US Department of Labor has prepared projections for 1970, 1980, 1985, 1990 and 2000 in an input–output framework, under the US Interagency Growth Project's US Economic Growth Model. The purpose of the model is to develop a more comprehensive and integrated framework for analyzing the implications of long-term growth for a number of problem areas, particularly problems of manpower utilization. Similarly, materials problems can be analyzed.

The US Economic Growth Model is characterized by a number of distinct but interrelated steps. The first of these is the development of a potential growth rate for the economy, which is derived by projecting the labor force to the target year and projecting the change in average hours worked. The next step is to calculate and distribute each of the major categories of expenditure into categories of demand, including final demand matching the input–output producing industry classification used.

Demand in the national income accounting system covers only final demand, that is, only that of the ultimate consumers. Thus, in order to include those industries whose products are not sold to ultimate consumers but are used up in the production process, input–output tables are utilized to translate final demand, such as personal consumption expenditure, government expenditure, investment, construction and exports into the output required from all industries and services. These include primary ferrous and nonferrous metals, chemicals, textiles, iron ores, copper ores, coal, crude petroleum, electricity, basic agricultural products, lumber and other materials as well as other sectors which do not usually sell a large fraction of their products to the ultimate consumer.

The basic input–output tables used in the US Economic Growth Model are those produced by the US Department of Commerce and these represent the technology and product mix for the year of the tables. A new set of projected input–output coefficients, or sales–purchase ratios, that would take into consideration changing technology and the anticipated changes in product mix due to different growth rates of product groups within the larger industry aggregates has been produced. These projected changes are related to the impact which they would have on an industry's purchases. Some changes are materials saving, others actually require larger inputs of materials; these factors are also introduced into the projections.

One criticism of input–output analysis is the fixity of its technical coefficients. There are several reasons why coefficients may change over time, but the two major ones are technological change (lower quantities or higher qualities of inputs used per unit of output) and technical substitution (relative changes in quantities of inputs used, as in a response to relative price changes). Fixed coefficients are likely to be most problematic when input–output analysis is used in forecasting. Still, a host of methodologies exist for projecting changes in technical coefficients.

The most rudimentary set of methods are limited *trend-line* approaches. The *ex ante* method (Fisher and Chilton 1975) utilizes expert opinion in projecting coefficients, and in short-term applications represents the self-fulfilling prophecies of engineers and plant managers. This method has the advantage of utilizing information from those at the scene of production, but must often reconcile divergent views among experts about the more distant future. The extrapolation method uses descriptive statistics or stochastic analysis of time-series data. In a disaggregated input–output context, its widespread application has been limited by lack of data. The heuristic approach also stops short of offering an explanation of the causation of coefficient changes, and uses simple correlation, analogy or metaphor. Examples are the application of national trends at the regional level and the logistic adoption function (Almon et al. 1974). Another basic method might be termed the mechanical method and includes wholesale updating techniques, such as the RAS method (Stone et al. 1963).

There are two shortcut *indexing* methods for dealing with technological change. The first is referred to as cost-engineering, which involves the application of fundamental engineering relationships and principles (Isard et al. 1959). Many of these principles, such as the fractional power law, can be applied to changes in the scale of operation on a year-to-year basis. Another method is the policy-cost index (Hannan and Rose 1988), which translates a policy goal into a measurable scale and calculates the cost of different levels of attainment, typically from corresponding microlevel data.

A third set of methods is more *behavioral* in nature. The best-practice method (Miernyk 1968, Carter 1970) is based on the fact that at any point in time there are different vintages of production processes in place. The shift from old processes to more productive new ones is essentially a form of technological change. The penetration–diffusion method (Ayres and Shapanka 1976) combines elements of the theory of the adoption of innovation and the logistic function of the heuristic method to more comprehensively explain the pace in terms of price competition, obsolescence and diffusion of innovation.

Another set of approaches involves more explicit *optimization.* This includes process analysis, which involves the consideration of a set of alternative techniques, but without ranking them as in the best-practice approach. The pseudo data approach (Klein 1974, Griffin 1977) uses the least-cost solutions of process analyses of different scale levels to provide the simulated data for the statistical estimation of a cost function. As in the case of process analysis, this approach has the advantage of utilizing engineering data to make up the deficiency of lack of time-series data in cases of new technology. The production-function approach (Hudson and Jorgenson 1974), on the other hand, typically uses time-series data, flexible, functional forms and the duality principle. It is even better suited to cases of technical substitution in response to changes in relative prices.

Given the demand projections and projected input–output tables, the necessary ingredients for estimating economic growth are present. The final step in the US Economic Growth Model is to develop employment estimates by industry. The BLS accomplishes this by a set of industry projections. Currently, the BLS is preparing similar projections for 2005 and later.

Using the BLS projected input–output tables, the impact analysis mentioned previously can be conducted. Unfortunately, the BLS input–output industry classifications are usually not detailed enough to cover many of the specific materials that may be of interest in the analysis. Nevertheless, BLS projections can be utilized as the point of departure for further analysis.

Since 1977, Professor Wassily Leontief and his associates at the Institute for Economic Analysis at New York University have been conducting consistent forecasting on production and consumption of basic minerals and primary and secondary metal processing materials industries within the framework of multiregional input–output models of the US and world economies. The study was designed to address the growing national concern regarding the adequacy of nonfuel minerals in the US economy to the year 2030 (Leontief et al. 1983).

The study was intended to answer a host of questions which would affect the demand for mineral resources. For example, what will be the effect on the output levels of the US mineral sectors, as well as on all other industrial sectors of the economy, if the reliance on foreign sources of supply is increased? How will the change in the rate of US population growth, and that of the rest of the world's population, affect demand for primary materials resources? How will increased substitution of aluminum and/or precast concrete for steel in construction, or plastic pipes for copper pipes, or plastic tubes for lead tubes affect the demands of the industries supplying the materials?

The team at New York University projected the revised United Nations world input–output model as well as its US input–output model to the year 2030 at ten-year intervals. With this New York University long-term, consistent forecasting model of the US and world economies, many types of impact analyses can be conducted on the demand and supply of minerals and other materials with different scenarios and assumptions for analyzing and evaluating policy options.

The US Forest Service developed IPASS "an **I**nteractive **P**olicy **A**nalysis **S**imulation **S**ystem that can be used to analyze the long-term economic and demographic effect of alternative forest resource management policies. The IPASS model is a dynamic analytical tool that forecasts growth and development of an economy. IPASS has potential use in applied research as well as policy analysis" (Olson et al. 1984).

IPASS is capable of forecasting a number of basic socioeconomic indicators, such as population, employment, income and investment on an annual basis. But IPASS requires a costly comprehensive data set on the economic and social situation for the initial year. Some savings can be realized by utilizing an input–output production module derived from the IMPLAN system.

4. Marketing Uses

The situations discussed so far involve evaluating government policies and operations, but input–output models also have many uses for business marketing analyses. Structural analysis has a number of implications for marketing research. Any single row in a base-year commodity flow or transactions table shows the detailed output distribution of the industry by its various immediate consumers. Such industrywide marketing information is valuable for any firm in the industry to plan marketing strategy as well as location of future plants and warehouses. The forward linkage coefficients available in output distribution tables can be used for evaluating the sales potential of an individual firm. For example, a businessman can compare the present marketing distribution position of his firm with the industrywide output distribution pattern and then recognize some of the probable areas of additional markets where the firm has yet to penetrate or exploit fully.

Furthermore, firms may want to rank and to

determine the strength of indirect demand potentials, which may be very useful in planning sales efforts for developing new markets as well as intensifying efforts in existing markets. The structure of the table of commodity flows records the end results of demand patterns that give the final result.

Moreover, firms in industry frequently know their customers but seldom are aware of the industries which purchase the products of their customers. It would be beneficial to know how the output of their major customers would directly affect the future sales of their own output. The input–output approach gives businessmen a better insight into such probable changes in the demand for the output of the industry that may result from new activities of some seemingly unrelated industries.

The indirect linkage of iron ore and steel industries to the demand requirements for automobiles is apparent to nearly everyone; however, the indirect linkage of the steel industry to other end products may be quite complex and difficult to trace from final demand components without the aid of the input–output tables. Such linkages would be useful for market research.

The production-requirement approach is used most often for backward analysis, which usually begins with the given final demand and then traces through the input–output matrix back through the processing industries and determines the production requirements of the supplying materials industry. Similarly, forward linkage analysis can ascertain the dependency of the output of one industry on any specific component of final demand.

Most of the materials industries have a common feature that distinguishes them from other industries in that their output distributions are principally shipped to the processing industries with very little shipped directly to final demands. This feature indicates that their respective output depends on two rather different factors: (a) the absolute production volume of the consuming industries and (b) the relative importance of the output of the supplying industry in meeting the direct requirement of the consuming industries. These two factors mean that the same changes in production volume of the different consuming industries may have very different effects on the demand for the output of the supplying industry.

An individual firm in a particular industry can use the input–output model not only to compare its own output and distribution pattern with that of the industry, but also to identify and measure the impact of its own operation within the economic system. IMPLAN models can be constructed and used to evaluate potential market opportunities for the product(s) of a firm within a region or regions. The use of the model can complement other market research techniques used as a supplementary tool to provide longer-range guidelines. Government agencies and some economic consulting firms make regular projections of gross national product and its components, personal consumption (household) expenditures, private investments, net exports and government purchases. Many firms subscribe to these sources as an alternative to making their own projections. However, these projections are for end-product demands and not intermediate demand. Now, with the interindustry model and the gross national product projections, it is possible for a firm to project its market potential and goal for the next five to ten years into the future. Given the range of projections of end products final demand, a range of intermediate demands may be established with a reduction in the risk in the decision making process.

Thus, there are many possible marketing uses of interindustry analysis for firms in the private sector, all of which could enable businessmen to formulate better informed production and investment programs and to make wiser managerial decisions.

5. Conclusion

The input–output model and its applications have been tested and are workable for guiding current decision makers on materials policy and analysis. The model and its applications represent a pragmatic and operational approach in economic model building. The applications and uses are mostly static but useful. Advances in research design, data collection and applications are pushing the state of the art further into dynamic types of input–output analysis.

The principal thrust of input–output analysis, particularly in the business world, is centered on long-range contingency input–output forecasting, which complements the more traditional methods. This is probably because of several inherent advantages of interindustry forecasting over other types: (a) it provides greater detail by specific industry and definite markets, (b) it closes the gap between macro- and microforecasting, (c) it can take into full account changes in technology and in the production mix methods as well as alter the output production mix in the industry, (d) it explicitly takes into full account all the indirect and derived demands within the given time period and (e) it provides an internally consistent framework of the economic structure that is fully integrated with the national income and products accounts for forecasting industry-by-industry sales.

As a word of caution, the input–output model in itself is not a self-sufficient forecasting model since it depends on the future state of technology and on exogenous projections of the mix of final demand components. The consistent projections of each producing sector output can be accepted as forecasts only if the future state of technology and the detailed mix of final demand components are given exogenously. Thus, the input–output model should not be used as a forecasting tool without fulfilling these necessary conditions.

Used in conjunction with modern computer technology, including personal computers, simulation for all major applications under various scenarios and assumptions, such as technology, economic growth rate, the level and components of government expenditures, tax and foreign trade policy, private consumption function and private investment can be performed readily with any interindustry model in conducting sound and consistent economic analysis. Input–output analysis can be a very powerful and versatile analytical tool, but, as with any model, it must be used carefully with a full understanding of its limitations and problems.

See also: Input–Output Analysis; Materials in the Industrial System

Bibliography

Almon C Jr et al. 1974 *1985 Interindustry Forecasts of the American Economy.* Heath, Lexington, MA

Alward G S 1989 *Micro-IMPLAN: Methods for Constructing Regional Economic Accounts.* US Forest Service

Alward G S, Palmer C J 1983 IMPLAN: an input–output analysis system for forest service planning. In: Seppala R, Row C, Morgan A (eds.) 1983 *Proc. 1st North American Conf. Forest Sector Modeling*, Williamsburg, VA

Augustinovics M 1970 Methods of international and intertemporal comparison of structure. In: Carter A P, Brody A (eds.) 1970 *Contributions to Input–Output Analysis.* North-Holland, Amsterdam

Auty R 1985 Materials intensity of GDP. *Resour. Policy* 11: 275–83

Ayres R U, Shapanka A 1976 Explicit technological substitution forecasts in long-range input–output models. *Technol. Forecast. Soc. Change* 9: 113–38

Carter A P 1970 *Structural Change in the American Economy.* Harvard University Press, Cambridge, MA

Chen C Y 1986 The optimal adjustment of mineral supply disruptions. *J Policy Model.* 8: 199–221

Chen C Y, Rose A 1991 Absolute and relative joint stability of input–output production and allocation coefficients. In: Peterson A W (ed.) 1991 *Advances in Input–Output Analysis.* Oxford University Press, New York

Chenery H B, Watanabe T 1958 International comparisons of the structure of production. *Econometrica* 26: 487–521

Davis H C, Cherniak H 1987 Interindustry approaches to the analysis of a supply disruption of a critical resource. *Resour. Policy* 13: 47–54

Devine P 1988 The effects of economic growth, technology change, consumption pattern change and foreign trade on domestic demand for primary metals, 1963–82. US Bureau of Mines, Washington, DC

Fisher W H, Chilton C H 1975 Development *ex ante* input–output flow and capital coefficients. In: Gossling W F (ed.) 1975 *Capital Coefficients and Dynamic Input–Output Models.* Input–Output, London

Ghosh A 1958 Input–output approach to an allocative system. *Economica* 25: 58–64

Griffin J M 1977 Long-run production modeling with pseudo-data: electric power generation. *Bell J. Econ.* 8: 112–27

Hannan M, Rose A 1988 The policy-cost index approach to modeling change in input–output coefficients. *Technol. Forecast. Soc. Change* 33: 13–22

Hudson E A, Jorgensen D 1974 US energy policy and economic growth, 1975–2000. *Bell J. Econ. Manage. Sci.* 5: 461–514

Isard W, Schooler E W, Vietorisz T 1959 *Industrial Complex Analysis.* MIT Press, Cambridge, MA

Klein L 1974 *A Textbook of Econometrics*, 2nd edn. Prentice-Hall, Englewood Cliffs, NJ

Leontief W W, Koo J, Nasar S, Sohn I 1982 *Techniques for Consistent Forecasting of Future Demand for Major Minerals Using an Input–Output Framework.* New York University, New York

Leontief W W, Koo J, Nasar S, Sohn I 1983 *The Future of Nonfuel Minerals in the US and World Economy.* Lexington, Lexington, MA

Lofting E M, Davis H C 1979 *The Input–Output Structure of US Mining Industries for 1958 and 1963: Transactions, Employment and Multipliers.* University of California, Riverside, CA

Lofting E M, Davis H C 1980 *State Input–Output Tables for 1977 Emphasizing the Mineral Industries: Arizona and Florida.* University of California, Riverside, CA

Lofting E M, Schink W O 1981 *The Input–Output Structure of the US Mineral Industries with RAS Update to 1977.* University of California, Riverside, CA

Miernyk W H 1968 Long-range forecasting with regional input–output model. *West. Econ. J.* 6(3): 165–76

Miller R, Blair P 1985 *Input–Output Analysis: Foundations and Extensions.* Prentice-Hall, Englewood Cliffs, NJ

Miyazawa K 1976 *Input–Output Analysis and the Structure of Income Distribution.* Springer, Berlin

Olson D, Schallau C, Maki W 1984 IPASS: An Interactive Policy Analysis Simulation System. General Technical Report PNW-170. US Forest Service

Oosterhaven J 1989 On the plausibility of the supply-driven input–output model. *J. Reg. Sci.* 28: 302–17

Opyrchal A M, Schantz R, Nelson D 1985 *The Importance of Mining and Minerals Processing in the Idaho Economy.* US Bureau of Mines, Washington, DC

Pyatt G, Round J (eds.) 1985 *Social Accounting Matrices: A Basis for Planning.* World Bank, Washington, DC

Rose A, Chen C Y 1990 Structural decomposition analysis of changes in material demand. Paper presented at the Annual Meeting of the AIME. Salt Lake City, UT

Rose A, Chen C Y 1991 Sources of change in energy use in the US 1972–1982. *Res. Energy* 13(1):1–21

Rose A, Stevens B, Davis G 1988 *Natural Resource Policy and Income Distribution.* Johns Hopkins University Press, Baltimore, MD

Skolka J 1977 Input–output anatomy of import elasticities. *Empirical Econ.* 2: 123–36

Stone R, Bates J, Bacharach M 1963 *Input–Output Relationships, 1954–1966.* Cambridge University Press, Cambridge

Wang K-L, Opyrchal A M 1981 *Economic Significance of the Florida Phosphate Industry: An Input–Output (I–O) Analysis.* Information Circular No. 8850. US Bureau of Mines, Washington, DC

K.-L. Wang
[US–China Council for International Exchange, Wheaton, Maryland, USA]

A. Rose
[Pennsylvania State University, University Park, Pennsylvania, USA]

Intensity of Use of Materials

The intensity of use of materials (i.e., the ratio of the amount of a material used by a nation in a year to the final goods and services produced by the nation over that year) has been applied increasingly in analytical studies relevant to world mineral resources. The quantitative core of this article rests upon data assembled through the mid-1970s (Malenbaum 1986). The article also seeks to take advantage of more recent views on the outlook for demand and supply of world metals as well as some modernization of the intensity of use concept and theory and its relation to technical innovation, the structure of economies, and market forces.

1. Concept and Measurement

The intensity of use (IU) of a material i in area j over time t is defined as

$$_{ij}\mathrm{IU}_t = {}_{ij}D_t/{}_j\mathrm{GDP}_t$$

where $_iD$ is the tonnage of material i used and GDP is the total gross domestic product, the total real value of final goods and services produced in a specific area and time period. Hence

$$_iD = \mathrm{GDP}\cdot{}_i\mathrm{IU} \equiv \mathrm{GDP}\cdot{}_iD/\mathrm{GDP}$$

The definition is tautological. The decomposition of $_iD$ for the important purposes of comparison over time and among regions must still cope with independent changes in GDP and D, given i, j and t.

With respect to GDP, modern economic growth theory places primary emphasis on the role of human quality inputs. Education, skills, attitudes and aspirations of the people constitute the major forces for progress. More quantity inputs, increases in such traditional factors as labor and capital of constant productivity are not primary contributors to dynamic economic and social change.

Nor can the seeming quantity of $_iD$ in itself be a determining force in GDP change. The apparent input–output aspect of the intensity of use notwithstanding, technological explanations of the intensity ratio find limited support. Specifically, intensity of use patterns appear to change with per capita product. Intensity increases with GDP/population as a "poor" region begins to develop; at a higher level of GDP/population intensity of use changes direction and declines with economic development. The underlying patterns of change in the two components of intensity of use are reflected in Figs. 1 and 2.

The precise definition and the specific values of D and GDP must conform to accepted practice of the market places of the world and of international statistical agencies. The metal or other material used is "apparent domestic consumption" which is domestic production plus imports minus exports, and is exclusive of materials traded in manufactured goods (e.g., in automobiles or their components) and exclusive of inventory changes of the material in the country during the year. Thus, the amount of a metal i used in time t (i.e., $_{ij}D_t$) refers to the weight of ore adjusted for metal concentration or, alternatively, the weight of the metal at a given stage of refining, with or without scrap. These alternatives are available in such regular international publications as the United Nations *Statistical Yearbook*, the US Bureau of Mines *Minerals Yearbook*, the UK *Metals Bulletin Handbook*, the German *Metal Statistics*, among others. Problems with the numerator of the intensity of use ratio are well known in the minerals field; differences in definition are constantly being narrowed among regions. Data for output, trade and use now tend to be delineated by the metal content of individual raw materials and are becoming standardized over most of the world.

For the denominator, the best measure is GDP, a magnitude that national and international agencies report by nation, by year and at a constant price level

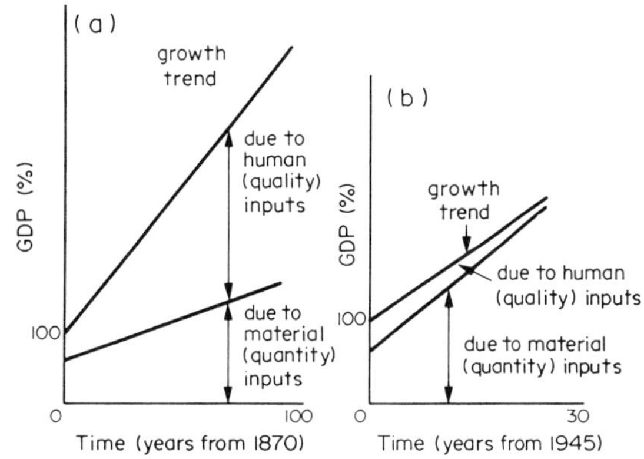

Figure 1
Human and material inputs in economic growth: (a) for rich countries and (b) for poor countries (after Malenbaum 1978)

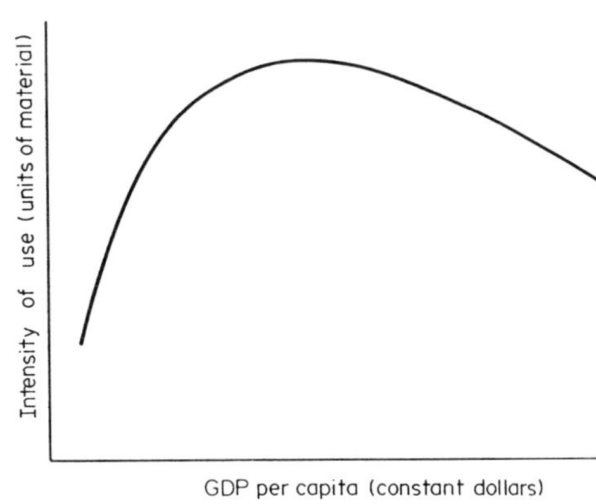

Figure 2
Typical intensity of use and GDP per capita (after Malenbaum 1978)

over time. The growing importance of comparability among nations assures persistent updating and improvement of national account data. Problems remain for standardizing comparisons of output among nations where factor endowments differ significantly; rich and poor nations have contrasting ratios of labor and of fixed capital supplies. This complicates comparisons between rich and poor nations and their supplies and demands for industrial materials. The present intensity of use measures rely primarily on UN data for real GDP, denominated for the world in US dollars and largely derived through converting national currencies by official exchange rates. Comprehensive alternative time series of international real GDP data are gradually becoming available.

Thus assembled, measures of intensity of use, ${}_{ij}D_t/{}_j\mathrm{GDP}_t$, have been computed for most nations for the years from about 1950. This article draws on analyses of the following materials: steel and iron ore; nickel, manganese ore, chromium ore, cobalt and tungsten; copper and aluminum; platinum metals; zinc and tin. Ten subdivisions of the world have been studied, each with a degree of homogeneity high enough to justify common assumptions about intensity of use and GDP changes (Malenbaum 1978):

(a) Western Europe—Organization for Economic Cooperation and Development (OECD) countries in Europe;

(b) Japan;

(c) other developed lands (ODL)—Australia, Canada, Israel, New Zealand and South Africa;

(d) the former USSR;

(e) Eastern Europe—former Soviet bloc countries, Albania and Yugoslavia;

(f) Africa (excluding South Africa);

(g) Asia (excluding Israel, Japan, China and related regions);

(h) Latin America;

(i) China (including Mongolia, North Korea and North Vietnam); and

(j) the USA (including Puerto Rico and offshore islands).

The poor nations, with over 75% of the world's population, consume some 10% of the annual world use of the twelve materials listed.

The measures give substance to the intensity concept: they differ significantly by region, time and material. Yet the historical record suggests clear patterns of behavior of intensity of use.

2. Intensity of Use Numbers

Table 1 presents intensity of use data for refined copper. The copper account is illustrative of the intensities of other industrial materials. Thus, in the USA copper intensity of use fell almost 30% from 1951–1955 to 1971–1975; from 1934–1938 the decline was almost 40%. Other rich nations which were important consumers reflect the reconstruction activities during the years after World War II, notably Japan and West Germany. Nonetheless, individual nations in this group also tend to have lower ratios in recent years than prevailed in the 1934–1938 period. For the rich countries, the decrease to the 1971–1975 level from the 1956–1960 level, at the peak of the Marshall Plan development years, was nearly 40%: the decrease was almost 10% from 1951–1955 when that program was only in its early phase. The underlying pattern is clear:

Table 1
Intensity of use: refined copper (metric tons per billion US dollars GDP, 1971 US prices)

Region	1934–1938	1951–1955	1956–1960	1961–1965	1966–1970	1971–1975
Western Europe	3307.0	2937.1	3315.1	3193.1	2675.9	2586.4
Japan	4116.0	2238.9	3067.3	3788.8	3998.1	3576.5
Other developed lands	1591.0	2314.0	2209.2	3003.7	2529.3	1943.8
USSR	1554.0	1839.4	1779.6	1906.0	1466.6	1776.8
Eastern Europe	NA	1697.2	1767.2	2008.3	2315.5	2378.7
Africa	NA	217.8	226.3	266.7	170.0	291.4
Asia	NA	452.2	644.1	812.4	416.6	550.7
Latin America	NA	1175.5	974.1	1121.3	1028.5	1388.5
China	NA	122.6	795.9	1309.5	1489.8	2019.4
USA	2724.0	2330.1	1966.1	2174.4	2006.8	1680.9
Totals						
World	NA	2116.2	3007.0	2294.5	2060.1	2000.6
Non-US world	NA	2009.9	3502.9	2349.3	2082.7	2126.9
Poor nations	NA	546.1	735.0	959.1	828.7	1134.8
Rich nations	NA	2384.1	3417.6	2521.4	2272.5	2160.0

Source: Malenbaum 1978, p. 108

modern industrial countries expand national product with diminishing inputs of refined copper per unit of real GDP.

Table 1 also shows the very different pattern emerging in the developing countries. In each region (especially China) the intensity of use for copper has expanded. For the poor nations together, the data for 1971–1975 relative to 1951–1955 show an expansion in intensity of almost 110%. However, because of the comparatively lower intensity levels that still pertain in the poor nations relative to the rich nations and because of the relatively small share of the total world gross product of the poor nations, the intensity changes in rich nations continue to dominate. Thus, refined copper per billion US dollars real GDP for the entire world in 1971–1975 was almost 6% lower than in 1951–1955. Hence a smaller tonnage of copper was used in producing a given real volume of GDP in the world of the 1970s than was required in the early 1950s. This prevailed despite major programs of traditionally copper-absorbing construction, electricity generation and transmission and communications expansion throughout the world during two or three decades of unparalleled economic growth.

The record for copper is not unique: Malenbaum (1978) provides the same information for all 12 of the materials specified above. Only for primary aluminum and the platinum group metals are the historical intensity patterns significantly different. Aluminum has obverse patterns over past decades as circumstances permitted extensive substitutions for other materials inputs, particularly for copper and steel. Rates of growth of intensity of use of aluminum are decreasing; the prospect is the gradual emergence in aluminum of the declining intensity of use pattern shown by copper. The situation for the platinum group is that apparent consumption data contain a relatively large final demand for platinum as jewelry, particularly in Japan but also in other Asian countries and in lesser degree in Africa. New demands for platinum began to emerge during the 1970s due to recognition by high technology industries of the unusual catalytic properties of platinum. These properties are of great importance in the production of plastics, rubber, pesticides and other products, and especially for their use for emission controls in engines. This growth in demand resulted in the emergence of alternative metals in the early 1980s, given the rapid price increases of platinum. Currently, platinum intensity of use gives evidence of a development similar to that in aluminum with the prospect of lower levels of intensity of use by the end of the 1990s.

Table 2 offers further insight into materials use in the years 1951–1975. For all materials and major regional components, comparison is made of change in materials use relative to GDP growth within this time interval, for the first decade, 1951–1960, and for the ten-year period 1966–1975. The elasticity concept measures relative changes of materials use that declines relative to growth in the output flows of total final goods and services, as against the absolute change measured by the intensity ratio. The table shows growth in materials that decline relative to the growth of GDP.

What forces account for this pervasive pattern of declining materials use relative to world growth in general and in rich lands particularly?

3. *Determinants of Intensity of Use Patterns*

The use of intermediate inputs in the process of generating the flow of final goods output is

Table 2
Income elasticities of demand for materials, 1951–1960 and 1966–1975

	World[a,b]		USA[a,b]		Non-US[a,b]		Poor nations[a,b]		Rich nations[a,b]	
Raw material	(1)	(2)	(1)	(2)	(1)	(2)	(1)	(2)	(1)	(2)
Crude steel	1.6	1.0	0	0.1	2.2	1.3	5.2	2.4	1.3	0.8
Iron ore	1.9	0.7	1.0	0.6	2.3	0.7	4.8	1.0	1.8	0.6
Nickel[c]	2.1	0.9	1.1	0.3	2.7	1.1	6.4[d]	2.6[d]	1.6[d]	0.8[d]
Manganese ore	1.5	0.6		0.4	1.6	0.6	1.0	0.9	1.3	0.5
Chrome ore	1.2	0.8	1.2	−0.5	1.2	1.0	1.5	1.9	1.2	0.8
Cobalt	0.6	0.6	−0.5	0.8	1.1	0.6	1.2	1.0	0.5	0.6
Tungsten	1.0	0.4	0.6	−0.1	1.0	0.4	0.8	0.7	1.1	0.3
Refined copper	3.1	0.9	−0.2	0	4.3	1.1	2.4	2.5	3.3	0.7
Primary aluminum	2.0	1.6	1.8	1.5	2.5	1.8	7.0	1.9	2.0	1.6
Platinum group[c]	2.1	2.1	1.9	0.9	2.3	2.7	13.8	1.0	1.8	2.2
Zinc	0.9	0.8	0.2	0.1	1.3	1.0	2.4	2.9	0.9	0.7
Tin	0.7	0.3	0.4	0.5	1.0	0.5	1.6	0.2	0.6	0.3

Source: Malenbaum 1979
a Column 1: % change 1956–1960 over 1951–1955 b Column 2: % change 1971–1975 over 1966–1970 c % change 1961–1965 over 1956–1960 (column 1) d Market economies only

appropriately identified with production functions and their theoretical framework. Steel inputs contribute to the new steel plant and to other capital creation or to an automobile and other goods for current use in a nation; this sequence pertains for the entire range of materials and the broad scope of final goods and services in the GDP. The input–output relationship also reflects technological changes that in time alter the coefficients of the production function. Research and development expenditures, whether in private or public entities, constantly press for such shifts. Innovations and inventions reflect the skills, knowledge and aspirations of man and society; there is a direct human contribution that serves to alter intensity of use through technological change.

The vast array of ratios for a single intermediate material (e.g., copper, Table 1) and the arrays cited for other materials which are so important in the elasticities of Table 2 reflect changes that can only be attributed in part to such shifts in production functions even in a dynamic context. A very important explanation of the intensity changes measured is the changing structure of the economy of a nation over a period of time. Certainly the prototype process of development, with movement from agricultural and extractive enterprise to a greater emphasis on industry and increasingly on services output, brings new materials intensity levels for the nation, even without any technological change in identified production processes. In addition to the technological and structural explanations for intensity change, adjustments are constantly and "normally" in process through market pressures towards the cost and efficiency goals of functioning economies. New national policies, changing ratios of capital and labor costs, as well as shifting prices for intermediate inputs and final products serve to bring substitutions in the production of a given product, among, for example, copper, aluminum and steel inputs, or among the ferroalloys. Such changes are apart from and beyond those attributable to new technology.

Structural and efficiency forces for intensity change join the technological forces in determining patterns for intensity measures. Present documentation is only suggestive of the relative importance of these forces, partly because they often operate simultaneously. Thus, market costs induce technological changes, and innovation or invention facilitates such structural shifts to increased manufacturing as against extractive enterprises. The literature on intensity measures suggests that structural forces have been predominant in the pattern since 1950, and that the influence of technological change may have been the least important of the three forces during those years.

Further research can yield more knowledge and thus enhance the value of the intensity of use concept for materials analysis and policy. Again, note that all these determinants are functions of the quality of humans: skills and knowledge, aspirations and commitment. Human attributes more than material conditions impart the critical flexibility to the use of materials in any nation. As is true for modern growth theory, the forces for activating intensity changes of materials as for serving economic expansion are human quality forces more than material quantity forces (Fig. 1).

With respect to the impressionistic representation of the intensity of use relationship of GDP/population (Fig. 2), the documentation for world development is available only in part. The entire curve is hard to measure for specific regions; poor nations reflect only the declining movement as their economies continue to progress. The data are revealing with respect to turning points so critical to the intensity concept. Thus the record permits some approximation to a single empirical pattern for intensity changes for each material and each region as real GDP per capita grows.

Figure 3 represents intensity of use of iron ore; again, parallel data are available for all the materials. The

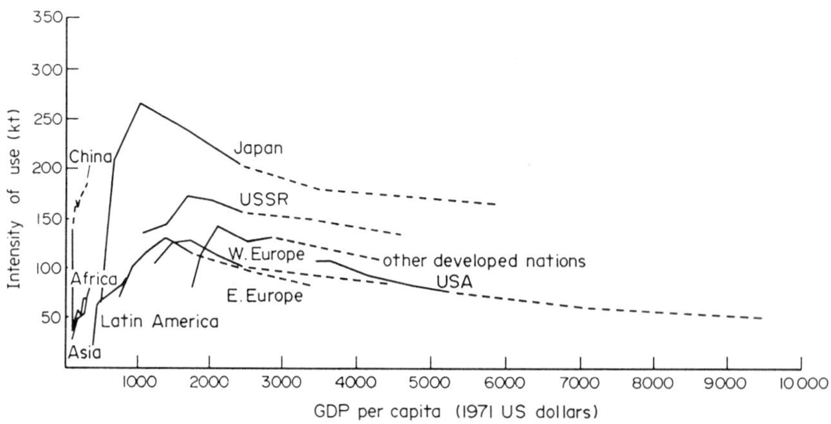

Figure 3
Iron ore intensity of use and GDP per capita: —— actual, --- projected (after Malenbaum 1978)

historical pattern (solid lines) reveals the downward slope of the inverted "U" of Fig. 2 in the case of the more developed regions. Also revealed is the predominantly upward path for the lands where per capita GDP is still relatively low and where structural adjustments are in progress. The data also permit interpretation of the regional differences in intensity levels at a point of time (e.g., the high figures for Japan and those for the USSR, as compared to the USA in the case of iron ore). The emerging patterns provide powerful tools that warrant further research.

4. *Relevance of Intensity of Use Ratios*

Scientific analyses in recent decades have increased awareness of the importance of materials supply and demand to the future of man and society. On the one hand all nations seek to expand their per capita GDP. Even conservative projections of world GDP imply use of industrial raw materials in tonnages two to three times the actual annual level of the early 1970s by the year 2000, and pressures for further such expansion exist. However, the possible depletion of the world's nonrenewable resources is also broadly accepted as a real threat to GDP and population growth, perhaps within the next century.

Research and analysis responsible for such propositions will continue to expand the world's capacity to live with alternatives in supply and demand of materials. In the early 1970s the tonnage of raw materials use in the year 2000 would have been projected at four to five times the actual 1970 tonnage levels. The idea of dynamic input–output relationships and the evidence of patterns of declining intensity of use of materials remain relatively unfamiliar. There is wide scope for additional research on other materials and for study directed towards a fuller understanding of the determining factors in intensity change. Perhaps these efforts will in time permit man and his organizations to decide the level of material inputs in a desired output and the required structure of GDP together with the real costs of such goals.

The evidence from past studies associates the factors that determine demand for materials with forces now increasingly accepted as central in the entire process of the economic development of a nation. They offer a demand prospect that counters the growth-limiting significance attributable to marked inelasticity of raw materials supply. Generating the desired level of GDP in the year 2000 with 50–60% less tonnage of materials input is a contribution to world materials supply. Capacities of man are substitutes for quantities of material; intensity of use considerations are fundamental to supply analysis.

Concern about materials exhaustion often encompasses concern about US dependence on foreign sources of supply. The evidence that patterns of materials intensity of use have increasingly made the country a relatively smaller user of the world's industrial raw materials is fundamental. This derivative of intensity of use patterns reflects the relative strength of US economic potential: the greater capacity to meet the needs of the population with smaller recourse to scarcer resources.

In addition, the outlook for declining intensities in the materials demand of the rich nations and for expanding intensities in demand of the poor nations offers complementarities in the future materials interests in the world. This should provide greater scope for solving critical interregional conflicts. Analysis of the occurrence, significance and burden of world materials supply shortages is an integral part of analysis of intensity of use patterns of demand.

5. *Concluding Observations*

This article opened with a reference to the article on intensity of use in the *Encyclopedia of Materials Science and Engineering* (Malenbaum 1986). The data recorded at that time stopped in the mid-1970s. Demand for important industrial metals had been expanding at rates that increased 5–10% annually for at least two decades into the early 1970s. In the following decade, however, the rate declined annually. GDP growth rates in the rich lands had been increasing by 5% or more each year through the mid-1970s. They grew at about half that rate until the mid-1980s (Tilton 1989). There is yet to emerge clear evidence of a resumption of the early growth experience of demand for key industrial metals and for a return of annual GDP growth to former levels.

The preceding four sections of this article were focused upon the role of the intensity of use of industrial metals and other products in the demand and supply of key inputs in the overall development process. Broadly, the theme highlighted in Fig. 1 was the key importance of human quality contributions to the development goals. The primary need here was for the poorer and less industrialized nations (Radetzky 1986) where considerable progress was being achieved. Still, the relatively limited scale of modern economic activity in these nations means that the world picture remains dominated by the industrial performance and other economic gains in the rich nations. Most of this article emphasizes the need to develop dynamic intensity of use impacts for an expanding world GDP.

No attention has yet been given here to recent writings on dematerialization (e.g., Herman et al. 1989) and on transmaterialization (e.g., Waddell and Labys 1988). Herman, Ardekani and Ausubel provide a strong case for dematerialization. Thus, Japan in 1984 consumed only 60% of the raw materials required for the same volume of output as in 1973. Furthermore, International Monetary Fund (IMF) estimates indicate that the amount of raw materials needed for one unit of industrial production is now no more than two-fifths of what it was in 1900, and this decline is accelerating. Waddell and Labys opt for a series of "normal replacement cycles" well illustrated in their study. Minerals demand experiences phases where traditional lower quality materials linked to mature industries undergo replacement periodically by higher quality, or technologically more appropriate materials. These replacement phases contrast to the once-and-for-all structural change of dematerialization. These publications point to the importance of further research on the possible relationships between inputs of materials and final outputs of goods and services. Both methods warrant study.

See also: Markets for Materials: Modelling; Materials in the Industrial System; Prices of Materials: Theory; Substitution of Materials: Economics

Bibliography

Brooks D B, Andrews P W 1974 Mineral resources, economic growth and world population. *Science* 185(4145): 13–19

Congressional Research Service 1980 *Emerging Issues in Science and Technology.* US Government Printing Office, Washington, DC

Considine T J 1986 Atrophy in metal demand: truth or consequences. *Mater. Soc.* 10(3): 529–38

Considine T J 1990 Recent trends in material consumption: the role of technology and economics. *Mater. Soc.* 14(2): 167–82

Fischman L L 1980 *World Mineral Trends and US Supply Problems.* Resources for the Future, Washington, DC

Herman R, Ardekani S A, Ausubel J A 1989 Dematerialization. In: Ausubel J, Sladovich H (eds.) 1989 *Technology and Environment.* National Academy Press, Washington, DC, pp. 50–69

Hibbard W R Jr 1986 Metals demand in the United States: an overview. *Mater. Soc.* 10(3): 251–8

Hyung Hwang K, Tilton J E 1990 Leapfrogging, consumer preferences, international trade and the intensity of metal use in less developed countries: the case of steel in Korea. *Resour. Policy* 16(3): 210–24

International Iron and Steel Institute Committee on Economic Studies 1972 *Projection 85: World Steel Demand.* IISI, Brussels

Landsberg H H 1976 Materials: some recent trends and issues. *Science* 637–41

Landsberg H H, Fischman L L, Fisher J 1963 *Resources in America's Future: Patterns of Requirements and Availabilities, 1960–2000.* Johns Hopkins University Press, Baltimore, MD

Leontief W W, Carter A P, Petri P A 1977 *The Future of the World Economy.* Oxford University Press, Oxford

Leontief W W, Koo J, Nasar S, Sohn I 1983 *The Future of Nonfuel Minerals in the US and World Economy: Input–Output Projections, 1980–2030.* Lexington, Lexington, MA

Malenbaum W 1973 World resources for the year 2000. *Ann. Am. Acad. Polit. Soc. Sci.* 408: 30–46

Malenbaum W 1975 Laws of demand for minerals. *Proc. Economic Committee, AIME.* American Institute of Mining, Metallurgical and Petroleum Engineers, New York, pp. 147–53

Malenbaum W 1978 *World Demand for Raw Materials in 1985 and 2000.* McGraw-Hill, New York

Malenbaum W 1979 Anatomy of world materials demand: its role in supply availability. *Miner. Congr. J.* 65(2): 59–64

Malenbaum W 1986 Intensity-of-use of materials. In: Bever M B (ed.) 1986 *Encyclopedia of Materials Science and Engineering.* Pergamon Press, Oxford, pp. 2355–61

Materials and Society 10(3): 237–541: papers from Metals Demand Conference, cosponsored by *Materials and Society* and the Department of Mineral Economics, Pennsylvania State University, May 19–22, 1986

Meadows D L, Meadows D H 1972 *The Limits to Growth.* Universe, New York

Radetzky M 1986 Issues and forecasting metal consumption in LDCs. *Mater. Soc.* 10(3): 491–511

Roberts M C 1986 An aggregate model of long term mineral requirements. *Mater. Soc.* 10(3): 303–28

Suslick S B, DeVerle P H 1990 Long-range metal consumption forecasts using innovative methods: the case of aluminium in Brazil to the year 2000. *Res. Policy* 16(3): 184–99

Tilton J E 1986 Atrophy in metal demand. *Mater. Soc.* 10(3): 241–3

Tilton J E 1986 Beyond intensity of use. *Mater. Soc.* 10(3): 245–50

Tilton J E 1989 The new view of minerals and economic growth. *Australian Economics Congress, 1988.* 253–78

Vogely W A 1986 Chairman's summary and conclusions: metals demand conference. *Mater. Soc.* 12(1): 59–86

Waddell L M, Labys W C 1988 Transmaterialization: technology and materials demand cycles. *Mater. Soc.* 12(1): 59–86

W. Malenbaum
[University of Pennsylvania, Philadelphia, Pennsylvania, USA]

L

Life-Cycle Analysis of Energy Consumption and Carbon Dioxide Emissions

Energy consumed in the production, fabrication, use and recycling of materials is becoming important in selecting materials because of concerns about supplies, costs and environmental impact, particularly global warming. Carbon dioxide is believed to be a prime agent in possible global warming and is evolved by most of the principal energy sources except hydro and nuclear power in the life cycle of materials. Other global warming emissions in the life cycle of materials should also be considered in a full assessment (e.g., radiatively active trace emissions with long residence times in the atmosphere, such as the perfluorinated carbons emitted in aluminum smelting).

Many of the fuels and energy technologies that have little or no possible global warming problems associated with them may not be affordable or widely available until well into the twenty-first century. During this interim period, significant reductions in carbon dioxide emissions could be achieved through energy conservation. Many energy conservation options involve changing what and how materials are used. Some of these changes also reduce life-cycle costs even with present value cost analysis. A prudent policy would be to implement these cost-effective changes while working on longer range answers to problems of possible global warming and meeting future energy needs.

Life-cycle analysis of energy usage and carbon dioxide emissions for material should include all the requirements in the manufacture, use and recycling of a product as shown in Fig. 1. The analysis should include energy and carbon dioxide emissions for raw materials and by-products and for producing and delivering the fuels. With recycled materials, particular care is required to employ production energies and emissions that reflect recycling of runaround, manufacturing and obsolete scrap.

Production energies and emissions vary widely for different materials. Judgments about suitability of materials from energy and emissions standpoints cannot be made directly from these data without considering the properties that determine how much of the material is required in a given end product. Differences in density, strength, stiffness, thermal and electrical conductivity and other properties among materials strongly affect the quantities required. Furthermore, the energy and emissions in using the end product can be many times the energy and emissions to produce it. Differences in efficiency in end use can cause the product with the greater production energy or emissions to have the lower life-cycle energy or emissions. Four examples illustrate these points.

1. Beverage Containers

For beverage containers, the principal steps affecting life-cycle energy and carbon dioxide emissions are producing the material and the container, filling and shipping the container and washing and refilling or recycling the container. The life-cycle values in Table 1 also depend strongly on recycling rates for cans and trippage for refillable glass containers, and to a lesser extent on container size. Three of the container materials—bimetal, glass and polyethylene terephthalate (PET)—have undergone rapid increases in recycling since the data of Table 1 were compiled. The lowest life-cycle energies and carbon dioxide emissions are obtained with refillable glass bottles at between five and ten trips per bottle. This type of container has the highest life-cycle values of those considered if discarded after only one trip.

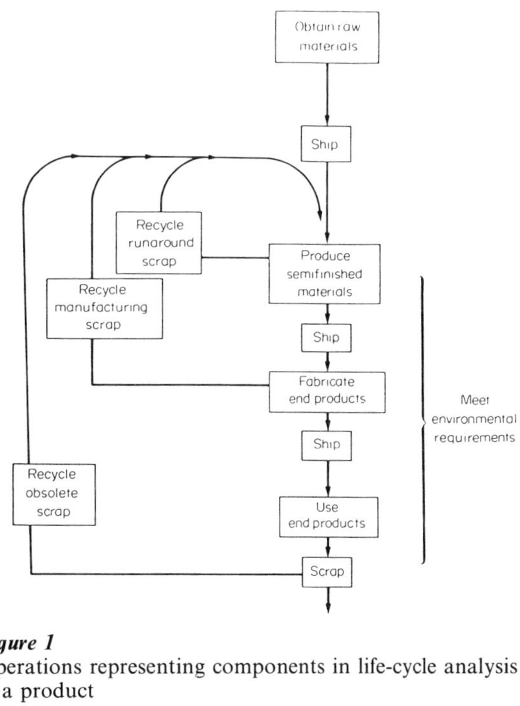

Figure 1
Operations representing components in life-cycle analysis of a product

Table 1
Life-cycle energy consumption and carbon dioxide emissions for soft drink containers in the USA

Container type	Percent recycle	Trips per container	Life-cycle energy per 0.355 l serving (kJ)[a]	Life-cycle carbon dioxide emissions per 0.355 l serving (g)
Aluminum two-piece can	0	1	5430	380
(0.355 l)	55.5[c]	1	3640[e]	234
Bimetal can (steel with aluminum top) (0.355 l)	4[b]	1	5270	508
Nonrefillable glass bottle (0.473 l)	2.4[b]	1	5670	306
Refillable glass bottle	0	1	10550	568
(0.473 l)	(80)	5[b]	2710	146
	(90)	10	1730	93
Polyethylene terephthalate (PET) bottle		1	2920	194
(2 l)		1	4750	316

a Includes energy to produce the material, fabricate, fill, package, ship and return for recycle or wash and refill b 1978 c 1982 d Being recycled into other products e Reflects effects of reducing can weight and energy conservation in production of aluminum

2. *Automotive Materials*

An important factor determining automotive fuel consumption is vehicle weight. Besides the direct effect of weight reduction on fuel consumption, an additional fuel saving is possible if the drive train capability is also reduced to just maintain performance as weight is reduced. The difference in the weights of the original and replacement parts and any supporting parts that are redesigned must be considered. For 1987 model US automobiles the fuel consumption weight coefficient for changes in vehicle weight with constant aerodynamic drag was 5.7×10^{-5} l kg^{-1} km^{-1} with drive train reoptimization. A 50% compounding of weight reduction in supporting parts is assumed.

The example chosen in Table 2 is automotive hoods. The aluminum version of this hood has the lowest life-cycle energy, even though its production energy is the

Table 2
Values used in computing life cycle energies and carbon dioxide emissions of automobile hoods

	Mild steel	High-strength steel	Aluminum alloy	Sheet molding compound
Weight (kg)	20.9	17.7	7.7	13.8
Energy to produce material and fabricate hood (MJ)	13300	1130	2150	1290
Energy to produce extra supporting parts (MJ)	420	319	0	196
Life-cycle energy with 50% weight compounding and drive train reoptimization (MJ)	12300	10100	5100	8000
Life-cycle energy with 75% recycling of aluminum and burning of plastic for fuel value (MJ)			4100	7700
Carbon dioxide emissions to produce and fabricate material for hood and supporting parts (kg)	173	143	150	105
Life-cycle carbon dioxide emissions to produce and operate (kg)	680	562	292	417
Life-cycle carbon dioxide emissions with 75% recycling of aluminum and burning of plastic for fuel value (kg)			221	431

highest of the hoods considered. Still lower life-cycle energies result with 75% recycling of aluminum and burning the plastic for fuel value. Recycling of plastics into engineered products such as automotive hoods is not envisioned. The carbon dioxide emissions generally parallel the energy values with adjustments for the different mix of energy sources for each material. Note that carbon dioxide emissions for production are eight to 20 times the weight of the hood and life-cycle emissions to produce and operate are 30 or more times the hood weight.

3. *Aerospace Materials*

Weight is even more important in aircraft than in automobiles for reducing lifetime energy and carbon dioxide emissions. The lifetime distance travelled for an aircraft is usually much greater than that for an automobile and the fuel needed to transport 1 kg of load 1 km is also higher. This leads to increased consideration of aerospace materials such as fibrous composites which allow reductions in weight because of their high specific strength and modulus. The energies and carbon dioxide emissions in Table 3 for the production of a conventional aluminum part and the equivalent part of graphite–epoxy composite all but disappear in the life-cycle values. The life-cycle energy consumption and carbon dioxide emissions of the composite part are each approximately 30% lower than for the conventional aluminum part.

4. *Electrical Conductors*

Electrical conductor size is governed by safety criteria. For economy the conductors are made as small as possible, yet large enough to dissipate their resistance heat through the insulation without overheating. If the conductor were larger, resistance heat would be lower at a given current. The larger conductor requires more energy for production, so the tradeoff between increasing the energy for production and lowering the resistance losses must be considered, as shown in Table 4. Increasing the size of a copper wire from No. 12 to No. 2 increases the energy for production tenfold but also reduces lifetime resistive losses tenfold. Since lifetime resistive losses with associated losses in a 33% efficient fossil-fuelled power plant are about 300 times the production energy for the conductor sizes currently used, the net result of increasing conductor size is to reduce life cycle energy and life-cycle carbon dioxide emissions by 85%. The larger conductor size in this example was chosen only to illustrate the point. Practical considerations such as space needs, ease of installation and costs also determine the conductor size to use.

Table 3
Aerospace materials: equivalent parts in aluminum and graphite–epoxy composites

	Aluminum	Graphite–epoxy composite
Weight of finished part (kg)	1	0.7
Ratio of purchased material weight to finished part weight	5	1.5
Production energy with recycling (MJ)	600	100
Lifetime fuel (l)	1140	800
Lifetime fuel (MJ)	47600	33400
Life-cycle energy (MJ)	48200	33500
Production carbon dioxide (kg)	42	6
Life-cycle carbon dioxide (kg)	2340	1620

Table 4
Effect of conductor size on life-cycle energies and carbon dioxide emissions for copper electrical conductors[a]

	Conductor size	
	No. 12	No. 2
Production Energy (MJ m^{-1})	3.3	34
Lifetime Resistance (MJ m^{-1})	942	101
Life-cycle (MJ m^{-1})	945	135
Production carbon dioxide (kg m^{-1})	0.20	2.0
Lifetime resistance carbon dioxide (kg m^{-1})	47.3	5.0
Life-cycle carbon dioxide (kg m^{-1})	47.5	7.0

a Current, 15 A; 10 h day^{-1}, 200 d $year^{-1}$, 40 year lifetime, 33% efficient fossil-fuelled power plant

See also: Economics of Materials, Energy and Growth; Energy Considerations in Metals Production

Bibliography

Altenpohl D G, Daugherty T S (eds.) 1980 *Materials in World Perspective.* Springer, Berlin

Cochran C N 1981 The energy balance of aluminum from production to application. *J. Met.* 33(7): 45–8

Heilman B 1989 Global warming. *Chem. Eng. News* 67: 25–44

Marland G, Rotty R M 1984 Carbon dioxide emissions from fossil fuels: a procedure for estimation and results for 1950–1982. *Tellus B* 36: 232–61

Payne P R, Doyle D W 1978 The fossil fuel cost of solar heating. *Proc. Intersoc. Energy Convers. Eng. Conf.* Society of Automotive Engineers, Warrendale, PA

Ramanathan V, Cicerone R J, Singh H B, Kiehl J T 1985 Trace gas trends and their potential role in climate change. *J. Geophys. Res.* 90: 5547–56

Research Triangle Institute 1976 *Energy and Economic Impacts of Mandatory Deposits*, PB-258 638. US Department of Commerce, Springfield, VA

Tester J W, Wood D O, Ferrari N A (eds.) 1991 Energy and the Environment in the 21st Century. MIT Press, Cambridge, MA
Verbraeck A 1976 *Proc. 9th Int. TNO Conf.: The Energy Accounting of Materials, Products, Processes and Services.* Netherlands Organization for Applied Scientific Research, The Hague, The Netherlands

C. N. Cochran
[Oakmont, Pennsylvania, USA]

Life-Cycle Costing

The cost of an item to the buyer includes not only the purchase price but also the cost of ownership for its lifetime and the cost of disposal—the total life-cycle cost. Durable, reliable and easily maintained items may have a lower life-cycle cost, despite a higher purchase price, than cheaper, more fallible and nonrepairable items. Materials properties and costs often determine the life-cycle cost characteristics of the design. An automobile fender of steel, for example, may be cheaper to manufacture than a fiberglass bonded fender, but repairs may be expensive and inferior for the steel fender and relatively inexpensive and satisfactory for the fiberglass fender. The choice between alternatives depends on the expected frequency of repairs and the buyer's assignment of the relative importance of first cost and the longer perspective of life-cycle cost.

1. Principles and Areas of Application

Formally, the life-cycle cost is the total cost of an item to the buyer including the development costs (if separately stated), the production costs (the usual purchase price for off-the-shelf items) and other investment costs, the operating and support costs, and the disposal costs.

In estimating or controlling life-cycle cost (LCC), the following four phases of a program are used.

(a) Research and development: those activities leading to the release of manufacturing plans and drawings.

(b) Investment or production phase: the production of the prime equipment and its support equipment, initial provisioning, the initial training of operational personnel, transportation to the user, initial tests or break-in, and facility construction (e.g., fuel storage tanks or repair stations). Acquisition of the elements of production (e.g., tooling, test equipment, etc.) is also included in this phase.

(c) Operations and support phase: the operational costs such as operators, fuel or utilities, maintenance replenishment provisioning, support equipment maintenance and replacement, replacement training, facility maintenance, and security. Modification or upgrading costs may be included in this phase or may be regarded as a new and distinct LCC program.

(d) Disposal phase: the disposal costs are usually relatively small and are often offset by the salvage value of the equipment. For some equipment, however, such as nuclear power plants, disposal costs are significant and should be estimated and controlled.

These later costs may be discounted to correct for the time value of money (Seldon 1979), and economic adjustment factors should correct for any expected monetary inflation (or deflation). The use of "constant dollars" is recommended for LCC analysis; if decision makers require it, results may be presented in both constant and "then-year" values (incorporating any predicted monetary inflation or deflation).

LCC analysis often provides guidance in choosing between alternatives in design, replacing aging equipment, making purchasing comparisons between various equipment, contractors or suppliers, choosing operations and support concepts, developing maintenance methods, and preparing logistics support plans. In design, for example, the selection of throwaway, nonrepairable or discard-on-failure items would be discouraged if the potential consumer could be shown the advantageous lower LCC of more durable designs. As increasing scarcity in time raises raw materials costs, more durable designs will have a more attractive LCC. Rising labor costs could negate the effect of materials cost increases; quantitative economic analysis can elucidate the controlling parameters.

LCC may provide financial insight for the evaluation of alternative mining methods or sites, the choice of processing facilities, buildings and energy costs, the choice of alternative materials for design and the choice of protective maintenance methods (e.g., finishes and washes).

2. Methods of LCC Analysis

LCC analysts often prepare cost estimates during the early phases of a project, as the most benefit can be gained before detailed designs or plans are available. Logistics plans and other operating support phase concepts are usually sparse or absent. In recent years, techniques have been developed for cost estimating with such rudimentary information. Even with these new techniques, often the LCC analyst must develop some exemplary plans for purposes of the baseline cost estimate; other personnel can then propose alternative plans, the cost of which will be compared with the baseline.

The cost estimates of the four phases (see Sect. 1) are combined to yield LCC and presented for analysis. The costs that comprise LCC must be combined and presented for analysis and review. The simplest model

Table 1
An example of LCC estimates phased over 11 years (in US dollars or other money units)

	Year							
LCC phase	1982	1983	1984	1985	1986	1987	1992	Total
Research and development	10	15	5					30
Investment		2	20	50	5			77
Operations and support					12	18	18	48
Disposal							1	1
Total	10	17	25	50	17	18	19	156

of LCC is:

$$\begin{aligned} \text{LCC} = {} & \text{Research and development} \\ & + \text{Investment} \\ & + \text{Operation and support} \\ & + \text{Disposal} \end{aligned}$$

Each of these four costs is time-phased and this variation with time is often portrayed as in Table 1. Analysts often portray LCC as a three-dimensional matrix (Fig. 1). Due to the volume of detail resulting from such a multidimensioned cost structure, computer techniques have been developed to ease the task of data manipulation, computation and structuring. The computer can generate requested output reports at whatever level of detail or generality is desired. Furthermore, the computer model can be used to convert constant dollars to inflated "real" dollars, or from one currency to another.

The computer mechanization portrayed utilizes cost-element values entered by the analyst. A further extension of the computer methodology would have the cost-estimating relationships built in to the computer program for some or all of the cost elements. For example, one of the operating and support costs might have a cost-estimating relationship as follows:

Operating and support maintenance costs per period = (number of maintenance men) × (labor rate per hour) × (hours per period)

Output formats of the LCC computer model should assist the analyst in understanding the details of cost contributions and show the decision makers the significant cost sensitivities of the program, as well as the overall costs. Various other formats and desirable characteristics of computer programs are discussed in the literature (Seldon 1979).

Computer software packages are available which can be used for performing LCC computations. The three-dimensional characteristics of the LCC can be approached utilizing personal computer spreadsheets which are actually two dimensional but can be arranged by the analyst into master and supplementary spreadsheets. This can be done whether the cost-estimating relations (CER) are exogenous to or contained within the model. True three-dimensional spreadsheets are becoming available. Several commercially available models simplify performance of LCC analysis and preparation of reports meeting industry and government standards. At one extreme is the PC version of the US Navy-developed Flex model (Zephyr Systems) in which the analyst inputs the CERs together with input supporting his hardware and logistics plan. At the other extreme are the Price-H/Price-HL models (Price Systems) combination which requires certain

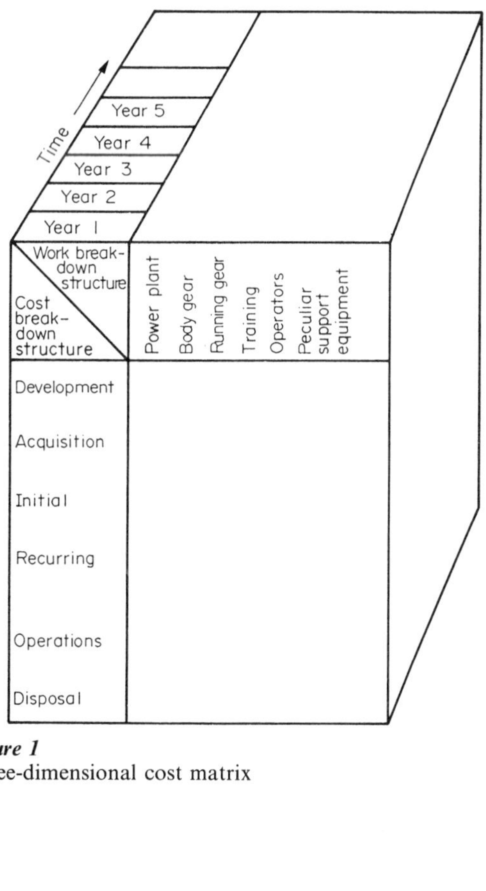

Figure 1
Three-dimensional cost matrix

hardware parameters and information concerning specific characteristics of logistic support but the actual CERs are proprietary and not available to the analyst for review or modification.

3. Determinants of Ownership Costs

Life-cycle costing focuses attention on the continuing ownership costs of an item (i.e., the operating and support costs). An automobile's ownership costs, for example, include operating costs (principally fuel) and support costs (principally maintenance). The calculation of operating costs, such as fuel and lubricants, is straightforward, even prior to having actual experience. Estimates of fuel consumption, for example, are routinely provided by design engineers. Maintenance costs, however, are more difficult to estimate.

Maintenance costs, particularly maintenance labor costs, usually dominate other support costs and should be given careful attention in a cost analysis. Most analysts divide maintenance costs into (a) scheduled or preventive maintenance, and (b) unscheduled or corrective maintenance (repair). Preventive maintenance should reduce the requirements for corrective maintenance and cost analysts can examine expenditures for preventive maintenance to determine whether a proper balance has been achieved for minimum total cost. Changes in the preventive maintenance schedule can often reduce the total maintenance cost. For this reason, the cost analysis is often developed in conjunction with a maintenance concept.

The reliability of an item, that is, how long it will operate before failure, determines how often corrective maintenance is required. The LCC analyst must, therefore, obtain estimates of the item's reliability (with a specified preventive maintenance schedule). The frequency of necessary repair actions should be estimated by specialized reliability personnel. They must identify the frequency of required repair actions. The predicted reliability will enable the analyst to estimate parameters such as maintenance labor required, maintenance material required, the facilities, maintenance personnel training, and maintenance tooling and test equipment. If maintenance engineering personnel (logistics planning engineers) are available, their help will improve the maintenance planning and, therefore, the cost estimates.

Durability, a measure of the life of an article, also affects a planning trade-off. For many items and materials, only a limited number of repairs can be made before the costs of satisfactory repairs begin to increase sharply. Furthermore, many mechanisms exhibit a wear-out phenomenon with the failure rate increasing with operating time. The general shape of the variation of failure rate with time holds true for many kinds of equipment (Fig. 2). The end of life or durability is represented by A in this figure. It should be noted that the initial failures B are significantly influenced by the quality control of the manufacturing process, and that the random failure rate C is determined principally by design.

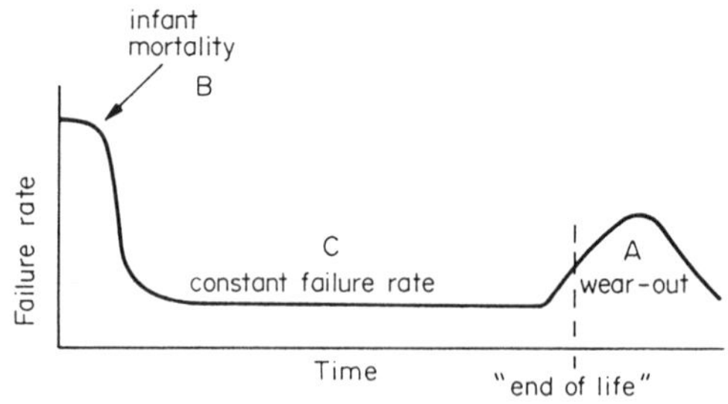

Figure 2
Diagram outlining the variation of failure rate with operating time

4. Other Operating and Support Costs

Replenishment provisioning identifies costs of replacing supplies and repair parts which are consumed during maintenance. Maintenance may be subdivided into prime equipment maintenance and support equipment maintenance, and further into scheduled and unscheduled maintenance. The finer breakdowns allow more precision in estimating.

The initial costs of training, such as technical manuals, training facilities, training of instructors and classroom aids, are accounted for in the investment phase, but during the operating and support phase the costs of replacement training due to attrition must be allocated. During the operating and support phase, costs will be incurred for:

(a) maintenance of support equipment,

(b) maintenance facility costs,

(c) crew pay and fringe costs,

(d) fuel and other provisions,

(e) utilities (e.g., electricity, gas, telephone, water and drainage), and

(f) taxes and rent depletion allowances.

Improved logistics planning provides a major benefit of the LCC analysis. Early in a program, analysis can alert the developer and the user to the cost implications of the design and suggest design changes and logistics plans to minimize the cost of product support.

5. Systems Engineering Using LCC with Cost–Performance Trade-Offs

An effective LCC program should reduce the LCC of the system under consideration. LCC can show desirable changes to the system configuration, design, or the way it is produced or used.

System engineering transforms an operational need (mission requirements) into a description of system performance parameters (design specifications) and a preferred system configuration. Design engineering develops a number of feasible design concepts that could perform to the design specification. Each alternative set of performance parameters and system configuration can be evaluated for LCC. Related pairs of performance parameters, such as service life and purchase price, can be varied over a possible range. The variation in LCC as these pairs are traded-off against one another, can be evaluated and the pair with the minimum resultant LCC can be selected. This process of examining various suitable alternatives and selecting the most desirable combination of performance and cost is known as the cost–performance trade-off process or cost–benefit analysis.

6. Marketing and Contractual Methods of Life-Cycle Costing

The buyer, the consumer and the user, both as individuals and as organizations, must be convinced that LCC is an important criterion in the choice between alternative courses of action. For relatively simple "off-the-shelf" items, the marketing task is to convince the consumer that the particular product has a lower LCC.

In buying large, complex, one-of-a-kind systems (e.g., an ore processing plant or military weapons systems) the LCC considerations become even more important. These large complex systems are procured as a design and construction package or as an architect and engineer package followed by a construction contract.

One cannot, from first principles, specify the LCC in such circumstances, particularly as the systems are often uniquely advanced. Therefore, the buyer must use contractual incentives that will motivate the contractor(s) towards a low LCC design and a low LCC product. Military systems procurement agencies have developed a number of contractual methods of motivating contractors toward lower LCC: (a) target logistics support cost, (b) reliability improvement warranty and (c) design to cost incentive (production costs as a surrogate for LCC).

Acknowledgement

The publishers wish to thank Mr E. I. Tinkham for having reviewed and updated this article.

See also: Cost Estimating

Bibliography

Brown R J, Yanuck R R 1980 *Life Cycle Costing: A Practical Guide for Energy Managers.* Fairmont, Atlanta, GA

Coble F W 1989 Teaming Price H and Price HL for design to cost. *Proc. 11th Ann. Conf. ISPA*

Earles M E 1978 *Formulas and Structures for Life Cycle Costing.* Eddins-Earles, Concord, MA

Headquarters US Army Materials Command 1987 *Logistics Support Analysis Techniques Guide*, AMC-P700-4. pp. 64, 116

Newbery C P 1989 Life cycle considerations for hardware systems. *J. Parametrics* 9(2)

Sackman H 1975 *Delphi Critique; Expert Opinion, Forecasting, and Group Process.* Lexington Books, Lexington, MA

Seldon M R 1979 *Life Cycle Costing: A Better Method of Government Procurement.* Westview, Boulder, CO

US Air Force Directorate of Computer Resources and Analysis 1988 *Project Officer's Guide to Life Cycle Cost (LCC) Models*

US Office of Management and Budget 1976 *Major Systems Acquisition*, Circular A-109. US Government Printing Office, Washington, DC

M. R. Seldon†
[Pomona, California, USA]

M

Markets for Materials

Marketing of basic commodities is quite different from marketing of manufactured goods or of services. The seller of a primary material must meet standard specifications of quality and size. Elements of style or design that are important in the case of manufactured goods are not relevant to basic commodities.

With the exception of a few bulky, low-value commodities (e.g., sand and gravel), prices of basic commodities are determined largely by the world balance between supply and demand. The centrally planned economies of the socialist countries have been an exception during the post-World War II period, but with the trend toward market economics that has developed in many of these countries from late 1989, they too will find their commodity markets influenced by global price trends.

Prices for a specific commodity in one market may differ somewhat from prices in other countries. These price differences are due to transport costs, import duties, quotas or other trade barriers. Nevertheless, the trend of prices tends to be similar worldwide. Thus it is rarely the case that the price of, say, timber is rising in Europe while it is falling in the USA.

In the marketing of manufactured goods promotional efforts are characteristically undertaken by individual sellars, emphasizing differences of style, performance and price. By contrast, much of the promotional effort for commodities is undertaken through industry associations that emphasize the advantage of a given commodity in competition with other commodities.

Thus, for example, in the manufacture of automotive parts in recent years there has been a fierce competitive marketing struggle between producers of stainless steel, aluminum, brass, zinc die castings and plastic moldings. Individual producers of these commodities have joined efforts through their trade associations to deal with the automotive industry. Decisions regarding these commodities are made on considerations of durability, fuel efficiency, weight and ease of manufacture.

1. Parties to Materials Contracts

In marketing commodities, sellers have used a wide variety of techniques. Each commodity has its own marketing mechanism; no two are exactly alike. Differences involve the pricing structure, the form in which the commodity is sold, length of the contracts, practices as to delivery of the goods and many other features.

A market may be defined as a "meeting of people for selling and buying." In the case of basic materials, the meeting may involve a crude product sold to a processor, a semiprocessed material offered to competing refiners or fabricators or the offering of a refined product to a broad group of industrial users. In some cases, merchants or commodity brokers may enter into such transactions as middlemen.

To illustrate, in the copper industry, ore from a small mine may be sold to a nearby concentrator, which in turn sells concentrates to a smelter that produces blister copper. Title to the blister may then be transferred to a refinery, which eventually sells a refined copper shape to a manufacturer or fabricator. Alternatively, in the case of a large integrated copper enterprise, mining, milling, smelting and refining may all be accomplished by the corporation before the first transfer of title takes place (and even in this case some copper producers own and operate fabricating facilities).

A parallel situation exists in the forest products field. The owner of a small area of wooded land may sell felled logs to a local sawmill, whereas there are many giant companies that not only harvest timber from their extensive land holdings and process this in their own mills, but manufacture and sell newsprint, plywood or finished paper products marketed under their own trade names.

2. Duration of Sales Contracts

An important marketing decision is the duration of the sales commitment made by the individual seller. In some cases, policy may be to market exclusively through spot sales for immediate delivery. Another seller may choose to sell its entire output under long-term contracts for a period of several years. These are the extremes—some sellers opt for annual contracts, others for monthly contracts. Many enter into forward sales engagements for a portion of their anticipated output, retaining a share to be disposed of on a spot basis.

Valid reasons exist for each of these alternative policies. A producer situated in an area where geographical proximity indicates an advantage in dealing with a single buyer may well elect to commit its entire output to that buyer for a protracted period. Typical examples of such arrangements are the mine-mouth contracts between a coal mine and a public utility owning a generating facility in the immediate vicinity; or sale of molten aluminium by a producer operating a reduction plant adjoining an automobile parts factory.

Essential to long-term arrangements is a price understanding that meets the requirements of both buyer and seller. In times of high inflation, such an understanding may prove difficult. Periods of pronounced fluctuations in foreign exchange rates create grave problems for long-term contracts between sellers located in one country and buyers located in a different country.

When materials are sold on spot or short-term contracts, these particular problems of price definition or currency fluctuation are largely avoided. However, in times of falling demand, a seller who relies on spot sales may have difficulty in disposing of his product (unless the commodity is traded on a futures market). If he cannot rapidly adjust his rate of production, serious problems will arise as unsold inventories mount.

For those commodities traded on futures markets producers are able to avoid inventory accumulation but this is almost always at the expense of putting severe pressure on price. Many commodities are not traded on futures markets (see *Trading of Materials on Commodity Exchange*).

3. *Elasticity of Demand and Supply*

Complicating the marketing of industrial materials are certain inherent factors.

(a) Demand is difficult to forecast.

(b) For most materials, demand is highly cyclical, responsive to every change in the economic climate. Some materials have exhibited strong secular growth trends in recent years (e.g., aluminium, plastics, nickel and fiber optics), but they are not immune to the business cycle.

(c) Consumer perceptions of materials availability may heavily influence market share. Thus, for example, political events that induced a temporary shortage of cobalt resulted in significant loss of market to substitute materials.

(d) For most materials, expansion of capacity involves long lead times and heavy capital investment. Thus supply tends to be inelastic in contrast with highly elastic demand. Price volatility that results from supply–demand imbalances in turn causes problems for financing of new capacity.

The conflict between relatively elastic demand, heavily influenced by trends in the world economy, and relatively inelastic supply is partially bridged for certain materials by the recycling of salvaged commodities.

Scrap is largely marketed on a spot basis. Collectors of scrap are extremely sensitive to alternative market possibilities for their offerings. The volume of scrap supplies is heavily dependent on economic conditions. New scrap, generated in the course of manufacturing operations, is available in larger volume when the economy is buoyant. Old scrap, salvaged from buildings, machinery and transportation equipment, will rise in volume because these items are more likely to be replaced in times of favorable business trends.

Thus, scrap supplies do tend to rise during an upturn in the business cycle and to dwindle during downturns—thereby cushioning somewhat the impact on the producers of primary materials. However, the response in scrap supplies is not immediate because of the time lag involved in collecting and processing salvaged items.

4. *Volatility of Market Prices*

On the whole, markets for industrial materials are characterized by recurring cycles of tight supplies and perceived shortages succeeded by periods of ample supplies and accumulating surpluses. The consequence is that prices of materials tend to be volatile—some would say erratic (see *Prices of Materials: Stabilization*; *Prices of Materials: Theory*).

During the 1980s, some observers believed that a fundamental change had occurred in the pattern of commodity prices. The sharp recession of 1981–1982 had caused severe drops in the prices of most industrial materials. During the subsequent pronounced economic recovery beginning in 1983, commodity prices continued to lag. An explanation then widely accepted was that there had been a permanent shift from product-based economies to service-based economies in the leading industrialized countries and that, consequently, materials demand would not keep pace with the general level of economic activity.

However, in late 1987 a sharp rise in prices of raw materials took place. Consuming industries had cut back their inventories substantially during the years 1983–1986. When demand rose, producer inventories began to drop also and by the end of 1987 for most commodities stocks were below normal levels, thereby causing the price rise.

As the 1990s began, observers could identify three potential sources of heavy demand for industrial raw materials. In the industrialized countries, due to the low level of capital investment during the 1980s, there was evidence of serious deterioration in the infrastructure of roads, bridges, utility plants and manufacturing capacity. In the developing countries that were newly industrializing, infrastructure was badly needed to provide the populations with improved living standards. In the centrally planned countries that were endeavoring to convert to market economies, most existing facilities were clearly obsolescent as well as deficient in respect of environmental controls.

While the potential for a period of rising consumption of materials exists, the capacity to finance projects that will meet needs in the three groups of countries is in question. Heavy debt burdens already incurred by governments and private corporations make the financing of projects difficult, if not unlikely.

Thus, one of the key elements in price determination—demand—is an imponderable. As to the second element—supply—the evidence suggests that fears previously expressed of potential shortages are not likely to materialize over the next several decades. Even with respect to minerals, which are nonrenewable resources, a high rate of discovery since World War II has materially expanded known reserves for virtually all mineral commodities. The third element—cost—is heavily influenced by monetary trends and political developments and has been a significant contributor to price volatility in recent years.

5. *Methods by Which Prices are Established*

Prices of some materials are established by producers and are referred to as administered prices. In these instances price changes tend to be tied primarily to cost factors. For other materials, prices are determined on open markets (i.e., commodity exchanges which are roughly equivalent to auction markets). The primary influence in such markets is the immediate balance between supply and demand. In still a third category are negotiated prices between buyer and seller. This category is mainly used for contracts involving bulk commodities sold on a long-term basis. Examples would include iron ore sold by an independent producer to a steel mill, bauxite sold to an alumina producer or coal sold to a mine-mouth generating plant owned by a public utility. In these cases the seller endeavors to obtain a fixed price, subject to escalation, that will cover production costs plus a return on investment. The buyer for his part commits himself only if the indicated price is comparable to or lower than the cost of alternative supplies.

In the short run, the factors influencing price differ somewhat among these three categories. In the long run, however, prices for all materials must take into account the three basic elements of supply, demand and cost of production. Supplies will only continue to be available if costs are covered. But prices that are established solely on the basis of cost, without reference to demand, will eventually result in mounting pressure in periods of poor demand and lead either to discounts from announced prices or capital-absorbing surpluses, or both.

The pervasive importance of price in materials markets cannot be overstated. In assessing the potential of a new project to produce materials, the first consideration in assessing feasibility is an assumption as to the probable future trend of prices. These then must be compared with projected operating and capital costs. This is as true for the cotton farmer or the timber harvester as it is for the developer of a mineral resource. In every case, the forecast of future prices must take into account the available evidence of global demand trends, supplies likely to be produced by competitors and the probable level of their costs.

Where the proposed investment is large, elaborate studies of markets are indicated. In those cases involving relatively small investments, the entrepreneur may choose to rely on a simple extrapolation of past trends. A common assumption is that current market conditions will persist. This rarely proves to be the case, however, and many projects have come to grief because of this assumption.

6. *Staffing to Sell Materials*

The size and scope of marketing staffs for producers of raw materials varies greatly. A relatively small producer may have no identified marketing staff. The cotton farmer, small logger or independent miner may choose to sell his output directly to a merchant or dealer based on his own business judgement, modified by such price information as may be gathered from the business section of the daily press.

On the other hand, the multimillion or multibillion dollar materials enterprises have marketing and sales departments of considerable size. But even for the largest materials producers, the fact is that they deal with a relatively limited number of buyers and hence do not require the large sales staffs characteristic of producers of manufactured goods.

Hence, marketing costs of the materials industries absorb a much smaller percentage of gross revenues than do the marketing costs of sellers of finished products. This is clearly evident in the annual reports of the large mineral companies in comparison with those of automobile, machinery or pharmaceutical companies.

See also: Economic Theory of Mineral Resources and Markets; Markets for Materials: Modelling; Markets for Materials: Principles and Public Policy Implications of Competition; Trading of Materials on Commodity Exchanges

Bibliography

Adams F G, Behrman J R 1876 *Econometric Models of World Agricultural Commodity Markets: Cocoa, Coffee, Tea, Wool, Cotton, Sugar, Wheat, Rice.* Ballinger, Cambridge, MA

Adams F G, Klein S A 1978 *Stabilizing World Commodity Markets: Analysis, Practice and Policy.* Heath, Lexington, MA

Cameron E N 1986 *At the Crossroads: The Mineral Problems of the United States.* Wiley, New York

Corey E R 1956 *The Development of Markets for New Materials: A Study of Building New End-Product Markets for Aluminum, Fibrous Glass and the Plastics.* Harvard Business School, Cambridge, MA

Fox W 1974 *Tin: The Working of a Commodity Agreement.* Mining Journal Books, London

Knoerr A (ed.) 1985 *Mineral Facts and Problems.* US Bureau of Mines, Washington, DC

Landsberg H H, Fischman L L, Fisher J L 1963 *Resources in*

America's Future: Patterns of Requirements and Availabilities, 1960–2000. Johns Hopkins University Press, Baltimore, MD
Malenbaum W 1978 *World Demand for Raw Materials in 1985 and 2000.* McGraw-Hill, New York
Manners G 1971 *The Changing World Markets for Iron Ore, 1950–1980: An Economic Geography.* Johns Hopkins University Press, Baltimore, MD
Mikesell R F 1979 *The World Copper Industry: Structure and Economic Analysis.* Johns Hopkins University Press, Baltimore, MD
Paarlberg R L 1988 *Fixing Farm Trade.* Council on Foreign Relations, New York
Peck M J (ed.) 1988 *World Aluminum Industry in a Changing Energy Era.* Resources for the Future, Washington, DC
Portney P R (ed.) 1982 *Current Issues in Natural Resource Policy.* Resources for the Future, Washington, DC
Strauss S D 1986 *Trouble in the Third Kingdom: The Minerals Industry in Transition.* Mining Journal Books, London
Vogely W A, Risser H E (eds.) 1976 *Economics of the Mineral Industries,* 3rd edn. American Institute of Mining, Metallurgical and Petroleum Engineers, New York

S. D. Strauss
[New Rochelle, New York, USA]

Markets for Materials: Modelling

This article is intended to provide an overview of the applications of the techniques of modelling to the markets for materials. It contains a brief consideration of the purposes of modelling and the modelling process, descriptions of alternative modelling techniques and a comparison of the strengths and weaknesses of these techniques with regard to their ability to deal with specific issues.

Materials industries are becoming increasingly complex and interrelated on both the supply and demand sides of the markets. On the demand side, the emergence of new and improved materials (e.g., high-strength steels, structural ceramics, filamentary composites) has increased the complexity of materials selection. In addition, new secondary processing and design techniques are enabling engineers to take better advantage of the properties of materials.

The technologies and economics of materials production and supply are also undergoing rapid changes. New processing technologies, fluctuations in the costs of factors of production and shifting political and economic policies among the industrialized nations are among those factors introducing critical uncertainties into the supply and availability of materials.

1. Purposes of Models

In such a dynamic and complex environment, industry and government policymakers often have need for a formal framework for projecting future events and/or estimating the consequences of alternative decisions. This need has led to the increased use of quantitative or other formal frameworks for analysis (e.g., models) of materials markets. Models, although unable to capture all of the intricate technological, economic and human relationships that govern the behavior of materials markets, are often used to narrow or explore the likely range of outcomes and the associated risks and uncertainties. Models of materials markets have been used for forecasting, investigating "what-if" questions and estimating costs and benefits of investment decisions and policies. Specific applications of models include:

(a) forecasting the production, consumption or prices of individual minerals or materials by product form or by end-use;
(b) analyzing market stabilization, that is, investigating control mechanisms or forms of market organization that lead to more stable demand, supply and price relationships;
(c) investigating the implications of disruptions in the supply of materials on future demand, supply and price relationships;
(d) investigating the impact of the introduction of a new material or processing technology on production costs and/or material properties and the implications for the competitive positions of firms, industries or geographic regions; and
(e) analyzing the effects of changes in final demands on the consumption of materials and the profitability of intermediate and primary processing operations.

2. The Modelling Process

A model of a materials market is a formal representation of the part of that system that is of interest. In its broadest sense a materials market model includes representations of the demand, supply, price and inventory components (Fig. 1), as well as the important economic, technological, political and social relationships that govern the behavior of the system. Of course, it is never possible to specify a model completely and precisely because of the limitations of the data and of our understanding.

Modelling of demand behavior normally begins with an interpretation of basic demand responses to prices and economic activity (i.e., the associated price and income elasticities). Price elasticities vary among different materials and among their diverse uses. They tend to be lower for materials without close substitutes. The degree to which demand responds to changes in economic activity depends on the intensity of materials use in the end products to which the materials serve as inputs. Those products with a relatively low income elasticity tend to experience proportionally smaller swings in shipments with fluctuations in the business cycle than those with higher elasticities.

The income elasticity of aluminum demand in the

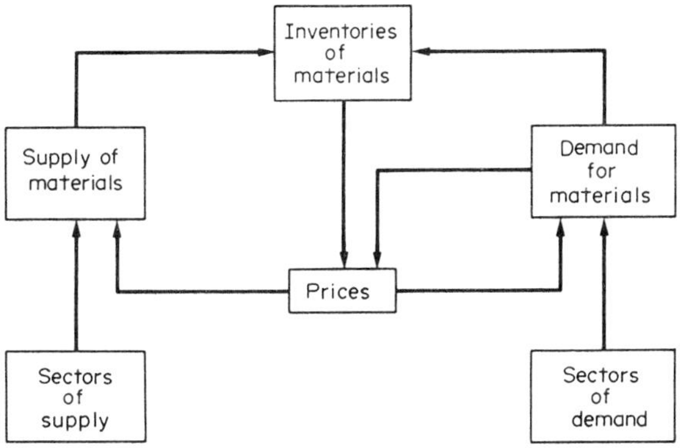

Figure 1
Basic elements of a materials market model

beverage container industry is about unity, while that for superalloys in jet engines is considerably greater. Thus, if other factors (e.g., relative prices) do not change, one would expect that changes in aluminum consumption would closely parallel changes in the aggregate economy while the pattern of superalloy consumption would exhibit fluctuations around the trend of jet-engine production.

Demand measurement is also influenced considerably by material substitution patterns. The variables of importance to intermaterials competition include relative prices (usually measured by the cross-price elasticity of demand), properties of materials (e.g., yield strength, specific modulus) and institutional barriers to use (e.g., standards and psychological factors such as resistance to change).

Modelling of supply behavior often involves consideration of geologic factors, resource development, the technology and costs of mining and processing and investment decisions. The production of a mineral commodity embodies the activities involved in finding it, in determining the size and physical characteristics of deposits and in developing the deposits to the point where mining and processing can begin. The first two of these activities are generally referred to as exploration. Exploration and development must be dealt with explicitly in a long-term model, where the location, size and grade of mineral deposits are expected over time. However, in a model with a short time horizon, or one in an industry where the characteristics of deposits (e.g., metal contents in the ore, depth of the mine) are unlikely to change considerably over time (e.g., the bauxite/aluminum industry), exploration and development may be considered as known variables or entered into the model as exogenous inputs.

The incorporation of these factors into a material-supply model involves consideration of important lags: (a) an implementation lag, which is the time lag between a change in price and the reaction by decision makers; (b) a technological or development lag, which is the time required to place new techniques or new mining capacity into full production; and (c) an exploration lag, which is the time between the decision to explore for new deposits and the utilization of the deposits in production.

Modelling inventory behavior is more important in the short run, as well as when inventories are greater or smaller than their average levels. The modelling of inventory behavior, particularly in international markets, is made difficult by both the outright lack of data and the disaggregation of data as categorized by the types of stockholding. Inventories can exist in the form of working stocks, buffer stocks or strategic stockpiles. They can be held by producers, consumers or intermediaries, each of whom may hold them for precautionary purposes or speculation.

Modelling price behavior is difficult because of the problem of ascertaining which of several prices actually represents the market equilibrating process. Although the spot price of a market with the largest volume of trading often serves this function, other prices also need to be taken into account—contract prices, forward prices, futures prices, administered prices, cost-plus prices and government-regulated prices. Prices are indicators not only of market development but also of complex adjustments taking place in the international system, such as inflation, exchange-rate alignments, speculation and military, natural or political crisis.

An important aspect of modelling price behavior is determining the nature of the market structure and the related adjustment mechanisms that move a market toward equilibrium. Price may not be the result of purely competitive adjustments. Consequently, market structures must be investigated to see if they reflect oligopolistic conditions. These appear to be more frequent when only one or a few large suppliers effectively control the market, as in the case of cobalt.

There are a number of steps involved in the modelling process. Although these steps are not well defined, they include:

(a) consideration of the modelling purpose,
(b) development of modelling methodology,
(c) specification of the model,
(d) model estimation,
(e) model validation, and
(f) model application.

Development of the modelling methodology and the specification of the model are often related. For instance, model specification usually involves consideration of such factors as

(a) the primary, intermediate or final levels of demand,
(b) temporal or spatial configuration,
(c) simultaneous or recursive causality,
(d) time span, and
(e) competitiveness of the market structure.

3. Modelling Techniques

The discussion of modelling techniques in this article is organized into three parts: econometric models, engineering models and input–output models. Although there is an overlap between methodologies, one can observe that econometric models relate more to classical materials demand, supply and price adjustments, while input–output models refer principally to the primary, intermediate and final utilization of materials throughout the economy. Engineering models often provide explicit details of processing and the transformation of specific materials inputs into manufactured goods. The chosen set of methodologies and the classification scheme are thus not unique. It simply comprises those methodologies that have been more popular and that can be organized into more-or-less distinct classes.

3.1 Econometric Models

In econometric modelling, the analyst hypothesizes a plausible mathematical relationship between a dependent variable of interest and one or more independent or explanatory variables. Using statistical techniques (e.g., regression) the analyst establishes the statistical validity of the hypotheses. In addition, the regression provides a best estimate of the quantitative relationship between dependent and explanatory variables.

The construction of econometric models requires that specific theories be employed to explain the complex relationships between cause and effect. Once the mathematical form of this relationship is defined, regression analysis provides estimates of the coefficients or parameters associated with each causal variable and yields the values of the errors resulting from the inability to explain completely the dependent variable.

Studying the demand for a given material in a specific market sector is a potential application for econometric modelling and will serve as an example to illustrate the technique.

Table 1
Historical demand, price and per capita income data in a hypothetical materials market

t	D_t	P_{t-1}	I_t
1989	85	80	95
1988	65	90	84
1987	85	73	87
1986	60	83	72
1985	73	75	75
1984	61	82	66
1983	61	77	59
1982	51	83	54
1981	49	94	51
1980	57	89	47

Two factors which typically affect the demand for a material are the real price of the material (real price is market adjusted to discount for inflation) and some measure of consumer wealth (e.g., per capita income, also discounted for inflation). One likely specification of the mathematical relationship between these variables is

$$D_t = A + BP_{t-1} + CI_t$$

where D_t is the demand at time t, P_{t-1} is the price at time $t-1$, I_t is the per capita income at time t and A, B and C are all coefficients to be estimated. Given the historical data in Table 1, a least-squares regression yields the following estimates for parameters A, B and C:

$$A = 78 \qquad (t = 46)$$
$$B = -0.63 \qquad (t = -2.1)$$
$$C = 0.56 \qquad (t = 4.6)$$
$$R^2 = 0.86$$

The R^2 statistic from the regression indicates that 86% of the behavior of demand can be explained by the two causal variables in the given mathematical formulation. The values in parentheses (t statistics) provide insight into the statistical significance of each of the coefficients and help to validate the hypothesized mathematical form of the demand equation.

The price of the material from the previous period was hypothesized to be one of the causal variables rather than the current price. Prices from previous periods, or lagged prices, are often used in econometric demand modelling since it is thought that demand does not usually respond instantaneously to changes in price. Additionally, all of the prices are expressed in constant dollars since the absolute magnitude of the price is not considered to be significant. Rather, it is the buying power of the dollar as it compares to the inflation-adjusted price that is thought to influence demand.

Modelling the demand for a single material in a specific application is of limited utility. Most econometric models cover a wider scope and delve to greater depths. However, a single regression equation illustrates the application of econometric modelling. It represents the least-common denominator from which all econometric models derive.

A full-blown econometric model might address the demand for a material in all market sectors. In doing so, the analyst must consider the effect that demand in one sector will have on demand in another. For instance, because there are limits on the available supply of any commodity, increased consumption in one sector must result in decreased consumption elsewhere. Also, the price of the commodity, considered to be exogenous in the one-equation model, may itself depend on the level of demand. Microeconomic theory states that as the demand goes up, so does the price. However, as price goes up new producers may be enticed into the market. This, in turn, will affect both the price and demand. Finally, the analyst must consider possible substitute materials and assess their impact on demand. Clearly, the demand for one material will decline if other functionally equivalent materials become less expensive. All of these complex interrelationships between variables can be effectively, if not easily, assessed with econometric modelling.

The major limitation of econometric modelling is that the hypothesized relationships are examined and assessed using historical data. If estimated econometric equations are to be useful for understanding the present or forecasting the future, it must be assumed that major changes in the relationships among the variables will not occur.

A second source of error in econometrics is known as specification error. Specification errors occur when the incorrect functional form of the regression equation is specified. In the above example, the equation was assumed to be linear (the effects of each independent variable were independent and additive). However, without a good understanding of the market being modelled, it is not clear that this is the correct relationship between the two causal variables. The analyst must have a clear understanding of the structure of the market to perform meaningful econometric modelling.

Finally, there are limitations on the data available to the analyst; a regression analysis is no better than the data on which it is built. For materials markets, aggregated data for most of the major materials classes are available. However, if a specific grade of material or a specific market segment is to be modelled, data may be unavailable.

3.2 Engineering Models

The study of engineering models should begin with the realization that such models do not possess a single unique specification or form. In the simplest form, an engineering model deals explicitly with engineering parameters and depicts the technical structure of a production process. This provides a linkage between the demand for finished products such as automobiles and appliances, and the derived demand for raw materials or other factor inputs such as energy and labor.

More advanced engineering models incorporate varying amounts of economic information (e.g., factor prices, labor productivities). Also, where there exists a spectrum of technically feasible inputs to the production process, advanced engineering models may incorporate optimization algorithms to select the best of these alternatives. For such models, a concise, mathematical definition of what constitutes "best" is required.

To illustrate the simplest form of engineering model, consider the production of raw steel from pig iron and steel scrap. As there are several technically feasible methods for producing steel, in this case the basic oxygen process (BOP) will be used for detailed examination.

To construct an engineering model of the BOP, the analyst must consider the following factors:

(a) the chemistry of the process,

(b) its thermodynamics,

(c) its kinetics, and

(d) the design and operation of the production facility.

Modelling the chemistry and thermodynamics of the process yields material and energy balances. The kinetics of the process define the rates of production and yield a measure of the size of the facility required to produce a given quantity. The design and operation of the production facility integrates information gleaned from the previous three sections and can incorporate economic information such as capital equipment requirements and costs, labor costs, insurance and maintenance.

A spreadsheet illustrating an engineering/economic model of a single basic oxygen furnace (BOF) appears in Table 2. In this spreadsheet, process inputs are divided into process materials, energy and others (including labor and overheads). The capital equipment costs (for a furnace that produces 200 000 t year^{-1}) are US\$11 million. Within each subdivision (for instance, process materials), engineering information is given in the units column. In this column, the quantity of an input required to produce one ton of product is presented. These values are determined through material and energy balances or, whenever possible, are taken from existing industrial practice. The cost of each input factor appears in the price per unit column. Except for entries in the "Others" category, multiplying the values of the "Units" and "US\$/Unit" columns yields the cost of a given input required for the production of a unit of output. This information is given in the third column.

Within the "Others" category, capital charges are computed by amortizing the initial investment of US\$11 million at the specified interest rate (12%) for the given number of years (10). These yearly payments are distributed evenly over the annual production volume of the furnace (200 000 t year^{-1}). For insurance and maintenance, the values are calculated as a percentage of the initial investment, again distributed over the annual production. Taxes are calculated as a percentage of other process costs (materials and energy).

The table represents one snapshot of the operating conditions that are technically feasible for the BOF. In a working engineering model, all of the complex relationships between process variables must be expressed parametrically. Relationships between vari-

Table 2
Basic oxygen furnace: process engineering and cost (production capacity = 200 000 tons per annum; initial investment = US$11 million)

Process inputs	Units	US$/Unit of material	US$/ton of steel produced
Process materials (basis: 1 ton hot steel)			
hot pig iron (short ton)	0.826	197.01	162.73
steel scrap (short ton)	0.32	90.47	28.95
oxygen (10^3 ft^3)	1.90	2.00	3.81
lime (short ton)	0.084	55.00	4.62
fluorspar (lb)	10.00	0.06	0.63
refractories (lb)	26.00	0.10	2.60
ferromanganese (lb)	13.50	0.20	2.67
silicomanganese (lb)	1.50	0.21	0.32
ferrosilicon (lb)	2.50	0.41	1.03
			207.36
Energy			
electricity (kWh)	30	0.0525	1.58
Others			
direct labor (man-hour)	0.407	21.00	8.55
capital charges	10 years	12.0%	9.73
taxes		1.2%	2.51
insurance		1.0%	0.55
maintenance		4.0%	2.20
			23.54
Cost of production (US$/short ton)			232.48

ables should be specified so that a change in one value induces the appropriate response in each coupled variable. For instance, it is possible to reduce the amount of steel scrap charged into the BOF. However, to maintain the same volume of output, such a change requires a concomitant increase in the amount of pig iron that is processed. Additionally, changing the amount of scrap affects other variables such as the oxygen requirements and the process productivity. In a complete engineering model, all of these complex interrelationships and constraints are explicitly specified.

Once constructed, it is possible to incorporate optimization techniques into engineering models. One of the most common of these involves the technique of linear programming. In optimization, the analyst generally specifies an "objective function," which consists of a mathematical representation of single or multiple attributes of interest. For example, an objective function might be specified in terms of the time required to produce one ton of steel or the cost of doing so. Optimization would involve maximizing output or minimizing cost. Optimization algorithms select the set of operating conditions which maximize or minimize the objective function, subject to the constraints on the input values or relationships among the process parameters (see *Process Analysis*).

3.3 Input–Output Models

Input–output analysis can be used to represent the structure of an economy for a detailed examination of the consumption and production of all goods and services. In most countries, the model is based on a square matrix (table) consisting of entries in value terms for the major intermediate (e.g., steel, chemicals, glass and lumber) and final (e.g., personal consumption, government consumption, investment and exports) consuming and producing sectors.

In each column, the quantities of goods (expressed in dollars) purchased by that sector in the production of its total annual output are entered, with each row entry indicating a different sector from which these inputs were consumed. For instance, assume that one billion dollars worth of coal was consumed in producing six billion dollars worth of steel. Under the column heading "steel production," at the row corresponding to "coal production," the value one billion would be entered. Similarly, the values of all other inputs required to produce six billion dollars of steel would be entered in the steel production column at their appropriate corresponding rows. The total of all these inputs plus the value added (e.g., labor, taxes, profits and other nonmaterial inputs) to these inputs in the conversion to steel must equal six billion dollars.

The values in each column of an input–output table constitute a shopping list of the goods and services required to produce the output of that column. As such, they reflect the production technology of that sector. One of the serious limitations of a static input–output model is that changes in the production technology inevitably lead to changes in the inputs that are required. These technological changes can be handled through use of dynamic input–output models, that is, a model in which the input structure is allowed to vary over time.

The rows of an input–output table describe the way the output of the sector corresponding to that row is distributed among the other intermediate producing sectors (including that sector itself) and the final demand sectors of the economy. In the steel example, the row labelled steel production contains entries of the dollar values of steel purchased by all of the other sectors for the production of their goods. Also, entries are made in the row for the purchase of steel (as steel) by the final demand sectors (e.g., government, exports).

In a properly constructed input–output table (specified in value units), the sum of the entries in each row must equal the sum of the entries in the corresponding column. The entries in the column equal the total purchases (outlays) of that sector. The entries in each row equal the total output of that sector purchased by all intermediate and final users. In any given year, total outlays must equal total output. Initial discrepancies can be attributed to changes in inventories or to profits (or losses) of the sectors.

Input–output analysis is useful for modelling materials markets because it provides a macroscopic overview of the economy, illuminating the relationship between the final demand for goods such as automobiles and the corresponding derived demand for products like steel. With input–output analysis, it is possible to trace the impact of changes in the macroscopic economy (for instance, changes in GNP), on the individual producing sectors (see *Input–Output Analysis*).

4. Model Hybridization and Linkage

Although models may cluster into broad groups that can be easily distinguished, a whole spectrum of modelling techniques exists and there is considerable overlap among the different methodologies. In particular, as the modelling process becomes more sophisticated there is a tendency for analysts to borrow skills from one another, so that modelling techniques begin to merge and become more difficult to classify. The principal ways in which mergers occur are through hybridization and linkage.

Many materials market models have been constructed that cannot be neatly characterized as belonging to a particular type. Often these models are hybrids of two or more techniques. For example, it is not uncommon for the demand side of a materials market model to be estimated econometrically. It might consist of regression equations that attempt to capture the determinants of consumption of a particular metal in each of several user sectors. However, the supply side of the same model might be principally engineering in character and might consist of detailed cost information on a deposit-by-deposit basis. Another example of a hybrid model should be a system dynamics model that makes use of price and income elasticities of demand that were estimated by statistical methods. Hybrid models are so common in practice that it is difficult to find an example of a model that is "pure" in the sense that it borrows nothing from other models.

Whereas hybridization involves constructing models that combine modelling methodologies, linkage involves combining two or more models in a systematic fashion. The constituent models that are linked can be constructed using the same or different techniques. One common practice is to link two or more commodity models when the respective commodities are substitutes in consumption. This can be accomplished by estimating production (or cost) functions for each using sector. Substitution will then occur among commodities as relative commodity prices vary.

Another form of linkage is to incorporate an input–output model into an econometric model. In this case, the principal economic relationships are estimated by statistical inference. The econometric structure is flexible in the sense that substitution among end-use groups is possible on a continuous basis. However, once the quantity demanded of each of the end-use groups is determined (on the basis of income and relative prices), detailed mineral consumption and other requirements can be derived through use of an input–output table. The income-generating implications of the detailed production mix can thus be fed back into the econometric model for either the sum or a subsequent time period.

Most materials market models must be linked to larger macroeconomic models in order to obtain forecasts of exogenous variables. This sort of linkage can be automated (i.e., the models can be run simultaneously on the computer) or it can consist of merely feeding macroeconomic forecasts into the smaller model. When the demand for materials is international in nature, a systematic view of the world economy is clearly essential to drive the model. The more detailed the model is, the more elaborate will be the macroeconomic forecasting infrastructure that it requires. In fact, the nature of the available macroeconomic forecasting infrastructure may be a key factor influencing the specification of the materials model that depends on it.

See also: Economic Theory of Mineral Resources and Markets; Markets for Materials

Bibliography

Burrows J C 1971 *Tungsten: An Industry Analysis*. Heath Lexington Books, Lexington, MA
Busch J V 1987 Technical Cost Modeling of Plastics Fabrication Processes. Ph.D. thesis, Massachusetts Institute of Technology
Busch J V, Field F R III 1989 Technical cost modeling. In: Rosato D V (ed.) 1989 *Blow Molding Handbook*. Hanser, Munich, Germany
Foley P T, Clark J P 1981 An economic/engineering analysis of cost–supply relationships. *Resour. Policy* 7: 171
Gordon R B, Koopmans T C, Nordhaus W D, Skinner B J 1987 *Toward a New Iron Age?: Quantitative Modeling of Resource Exhaustion*. Harvard University Press, Cambridge, MA
Koch G S Jr 1989 Statistical and computer models in mining and exploration. In: Carr D D, Herz N (eds.) 1989 *Concise Encyclopedia of Mineral Resources*. Pergamon, Oxford, pp. 339–43
Labys W C 1973 *Dynamic Commodity Models: Specification, Estimation and Simulation*. Heath Lexington Books, Lexington, MA
Labys W C 1977 Minerals commodity modeling: the state of the art. *Proc. Mineral Economics Symp. Minerals Policies in Transition*. American Institute of Mechanical Engineers, Washington, DC, pp. 80–106
National Research Council Committee on Nonfuel Mineral Demand Relationships 1982 *Mineral Demand Modeling*. National Academy Press, Washington, DC
National Research Council, National Materials Advisory Board, Commission on Engineering and Technical Systems 1982 *Analytical Techniques for Studying Substitution Among Materials*. National Academy Press, Washington, DC
Office of Technology Assessment 1985 *Strategic Materials: Technologies to Reduce U.S. Import Vulnerability*. Report No. OTA-ITE-248. OTA, Washington, DC
Ray W H, Szekely J 1973 *Process Optimization: With Application in Metallurgy and Chemical Engineering*. Wiley, New York
Sparrow F, Soyster A 1980 Process models of minerals industries. *Proc. Council of Economics AIME*. American Institute of Mining, Metallurgical, and Petroleum Engineers, New York, pp. 93–101
US Department of the Interior, Bureau of Mines 1988 *Cost Estimating System Handbook*. US DoI, Washington, DC
Vogely W A (ed.) 1976 *Mineral Materials Modeling*. Johns Hopkins University Press, Baltimore, MD

J. P. Clark
[Massachusetts Institute of Technology, Cambridge, Massachusetts, USA]

J. V. Busch
[IBIS Associates, Wellesley, Massachusetts, USA]

Markets for Materials: Principles and Public Policy Implications of Competition

Economic competition is the vigorous rivalry among firms for sales. Some disagreement prevails among lawyers, politicians, businessmen and economists about desirability of strong competition. However, greater competition is more often considered preferable. The more intense debate is about how strongly competitive actual markets are and the proper public policy posture to adopt.

Mineral materials, particularly metals, have often been considered inadequately competitive industries. The position of International Nickel in its industry was the only practical example that came close to resembling the textbook model of pure monopoly—the firm without competition. Other mineral industries such as those for steel, aluminum and diamonds have often been cited as examples of cooperating oligopolies—industries where a few sellers tacitly or openly work together to limit competition.

These visions were always imprecise, and the industries were recognized to resemble only approximately the extreme cases of market control postulated in economic theory. Increasingly, the vision of market imperfection has become less correct, and it is now recognized that competitive forces are stronger than many industry critics would admit.

1. Definition and Desirability of Competition

Competition is defined in economic theory as the situation in which firms operate as if they have no control over price and thus are termed price takers. Such firms are so small a part of the market that any change in their output has no noticeable effect on prices and thus they respond passively to the prevailing prices. Farmers selling in an organized exchange such as the Chicago Board of Trade most clearly correspond to such price takers, but, as discussed below, so may firms in industries with different procedures for price setting.

To understand the virtue of competition, it is essential also to understand the concept of economic efficiency (i.e., the assurance that every possible socially beneficial transaction takes place). The benefits are improved patterns of production and its utilization. Specialization in one or a few activities and trade allows disposal of less valuable goods for goods that others possess that are more valuable to the purchaser than to the vendor. An efficient specialization (and trading) process makes people more productive and improves their use of the benefits of their productivity through raising consumption levels.

The competitive process facilitates creation of a market order far more complex than any one person or small group could hope to understand. Millions of specialized people, each armed with the particular facts needed for their activities, separately produce goods desired by others. Market prices act as the signalling devices. Sellers and buyers learn where good opportunities lie by observing what prices are charged. The system then works through the market reconciliation of these separate actions. This is what Adam Smith meant by the "invisible hand" (Smith 1979). Others (e.g., Hayek) have described it as the result of human action but not human design.

Competition ensures that all available opportunities are actually taken. Substantial neglected opportunities for profitable trade will cause market forces to bid up the prices of the most desired resources and force down the prices of the least desired ones. Price taking behavior is critical to ensuring this result. By definition, a price taker accepts prevailing prices as given and acts accordingly. Price taking arises because the effects of the actions of a firm are so negligible. The price taker then acts by moving into those areas that, at the given prices, appear profitable and out of areas that look unprofitable. (While no one price taker greatly affects or benefits from such actions, the collective impact of response by many price takers is the lowering of prices in areas entered and raising of prices in areas left.) In contrast, a firm with perceptible control over prices does recognize the impacts and restrains itself in shifting resources into better uses.

This competitive process is by no means a guarantee of economic utopia. It cannot even ensure economic efficiency. There are, for example, critical commodities such as national defense and an efficient level of environmental quality that cannot be efficiently secured in the private market. However, the difficulties that prevent the rise of private markets also hinder public decision making (see Coase 1988).

2. *Fairness*

A further problem is that the benefits are determined by the preexisting patterns of resource ownership. A principal objection to unrestrained competition is that the resulting income distribution will prove in some sense to be undesirable. Questions are perennially raised about whether this outcome should be altered. Here, too, answers are hard to come by since fairness is a subjective, controversial concept.

The vast literature on fairness indicates that a balance must be struck subjectively among desires to assist the poor, needs to maintain economic incentives and individuals' rights to enjoy the fruits of their own labor. More importantly, the impacts of changing market conditions on different people are impossible to determine.

When prices are forced up, buyers complain; falling prices cause producer complaints. Vigorous competition is supported only until shifts in markets cause competitive forces to produce a deterioration in someone's position. The potential losers then seek and too often actually get political intervention to thwart the workings of the market.

Such protection may, in principle, seem fair if it protects the weak against the strong, but frequently the politically and economically powerful are the gainers. Given the uncertainties of effects and the inordinate misuse of market regulations to protect the powerful, regulation of competition should concentrate on increasing efficiency. Separate, direct assistance programs should be given to the poor.

3. *Feasibility of Competition—Economies of Scale and Other Problems*

A basic concern about competition is that an inconsistency can arise between producing a given output at the lowest possible cost and having enough firms so that none has control over prices. Cost minimization may require such large firm sizes relative to the total extent of the market that only a few low-cost firms can exist. What constitutes the lowest cost scale of production in different industries is not adequately measured. Therefore, judging accurately whether any extant firm is above or below the most efficient size is not possible. Evidence is lacking about how many firms can operate, so caution is needed when evaluating existing market structures.

Hidden in the debate are two radically different visions about the nature of economic reality. One side sees the economy beset by numerous difficulties to establishing firms, and that the behavior of firms often has undesirable effects. Trading between firms and customers is difficult, but market-sharing agreements among rivals are more readily established or more deserving of concern. Governments should stimulate competition. The other side sees greater ease of entry and transacting or believes that the adaptability of markets is more critical. The latter view also suggests that rivalry for markets is the natural condition, and that this promotes dealing with customers and discourages cooperating with rivals. It is also feared that government tends to reduce rather than promote competition.

Where only a few firms operate, their behavior is not readily predicted. The firms would like to control outputs to secure the highest possible profits. Even if no legal barriers existed to such arrangements, they are hard to establish and maintain. Market rigging requires the creation and continuation of an industry consensus. Accord must exist about both the preferable industry total and how best to divide that total among firms. Each firm must both accept that consensus and be confident that all other firms also are cooperatively inclined. Historically, hints of breakdown in painfully constructed accords have led to panic reactions and a quick breakdown. The implications of these problems can differ widely from case to case. Economic theorists have produced many elaborate mathematical analyses of critical cases (see Tirole (1988) and Schmalensee and Willig (1989) for advanced discussions with extensive bibliographies and Scherer and Ross (1990) and Carlton and Perloff (1990) for simpler reviews).

In addition, a vast literature has arisen discussing how the threat of creating new firms can further restrain the exercise of monopoly power. Bain (1949), who pioneered the approach, created the impression that such threats often had limited impacts. Later writers such as McGee (1979) and Easterbrook (1981) have argued that the threat often would force purely competitive behavior.

The views are as varied as the views on competition generally. Those who believe competitive forces are vigorous similarly believe that deviations from competitive pricing are small. Those at the other extreme seem to feel that many industries consist of oligopolists who face few challenges.

In between are those who believe that various amounts of potential competition face existing firms. Here, the poles range from one where the threat produces competitive behavior to another in which a few potential rivals must be dealt with. It is only with cases near the latter pole in which entry threat models become interesting.

Existing firms have options that include permanently setting prices low enough to deter entry, letting entry occur and allowing the entrants to share in industry profits or trying to scare the entrants out by using price cuts after entry.

During the 1970s, an extensive debate arose on the relationship of efforts to frighten entrants to the classic accusations that companies often engaged in "predatory" pricing to ruin rivals. Possibly less severe price cuts might be required to scare out an entrant than to chase away established firms. Predatory pricing usually refers to selling at prices that are lower than (marginal) costs in the hope that existing rivals will be permanently driven out of existence.

Areeda and Turner (1975) of Harvard Law School started the debate by proposing simple rules for detecting predation. Numerous economists criticized Areeda and Turner for being too lenient about predation. Successive critics of Areeda and Turner proposed increasingly more complex alternative enforcement rules starting with tests based on output and future price behavior, venturing into proposals to assess the motives for intervention and ending up suggesting "miniature" market analyses. However, McGee (1979), in turn, criticized the others for exaggerating the threat of predation and proposing unrealistic enforcement. McGee pointed out that the most elaborate of these (Joskow and Klevorick 1979) amounted to a complete investigation of the monopoly power of a firm.

The issues are whether predation occurs often and, if it does, whether it is effective. McGee's basic points are that the strategy is not likely to work and it is a more expensive way to destroy competition than making a deal. Predation seems more an excuse for uncompetitive firms to secure relief than a serious problem.

Predation can only work very rarely. At most, the firms can be bankrupted. The physical facilities are likely to remain in existence and, should the predator try to raise prices, the facilities could be reopened at a sufficiently low cost to prevent substantial price increases. Thus, predation is often a dubious charge. Efforts to limit present competition on the grounds that predation prevails are likely to involve unwise sacrifice of low current prices. The expected payoff in prevention of future price rises is unlikely to occur. Thus, costs are incurred for no gain.

However, what actually occurs theoretically with limiting entry and predation can differ radically with prevailing conditions. However, McGee's complaints about the dubious practical relevance of elaborate theories about cheaply thwarting competition have not been adequately refuted. Clear evidence has not emerged that desirable entries often are prevented by existing firms or by widespread use of predation. The concept of discouraging entry seems particularly inapplicable to the mineral industries of the 1990s.

Baumol et al. (1982) created the concept of contestability to provide measures of the relationship between production technology and ease of entry. They show that when capital is mobile and flexible, the industry will act as if it were highly competitive no matter how few firms actually operate. The extension of the impeccable theorizing to the empirical conclusion that such pressures often occur made the analysis controversial.

A further, even more controversial idea of Demsetz (1968) is that if the rights to operate firms with market power were open to competitive bidding, competitive results could be produced. Demsetz's critics concentrate on the problems of monitoring performance. However, bidding could be relied upon if the fear of better offers forced maintenance of good performance. One critical question is whether the market is sufficiently contestable to ensure competitive behavior. Another is whether the biases that produce unsatisfactory regulation will lead to imposing inefficient requirements on the franchise holder.

4. Innovation

Market participants do not just respond to profit opportunities; a vital part of the competitive process is continually stimulating creation of new opportunities. A powerful motive for introducing new ideas is the resulting ability to be more profitable than those who retain old ways. The full competitive process involves the continual disruption of the status quo by the introduction of new ideas and the diffusion of the benefits of these ideas by the ultimate spread of competition among the innovator and its imitators.

A further complication is the wide range of possible responses to the discovery of better business techniques. Ideally, the concepts should spread quickly through the economy and produce a perceptible reduction in market prices. For many reasons, ranging from the difficulties of transmitting knowledge to monopolization of ideas, the process of diffusion may be slower than would be efficient. At best, the adoption of new ideas will take time because of the inherent lags in learning about and adopting new methods.

Therefore, complaints often heard about lagging innovation are difficult to evaluate. There is uncertainty about what the most efficient pace should be and hence lack of confidence about whether the actual pace is really too slow (or too fast). Difficulties arise both in the appraisal of initially developing innovations and in their diffusion throughout the economy.

5. *Problems of Restricting Competition*

Restrictions on competition usually produce inefficiency. This is true, for example, of numerous public policies limiting competitive pressures on firms and of private price-fixing accords. As a result, the US Supreme Court decided in the early twentieth century that all price-fixing agreements violated the Sherman Antitrust Act. However, the flat ban clearly extends only to explicit agreements specifically concerned with prices competing firms will charge. Considerable difficulty arises in deciding the impact of other cooperative practices. Some may promote price fixing, others may simply better inform buyers and sellers and promote more vigorous competition.

Since the 1970s, deregulation of industry has been a popular policy concept in the USA. Decontrol is advocated to remove the protection these rules give some of the covered industries (e.g., airlines and trucking). Similarly, for over two centuries, economics texts have stressed that free trade among nations is a powerful force for strengthening competition and have thus advocated reductions in trade barriers.

6. *Concepts for Measuring Competition*

The problems of estimating the vigor of competition are formidable. The best estimations emerge directly from the definition of competition. Price taking technically means an infinite change in demands. Selling for less would lead to all customers trying to buy from the price cutter; trying to sell at higher price would cause a total loss of sales. The responsiveness of sales to price is conventionally measured by elasticity E. Elasticity is the ratio of the percentage change in output Q to the percentage change in price P (or its absolute value). Thus

$$E = \frac{dQ/Q}{dP/P} = \frac{P\,dQ}{Q\,dP}$$

Demand is elastic ($|E| > 1$) if $|dQ/Q| > dP/P$ (i.e., if the absolute value of the percentage decline in quantity produced by price increase is greater than the percentage price rise). The infinite decrease in Q produced by a price increase by a price taker is the highest possible elasticity. Less elasticity means less sensitivity and more control over price. An elasticity so high that output restriction is unprofitable is the most reliable measure of strong competition.

An alternative but closely related measure is the Lerner monopoly index $(P - \mathrm{MC})/P$, where MC is the marginal cost. It is useful because a zero level—possible only if $P = \mathrm{MC}$—is identified in economic theory with efficiency. Discussions of the index show that for profit-maximizing firms, it is the reciprocal of the elasticity of demand.

The index has drawbacks for policy application. It only shows the percentage loss at the prevailing output level. To see whether the loss makes it worth investing in correction, the absolute loss at the actual output level and on the output eliminated by monopoly must be known.

7. *Practice of Appraisal*

The available data and analytical tools are inadequate to provide good measures either of demand elasticity for the firm or of the Lerner index. As a result, efforts are made to approximate the measurements. A proper approximation would be to consider explicitly the forces that make for higher elasticity and lower Lerner ratios, examine the extent to which they are present and make a presumption about the probable level of monopoly power.

The basic argument is that the more rivals exist and the less the market role of each, the higher the elasticity is likely to be. Two concerns in measuring rivalry are considering the extent to which different commodities compete with each other and then, given the definition of the relevant competing products, determining the geographic scope of competition among suppliers of the market. The first issue involves intermaterial substitutability: the extent to which, for example, different metals compete with each other or metals compete with plastics or glass. Necessarily, any one rival material cannot fully replace another. The total effect of rivalry from many other materials may more than offset the problem of the existence of only a limited number of producers of any one commodity.

In any case, the more easily that rival materials can be transported, the broader the geographic extent of the markets. The possibilities range from quintessentially local markets (such as for personal services) to the worldwide markets that are typical for materials. The regional extent of markets is more obvious since the ability to compete will be reflected in trade movements. These, however, necessarily will be one sided. Goods move from low-cost areas to high-cost areas so linking will normally take the form of one area supplying another. Elzinger and Hogarty (1973) proposed using the existence of trade flows as a measure of linkage. A popular though unsatisfactory alternative method is to examine mechanically the structure of an industry (see *Structure of Materials Industries*).

8. *Fears Over Competition in Materials*

Metals industries have often been viewed as the most obvious example of imperfect competition. Given its size and organization, the steel industry was a favorite example. The creation of US Steel to control at the time about two-thirds of the industry and the apparent practice of US Steel and other producers to adopt restrained competitive behavior created considerable suspicions of imperfect competition. The many critics of the industry argued that it tacitly agreed to limit competition in order to enjoy collective monopoly power. At least, US Steel and a few of the largest companies were willing to restrict outputs and raise prices. A fringe of competitors took advantage of this restraint to expand output and so eroded US Steel's market share.

The 1920 US Supreme Court decision that US Steel was not guilty of violating the Sherman Antitrust Act prohibition of monopolization only intensified the debate. Critics of the Supreme Court have alleged that it either misunderstood the facts or took too narrow a view of what was illegal. The court seemed unconcerned about possible informal cooperation (through actions taken independently but with recognition of what others would do). Such an attitude could have arisen from the Justices' unawareness that independent actions of firms with control over price can have the same result as formal market rigging accords. Alternatively, the Court may have felt that the legal restrictions did not extend to the (obviously difficult to measure and prove) informal approach to output restriction. Of course, another possibility is that US Steel and the other firms actually had little market power.

Later court decisions ruled that the Aluminum Company of America (Alcoa) had excessive market power, that US Steel could acquire a small steel fabricator in the West, and that acquisition of Youngstown by Bethlehem Steel would lessen competition.

The Alcoa case extended from 1937 to 1950. In the meantime, World War II led to the building of government-owned plants that were sold after the war to new companies that provided competition to Alcoa. Therefore, the implementation of the finding that Alcoa was a monopoly was limited to severing the tie between Alcoa and Canada's Aluminium Limited.

The Alcoa decision went to great lengths to differentiate from the US Steel case by asserting that Alcoa had a larger market share and was more aggressive in keeping out competitors. Critics have noted that the market share definition used ignored substitution from copper and, for that matter, the availability of scrap aluminum.

9. Increasing International Competition

Whatever the courts have held, economic and technical realities have greatly increased competition in materials markets. The best known aspect of the change is the development of numerous new technologies to increase the versatility of materials. Another important development has been improvement in transportation that has facilitated international trade. An associated development has been expansion of ore extraction in countries that are not major processors or consumers and which largely export ore. Further boosts, on balance, have been provided by lessening of trade barriers, although reversion to protectionism still occurs. Still another important influence has been the efforts of companies previously in materials production and of outsiders to diversify. This has created new participants in production of many materials.

Steel remains the most widely used metal in the world, and the number of leading companies has shrunk through various mergers, particularly in Europe, and shutdowns, particularly in the USA. Nevertheless, only a few diehard believers in the severe imperfections of competition would claim that the industry is inadequately competitive. The industry itself—especially in the USA, Western Europe and now Japan—complains that too much competition exists. Steel output growth has been hindered by numerous forces, such as the sluggishness of demand in the heavy goods markets on which the industry depends, the development of technologies to reduce the amount of steel needed in specific applications and the rise of substitute materials. Additionally, a world steel market has developed in which imports are a major influence on the USA and Western Europe.

A major element in the rise of international competition is the expansion of output and exports by Japanese producers. In this case, and others involving aggressive new competition from foreign countries, charges are frequently made (but never substantiated) that the Japanese and other new competitors somehow have an unfair advantage.

Questions arise about whether US public policy overly inhibits the competitive strength of the US industry. Some suggest that the habit of US presidents of pressuring the steel industry to hold down prices has caused undesirable losses, but others suggest these efforts were ineffective anyway. The industry complains that it is being hindered by health, safety and environmental regulations. The regulations, in principle, are serving desirable social functions, so the complaints of the steel industry are valid only to the extent that the regulations impose costs greater than the resulting health, safety and environmental benefits.

Important changes in steel industry economics have influenced the rise of Japan as a steel producer. In particular, the increased ease of ocean transportation and the resulting stimulus to developing raw materials for export has benefited Japan at the expense of the USA and Western Europe. A major aspect of the development has been the rise of substantial iron ore and metallurgical coal production in Australia. Australia largely exports both minerals. The significant distances between the coal and iron ore deposits are such that shipping to Japan is not notably more costly than shipping one near the other. Also there are advantages in locating steel mills close to consumers. This is a typical example of how mineral occurrence interacts with transportation economics and affects the location of industrial activity.

Other countries such as South Korea, Taiwan and Brazil have followed Japan in developing steel industries with capacity intended to serve export markets.

Numerous changes in the technologies in steel-using industries (e.g., construction, automobiles, home appliances and packaging) have affected competition. In construction, steel-reinforced concrete has become an important alternative to using all-steel beam construction. In this and other sectors, considerable effort has been devoted to introducing other metals and plastics in various applications. Similarly, the competition between usage of cans, bottles and cartons has intensified. Aluminum has become an alternative in some applications to tin-plated steel; plastics have been widely used by themselves or in conjunction with other materials as plastic pouches for frozen foods.

Similarly, international substitution, reduced requirements and increased international trade have affected many other materials. For example, it is even clearer than at the time of the Alcoa decision that aluminum competes heavily with many materials. The interaction with the steel industry and its rivals has already been noted. A critical and long extant rivalry is with the copper industry. A major element in the progress of the aluminum industry has been the ability to compete with the copper industry for uses such as electrical wire and cable.

10. International Competition and Public Policy in Materials

The steel example is a classic case of how technological progress by easing substitution among resources and facilitating international trade can greatly increase competition in an industry. It is also a graphic illustration of the uneasy status which vigorous competition has among policy makers. Domestic steel producers in the USA and Western Europe and the powerful labor unions of their workers have reacted predictably to the rise of competition from Japan and elsewhere. Those crowded by competition claim that the rivals are competing on an artificial basis. This has many definitions but two favorites are that foreign governments are subsidizing the industry or that predatory pricing as defined and criticized above is being undertaken.

The foreign subsidy argument is suspect because the charges are often unfounded. In any case, the subsidies are a gift from the foreign powers and the importing countries are worse off if they intervene to prevent acceptance of the favor. Predation is a rarely substantiated charge.

Despite these considerations, governments can and often have intervened to limit international trade. In 1983, the USA had complex informal limited protection for steel; in previous years, lead and zinc, petroleum and uranium industries have been aided. The aid is usually justified by concerns of predation, foreign subsidies and the risk involved in depending on foreign concerns. This last argument is applied whether the supplier is a stable ally or an unstable hostile country.

Some observers contend that society should cushion the adaptation of industry to changing competitive circumstances. Difficult questions arise about the subjective matters of what changes should be compensated for, what form the compensation should take and the amounts of aid that should be provided.

A risk-free society is an impossibility. Moreover, risk taking is often worthwhile because of the benefits. While it is recognized that protection from particularly dire outcomes such as starvation is appropriate, controversy prevails about how far aid should be extended. Considerable doubt arises over aiding failing businesses or even workers who must move from usually highly paid jobs to lower paid ones. Further concern relates to whether losses to imports deserve treatment different from losses to domestic competition.

Still further issues arise about how well aid programs work. Actual programs often give enough aid to prevent the radical changes that are desirable to promote efficiency but still do not fully compensate the losses.

See also: Cartels and Producer Associations; Economic Theory of Mineral Resources and Markets; Markets for Materials; Prices of Materials: Theory; Structure of Metals Industries

Bibliography

Areeda P 1981 *Antitrust Analysis: Problems, Text, Cases*, 3rd edn. Little, Brown & Co, Boston, MA

Areeda P, Turner D F 1975 Predatory pricing and related practices under section 2 of the Sherman Act. *Harv. Law Rev.* 88: 697–733

Areeda P, Turner D F 1976 Scherer on predatory pricing: a reply. *Harv. Law Rev.* 89: 891–900

Areeda P, Turner D F 1978 Williamson on predatory pricing. *Yale Law J.* 87(7): 1337–52

Bain J S 1949 A note on pricing in monopoly and oligopoly. *Amer. Econ. Rev.* 39(1): 448–64

Barnett D F, Crandall R W 1986 *Up from the Ashes.* Brookings Institution, Washington, DC

Barnett D F, Schorsch L 1983 *Steel Upheaval in a Basic Industry.* Ballinger, Cambridge, MA

Baumol W J 1977 Quasi-permanence of price reductions: a policy for prevention of predation. *Yale Law J.* 89(1): 1–26

Baumol W J, Panzar J C, Willig R D 1982 *Contestable Markets and the Theory of Industry Structure.* Harcourt Brace Jovanovich, New York

Bork R H 1978 *The Antitrust Paradox: A Policy at War with Itself.* Basic Books, New York

Carlton D W, Perloff J M 1990 *Modern Industrial Organization.* Foresman/Little, Brown, Glenview, IL

Coase R H 1988 *The Firm, the Market and the Law.* University of Chicago Press, Chicago, IL
Crandall R W 1981 *The US Steel Industry in Recurrent Crisis: Policy Options in a Competitive World.* Brookings Institution, Washington, DC
Demsetz H 1968 Why regulate utilities? *J. Law Econ.* 11: 55–65
Duke R M, Johnson R J, Mueller H, Qualls P D, Raush C T Jr, Tarr D G 1977 *The United States Steel Industry and Its International Rivals: Trends and Factors Determining International Competitiveness.* US Government Printing Office, Washington, DC
Easterbrook F H 1981 Predatory strategies and counter-strategies. *Univ. Chicago Law Rev.* 48: 263–337
Elzinger K, Hogarty T F 1973 The problem of geographic market delineation in antimerger suits. *Antitrust Bull.* 18(1): 45–81
Fisher F M, McGowan J J, Greenwood J 1983 *Folded, Spindled and Mutilated Economic Analysis and US v IBM.* MIT Press, Cambridge, MA
Friedman D D, 1990 *Price Theory: An Intermediate Text.* South-Western, Cincinnati, OH
Gordon R L 1985 The production of mineral commodities. In: Vogely W A (ed.) 1985 *Economics of the Mineral Industries,* 4th edn. American Institute of Mining, Metallurgical and Petroleum Engineers, New York, pp. 99–159
Hayek F A 1948 *Individualism and Economic Order.* University of Chicago Press, Chicago, IL
Hayek F A 1960 *The Constitution of Liberty.* University of Chicago Press, Chicago, IL
Hayek F A 1973–1979 *Law, Legislation and Liberty,* Vols. 1–3. University of Chicago Press, Chicago, IL
Hayek F A 1989 *The Fatal Conceit: The Errors of Socialism.* University of Chicago Press, Chicago, IL
Howell T R, Noellert W A, Kreier J G, Wolff A W 1988 *Steel and the State: Government Intervention and Steel's Structural Crisis.* Westview, Boulder, CO
Joskow P, Klevorick A K 1979 A framework for analyzing predatory pricing policy. *Yale Law J.* 89(2): 213–70
Krouse C G 1990 *Theory of Industrial Competition.* Blackwell, Cambridge, MA
Labys W C 1980 *Market Structure, Bargaining Power, and Resource Price Formation.* Heath, Lexington, MA
Lerner A P 1943 The concept of monopoly and the measurement of monopoly power. *Rev. Econ. Stud.* 18: 157–75
Lima J G H 1991 *Restructuring the US Steel Industry: Semi-Finished Steel-Imports, International Integration, and US Adaption.* Westview, Boulder, CO
Mariger R 1978 Predatory price cutting: the Standard Oil of New Jersey case revisited. *Explor. Econ. Hist.* 15: 3412–67
McGee J S 1958 Predatory price cutting: the Standard Oil (NJ) case. *J. Law Econ.* 1: 137–69
McGee J S 1979 Predatory pricing revisited. *J. Law Econ.* 23: 289–329
McGee J S 1988 *Industrial Organization.* Prentice-Hall, Englewood Cliffs, NJ
Neale A D, Goyder D G 1980 *The Antitrust Laws of the United States of America: A Study of Competition Enforced by Law,* 3rd edn. Cambridge University Press, Cambridge
Posner R A 1976 *Antitrust Law, An Economic Perspective.* University of Chicago Press, Chicago, IL
Posner R A 1986 *Economic Analysis of Law,* 3rd edn. Little, Brown, Boston, MA
Posner R A, Easterbrook F H 1981 *Antitrust: Cases, Economic Notes, and Other Materials.* West, St Paul, MN
Salop S C (ed.) 1981 *Strategy, Predation, and Antitrust Analysis.* US Federal Trade Commission, Washington, DC
Scherer F M 1976a Predatory pricing and the Sherman Act: a comment. *Harv. Law Rev.* 89: 869–90
Scherer F M 1976b Some last words on predatory pricing. *Harv. Law Rev.* 89: 901–3
Scherer F M, Ross D 1990 *Industrial Market Structure and Economic Performance,* 3rd edn. Houghton-Mifflin, Boston, MA
Schmalensee R, Willig R D (eds.) 1989 *Handbook of Industrial Organization.* North-Holland, Amsterdam
Smith A 1979 *An Inquiry into the Nature and Causes of the Wealth of Nations.* Oxford University Press, Oxford and Liberty Classics, Indianapolis, IN
Spulber D F 1989 *Regulation and Markets.* MIT Press, Cambridge, MA
Stigler G J 1968 *The Organization of Industry.* Irwin, Homewood, IL
Stiglitz J E, Mathewson G F (eds.) 1986 *New Developments in the Analysis of Market Structure.* MIT Press, Cambridge, MA
Tirole J 1988 *The Theory of Industrial Organization.* MIT Press, Cambridge, MA
Williamson O E 1967 *The Economics of Discretionary Behavior: Managerial Objectives in the Theory of the Firm.* Markham, Chicago, IL
Williamson O E 1975 *Markets and Hierarchies: Analysis and Antitrust Implications.* Free Press, New York
Williamson O E 1977 Predatory pricing: a strategic and welfare analysis. *Yale Law J.* 87: 284–340
Williamson O E 1979 Williamson on predatory pricing II. *Yale Law J.* 88: 1183–200
Williamson O E 1985 *The Economic Institutions of Capitalism: Firms, Markets, Relational Contracting.* Free Press, New York

R. L. Gordon
[Pennsylvania State University, University Park, Pennsylvania, USA]

Materials and Innovation

The notable innovations in materials, and those mostly discussed in this article, are innovations that enable other technological innovations to be made or those that replace a material in current use. Examples are carbon filaments for incandescent electric lamps and nylon, which replaced silk. The major part of the work of materials scientists and engineers, however, is concerned with producing less dramatic changes. Once a new product has been proved to be technologically feasible, an enormous amount of work is still required to adjust processing procedures and compositions to yield acceptable and reproducible properties. For most materials, optimization continues for years, providing a continuously improving product. Innovations in processing that yield essentially the same product can also have great impact; for example, the process of floating hot glass sheets on molten tin to make them flat revolutionized the flat glass industry.

Early innovations were made by discovery and empirical development, but at the end of the nineteenth century, progress in chemistry facilitated the introduction of a number of new materials, ending an unproductive period of many centuries. Since then, science has played an ever-increasing role in materials innovation. Contributions from other sciences aided the development of "metals science" after the end of World War II. This soon expanded to "materials science," a loose federation of the fields of metallurgy, polymer science and engineering, electronic materials and ceramics.

1. The Course of Materials Innovation

Innovations in materials have not been made at a uniform rate. The seven metals of the ancients (gold, silver, tin, lead, copper, mercury and iron) were known before 1500 BC. Zinc metal was added around 1500 AD, but except for that, no new major engineering materials were introduced until the latter part of the nineteenth century. As is shown in Table 1, among the earliest of these recent materials innovations were high-temperature resisting materials for the production of light: carbon filaments and later slender yttria-doped ionically conducting zirconium oxide rods for electric lamps, as well as ceria-doped thorium oxide for gas mantles. Early in the twentieth century, plentiful electrical power at Niagara Falls permitted electrochemists to develop processes to make synthetic graphite, silicon carbide and synthetic aluminum oxide abrasives. Following this, new materials were introduced at a steady rate, as shown in Table 1.

The demands for armor and weapons during World War II caused a great spurt in the rate of appearance of new materials that continued until about 1960 (see Table 2). The special materials required by the atomic bomb and civilian nuclear energy, substantially all the materials used in modern electronics, especially silicon itself, the great majority of modern engineering plastics, glassfiber–plastic composites and new nonsilicate ceramics such as ferrites, ceramic piezoelectric materials, synthetic diamonds and pore-free polycrystalline alumina were all introduced in these two enormously productive decades. Since then the rate of appearance of new materials has decreased. As shown in Table 3, important ones have appeared, and great improvements have been made in those introduced earlier, but the introduction rate of the earlier period has not been matched.

2. The Rise of Materials Science

Shortly after the start of World War II, the needs of such wartime developments as the atomic bomb, radar and new armor caused a surge in materials development work. The need to solve certain problems quickly led to an interdisciplinary approach and chemists, physicists and crystallographers worked with metallurgists on the development of new materials. The association introduced the newcomers to a variety of other unsolved problems in metallurgy and this in turn led to the innovative idea that a "science of metals" was possible. From that time, and continuing well into the 1960s, there was a major effort to understand the behavior of materials and to "make metallurgy

Table 1
Early innovations in materials

Material	Application	Decade of introduction
Aluminum, Hall/Héroult process	applications requiring light weight	1880–1890
Cellulose nitrate, camphor plasticized (celluloid)	billiard balls, collars, molded shapes	1880–1890
Carbon filament	electric lamps	1880–1890
Thorium oxide + 0.5% CeO_2 (Welsbach mantle)	gas mantles	1890–1900
Zirconium oxide + 15% Y_2O_3 (Nernst glower)	lamp filaments	1900–1910
Phenol–formaldehyde resins	moldable plastics	1900–1910
Synthetic graphite	electrodes	1900–1910
Galena	radio detectors (crystal sets)	1910–1920
Silicon carbide	abrasives	1910–1920
Ductile tungsten	lamp filaments	1910–1920
18–8 stainless steel	rust-resistant steels	1910–1920
Age-hardenable aluminum	light structural alloys	1920–1930
Polystyrene	packaging, capacitors	1920–1930
Rayon	fabric fibers	1920–1930
Tungsten carbide (cobalt cemented)	tool materials	1920–1930
Alnico	hard magnetic alloys	1920–1930
Acrylic resins	glazing, shapes	1930–1940
Nylon	fibers, stockings	1930–1940

Table 2
Materials innovations, World War II to 1960

Material	Application	Decade of introduction
Beryllium oxide	neutron moderators	1940–1950
Ferrites	high frequency cores	1940–1950
Ferroelectrics (e.g., $BaTiO_3$)	capacitors, electromechanical transducers	1940–1950
Glassfiber–plastic composites	fishing rods, structures	1940–1950
Germanium	semiconductors	1940–1950
Polytetrafluoroethylene	nonstick pans	1940–1950
Polycarbonate plastics	tough glazing	1940–1950
Polyethylene plastics	fibers, sheet, bottles	1940–1950
Plutonium	nuclear bombs	1940–1950
Silicones	fluids, elastomers, shapes	1940–1950
Uranium	nuclear bombs, reactors	1940–1950
Alumina, pore-free	envelopes for high-pressure sodium vapor lamps	1950–1960
Crystallizable glasses (glass ceramics)	moderately heat-resistant ceramics	1950–1960
Diamonds (high pressure)	superabrasives	1950–1960
Epoxy	adhesives, electrical insulation	1950–1960
Polyimides	high-temperature films, molded parts	1950–1960
Polyphenylene oxide (Noryl)	impact-resistant molded parts	1950–1960
Silicon (semiconductor grade)	semiconductors	1950–1960
Titanium	airframe structural components	1950–1960

scientific." The high mobility of people in the immediate postwar period permitted the assembly of a number of interdisciplinary groups to work on metals at several locations. Out of this came the field of metals science and, eventually, materials science.

The materials-producing industries lent support for these new activities, but a new development was that a large part of the support came from the materials-using industries interested in future technological developments likely to be limited by available materials, and from government agencies, particularly military agencies and those concerned with nuclear energy. Guidance for the development of required new materials was expected to come from the new science.

A new idea in materials science was to study materials properties and behavior independently of any specific material. This approach avoided the complexity of real materials and investigated elementary phenomena in the simplest possible systems. Single crystals, high-purity materials, model systems such as alkali halides and purely mathematical or geometrical models were used instead of complex alloys satisfactory for industrial applications. Understanding of the phenomena on an atomic basis was sought and usually achieved.

Materials science covers an enormous range of topics. Theoretical understanding of mechanical, electrical, magnetic and other properties in a wide range of materials has been gained. A consequence of this is to allow the definition of the theoretical limits of many of these properties. For example, the maximum possible values of flow and fracture stress and of

Table 3
Materials innovations, 1960 to 1990

Material	Application	Decade of introduction
Uranium dioxide	nuclear fuel	1960–1970
Zirconium (hafnium-free)	nuclear reactor components	1960–1970
Graphite filaments	high strength filaments	1960–1970
Gallium arsenide	semiconductors	1960–1970
Polypropylene	filaments, fibers, bottles	1960–1970
Low-absorption glass	fiberoptic communications	1970–1980
Zinc oxide–bismuth oxide varistor	electrical surge protection	1970–1980
Diamond (chemical vapor deposited)	various	1980–1990
Amorphous metal	transformer cores	1980–1990

magnetic saturation and certain electrical properties have been stated with precision. The mechanism and kinetics of reactions occurring during phase transformations and heat treatment have been established for many systems, their role in determining microstructure demonstrated and improved understanding of the relationship between microstructure and physical properties obtained.

Research aimed at such new knowledge requires financial support. In the USA, the government has been a strong supporter of materials research; this has been valuable for basic research, but biases it toward the special and frequently expensive materials needed for military applications and has reduced the driving force for the development of materials for industry and commerce.

3. Recent Materials Innovations

3.1 Metals

Most recent innovations in metals have resulted from research prompted by the needs of other technological areas. A World War II example was the development of techniques for the reduction and shaping of uranium and plutonium for the atomic bomb. This involved close collaboration among metallurgists, physicists and chemists and set the stage for many further interdisciplinary developments. The nuclear materials effort continued after the war with the development of hafnium-free zirconium as a fuel cladding material and of beryllium as a neutron moderator in nuclear reactors.

The performance of gas turbines is limited by the high-temperature resistance of available materials. Since their introduction immediately after World War II, a major metallurgical concern has been the production of superalloys that will resist ever higher stresses for longer times at higher temperatures. Success has been achieved, but it has required great metallurgical ingenuity, mainly by alloy development work carried out by engine manufacturers or with close collaboration between them and materials producers. A recent approach, called rapid solidification technology (RST), involves rapidly cooling tiny liquid droplets to form a fine grained powder. This is pressed to approximately final shape ("near net shape"), sintered and finally consolidated by hot isostatic pressing (HIP).

The development of techniques for making commercially useful superconducting wire has made magnets with a field of 15 T a reality; these in turn have enabled the development of (nuclear) magnetic resonance imaging (MRI) devices.

Alloys of iron containing boron or phosphorus can be rapidly solidified to produce an amorphous or glassy product that is magnetically soft; transformers made with this material have lower losses than those made with the more conventional crystalline silicon iron. There have been important innovations in hard magnetic materials as well (Livingston 1990). In 1950, most permanent magnets were made from the alloy alnico. At that time, the magnetically hard ferrite (oxide) magnets were introduced and by 1960 they had surpassed alnico in tonnage. In 1970, the very high coercive force Co_5Sm magnets appeared, as did the serendipitously discovered $Fe_{14}Nd_2B$ magnets with energy products of over $400\,kJ\,m^{-3}$. The rare-earth magnets now hold a large share of the magnet market.

3.2 Ceramics

The first of the modern single phase, nonsilicate ceramics were the soft magnetic ferrites, developed in the Netherlands in the late 1940s. They were called ceramics because they were made by sintering powders, although they had little in common with the silicates for which the term had previously been used. Other new ceramic materials followed that have found most important places in modern technology. Examples of these are ferroelectric materials based upon barium titanate, varistor materials made from bismuth-doped zinc oxide and pore-free, translucent aluminum oxide that made possible the exceedingly efficient, high-pressure sodium vapor discharge lamp now used almost exclusively for outdoor lighting. The ability to make uranium dioxide into fuel pellets with highly reproducible dimensions and controlled density has been essential to the development of modern nuclear reactors. Zirconium oxide is used as the electrolyte in the cell that determines the partial pressure of oxygen in the exhaust of fuel-injected engines and provides information for automatic fuel control.

A major processing innovation in ceramics occurred in the production of chip carriers for integrated circuit chips. Silicate-bonded alumina ceramic plates with circuits printed on them in tungsten powder ink are combined into a complex multilayer monolithic piece to which chips can be attached. The material used was old; the innovation was the development of a process for making the desired product with high dimensional accuracy.

3.3 Electronic Materials

For electronic materials, an early barrier to development of transistors was the need for high-purity silicon single crystals. In 1952, Pfann developed zone refining, which led to a plentiful supply of high-purity, high-perfection silicon single crystals. In many ways, silicon technology has been a model example of materials science and engineering. The properties to be controlled are electrical and their relation to material composition and processing are well understood, so that devices can be designed theoretically, and when made, they operate as planned. The goal of processing is to create *p–n* junctions in desired locations by diffusing various elements into the silicon. This is now accomplished with almost incredible skill at dimensions which are at the limit of resolving power of the optical microscope.

Certain semiconductors composed of group III and group V elements (chiefly gallium arsenide and related types) have higher carrier mobilities and therefore an inherent advantage for some high-speed, high-frequency and some high-power applications. They are used in some applications and if the difficulties of processing can be dealt with, they may further replace silicon, but processing difficulties, coupled with the fact that they lack the protective natural oxide of silicon, are sufficiently serious that silicon will retain its prominent position for the foreseeable future.

3.4 Polymer-Based Materials

Metals, advanced ceramics and most of the materials for electronics applications share common factors: they are inorganic materials synthesized or won from their ores by high-temperature processes, and their final fabrication usually involves high-temperature treatment. The polymer-based materials have a different heritage. They are products of the organic chemistry laboratory and, if not synthesized from the atom up, at least they are made from precursors selected to produce the desired final molecular structure in the most efficient and least expensive way. A myriad of different properties have been produced in polymers, and it seems likely that more interesting ones will be produced in the future.

Although there were important polymer developments before World War II, the field has blossomed since then, and by the late 1970s the annual volume of plastics produced exceeded that for metals. A few basic characteristics account for their explosive growth. Petroleum is still plentiful and relatively inexpensive. It can be used to make plastics of widely different properties that can be formed into complex and attractively finished shapes more inexpensively than most other materials.

4. The Role of Materials Science in Innovation

Most commonly, workers in materials science have used simple materials, those that are very pure, single crystals or even model materials, in order to obtain an understanding of materials properties. The real materials of engineering are usually complex and the understanding of materials science can only qualitatively predict properties from an approximate knowledge of composition and structure or qualitatively predict the effect of a treatment on properties. As a result, discovery plays an important role in the appearance of new materials and empiricism plays a similar role in optimizing processing to further improve properties. Once a new specimen is made, its structure determined and its properties measured, materials science can play a critical role in explaining the result obtained and in suggesting an appropriate next procedure to try.

It is sometimes claimed that a materials scientist can synthesize a material with desired properties from the atom up, but that is rarely possible. In very definite systems, particularly those based on silicon single crystals in which properties are modified by incorporating dopants by diffusion or by depositing thin layers on the surface, very exact prediction is possible because the system is simple, well understood and the processing susceptible to exact control. Such simplicity is not generally encountered.

5. Future Innovation in Materials

Three of the currently most sought after improvements in materials are higher temperature resistance (by engine manufacturers), improved strength–weight ratio (by aerospace and automobile manufacturers) and shorter switching times (for computer and telecommunications equipment manufacturers). Together with better corrosion resistance, which everybody wants, these items head the list of needs elicited in a poll of eight American industries (*Materials Science and Engineering for the 1990's* 1989). Research aimed at such materials is active now and will continue, but major improvements will be difficult. Materials have already been devised that meet these goals but in all cases there are deficiencies that prevent their general use. Silicon carbide and silicon nitride ceramics can be used 400 °C to 600 °C above feasible temperatures for superalloys and have been used in some parts of small heat engines (Sheppard 1990). The problem is that they are brittle. The toughest of the high-temperature ceramics absorb only about one-tenth of the energy absorbed by the most brittle superalloys used. Materials science provides confirmation that ceramics cannot be made ductile. Although relatively small pieces can be made with sufficiently small defects to display high strength, the probability of finding a critical size flaw in larger pieces is so great that it probably rules out the use of ceramics for those high-temperature applications where high tensile strength is required.

The other materials face similar kinds of problems. Gallium arsenide and related compounds have high electron mobility and can provide more rapid switching, and are now used in many special applications. The difficulties of manufacture and the lack of a natural protective oxide coating impede their more general use.

Glassfiber–plastic composites have been enormously successful, and carbon fiber–plastic composites have been moderately successful in a number of special applications. Other fiber-reinforced products have received less acceptance. It is easy to make fiber-reinforced products strong in one or two dimensions, but it is most difficult to arrange fibers so that they give strength in three dimensions.

The demand for new materials will continue, and from time to time, truly new materials with higher temperature resistance, better electrical properties or higher strength–weight ratios may be developed, but the enormous innovation rate of the 1940s and 1950s, or even that of the 1960s and 1970s will not be matched. In these, and other well researched areas, improvements will be gradual. If the intrinsic property of the material cannot be improved, it will be made more acceptable for the job at hand by ingenious methods such as the protection of superalloy turbine blades from high-temperature gases by high-melting plasma-sprayed coatings and gas cooling through internal channels.

Materials processing will receive more attention for several reasons. First, new products can be made from relatively old materials by precise, well controlled processing. Two current examples are semiconductor chip carriers and multilayer capacitors which are made by precisely assembling tape-cast plates of conventional silicate-bonded alumina ceramic (Mistler 1990).

Second, with the end of the cold war, military demands for materials with the most extreme properties regardless of cost will lessen, and those properties, particularly cost, that are important for civilian applications will become more important. The major part of the cost of most materials can be attributed to processing and processing improvements will be needed to reduce the cost of some of the esoteric high-cost materials used in military devices, so that they can become available to the general population.

Finally, environmental considerations are becoming ever more important. They will impose increasing restrictions on manufacturing processes and will require great ingenuity by process engineers to meet them.

Corrosion resistance is an area of such great importance that methods of improving it in materials will receive much more attention. Many aspects of corrosion are well understood, but problems remain.

There are still great opportunities to discover new compositions of matter that have as yet undetermined properties, some of which will be technologically important and others that may be primarily of academic interest. *Science*, February 9, 1990 presented a series of articles describing progress in several of these areas (Cava 1990, DiSalvo 1990, Dye 1990, Greene et al. 1990, Stucky and McDougall 1990, Yarborough and Messier 1990).

See also: Productivity and Technological Improvements: Managerial Evaluations; Substitution of Materials: Economics; Substitution of Materials: Technology

Bibliography

Barraclough K C 1986 Metallurgy, history of. In: Bever M B (ed.) 1986 *Encyclopedia of Materials Science and Engineering.* Pergamon, Oxford, pp. 3009–26

Cava R J 1990 Structural chemistry and the local charge picture of copper oxide superconductors. *Science* 247: 656–62

Chrétien J F, Clavel M, Quarante D 1986 Materials and product design. In: Bever M B (ed.) 1986 *Encyclopedia of Materials Science and Engineering.* Pergamon, Oxford, pp. 2758–65

DiSalvo F J 1990 Solid state chemistry: a rediscovered chemical frontier. *Science* 247: 649–55

Dye J L 1990 Electrides, ionic salts with electrons as anions. *Science* 247: 663–8

Greene B I, Orenstein J, Schmitt-Rink S 1990 All-optical nonlinearities in organics. *Science* 247: 679–87

Livingston J D 1990 The history of permanent magnet materials. *J. Metals* 42(2): 30–4

Mark H 1986 Polymer science and engineering: historical development. In: Bever M B (ed.) 1986 *Encyclopedia of Materials Science and Engineering.* Pergamon, Oxford, pp. 3710–15

Materials Science and Engineering for the 1990's 1989 National Academy Press, Washington, DC, p. 39

Mistler R E 1990 Tape casting: the basic method for meeting the needs of the electronics industry. *Am. Ceram. Soc. Bull.* 69: 1022–6

Sheppard L M 1990 Automotive performance accelerates with ceramics. *Am. Ceram. Soc. Bull.* 69: 1012–21

Stucky G D, McDougall J E 1990 Quantum confinement and host/guest chemistry. *Science* 247: 669–78

Yarborough W A, Messier R 1990 Current issues and problems in chemical vapor deposition of diamond. *Science* 247: 688–96

J. E. Burke
[Burnt Hills, New York, USA]

Materials and the Environment

For the purpose of this article the term material encompasses any substance used by man without significant chemical conversion. The notion of a materials cycle is actually based on an analogy with natural biogeochemical cycles involving water, carbon and nitrogen. However, the latter are true cycles, whereas the so-called materials cycle is incomplete except in the very crude sense that materials are extracted from the environment, processed, used and finally disposed of as waste. It is important to bear in mind that materials are normally returned to the environment in a form very different from that in which they were extracted. A graphic characterization of the situation is given by Fig. 1.

The materials cycle, in general, can be visualized in terms of a system of compartments, or "stocks," linked by "flows." A "closed" cycle exists if the flows into each compartment are balanced by flows out of each compartment, so that the stocks in each compartment remain relatively constant. If one or more stocks are growing and others are decreasing the cycle is "open."

Examples of closed cycles in nature are the hydrological cycle, the carbon–oxygen cycle and the nitrogen, phosphorus and sulfur cycles. The hydrological cycle is purely physical: water is transformed

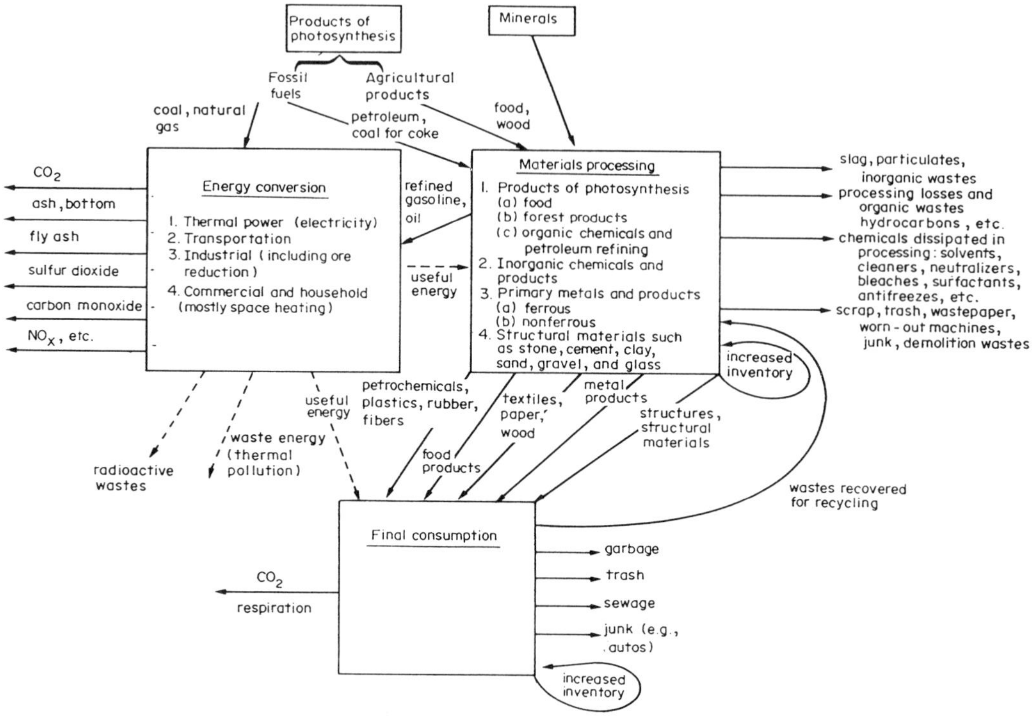

Figure 1
Materials flow in the economy (after Ayres and Kneese 1969)

from liquid to solid or vapor, and transported between geographical locations by ocean currents, rivers or air currents. The other cycles involve biochemical reactions, of which the most important are the carbon fixation process in green plants (photosynthesis), the complementary aerobic respiration process, the anaerobic fermentation and methanation processes and the nitrogen fixation process carried out by certain soil organisms. In each case there is a relatively constant stock of the element (in a characteristic chemical compound) in each environmental compartment, with balanced flows (transformations) into and out of the compartments.

By contrast, the materials cycle in human activity is open; high quality (low-entropy) raw materials—metals and fossil fuels—are extracted from the earth, processed, and discarded in degraded (high-entropy) form. However, these stockpiles are finite and will eventually be exhausted. Moreover, the environment's capacity to assimilate degraded wastes is also limited. The phrase "closing the materials cycle" conveys the idea that an unbalanced system is inherently unsustainable in the long run. To achieve balance, the extent of materials degradation through processing and use must be minimized, production and consumption wastes must be recovered (in as useful a form as possible), and extractive processes must be replaced by reconditioning, remanufacturing and materials recovery/recycling processes.

The idea of recycling low-grade wastes is intellectually appealing but attractive opportunities for recycling are currently limited. In practice each waste flow in the economy must be regarded as a potential source of materials on an equal footing with virgin sources. If a waste stream is easily recovered and can be processed at moderate cost compared with natural sources it will generally be exploited. Not surprisingly, the more economically valuable a material is, the more likely it is to be recycled. Thus, it pays to go to considerable lengths to recover platinum from industrial catalysts or silver from film processing wastes but, to date, it is cheaper to extract aluminum from bauxite ore than from fly ash. Similarly, it is cheaper to extract sulfur from natural deposits via the Frasch process than to recover it from combustion products, even when the flue-gas desulfurization is subsidized.

It is clear that materials recycling will gradually become more widespread as high-quality virgin sources are exhausted or as increasingly strict environmental regulations force processing of combustion products or waterborne waste streams to remove hazardous substances. Nevertheless it is most likely that significant quantities of materials in forms that are not economically recoverable (such as fuels, pigments, surface coatings, flocculants, water softening agents, detergents, antifreeze, fuel or plastic additives, lubricants and solvents) will be discarded for many decades to come unless environmental controls are tightened considerably. Fuels are, of course, sources of energy.

The other dissipative uses are all examples of what may be termed labor-saving, capital-saving or energy-saving—but nonstructural—materials. For example, lubricants cut down on friction enabling a machine to function with less energy input and to last longer, detergents save scrubbing, and so on (Ayres 1978). In nonstructural uses, the materials are physically or chemically transformed by use, usually into low-grade dispersed forms. Since it is unlikely that an advanced economy will dispense with these categories of uses of materials it is not feasible in principle to recover a high percentage of many materials.

The discharge of wastes to the environment creates a variety of possibilities for adverse effects on man, both direct and indirect. Major examples include:

(a) disturbance to land (from mining, materials processing, construction, or waste disposal);

(b) contamination of soil, ground or surface water used by humans, livestock or for irrigation (by acids, sludges or toxic industrial wastes), with secondary effects on fish and birds;

(c) contamination of air by toxic or irritating combustion products (e.g., total solid particulates, oxides of sulfur and nitrogen) with direct effects on health;

(d) disturbance of fresh water ecosystems by eutrophication or acid-rain deposition;

(e) disturbance of ocean ecosystems due, for example, to oil spills, ocean dumping and ocean mining; and

(f) climatic disturbance due to buildup of CO_2, methane, CFCs, N_2O and/or other chemical pollutants in the atmosphere.

Further discussion can be most usefully carried out by separately considering three major categories of materials: nonfuel minerals, fuels, and plastics (including synthetics).

1. Minerals

It is convenient to divide the minerals category into metals and nonmetallic minerals. Metals are worth considering separately because they are mainly used in structural applications (i.e., embodied in products), and consequently recovery and reuse is often feasible. The major tonnage metals are: iron (and steel), aluminum, copper, zinc, lead and nickel. World production of metals and ores for the years 1988 is summarized in Table 1.

Two important points to emphasize at the outset are that the first two metals listed (iron and aluminum) are obtained almost exclusively from oxide ores and that their compounds are comparatively benign. Environmental problems associated with their extraction and processing are primarily due to the large scale of operations and the need to separate and dispose of significant quantities of economically useless or mildly hazardous materials. The quantities of solid mining and milling wastes produced by the mining of metal ores are typically much greater than the quantities of contained metals (Table 2). This is especially true for copper, gold, uranium and platinum group metals, which involve very large tonnages of raw materials relative to the tonnage of final product.

Environmental controls are primarily needed, in the case of iron/steel facilities, to minimize the discharge of

Table 1
World production of metal ores (1988) (10^3 t)

	Gross weight	Metal content (%)	Net weight	Mine and mill waste
Aluminum	97660	37.30	36400	61220
Chromium	14000	32.50	4207	10000
Copper	>1750000[a]	0.50	8450	>1740000
Gold (1979)	170000	0.0007	1.19	170000
Iron	953000	59.00	565400	390000
Lead	101000[b]	3.33	3380	98000
Manganese	30500	28.00	8540	22000
Nickel	80000[c]	1.00	835	79000
Platinum group (1979)	30000	0.0007	0.20	30000
Uranium (1978)	1900000	0.002[d]	36	1900000
Zinc	220000[b]	3.22	7115	213000

a Extrapolated from US data for 1978 (0.5% grade ore). For surface mines overburden averages 2.5 t per tonne of ore in addition b Total, including overburden, extrapolated from US data c Based on US Bureau of Mines estimates of the average grade of total known reserves. Sulfide ores currently being mined are richer, however d Barney (1980)

Table 2
World production of nonmetallic minerals (10^3 t)

	Production (net)	Mine and mill waste	Land utilized (km^2)
Crushed stone[a] (1978)	2768000	635	935
Sand and gravel (1979)	7485000		1400
Clays (1978)	520000	420	290
Phosphate rock[b] (1988)	53156	110000	110
Gypsum[c] (1979)	64920		
Salt (1979)	165140		

a Including limestone b P_2O_5 content c Mine production; excludes by-product gypsum

toxic gases from coke ovens and furnaces, and waterborne acidic wastes or sludges from pickling sheet or strip. In the case of aluminum smelting, the most serious environmental problem is to prevent the escape of fluorine from the electrolytic cells. (Fluorine originates in the breakdown of the molten cryolite bath, not in the aluminum ore.) A minor problem is the disposal of so-called red mud from bauxite processing, although recovery of iron and possibly silicon from bauxite wastes may ultimately be feasible.

Environmental problems associated with end-use consumption and disposal of iron/steel and aluminum are also primarily due to the wide variety of minor uses. Scrap (e.g., from demolished structures, pipelines, worn-out rails or rolling stock, obsolete machinery and junk cars) is normally recovered and recycled via electric furnaces, but it is much harder to recover metal economically from small items, such as containers, wire products, foil, fasteners (e.g., nails), razor blades, mattresses and springs. These tend to be mixed with other kinds of waste, as in household refuse, or are scattered over the landscape as litter. In this context, aluminum cans and bottle caps constitute by far the most serious problem, since aluminum is highly corrosion resistant, whereas small iron or carbon steel objects eventually rust away.

The other major tonnage metals (copper, zinc, lead, nickel) are geochemically quite different from iron and aluminum. Firstly, being chalcophiles, they are typically found in nature as sulfides rather than oxides, although nickel is mined in both forms. The sulfides are commonly first converted to oxides by the simple process of roasting: the sulfur is oxidized and driven off as SO_2. Copper, zinc, lead and nickel smelters are major sources of air pollution despite recent efforts to improve environmental controls in some countries. For example, the large copper–nickel smelters in the Sudbury district of Ontario emit 2.7 Mt of sulfur oxides annually causing severe damage to forest over nearly 2000 km^2 and lesser damage over a much larger area.

A second significant characteristic of these metals is that copper, zinc, lead and nickel are typically produced with coproducts or by-products. Major coproduct groups are shown in Table 3. Thus, a copper, zinc, lead or nickel smelting operation will normally yield tailings, slags, ashes, flue dust or sludges that are rich in by-products (such as antimony, arsenic, bismuth, cadmium, selenium or silver) which may or may not be desirable. Several of these by-products are not worth enough to justify highly efficient recovery methods, so significant quantities of these metals are discarded or lost in refining operations (see *By-Products and Coproducts*).

A third point is that many compounds of the most important metals other than iron and aluminum, including copper, zinc, lead, nickel and other transition metals (chromium, manganese, cobalt, vanadium) as well as mercury, are quite toxic to animals and plants. The extreme toxicity of lead, arsenic, cadmium and mercury is well known, but some salts of copper, zinc, chromium, bismuth and thallium are also toxic enough

Table 3
By-product and coproduct groups

By-product groups

Copper	Zinc	Lead	
antimony	antimony	antimony	
arsenic	cadmium	bismuth	Platinum
cobalt	gallium	selenium	iridium
gold	germanium	silver	osmium
rhenium	indium	tellurium	palladium
selenium	silver	Nickel	rhodium
silver	thallium	cobalt	ruthenium
tellurium	Zirconium	manganese	Lithium
Niobium	hafnium		rubidium
tantalum			

Coproduct groups

Brine	Rare earths
sodium	yttrium
potassium	europium
boron	erbium
magnesium	terbium
chlorine	dysprosium
bromine	ytterbium

to have found agricultural uses as insecticides, fungicides and rodenticides.

Clearly, nonferrous metal extraction, processing, and use all create possibilities of significant environmental damage. The cases of zinc and its by-product cadmium are typical. Figure 2 shows schematically the magnitudes of various flows in zinc-refining operations. Significant fractions of zinc and cadmium are lost in processing and much of this may be accessible to biological organisms. Major uses of zinc are shown in Fig. 3.

Environmental hazards are also associated with end uses of some metals, especially where the metal is toxic and the use is nonstructural. In the case of cadmium, for instance, almost all of the present uses are inherently dissipative. These include electroplating, pigments, and stabilizers for PVC. A flow chart for cadmium is shown in Fig. 4.

Taking into account extraction, processing and end-use, human activities have sharply increased the natural rates of mobilization of many metals, including some of the more toxic ones. Comparisons of worldwide natural and man-induced rates of discharge into the oceans are given in Table 4. Estimated emissions to the atmosphere are shown in Table 5.

Nonmetallic minerals such as stone, gravel, clay and sand (Table 2) are mined in larger quantities than metal ores, although processing losses are smaller and these materials are somewhat more benign. They are used mostly in construction.

Environmental impacts arise from mining itself, which is mostly on the surface, and significant land disturbances, dust and erosion which often accompany such activity. Processing of phosphate rock (to produce fertilizer) also results in significant further environmental impacts of several kinds, including toxic fluoride emissions to the atmosphere and accumulation of mildly radioactive tailings. It is noteworthy that erosion rates for disturbed land, such as active strip mines, is up to ten times greater than for cropland and perhaps 200 times greater than for natural grassland.

2. Fuels

Fossil fuels (petroleum, coal, natural gas) are used in tonnages comparable to stone, clay and sand (see Table 6). Land disturbance from source mining of coal is a major environmental effect. Cleaning and washing of coal also creates large piles of waste material which can later cause problems, such as mudslides and acid run-off.

However, the major environmental problems associated with fossil fuels are attributable to the combustion process and its waste products. Most air pollution is attributable to combustion of fossil fuels, the contaminants being: carbon dioxide, carbon monoxide, unburned or partially burned hydrocarbons, total solid particulates, oxides of nitrogen and oxides of sulfur. Recent emissions are given in Table 7.

Because of the vast quantities involved some of these effluents, notably oxides of sulfur and nitrogen and total solid particulates, can cause respiratory and other health problems in polluted areas. Unburned hydrocarbons, especially in the presence of ultraviolet radiation, can produce a mix of highly reactive, irritating and carcinogenic compounds called smog. Moreover, oxides of sulfur and nitrogen combine with

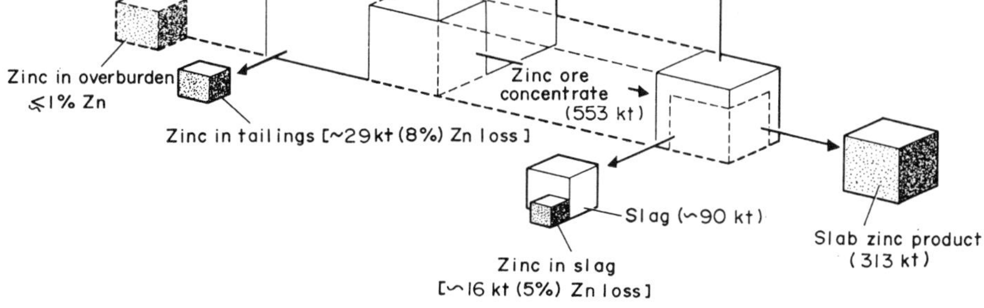

Figure 2
Zinc flow in mining and processing of zinc ores in the USA (1968) (after Fulkerson et al. 1972)

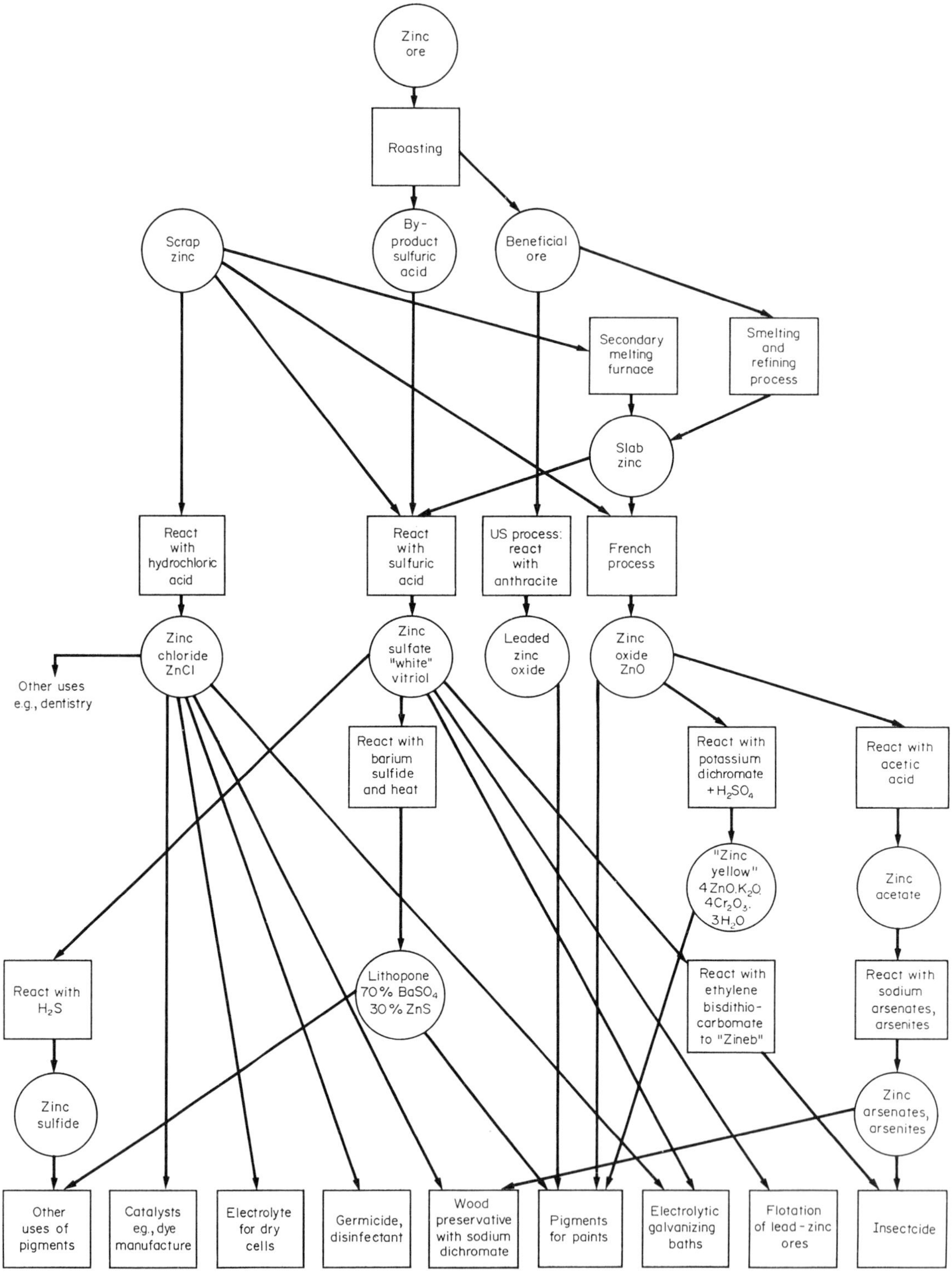

Figure 3
Zinc process product flow (after Ayres et al. 1988)

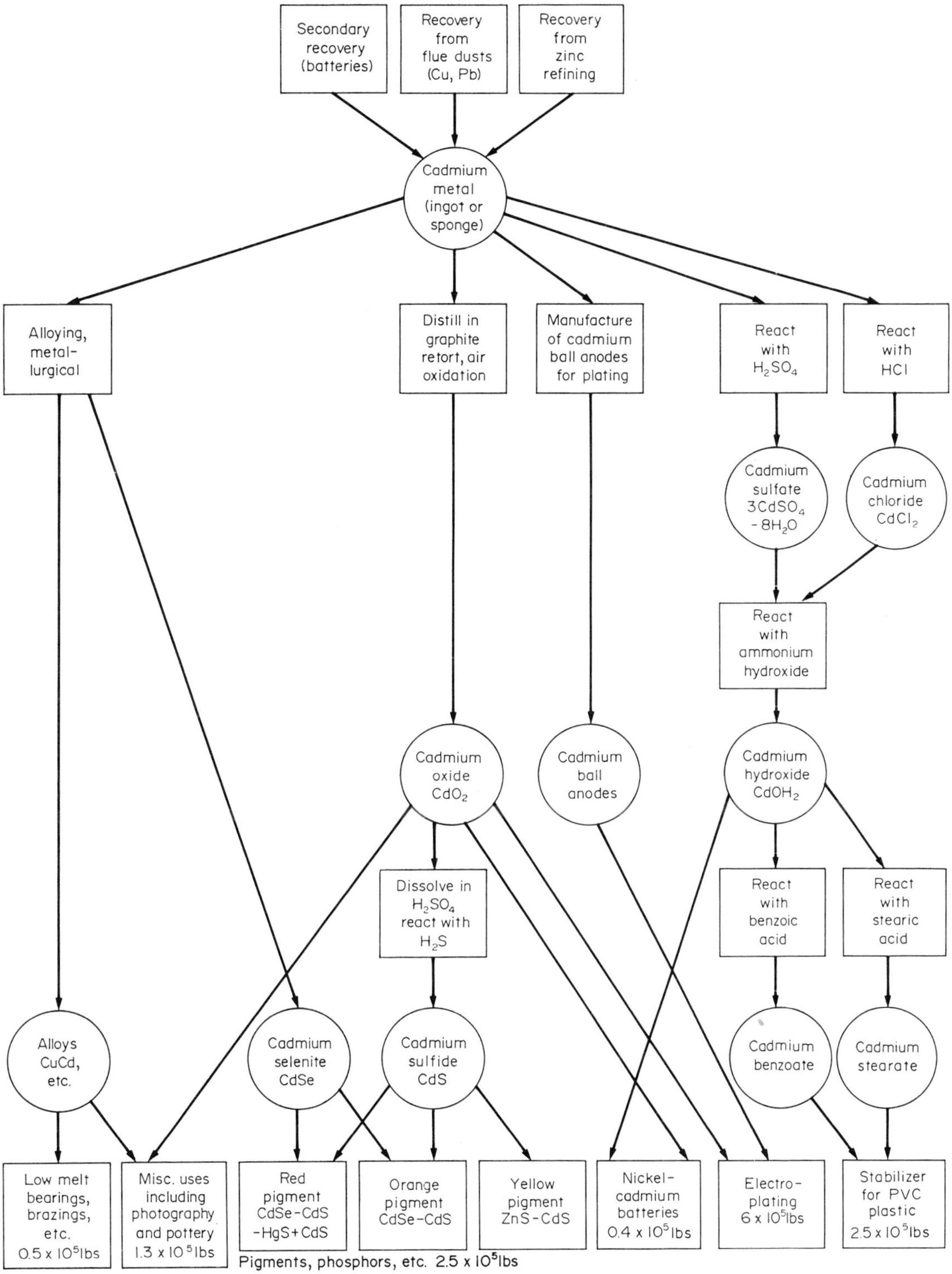

Figure 4
Cadmium process product flow (after Ayres et al. 1988)

Table 4
Estimates of annual discharges of metals into the oceans (kt)

Metal	By geological processes (in rivers)	By man (in mining)
Iron	25000	319000
Manganese	440	1600
Copper	375	4460
Zinc	370	3930
Nickel	300	358
Lead	180	2330
Molybdenum	13	57
Silver	5	7
Mercury	3	7
Tin	1.5	166
Antimony	1.3	40

Source: Waldichuk (1977)

atmospheric moisture and return to earth as acid rain, downwind of major power plants or smelters. This process slowly but irreversibly damages conifer forests and freshwater lakes. Airborne sulfate particles from anthropogenic sources (SO_x) are now estimated to be comparable in mass to natural sources (i.e., 130–200 Mt year^{-1}) (Robinson 1977).

It is important to note that atmospheric emissions and other wastes from the combination of natural gas and petroleum products are relatively minor compared with the wastes that are associated with use of coal. Natural gas has little sulfur to begin with, and it is easily removed. In the case of petroleum, refining automatically removes the sulfur from most distillate products (and concentrates it in residual fuels, tar and coke) and total sulfur removal is approaching feasibility.

Table 6
World fossil fuel consumption (10^6 t)

Energy	1970	1980	1985	1988
Coal, hard	2134.3	2728.0	3162.0	3451.0
Brown coal and lignite	790.8	1004.0	1169.0	1223.0
Crude petroleum	2275.0	2982.4	2654.0	2924.0
Natural gas	580.7	807.6	920.0	1022.0
Total	5780.8	7522.0	7905.0	8620.0

Source: UNSO (1988)

The biggest difference between natural gas/petroleum fuels and coal is their ash. Gas is ash-free and even residual oil has an insignificant ash content, but coal ash content varies from a few percent to 20% or more. In conventional combustion most of this is bottom ash, which becomes a solid waste disposal problem because of the large amounts of coal involved. However, around 10% of the ash goes up the stack as fly ash and, even with efficient electrostatic precipitators, a small percentage is lost to the environment as solid particulates. Apart from the undesirable physiological effects of small particles, fly ash contains both toxic and radioactive constituents, notably zinc, vanadium, nickel, chromium and manganese (Table 8).

Assuming the world's energy needs are met increasingly from coal in the future as natural gas and petroleum become relatively scarcer, the environmental health problems associated with coal combustion—

Table 5
Worldwide atmospheric emissions of trace metals (10^3 t year^{-1})

Element	Energy production	Smelting, refining and mining	Manufacturing processes	Commercial uses, waste incineration and transportation	Total anthropogenic contributions	Total contributions by natural activities
Antimony	1.3	1.5		0.7	3.5	2.6
Arsenic	2.2	12.4	2.0	2.3	19.0	12.0
Cadmium	0.8	5.4	0.6	0.8	7.6	1.4
Chromium	12.7		17.0	0.8	31.0	43.0
Copper	8.0	23.6	2.0	1.6	35.0	6.1
Lead	12.7	49.1	15.7	254.9	332.0	28.0
Manganese	12.1	3.2	14.7	8.3	38.0	12.0
Mercury	2.3	0.1		1.2	3.6	317.0
Nickel	42.0	4.8	4.5	0.4	52.0	2.5
Selenium	3.9	2.3		0.1	6.3	3.0
Thallium	1.1		4.0		5.1	29.0
Tin	3.3	1.1		0.8	5.1	10.0
Vanadium	84.0	0.1	0.7	1.2	86.0	28.0
Zinc	16.8	72.5	33.4	9.2	132.0	45.0

Source: Nriagu (1990)

Table 7
Atmospheric emissions due to fossil fuel use, 1987 (10^6 t)

	US and Canada	Europe	USSR	Asia[a]	Total world
Carbon dioxide (1987)	4877	4400	3927	5206	20900
Methane[b] (1987)	26.5	8.9	6.3	13.8	69
Carbon monoxide (1982–4)	76.0	>36	NA	NA	300–1600
Sulfur dioxide (1982–4)	24.86	>25	28 (1979)	>16	160–180
Oxides of nitrogen (1982–4)	21.37	>14	NA	NA	75
Particulates[c] (1982–4)	7.6	>15	NA	13.7[d]	~300
Lead (1982–4)	0.0584	NA	NA	NA	NA
Lead (1985)	0.0216	NA	NA	NA	~0.45

Source: WRI (1991)
a Including Japan and China b From coal mining and gas leaks c Including sulfate particulates d China only

especially fly-ash emissions—will become more significant.

Finally, the buildup of atmospheric carbon dioxide from massive fossil-fuel combustion appears to be capable of affecting the heat balance of the earth itself via the so-called greenhouse effect. Although scientific uncertainties remain, evidence seems to be accumulating to support the hypothesis that CO_2 buildup will result in a general warming of the earth's surface, particularly in the polar regions. This, in turn, is likely to cause northward shifts in the average locations of the jet streams (which control precipitation patterns). The resulting climatic impact might be favorable for some sparsely populated northern countries (e.g., Canada, Siberia) but could be quite unfavorable for the more heavily populated countries in the temperate zone.

Table 8
Trace elements in coal

Element	Concentration (ppm)	Total amount (t year^{-1})	Percentage of 1968 US consumption of the metal
As	10	4500	21
B	100	45000	57
Ba	60	27000	4
Be	0.3	14000	46
Cd	0.1–0.5	45–230	1–4
Co	5	2300	36
Cu	15	6800	0.5
Cr	30	14000	3
Pb	5	2300	0.3
Mn	30	14000	2
Hg	0.1–0.5	45–230	2–11
Mo	1	450	0.2
Ni	50	23000	16
V	70	32000	610
Zn	100	45000	3

Source: Fulkerson et al. (1972)

In addition, it has been suggested that a few degrees of general warming might trigger a fairly rapid melting of floating ice in the Arctic and Antarctic oceans. Although this appears unlikely, the effect would be to cause the sea level to rise by up to 10 m. A rise of sea level of even a few meters would cause drastic changes in the map of the world. Many island groups would disappear under water (the Bahamas, for instance), and offshore barrier islands would be swept away (including most of Long Island). Highly productive coastal and delta areas such as the Nile (Egypt), Pearl, Yellow, Yangtze (China), Mekong (Vietnam), Irrawaddy (Burma), Brahmaputra (Bangladesh), Indus (Pakistan) and Mississippi (USA) could be inundated. Many of the world's largest cities, including New York, Los Angeles, Tokyo, Osaka, London, Hamburg, Amsterdam, Rotterdam, Alexandria, Bombay, Calcutta, Djakarta, Shanghai, Canton, Hong Kong and Singapore are close to sea level. Thus, it is not exaggerating to note that if significant ice melting should occur, the effects could be catastrophic.

3. *Packaging Materials and Synthetics*

The convention adopted in this Concise Encyclopedia is to consider materials as substances that are used without significant chemical conversion. Packaging materials, synthetic fibers, plastics and rubber fall into this category. In the interest of completeness, therefore, it is worthwhile to point out that synthetic polymers—mostly plastics and rubber—are made largely from derivatives of petroleum. Petrochemicals already account for perhaps 7% of world petroleum consumption, and this percentage can be expected to increase. Thus, use of synthetics is inseparable from extraction and refining of petroleum, though in the future, chemicals may also be manufactured from coal.

The three most important industrial polymers are polyethylene, polystyrene and polyvinyl chloride (PVC). There are, of course, a host of other resins including polyesters, phenolics and acrylics. Neglecting environmental problems associated with the paper and chemical industries there remains a significant problem of disposal. The physical volumes involved are equal to or greater than the volume of metals produced each year. Most plastics are not biodegradable: they must be disposed of in landfills or by incineration. Incineration of chlorinated plastics in particular is environmentally hazardous because of the emission of toxic or carcinogenic compounds. Tires are extremely difficult to incinerate. Nondegradable litter, especially polystyrene foams and polyethylene films, is an increasing problem in many public places, such as beaches.

Paper remains an important material in the economy. It is, of course, made from woodpulp by a process in which one component of the wood (the lignin) is chemically dissolved and removed. In older pulp and paper mills, this waste material was simply discarded into the nearest river. There it created biological oxygen demand (BOD), removing the dissolved oxygen and killing organisms (such as fish) that must have oxygen to survive. In recent years pulp and paper mills in most countries have substantially reduced their waterborne waste output, under pressure of environmental regulations (see *Paper and Paperboard: Environmental Considerations*).

4. Policy Issues

It is generally acknowledged that the environment is a resource. The fact that it has been available in the past for use as a "sink" for the disposal of industrial and consumption wastes at zero (or very low) cost is a consequence of its indivisibility and the fact that it cannot be contained and privately owned or exchanged in a market place. It is, in short, a public good, outside the market pricing mechanism. It is an axiom of economics that (in the absence of cultural restraints) the use of unpriced public goods will be excessive unless regulated by public policy. This is the nub of the problem.

Environmental services to man can be said to include the biosphere (photosynthesis, the water, carbon and nitrogen cycles, biodegradation—decay—of organic materials), the maintenance of oxygen in the atmosphere (the atmosphere of the earth prior to the evolutionary development of the blue–green algae did not contain free oxygen) and the ozone layer in the stratosphere, and a host of local "public service" functions. There is no question that these environmental functions (and others) are interfered with, for example, by large-scale mobilization of toxic materials, erosion, silting, buildup of CO_2 and particulates in the atmosphere. Unfortunately, much remains to be learned about the specific mechanisms involved and their quantitative impact. Research is therefore the first prerequisite of intelligent environmental policy.

In the absence of the scientific knowledge that would permit us to discriminate between minor and major problems (e.g., between reversible and irreversible disturbances), the safest strategy is to limit the mobilization of materials by human activities to levels significantly below the levels found in natural processes. By this criterion, the processing and disposal of heavy toxic metals into surface waters is already excessive. Similarly, combustion and other processes are injecting certain materials into the atmosphere at rates comparable with or greater than natural processes.

Policy instruments for abating these undesirable flows include emissions regulations applicable to particular sources (e.g., mines or smelters) or on the use of particular substances (e.g., phosphates in detergents, chlorofluorocarbon propellants). More broadly, however, policies tending to promote conservation of scarce resources will also tend to reduce environmental problems. It is important to note that many relevant existing policy instruments have had the opposite effect of encouraging excessive use of materials, especially virgin materials. In the USA, the mineral depletion allowance (recently reduced, but still in effect) and freight rates favoring transportation of mine products as compared with recycling scrap, are two examples. Similarly regulations, or indirect subsidies, tending to reduce consumers' energy costs (e.g., for gasoline or home heating) also tend to encourage excessive use of energy and its associated environmental problems.

Severance taxes on mineral extraction, on the other hand, would tend to discourage use of virgin resources and encourage conservation and recycling. Similarly, effluent taxes or fees would tend to discourage the discharges of recoverable wastes into surface waters or the atmosphere. This approach is generally preferred by economists over the more common regulatory approach ("standard-setting") because it allows greater discretion to the waste generator. It is frequently opposed by environmentalists on the grounds that imposition of effluent charges would not guarantee meeting desired environmental standards. If a given fee does not yield the desired result, of course the answer would be for the public authority to raise the effluent charge until it is effective. In the long run, the price of virgin materials will rise as the high-quality reserves are exhausted. This will also induce conservationist activity, though not necessarily optimum recovery of toxic by-products. A combination of approaches will certainly be needed.

An alternative policy approach that has recently received increasing attention is the use of tradeable emissions permits. The initial allocation of permits would have to be based on some political process, taking into consideration both relative contribution to the pollution problem and relative ability to ameliorate it. After the initial allocation, those able to reduce their emissions would be able to sell excess permits to those unable to do so. In principle, the tradeable permit idea would be applicable to almost any type of emission, and could even be implemented internationally. In the

providing a straightforward and equitable mechanism for transferring resources from the industrially developed countries (which generally contribute more than their share of emissions) to the less-developed countries, which are chronically short of hard currency for development.

See also: Paper and Paperboard: Environmental Considerations; Recycling of Demolition Wastes; Recycling of Metals; Residuals Management

Bibliography

Ayres R U 1978 *Resources, Environment and Economics: Applications of the Materials/Energy Balance Principle.* Wiley, New York

Ayres R U, Ayres L W, McCurley J, Small M, Tarr J, Widgery R C 1988 *An Historical Reconstruction of Major Pollutant Levels in the Hudson–Raritan Basin 1880–1980*, NOAA Research Memorandum, National Oceanic and Atmospheric Administration, Rockville, MD

Ayres R U, Kneese A V 1969 Production, consumption and externalities. *Am. Econ. Rev.* 59(3): 282–97

Barney G O 1980 *The Global 2000 Report to the President: Entering the Twenty-First Century.* Pergamon, New York

Fulkerson W, Goeller H E, Phelps D H 1972 Environmental sources, sinks and chemical forms of heavy metals. *American Association for the Advancement of Science, Selected Symposium.* Westview, Boulder, CO

National Academy of Sciences 1973 *Materials and Man's Needs*, Report of the Committee on the Survey of Materials Science and Engineering (COSMAT) 1973. National Academy of Sciences, Washington, DC

National Academy of Sciences 1975 *Mineral Resources and the Environment*, Report of the Committee on Mineral Resources and the Environment (COMRATE) 1975. National Academy of Sciences, Washington, DC

Nriagu J O 1990 Global metal pollution—poisoning the biosphere. *Environment* 32(7): 7–32

Robinson G D 1977 Effluents of energy production: particulates. In: *Energy and Climate*, Report of National Academy of Sciences Geophysics Study Committee. National Academy of Sciences, Washington, DC

United Nations Statistical Office 1988 *Industrial Statistics.* UNSO, New York

Waldichuk M 1977 *Global Marine Pollution: An Overview.* United Nations Educational, Scientific and Cultural Organization, Paris

R. U. Ayres
[INSEAD, Fontainebleau, France]

Materials Handling: Automation, Unit Materials

Basic materials handling equipment has changed only slightly in the past several years. The substantive change in materials handling has been the application of more computer-based controls and systems to this equipment.

Automated materials handling, in a sense, has become an industry within an industry, a relatively small, high-technology island that has generated higher growth rates than the total industry by providing equipment that increases the productivity and efficiency not just of the materials handling function of a facility but of the entire facility.

In an automated materials handling environment, the coordination of the physical storage and movement of materials takes on the added dimension of coordinating these activities with other computer-controlled activities.

(a) Materials storage and movement are coordinated with computer-based management systems such as production scheduling, that is, the production scheduling computer generates the pick list for components required by manufacturing; the components are removed from storage electronically and delivered to the appropriate production location under electronic control. At the same time, inventory control computers are automatically updated, components are reordered if necessary, and purchase orders and invoices are generated electronically.

(b) Automated materials handling equipment communicates and is coordinated with other computer-controlled equipment, if required, such as robots that will then become part of the materials movement process. Automated guided vehicles (AGVs) can carry robot arms when necessary. Packaging equipment, palletizers, depalletizers, bundling and unbundling equipment can all be interfaced with automated materials handling systems to streamline the incoming/outgoing material functions.

(c) Materials storage and movement are coordinated with automatic identification systems, including bar code and laser-based systems that can communicate current materials handling activity to other facilities, to higher management or to vehicles. This information can be used to direct or redirect vehicles or to signal the status of a particular materials handling operation.

Integration is key to implementing automated materials handling. The advances in computers, microprocessor-based controls and programmable logic controllers have enabled sophisticated automated materials handling systems to exist. Sensor technology, likewise, has enabled more unmanned operation of both storage and transport systems.

In many industrial environments, control systems will be hierarchical, that is, there will be layers of controls, with the lower levels collecting data and reporting to supervisory control systems, which, in turn, can transmit summary data to management. Microprocessor-based controls are now built into much materials handling equipment that previously operated as stand-alone equipment; these devices form the lowest level of automated materials handling control; programmable logic controllers are widely used on AGVs, automated storage and retrieval systems (AS/RS) and conveyor systems which require more complex control and sequencing with other functions.

The development and application of new sensors has been spurred by the widespread trend of adding diagnostic capabilities at various points in the material flow. For example, sensors on a conveyor can determine the arrival of a product or component and effect the transfer of that product or component to an appropriate workstation. The type of sensors that will be in most demand in the future are the smart sensors, that is, those that have the capability to convert an analog signal to digital format, are equipped with logic signal conditioning circuitry and, in some cases, have memory capability.

This article will first describe the individual computer-based controls that are used in automated materials handling systems, then describe how these controls are used with specific materials handling equipment.

1. Information and Control Systems

Yesterday's materials handling system has evolved into today's materials management system. Materials management itself is now as much of an information management system as it is a physical materials management system. Computers have been the obvious tools to use for this information management task for many years, and decreases in hardware costs and increases in software capability now make it much easier to reduce the paperwork associated with tracking material through a facility or network of facilities without an overwhelming capital expenditure.

The foundation of any computer-based information system is the database—the collection of data that keeps track of what material is expected, where material already in-house is, either in storage or in processing, and when material is expected to be sent out. It will contain the production or processing schedules, costs, personnel and machine loading. Helping to keep the data in the database accurate and timely are an assortment of automatic identification devices and the on-board equipment controllers, including microcomputers and programmable logic controllers. But, it is the materials management software that coordinates the activities of all the data collection activities in a facility and manages, directly or indirectly, all the materials handling operations. Figure 1 illustrates a typical materials handling control hierarchy.

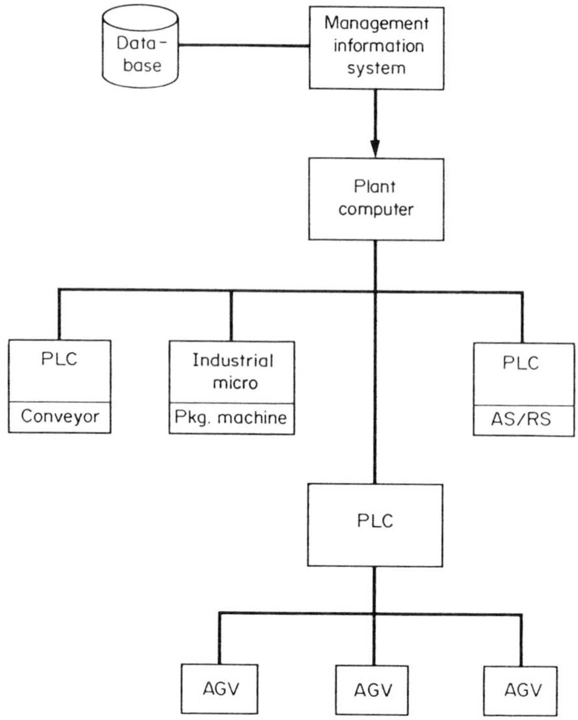

Figure 1
Typical plant information flow

1.1 Materials Management Software

Materials management software is available in a number of different forms. It is available to run on large mainframe computers, minicomputers and most recently on personal computers. All three types of systems serve the same function.

The most generic software for materials management are MRP II, or manufacturing resource planning, and just-in-time (JIT).

(*a*) *Manufacturing resource planning (MRP II) systems.* These systems have been called the central nervous system of a manufacturing enterprise. Contained within these systems are the software modules that plan and schedule operations, allow production and materials schedules to be explored for better alternatives, monitor operations against the plan, allow operating results to be projected, and tie financial reporting to operating figures.

Virtually all modern materials management software systems are modular, that is, they are comprised of separate modules that perform different functions and the modules can be purchased and configured as required for any given operation. Modules include such functions as receiving and shipping, purchasing, location control, scheduling, shop floor control, pick line replenishment, order processing, invoicing and reporting.

The importance of such systems is obvious; through the data they generate, collect and manage, MRP II systems establish and maintain links between the shop floor, the engineering department and the front office.

MRP II systems interact with automated materials handling systems in two ways:

(i) because MRP II systems determine the production schedule from business forecasts they can also be used to generate lists of components required to satisfy that production schedule—more sophisticated systems can sequence removal of these components from an AS/RS; and

(ii) sensors on components or finished goods moving through a production facility can communicate status to higher level control systems—ultimately, these systems can notify MRP II systems modules that can generate delivery and invoicing paperwork.

(*b*) *Just-in-time (JIT) systems.* These systems, with technology related to MRP II, are an adaptation of the Japanese Kanban approach. Their basic tenet is to keep inventory, particularly work-in-process (WIP) inventory, and inventory costs at a minimum. Purchased parts or raw materials are delivered direct to the production line, several times a day if necessary. Internally, one production area makes only as many components or subassemblies as the next portion of the production operation can use that day or that shift. The system turns inventory into products as quickly as possible.

Obvious implications for automated materials handling involve closer tracking of materials throughout a facility and closer electronic links with suppliers of raw materials or components. In a JIT environment, materials handling activities are less concerned with moving materials to and from warehouses or storage systems than they are in moving materials from one place to another within the manufacturing area, for example, from delivery dock to workstation or from one workstation to another.

(*c*) *Simulation software.* In addition to computer-assisted materials management software, the use of software simulation packages to model materials handling and other automated systems is growing dramatically. User organizations have recognized that the best way to minimize risks and uncover problems before they become expensive is by software simulation. Simulation via computer model attempts to simulate the materials handling system as it would operate on the floor. These computer models can be used to (a) establish the scope and size of different systems, (b) evaluate different approaches during the design of a materials handling system, (c) test and debug components of the system during development, and (d) define the type and quantity of equipment needed for the materials handling application.

1.2 Control Systems

Control systems used in the materials handling industry for control of the various types of machinery vary. The most common type of control is the programmable logic controller (PLC). Industrial microcomputers are also used. In general, the PLC is used on materials handling equipment where a number of functions must be monitored simultaneously, such as AGVs, and microcomputers are used in simpler control situations, such as in packaging machinery.

(*a*) *Programmable logic controllers (PLCs).* These were developed in the early 1970s to replace hard-wired relay logic. Since that time, the devices have grown in complexity and capability. Currently, PLCs are available that perform the simple sequencing and direct control tasks they were first applied to, and PLCs are available that perform complex data acquisition, decision logic, distributed control and manufacturing cell control. Because the PLC is so versatile, it has become a basic automation building block in both discrete and process industries.

PLCs of all complexities comprise certain common elements. These include input/output interfaces, a processor, a programming language and a programming device. In operation, a PLC monitors and controls the status of its environment by examining the input signals it receives from the devices and/or sensors to which it is connected. Based on its preprogrammed logic, the PLC can then instruct those devices to implement the appropriate functions.

The program logic for PLCs is stored in the device memory via a programming device that was traditionally a hand-held device supplied by the equipment vendor. More recently, personal computers are being used to program PLCs. The programming language used by PLCs traditionally was similar to the ladder logic language it replaced; however, as the devices grow more sophisticated, languages are moving closer to computer programming languages.

(*b*) *Industrial microcomputers.* These are, in fact, not a single entity, but a variety of computing solutions designed for industrial applications.

Embedded computers are board-level microprocessors that are embedded in the machinery. These devices interface with their surroundings through smart sensors and control the operation of the equipment. They are the simplest industrial computer systems, preprogrammed by the equipment supplier, and they have a limited range of functions. If these microprocessors malfunction, they are replaced rather than repaired.

More sophisticated industrial microcomputers are computer systems that are designed for the industrial environment. Their cabinets are specially sealed and hardened to withstand the dirt, dust, heat, electrical noise and even fluids that might be encountered in an industrial environment. They feature and require reliability, ease of installation, and maintenance-free operations for long periods of time. Moreover, they must operate in real time in most cases.

The industrial microcomputer is similar in these respects to the PLC, but it operates with more computerlike programming languages and software packages. As both types of control develop, the difference between the two becomes less distinct.

1.3 Connecting the Hierarchy

In most operations today, a variety of types of hardware and software must work together. Communication between systems is a problem, especially if those systems run on different hardware.

Most PLC vendors have developed their own proprietary networks which allow their products to communicate with each other and with the communications protocols of higher-level computers (on a selected basis, e.g., Digital Equipment Corporation (DEC) computers or IBM computers).

There are many initiatives under way to simplify the task of data communications—MAP (the manufacturing automation protocol), is the most universal effort to date.

1.4 Trends

There are certain hardware/software capabilities that are emerging as standards. Menu-driven systems are replacing command-driven systems. Pop-up menus and graphics windows are common features of much of the software being written today. These types of features make computer systems easier to use and therefore much more effective.

Artificial intelligence or expert systems is another area of computer technology with potential value for materials handling. Such systems are just beginning to be used in manufacturing operations to determine the best way to machine a part or to plan a production process. Because these systems capture the intelligence of an operation, they could conceivably help design the material flow through a facility.

2. *Automatic Identification Systems*

The entire field of automatic identification is, in fact, just a subset of broader automation technologies. The data collected through bar codes, optical characters and voice recognition are only valuable if they improve the decision-making process by providing timely information on the identity, status and location of products or people so that appropriate action may be taken.

The advantages of all automatic identification technologies lie in their ability to capture data with greater accuracy than manual methods. This, in turn, contributes to improved operations management, improved quality control, improved customer service and, perhaps, labor savings.

Automatic identification systems act as the senses of computer control in that they provide the computer with information on parts and products moving through manufacturing and distribution operations.

Table 1
Principal features of automatic identification techniques

Technology	Advantages	Disadvantages
Bar code	high speed moderate cost	line of sight required not human readable
OCR	low cost human readable	requires contact line of sight required
Voice	adaptable flexible	moderate speed high cost requires quiet workplace
RF/SAW	read/write capability high speed line of sight not required	high cost significant integration effort

Automatic identification systems can supply this information in real time and often without the need for human input.

Automatic identification bridges the gap between the physical movement of the material and the information flow that is required to manage the physical operation. Not only does automatic identification reduce the cost of data entry, but it significantly increases the accuracy of such data.

The three major categories of automatic identification are bar coding systems, optical character recognition systems and voice recognition systems. Table 1 describes the principal features of each.

2.1 Bar Coding Systems

Of all the automatic identification techniques available, bar coding is the most widely used. Bar coding technology involves bar code printers, bar code scanners that decode the scanned image, computer technology to perform whatever action is required, and the bar code symbology itself.

Bar code symbologies are the machine-readable representations of data, that is, the particular pattern of wide and thin lines. The awareness by a number of industries of the benefits of bar code technology and the benefits of a common code has prompted those industries to standardize codes that suit their particular needs.

Bar code scanners operate by illuminating an area of the bar code symbol with light generated by an internal light source and then collecting a portion of the reflected light with a photodetector. The photodetector generates an analog signal, which is amplified and transmitted to an analog-to-digital converter before the signal is sent to a microprocessor for decoding. Most scanners include the analog-to-digital converter.

A number of types of bar code scanners are available

to suit particular requirements. These include wand scanners, fixed beam scanners, moving beam scanners and portable laser scanners.

Bar code printers are the specialized printers that create the bar code pattern on labels or other documentation.

In a working environment, bar code labels are either purchased preprinted or printed on-site. On-site printing enables the user to change variable information, label layouts and comply with a number of bar code industry requirements. It also allows the user to print small quantities of labels and print them quickly as demands dictate.

The choice among these technologies must be made in light of volume, flexibility, resolution and cost.

The computer technology involved in bar code systems consists of the microprocessors that decode the bar code symbol and transmit the data to a higher-level computer for decision making and the higher-level computer that takes some action based on the information it receives.

2.2 Optical Character Recognition

Optical character recognition (OCR) is the processing of human-readable numbers or letters that are scanned with a light source. When the pattern is recognized by the reader, it converts that pattern into electrical impulses for transmission to a computer. Popular applications of OCR include check and document marking, along with merchandise tickets and labels in retail stores. In manufacturing, work orders and routing slips are being printed in OCR and scanned at workstations for tracking parts and assemblies.

2.3 Voice Recognition

Voice recognition devices are those products that facilitate the interaction between man and machine by the use of voice input. Many of the early successful voice systems have been specialized systems. These include a system designed for airlines to assist baggage handlers in the movement of baggage to and from planes and a quality assurance system designed for automotive assembly lines.

Voice recognition systems have the advantage of being very adaptive because they use a person for the data collection. The person interprets what is seen and expresses it verbally. However, that advantage can also be a problem. The system depends on the person's ability to verbalize correctly. Other problems exist also. Most systems are only 99% accurate when used in a quiet environment.

2.4 Emerging Automatic Identification Technologies

Three emerging automatic identification technologies have been the subject of extensive recent development: machine vision, radio frequency (RF) and surface acoustical wave (SAW).

The object recognition speed of vision systems has increased dramatically in recent years. In addition, video cameras are becoming much more compact and rugged so they can fit more easily into an industrial environment. Further reductions in the cost of vision systems will also increase their application.

Both RF and SAW will play an increasing role in factories and warehouses as the result of newer developments. Because RF and SAW codes do not have to be seen to transmit data, RF and SAW are suitable for heavy industrial operations, such as heat treat ovens, that would destroy conventional labels. The two techniques also fit processes or environments in which a tag containing the chip may be coated with dirt, dust or grease, and those that require the tag to be out of the line of sight of the reader.

The control center of both technologies is a programmable chip. While some RF chips are passive, others have read/write capabilities, allowing data to be changed, added to or deleted. Such chips support local decision making in decentralized, distributed control of materials handling equipment. Mounted on an AGV, for example, a read/write tag can direct a subassembly to various stations, direct manufacturing steps, signal a computer when the manufacturing process is complete and send the vehicle on its way.

3. Controls for Specific Materials Handling Equipment

3.1 Unitization/Deunitization Equipment

Packaging equipment is that equipment which unitizes material so that many individual items can be handled as one. Conversely, the category also covers the equipment that deunitizes or unbundles material so that it may be sorted or distributed. Packaging equipment consists of palletizing and depalletizing equipment and bundling and unbundling machinery, including shrink wrap and strapping machinery.

Packaging machinery serves three functions:

(a) containment of the material or product being handled;

(b) protection of that material from damage or other hazards; and

(c) communication or identification of the packaged product, its source and/or destination.

In many cases, this equipment is interfaced with other automated materials handling systems, conveyors, AGVs, feeders and so forth. It is also a function of the packaging system to facilitate the interaction between the package and the people or other equipment that come into contact with it.

The current trend in packaging equipment of all types is towards better control and more integration with other plant or facility operations. With the addition of computer-based controls such as programmable logic controllers, packaging equipment can be truly part of a materials handling strategy. It can communicate with other equipment, for example the conveyors or diverters that feed cartons to it or receive its output. Automatic identification systems, for example bar codes, optical scanners or photosensors, can identify materials to be packaged or palletized *en route* and communicate to the packaging equipment the dimensions of the material, the desired stacking height, the stacking pattern, or whatever other information is required for efficient material flow.

Often the total operation—conveyors, AGVs, palletizers, and so on—is monitored from a central location that feeds information to the individual machine controls and receives information from these controls as the operation progresses. Central control stations of this sort may use graphics displays to give the operator a "view" of the process.

3.2 Materials Movement Systems

(*a*) *Automated guided vehicles* (*AGVs*). These are driverless trucks that operate under computer control. Most AGVs in use today follow a path defined either by a wire embedded in the plant floor or by a reflective strip of paint or tape. Because of the computer control, they are able to interface automatically with a variety of other equipment, including machine tools, production machinery, AS/RS and robots. Guided vehicles can perform a variety of tasks in an industrial environment. Their principal advantage is that AGVs are able to move over "flexible" routes.

Controls for AGV systems vary. They range from on-board controls to central computer systems. On-board controls may consist of switches or pushbuttons which correspond to plant locations or workstations to which the vehicle is being directed. On-board controls may be microprocessor-based and able to handle vehicle positioning, acceleration and deceleration, activating lifting devices, and so forth. These devices can operate alone or be directed by a floor controller or a central computer.

Floor controllers serve as an interface between the vehicle and a central control computer. These controllers, generally a cluster of microprocessors, generate guidepath frequencies, process communication between the vehicle and the central computer, track individual vehicles, detect and compensate for errors, and may control other factory floor equipment.

The central control computer may be a programmable logic controller, a minicomputer or a mainframe, depending on the complexity of the system and how many other functions the computer performs. The central control will probably keep track of all the vehicles in the system and correlate their activity with other activities in the plant, such as AS/RS, or a material tracking system. Graphic displays at this central location may give an overview of the plant and all the material transport systems in use. Vehicle alarm conditions can be monitored at the central location.

Inertial guidance has been employed in some new AGVs. Self-navigation capabilities, relying on sensory perception and sophisticated control laws, represent the most important control capability. This technology will greatly increase system flexibility.

(*b*) *Conveyors.* Once thought of as a mature technology, conveyors are assuming new functions in today's integrated environment. At one time, conveyors principally provided transportation; now, in addition, they serve as input/output stations, accumulators for staging work-in-process, and stations for sizing, weighing and identifying loads.

One of the major trends in the use of powered conveyors is towards having them work with other types of materials handling equipment under computer control. Guided vehicles, for example, can pick up material from or deliver it to conveyors.

Organizations are beginning to interface robots with conveyors. In this application, accurate positioning is important. Unless the robot is equipped with vision so that it can adjust to variations in material position, special positioning devices may have to be built into the conveyor system. In most cases, robotic vision systems would be prohibitive for bulk materials handling applications, both in cost and in complexity.

Along with changes in the way conveyors are used, there have been great changes in the way conveyors are controlled. The programmable logic controller (PLC) has had the biggest impact on conveyor control. Even the simplest or shortest conveyor systems can have PLC control.

Computer-based controls such as programmable logic controllers are necessary to the integration of conveyor systems with other materials handling equipment. The ability of a conveyor controller to communicate electronically with an AGV control or a robot control can improve the efficiency of the material transfer.

(*c*) *Automated monorail carrier systems.* These are among the most versatile type of overhead materials handling systems. These overhead systems serve a broad range of applications in the transportation, processing and storage of materials that cannot be effectively and economically handled by other types of conveyors.

Automated monorail controls are similar to those used for AGVs. Often control systems are hierarchical, that is, with on-board microprocessors to direct the sequencing and drive of the trolleys, and programmable logic controllers to manage traffic control.

Host computers can be used to integrate the operation of the automated monorail system with other automated systems in the facility.

The technology is fully developed. The control systems permit the carriers to be sent from one point to another and can be programmed for sequenced stops and for particular functions at those stops.

Accessories include lift sections for travelling between floors or rail levels, turntables for track changes, and loops and twist sections for turning the carriers or load upside down or sideways.

3.3 Materials Storage and Retrieval Systems

(*a*) *Automated storage and retrieval systems (AS/RS).* These are essentially storage racks into which parts are loaded and from which parts are unloaded automatically. They typically consist of input/output staging areas, storage/retrieval (S/R) machines which shuttle loads in and out of storage and the storage racks themselves.

Control of an AS/RS can vary widely. A manual input station could control storage and retrieval through prepunched cards or pushbuttons.

In a more sophisticated system, AS/RS controls can be a combination of computers, microprocessor controls and programmable controllers, working together under a supervisory control which is normally a computer. This type of system not only controls the moving of loads into and out of storage, but can provide a host of data to other computers for managing inventory, for coordinating other plant activities, or for management information systems (MISs).

Many manufacturing companies are exchanging data between their manufacturing resource planning (MRP II) systems and AS/RS; the MRP II system generates a pick list which is electronically transmitted to the AS/RS supervisory computer to enable the automatic retrieval of the necessary parts and materials for the current production cycle. The AS/RS computer tells the MRP II system when the picking has been successfully completed so that inventory records may be updated.

This type of feedback is not limited to manufacturing operations; timely information on the whereabouts of materials as they move through various stages of processing is equally useful in a distribution environment.

AS/RS equipment is available from manufacturers complete with control systems that allow data to be uploaded from the storage/retrieval microprocessor to a host computer's inventory control or planning system. Such systems also allow sequencing instructions to come from the host. When combined with these sophisticated control schemes, AS/RS become vital links in integrated business or industrial environments.

Because the AS/RS can be linked to a plant or facility management information system, the storage area can also play a part in facilitating inspection. Since the computer's data include the storage location of all materials in the AS/RS area, the AS/RS can automatically segregate a certain segment or lot of material while certain items in that segment or lot are inspected. This kind of operation keeps congestion off the shipping docks and automatically guards against release of faulty or incorrectly labelled materials to processing.

(*b*) *Carousels.* These are used to store small parts or work-in-process in totes or other containers which rotate vertically or horizontally to present the selected container to the operator. They are similar to mini load AS/RS and are an alternative technology to them.

Carousels are normally controlled by microcomputers, often personal computers, that can be linked to a host computer. In manufacturing situations in which carousels are implemented as part of a control hierarchy, the host computer would contain the production schedules and download the picking requirements to the microcomputers. Information downloaded from the host will cycle the carousel through the picking sequence, that is, the picker need not enter the part number for each bin to be picked. Once the picking sequence has begun, the carousel will automatically rotate to the required bins. The micros, in turn, would transfer data relating to the picked parts back to the mainframe.

An emerging trend is the use of carousels interfaced with computers, voice and vision systems to form systems that will make the operator's job easier or eliminate it entirely.

See also: Materials Handling: Bulk Materials; Materials Handling: Unit Materials; Materials Requirements Planning

Bibliography

Apple J M 1972 *Materials Handling Systems Design.* Wiley, New York

Eastman R M 1987 *Materials Handling.* Dekker, New York

Hollier R H (ed.) 1986 *Proc. 3rd Int. Conf. Automated Materials Handling.* IFS (Publications), Birmingham, UK

Kulwiec R A 1985 *Material Handling Handbook*, 2nd edn. Wiley, New York

Orlicky J 1975 *Material Requirements Planning.* McGraw-Hill, New York

Williams G W (ed.) 1990 *Factory Automation Systems*, Datapro Manufacturing Automation Series. McGraw-Hill, New York

This subject area is also covered by the journals *Material Handling Engineering*, published by Penton Publishing, Cleveland, OH, and *Modern Materials Handling*, published by Cahners, Newton, MA

A. M. Greene
[Arthur D. Little, Cambridge, Massachusetts, USA]

Materials Handling: Bulk Materials

Materials handling, as defined by the Conveyor Equipment Manufacturers' Association (CEMA), is the art and science involving the movement, packaging and storage of substances in any form. This article deals specifically with the handling of bulk materials or those substances whose physical characteristics permit them to be classified as free flowing. Whether the substance is flour, iron ore, snow or waste material, it must be transported in an efficient and economic manner.

The bulk materials handling systems commonly in use include feeding devices, conveyors, large heavy-duty equipment and slurry pipelines. Material characteristics have a large impact on the appropriate system selection and design.

1. Handling Systems

In selecting the most suitable materials handling system there are typically three major considerations: the origin and destination of the material, the volume to be transported and the characteristics of the material itself. Other considerations, such as environmental issues, should also be taken into account.

1.1 Feeding Devices

Most materials handling systems, particularly conveyor belts, are highly dependent on an even flow rate for effective handling. To avoid overflows or underflows and to ensure a constant flow rate, feeders are typically used. The type of feeder required depends on the material to be loaded, the required flow rate and the storage techniques. Typically, a surge bin or hopper of some sort is associated with the feeder. Some of the more typical feeders are described in Table 1; Fig. 1 illustrates these feeders.

1.2 Conveyors

While the purpose of the feeders is to load material at a constant rate from the storage facility to the handling system, conveyors take the basic principles of feeders and apply them to long-distance transportation. The most common and most popular conveyor system currently in use is the belt conveyor.

The popularity of belt conveyors is due to their high capacity, high reliability and low power requirements. Although high capital costs are incurred, their low operating costs make them one of the most cost-effective bulk materials handling systems available. In addition, they can be used to transport a wide variety of materials over both short and long distances. In recent years, overland belt conveyors have been installed where previously railroad spurs or truck hauling systems would have been used. Most of these overland conveyors are under 10 km long, although there are some as long as 20 km.

Aside from the aforementioned materials handling system design criteria, other parameters for conveyor-belt design and selection must also be considered. Power requirements, for example, are of considerable importance. The horsepower required to drive a belt must be able to overcome several power draws. This power is required to lift or lower the load, overcome the resistance of the conveyor parts, accelerate the material load and accelerate the complete conveyor load.

A belt-conveyor system is comprised of several components: the belting; the support structure; the drive, idler and take-up pulleys; and safety devices. According to the Conveyor Belt Selection Guide (Uniroyal 1978), there are nine steps for properly designing a belt conveyor system: (a) define material type; (b) select a minimum belt width; (c) compare the capacity, speed and belt width; (d) select carrier spacing; (e) calculate

Table 1
Feeder devices and their applications

Feeder device	Operating mechanism	Applications
Rotary table	A rotating disk centered beneath a hopper, blade or plough discharges off the edge of the disk	Suitable for materials that tend to arch over restricted openings
Rotary vane	A series of rotating vanes in an enclosed case directly mounted below a chute	Suitable for fine-grained materials
Reciprocating plate	A hopper discharging into a plate which has a reciprocating action imparted on it. Flow rate is adjusted by the angle of the plate and frequency of the reciprocating action	All sizes of materials
Vibrating	A vibrating motion applied to a pan or trough. Flow rate is adjusted by the vibration level	All sizes of materials but not suitable for sticky materials
Bar flight	Bars mounted between two strands of chains which scrape material along the bottom of a trough	Good for small-lump material
Belt	Short conveyor belts. The belts are typically not troughed	Applicable to most materials, except those of extreme temperatures
Apron feeders	Metal plates or pans mounted between two chains moving the material	Used for large abrasive materials
Screw	Uses a rotating, helically formed flight to move materials. The rpm of the screw assembly adjusts the flow rate	Applicable to most materials except those of extremely fine or extremely coarse size

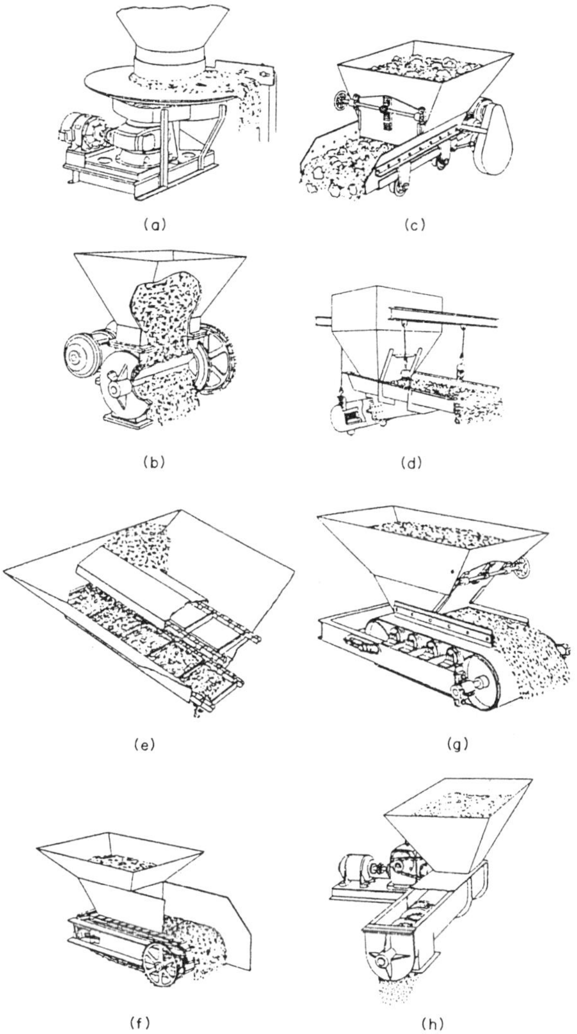

Figure 1
Types of feeder devices: (a) rotary table feeder, (b) rotary vane feeder, (c) reciprocating plate feeder, (d) vibrating feeder, (e) bar flight feeder, (f) belt feeder, (g) apron feeder, and (h) screw feeder

the horsepower; (f) compare the belt tensions; (g) select the belt; (h) select pulley diameters; and (i) select pulley shaft diameters.

In recent years, belt conveyors have made many sophisticated advances, most noticeably in the area of computer control and monitoring. Belt lines can now be automatically shut down or restarted by these monitoring devices. Overload conditions, overheating of motors and fire detection are some of the parameters that can be monitored.

Belt conveyors, reciprocating conveyors, apron conveyors, screw conveyors and bar flight conveyors are all based on the same principles as the feeder devices with similar names described in Table 1. Table 2 describes several other conveyor types of interest.

1.3 Heavy-Duty Transporting Equipment

The conveyor systems previously described deal with a continuous flow of material. There are, however, many cases where a continuous flow is not practical or possible. Surface mining or civil construction projects requiring considerable earth moving are typical examples where large volumes of materials are to be intermittently handled. Another example of when conveyor systems become impracticable is when the shipping distance is great and many physical obstacles exist such that rail or ship haulage is the only means of practical transportation. Various types of large-scale materials handling equipment are described below.

Locomotive and rail car haulage played a major role in the development of the industrial revolution. The locomotive still contributes greatly to the daily shipping of many commodities. A train consisting of 100–150 rail cars weighing approximately 100 t each is quite common in the USA. Locomotives can be powered by several different sources, including battery, diesel, compressed air, trolley, or some combination of these such as a trolley–battery combination.

Unit trains have gained in popularity in recent years. A unit train is a dedicated set of cars and a locomotive to haul materials. The locomotive and cars usually remain coupled to each other whether the cars are loaded or not. Unloading is by either rotary-dump or bottom-dump rail cars. The bottom-dump rail cars can

Table 2
Conveyor devices and their applications

Conveyor device	Operating mechanism	Applications
Wide chain drag	Uses an endless chain to drag material along a trough	All sizes are suitable
Bucket	Uses buckets attached to a chain to move material either horizontally or vertically	Suitable for all sizes except extremely lumpy
Weight larries	An assembly typically rail mounted from the ceiling comprised of a scale, hopper, chute and gate. Transports materials from overhead storage bins to one or several discharge bins	Applicable to most bulk materials

be activated by several techniques. The first method is to use an electric motor installed on the cars themselves to open the bottoms. The second method is to have the doors open and close automatically on entering the dump area by engaging a gear train to the bottom door assembly which is driven from outside the car itself. Rotary-dump cars also have two different dumping techniques. One method uses an outside motor to rotate the rail car into the dumping position. The center of rotation is the swivel couplings used to connect the cars thus permitting unloading without disconnecting the cars. The second method, gravity-type rotary dumps, utilizes a counterweight instead of a motor to rotate the cars. When a loaded car is put into the dumping unit, the counterweight makes the dumping unit top heavy, thus causing the rotation. Once the car is empty, imbalance is again created and the unit returns to its upright position. The disadvantage to the gravity dump system is that the rail cars must be uncoupled before dumping.

There are four steps involved in the selection of the locomotive for a specific application. The required values to be calculated are: (a) the drawbar pull (the effort exerted by the locomotive on the train); (b) the available tractive effort; (c) the locomotive weight; and (d) the required braking effort.

Hoists are used to move intermittent loads within a fixed area. They can be used for transporting loads along a vertical or inclined path and can be either of the small portable class used in erection work or the large stationary class used in the mining industry. Large hoists typically operate at high rope speeds. For example, a 3 m diameter drum usually operates at rope speeds greater than 13 m s^{-1}.

There are two basic hoist types, drum and friction. A drum hoist acts as a spool for wire rope to be wrapped on to raise the load, and fed off to lower the load. A friction hoist, on the other hand, merely has a continuous wire rope pass over the drum and uses the tension differential from the rope's entry point on the drum to the exit point on the drum to move the load.

Ships and *barges* of many shapes and sizes have long been used as a means of transporting goods from one area to another. In recent years, much interest has been generated by the coal loading terminals and other new port facilities, which have an increase in storage capacity and can load larger ships at a quicker rate. Some new port facilities can load ocean ships with a 152 cm (60 in) belt at rates as high as 1.6–2.0 kt h^{-1}.

In general, waterways attract heavy industrialization because of the inexpensive freight rates typical with barging. The typical barge loading system is the elevating boom type, which has capacities in excess of 1.5 kt h^{-1} and often up to 3 kt h^{-1}. The barge unloading system typically utilizes cranes and clamshells.

Haulage and *industrial trucks* are commonly used when material is to be intermittently transported over various paths and maneuverability is of prime importance. Industrial trucks include such types as forklifts, tractor trailers and platform trucks usually associated with unit materials handling. Haulage trucks, however, are commonly used for transporting gravel to construction sites or hauling ore and waste rock in surface mines. Such haulage trucks can be for off-highway or on-highway use, depending on size. Off-highway haulage trucks usually range from 35 t to 350 t in capacity, with the 170 t trucks being the most popular in the large mining operations. Off-highway truck transportation is typically limited to moderate haulage distances for cost reasons.

Haulage trucks may be of the bottom dump, rear dump or side dump types, with the rear dump being the most common. Certain considerations are important for the appropriate selection of a haulage truck. For example, the load weight, the material's characteristics, the haul distance, the grade and rolling resistance of the road, and the desired production or handling rates all affect truck selection.

Earth-moving equipment pushes, pulls, or displaces large volumes of earth for such projects as dam construction, surface mining and agricultural development. Equipment such as backhoes, tractors (wheeled or tracked), powershovels, draglines, front-end loaders, graders and scrapers are commonly used. The selection of the equipment depends not only on the type of work to be done, the material to be moved, and the rate at which it must be moved, but also the equipment's reliability and parts availability. This is true for much of the equipment described previously; however, due to the harsh operating conditions for earth-moving equipment, the reliability and parts availability are of particular concern.

Most earth-moving equipment is diesel powered; however, some items such as draglines and powershovels are electrically powered. For certain pieces of equipment (i.e., tractors and graders) the primary equipment specifications are weight and power. These two criteria identify the amount of force the machine can exert. For the loading and digging equipment, the bucket size, cycle time, maneuverability, operating reach and other specifications become important.

Dredges are large floating devices used for moving materials located below the surface of the water. Dredges are commonly used to maintain the proper water depths in rivers and harbors, and in recovery of underwater mineral deposits. There are various types of dredges, including draglines, clamshells, bucket lines and hydraulic systems. The bucket-line dredge is the most common dredge and utilizes buckets mounted on a continuous chain. The buckets discharge into hoppers; this is usually followed by some preliminary sizing and beneficiation if the dredge is being used for mining purposes. The dragline and clamshell types, unlike the continuous flow of material provided by the bucket line, provide an intermittent or repetitive flow. Hydraulic systems transport materials in slurry or pulp form and provide a continuous flow. Hydraulic dredges are more commonly associated with nonmining applications such as deepening and widening of rivers or harbors. Cutterheads are often used to break the material prior to handling and the typical mode of gathering the material is suction.

1.4 Slurry Pipeline Transportation

The transport of solids by fluid techniques has been practised for a very long time. Favorable economics of slurry pipelines when compared to other heavy-duty transporting equipment have resulted in numerous long-distance slurry pipeline projects since the late 1950s, as shown in Table 3.

Transporting a mixture of solids and liquids in pipes differs from pumping homogeneous liquids in a number of ways. Characterizing slurry flow is more complex than liquid flow for two reasons. Firstly, with a two-phase system, the properties of the solid particles are superimposed on the liquid and, in addition, the properties of the mixture may resemble neither those of the liquid nor those of the solid. Secondly, a range of slurry behavior is possible depending on the range of conditions dictated by the application. This will always be between the two extremes of flow: homogeneous flow and heterogeneous flow. Homogeneous flow is where the solids are uniformly distributed throughout the liquid medium. Systems which approach this extreme are those with high slurry concentrations and fine particle sizes which make particle inertia effects relatively minor. Heterogeneous flow is where the solids are not uniformly distributed resulting in pronounced concentration gradients. These systems have coarser particle size distributions and lower solids contents. Particle inertia effects become significant and the two phases often can retain their separate identities to a large extent.

As illustrated in Fig. 2, the pipeline is but one step in a multistage transportation system designed to move a commodity from one point to another. The entire system often involves specialized equipment and related procedures for handling, storing, loading, transporting and discharging bulk granular material as slurries. System design is highly application specific, but must consider such variables as specific gravity of solids, particle size distribution, particle–particle and particle–liquid interaction, and hardness and abrasiveness of the material.

Future developments in slurry technology will focus not just on long-haul schemes, but will also include short pipeline systems at port facilities for loading and unloading ships. Recent experimentation has demonstrated the feasibility of handling slurries which are up to 80 wt% solids for such applications.

2. *Material Characteristics*

The characteristics of the material to be handled have a large impact on the selection and design of the handling systems. These characteristics also vary with each specific application. For example, such environmental conditions as temperature and humidity may have an effect on the material's moisture content, volume, or other parameters.

Table 3
Slurry pipeline operations (adapted from Wasp et al. 1977)

Material deposits	Length (km)	Pipe diameter (cm)	Capacity (Mt year^{-1})	Year of construction
Coal				
Consolidation	174	25.4	1.3	1957
Black Mesa	439	45.7	4.8	1970
Etsi	1667	96.5	25	1979
Alton	290	61.0	10	1981
Iron concentrate				
Savage River	85	22.9	2.25	1967
Waipipi (iron sands)	10	20.3 and 30.5	1.0	1971
Peña Colorada	45	20.3	1.8	1974
Las Truchas	27	25.4	1.5	1976
Sierra Grande	32	20.3	2.1	1976
Samarco	407	50.8	12	1977
Copper concentrate				
Bougainville	27	15.2	1.0	1972
West Irian	111	10.2	0.3	1972
Pinto Valley	18	10.2	0.4	1974
Limestone				
Rugby	92	25.4	1.7	1964
Calaveras	27	17.8	1.5	1971

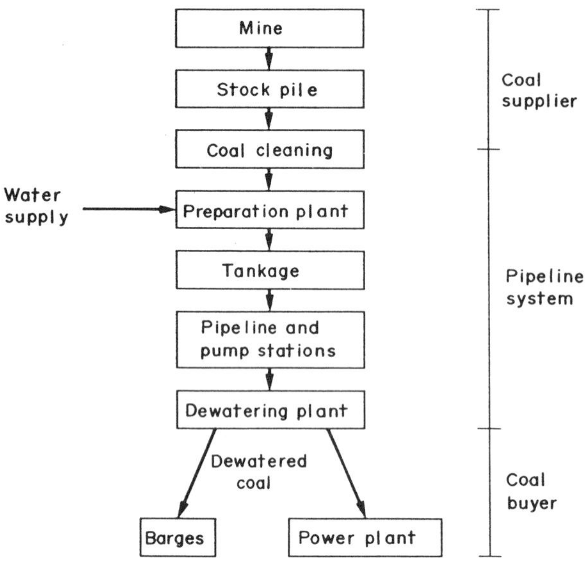

Figure 2
Coal transportation system: mine to power plant

The characteristics of concern are: size consist, bulk density, specific gravity, abrasiveness, flowability, friability hardness and corrosive action. If belt conveyors are being considered, then the angle of repose, angle of maximum inclination, angle of internal friction and angle of external friction must also be considered.

Size consist is the size ranges for bulk materials. The categories into which they are broken are shown in Table 4. The size of the material is of major importance in selecting the proper conveyor: for example, lumpy materials cannot be handled by a screw conveyor.

Bulk density is the weight per unit volume of the material and is given as either loose, vibrated or packed. The vibrated bulk density is a measure of the material once it has been compacted by a vibrating motion. A measure of packed density is attained by applying a set pressure to material in a form.

Specific gravity is the ratio of the mass of a solid or liquid to the mass of an equal volume of distilled water at 4 °C. The specific gravity is used only rarely when lumpy materials must be handled. More commonly, the loose bulk density is more applicable in the selection of handling systems.

Abrasiveness is the ability of a material to wear away by friction the surfaces of the materials with which it comes into contact. The abrasiveness of a material is a function of four of its other characteristics: size, shape, hardness and bulk density. The levels of abrasiveness as defined by CEMA are mildly abrasive, moderately abrasive and extremely abrasive.

Flowability is the level to which a bulk material will move when some outside force is applied. It is usually an indication as to the level of particle freedom in the material.

Friability is an indicator of the material's ability to crumble or break apart.

The corrosive action of a material is of major importance in the selection of an adequate handling system. Improper consideration of this property could lead to costly replacement and maintenance costs. The corrosive action of a material may be related to its chemical composition or to its moisture content.

The angle of repose is the angle between the slope of a freely formed pile and the horizontal. Depending on the particle size and piling method, the angle of repose may vary for the same material.

The angle of maximum inclination is the maximum angle at which a conveyor belt can transport material upwards. This angle varies with the material, the belt speed and the troughing angle.

The angle of internal friction is the angle between the normal stress axis and yield locus of material. This angle is an indication of the tendency of the materials to slide or roll among themselves.

The angle of external friction is the angle between the normal stress axis and the wall yield locus. This angle is an indication of the level to which material will slide on another surface.

Aside from the characteristics mentioned above, there are other properties which should be considered in the handling of materials, for example, whether a material is explosive, whether it deteriorates in storage or whether it packs under pressure.

There are many tables and charts published that supply bulk material characteristics for a wide range of materials. *Belt Conveyors for Bulk Materials* prepared by CEMA has an extensive listing of various materials and their characteristics. Many of the properties previously described are determined by sampling and testing and most of the results from such testing have also been put in tabular form.

See also: Materials Handling: Automation, Unit Materials; Materials Handling: Unit Materials

Table 4
Size classification for bulk materials

Category	Size
Very fine	No. 40 sieve ($\leq$0.041 cm)
Fine	No. 6 sieve ($\leq$0.335 cm)
Granular	$\leq$1.27 cm
	$\leq$7.62 cm
	$\leq$17.78 cm
Lumpy	$\leq$40.64 cm
	$>$40.64 cm to be specified
Irregular	stringy, fibrous, cylindrical, slabs

Bibliography

Anon 1983 Slurry technology advances. *Min. J. (London)* 300: 408

Caterpillar Performance Handbook, 10th edn., 1979 Caterpillar Tractor, Peoria, IL
Conveyor Equipment Manufacturers' Association 1979 *Belt Conveyors for Bulk Materials*, 3rd edn. CBI Publishing, Boston, MA
Conveyor Equipment Manufacturers' Association 1980 *Classification and Definitions of Bulk Materials.* CEMA, Washington, DC
Conveyor Equipment Manufacturers' Association 1988 *Conveyor Terms and Definitions.* CEMA, Washington, DC
Cummins A B, Given I A (eds.) 1973 *SME Mining Engineering Handbook.* Port City Press, Baltimore, MD
Hawkins H V, Behnke E R, Manney C J, Parsley H V, Braker E A, Foster R J 1967 Materials handling. In: Baumeister T, Marks L S (eds.) 1967 *Standard Handbook for Mechanical Engineers*, 7th edn. McGraw-Hill, New York, Chap. 10
International Journal of Storage and Handling of Bulk Materials (October 1989): issue devoted to Bulk Solids Handling
Knill B (ed.) 1973–1974 *Material Handling Engineering Handbook and Directory.* Industrial Book Publishing, Wallingford, CT, pp. A5–156
Marconaflo 1981 *Marconaflo Slurry Systems.* Marconaflo Division, McNally Group, Salt Lake City, UT
Pfleider E P (ed.) 1968 *Surface Mining.* American Institute of Mining, Metallurgical and Petroleum Engineers, New York
Raymus G J, Steymann E H 1973 Handling of bulk and packaged solids. In: Perry R H, Chilton C H (eds.) 1973 *Chemical Engineers' Handbook*, 5th edn. McGraw-Hill, Kogakusha, Tokyo, Chap. 7
Uniroyal 1978 *Conveyor Belt Selection Guide*, Catalog 120 revised. Uniroyal Industrial Products, Middlebury, CT
Wasp E J, Kenny J P, Ganghi R L 1977 *Solid–Liquid Flow Slurry Pipeline Transportation.* Gulf Publishing, Houston, TX

W. S. Kelly
[Cyprus Minerals, Englewood, Colorado, USA]

M. J. Laurila
[Process Technology, Calumet, Michigan, USA]

Materials Handling: Unit Materials

The Materials Handling Division of the American Society of Mechanical Engineers defines materials handling as "the art and science involving the moving, packing, and storing of substances in any form." For unit materials handling, the substance is a unit; these substances are manufactured or natural materials with macro dimensions and cannot be readily fluidized. For bulk materials handling, the substances usually can be fluidized and include grains, gravel, cement and other molecular or aggregate substances. However, bags or containers of grains, gravel, cement or flour would be considered units. Bulk materials handling is detailed elsewhere (see *Materials Handling: Bulk Materials*).

In unit materials handling, the concept of a unit may be pervasive, as provided by the following illustration. The primary unit may be a single egg. A secondary unit could be the carton of a dozen eggs, which is also considered a unit in materials handling. The tertiary unit would then be 24 egg cartons in a corrugated shipping container called a case. Continuing in the macro scale, the next unit could be a 40 × 48 in (1016 × 1219 mm) pallet holding 30 egg cases; the next unit could be an 8 × 8 × 20 ft (2438 × 2438 × 6096 mm) container holding as many as 24 pallets if they were double stacked, or 207 360 primary units, or eggs.

The fundamental principle in unit materials handling is the concept of unitization and its converse, deunitization. These basic principles are the foundation on which economic unit materials handling is founded. Materials handling does not add any intrinsic value to the units being moved, but only to the utility of the new time or place to which the unit has been moved. Packaging is discussed here only so far as it relates to the principles of unitization, deunitization and unit movement; it is not treated in any detail as a separate technology from materials handling or as an art form. Storage is treated in discussions of storage and retrieval systems and also in the whole area of sorting, order filling and order assembly systems. These latter operational categories are frequently included in an overall category, namely, warehousing.

1. Unitization

Unitization is the process of assembling and/or packaging units of material into a relatively standard macro configuration that includes two or more units. Typical examples of unitization are palletizing and container-loading operations—whether a small corrugated shipping container or a large steel or aluminum freight container. Frequently, unitizing is considered a packaging operation.

The initial unitizing of bulk materials—whether liquids or solids—is the subject of bulk materials handling. Following this initial unitization, the units—whether bottles, jars or boxes—are further unitized through orienting and accumulating the units for multipackaging, such as wrapping or film packaging and shrinking the film, or banding or bundling. Other forms of unitizing involve semiautomatic or automatic casing machines. The semiautomatic machines employ manual unit loading of the primary package—whether a bottle or a multipack. The automatic casing machines integrate the casemaking, case loading, case closing and sealing operations.

The next step in unitizing is palletizing, which can be manual, semiautomatic or fully automatic (see Fig. 1). For the semiautomatic palletizer, the operator arranges the case or bag units in each tier by manually orienting them into a distinct pattern. Then the tiers are deposited on the pallet automatically. For the fully automatic palletizer, the pallets or slip-sheets are delivered automatically. The tier patterns are programmed into the palletizer, which then automatically

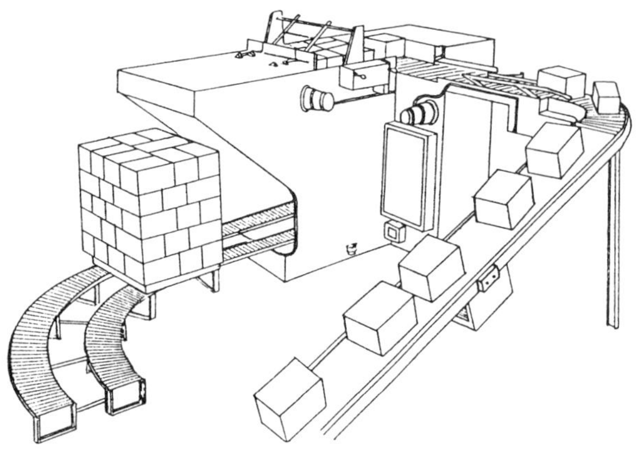

Figure 1
A palletizer

arranges the cases or bags for each tier. When the last tier is deposited, the pallet may be automatically indexed to a second station, which could be an automatic strapper or a stretch or shrink film wrapper to secure the load.

Besides the case and bag palletizers, there are also bulk palletizers on which bottles or cans are accumulated on a conveyor belt and then swept onto a tier sheet on top of a previous layer of bottles or cans. The tier sheet is then lowered onto the pallet. The complete unit load is then either strapped or shrink-filmed to contain the individual cans or bottles.

The next level of unitization is the semiautomated or automated loading of freight containers or semitrailers for long-distance movement of the unit load. Semiautomatic loaders normally use a forklift for handling the unit pallet load or the unit slip-sheet load. For slip-sheet operation, the forklift is equipped with a pusher to push the load off the fork or with push–pulls to pull the unit slip-sheeted load onto the forks and then push it off into the container or semitrailer.

For the automated unitizing of the load for the freight container or the semitrailer, the unit pallet or slip-sheet loads are assembled progressively onto a roller mat conveyor (Fig. 2). The loads are then compacted on the conveyor, which moves forward to deliver the load into the container or semitrailer. A hydraulic ram is part of the roller mat subsystem, which first holds and then pushes the load forward to compact

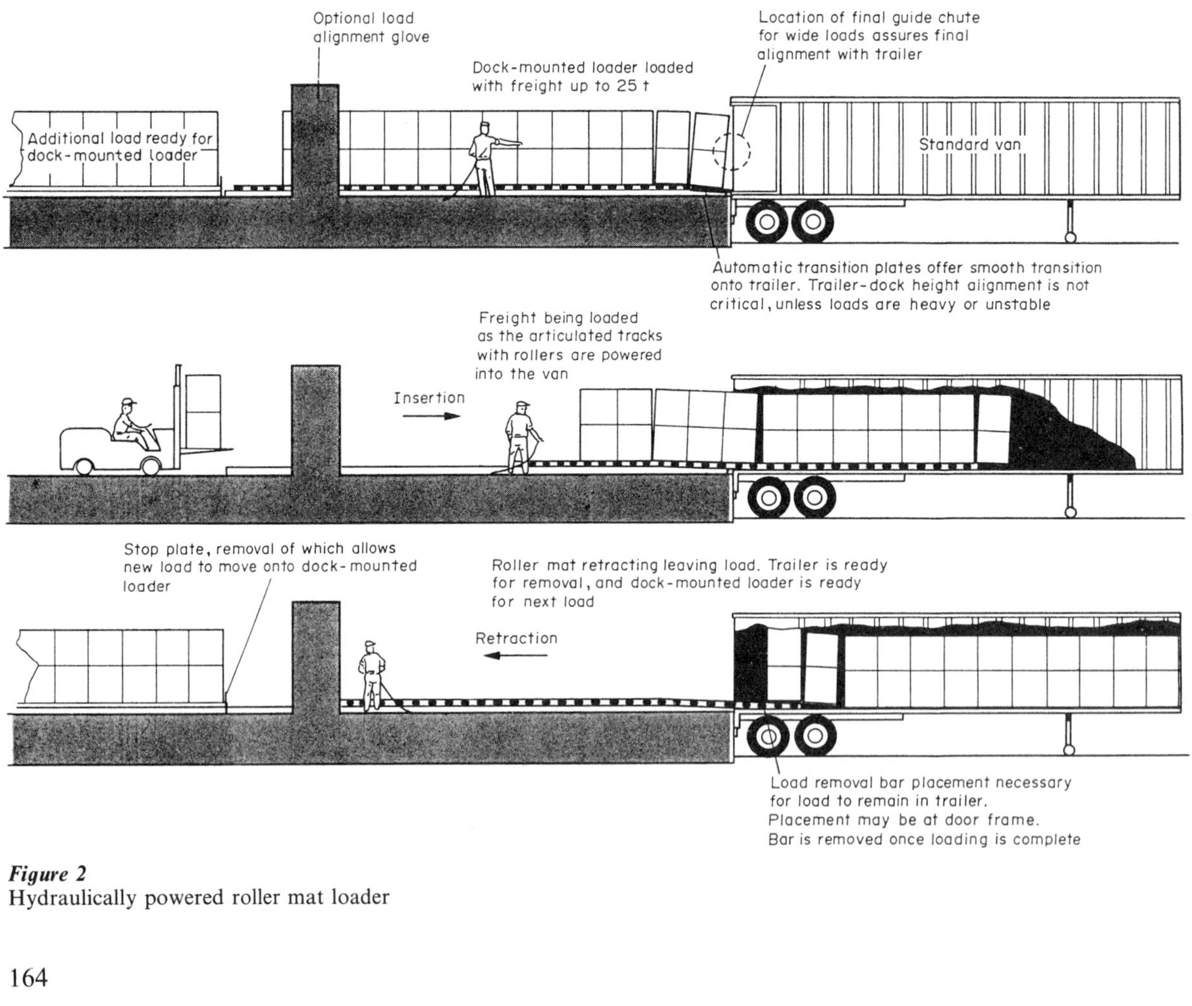

Figure 2
Hydraulically powered roller mat loader

it longitudinally while the roller mat conveyor is being extracted from beneath the pallet or slip-sheet units that are being deposited in the container or semitrailer.

Unitizing in the service industries and in order fulfillment remains mostly manual. For instance, unitizing is strictly manual in container loading of baggage or cargo in LD3 or other containers used in wide-body aircraft. In order fulfillment, such as in mail order distribution centers, and in most nonfood industries where multiple products are shipped as unit loads, unitizing is mostly manual. In the transportation industries, however, all the container or pallet unitized loads are handled semiautomatically under manual controls.

2. *Deunitization*

Deunitization is the converse of unitization. Typical automation of deunitization is the automated depalletizing of product loads or the automated emptying of containers. Deunitizing usually involves separating each unit being unloaded, inserting it into the material flow, and possibly orienting the unit. The orientation process, such as for facing or address reading, is mostly manual.

Deunitizing is fully automated in some industries. In the beverage industry, for example, bottles and cans are automatically unloaded from unit containers or unit bulk loads, unscrambled and separated for filling. In advanced grocery warehousing, unitized pallet loads are automatically depalletized, and the case unit is individually conveyed under orientation control to mechanized picking lines or to flow racks for manual picking.

Other examples of automated deunitizing are automatic container unloading in bulk mail facilities and parcel post activities, and automated mail cart unloading at airline facilities at international airports to separate the mail bags and parcels for mechanized sorting. However, in most of the service industries and in industries in which the unit products are fragile, controlled unloading and deunitization remain mostly manual.

3. *Unit Materials Movement Systems*

A unit materials movement system is defined as the movement through space and time of one unit of material, whether the unit is a single, discrete object or a unitized load of discrete objects. Unfortunately, there is no logical classification of movement systems. Classification tends to be arbitrary and has, at most, only historical significance. Carson et al. (1972) divide movement systems into three types—conveyors, cranes and hoists, and industrial trucks. Table 1 presents the general characteristics of these movement systems. Conveyors move uniform loads continuously from point to point over fixed paths where the primary function is transportation. Cranes and hoists are used to move varying loads intermittently to any point within a fixed area. Industrial trucks move mixed or uniform loads intermittently over various paths with suitable surfaces where the primary function is maneuvering.

A more basic classification of unit materials movement systems may be explored by dividing the system into its two components—the guideway and the unit materials carrier. The guideway and carrier may be synonymous in the case of a chute or slide, or may be distinctly different entities as is the case for conveyor or industrial truck movement systems. For a slider-bed belt conveyor, the slider bed and its frame would be the guideway; the belt would be the unit materials carrier. For a monorail, the rail and its structure would be the guideway; the chain and its attachments would be the carrier. For an industrial truck, the aisles of the factory and warehouse and the roadways outside the factory would be the guideway, and the truck itself would be the unit material carrier. For a bridge crane, the runway structure and bridge would be the guideway; the trolley and hook and hoisting mechanism would be the carrier. Hence, part of the guideway is static and part dynamic.

Figure 3 shows a matrix of the basic system characteristics of the carrier path—whether the path is fixed or flexible; whether the guideway is static and nonpowered, static and powered or dynamic and powered; and whether the carrier is static and nonpowered, dynamic and nonpowered or dynamic and powered. Conveyors generally have the characteristics of a fixed carrier path system, and industrial trucks, hoists and cranes have the characteristics of a flexible carrier path system.

The choice of a unit movement module or system is dependent principally on the physical characteristics of the unit of material and the economy that can be achieved in its movement. Hence, unitization usually occurs before movement. Discrete units, rather than unit loads, are moved primarily between continuous and/or automated factory operations, and a conveyor system is usually chosen. Unit loads are more often moved by industrial trucks than fixed path conveyors for factory batch operations and warehousing.

Tables 2 and 3 illustrate the types of movement systems commonly used in fixed as well as flexible unit materials carrier path systems. Transportation systems such as motor vehicles, railroad cars, marine vehicles or air transports are not considered as unit materials movement systems here.

4. *Unit Materials Storage and Retrieval Systems*

Storage and retrieval are defined as the input of a unit of material, or a unit load of material, into a known stored position from which it can be withdrawn at

Table 1
General characteristics of basic material handling equipment types (Carson et al. 1972)

Task and equipment characteristics	Conveyors	Cranes and hoists	Industrial trucks
Material			
volume	high	low, medium	low, medium, relatively high
type	individual item, unit load, bulk	individual item, unit load, variety	individual item, unit load, variety
shape	regular, uniform, irregular	irregular	regular, uniform
size	uniform	mixed, variable	mixed, or uniform
weight	low, medium, heavy, uniform	heavy	medium, heavy
Move			
distance	any, relatively unlimited	moderate, within area	moderate (75–100 m)
rate, speed	uniform, variable	variable, irregular	variable
frequency	continuous	intermittent, irregular	intermittent
origin, destination	fixed	may vary	may vary
area covered	point to point	confined to area within rails	variable
sequence	fixed	may vary	may vary
path	mechanical, fixed point to fixed point	may vary	may vary
route	fixed area to area	variable, no path	variable, but over defined path
location	indoors, outdoors	indoors, outdoors	indoors, outdoors
cross traffic	problems in bypassing	can bypass, no effect	can bypass, maneuver, no effect
primary function	transport, process/store in move	lift and carry, position	stack, maneuver, carry, load, unload
% transport in operation	should be high	should be low	should be low
Method			
load support	none, or in containers	suspension; pallet, skid, none	from beneath; pallet, skid, container
load/unload	automatic, manual, designated points	manual, self, any point	self; any point on available path
operator to accompany load	no	may or may not, usually does	usually does; may be remote
Building characteristics			
cost of floor space	low, medium	high	medium, high
clear height	if enough, conveyor can go overhead	high	low, medium, high
floor load capacity	depends on type of conveyor and material	depends on activity	medium, high
running surfaces	not applicable	not applicable	must be suitable
aisles	not applicable	not applicable	must be sufficient
congested areas	fair	good	poor

command and moved to a known position with access outside the storage and retrieval system.

Historically, the input and retrieval of the material has been manual. For manual movement into and out of storage, the storage equipment has typically been shelving, racks and drawers. Storage and retrieval became mechanized with the introduction of the industrial truck (i.e., the forklift truck with its many lift attachments for specialized unit materials and unit loads).

To improve the economy of storage, storage space has been increased by utilizing an extended height dimension. Modern mechanized and automated storage and retrieval systems use racks that are structurally integrated into the building with heights as great as 50 m.

The storage and retrieval machine, which may be manually operated or under computer control, is a stacker crane. These systems can handle palletized loads, almost any type of container, or individual units of material ranging from rolls to complex shapes. The individual units of material range from a miniload of less than 225 kg to a maxiload of several thousand kilograms.

These automated storage and retrieval systems use guidance systems with multidegree freedom of movement so that the storage and retrieval machine can retrieve a unit of material or unit load in a minimum of time by employing the most effective combination of x, y and z movements. The actual input–output transfer mechanism is usually a separate, manually controlled or automated device on the storage and retrieval machine and is actuated only when the storage and retrieval machine is lined up with a slot position.

Other storage and retrieval systems use live, rather than static, storage of the unit material or unit load. In these instances, storage is normally in a carousel type of configuration consisting of bins or drawers. The carousel shape is designed to save as much space as possible. Practically all storage and retrieval systems employ random access to the slot positions—whether the slots are dynamic or static.

5. Unit Materials Sorting, Order Picking and Order Assembly Systems

Sorting is defined as isolating and selecting a unit item or unit load from a dynamic stream of items or unit loads. Order picking is defined as selecting and retrieving a unit item or unit load from, usually, a static storage position. The storage itself, however, can be a combination of static and dynamic—for example, a flow rack with the unit repositioning itself in the replenishment process for picking. Order assembly is the process of accumulating an order, which may consist of many unit items for unitizing and shipment.

Automated sorting involves reading a bar code on the unit item or unit load and removing the item or load from the moving stream. Other systems use voice recognition or manual keyboard encoding to provide control. The most common mechanical devices for sorting are diverters, which tend to lose the orientation of the unit item or unit load when removing the item from its dynamic stream, such as on a belt conveyor. The more advanced mechanical devices are pulling or

Basic system characteristics of carrier path	Guideway characteristics	Unit materials carrier characteristics		
		Static	Dynamic	
		Nonpowered	Nonpowered	Powered
Fixed	Static, nonpowered	Chute Slide Deadplate	Conveyors gravity or manual (roller, wheel) monorail Dumbwaiters (manual)	Automatic guided vehicles (AGVs) Shuttle cars
	Static, powered		Conveyors belt (slider bed, roller bed), slat, live roller, air film, pneumatic tube, monorail (powered), car, tow, apron, drag, flat top, arm, process/plastic belting Fixed hoists Elevators Dumbwaiters (powered)	
	Dynamic, powered		Conveyors arcing, telescoping, mobile, switching, sorting	
Flexible or free	Static, nonpowered		Industrial trucks (manual) Hoists (manual) Lift tables (manual)	Industrial trucks (powered) Mobile cranes Jib cranes Hoists (powered) Lift tables (powered) Derricks
	Static, powered or combination static and dynamic, powered			Bridge cranes Gantry cranes Climbing cranes
	Dynamic, powered			Robots Automatic guided vehicles under computer guidance control

Figure 3
Unit materials movement system or module classification according to basic characteristics of the carrier path

Table 2
Fixed carrier path unit materials movement systems and their applications

GUIDEWAY, STATIC AND NONPOWERED; CARRIER, STATIC AND NONPOWERED

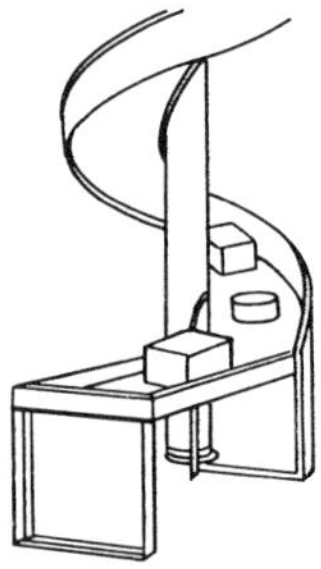

Chute: A trough through which unit materials are directed and lowered by gravity. The trough may be open or enclosed, straight or curved (e.g., a spiral). *Applications*: Any unit material that is not fragile. *Speed*: 0.4 m s^{-1} max

Slide: A sloped surface to permit free flow of unit material, such as packages, without particular confinement or restraint at point of discharge. Similar to a straight chute without restraining sides or top. *Applications*: Any unit material that is not fragile. *Speed*: 0.4 m s^{-1} max

Deadplate: A metal or other hard surface onto which packages come to a stop following conveying, such as by a chute or slide or other conveyor. *Applications*: Any unit material that is not fragile. *Speed*: 0.4 m s^{-1} max

GUIDEWAY, STATIC AND POWERED; CARRIER, DYNAMIC AND NONPOWERED

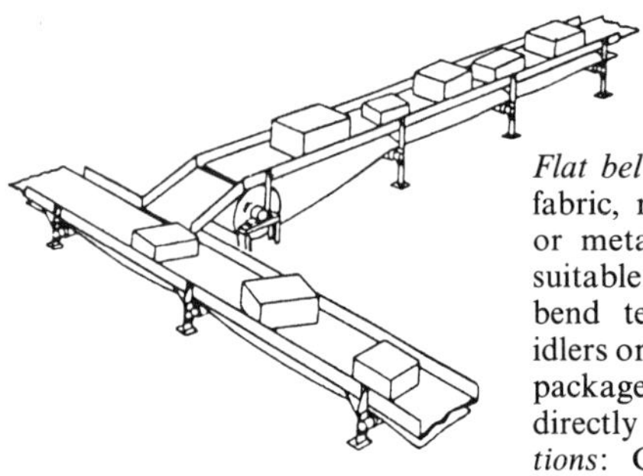

Flat belt conveyor: An endless fabric, rubber, plastic, leather or metal belt operating over suitable drive, tail end and bend terminals and overbelt idlers or slider bed for handling packages or objects placed directly on the belt. *Applications*: Carrying objects, cartons, bags used in manual and automated sorting applications. *Speed*: 4.0 m s^{-1} max

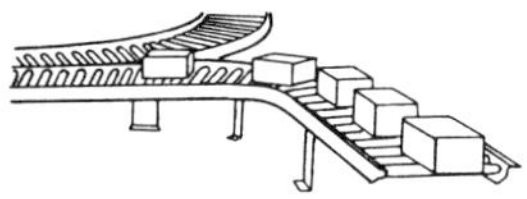

Roller conveyor: A series of rollers supported in a frame over which objects are advanced manually, by gravity or by power. *Applications*: Almost any object, either the unit material or unit load which has a relatively flat, rigid or semirigid surface which contacts two or preferably three or more rollers. *Speed*: Limited to inertia considerations of unit being moved

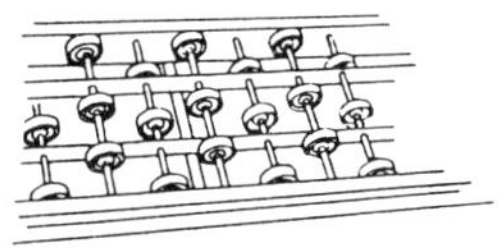

Wheel conveyor: A series of wheels supported in a frame over which objects are moved manually or by gravity—usually skatelike wheels. *Applications*: Used for moving relatively light unit loads; frequently used for truck loading or unloading because of its portability. Ideal for curves because of its simple construction. *Speed*: 0.4 m s^{-1} max

Flat top conveyor: A chain conveyor with specially designed links or slats which form a continuous top and are supported with chain rollers. *Applications*: Bottles, cans, jars. Two parallel flat top chain conveyors are employed for moving cars and trucks through final assembly inspection and assembly operations. *Speed*: 1.0 m s^{-1}max

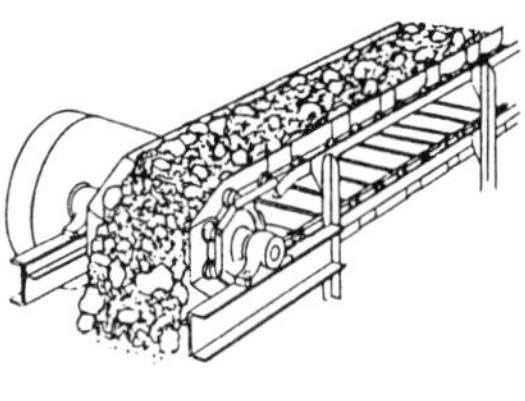

Apron conveyor: A series of pans or plates attached to a chain or pivotally attached to one another to form a conveying medium, the apron being the carrier. *Applications*: Hot forgings and castings, turnings and chips, greasy, oily materials, materials that require draining. *Speed*: 0.5 m s^{-1} max

Drag conveyor: A type of conveyor having one or more endless chains which drag unit materials through a trough. *Applications*: Abrasive unit materials, hot metals, logs, lumber, vehicles. *Speed*: 0.1 m s^{-1} max

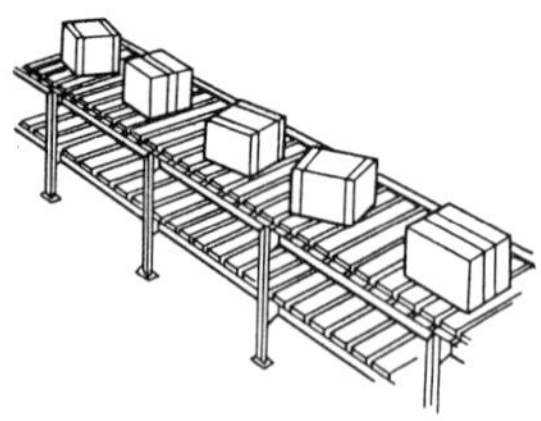

Slat conveyor: A conveyor employing one or more endless chains to which nonoverlapping, noninterlocking space slats are attached. *Applications*: Hot, wet or heavy unit materials or unit loads. *Speed*: 0 to 0.4 m s^{-1}

Chain conveyor: A series of links pivotally joined together to form a medium for conveying or transmitting motion or power. General classes of chain common to the conveyor are detachable pintle, combination, roller, rivetless, coil, inverted tooth and bar link chains. *Applications*: Unit materials that are heavy, greasy or hot. *Speed*: 1.0 m s^{-1} max

Table 2 (*continued*)

GUIDEWAY, STATIC AND POWERED; CARRIER, DYNAMIC AND NONPOWERED (*continued*)

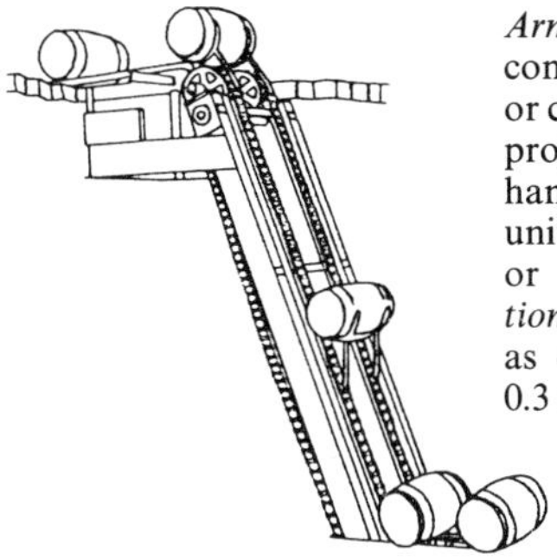

Arm conveyor: A conveyor consisting of an endless chain or chains to which are attached projecting arms or shells for handling packages or other unit materials in a vertical or inclined path. *Applications*: Large containers, such as drums and barrels. *Speed*: 0.3 m s^{-1} max

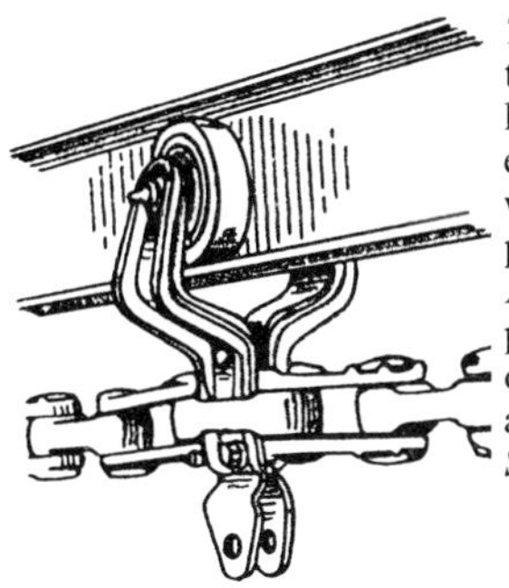

Trolley conveyor: A series of trolleys from or within an overhead track and connected by an endless chain or other linkage with unit materials or loads suspended from the trolleys. *Applications*: Paint lines, intraplant movement of materials; other continuous processing, such as baking, cooling, degreasing. *Speed*: 1.0 m s^{-1} max

Tow conveyor: An endless chain supported by trolleys from an overhead track or running in a track at the floor with means for towing floor-supported trucks, dollies or carts. *Speed*: 1.0 m s^{-1} max

Air film conveyor: A moving cushion of air with both vertical and horizontal components propels relatively light unit materials along a guideway utilizing low-pressure air. *Applications*: Tissue products, cans, shipping containers. *Speed*: 2.5 m s^{-1} max

Dumbwaiter: A hoisting and lowering mechanism with a car of limited capacity and size which moves and guides in a substantially vertical direction and is used exclusively for carrying unit materials. *Applications*: Any unit material that is within the weight and space limits of the car. *Speed*: 0.8 m s^{-1} max

Elevator: A hoisting and lowering mechanism equipped with a car or a platform which moves and guides in a substantially vertical direction and which serves two or more floors of a building or structure. *Speed*: 0 to 7.6 m s^{-1} depending on type of structure

Car-type conveyor: A series of cars attached to and propelled by an endless chain or other linkages running on the horizontal or a slight incline. *Applications*: Large or heavy unit materials, foundry production lines, assembly operations. *Speed*: 1.0 m s^{-1} max

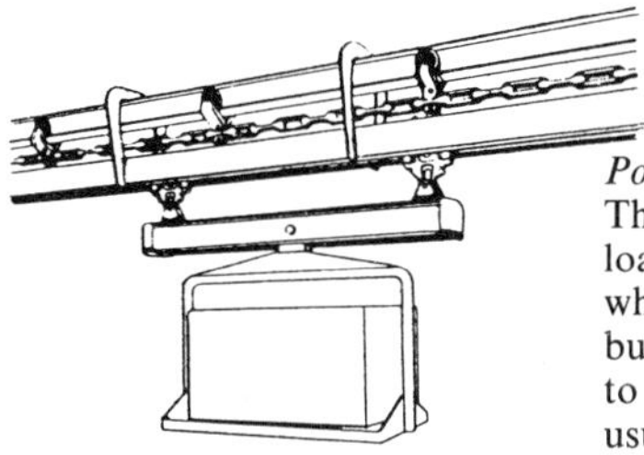

Power and free conveying: The unit material or unit load is carried on a trolley which is chain propelled but may be switched at will to a second track, which is usually a free monorail track. The power track usually has pushers which engage the trolleys on the free track. *Applications*: Temporary or float storage on production, assembly and test lines; routing and sorting. *Speed*: 1.0 m s^{-1} max

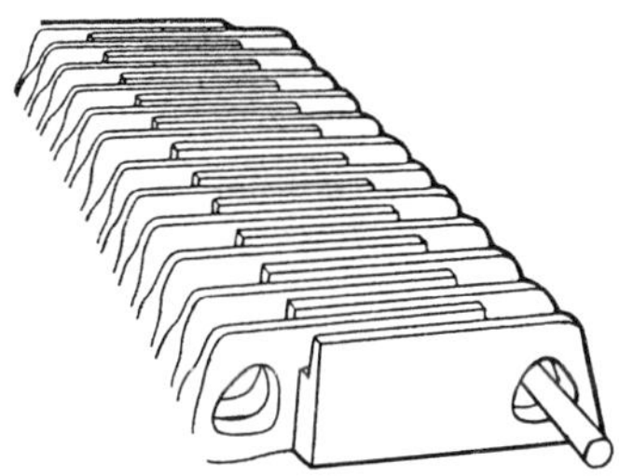

Process/plastic belting: An endless plastic belt made of individual modules operating over a suitable drive supported normally by plastic wear strips for handling and processing food and food beverage containers. *Applications*: Food industry. *Speed*: 0.5 m s^{-1} max

Pneumatic tube conveyor: A tube guideway system through which carriers with dynamic seal capabilities are propelled by positive pressure air throughout the system. *Applications*: Document transfer in banking and security industries, document and medicine transfer in hospitals. *Speed*: 2.5 m s^{-1} max

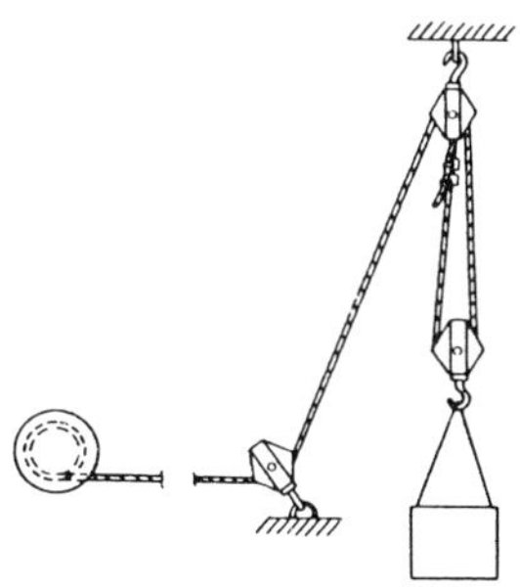

Fixed hoist: A lifting device with a load attached to the lower block, which is in turn supported by ropes or parts of lines suspended from the upper block. For two ropes, each rope must carry half the weight of the load, thus giving the system a mechanical advantage of two. Mechanical advantage is governed by the number of ropes supporting the load. *Applications*: Any intermittent lifting task; mine hoists. *Speed*: 2.5 m s^{-1} max

Table 2 *(continued)*

GUIDEWAY, DYNAMIC AND POWERED; CONVEYOR

Mobile: A conveyor supported on structures which are movable under their own power. *Applications*: Loading and unloading unit material in air transportation. Other intermittent unit materials applications. *Speed*: Same as belt, roller, etc.

Telescoping: A type of conveyor, typically belt or roller, the length of which may be varied by telescoping frame members. The telescoping capability is normally powered. *Applications*: Loading and unloading containers, vans, trucks. *Speed*: Same as belt, roller, etc.

Sorting: A conveyor which receives mixed unit loads and discharges them to segregated spaces or conveyors in response to an automatic dispatch control. *Applications*: Order fulfillment, mail order, postal service, parcel post, parcel services. *Speed*: Same as belt, roller, etc.

Arcing: A powered conveyor, typically belt or roller, that is hinged at its input end and is powered to move in an arc, typically to match the speed and deliver packages or other units of material to a dynamic sorting conveyor such as a tray sorter. *Applications*: Mail order, parcel sorting postal service. *Speed*: Same as belt, roller, etc.

Switching: A conveyor which has sections of dynamic guideway that can switch the unit material or the container or carrier holding the unit material from one guideway to another at a converging or diverging section. *Applications*: Sorting and destination-controlled movement systems (e.g., outbound baggage systems). *Speed*: Same as belt, roller, etc.

GUIDEWAY, STATIC AND NONPOWERED; CARRIER, DYNAMIC AND POWERED

Automatic guided vehicles (*AGVs*): A driverless industrial truck following a fixed path guideway. The guidance is usually based on a wire embedded in the floor, or an optical surface on the floor. The optical pathway normally requires a relatively clean environment. The AGV interfaces with pickup and delivery, as well as work stations. The AGV also has capabilities of towing trains throughout a facility, and usually has switching capabilities. *Applications*: Automated factory, warehousing, assembly operations. *Speed*: 1.0 m s^{-1} max

Shuttle cars: A driverless vehicle following a fixed path, usually a railsystem, capable of forward and reverse movement between fixed stations for automatic loading and unloading of the work or units that it carries. There is no switching involved. *Applications*: Unit materials movement between fixed locations, assembly operations, fixed on-line production facilities, transfer operations. *Speed*: 1.0 m s^{-1} max

pushing devices that preserve the orientation of the unit item. The most rapid system, however, as far as direct sorting from multiple input channels is concerned, is the tilt tray or gull wing sorter, which drops the item from the tray into a velocity-controlled chute at the designated sorting point. The tilt tray or gull wing sorter moves like a carousel in a closed loop between the input channels and the destination sorting chutes. Multiple-layer gull wing tray devices can sort at rates up to 600–800 unit items per minute. The pulling devices are usually limited to about 80 items per minute.

Order picking is usually a manual operation based on computerized routing of the information to minimize travel time of the order picker and the movement device. Automated order picking involves oriented packaging or other ordered arrangements of the unit item so that pick-and-place or other escapement or gate-type devices can automatically pick one or more of the unit items from the ordered arrangement of these items. Random selection or picking of a unit item from an unordered collection of unit items is not automated and requires a manual order picking operation.

Order assembly is normally accomplished by picking a specific order over a specified period of time and assembling it through a conveyor system. Alternatively, order assembly can be accomplished progressively by using a common container that travels between order picking stations, automatically picks the items for a specific order, and then delivers the assembled order to a packing (unitization) station.

The trend has been to automate the sorting operation by having a laser read the bar code, or by using optical character recognition (OCR) identification of the unit item to be sorted. This order picking process requires orienting the item so that the bar code or OCR can be read. Order assembly requires a designated time frame or a controlled space that can be progressively filled. The trend is toward extending the randomness of the unit item's position as the reliability of sorting, picking and order assembly is

Table 3
Flexible carrier path unit material movement systems and their applications

GUIDEWAY, STATIC AND NONPOWERED; CARRIER, DYNAMIC, NONPOWERED

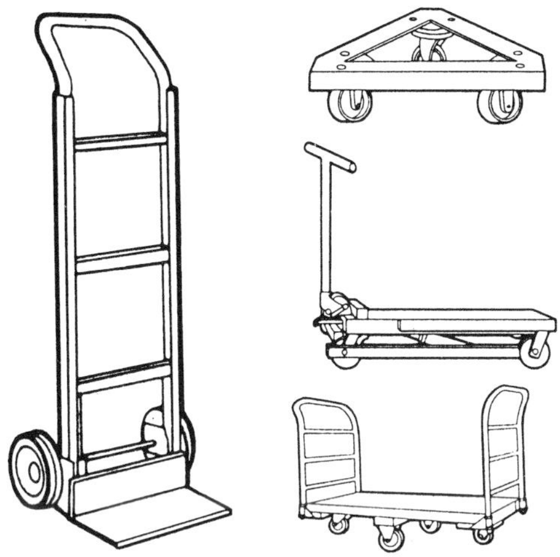

Industrial trucks, manual: A hand-operated vehicle used for the movement of unit materials or unit loads intermittently over flexible paths having suitable running surfaces and clearance. Manual industrial trucks may be classified by the number of wheels. A single-wheel hand truck would be typically a wheelbarrow. Two-wheel hand trucks take a number of forms depending on the unit materials or unit loads being handled. The dolly is a small, low-platform-type load carrier with one or more rollers, casters or wheels. The four-wheel hand truck is a rectangular load-carrying platform with four to six wheels for manual pushing, usually by means of a rack or handle at one or both ends; some have two larger wheels at the center of the platform also for easy maneuverability. The hand lift truck is essentially a wheeled platform that can be rolled under a pallet or skid and equipped with a lifting device designed to raise loads just high enough to clear the floor and permit manually moving the load. *Applications*: Unit materials and unit loads in warehousing, all industry, farming, etc. *Speed*: Low (walking pace), 1.5 m s^{-1} max

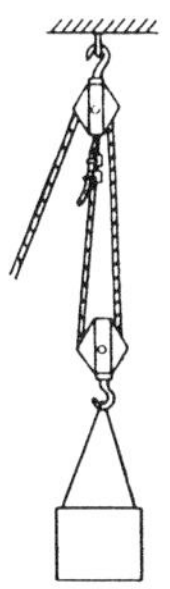

Hoist, manual: A device for lifting or lowering unit materials suspended from a hook on the end of a retractable chain or cable, usually supported from overhead by a hook or travelling on a track. *Applications*: Used for relatively light loads and to service a workshop. *Speed*: 0.3 m s^{-1} max

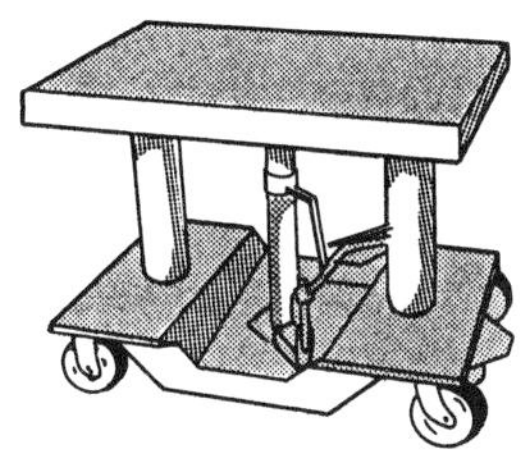

Lift table, manual: A wheeled, mobile table that can be elevated or lowered manually by a mechanical or hydraulic mechanism. *Applications*: Transferring unit materials, either work or tools machine to machine or machine to or from storage. *Speed*: 0.3 m s^{-1} max

GUIDEWAY, STATIC AND NONPOWERED; CARRIER, DYNAMIC AND POWERED

Lift truck: A self-loading, counterbalanced, self-propelled wheeled vehicle carrying an operator, designed to carry a load on forks or other attachments, fastened to a telescoping mast, which is mounted ahead of the vehicle to permit lifting and stacking of loads. *Applications*: Used for lifting, lowering, moving, stacking, loading, unloading and maneuvering unit materials and unit loads. *Speed*: 9.0 m s^{-1} max

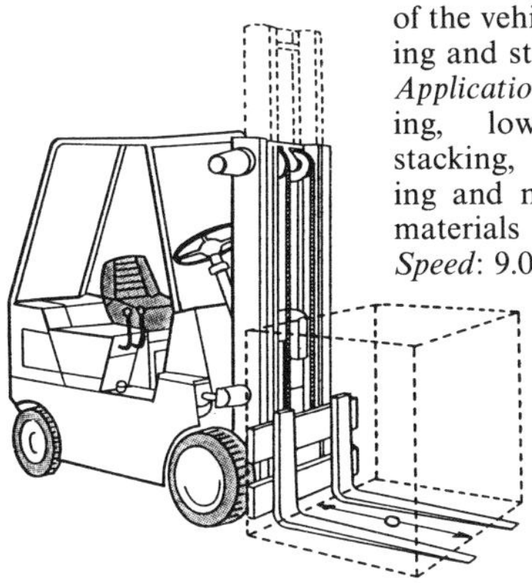

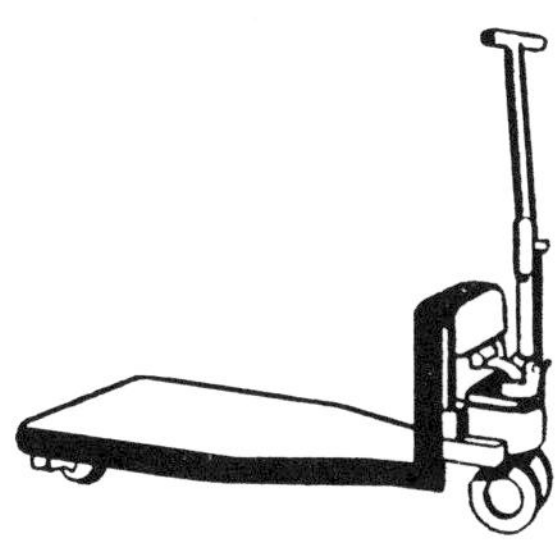

Platform lift truck: A truck designed to handle loads on skids by means of an elevating platform which lifts the skid. *Applications*: Low-lift hauling, handling work between production areas. *Speed*: 4.0 m s^{-1} max

Straddle carrier: A truck whose inverted frame has four wheels mounted on legs, which permits the vehicle to straddle a unit load or unit of material, pick it up with hydraulically operated load-carrying shoes mounted within its frame and then move the load. *Applications*: Used for long unit loads, fast movement of large loads. *Speed*: 25.0 m s^{-1} max

Table 3 *(continued)*

GUIDEWAY, STATIC AND NONPOWERED; CARRIER, DYNAMIC AND POWERED *(continued)*

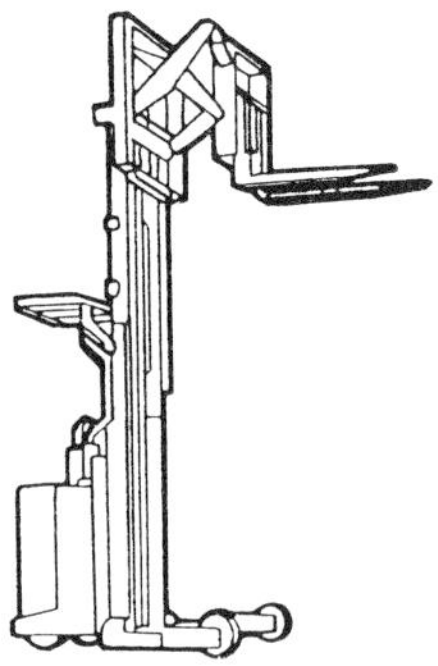

Reach truck: A variation of the straddle truck in which the forks reach out for the load on a pantograph-type device, which permits the forks to travel forward to engage the unit load, lift it and then retract it to the mast for travelling. *Applications*: Used for warehousing, storage in narrow aisles and tight quarters. *Speed*: 2.0 m s^{-1} max

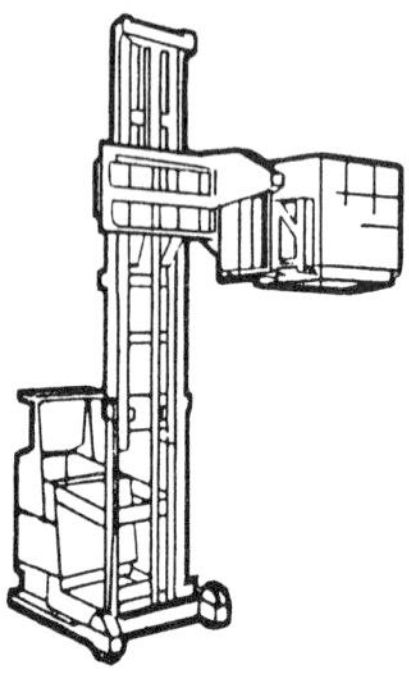

Side loader truck: A four-wheel truck which picks up by means of a mast, with forks fully mounted in a bay at the center of the truck chassis. This arrangement permits the mass to travel transversely across the chassis, and in the extreme outboard position lift loads within reach of the forks. The load is then retracted and placed on the chassis or deck for carrying. *Applications*: Handling long loads—pipe, lumber, sheet metal, rock laths. *Speed*: 2.0 m s^{-1} max

Mobile crane: A crane that can move freely about the job site under its own power without being restricted to a predetermined travel path. A crane is a device for raising, lowering and/or moving unit loads laterally through the use of a hoisting mechanism employing ropes but whose hoisting engine is an integral part of the machine. Mobile cranes may have crawler bases, wheel bases or truck bases. Mobile cranes may be of the latticed boom type, the telescoping boom type or the tower type having a cantilevered boom that is counterweighted or counterforced. *Applications*: Construction, shipbuilding, maintenance. *Speed*: 5.0 m s^{-1} max

Platform trucks: A fixed level, nonelevating, load-carrying truck which supports the load on a platform. *Applications*: Used for heavy loads, typically in maintenance work, carrying tools as well as work to the maintenance site

Tractor trailer train: A handling system consisting of a three- or four-wheel self-propelled vehicle designed for pulling loaded carts or trailers. They may be either a rider type or a walkie type. *Applications*: Used in warehousing and receiving and shipping for large volumes of unit materials over distances of 100–125 m. *Speed*: 6.0 m s^{-1} max

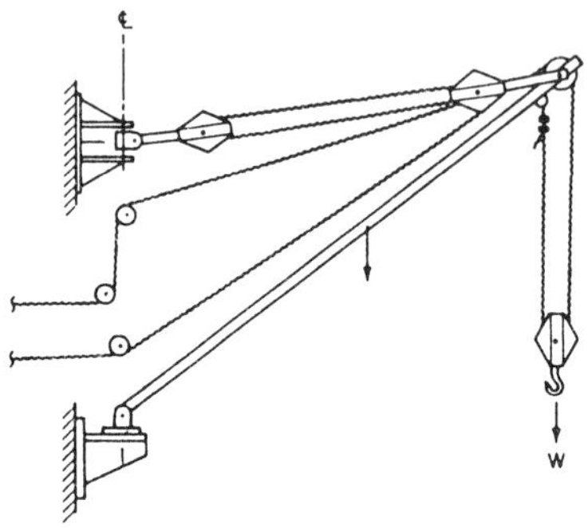

Derrick: (The name is derived from a London hangman of about 1600, denoting the similarity between a gallows and lifting devices in use at that time). A device for raising, lowering and/or moving loads laterally through use of a hoisting mechanism employing ropes, but whose hoisting engine is not an integral part of the device. Types of derricks include the Chicago boom derrick which can be mounted on a building frame during or after construction. A guy derrick is a Chicago boom derrick mounted on a tower (the mast) which is held vertical by six or more guy ropes. Other derricks, detailed by Shapiro (1980), include gin pole, stiffleg, gallows, A-frame, basket, shaft hoist, cathead and flying strut types. *Applications*: Construction, shipbuilding, maintenance. *Speed*: 5.0 m s^{-1} max

Pedestal-mounted crane: Pedestal-mounted tower cranes are of two principal types. The first is a saddle jib or hammerhead jib crane. The jib is basically a horizontal boom that is cantilevered and counterweighted. The saddle jib has a trolley capable of travelling the length of the jib. The trolley includes the upper block and supports the load block suspended from it. The other type is a luffing jib. The jib is mounted to the tower with a pivot and a rope system for raising, lowering and pivoting the jib. *Applications*: Construction, shipbuilding, maintenance. *Speed*: 5.0 m s^{-1} max

GUIDEWAY, STATIC AND POWERED; CARRIER, DYNAMIC AND POWERED

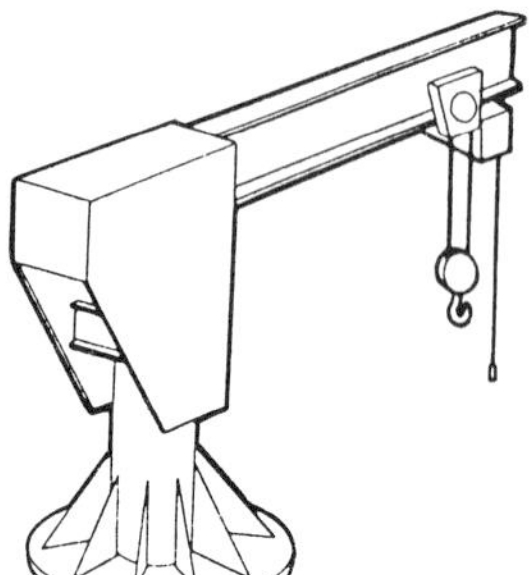

Jib crane: A lifting device travelling on a horizontal boom mounted on a column or mast which is attached to the floor, walls or posts, or mounted on rails. *Applications*: Serving a workplace. *Speed*: 0.5 m s^{-1} max

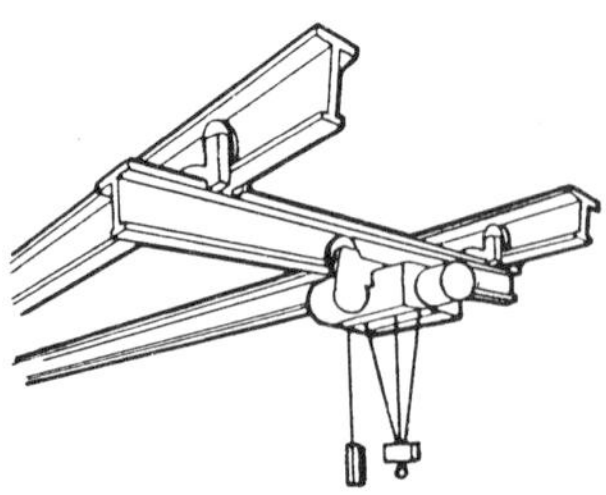

Gantry crane: A crane commonly of the bridge type, which is supported by a structure spanning an intervening space on legs travelling on tracks or two legs mounted on wheels for aisle or road movement. *Applications*: Lifting and moving large, heavy objects; foundries, steel mills, shipbuilding, yard storage. *Speed*: 0.5 m s^{-1} max

Table 3 *(continued)*

GUIDEWAY, STATIC AND POWERED; CARRIER, DYNAMIC AND POWERED *(continued)*

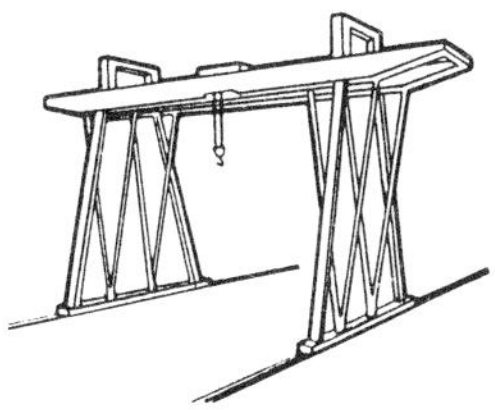

Bridge crane: A lifting device mounted on a bridge consisting of one or two horizontal girders, which are supported at each end by trucks riding on runways installed at right angles to the bridge. Runways are installed on building columns, overhead trusses or frames. Lifting device moves along bridge while bridge moves along runway. *Applications*: Lifting and moving large, heavy objects, foundries, steel mills, shipbuilding, yard storage. *Speed*: Bridge, 2.5 m s^{-1} max; hoist, 0.5 m s^{-1} max

Climbing crane: Tower cranes identical to free-standing cranes from the slewing circle up, but otherwise entirely different. The whole crane, including a fixed length of mast, lifts itself up on the structure as the work of construction progresses. The crane normally climbs through an elevator shaft constructed to accommodate it. The crane is normally lifted by hydraulic jacks and then the lifted structure is secured by dogs. *Speed*: 5.0 m s^{-1} max

GUIDEWAY, STATIC AND POWERED; CARRIER, DYNAMIC AND POWERED

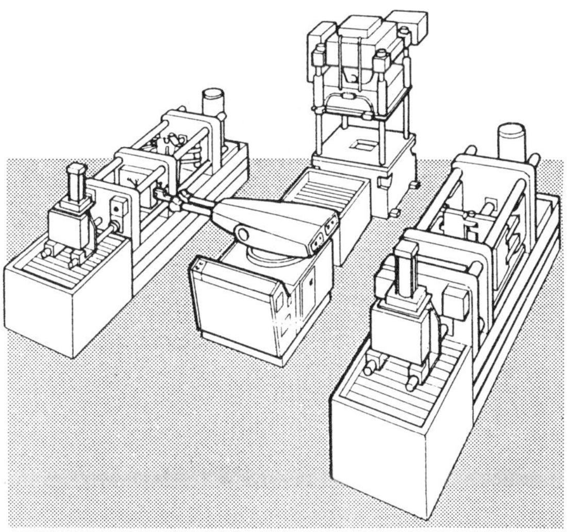

Robot in die casting installation

Robots: (Guideway and carrier are synonymous.) A robot is a reprogrammable, multifunctioned manipulator designed to move material, parts, tools or specialized devices through variable program motions for the performance of a variety of tasks. All robots consist of two major component systems. First, there are moving parts, chiefly comprising the arm, wrist and hand elements. Then there is the control system ranging from a series of adjustable mechanical stops or limit switches, to computer-type controls which give the robot a programmable memory. This memory allows the robot drives to follow a path that is accurately defined all along its length by a series of continuous coordinates. It can also be coupled with another computer or machine control system to synchronize the robot for the most efficient and safe movement or production operation possible. *Applications*: Feeding and removing work from machines; assembly operations. *Speed*: 1.0 m s^{-1} max

Limited sequence robots: Typically these robots use a system of mechanical stops and limit switches to control the movements of arm and hand

Playback robots with point-to-point control: The methods for achieving positional control of each limb rely on the position of some form of servomechanism

Playback robots with continuous path control: The robot has a memory size sufficient to allow path control that is, to all intents and purposes, continuous

improved. This random positioning will provide greater flexibility for the sorting, picking and assembly systems.

Depending on the amount of information required and the shape, density and fragility of the unit item, automated sorting has proved to be much faster than manual sorting—from several times as fast for large, fragile and bulky unit items to orders of magnitude faster for small, more uniform unit items such as small parcels, rolls and letters. Automation is the solution to high-volume sorting problems as long as the unit items to be sorted and assembled into unit loads are within specified size and weight limits (i.e, the parameters for reliable performance of the system). If the system is required to handle outsized unit items, automation will generally fail; for success, presorting is necessary to cull the outsize, low-volume items that will not fit within the system. For most automated materials handling systems, buffers or floats involving a queue of units are necessary and are an economic solution to variations in supply and demand between the operations of the system. There has, however, been a trend toward reduced queue length of buffers in automated systems as system reliability has been improved.

See also: Materials Handling: Automation, Unit Materials; Materials Handling: Bulk Materials

Bibliography

Apple J M 1972 *Material Handling Systems Design.* Ronald Press, New York, pp. 471–531
Bowman D 1972 *Lift Trucks; A Practical Guide for Buyers and Users.* Cahners, Boston, MA, p. 7
Carson G B, Bolz H A, Young H H (eds.) 1972 *Production Handbook*, 3rd edn. Ronald Press, New York
Eastman R M 1987 *Materials Handling.* Dekker, New York
Engelberger J F 1980 *Robotics in Practice: Management and Applications of Industrial Robots.* AMACOM, New York
Hollier R H (ed.) 1986 *Proc. 3rd Int. Conf. Automated Materials Handling.* IFS (Publications), Birmingham, UK
Industrial Engineering Handbook, 2nd edn., 1963. McGraw-Hill, New York
Kulwiec R A 1985 *Material Handling Handbook*, 2nd edn. Wiley, New York
Marks' Standard Handbook for Mechanical Engineers, 8th edn., 1978. McGraw-Hill, New York
Material Handling Engineering Directory and Handbook 1974. Industrial Publishing, Cleveland, OH
Shapiro H I 1980 *Cranes and Derricks.* McGraw-Hill, New York, pp. 79, 293

This subject area is also covered by the journals *Material Handling Engineering*, published by Penton Publishing, Cleveland, OH, and *Modern Materials Handling*, published by Cahners, Newton, MA

R. H. Bode
[Arthur D. Little, Cambridge, Massachusetts, USA]

Materials in the Industrial System

The notion of "materials" in an industrial system is rooted in a model of staged processes of production. In this conception, certain processes such as agriculture and mining extract a raw material from the natural environment and subsequent processes refine and transform it into materials. At a still later "fabricating" stage, the refined material is converted into a finished product. For example, iron ore is mined (extractive stage), converted to steel (materials stage) and cast, stamped and machined into components that are assembled into automobiles (fabricating stage); hydrocarbon feedstocks are mined (extracted), converted into plastics (materials) and molded (fabricated) into, say, plastic garden hose; wood is cut (extracted), milled and/or converted to plywood, chipboard or other building stock (materials), and then fabricated into buildings and boats.

This article is concerned with the sequence of producing and processing materials in the context of the overall industrial system. The system is represented by an economy-wide input–output table. In this general context some major industrial activities, particularly the delivery of energy and services, fall outside the materials processing sequence. Furthermore some commodities, such as petroleum, are raw materials for engineering materials as well as for fuels and lubricants.

The end products of the industrial system vary in the kinds and amounts of materials that they require. Over time, a decrease is observed in the total volume of materials used to deliver most end products. However, changes affecting individual materials and products may not be apparent, because several products are subsumed in a single category of an industrial classification scheme. Increases in the use of one material may offset decreases in the use of another when the two are aggregated into a single materials supplying sector.

Technological change expands the variety of materials that man can produce. Highly specialized new materials are developed to meet the needs of particular fabrication processes or to improve particular products. In addition, the range of choice of materials that may be used to produce many individual products is becoming wider. These changes, in turn, modify the interindustrial linkages represented in an input–output table.

1. Materials in a Staged Productive Sequence

Figure 1 shows the model of staged processes of production schematically. The diagram is in the form of a simple input–output table. Sectors listed along the left send their products to sectors listed across the top. Each sector represents a stage of processing. A check signifies a transaction. The check in the top row of the diagram indicates that the extractive sector sends its outputs to materials producers or, looking at it from the point of view of the receiving sector, materials producers receive their inputs from the extractive sector. Fabricators are shown as delivering finished products to final users. This hypothetical scheme is strictly staged; later stages depend on earlier stages for their inputs, but not vice versa.

The above model gives a plausible approximation of the role of materials in an idealized industrial scheme. However, it has very significant limitations in representing a real economy. First, it does not deal with nonmaterials inputs into the various processes. Fuels and services (as well as labor and capital, which are not considered in this scheme) are used by virtually all sectors. But fuel and service sectors are not fabricators. These sectors do not consume significant amounts of materials on current account. Figure 2 incorporates fuels and services as "general inputs" into the industrial scheme.

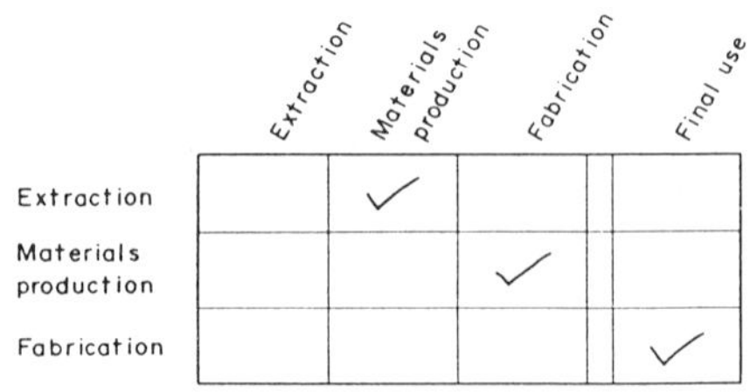

Figure 1
Staged production

	Extraction	Materials production	Fabrication	Fuel	Services	Final use
Extraction		✓		✓		
Materials production			✓			
Fabrication						✓
Fuels	✓	✓	✓	✓	✓	✓
Services	✓	✓	✓	✓	✓	✓

Figure 2
Staged production including general sectors

	Extraction	Materials production	Fabrication	Final use
Extraction		✓		
Materials production			✓	
Fabrication	×	×	×	✓

Figure 3
Staged production with capital flows: √ represents transactions on current account; × represents transactions on capital account

Figure 2 shows all sectors drawing inputs from the general sectors while fabricators alone purchase materials. The expanded model of Fig. 2 still does not deal with some very important features of the real world. Petroleum is extracted and refined to produce both fuel and materials. Wood is also used both as a material and as a fuel. Grains are grown (extracted) and milled (refined), but they do not normally produce materials that are to be fabricated. Thus, many productive sequences do not correspond to the simple materials sequence prototypes illustrated in Figs. 1 and 2.

A further complication is associated with the accounting distinction between goods on capital and current account. In Figs. 1 and 2, fabricated goods are shown as sold only to final users. Actually some fabricated goods, particularly containers and spare parts, are used as inputs on current account in the extractive and materials sectors. These transactions have been ignored because they represent a minor fraction of fabricated products.

The great bulk of fabricated goods comprises durables, machines and buildings that are purchased on capital account. According to the conventions of national income accounting, sales of goods on capital account are viewed as sales to final users. In reality, of course, durable goods make an essential contribution to the productive process in all sectors. They are segregated from current account purchases in an input–output table, and in the national accounts in general, because these goods are not "used up" rapidly in production. Hence their purchase is postponable and likely to occur irregularly over time. Nevertheless, most sectors require capital goods. If purchases of durable goods on capital account are taken into consideration, Fig. 1 must be modified to look like Fig. 3. The introduction of capital items obscures the stagelike structure of intersectoral flows that characterizes current account transactions. Only fabricators buy materials directly, but all sectors depend on materials indirectly, because they use fabricated products.

There are, of course, many kinds of materials and many different fabricating processes. Furthermore, technological change results in a proliferation of both. Thus, the extractive and materials sectors in Figs. 1–3 are really aggregates of many subsectors. The presence of such subsectors can be symbolized by subdividing each stage into two parts, as shown in Fig. 4. A realistic representation would require many more subsectoral divisions.

If each material were sold only to a particular specialized fabricator, the pattern of checks in the table would continue to fall along a single diagonal line as in Figs. 1 and 2. This special case is illustrated by the entries of checks in Fig. 4. Actually, many real-world products can be fabricated from alternative materials or from combinations of materials, and individual materials can be fabricated into numerous different products. Hence, these options appear as off-diagonal entries, as illustrated by the O's in Fig. 4.

Recycling permits the substitution of scrap for extractive products in making materials. Depending on where the scrap originates, recycling will warrant off-diagonal entries in the materials columns in addition to or instead of the entries signifying the flow of extractive products into materials.

Figures 1–4 are based on hypothetical data. Were we to depict actual transactions among real plants or firms, rather than hypothetical transactions among abstract units, we would find the picture still more complicated. Integrated establishments can encompass activities

	Extraction I	Extraction II	Materials production I	Materials production II	Fabrication I	Fabrication II	Final use
Extraction I			✓				
Extraction II				✓			
Materials production I					✓	O	
Materials production II					O	✓	
Fabrication I							✓
Fabrication II							✓

Figure 4
Staged production with two sectors at each stage: O represents transactions added to the system when particular products can be fabricated from a variety of materials

spanning many stages of production. Individual establishments (the economic units of the US Census of Manufactures) may purchase extractive products, refine them into materials and fabricate them, thus combining several stages. For such establishments, interplant transactions would furnish no record of separate stages. Statistics of transactions among sectors are likely to combine establishments with different degrees of integration into a single aggregate sector.

Given all these considerations, it is not surprising that an empirically based input–output table is not faithful to the prototype of Fig. 2 where the full output of each stage becomes input into the next. However, the US tables do approximate this pattern over the span of materials and fabricating sectors (see, for example, *Scientific American* 1981).

2. Materials Requirements for Various End Products

Some products use more materials than others. It is useful to estimate the materials intensity of different final products (see *Intensity of Use of Materials*; Herman et al. 1989) so as to gauge how changing conditions will affect total materials requirements. What does a slowdown of economic growth mean for the materials sectors? How would a shift in consumer expenditures from automobiles to entertainment affect them? How can needs for individual materials be anticipated? Questions like these make it important to be able to measure the materials content of specific end products.

In many instances, materials remain identifiable after they are fabricated. The various kinds of metal in an automobile can be observed directly and even measured by weighing the components. Indeed, the weight of the various materials embodied in US automobiles are published regularly (*Ward's Automotive Yearbook* 1992). The "material content" of an automobile so measured underestimates the total amount of material used to make it, since it takes no account of scrap and waste generated in the course of fabrication. Total materials use could be observed by monitoring the fabrication of each component and measuring the amounts of materials entering the productive process at each stage.

The total amount of a given material required to produce a given final product can also be estimated on the basis of the information in a published input–output table. Assume that sector i produces a material while sector j produces a fabricated product. Then a given direct coefficient a_{ij} specifies the amount of material i directly required to produce a unit of good j. (Normally i and j are measured in "dollars' worth" units.) If A is a matrix of direct coefficients and I is the identity matrix, an element of the Leontief inverse matrix $(I-A)^{-1}$, A_{ij}, measures the total amount of material i that is required to deliver a unit of j to final use. A_{ij} will normally be greater than a_{ij}, since it includes indirect as well as direct requirements of material i to produce j. Thus, if i is steel and j is automobiles, a_{ij} will measure steel directly purchased by the auto sector, while A_{ij} will include both the direct purchase of steel by the auto sector and steel purchased to produce the nuts and bolts that are, in turn, sold to the auto sector (Leontief 1986) (see *Input–Output Analysis*).

Table 1 presents information on the materials intensity of the goods produced by various sectors as measured by the Leontief inverse of the US input–output table for 1982. Each element of Table 1 indicates the dollar amount of a particular material used, directly or indirectly (i.e., anywhere in the system), to produce a dollar's worth of the sector's output. Materials are enumerated along the left side of the table; the goods being produced are enumerated across the top. The number 0.085 in the upper left-hand corner of the table, for example, indicates that 8.5 cents' worth of wood were used, directly and indirectly, for every dollar's worth of new construction put in place in 1982.

According to the information in Table 1, new construction and household furniture are very "wood-intensive." All other types of products require less than two cents' worth of wood for every dollar of final product. Only construction calls for significant amounts of stone and clay products, and glass requirements are relatively small for all sectors. On the other hand, most of the products enumerated in Table 1 require significant amounts of ferrous and nonferrous metals. Automobiles and household appliances draw very heavily on steel. Electrical machinery and aircraft draw more heavily on nonferrous than on ferrous metals.

The column sums given in the bottom row of the table measure the total value of all the enumerated materials per dollar of output of each sector. Note that the ordnance and computer and office machine sectors are the least materials-intensive of all, followed by aircraft and instruments. Since all of these four product lines are rapid growers relative to the rest of the economy, there should be some tendency for the materials intensity of the economy as a whole to decrease.

The US product mix is changing in such a way as to relax emphasis on materials requirements. The relative growth of services should reinforce this tendency.

Table 2 presents the direct materials requirements corresponding to the total requirements cited in Table 1. In most cases, direct purchases are much smaller than the corresponding total purchases. This is not surprising. There is a very complex division of labor among plants. Fabricators are likely to buy parts and components from other fabricators, who in turn buy the materials. The latter may also buy semifabricated components rather than direct materials, and so on.

Input–output information is very useful for the analysis of materials intensity, but several caveats

Table 1
Total amounts ($ per $) of specific materials required to produce various fabricated goods in the USA, 1982

	New construction	Ordnance	Household furniture	Office, computing, accounting machines	Electric industrial equipment	Household appliances	Motor vehicles	Aircraft	Scientific and controlling instruments
Wood	0.085	0.005	0.161	0.005	0.006	0.015	0.006	0.003	0.008
Plastics	0.009	0.008	0.042	0.019	0.012	0.038	0.021	0.008	0.028
Glass	0.003	0.002	0.007	0.004	0.002	0.009	0.013	0.002	0.007
Stone and clay	0.069	0.005	0.009	0.008	0.013	0.013	0.009	0.006	0.007
Ferrous metals	0.067	0.045	0.046	0.033	0.081	0.122	0.143	0.048	0.041
Nonferrous metals	0.042	0.051	0.026	0.056	0.098	0.089	0.059	0.077	0.065
Total	0.275	0.116	0.291	0.125	0.212	0.286	0.251	0.144	0.156

Source: Interindustry Economics Division (1988)

Table 2
Direct requirements ($ per $) of specific materials to produce various fabricated goods in the USA, 1982

	New construction	Ordnance	Household furniture	Office, computing, accounting machines	Electric industrial equipment	Household appliances	Motor vehicles	Aircraft	Scientific and controlling instruments
Wood	0.055	0.001	0.109	0.000	0.001	0.006	0.001	0.000	0.001
Plastics	0.000	0.001	0.004	0.002	0.003	0.018	0.001	0.001	0.008
Glass	0.001	0.000	0.005	0.000	0.000	0.007	0.008	0.000	0.005
Stone and clay	0.057	0.001	0.004	0.001	0.008	0.007	0.003	0.001	0.003
Ferrous metals	0.028	0.021	0.021	0.006	0.047	0.075	0.061	0.022	0.015
Nonferrous metals	0.012	0.019	0.007	0.013	0.051	0.041	0.015	0.033	0.031
Total	0.153	0.043	0.150	0.022	0.110	0.154	0.089	0.057	0.063

Source: Interindustry Economics Division (1988)

must be entered. First, there are serious drawbacks to relying on value measures of materials inputs. The specified materials have very different unit prices and these prices change over time. Furthermore, the physical properties of a unit of each material are very different. Thus, one cent's worth of steel and one cent's worth of plastics are not generally interchangeable. In particular, there should be no presumption that it would take a dollar's worth of any material to replace a dollar's worth of another, should the substitution be feasible.

Second, the stability of the input–output coefficients cannot be guaranteed. These ratios do tend to change with relative prices and with the introduction of new processes. Hence, generalizations about materials intensity for a given year must be extrapolated with caution. Third, as was noted above, requirements for materials measured in this way do not include expenditures on capital account. The materials required to expand capacity when a given sector's output goes up are not tied to that output in standard input–output computations. Instead, capital goods are themselves counted as final goods. Hence the measures described here tend to underestimate total materials requirements for specific final goods. More sophisticated computations can be made to allow for the materials content of capital goods in the production of other products. However, these computations require information that is not readily available.

When the materials embodied in capital goods are taken into account, certain sectors that would appear to generate only minimal demands for materials present more significant requirements. Hotel services, for example, require very little materials on current account; however, hotel buildings embody a lot of materials, and buildings are essential to the service. If there is any excess capacity, hotel services can be increased without any additional building. Under these circumstances, an increase in hotel services does not require additional materials. If capacity must be expanded, a large demand for materials will be generated.

Every input–output sector represents an aggregate of many establishments, and materials requirements are likely to vary within as well as among sectors. The coefficients for a sector characterize average conditions but do not necessarily apply to any given product or plant. This cautions against the use of input–output information to estimate requirements of strategic materials.

By definition, strategic materials are essential because they lack ready substitutes. Requirements estimated on the basis of sectoral average use tend to understate the needs generated by an increase in special products alone.

Furthermore, requirements for strategic materials for a particular product may have to be traced upstream in the productive process, to components made of special materials. Thus, the aircraft industry may buy bolts of a special steel. The input–output table does not distinguish between bolts suitable for autos and those suitable for aircraft. Instead, the average material needs of the bolt sector are assumed to characterize all bolts. An increase in the output of aircraft generates a larger, and an increase in autos a smaller, demand for special materials than would emerge from an aggregated input–output computation. Special methods have been proposed (Noble and Ayres 1978) to avert this difficulty. Given the problem, it is advisable to consult direct technical information rather than to rely on traditional input–output methods to estimate requirements for strategic materials.

3. Changes in Materials Intensity over Time

Authorities tend to agree that the economy is becoming less materials-intensive with time, although measurement problems complicate the task of demonstrating this proposition (Carter 1970). Radcliffe (1976) shows that the use of most materials increases no more rapidly than population in the long run; GNP growth rates do continue to exceed population. Three important trends contribute to the economy-wide trend of falling materials intensity:

(a) increases in the relative importance of services and of products of low materials intensity in the nation's product mix;

(b) a tendency toward reduction in the size and weight required for a given level of performance for many products; and

(c) use of materials that perform a given function at lower cost—particularly an increased use of plastics relative to metals.

The graphs in Figs. 5–7 illustrate the trends in specific industries that underlie the economy-wide decline in materials intensity. Figure 5 shows the change in relative importance of different kinds of durable goods produced by the American economy between 1972 and 1977. Each line in the graph traces changes over time in the ratio of purchases of a particular durable good to the sum of all durable goods purchases. The reference year is 1972 (i.e., all ratios are unity in that year). The rapidly rising solid line shows that office computing and accounting machines rose by 50% in relative importance between 1972 and 1977. Note that the two materials-intensive products—special industry machinery and transportation equipment—declined, while the less materials-intensive products rose in importance during this period. Similar trends have been apparent throughout the entire postwar period (Carter 1970).

Figure 6 shows how input coefficients for each general input, averaged across all consuming sectors, changed over time. The average input coefficient for business services increased by ~30% between 1972 and 1977. Figure 7 shows how average input

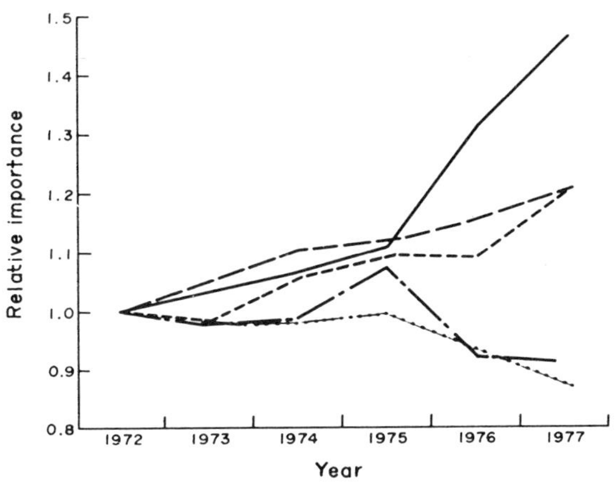

Figure 5
Changes over time in the relative importance of particular durable goods: ——, office, computers; — — —, optical; – – –, instruments; — - —, special industrial machinery; ·····, transport equipment

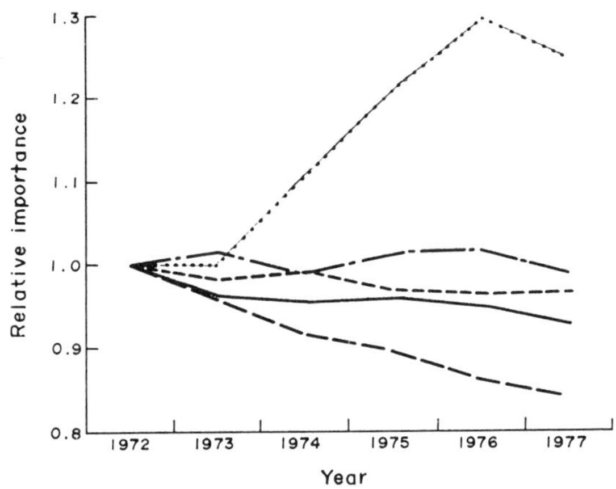

Figure 7
Changes over time in input coefficients for steel into various fabricating sectors: ——, plumbing and heating; — — —, agricultural machinery; – – –, household appliances; ·····, mobile homes; — - —, automobiles

coefficients for steel into most end markets declined between 1972 and 1977. These observations are also continuations of trends observed for earlier periods.

Scholars with an engineering perspective use the term dematerialization for the apparent decline in materials intensity over time (Herman et al. 1989). In accord with the findings reported in this article, they cite good reasons, such as miniaturization and the growing relative importance of services, why dematerialization is likely to continue. Serious environmental problems associated with materials production and use suggest that, as income and population increase, the world industrial system can only be sustained if materials intensity continues to fall substantially, along with major increases in recycling (Ayres 1989).

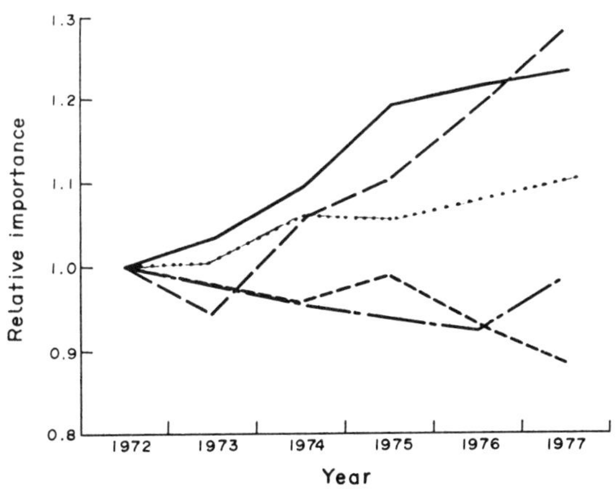

Figure 6
Changes over time in coefficients for selected general inputs averaged across all consuming sectors: ——, communications; — — —, petroleum products; ·····, business services; — - —, printing and publishing; – – –, utilities

4. *Diversification of the Material Base for Various End Products*

For the period from World War II to around 1960, it was shown that the traditional dominance of particular materials in each end use market was breaking down as improved varieties of synthetics and of traditional materials increased fabrication options. In particular, steel and copper were losing ground to various other metals and to plastics in many end markets (Carter 1970). The evidence presented in Sect. 3 indicates that these two trends have continued. Despite rising costs of feedstocks, plastics can still hold their own in competition with metals. As synthetic materials improve, the number of technical options increases in many production processes. The specialization of a single material to a single end use, as shown by the checks in Fig. 2, is replaced increasingly by the diversified market situations characteristic of multiple alternatives, as illustrated by the addition of the O's in Fig. 4. Good canoes are made of wood, metal or fiberglass or of some combination thereof. Similar statements can be made about refrigerators and prefabricated bathrooms.

On the other hand, technologically sophisticated products like aircraft and computers have depended on current developments in the materials sciences. Thus, while specific materials are competing more actively for shares in each others' markets, there are also significant instances of a stricter specialization calling for particular materials designed to support special technical functions. On the one hand, science continues to increase the range of options among materials to be used in manufacturing a given product. On the other hand, scientists and engineers are called upon to design new materials that satisfy the exacting demands of emerging products. There is no basis for predicting which tendency will be dominant in the near future.

See also: Input–Output Analysis; Intensity of Use of Materials; Materials Requirements Planning; Process Analysis

Bibliography

Abelson P H, Hammond A L (eds.) 1976 *Materials: Renewable and Non-Renewable Resources.* American Association for the Advancement of Science, Washington, DC

Alexander W O 1967 The competition of materials. *Sci. Am.* 217(3): 255–66

Ayres R U 1989 Industrial metabolism. In: Ausubel J H, Sladovich H E (eds.) 1989 *Technology and Environment.* National Academy Press, Washington, DC, pp. 23–49

Carter A P 1970 *Structural Change in the American Economy.* Harvard University Press, Cambridge, MA

Carter A P 1980 Changes in input–output structure since 1972. *Inter-Industry Review.* Data Resources, Lexington, MA

Herman R, Ardekani S A, Ausubel J H 1989 Dematerialization. In: Ausubel J H, Sladovich H E (eds.) 1989 *Technology and Environment.* National Academy Press, Washington, DC, pp. 50–69

Interindustry Economics Division 1988 Annual input–output accounts of the US economy, 1982. *Surv. Curr. Bus.* 68(4): 31–46

Leontief W 1986 *Input–Output Economics*, 2nd edn. Oxford University Press, New York

Noble S B, Ayres R U 1978 Materials/energy accounting and forecasting models. In: Ayres R U (ed.) 1978 *Resources, Environment and Economics: Applications of the Materials/Energy Balance Principle.* Wiley, New York, pp. 111–12, 114

Radcliffe S V 1976 World changes and chances: some new perspectives for materials. In: Abelson and Hammond 1976, pp. 24–31

Scientific American 1981 *The Scientific American Input–Output Table for 1972.* Scientific American, New York

Simpson D, Tsukui J 1965 The fundamental structure of input–output tables, an international comparison. *Rev. Econ. Statist.* 47: 434–46

Ward's Automotive Yearbook 1992. Ward's Communications, Detroit, MI

A. P. Carter
[Brandeis University, Waltham, Massachusetts, USA]

Materials Requirements Planning

Materials requirements planning (MRP) is a technique, or a set of systematic procedures, for managing inventory and controlling replenishment of component materials in a manufacturing operation. The MRP procedure requires as input a master schedule for finished products and a bill of material. It calculates time-phased requirements for component items and converts these requirements into replenishment orders. The procedure is repeated routinely as updated information is received in order to adjust previous results and to extend planned orders into the future.

MRP is often discussed as if it were an independent system but, in reality, it is one component module in a total manufacturing control system. The increase in popularity of the MRP technique, which began in the 1960s, is a direct result of the increased computation speed and memory capacity of computers. The technique has been in existence for many years but its application was previously limited because of its extensive data manipulation and record keeping requirements.

This article explains the MRP methodology and the role of MRP in a total manufacturing control system, commonly called manufacturing resource planning or MRP II systems. It explains different types of MRP systems and the relationship of MRP to the just-in-time (JIT) concept. Some features of MRP systems are discussed including alternative lot-sizing models for manufacturing and purchase orders and the concept of safety stock requirements. Finally, the weaknesses and benefits of MRP are discussed.

1. Methodology

Differentiation between dependent and independent demand is a critical concept in MRP methodology. Demand is independent when it cannot be derived from the demand for any other item within the scope of the system. Demand for finished end-items and service demand for spare parts are examples of independent demand. Demand is dependent when it can be derived from the demand for another item. The demand for subassemblies, parts and raw materials is dependent. A part which is a component of an active end-item and also has service usage as a spare part has both dependent and independent demand.

MRP applies to dependent demand items and generates time-phased requirements that serve as a forecast of future demand for component items. The MRP procedure consists of the following four basic steps.

(a) *Netting* subtracts the on-hand and on-order quantities from gross requirements to give net requirements. An example is shown in Table 1 for a typical dependent demand item. This item has current inventory on-hand of 15 units and an open replenishment order for 120 units that is scheduled to be received in period 1. The net requirement for this item is zero in the first two periods and is less than the gross requirement in the third period. After the third period, the gross and net requirements are equal.

Table 1
Netting

	Time period						
	Current	1	2	3	4	5	6
Gross requirements		50	75	90	100	35	100
Scheduled receipts		120	0	0	0	0	0
Projected on-hand	15	85	10	−80	−180	−215	−315
Net requirements		0	0	80	100	35	100

(b) *Lotting* determines individual order, or batch, quantities for replenishment to satisfy the time-phased net requirements. There are many alternative lot-sizing models (these are discussed in Sect. 6). Some of the models involve economic balancing of setup and inventory carrying costs and others follow simpler rules such as using requirements for a fixed number of periods. To demonstrate the concept of lotting, the fixed period requirements rule is used in an example. Lot quantities are set equal to the total net requirements in two period increments. Table 2 shows the resulting planned lot quantities. The first planned lot is the sum of the net requirements in periods 3 and 4 and the second lot is the sum of the net requirements in periods 5 and 6.

(c) *Offsetting* determines the appropriate time for release of planned orders so that they will be received or completed in time to satisfy demand. The planned release date for an order is determined by subtracting the lead time from the date of the earliest net requirement that the order is intended to satisfy. Continuing the example in Tables 1 and 2 and assuming that the replenishment lead time is two periods, the planned order releases should occur in periods 1 and 3 as shown in Table 3.

(d) *Explosion* calculates gross requirements for the next lower level component(s) that will be required to produce the planned orders for manufactured parts or assemblies. Explosion considers the quantity of each component required to make the higher level item and also adjusts for shrinkage, or yield loss, of components in the manufacturing process. The "quantity per" factor for the component is multiplied by the planned order quantity for the parent item and then, if appropriate, divided by the yield fraction to give the gross requirement for the component. For instance, assuming that each unit of the example item is made up of one unit of component A and three units of component B and that there is a 10% shrinkage loss in assembling component A into the parent item, calculation of gross requirements for the two components is shown in Table 4.

Table 2
Lotting

	Time period					
	1	2	3	4	5	6
Net requirements	0	0	80	100	35	100
Lot quantities			180		135	

The process of determining net requirements, lot sizing and offsetting of planned orders is next performed for the components, A and B. If A and B are also manufactured items, their planned orders are exploded into gross requirements for their components. This procedure is repeated until planned orders are generated for purchased parts and materials.

In an ongoing situation, each MRP planning cycle re-evaluates and adjusts the plans generated in the previous cycle. Changes may occur between update cycles in the gross requirements for end-items, in the inventory on-hand quantities and in the quantity or expected receipt date for open orders. The MRP system will revise planned (unreleased) orders and will generate messages giving instructions to adjust released orders, when appropriate. It is this ability to re-evaluate plans in light of changing conditions and to signal appropriate corrective actions that makes MRP such a valuable tool in manufacturing control.

2. Role of MRP in a Total Manufacturing Control System

Total manufacturing control systems have come to be called manufacturing resource planning or MRP II systems. A diagram of a typical manufacturing resource planning system is shown in Fig. 1.

It is not necessary for all component modules in an MRP II system to be computerized. Moreover, the detailed procedures in the various modules may differ depending on the characteristics of the products, the manufacturing process and the company.

Table 3
Offsetting

	Time period						
	Current	1	2	3	4	5	6
Gross requirements		50	75	90	100	35	100
Scheduled/planned receipts		120	0	180	0	135	0
Projected on-hand	15	85	10	100	0	100	0
Planned order releases		180		135			

MRP is a central element in an MRP II system. The effectiveness of MRP depends on good input information and on the ability of other component modules to utilize and process the output from MRP.

One input requirement of MRP is an accurate bill of materials or product structure. This file records the component parts of all manufactured parts. It is used for the explosion of dependent demand requirements. Another input requirement is an item master file which contains parameters for each item such as the lead time and lot size and also records the inventory on-hand status and the open replenishment order balance. The status of individual replenishment orders is usually maintained in separate work-in-process and open purchase order files.

Another critical input to an MRP system is the master schedule. A master schedule is usually developed in two steps. The first step, called production planning, establishes the aggregate production rate for the critical work centers in the manufacturing process. The production plan specifies the manpower level and/or the number of shifts per day that each of the critical work centers should be operated in each time period of the planning horizon. Another result of this step is the planned production output and inventory level for aggregate product categories in each time period.

Master scheduling determines the production quantities for individual end-items in each time period and ensures that the sum of the individual product schedule quantities matches the planned aggregate production rates. The master schedule quantities are calculated to satisfy the forecast of independent demand for end-items. In situations in which finished products are custom assembled to customer order, the master schedule is usually prepared in terms of major subassemblies. A separate assembly schedule is prepared containing the customer orders for finished products. The horizon for the assembly schedule is much shorter than the horizon for the master schedule.

The length of the planning horizon for the master schedule is an important consideration. The schedule horizon should be at least as long as the total cumulative manufacturing and purchasing lead times for the end-items being produced (see Fig. 2). The schedule horizon is usually divided into a firm portion and a tentative portion which has significance for

Table 4
Explosion

	Time periods	
	1	3
Planned orders for product X	180	135
Exploded gross requirements		
Component A	$200 = \dfrac{180 \times 1}{0.90}$	$150 = \dfrac{135 \times 1}{0.90}$
Component B	$540 = \dfrac{180 \times 3}{1.0}$	$405 = \dfrac{135 \times 3}{1.0}$

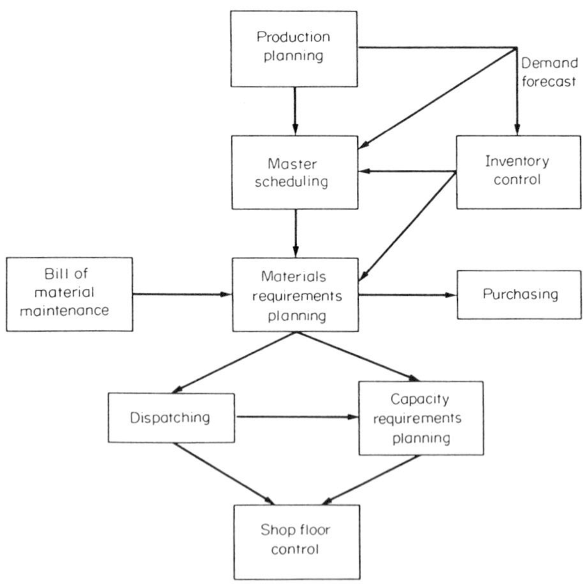

Figure 1
Manufacturing resource planning system

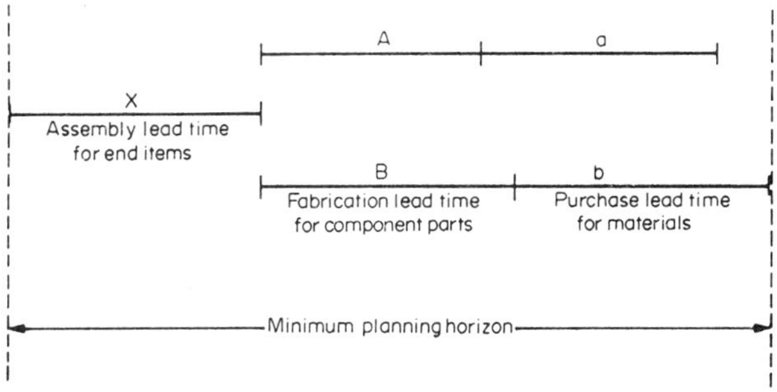

Figure 2
Planning horizon

setting safety stock requirements (see Sect. 7). Planned orders within the firm schedule horizon are not changed except in special circumstances.

The production planning and master scheduling modules are very important because they must ensure that the schedule is feasible from a capacity standpoint. Often, master scheduling includes a step called "rough-cut capacity planning," which calculates an approximate projection of the future workload on critical work centers based on the current master schedule. The MRP procedure does not check for capacity feasibility.

On the output side, MRP provides information needed in order to prioritize and expedite open released orders (both manufacturing and purchase orders) and to schedule and dispatch planned future orders. The primary outputs of an MRP system are the following:

(a) order release notices, which call for the release of planned orders;
(b) rescheduling notices, which call for changes in open order due dates and quantities (quantity changes are usually limited to purchase orders only);
(c) cancellation notices, which call for cancellation of open orders;
(d) exception messages, reporting errors and conditions requiring review;
(e) planned orders scheduled for release in the future; and
(f) time-phased requirement, order and inventory profiles.

The capacity requirements planning module projects the time-phased load in man-hours and/or machine-hours on designated work centers or stations based on both released and planned manufacturing orders. Projected temporary overloads and idle capacity are reported and may initiate reassignment of labor between work centers or adjustments to the schedule.

Shop floor control generates job queue lists in priority sequence for each work center. Many different priority rules may be used to sequence the job queue, such as earliest order due date, earliest operation scheduled start time, shortest operation run time, and so on. Orders that are behind schedule are usually highlighted in an exception report.

A recent development in manufacturing control systems is to replace the capacity requirements planning and shop floor control modules with a finite capacity scheduling module. This module creates an explicit schedule for the manufacturing orders on specific work stations. The schedule specifies expected start and end times for each order operation and ensures that there is capacity available on the required equipment. Finite capacity scheduling may also consider the availability of other resources such as labor, tooling and materials handling in preparing the detailed schedule. Usually, the system updates the schedule to reflect current status at fairly frequent periodic intervals or when an exception condition, such as a machine breakdown, occurs. The scheduling module typically utilizes a combination of automatic rules and a user interface for manual schedule adjustment.

The purchasing module provides the capability to process and print purchase orders, tracks open purchase orders and generates messages for follow-up on orders that are past due and possibly at appropriate times before scheduled delivery. The system also records total vendor status and history information for use in negotiating with suppliers.

3. Repetitive vs Job Shop Manufacturing Environments

MRP and MRP II systems originated in job shop manufacturing companies. When these systems became popular, it was discovered that the classic design does not work very well when implemented in a repetitive manufacturing environment. Some of the significant contrasting characteristics of the two types of manufacturing environment are shown in Table 5.

The classic job shop MRP II system results in too much paperwork to be efficient in a repetitive manufacturing environment. For example, a shop order

Table 5
Characteristics of job shop and repetitive manufacturing environments

Job shop manufacturing	Repetitive manufacturing
Functional work centers	Flow shop or assembly line
Large number of alternative process routings	One basic routing with possible optional operations
Discrete batch orders	Long continuous runs, possibly consisting of a mix of variations of a basic product

is printed for each separate product batch and an issue requisition is printed for each component part for each batch. Therefore, a number of changes have been made to the standard MRP II logic for application in a repetitive manufacturing environment.

(a) The master schedule is expressed as a daily production rate rather than discrete batch orders. If appropriate, a planning bill of material is used to specify the mix of variations of a basic product model (for example, the mix of colors for a particular product model may be 50% white, 30% red and 20% blue).

(b) No order paperwork is generated.

(c) Material usage is recorded for all component parts when a subassembly or assembly is reported as complete by exploding the bill of material rather than generating stockroom requisitions and recording component usage when the material is issued from the stockroom. This procedure is called "backflushing."

(d) Scrap is reported as a general inventory adjustment rather than recorded against a specific production order.

Most MRP II package systems now include special modules for use in a repetitive manufacturing environment.

4. *Regenerative vs Net-Change MRP Systems*

The early computerized MRP systems used the regenerative approach in which requirements are exploded for all items in the master schedule all the way down to purchased parts and materials each time the MRP system is run. Since computer processing time for a regenerative MRP run can be quite long, the system is usually not run very often. A weekly or biweekly replanning cycle is common. The volume of output from each run is large. Actions are initiated and exception reports generated only when the MRP system is run. Thus, while it is possible to update job queue lists or revise schedules more frequently, the priorities and due dates used in these other system modules will remain unchanged until the next MRP run even though events such as scrap, inventory adjustments and machine breakdowns may have occurred in the interim.

Net-change MRP systems are a more recent development. The basic concept is to make a limited explosion only for items affected when a change is recorded in the master schedule or in the inventory on-hand quantity or open order status for any item. A net-change system can be on-line with change transactions processed throughout the day, but it is more common to make daily or once-a-shift batch runs. The key characteristic of a net-change MRP system is that the net, lot and offset processing is performed only when there is a change affecting the requirements, on-hand or on-order status for an item and explosions are made only for these items and their components.

The primary advantage of a net-change MRP system is that it is more readily updated, corrective actions in response to reschedule messages can be initiated sooner and priorities always reflect (almost) current conditions. Another benefit is that the manual workload to review outputs, initiate corrective actions and follow up on exceptions is spread more evenly over time.

The disadvantages of net-change systems are as follows.

(a) They are more costly to develop and operate.

(b) Errors tend to accumulate and perpetuate. If an error is made in creating the master schedule, it will remain until it is detected manually or is aged out of the schedule. In a regenerative system, a complete new master schedule is supposed to be generated each cycle.

(c) Net-change systems can be overly sensitive, issuing frequent minor and counteracting directions. This can be overcome by including tolerances and damping factors in the design so that only important actions are initiated.

5. *Relationship of MRP to Just-in-Time*

The relationship of MRP to the new concept of just-in-time (JIT) is often not well understood. MRP is a planning tool as explained earlier in this article. JIT is a philosophy of manufacturing management that strives to eliminate waste in manufacturing operations and improve quality and productivity. Waste is defined as any activity that does not add value to the end product. One element of waste that the JIT philosophy attempts to eliminate is in-process inventory. Thus, one of the important concepts of JIT is that materials should be delivered to work stations just in time to perform scheduled operations.

MRP is a tool that can be used to support the JIT philosophy. If lead times are set appropriately, MRP and the other planning components of an MRP II system can plan requirements for purchased materials (e.g., for blanket purchase orders) as well as plan manufacturing capacity to support JIT production.

There is a conflict between JIT and MRP II at the execution level of the system, that is, in the way in which production and purchase orders are created and controlled. MRP II is referred to as a "push" system because the orders are released and pushed to completion based on projected need dates determined in MRP netting and lotting. Under the JIT philosophy, material is "pulled" through each stage when a need is created at the next stage. This is referred to as a "kanban" system.

The applicability of kanban depends on the characteristics of the manufacturing operation (Karmarkar 1989). It works best in high-volume, repetitive manufacturing with stable product demand. Kanban does not work well when demand for products is erratic or products are custom engineered although other aspects of JIT such as setup time reduction, minimal work-in-process and total quality control are still applicable.

The special features included in an MRP II system for a repetitive manufacturing environment are also applicable to JIT. The primary change required in an MRP II system to operate in a JIT environment is to eliminate the generation of order release notices and job queue lists (Raeder 1987; Discenza and McFadden 1988).

6. Lot Sizing for Dependent Demand Items

There are many alternative models for determining the manufacturing lot quantity or purchase order quantity for dependent demand items in an MRP system (Orlicky 1975 Chap. 6, Silver and Peterson 1985 Chap. 8). The most common models are listed in Table 6 along with a brief explanation of the methodology used by each model and the conditions under which the models are recommended for use.

The models are classified into three categories: naive, static and dynamic. The fixed order quantity, lot for lot and fixed period requirements models are classified as naive because they are based on judgement rather than a formula that attempts to minimize setup and inventory carrying costs. The economic order quantity and economic time supply models are classified as static models because they both utilize a formula to determine a value that minimizes the sum of setup and inventory carrying cost, but the model includes an assumption of constant demand and the results are updated externally to the MRP module, usually infrequently. In the dynamic models, the lot quantity is recalculated each time a result is required, based on the current time-phased requirements. Thus, when a dynamic model is used, the order quantity and/or the order interval may vary each time they are recalculated.

One desirable characteristic for the order quantity of a dependent item is that the quantity corresponds to the total requirements in an integer number of periods so that the inventory will be depleted to zero (or the safety stock quantity) at the end of a period. This results in a lower average level of inventory investment. This benefit is greatest for low volume and intermittent demand items that have frequent periods with zero demand. With the fixed order quantity and economic

Table 6
Common lot-sizing models for dependent demand items

Model	Type	Methodology	Recommended use
Fixed order quantity	Naive	Judgement or rules of thumb	Not recommended
Lot for lot	Naive	Single period net requirements	Infrequently used items and items with low setup cost
Fixed period requirements	Naive	Judgement or rules of thumb	Initial implementation
Economic order quantity	Static	Optimization; minimize sum of setup and inventory carrying cost assuming constant demand	Not recommended
Economic time supply	Static	Optimization; minimize sum of setup and inventory carrying cost assuming constant demand	Satisfactory in many applications
Least unit cost (LUC)	Dynamic	Heuristic; minimize total setup and inventory carrying cost per unit produced	Satisfactory in many applications
Least total cost (LTC)	Dynamic	Heuristic; minimize total setup and inventory carrying cost	Satisfactory in many applications
Part-period balancing	Dynamic	Heuristic; same as LTC but with a look ahead/look back adjustment	Satisfactory in many applications
Silver–Meal algorithm	Dynamic	Heuristic; minimize total setup and inventory carrying cost per unit time	Satisfactory in many applications
Wagner–Whiten algorithm	Dynamic	Optimization; dynamic programming approach to minimize total setup and inventory carrying cost	Too complex for most practical applications

order quantity models, some remnant inventory will remain at the end of every period except in rare instances. Therefore, these rules are not recommended for most items in MRP systems.

The four dynamic models that use heuristic rules are similar but use a slightly different methodology. No one rule consistently gives better results than the others. The results depend on the specific pattern of time-phased requirements and other parameter values. The Wagner–Whiten algorithm gives theoretically optimum (minimum cost) lot quantities but it is the most complex of all the models.

The dynamic lot-sizing models give a lower cost solution than any of the naive or static models, at least theoretically. However, since actual requirements may vary from the planned values used in the calculation and the cost parameters are only approximate, the theoretical advantage of these models may not be real. Moreover, all of the dynamic models require substantially more computation time than the other models because the lot quantities are recalculated every time a value is required. Thus, there is a tradeoff between the cost of additional computer processing time for the MRP system and the possible savings in setup and inventory costs. In practice, the dynamic models are rarely used, although this is probably more as a result of their complexity than unsatisfactory cost–benefit performance.

It is common to include several different lot-sizing models in an MRP system and to use different models for different items. A good approach is to gradually refine the lot-sizing models used in an MRP system. The system can be initially implemented using only naive lot-sizing models, gradually switching selected items to static or dynamic models at a pace that allows the user to evaluate the results and determine whether the changes are beneficial.

In many situations, it is desirable to adjust the calculated lot quantity to make the result more practical. Three types of adjustment are common.

(a) *Minimum and maximum limits.* A maximum limit is used to avoid producing a long time supply quantity when there is a possibility of obsolescence; a minimum is used to avoid a setup for a very small run quantity. Limits can usually be specified in absolute quantities or time supply values.

(b) *Scrap allowance.* The order quantity is increased to compensate for scrap loss in the manufacturing process. The allowance is usually expressed as a fraction of the lot quantity.

(c) *Basic ordering unit.* The order quantity is rounded to the next higher integer multiple of the specified ordering unit. This is often used when a standard unit of measure of a component material is converted to a fixed number of units of the parent item (e.g., a standard size sheet metal sheet may produce four stampings of a particular part).

7. Safety Stock for Dependent Demand Items

The purpose of maintaining safety stock inventory is to ensure material availability with a desired degree of (statistical) confidence when there is uncertainty in supply and demand.

For independent demand items (end-items), it is easy to obtain a measure of demand uncertainty by calculating the mean absolute deviation or standard deviation of forecast error in the forecasting system. This measure of demand uncertainty is used in one of several alternative customer service models to calculate safety stock requirements (Brown 1967 pp. 167–8, Smith 1984). Supply uncertainty is usually relatively unimportant, so it can be ignored. If supply uncertainty is significant, a measure of lead time variability can be included in the safety stock calculation but the model requires the assumption that lead time variability is statistically independent of demand variability. This assumption is often not valid but the results are usually satisfactory.

In the case of dependent demand items, there is no readily available measure of demand uncertainty and supply uncertainty is likely to be just as important as demand uncertainty. Standard MRP systems do not calculate a measure of demand uncertainty. Moreover, if the MRP update cycle is run fairly frequently and exception messages and priorities are used to expedite replenishment orders, the replenishment lead time will tend to decrease when demand is higher than expected. Thus, there is usually a strong correlation between supply and demand uncertainty for dependent demand items in an MRP environment.

It is possible to reduce the demand uncertainty for dependent demand items by increasing the firm schedule horizon of the master schedule; however, this increases the safety stock requirement for end-items. Since it is cheaper to maintain safety stock inventory in raw material and semifinished form rather than in finished goods, it is usually undesirable to extend the master schedule firm schedule horizon in order to avoid carrying safety stock for dependent items.

There are two ways to provide protection against uncertainty for dependent demand items: safety stock and safety lead time. Safety stock is a minimum level of inventory that is considered to be required at all times when calculating net requirements. Safety lead time is provided by specifying a replenishment completion time that is earlier than the expected need time. It has been shown that the safety lead time method results in a lower average inventory when the primary uncertainty is in timing, either demand timing or replenishment arrival time, and the safety stock method is better when the primary uncertainty is in quantity. Thus, safety lead time is the better method for items that are demanded and replenished infrequently and safety stock is the better method for high usage items.

While analytical approaches have been formulated

(Etienne 1987), the common method of handling safety stock for dependent demand items in actual practice is to specify values using an arbitrary rule, such as a given number of weeks' supply, and adjusting these values based on empirical performance.

8. Pegged Requirements

Pegging of requirements is a feature sometimes included in MRP systems that provides the capability to relate the gross requirements for a dependent demand item back to manufacturing orders, customer orders and independent demand forecasts for higher level items. Single-level pegging, which is most common, provides connector links from gross requirements for a dependent demand component to the next higher level (parent) item only. With single-level pegging, if a component is several levels down in the bill of materials, it is necessary to trace through the pegging report level by level to determine the end-item that is the source of a requirement. This can take time and leads to the desire for what is called "end-item pegging" or "full pegging." Full pegging identifies the source at the master schedule level for component requirements at all levels in the bill of materials.

Single-level pegging is fairly simple to include in an MRP system but full pegging is difficult and expensive to include. It is often impossible to provide unambiguous pegging references because safety stock requirements, scrap adjustments and other lot size adjustments tend to obscure the connection between uses and requirements; this problem increases with the number of levels between the requirement and its source. Full pegging works best when products are made to order and there is little commonality of usage for components.

9. Firm Planned Orders

A firm planned order capability is another optional feature that is often included in MRP systems. Normally, MRP systems will freely adjust order quantities and the scheduled release dates on planned orders as conditions change from one planning cycle to the next. The firm planned order feature provides the ability to freeze both the quantity and the scheduled release date on selected planned orders. It also prohibits the system from generating another planned order for the same item with the same release date. If it is not possible to satisfy requirements because firm planned orders block creation of an additional planned order in time to meet the requirement, an exception message is generated.

Firm planned orders are useful to plan around capacity limitations (e.g., overloaded equipment, planned maintenance and operator vacations) and outside vendor supply constraints.

10. Weaknesses

MRP systems are frequently criticized for two weaknesses:

(a) the inability to recognize capacity or other resource constraints in backscheduling to determine need dates for component materials; and

(b) the use of fixed lead times.

These weaknesses are related and can be largely overcome by effective utilization of other modules in the MRP II system.

Capacity constraints should be considered by rough-cut capacity planning in the master scheduling module. Unfortunately, this check is only approximate and depends on manual procedures to close the loop; that is, rough-cut capacity planning may detect overloaded work centers but the master schedule must be adjusted manually to make the schedule feasible to avoid problems later.

A finite capacity scheduling module will ensure feasibility of the detailed schedule but the detailed schedule may not be consistent with the master schedule if the master schedule contains capacity infeasibilities. If the master schedule and detailed schedule are not consistent, then the schedules for multiple components of the same parent may not be synchronized.

The assumption of fixed lead times in MRP calculations is also addressed in finite capacity scheduling. The elapsed time for each operation in each work center is determined from the schedule rather than using fixed time allowances. Again, deviations between the fixed lead times used in the MRP calculations and scheduled operation durations will result in inconsistencies between the detailed schedule and the master schedule and associated MRP results.

Thus, coupling MRP with finite capacity scheduling obviates the weaknesses of MRP but may cause inconsistencies between the detailed schedule and the master schedule. It is important to close the loop and bring these two schedules into agreement to maximize the benefits from the MRP II system. This can only be done manually.

11. Benefits

Four major benefits are attributed to MRP systems, although the extent to which these benefits can be realized depends on the other elements of the MRP II (total manufacturing control) system.

(a) *Reduced inventory investment.* This is a real measurable benefit and the savings can be dramatic. One reason that an MRP system reduces inventory investment is that replenishment orders for component parts and materials can be timed so that they are received just before the components are needed to make higher level items. Without MRP, the usual practice is to release a replenishment order whenever the inventory quantity for component items reaches a predetermined level (reorder point) and the material may remain in inventory for a long time before it is used. Another reason for lower inventory is that the order quantities can be matched to the pattern of requirements, which is usually erratic for dependent items. Using a lot-sizing rule that sets the order quantity equal to the requirements of an integer number of periods results in lower inventory, as explained earlier. A third reason for lower inventory is that safety stock requirements for dependent demand items can be smaller because the explosion of requirements from planned orders for parent items provides a more accurate forecast than independent forecasting techniques. A large part of demand variability is not uncertain. Finally, in-process inventory will often decrease because lead times can be reduced as a result of the tighter monitoring and control of replenishment orders.

(b) *Better order prioritization.* MRP coordinates need dates for all components of a parent item. When these need dates are utilized in an effective shop floor control system, it ensures that one component is not expedited to completion only to find that another component that is required in order to proceed is not available. Also, MRP can de-expedite or give a low priority to orders if it is found that they are not needed as soon as expected when they were released.

(c) *Early warning signals.* The ability to trace requirements to pegged orders from level to level provides an early indication of difficulty in meeting end-item commitments due to delays that occur to component orders several levels down in the product structure. If an entire batch of a part has to be scrapped and restarted due to a quality problem, it is possible to determine the specific master schedule entries that are affected and decide whether it is necessary to change the master schedule.

(d) *Better quality of life.* A commonly reported benefit of successful MRP-based manufacturing control systems is a better quality of life due to more orderly manufacturing operations. There are fewer crises as most problems are anticipated and avoided or minimized. Management has more time to handle the problems that do arise and can spend time on activities such as training and vendor selection. This can result in reduced setup costs, improved quality performance, improved productivity and lower purchased material costs.

See also: Risk Management

Bibliography

Anderson J C, Schroeder R G, Tupy S E, White E M 1981 *Material Requirements Planning: A Study of Implementation and Practice.* American Production and Inventory Control Society, Falls Church, VA

Brown R G 1967 *Decision Rules for Inventory Management.* Holt, Rinehart and Winston, New York

Discenza R, McFadden F R 1988 The integration of MRP II and JIT through software unification. *Prod. Invent. Manage.* 29(5): 49–53

Etienne E C 1987 Choosing optimal buffering strategies for dealing with uncertainty. *J. Oper. Manage.* 7(2)

Karmarkar U 1989 Getting control of just-in-time. *Harvard Bus. Rev.* 67(5): 122–31

Orlicky J A 1975 *Material Requirements Planning: The New Way of Life in Production and Inventory Planning.* McGraw-Hill, New York

Raeder R H 1987 Toward the achievement of manufacturing excellence with a JIT modified MRP II system. *APICS 30th Annual International Conference Proceedings*, St Louis, MO, pp. 386–9

Sheridan J H 1989 MRP II, still a sound control strategy? *Ind. Week* 238(13): 39–45

Silver E A, Peterson R 1985 *Decision Systems for Inventory Management and Production Planning*, 2nd edn. Wiley, New York

Smith D J 1978 Material requirements planning. In: Hax A C (ed.) 1978 *Studies in Operations Management.* North-Holland, Amsterdam

Smith D J 1984 Choosing cost-effective customer service levels. *APICS Congress for Progress IX Proceedings*, Philadelphia, PA, pp. 118–27

Vollman T E, Berry W L, Whybark D C 1988 *Manufacturing Planning and Control Systems*, 2nd edn. Dow Jones–Irwin, Homewood, IL

Wight O W 1981 *MRP II: Unlocking America's Productivity Potential.* CBI, Boston, MA

D. J. Smith
[ITP Systems, Paxton, Massachusetts, USA]

Mineral Resources: Definitions, Uses, Classifications and Future Availability

Minerals are directly or indirectly the source of most of the materials used by industrial society. Their increasing use has made possible the rise in standard of living in both developed and developing countries. With increased use, concern and controversy have developed about the adequacy of resources to meet future needs. Much of both the concern and controversy arises from a lack of understanding of what mineral resources are and of how supplies are developed.

1. Definitions

The term mineral has several connotations. Used by the mineralogist, it may refer to a naturally occurring solid inorganic substance with distinctive physical properties and a composition that can be described by a chemical formula. Used by the mineral economist, it may refer to a commercially traded product derived from a naturally occurring, nonliving, organic or inorganic, solid, liquid or gaseous substance. Used to describe resources, it refers to naturally occurring, nonliving, organic or inorganic, solid, liquid or gaseous substances, that have a known or potential use. Reference to the mineral resources of a country or region, for example, would be likely to be all-encompassing and include its water resources, its energy resources, such as coal, gas, petroleum, uranium and geothermal energy, and its potentially usable deposits of metallic and nonmetallic minerals, such as deposits of ferrous, base and precious metals, phosphate rock, saline minerals, sand, gravel and building stone.

In reporting specific aspects about mineral resources, the scope of the subject is generally limited to substances that are commercially traded. Water resources are thus treated separately, except for brines containing commercially recoverable dissolved salts. Air is not commonly thought of as a mineral resource, but gases recovered commercially from the atmosphere—argon, nitrogen and oxygen, for example—are counted as minerals in production statistics.

The term ore refers to solid substances currently recoverable at a profit. Generally it applies to explored and developed deposits of metallic minerals but is sometimes used to describe producible deposits of a few nonmetallic minerals such as phosphates or potash.

An important part of the concept of mineral resources is that they have a known or potential use. They thus do not represent permanently fixed quantities but tend to be extended over time as new uses are found for naturally occurring substances and as technologies are developed to recover valuable products from deposits once considered too lean, refractory or inaccessible to have any potential use. Mineral resources to prehistoric man would have thus been limited to flint, salt, water and rocks that could be used for tools, weapons or crude construction. By the end of the nineteenth century about 30 chemical elements were in commercial use; aluminum was known only in the laboratory and deposits rich in that element would not have been considered a resource. Today about 80 of the elements and 135 mineral commodities are in commercial use in the USA and mineral deposits, such as taconite (a hard, highly siliceous iron-bearing rock), that were once considered worthless are now in commercial production.

Mineral resources (excluding water) have been described as nonrenewable resources, in contrast to renewable resources such as surface water and timber, which are or can be replenished naturally or artificially. The geological processes by which most mineral deposits are formed take so long for completion that they cannot be thought to replenish deposits extracted from the ground and dispersed by use. However, it is important to recognize that mineral resources are extendable as the result of advancing technology that develops uses for substances not usable before, permits concealed deposits to be discovered, improves recovery efficiency or makes it possible to mine and process deposits once considered worthless. The life of mineral resources can also be extended by increased efficiency of use. For example, the card-size calculator weighing 35 g and costing about US$5 in 1990 will perform more functions than the calculating machine of 1950 weighing 15 kg and costing US$500 or more.

2. *Uses*

Some mineral products are used in the form in which they are mined, with only shaping, crushing, cleaning or other treatment that does not change their composition. Examples include sand, gravel, crushed stone, metal, building stones, coal in most of its uses, natural gas, gemstones and some abrasives. Most minerals are processed, however, to yield an artificial product—a metal, chemical or other material from which most of the things used by industrial society are made. Minerals provide the materials from which labor-saving machines, lighting, heating and cooling systems and transportation and communication networks are built and powered. Processed and utilized with the aid of inanimate energy, minerals are the source of most chemicals, drugs, paints, ceramics, synthetic textiles, plastics and metal products. They are the raw materials from which most homes, buildings, dams and other engineering works are constructed. Also, converted to fertilizers, insecticides, farm machinery and inanimate energy, they are key elements in the high productivity of modern agriculture.

It is the extensive use of mineral resources, in fact, that forms the physical basis of the high standard of living enjoyed by industrial societies. Strong positive correlation is found, for example, between per capita national income and per capita consumption of energy and key mineral commodities in a given country over time and among the countries of the world today. Key though they are to affluence, their value as raw materials is relatively small in the gross national product of industrialized countries—for example, only about 6% of it in the USA in 1990.

Along with the affluence developing from the high use of mineral resources have come two other phenomena. One is the growth in world population, made possible mainly by increased food production, improved sanitation and control of infectious diseases. The other is environmental strain related mainly to air, water and land pollution arising from industrial society's high use of chemicals and fuels and, to a lesser extent, to mining. If unmitigated, these phenomena could cancel the social gains from the advance of mineral technology. Two of the most important challenges of the future are to achieve population stability and to avoid environmental pollution that would reduce the carrying capacity of the land or affect the long-term stability of the global ecological system (Brooks and Andrews 1974).

3. Classification

For many commercial and governmental purposes it is desirable to know the magnitude of specific mineral resources—at a specific mine site, in a region, in a country and in the world at large. Over the years, diverse terms and classifications have been introduced in recognition that estimates of mineral resources must be labelled to indicate the degree of certainty involved. None of these terms or classifications has achieved universality in use, but the terms *proved*, *probable* and *possible*, and *measured*, *indicated* and *inferred* are illustrative of attempts to indicate degrees of certainty in resource estimates. For many years, such estimates were made only of known deposits believed to be recoverable at a profit, but, as advancing technology has lowered cutoff grades and shown the power of exploration in discovering new deposits, and as mineral shortages have developed or threatened to do so, interest has grown in the quantitative assessment of deposits that are as yet undiscovered or that are not profitably producible now but that may come within economic reach in the future.

Figure 1 shows a classification based on the degree of certainty of the estimate (increasing from right to left) and the economic feasibility of recovery (increasing from bottom to top). A similar system was adopted by the US Bureau of Mines and the US Geological Survey in 1976 and modified somewhat in 1980 (see also Wood et al. 1983, McKelvey and Masters 1984). A group of experts convened by the Secretary General of the United Nations in 1979 recommended an almost identical system, using letters and numbers for the various categories instead of names that may not be translatable into certain other languages. Although there is not yet universal agreement on a classification system for mineral resources, publication of diagrams similar to Fig. 1, drawn from earlier versions of it (e.g., McKelvey 1972, Schanz 1975) in many other countries indicates that its principles are widely accepted, even though the system has some shortcomings (Zwartendyk 1981, Harris and Skinner 1982). The following discussion relates mainly to Fig. 1, but it is the principles of classification that are being emphasized, rather than the terminology or the exact definition of the boundaries of individual categories.

The most important feature of the classification is to distinguish between identified deposits that are recoverable with existing technology and at current prices from those that are as yet undiscovered or, if known, are not yet within economic reach. In this classification the term reserves is used only for identified deposits that could be produced commercially at the time the estimate is made; the term resources is used for subeconomic and undiscovered

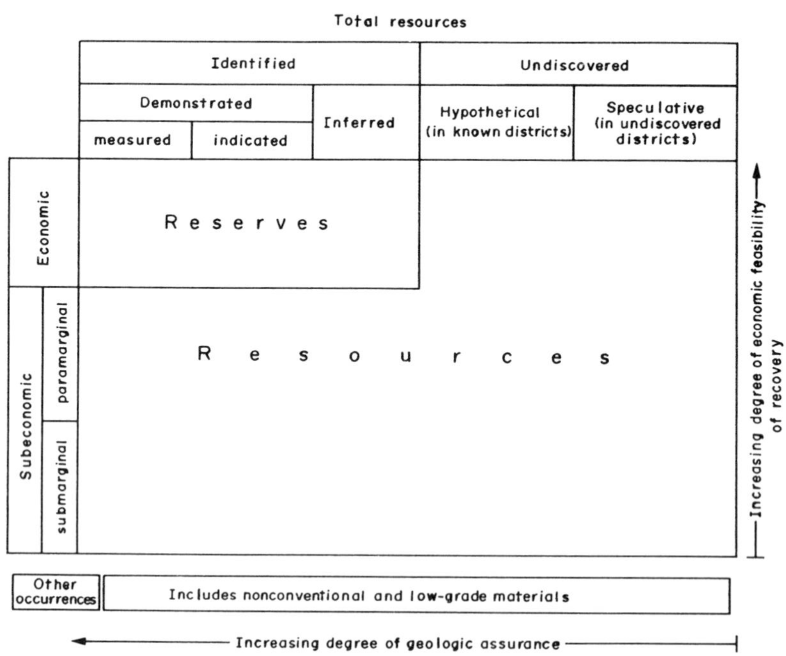

Figure 1
Classification of mineral resources

categories and for the totality of all categories for either one or all minerals. Although resources may include undiscovered deposits that, even if found, are far from being economically recoverable, they include only those deposits conceived to be recoverable at some time in the future (in the United Nations recommendations the future is defined as within the next two or three decades). Estimates of deposits not conceived to have such a potential would be classed as *other occurrences*.

Another important distinction for some minerals, not self-evident in Fig. 1 but stressed in the United Nations report and to a lesser extent in the 1980 US Bureau of Mines–Geological Survey system, is between recoverable and *in situ* quantities. In Fig. 1, reserves would include only currently recoverable amounts, but some of the remainder might be shown in one or more of the subeconomic categories. Until recently, for example, crude oil recovery averaged only around 30% of that originally in place; with higher prices and advancing technology it is reasonable to expect that the percentage would increase, perhaps to 60% or so, and that part of the oil originally in place could be classed as a paramarginal and part as a subeconomic resource. In the system recommended by the United Nations expert group, estimates of *in situ* versus recoverable would be reported separately, as they would to some extent in the 1980 US Geological Survey scheme. However the distinction is made, it is an important one for those minerals for which current extraction practice leaves a substantial portion in the ground. At least some of the remainder is a resource, and an estimate of its amount establishes a target for advance in extractive technology or an indication of amounts that might become available with increased prices.

A good indication of the uncertainty that exists in estimates, even where discoveries have been made and some exploratory drilling completed, is seen in the report of the National Research Council (NRC) (1976) on gas reserve estimation of offshore producible shut-in leases in the Gulf of Mexico; that is, leases on which a commercial discovery of oil or gas has been made but that are not yet in production. Because of concern that outer continental shelf Federal leases were deliberately shut in to raise gas prices, in 1974 the Federal Power Commission (FPC) estimated the proved and probable reserves of the 168 leases then classified as producible shut-in. The results, showing proved reserves of 130×10^9 m^3 and probable reserves of 93×10^9 m^3 of gas, were challenged as being much too conservative.

Accordingly, the NRC was asked to make an independent estimate, using the same data from US Geological Survey files used by the FPC. The NRC selected 33 as a representative sample of the 168 leases and commissioned two consulting firms to make independent estimates of the gas reserves. Their results indicate that rather than being too low, the FPC estimates may have been much too high. In the sample, the FPC's estimate of both proved and probable reserves was 42×10^9 m^3, compared with "best" estimates of the two consultants of 18×10^9 m^3 and 13×10^9 m^3 (the consultants also made high and low estimates which were 18–28% lower and 32–45% higher than their best estimates). Estimates of reserves in some individual reservoirs differed by more than an order of magnitude and, with few exceptions, the FPC was generally the highest, consultant A generally the lowest and consultant B generally in between.

Because many of the reservoirs are structural traps associated with salt domes, differences in interpretation of the extent of reservoirs at this early stage of exploration are understandable, but that such differences were not random indicates that differences in philosophy—possibly born from experience or the lack of it—played a significant part. In any case, the wide range in estimates of reserves in discovered and partly explored fields in the Gulf of Mexico made by experts using the same data is indicative of the great uncertainty that necessarily accompanies estimates of reserves, to say nothing of undiscovered resources. They are fittingly described as estimates, with the lack of precision implied by that term.

In the light of such uncertainty, it may be asked if such estimates serve any useful purpose. Obviously they must be used with the understanding that they are only estimates, subject to a wide margin of error, and that they need to be continually updated as new information is acquired. But taken with that understanding, they show with respect to reserves that there are, or are not, known deposits that can be mined with existing technology at current prices; they form the basis for commercial sale and purchase of mineral properties, guide plans for mineral development and production and provide governments with information bearing on the design of near-term supply policies. Estimates of identified subeconomic resources show whether there are known deposits large enough to justify technological research and development—valuable information for both industry and government. Estimates of undiscovered resources show whether there are targets for additional exploration—information as essential to government in formulating policy as it is to industry in laying plans to develop future supply.

Estimates in the framework of the classification illustrated in Fig. 1 help to emphasize that the development of mineral resources is a dynamic process (Schanz 1977) and that the magnitude of usable resources, although limited by a finite earth (a point that some would not concede, now that space exploration has begun), is not fixed but increases with advancing technology and exploration and, of course, decreases with increasing production. The mineral deposits being produced today were once in the category of undiscovered subeconomic resources. For most minerals, the measured reserves of tomorrow will come from the now subeconomic or undiscovered categories; for some they may come from today's "other occurrence" category and for still others they may come from unknown types of deposits that would not even be listed in 1984 in the "other occurrence" category (DeYoung 1978).

4. Future Availability

Concern about the adequacy of mineral resources to meet future national and world needs has mounted from time to time, receiving initial prominence in the USA with the Governor's conference on conservation convened by President Theodore Roosevelt in 1908. In his opening address to the conference (see Nolan 1958) he said:

> "I have asked you to come together now because the enormous consumption of these resources, and the threat of imminent exhaustion of some of them, due to reckless and wasteful use, once more calls for common effort, common action. We want to take action that will prevent the advent of a woodless age and defer as long as possible the advent of an ironless age."

Many others since then have expressed similar concerns. In the widely publicized study sponsored by the Club of Rome (Meadows et al. 1972) it was concluded that world resources would be exhausted in less than 100 years if growth in consumption were to continue at current rates. Skinner (1976), foreseeing the dawn of a second iron age to be "much closer than most of us suspect" concluded that by the end of the next century "mining of scarce metals will be increasingly a memory." Concerned about the prospects of shortages of several metals, some believe that a "resource war" designed to capture supplies of metals that come largely from a few African countries is imminent if not already under way (Council on Economics and National Security 1980).

Others, however, see no signs of impending mineral scarcity. Barnett and Morse (1963), for example, found that the labor cost per unit of output in the USA declined rather steadily from 1870 to 1957, as did the unit costs of minerals compared with other extractive goods, and Barnett (1979) reports that these trends continued through 1970 and seem to apply throughout most of the world. Simon (1981), expressing views with which many disagree, recently stated that "over the course of history, up to this very moment, copper and other minerals have been becoming not more scarce but more abundant" and concluded that "with knowledge, imagination, and enterprise, we and our descendants can muster from the earth all the mineral raw materials that we need and desire, at prices that grow smaller relative to other prices and to our total income." Table 1 shows that, except for the last three years, prices and average grade for copper in the USA declined over the period 1900–1986; reserves increased throughout the entire period. Similar trends have been shown by many other minerals as advancing technology and exploration have transformed undiscovered and submarginal resources into measured reserves. But what about the future?

In thinking about future mineral availability it is useful to differentiate several kinds of availability (Sheldon 1977, Brobst 1979). One is the geological

Table 1
Changes in copper price, grade of ore mined and US ore reserve estimates, 1900–1990

Year	US producer's price refined and delivered ($\$ lb^{-1}$) Current	1987	Average US mine recoverable yield (% Cu)	Demonstrated ore reserve base (10^6 t)
1900	0.16	2.31	NA	NA
1910	0.13	1.87	1.88	NA
1920	0.18	1.09	1.63	NA
1930	0.13	1.09	1.43	17
1940	0.12	1.05	1.20	22
1950	0.22	1.06	0.89	27
1960	0.32	1.23	0.73	30
1965	0.35	1.23	0.70	30
1970	0.58	1.63	0.59	78
1975	0.64	1.27	0.47	84
1980	0.01	1.39	0.48	92
1984	0.67	0.73	0.58	90
1986	0.66	0.68	0.60	90
1988	1.21	1.17	0.60	90
1989	1.31	1.22	0.62	90
1990	1.23		0.61	90

Source: Janice Jolly, US Bureau of Mines (1991)

availability of the mineral or the mineral resource as it is currently conceived. Another is technological availability. For example, the availability of energy from fusion of deuterium is currently limited by technology. If there should be a technological breakthrough, economic availability may become limiting, at least at first. In the 1980 revised mineral classification of the US Bureau of Mines and US Geological Survey still another kind of availability—legal—is recognized; restricted reserves or resources are defined as deposits that meet all the requirements of the definitions of reserves and resources but are restricted from extraction by laws, regulations or governmental policy.

Also, of course, there is marketplace availability, partly a function of the other limiting factors but also of the extent to which exploration and development have resulted in productive capacity adequate to meet demand. The USA was not short of economically producible petroleum resources during the Arab embargo of 1973 and the Gulf crisis of 1990–91 but it was short of the productive capacity to meet all its own needs. Availability limited by technology, economics, production capacity or legal restriction is subject to change over time, but geological availability is finite, even though the magnitude of the deposits of a mineral or of its resources are unknown or at least uncertain.

In examining the future availability of minerals, the rate of consumption as well as the magnitude of unmined mineral resources must be considered. Meadows et al. (1972) assumed a fixed 250-year supply of minerals if used at present levels; their estimate of a 100-year lifetime of reserves if used at exponentially increasing rates of consumption thus has no merit as a prediction (which they did not claim it to be). But they were right in emphasizing the impossibility of continued exponential growth in mineral consumption. For example, a billion-year supply of anything if used at current levels would be exhausted in only 584 years if used at an exponentially increasing rate of 3% a year (which is a modest rate of increase compared with that of many minerals in recent decades). A 2-billion-year supply would last only one more doubling time—24 years—at the 3% annual rate of increase. Described in another equally disturbing way, no matter how many doubling times it takes to use, at an exponentially increasing rate, 25% of a fixed supply, the remainder would be exhausted in only two more doubling times. Exponential growth would never continue unabated to the point of exhaustion, of course, but by the time market forces signalled an impending scarcity, maneuvering room for adjustment might be uncomfortably small.

There appears, however, to be a tendency for the rate of increase in consumption to slacken, once initial uses are satisfied and recycling begins. In the USA, for example, the rate of annual increase in primary mineral demand for 24 commodities has been declining since the 1960s and for 33 more it averaged less in the 1970–76 period than in the 1960s. In the period 1976–90 the declining trend continued due to increased use of secondary sources (scrap) and competition from advanced materials. Of 24 nonferrous metals in that period, the primary supply demand decreased or was essentially flat for 14 metals (US Bureau of Mines 1987, 1991). Nevertheless, exponential growth in consumption (primary plus scrap) is continuing for most minerals.

Mineral resources are irregularly distributed over the earth's surface and no country has domestic supplies of all the minerals it needs. The situation and outlook with respect to mineral availability thus varies from country to country and commodity to commodity. The discussion that follows focuses on mineral trends in the USA, many of which are representative of those in other industrialized countries, and in the world at large, the mineral trends of which are strongly influenced by the industrialized countries on the demand side and by the developing countries on the supply side.

As shown by data assembled on the nonfuel minerals by the US Bureau of Mines (1991), US primary production of raw nonfuel minerals increased in all but three of the last 30 years. US production was greater in 1979 than in 1958 (US Bureau of Mines 1979, 1981) for 42 mineral commodities and less for 18 others, reflecting declining availability of competitive sources for some, such as cobalt, fluorite, gold, manganese and platinum, and increasing costs for others, such as dimension stone. In recent years (1986–90), US production continued to decline in cobalt (nothing since 1971), fluorite and manganese (nothing since 1970) and has remained about constant in dimension stone. Production has increased markedly in platinum since 1987, when a mine in Montana came on stream, and in gold. Taking price in constant dollars as a related measure of availability, 1979 prices were lower than those in 1958 for 42 commodities and higher for 40 others. Because the improved technology resulting in decreasing prices has been related to the increased use of low-cost energy in the production process for many minerals, the large increase in energy costs since 1973 led to some reversals in a previously declining trend. Thus, the price was lower in 1973 than in 1958 for 20 of the 40 commodities for which 1979 prices were higher than 1958 prices. For some 18 commodities, however, the price increase over the period was relatively continuous, reflecting for the most part diminishing availability coupled with increasing demand.

World production of all mineral commodities but six has been steadily increasing and imports have thus been offsetting diminishing domestic supplies in the industrialized countries. The six for which 1979 production was less than in 1968 are arsenic, beryllium, gold, indium, mercury and thallium. Gold production peaked in 1970, when its price in 1978 dollars was US\$60.59 a troy ounce, and the continued increase in its price subsequently did not reverse the decline in world production through 1979. However, gold production increased for the period 1980–1990 in both the world and the USA. The USA was the world's second largest producer of gold in 1990, after South Africa, and for the first time in about 50 years produced more than the USSR, the previous second largest producer (US Bureau of Mines 1991). World production of beryllium peaked in 1984 and of mercury in 1979 but both have since declined.

Attempting to assess mineral availability for the remainder of the century, the US Bureau of Mines (1979) forecast a decline in US demand during the period 1976–2000 for only mercury, thallium, sheet mica and barite and an annual increase of 2% or more per year for 64 of the 84 commodities projected. However, although the domestic demand for mercury had declined as of 1990 compared with previous years (US Bureau of Mines 1991), increases were recorded in thallium, because of its use in superconductor compounds, and in barite, because of its use as a heavy medium in petroleum drilling; the demand for sheet mica was unchanged. It was estimated that primary US production will supply 90% or more of the cumulative demand for 32 commodities and <10% for 20 others. In addition to its estimate of year 2000 production, the Bureau calculated what 2000 production would be if based on a projection of the past 20-year trend; for 23 commodities this "projected" 2000 production would be less than half of the Bureau's "estimated" year 2000 production and for nine others it would be <100% of the estimated production, indicating anticipation of developments that will accelerate domestic production of many minerals over what would be expected from the trend of the last 20 years.

The Bureau also compared these demand estimates with US reserves and identified resources. For the USA, the ratio of reserves to 1976–2000 cumulative demand is 0.1 or less for 18 commodities, between 0.1 and 1 for 15 others, >1 for 36 and "adequate" for four more. For the 38 commodities for which there are estimates of identified resources, the ratio to cumulative demand is >1 for all but 10. Presently undiscovered resources will surely be found for many minerals and will add to their future supply.

For the world as a whole, the Bureau of Mines estimates that the ratio of reserves to 1976–2000 cumulative demand is 0.1 or less for only fluorine, 0.1–1 for 12 commodities, >1 for 61 others and "adequate" for five more. The ratio of world identified resources to cumulative demand is <1 for only two commodities: fluorine and asbestos.

Turning to the future availability of energy resources, the changing situation and prospects for future supplies of energy are illustrative of future availability problems and possible solutions for mineral resources in general. The shortage of crude oil in 1973 as a result of the Arab embargo on exports to a world that had become dependent on them to a significant extent, followed by US shortages of natural gas during the winter of 1976–77, were strong shocks that have had a wide range of consequences. World oil prices, controlled largely by the Organization of Petroleum Exporting Countries cartel, skyrocketed to levels many times those prevailing in 1973, resulting in sharp increases in the prices of consumer products such as gasoline.

Widened recognition of the industrial society's dependence on energy and broadened realization that low-cost conventional reserves of oil and gas are limited led the US government to develop policies that would encourage conservation of energy and its more efficient use on the one hand and encourage exploration and the development of alternate sources of energy on the other. Crude-oil prices established after the embargo were decontrolled, and decontrol of natural-gas prices is in process. Several significant trends are developing as a result of these changes. Energy consumption per unit of gross national product was 20% less on a constant-dollar basis in 1982 than in 1973 (US Department of Energy 1984a). Industry consumed 28% less energy in 1982 per unit of output than in 1973, but its output increased; its energy consumption per unit of output has decreased by around 3.5% a year since 1970. Consumption of both oil and gas began to decline as people drove at slower speeds, shifted to improved-gas-mileage automobiles, added insulation to their homes and took other conservation measures, and in 1978 oil imports also began to decline. Oil consumption decreased 21% in the period 1979–83 (US Department of Energy 1984b), and in 1983 oil imports were down to 36% of the 1973 level. This trend has been reversed in recent years with the Federal government no longer promoting conservation measures.

Higher prices have also stimulated oil and gas exploration. More than twice as many rotary rigs were active in the USA in 1979 than in 1971, completing 1.88 times as many wells, drilling 1.90 as much footage and finding 1.45 times as much liquid hydrocarbons and 2.38 times as much natural gas as in the low-discovery years of the early 1970s. Whereas production of crude oil had begun to decline after apparently peaking in 1970, it began to increase again in 1977. Even though the increases are modest, that trend is still continuing—a reversal similar to that which began in 1963 for the production of lead in the USA (Harris 1977). Research and development to improve the percentage of oil recovered from the ground has been stepped up, leading to a reversal in the production decline in California in 1976 and promising to make available extensive heavy crude oil resources elsewhere.

Production of energy in the USA has also begun to increase from other sources—coal, hydropower, geothermal energy, oil shale, biomass and solar. Plans have been announced for the development of tar sand production in Texas and construction of a peat-supplied 120 MW power plant in North Carolina. Research and development are under way on the recovery of natural gas as well as geothermal energy from the extensive geopressured reservoirs of the Gulf

Coast and on the recovery of methane occluded in coal. Except for coal and hydropower, these sources in 1981 were not making significant contributions nationally, but they have the potential for doing so.

In spite of public uncertainty about its safety, even nuclear energy has yielded significantly increased production in recent years. As of 1973, 40 plants with a total of 19 843 MW capacity had been licensed by the Nuclear Regulatory Commission; as of the end of 1983, 81 plants were licensed with a total capacity of 64 058 MW (US Department of Energy 1984a,b). According to the Office of Technology Assessment, however, as reported by *Chemical and Engineering News* on 13 February, 1984, "Without significant changes in technology, management and level of public acceptance, nuclear power in the US is unlikely to be expanded beyond the reactors already under construction."

The recent developments in the energy field illustrate how mineral resources can be created and extended not only by advancing technology and exploration but also by political, social and economic ingenuity. As shortages develop, price increases (if permitted) encourage exploration, improved recovery technology, technically more efficient use, conservation in the sense of curtailing unnecessary use and avoiding waste, recycling and substitution by a more abundant material—any or all of which extend supplies. The limits of these processes cannot be thought to have been reached for any mineral resource. Even for fluorine, for which the US Bureau of Mines estimated the ratio of world identified resources to world cumulative demand to the year 2000 to be only 0.8, some relief is already in sight, for the Bureau did not include in its estimates fluorine in identified resources of phosphate rock, in which it is present in amounts of 3–4% and from which its by-product recovery has begun in recent years.

Because of the near impossibility of predicting the result of the interplay of the factors that extend resources or accelerate their decline, it is impossible to be definitive about the lifetime of mineral resources beyond the certainty that exponential growth in consumption cannot continue indefinitely. Reserves for all minerals can be extended for a time, perhaps a very long time for some minerals, particularly if some of the very large potential resources of energy—such as dissolved gas in geopressured reservoirs, tar sands, deuterium in natural waters for fusion—can be brought within technological and economic reach and if the environmental degradation that might deter their development can be avoided. Still uncertain, for example, is whether or not continued combustion of fossil fuels will raise the atmospheric temperature as a result of the greenhouse effect caused by increased CO_2 content. The climatic effects of such a development might be so adverse as to necessitate curtailment rather than expansion in the use of fossil fuels.

Even though economic tests do not indicate developing scarcity of many minerals on a world scale, there are many signs of regional and national depletion, especially in industrialized countries. The USA, for example, has shifted from being a net exporter to a net importer of many minerals, including oil, not necessarily because all ultimately producible deposits of those minerals have been exhausted but because some of the remaining deposits are economically not competitive with foreign sources or because they are not adequate to meet expanded needs. Producible deposits of nearly all minerals are becoming more difficult to find; this is especially true for some of those—titanium, for example—for which production has fallen in spite of rising prices. New discoveries are still being made, however, and ultimately producible resources should not be thought to be limited to those now identified.

5. Summary

Mineral resources are naturally occurring solid, liquid and gaseous substances useful or potentially useful to man. They are the source of most of the materials and energy used by the industrial society. Historically, the per capita gross national product of an industrial society has increased in nearly direct proportion to increase in their consumption. Mineral resources include identified deposits producible under current economic and technological conditions (termed *reserves*), undiscovered deposits of the same quality (*undiscovered economic resources*) and known and unknown deposits that cannot yet be produced profitably (*identified* and *undiscovered subeconomic resources*).

Estimates of the magnitude of mineral resources have low precision and may be expected to change over time as the result of further geological exploration and technological advance that transform undiscovered and subeconomic resources into reserves. In assessing the future availability of mineral resources, account must be taken of rate of consumption. During the twentieth century, mineral resource output in the USA has increased at an annual rate of about 3.7% per year, and in its forecasts of demand to the year 2000 the US Bureau of Mines has assumed growth rates for most nonfuel minerals in the range of 2–6%. One of the few certainties about the future availability of mineral resources is that none is adequate for long-continued, exponential growth in consumption. In the USA, higher prices for crude oil began in the late 1970s to reduce consumption through more efficient and conservative use while at the same time stimulating actions that enlarge supply.

The USA has already become a net importer of many minerals. Estimated reserves are lower than the forecast cumulative demand to the year 2000 for nearly half of the commodities for which estimates are available, and estimated identified resources are insufficient for about a quarter of them. For the world at large, estimates of reserves and identified resources are larger than the forecast cumulative demand to the year 2000 for nearly all commodities. Production of a few has begun to decline in spite of rising prices, indicating that profitably producible deposits are becoming extremely difficult to find. Even so, new discoveries, as of gold are still being made and the limits of efforts to extend supply through exploration, improvement in extractive technology, increase in efficiency of use, reduction of waste and conservative use, recycling and substitution have not been reached for any resource. Geological availability—unknown and perhaps unknowable in any but the most approximate sense—of a mineral is finite, but mineral resources are created by technological ingenuity and their life can be much extended by scientific, technological, political, economic and sociological creativity.

Acknowledgement

The publishers wish to thank Professor N. Herz for having reviewed and updated this article.

See also: Availability of Materials; Resource Appraisal; Wood: Future Availability

Bibliography

Barnett H J 1979 Scarcity and growth revisited. In: Smith V K (ed.) 1979 *Scarcity and Growth Reconsidered, Resources for the Future.* Johns Hopkins University Press, Baltimore, MD, pp. 163–217

Barnett H J, Morse C 1963 *Scarcity and Growth, Resources for the Future.* Johns Hopkins University Press, Baltimore, MD, pp. 1–288

Brobst D A 1979 Fundamental concepts for the analysis of resource availability. In: Smith V K (ed.) 1979 *Scarcity and Growth Reconsidered, Resources for the Future.* Johns Hopkins University Press, Baltimore, MD, pp. 106–42

Brooks D B, Andrews P W 1974 Mineral resources, economic growth, and world population. *Science* 185: 13–19

Council on Economics and National Security 1980 *The Resource War and the U.S. Business Community: The Case for a Council on Economics and National Security.* National Strategy Information Center, Washington, DC, pp. 1–42

DeYoung J H Jr 1978 *Mineral Resources Map of the Chandalar Quadrangle, Alaska,* US Geological Survey Map MF 878-B. USGS, Arlington, VA

Group of Experts on Definitions and Terminology for Mineral Resources 1979 *The International Classification of Mineral Resources,* Secretariat Economic Report No. 1. United Nations, New York, pp. 1–8

Harris D P 1977 Conventional crude oil resources of the United States: recent estimates, methods for estimation and policy considerations. *Mater. Soc.* 1: 263–86

Harris D P, Skinner B J 1982 The assessment of long-term supplies of minerals. In: Smith V K, Krutilla J V (eds.) 1982 *Explorations in Natural Resource Economics.* Johns Hopkins University Press, Baltimore, MD, pp. 247–326

McKelvey V E 1972 Mineral resource estimates and public policy. *Am. Sci.* 60: 32–40

McKelvey V E, Masters C D 1984 Undiscovered oil and gas resources: procedures and problems of estimation. *Proc. 27th Int. Geological Congr.,* Vol. 13. VNU Science Press, pp. 335–52

Meadows D H, Meadows D L, Randers J, Behrens W W III 1972 *The Limits to Growth.* Universe Books, New York, pp. 1–205

Miller B M, Thomsen H L, Dalton G L, Coury A B, Hendricks T A, Lennartz F E, Powers R B, Sable E G, Varnes K L 1975 *Geological Estimates of Undiscovered Recoverable Oil and Gas Resources in the United States,* US Geological Survey Circular 725. USGS, Arlington, VA, pp. 1–78

National Research Council, Panel on Gas Reserve Estimation, Board on Mineral Resources 1976 *Gas Reserves Estimation of Offshore Producible Shut-in Leases in the Gulf of Mexico.* National Academy of Sciences, Washington, DC, pp. 1–170

Nolan T B 1958 The inexhaustible resource of technology. *Perspectives on Conservation, Resources for the Future.* Johns Hopkins University Press, Baltimore, MD, pp. 49–66

Peters W C 1989 Mineral deposits: examination and evaluation. In: Carr D D, Herz N (eds.) 1989 *Concise Encyclopedia of Mineral Resources.* Pergamon, Oxford, pp. 204–7

Schanz J J Jr 1975 Resource Terminology: an examination of concepts and terms and recommendations for improvement. *Resources for the Future.* Johns Hopkins University Press, Baltimore, MD, pp. 1–116

Schanz J J Jr 1977 United States minerals—a perspective. *Min. Cong. J.* 63(2): 26–9

Sheldon R P 1977 Factors limiting the future primary mineral supply system. In: Marios M (ed.) 1977 *Towards a Plan of Action for Mankind, Long Range Mineral Resources and Growth,* Vol. 1. Pergamon, Oxford, pp. 63–6

Simon J L 1981 The scarcity of raw materials. *Atl. Mon.* June: 33–41

Skinner B J 1976 A second iron age ahead? *Am. Sci.* 64: 258–69

Tooker E W 1989 Ores of the future. In: Carr D D, Herz N (eds.) 1989 *Concise Encyclopedia of Mineral Resources.* Pergamon, Oxford, pp. 241–5

US Bureau of Mines 1979 *Mineral Trends and Forecasts.* US Government Printing Office, Washington, DC, pp. 1–25

US Bureau of Mines 1981 *Mineral Facts and Problems,* US Bureau of Mines Bulletin 671. USBM, Washington, DC

US Bureau of Mines 1983 *Minerals Yearbook,* Vol. 1. USBM, Washington, DC

US Bureau of Mines 1984 *Mineral Commodity Summaries.* USBM, Washington, DC

US Bureau of Mines 1987 *Mineral Industry Surveys: Nonferrous Metals Supply/Demand Data 1975–1985.* USBM, Washington, DC

US Bureau of Mines 1991 *Mineral Commodity Summaries.* USBM, Washington, DC

US Bureau of Mines, US Geological Survey 1976 Coal resource classification system of the US Bureau of Mines and US Geological Survey. *Geol. Surv. Bull. (US)* 1450-B: 1–7

US Bureau of Mines, US Geological Survey 1980 Principles of a reserve/resource classification for minerals. *Geol. Surv. Circ. (US)* 831: 1–5

US Department of Energy 1984a *Annual Report of Energy Conservation Indicators for 1982,* DOE/EIA-0441 (82). US Government Printing Office, Washington, DC

US Department of Energy 1984b *Monthly Energy Review* for December 1983. US Government Printing Office, Washington, DC

Wood G H Jr, Kehn T M, Carter M D, Culbertson W C 1983 Coal resource classification of the US Geological Survey. *Geol. Surv. Circ. (US)* 891: 1–65

Zwartendyk J 1981 Economic issues in mineral resource adequacy and in the long-term supply of minerals. *Econ. Geol.* 76: 999–1006

V. E. McKelvey[†]
[St Cloud, Florida, USA]

Multiattribute Utility Analysis

Multiattribute utility analysis (MAUA) has emerged as a powerful tool for materials selection and evaluation. An operations research technique, MAUA has been used in a wide range of engineering areas, of which materials science and engineering is one of the more recent. Utility analysis affords a rational method of materials selection which avoids many of the fundamental logical difficulties of many widely used alternative approaches. Furthermore, its use extends beyond the problem of materials selection into the more speculative areas of materials competitiveness and market demand.

1. Materials Selection

The selection of a material system for an engineering application requires the designer to find the best match between the technical and economic requirements of the application and the performance and manufacturing characteristics of the available material alternatives. The identification of this best match is frequently a difficult problem, relying on the designer's accumulated experience, expertise and judgment.

Materials selection has emerged as a major problem for engineers. The available set of materials, rapidly growing both in type and number, has vastly expanded the number of possibilities meriting serious consideration for many engineering applications. At one time, an engineer could rely on engineering handbooks, journals and experience to select the appropriate material for an application. Today, however, engineers are forced to look for systematic techniques for managing and analyzing engineering data on the growing array of materials.

At the same time, the suppliers of materials face an equally perplexing problem. While advances in materials science continue to expand the horizons of material performance, the material developer finds it increasingly difficult to establish what constitutes a desirable material. By its nature, the designer's process of material selection happens outside the ken of the suppliers of the material alternatives. Because the suppliers are frequently isolated from this process, they receive little insight into the rationale for the success or failure of their offerings. The lack of this information not only limits their ability to identify and rectify the limitations of their current offerings, but it also makes it very difficult to ascertain what success new offerings will have.

These difficulties are a consequence not only of the ever-increasing number of material alternatives available, but also of the very nature of the materials selection problem. While several analytical tools have been developed to cope with the former problem, notably screening and indexing techniques (Clauser et al. 1965, Sharp 1966, Verink 1966, Hanley and Hobson 1973, Kusy 1976, Alexander and Appoo 1977, Rosen and Bennett 1979, Miaw and Wilson 1982), none of them successfully treats the latter (Field and de Neufville 1988, de Neufville 1990).

When evaluating competing material alternatives, the analyst is called upon to identify the "best" alternative for the application under consideration. The identification of this best alternative is quickly complicated by two elements of the materials selection decision. First, the indices of performance of a particular material alternative are incommensurable on any objective basis (e.g., yield strength and tensile modulus). While the application of design constraints and engineering relations can occasionally yield analytical relationships which can be exploited for design purposes (e.g., thickness, modulus and curvature relations for calculating plate stiffnesses), there are no global relationships which are able to transform complex measures of performance, like cost and manufacturability, into analytical design relations.

Second, there are no such global analytical relationships because, by their very nature, they cannot incorporate the essence of the design process, which is the use of engineering judgment to develop strategies for solving multiobjective problems. For example, all automobile body panel designers might agree on two facts, all other factors being equal: (a) lower cost body panels are always preferred, and (b) lower weight body panels are always preferred. If so, it is easy to predict that US\$50, 25 lb (11 kg) panels will always be preferred over US\$100, 40 lb (18 kg) panels (all other things being equal).

On the other hand, it is a very different matter when the alternatives are a US\$100, 25 lb panel and a US\$50, 40 lb panel. If asked to identify the better of the two alternatives, the usual answer is "it depends." The reason for this response is that the simple decision rules are deficient when there is no single alternative having attributes no worse than the best offered by any other alternative; there is no "dominant" alternative. Rather, the choice must be made from members of the "nondominated" set of alternatives. And the selection of the "best" from this set of alternatives cannot be made on the basis of objective rules; subjective judgments by the designer will determine the preferred alternative.

Returning to the example given, a designer developing a low-performance, high-volume automobile on a tight production budget would probably select the US\$50, 40 lb panel. On the other hand, the same designer working on a high-performance, low-volume application might lean towards the US\$100, 25 lb panel. There is no inconsistency here; rather, the designer's preferences (which might themselves be driven by product strategies, corporate policies, government regulations, etc.) enable him to trade-off these incommensurable characteristics in order to select the alternative that best meets his design goals. The effectiveness of the selection will depend, in turn, on the preferences of the customers for the price–performance combination that is chosen to provide.

2. *Multiattribute Utility Analysis*

Efforts to understand how this choice takes place, and to help people make the most effective use of the information that they have to make these choices, have led to the development of a branch of operations research known as decision analysis. Within this field, researchers into multiobjective problem solving have formulated the theory of utility analysis, which has been applied to great effect in many engineering fields.

A utility function is a mapping of a multidimensional attribute space into a single-dimensioned preference space. A simple attribute space (cost and weight) was considered in Sect. 1. The critical elements of this attribute space are that its dimensions correspond to the performance attributes that underlie the decision being studied, and that the limits of this space are well defined at the outset of the analysis. Based on repeated evaluations of carefully constructed decision problems, an analyst can define a mathematical mapping of this performance space into a single dimension of preference, which establishes an ordering to all points in the attribute space. The defining characteristics of any utility function, given two alternatives, A and B, and a utility function $U(x)$, are

$U(A) > U(B)$ if A is preferred to B; and

$U(A) = U(B)$ if A and B are equally preferred

While there are a wide range of potential transformations which can meet these criteria, decision analysis focuses on a subset of utility functions, classified as von Neumann–Morgenstern utility functions. Such utility functions are characterized by the fact that they are consistent under probabilistic expectation; that is, if an individual is indifferent between a given alternative A and a situation in which there is a probability p that alternative B will obtain and a probability $1-p$ that alternative C will obtain, then $U(A) = pU(B) + (1-p)U(C)$. Graphically this is represented as a lottery of the form

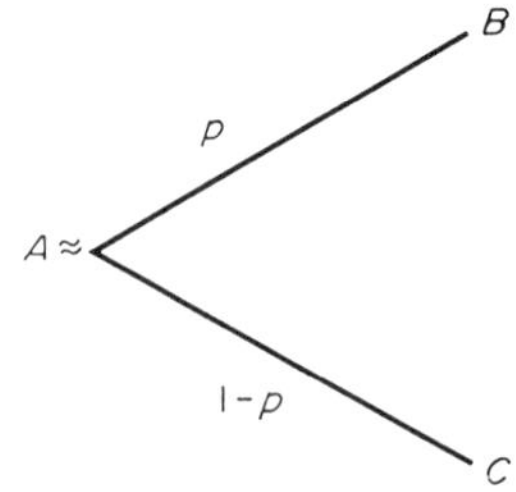

This characteristic of von Neumann–Morgenstern utility yields a powerful ability; utility functions of this form can be successively approximated through repeated inquiries under which individuals are asked to make choices under uncertainty. For example, by offering a subject a choice between \$10 and a 50:50 chance of winning \$$X$ (versus a \$0 outcome) and by varying X, it should be possible to establish the relative utilities of \$10, \$0 and the value of X that makes the subject just indifferent between the alternatives. Persistent questioning could yield a fairly detailed mapping of dollars onto utility space, limited only by the subject's patience and the questioner's persistence. Such a utility function could be used to predict the subject's choice in any situation falling within the domain of the measured space.

It is important to recognize that the scale of measurement is defined by the analyst at the outset of the analysis; extrapolation of utility functions outside the boundaries of measurement cannot be reliably accomplished. However, within the measurement boundaries, fairly detailed definition of the utility function can be accomplished with relatively few inquiries.

While such single-attribute utility functions do have their uses, it is through the extension of this concept to multiple dimensions that it becomes literally possible to compare apples with oranges, and then suggest which is better in the eye of a particular beholder. The extension of the von Neumann–Morgenstern framework to multiattribute utility functions resolves the question of interdimensional incommensurability through the development of utility functions which treat several attributes simultaneously, without a geometric increase in measurement complexity.

This technique has been widely used within the engineering community with considerable success. During the 1980s, the use of this technique was extended to the analysis and evaluation of problems of materials selection. The problem by its very nature parallels the problems for which multiattribute utility analysis was developed to address, and it has been successfully applied to the study of this class of problems.

A feature of utility functions which makes them particularly valuable in these analyses is the fact that utility is not a simple ordinal measure; rather, utility is a cardinal measure over the range of measurement (Keeney and Raiffa 1976). Thus, an inverse utility function is meaningful, and it can be used to identify the degree to which individual attributes of a decision alternative should change to achieve a desired level of utility. This fact suggests that it is possible, through the use of this technique, to go beyond simple ordering of a complex set of alternatives and to actually quantify the degree to which one alternative surpasses the others in terms of performance attributes such as cost and weight.

2.1 Application

Materials selection using utility analysis proceeds as do many of the standard techniques for materials selection, except that the criterion of choice is different. Instead of maximizing a weighted index, overall utility is maximized.

The first step is to identify the performance characteristics which are important for a particular application. There are, of course, a wide range of performance characteristics which go into the design of any engineering component. They may be gleaned from the relevant engineering literature and texts, professional societies and standards, and current industrial practice. The number of criteria can be considerably more than the limiting number of six cited. However, it is rarely necessary to treat all of these criteria in a multiattribute utility analysis. In most cases, the majority of these criteria are "binary" and do not require the extensive treatment that utility analysis entails. Binary characteristics are those for which there are clearly defined limits of performance; whose performance below a particular level is completely unacceptable and above which is not particularly valued by the decision maker. There are usually many such characteristics in any engineering design/materials selection problem. While in isolated cases engineering design may take into account and balance the multitude of performance criteria under consideration, most engineering designs set simple performance limits for most characteristics and concentrate on a subset of these for the bulk of their engineering decision making.

The second step is to construct the questionnaire for measuring utility. This is a standard procedure (Keeney and Raiffa 1976). Briefly, these questionnaires revolve around asking the subject to make pairwise choices between two alternatives offering both different levels of engineering performance and different levels of uncertainty. Two classical scenarios involve a materials acceptance scenario (i.e., choose between the available shipment of materials, which offers an uncertain level of performance, or demand a shipment that precisely offers some intermediate level of performance) or a research funds allocation problem (i.e., choose between two research programs, one of which may offer better ultimate material performance, although with less certainty).

The critical feature of the questionnaire is that it must be representative of actual decisions made by the interview subject. If the scenario presented is too unrealistic, the analyst cannot be assured of the validity of his data, nor can he be assured that his interview subject will treat the effort with any degree of seriousness. As a consequence, the development of these questionnaires calls for considerable insight and ingenuity on the part of the analyst.

Finally, the procedure calculates the overall utility of all the characteristics of a material. These results can be used in several ways, each offering powerful insight into the way in which materials are selected. First of all, the materials can be ranked according to their utility. Additionally, and as a special advantage of the utility method, the utility can indicate by how much a performance characteristic must change in order for the decision maker to be just indifferent between any one alternative and another. With such information, it is possible to define engineering and design objectives in a fashion much more meaningfully, by directly taking into account the preferences of the target market.

A third form of study focuses on the analysis of the form of the generated utility function itself. The shape of the function, particularly its curvatures and inflections, reveal how the relative importance of individual characteristics change as different levels of importance are achieved, indicating critical decision thresholds.

For example, consider Figs. 1 and 2. In each case, the figures present a contour map of a utility function for cost and tensile modulus. The contours show combinations of cost and modulus which yield the same utility value; these "isoutility" lines therefore are composed of combinations of cost and modulus which are equally preferred by the subject.

In Fig. 1, the isoutility lines are almost vertical. This shape indicates that, over the range represented here, changes in modulus have a relatively small effect on

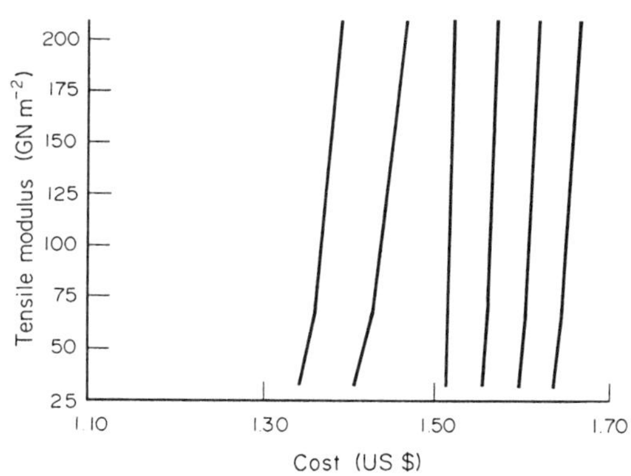

Figure 1
Cost vs tensile modulus isoutility map: cost dominant

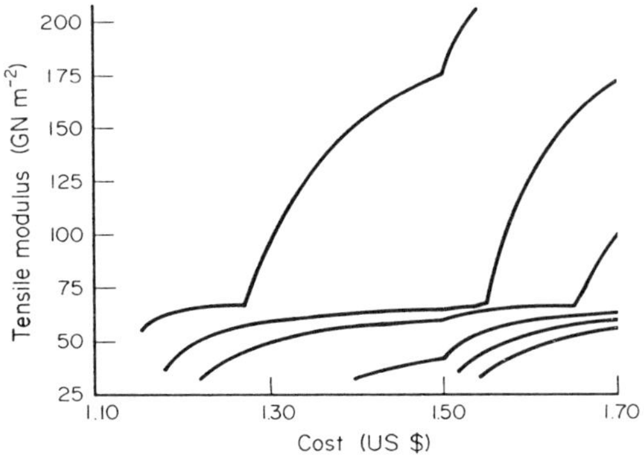

Figure 2
Cost vs tensile modulus isoutility map: neither attribute dominant

the level of utility achieved, while changes in cost have a very large effect. Therefore, cost is relatively more important to this decision maker and is likely to be the determining factor in materials selections, provided some minimum level of performance is achieved. Certainly, these curves indicate that he would be unwilling to pay a premium to achieve better than nominal performance.

In Fig. 2, the shape of the isoutility curves is more complex (and more typical). The slopes of the curves vary over the entire range of cost and modulus, suggesting that the relative importance of the attributes depends on the levels available. Furthermore, note the bunching of the isoutility curves in the vicinity of 70 GN m^{-2}. This appearance suggests that below 70 GN m^{-2} increasing strength is far more important than reducing cost, while above 70 GN m^{-2} the reverse is true.

Such insights are particularly valuable, not only in evaluating the competitive position of material alternatives, but also in speculating on the effects of potential changes in material performance on competitiveness. Such evaluations can be used to establish research objectives or to devise material marketing strategies.

2.2 Limitations

While utility analysis offers new avenues in materials selection techniques, it does have a number of limitations. First, for materials selection problems involving more than six important characteristics, the interview process takes too long to be feasible for most situations. There are real limitations on the time one can expect to get from decision makers in these areas and there are similar limitations on the time they can be expected to concentrate on a series of difficult questions.

The technique can thus be expensive. It is not only time consuming to develop and administer the questionnaire, but also to identify the individuals who actually select materials. These individuals are rarely so apparent that they can be picked out of an organizational chart, and a familiarity with the engineering practice and organizations of the industry being analyzed is a necessary part of the analysis.

Finally, the administration of the questionnaire requires an interviewer sensitive to the interviewees' reactions as well as to the needs of the analyst. There is a real skill which must be developed before the results of an interview can be reliably employed. A familiarity with the engineering issues that the decision maker faces is vital to the development of a questionnaire comprising the give and take which is the hallmark of a good interview.

3. *Materials Analysis Applications*

To illustrate the wide range of uses to which MAUA can be applied, two analyses drawn from recent work will be discussed in the following sections. The first will present a straightforward application of MAUA to evaluating a set of available material alternatives, while the second will undertake a more speculative venture into estimating competitive pricing of a new material.

3.1 Case 1—Automobile Camshaft Materials

In the study by Nallicheri et al. (1990), material and processing alternatives for making camshafts were analyzed, focusing on five process alternatives: nodular casting, steel forging, microalloy forging, a hybrid process (segmented camshaft) using a hollow steel tube with forged lobes held in place by hydraulic expansion of the tube, and the fully machined camshaft. While nodular cast camshafts are the predominant form used today, there are forces driving automakers to consider material and process alternatives. These forces are reduced weight, increased surface strength, increased stiffness and, of course, reduced cost.

Based on these concerns, an analysis of the competitive position of the five alternatives was undertaken using MAUA. Following an extensive literature review and discussions with camshaft designers, the attributes presented in Table 1 were chosen for detailed study.

Only physical attributes which could be traded off against one another or against cost were considered. By no means should these properties be considered exhaustive; rather, these were characteristics that merited the type of analysis that MAUA makes possible because there were indications that these characteristics could be traded off against one another. Other equally important characteristics were not included because simple "go/no go" materials selection criteria based on fixed performance targets had been established.

Responses indicated that, while the camshaft ought to meet strength specifications, designers do not seem

Table 1
Attributes analyzed for the camshaft

Attributes	Typical (cast)	Best	Worst
Part cost (US$)	12.0	5.0	40.0
Part weight (kg)	2.7	1.4	9.1
Material modulus (GN m^{-2})	145	275	105
Compressive yield strength (MN m^{-2})	1590	2070	1040

Table 2
Matrix of attributes for the alternatives of concern

	Cost (US$)	Weight (kg)	Modulus (GN m^{-2})	Compressive strength (MN m^{-2})
Nodular casting	10.71	2.7	145	1590
Steel	15.48	2.7	200	2030
Microalloy	13.64	2.7	200	2030
Assembled	15.70	1.5	200	2030
Fully machined	17.57	2.7	200	2030

to attach any value to exceeding those specifications. The prime design issue is increasing the stress-carrying capacity of the lobes. The modulus is a key contributor to the stress-carrying capacity of the cam lobes, through the Hertzian stress equation (Turkish 1946). While this relationship suggests that optimal surface stress performance is a complex function of both material modulus and compressive strength, subsequent analysis of designer preferences suggested that the individual characteristics also were accorded value individually, and they were thus assessed independently.

The performance of the five alternatives (under specific conditions—see Nallicheri et al. (1990) for more details) is presented in Table 2. The results of the utility analysis yielded a ranking for the five alternatives, which is shown in Table 3. Steel has a higher utility than nodular iron primarily because of its higher modulus and higher compressive strength. Although steel forgings are more expensive than castings, they still have a higher utility due to an overriding preference for high modulus and compressive strength. Apart from the enhanced modulus over the cast cams, the assembled camshaft also offers a reduction in weight. Thus, depending on the needs of the engine, it may be worthwhile to select this route, even though it is more expensive than the steel forging.

The importance of compressive strength is further illustrated in Fig. 3, showing an isoutility map in cost and compressive strength space (all other attributes held fixed). The notable feature of this figure is the fact that, over most of the domain plotted, cost is the dominant figure of merit, having the greatest influence on utility. However, there is a clear inflection point in the vicinity of 1500 MN m^{-2}, indicating that 1500 MN m^{-2} is an important threshold of performance.

These results were essentially confirmed in follow-up discussions with automobile camshaft designers, who are definitely moving in the direction of higher performance materials for this application. The results strongly support the contention that forged camshafts, using either microalloy steel or powder forged steel, are going to become increasingly important in camshaft design. Furthermore, the results suggest that cost improvement, through manufacturing innovations like near net shape processing, are generally of greater interest to these automobile part designers than major performance improvements.

This case shows how multiattribute utility analysis can be used to analyze a complex material choice problem, where the performance of in-

Table 3
Ranking of alternatives based on MAUA results

Alternative	Utility ranking
Assembled	1
Microalloy steel forged camshafts	2
Steel forged camshafts	3
Nodular cast camshafts	4
Fully machined camshaft	5

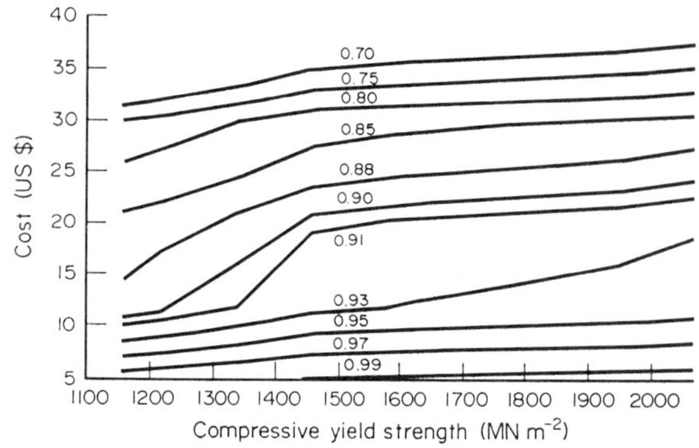

Figure 3
Cost vs compressive strength: camshaft case study

dividual alternatives varies in a complex fashion. MAUA provides the analyst with a method for directly assessing the value of varying levels of performance and relating this value to the overall value of alternatives with large differences in performance in several dimensions. Beyond this ability, analysis of the utility function itself yields insight into elements of the designer's thinking process by revealing critical levels of performance which have particular value to the designer. This information can be used to target particular markets with new material offerings as well as to direct material development efforts.

3.2 Case 2—Computer Backplane Printed Circuit Board Materials

In the study by Ng and Field (1983), a novel application of MAUA, linking it with technical cost modelling, was discussed. The problem centered on selection of an appropriate polymer–reinforcement blend (frequently referred to as laminates) for use in a specialized printed circuit board application, the computer backplane.

The backplane, sometimes called backpanel, provides interconnections for the daughter boards (PCBs). This application was chosen for the following reasons.

(a) As the desired speed of computers increases, signal delay at the board level (a function of physical characteristics of the laminate materials and the design of the circuit board) becomes more important to the overall performance of the computer. As a consequence, there is a pressing need to improve the performance of PCBs through improved design, manufacturing changes and careful selection of materials.

(b) The many new materials introduced into the PCB market offer a wide range of properties and cost. To date, despite strong marketing efforts and promises of improved performance, only three materials (epoxy, bismaleimide triazine (BT) and polyimide) are used, in conjunction with glass reinforcing fibers, in the commercial industry (excluding prototype and microwave applications). Of the three materials, epoxy–glass dominates the industry. Although numerous reports have predicted growth for other materials, there is still no clear material winner at this time (Murray 1987).

Because MAUA evaluates alternatives on the basis of their performance characteristics, identification of those characteristics which define and determine the relative competitiveness of these alternatives is critical. Through the examination of engineering literature and discussions with PCB industry engineers, an exhaustive list of performance characteristics was made. After screening the attributes for binary, insignificant, aggregable and intangible attributes, and consulting with industry experts, the scope of the MAUA was restricted to three attributes.

(a) *Dielectric constant.* This is an important attribute since it is a measure of signal delay of the PCBs. A lower dielectric constant is always preferred.

(b) *Glass-transition temperature (T_g).* This attribute is important because it is an indication of the thermal stability of the PCBs. A high T_g results in a lower variation in the coefficient of thermal expansion (CTE) during temperature cycling.

(c) *Cost of bare board (backplane).* The cost of backplane is an important factor in material selection. In this case study, the cost of a bare board was estimated using a PCB cost model developed for this purpose, and the cost estimates were verified by the interviewees. The cost of bare board is a preferred attribute in this case because PCBs have very high value added in manufacturing. Therefore, using raw material cost alone can be very misleading.

Although polyimide, epoxy and BT are the most commonly applied materials, a fourth material, cyanate ester (CE), was also considered a potential candidate for these applications, although an actual

Table 4
Average properties of laminates

Attributes	Epoxy	BT	Polyimide	CE
Glass-transition temperature, T_g (°C)	125	181	260	245
Dielectric constant, E (1 MHz)	4.5	4.1	4.3	3.5
Cost (US$)				
8 layers	330	430	571	688
14 layers	690	860	1200	1400

Table 5
Results of MAUA—relative rankings

Industry	Epoxy–glass	BT	Polyimide	CE
Computer vendors	4	2[a]	3	1
Communications	1[a]	2	3	4

a Material with the highest utility available in 1989

market in the material did not exist. In fact, only prototype laminates were being made available. Based on technical data and technical cost estimates, the matrix of performance characteristics shown in Table 4 was developed for the four alternatives. Several representatives of computer and communications companies were contacted, and their utility functions were developed and applied to these data. The aggregated utility results derived from these data are presented in Table 5.

As the results show, for the computer vendors the BT blend has the highest utility of the available materials. The utility of polyimide actually followed BT closely, while that of epoxy–glass trailed by a much larger margin. In fact, if the dielectric constant of polyimide were 4.2 (as claimed by some suppliers) instead of 4.3, its utility would be higher than that of BT. Thus, it can be concluded that both polyimide and BT are very attractive and their use will continue to grow in this application until a better material becomes available.

On the other hand, the communications industry results reflected a different set of attitudes. In this industry, cost is the main concern. Therefore, inexpensive epoxy–glass is preferred. Information from departments at AT&T confirmed the utility results, although the PCBs designed for more critical applications have already moved to the higher-T_g materials. However, it is estimated that more than 99% of all the PCBs used by AT&T are epoxy–glass.

These MAUA results indicated that, while the computer industry favors high-T_g materials such as polyimide and BT, their "preferred" material (were it available) would be CE. However, the CE price used in the analysis reflects a rough approximation of the possible laminate price, rather than an actual price. The actual potential of CE in the commercial market depends on how the material is priced and its manufacturing requirements since, to date, no supplier has committed to commercial pricing or volume production. However, utility analysis, in conjunction with cost modelling, can be used to establish at how high a price a CE laminate vendor can offer the material and remain competitive.

Figure 4 shows how the utility of each of the material alternatives varies as the final cost of the baseline circuit board varies, all other physical characteristics remaining fixed. Where market price data exist, the point corresponding to the actual production cost is indicated. Note that these results suggest that if the CE board were to cost less than US$2200, its utility would exceed that of polyimide, which is the alternative having the highest utility.

Through the use of a PCB cost model, relating finished PCB cost to process yield and material cost, it is possible to take this analysis further. Given a targeted bare board price, it is possible to evaluate the cost of the laminate and prepreg that would yield a board at that price. While there is a general consensus that CE laminates will be priced at the level of polyimide laminates, speculation on the production yield is less unanimous. By making assumptions regarding yield and processing parameters (as recommended by suppliers, but *not* verified by users in practice, since the material is not available for production), it is possible to establish the combinations of new material price and processing yield required to achieve the necessary bare board price. A graph showing the results of this analysis is presented in Fig. 5.

Most PCB manufacturers would be unwilling to produce at less than 60%, while 90% yield is a credible upper limit for 14-layer board yields in epoxy–glass. As Fig. 5 indicates, almost all combinations of yield and price would in fact make CE a viable competitor in the high-performance end of the marketplace. Whether this competition would occur at the expense of other high-performance materials,

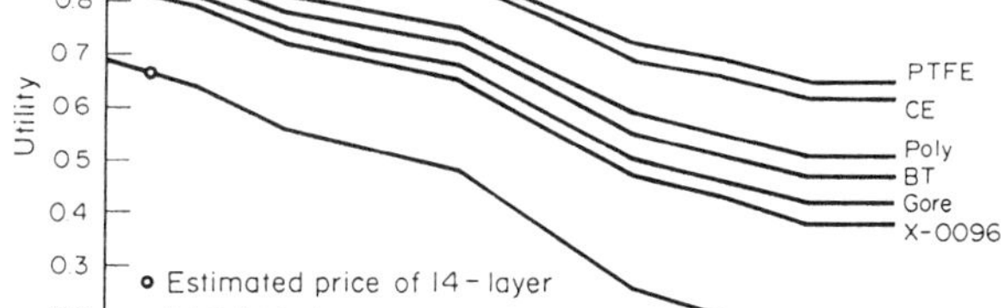

Figure 4
Effect of cost on utility

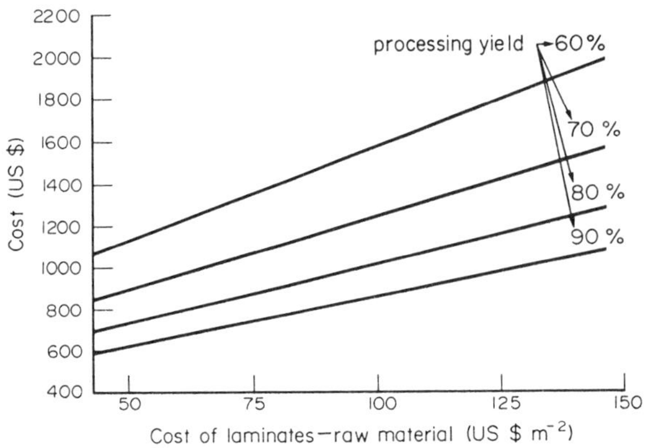

Figure 5
Estimated cost of CE PCBs as a function of yield

or whether CE would open new markets will depend on how the material is ultimately marketed. However, the results of the analysis clearly demonstrate that CE offers a realistic and credible alternative to the materials employed in this market.

4. Summary

The utility analysis approach to materials selection yields results that cannot be matched by any of the currently available techniques. By constructing a formal representation of a materials decision maker's preferences, it is possible to obtain both accurate and flexible representations of the materials selection process.

The analysis is neither sensitive to arbitrary choices of normalizing constants, nor limited to linear representations of materials preference, as the other techniques assume. Furthermore, the observed preference structure can be applied not only to the problem of assessing the relative suitability of materials for a particular application, but also to the question of the degree of difference between alternatives, the relative importance of performance characteristics and the way in which various performance characteristics are factored into the materials selection decision.

The advantages of the utility analysis approach do come at a price: it places some burden of time and energy on the engineers and decision makers. Much of this can be alleviated, however, through careful application of the underlying theory as well as the use of interactive computer programs for assessing utility and performing the analysis.

Bibliography

Alexander W O, Appoo P M 1977 Materials selection—the total concept. *Des. Eng.* (November): 59

Clauser H R, Fabian R J, Mock J A, 1965 How materials are selected. *Mater. Des. Eng.* 62: 109

de Neufville R 1990 *Applied Systems Analysis; Engineering Planning and Technology Management.* McGraw-Hill, New York

Field F R, de Neufville R 1988 Materials selection—maximizing overall utility. *Met. Mater.* (June)

Hanley D P, Hobson E 1973 Computerized materials selection. *J. Eng. Mater. Technol.* 95: 197–201

Keeney R, Raiffa H 1976 *Decisions with Multiple Objectives: Preferences and Value Tradeoffs.* Wiley, New York

Kusy P F 1976 Plastic material selection guide. Society of Automotive Engineers Technical Paper No. 760663. SAE, Warrendale, PA

Miaw D C, Wilson W R D 1982 Uses of merit in computer-aided material and manufacturing process design. *J. Mech. Des.* 104 (October): 806

Murray J 1987 Today's substrate. *Circuits Manuf.* (November)

Nallicheri N V, Clark J P, Field F R III 1990 An economic assessment of alternative manufacturing processes for the camshaft. Society of Automotive Engineers Technical Paper No. 901741. SAE, Warrendale, PA

Ng L H, Field F R III 1989 *Materials for Printed Circuit Boards: Past Usage and Future Prospects, Materials and Society*, Vol 13. New York

Rosen H J, Bennett P M 1979 *Construction Materials Evaluation and Selection—A Systematic Approach.* Wiley, New York

Sharp H J (ed.) 1966 *Engineering Materials—Selection and Value Analysis.* Elsevier, New York

Turkish M C 1946 *Valve Gear Design.* Waverly, Baltimore, MD

Verink E D Jr, (ed.) 1966 *Methods of Materials Selection.* Gordon and Breach, New York

J. P. Clark and F. R. Field III
[Massachusetts Institute of Technology, Cambridge, Massachusetts, USA]

N

Natural Fibers in the World Economy

Among the many vegetable and animal fibers produced in the world, those that are most important commercially as textile fibers are cotton, jute, sisal, flax and wool. These five fibers have been selected here to illustrate the various global economic forces affecting production, consumption and trade of natural fibers.

Natural fibers are produced seasonally and are subject to vagaries of weather, insects, diseases and other factors. Thus, the supply and quality of these fibers can be quite variable from year to year. Additionally, these fibers must compete with other agricultural commodities for land and other resources as well as competing with manmade fibers in all important end uses. To be viable economically, a fiber must be available in sufficient quantity and quality and at a competitive price. These requirements have often favored manmade fibers. All natural fibers have lost markets to manmade fibers since the 1960s, but the sharp losses of earlier years have ceased. During the early 1970s natural fibers were rapidly losing market share primarily because of the new performance characteristics of manmade fibers. Since 1980, however, the competitive losses of natural fibers have bottomed out, blend levels have stabilized in a number of important product markets, and a strong consumer preference for natural fibers is boosting global consumption.

World textile fiber production, including cotton, wool, silk, rayon, acetate and noncellulosic fibers except olefin, has more than doubled since 1965, reaching about 37×10^6 t in 1988—an average annual growth of about 4.8% during the period. World population grew by only about 2% annually, which implies that a sharp expansion in per capita fiber use occurred during the 1970s and 1980s. Marked shifts have taken place in the composition of fiber use. Cotton production increased 55% during 1965–1988, but its share of the world fiber market declined from over 61% in 1965 to about 49% in 1988 (see Fig. 1 and Table 1). The development of polyester fibers in the late 1950s brought about intense competition with cotton, rayon and acetate, and was instrumental in cotton's loss of market share. Noncellulosic fibers (primarily polyester) accounted for 38% of the world fiber market in 1988, cellulosic manmade fibers (rayon and acetate) about 8% and wool and other natural fibers accounted for the remaining 5% of the fiber market.

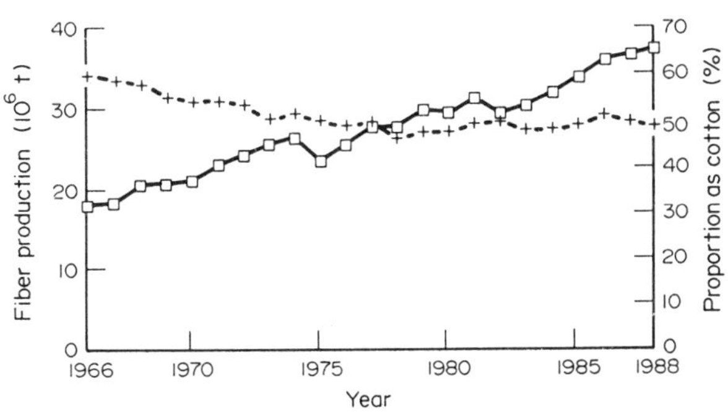

Figure 1
World fiber production and cotton's share: □, fiber production; +, cotton's share

1. Cotton

Cotton is the single most important textile fiber in the world. Its share of world fiber production has remained around 48–50% since the mid-1970s. It is an important source of foreign exchange for many cotton exporters such as the USA, the former USSR, Egypt, Sudan, Pakistan, Turkey and Central American countries. Millions of people around the world are either wholly or partly dependent on cotton for their employment and incomes, ranging from farm supply industries and producers to retailers of final products. Its importance

Table 1
World fiber production and cotton's share

Year	Production (10^6 t)	Cotton's share (%)
1966	18.1	59.7
1967	18.3	58.4
1968	20.6	57.5
1969	20.7	54.9
1970	21.1	53.9
1971	23.1	53.9
1972	24.4	53.1
1973	25.7	50.4
1974	26.5	51.5
1975	23.7	49.9
1976	25.6	48.9
1977	27.8	49.8
1978	27.8	46.3
1979	29.7	47.4
1980	29.4	47.6
1981	31.1	49.3
1982	29.4	49.8
1983	30.3	47.9
1984	31.8	48.1
1985	33.7	49.1
1986	35.9	51.0
1987	36.5	49.7
1988	37.2	48.9

is best understood in relation to the world textile fiber situation.

1.1 Production

World cotton production has increased steadily since 1965, rising from 54.4 × 10^6 bales in 1965 to 84.3 × 10^6 bales in 1988 (Table 2). Most of the secular production increase has come from higher yields per hectare, as the harvested area has remained fairly stable at about 30–34 × 10^6 ha. World yields averaged about 373 kg ha^{-1} during 1965–1975, compared with 535 kg ha^{-1} during 1985–1988. Sharply higher yields in China, Pakistan and the USA have been responsible for most of the recent growth in world yields.

The USA accounted for about 27% of world cotton production in 1965, compared with about 18% by 1988. China, Pakistan and India have greatly increased their cotton production since 1965, and are accounting for a large share of global cotton output. China's cotton production in 1988 jumped to 23% of the world total in response to a rapidly developing textile industry and increased emphasis on the crop by the government. China now ranks as the world's largest producer and consumer of cotton fiber.

1.2 Consumption

Total consumption of cotton increased by about 31 × 10^6 bales over the last two decades, from 52.1 × 10^6 bales in 1965 to about 83 × 10^6 in 1988 (see Fig. 2 and Tables 3 and 4). The largest expansion in mill use has taken place in the developing countries of Asia and Africa, especially China, South Korea and Taiwan. The USA and a number of European countries, on the other hand, have experienced long-term declines in total mill consumption of cotton. However, as mentioned earlier, this trend appears to be bottoming out or even reversing in recent years, particularly in the USA. The growing importance of cotton consumption in Asia (excluding the former USSR) is indicated by its share of world use—about 57% in 1988. Asian cotton use has expanded to fill the needs of a growing population, but also to produce textile and apparel products for export, yielding an important source of foreign exchange.

Cotton is used primarily in clothing and home

Table 2
Cotton: area harvested, yield and production for selected countries and world

Year	Brazil	China	India	Pakistan	USA	USSR	World
			Area (1000 ha)				
1965	2226	4775	7932	1568	5510	2428	33067
1970	2469	4997	7689	1748	4514	2752	31885
1975	1815	4955	7446	1851	3560	2922	29992
1980	2015	4920	7823	2108	5348	3147	32230
1985	2290	5140	7581	2366	4140	3316	31806
1986	2165	4306	7280	2502	3427	3475	29910
1987	2312	4844	7400	2568	4061	3527	32199
1988	2285	5577	7700	2440	4833	3432	34264
			Yield (kg ha^{-1})				
1965	244	347	126	266	590	790	362
1970	199	458	125	311	492	852	376
1975	218	479	155	267	508	865	380
1980	309	549	169	341	453	934	438
1985	362	805	240	522	706	839	545
1986	303	824	222	527	618	762	513
1987	327	876	209	573	791	700	545
1988	314	753	232	592	694	806	535
			Production (1000 bales[a])				
1965	2499	7600	4600	1915	14938	8810	54435
1970	2251	10500	4400	2500	8204	10770	55035
1975	1815	10900	5300	2269	8302	11610	54061
1980	2857	12400	9090	3300	11122	13498	64815
1985	3812	19000	8355	5669	13432	12777	79562
1986	3008	16300	7418	6058	9731	12157	70434
1987	3468	19500	7105	6764	14760	11345	80607
1988	3300	19300	8200	6637	15411	12700	84253

Source: US Department of Agriculture (1990 and earlier issues)
a 1 bale = 218 kg (480 lb) net

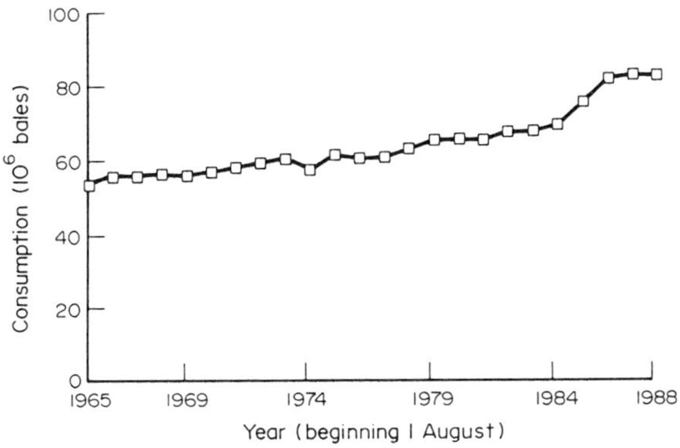

Figure 2
World cotton consumption

Table 3
World cotton consumption

Year	Million bales
1965	54.0
1966	56.0
1967	56.1
1968	56.5
1969	56.2
1970	57.3
1971	58.6
1972	59.8
1973	60.9
1974	57.9
1975	61.9
1976	60.9
1977	61.2
1978	63.5
1979	66.2
1980	66.1
1981	66.1
1982	68.3
1983	68.3
1984	69.9
1985	75.8
1986	82.5
1987	83.4
1988	82.9

furnishings, with lesser amounts used in industrial products. While no worldwide data are available, it is estimated that about 60% of all cotton used goes into clothing, 30% into home furnishings, and the remaining 10% into industrial products. Within each of the three major product markets, cotton fiber continues to compete with manmade fibers in specific applications. Overall, however, it is estimated that of all fiber used in clothing, 55% is cotton, 44% manmade fibers and about 1% wool. In the home furnishings market, cotton's share is about 25%, while in industrial applications, cotton accounts for about 18% of all fibers used.

1.3 Trade

World cotton trade has experienced slow but steady growth since the early 1960s. Beginning in 1986, however, trade jumped sharply to almost 26×10^6 bales as large world supplies and lower prices encouraged greater mill consumption (Table 5). Improved world economic conditions enabled world cotton trade to expand.

Most of the growth in imports occurred in the developing nations where cotton textile manufacturing expanded to meet growth in domestic demand, and also growth in cotton textile exports. Asian nations have accounted for an increasing share of world cotton imports since 1965 (Table 6). Except for China, most Asian nations produce no cotton and must import all of their cotton requirements. China's imports increased sharply during 1970–1980 reaching 17% of world imports in 1980. After 1980, cotton production in China grew rapidly and imports dropped to nearly zero. In

Table 4
Cotton consumption in specified countries and world

Year	China	India	Japan	South Korea	USA	USSR	World
				(1000 bales[a])			
1965	7900	5000	3200	340	9596	6950	52121
1970	10500	5250	3508	550	8204	8170	57300
1975	11500	6150	3166	913	7250	8900	61705
1980	15100	6306	3295	1447	5891	9150	66118
1985	18400	7191	3146	1707	6399	9400	75811
1986	20200	7847	3445	1844	7452	9350	82519
1987	20000	7612	3477	1995	7617	9300	83352
1988	20500	7960	3200	1900	7300	9300	82930

Source: US Department of Agriculture (1990 and earlier issues)
a 1 bale = 218 kg (480 lb) net

Table 5
World cotton supply and utilization

Year	Beginning stocks	Production	Imports	Total supply	Consumption	Loss[a]	Exports	Ending stocks
				(1000 bales[b])				
1965	28755	57110	17127	102992	54013	−260	16946	32293
1970	23316	55304	18872	97493	57295	74	17748	22375
1975	33435	54195	19530	107160	61907	145	19093	26014
1980	21132	64978	20691	106802	66118	530	19699	20454
1985	42437	79562	21474	143474	75811	227	20260	47175
1986	47175	70434	25691	143301	82519	295	25958	34529
1987	34529	80607	24207	139344	83352	243	23523	32226
1988	32226	84253	24907	141386	82930	328	24615	33512

Source: US Department of Agriculture (1990 and earlier issues)
a Unaccounted difference between total supply and total use b 1 bale = 218 kg (480 lb) net

recent years, however, China has come back into the import market, supplementing lower cotton supplies because of production shortfalls. South Korea and Taiwan have developed extensive textile economies, currently accounting for a combined total of almost 15% of world trade in cotton. Japan remains the largest cotton importing nation, accounting for total purchases of over 3×10^6 bales in 1988. Japan's mills, as well as those in Western European nations, have gradually shifted from cotton to manmade fibers, and face increased competition from imported cotton textiles. These developments are consistent with the historical shift of cotton manufacturing capacity from the developed nations to the developing nations.

The major exporting nations include the USA, the countries of the former USSR, Pakistan, Sudan, Egypt,

Table 6
Cotton trade for selected countries and world

	Exports (1000 bales[a])						
Year	Egypt	Pakistan	Sudan	Turkey	USA	USSR	World
1965	1575	492	570	959	3035	2240	16946
1970	1397	473	1049	1124	3897	2450	17748
1975	775	418	1097	2163	3311	3890	19093
1980	749	1490	426	1028	5926	4070	19699
1985	837	3148	499	322	1960	3170	20260
1986	586	2884	820	510	6684	3462	25958
1987	428	2383	825	198	6582	3417	23523
1988	280	3500	675	450	5900	3500	24615
	Imports (1000 bales[a])						
Year	China	Italy	Japan	South Korea	Taiwan	W. Germany	World
1965	500	1013	3078	327	305	1250	17127
1970	500	816	3669	557	735	1084	18872
1975	900	886	3220	1013	1024	1040	19530
1980	3550	870	3207	1527	965	724	20691
1985	1	1194	3054	1682	1534	987	21474
1986	16	1539	3688	1901	2357	1376	25691
1987	86	1435	3431	1960	1593	1135	24207
1988	1800	1275	3200	1900	1700	1050	24907

Source: US Department of Agriculture (1989b and earlier issues)
a 1 bale = 218 kg (480 lb) net

Turkey and Mexico. In recent years, Australia has developed as a major exporter of cotton, and China has periodically offered large volumes of cotton for sale. The USA has traditionally maintained the largest share of the world export market, accounting for nearly 24% in 1988. Market shares of other exporters, such as Pakistan and China, have varied widely from year to year, while countries such as Egypt and Turkey have been experiencing declining exports and market share.

1.4 Market Structure

In the USA and throughout the world, vegetable fibers are typically produced by many small firms which, individually, have no measurable effect on market prices. In the USA, the complex forces of supply and demand are reflected through the individual decisions of thousands of producers and buyers.

In the USA, the number of farms harvesting cotton declined dramatically between 1969 and 1987 from nearly 200 000 to 43 000. Farms continue to increase in size in response to economic and technological forces. However, the increasing size and concentration of cotton production on fewer farms does not seem to threaten the traditional family farm business organization. The corporate form of organization, although increasing, is undertaken by farm operators chiefly to take advantage of tax policies, limited liability or transfer provisions. Recent levels of profitability in cotton production, as in other major sectors of crop production, are not attracting a substantial influx of capital investment by nonfarm corporations.

The physical marketing of cotton begins when the producer delivers the fiber and attached seed to the gin, where the seed is removed and the fiber is baled. The baled fiber typically moves to warehouses for further compression and storage. Some cotton fiber moves directly from gins to textile mills without storage. In the late 1980s, approximately 5800 textile mills in the USA used cotton to some extent, ranging from spinning and weaving mills to padding and upholstery mills. Manufacturing of finished apparel products took place at about 16 400 establishments—chiefly cutters and garment makers.

Market structure in other countries, especially in the centrally planned economies, has differed markedly from that in the USA. For example, in China, economic decision making on the 2100 state farms is fully controlled by the government (US Department of Agriculture 1982). On the other hand, the production teams of the commune system are given more autonomy to make economic decisions, provided that the State's policies and production plans are met. In 1980 there were approximately 5.2×10^6 production teams on 53 348 communes. Although China's long-range policy favors state-owned farms, they contributed less than 4% of total farm output, compared with nearly 80% produced by communes and about 17% by households on private plots. About half of the state-owned farms specialized in grain and cotton.

1.5 Competitive Factors

Cotton competes in world markets not only with the production of cotton from several exporting countries but also with manmade fibers. Raw cotton prices generally exceeded those of rayon and polyester staple fibers throughout the 1970s. Following the relatively high cotton prices of 1980 and subsequent price increases for raw materials used to produce manmade fibers, mill-delivered prices of raw cotton have compared favorably with rayon and polyester staple (Table 7). Although price competition is only one of many determinants of cotton–manmade fiber substitution, it is important in the maintenance of market share. In contrast to natural fiber production, manmade fiber production is dominated by a few large firms. This

***Table** 7*
Prices of cotton and manmade fibers in the USA

Year	Cotton[a] (US\$ kg^{-1})	Rayon staple[b] (US\$ kg^{-1})	Polyester staple[c] (US\$ kg^{-1})	Price ratios	
				cotton rayon^{-1}	cotton polyester^{-1}
1965	0.62	0.63	1.96	0.98	0.31
1970	0.66	0.57	0.93	1.16	0.71
1975	1.21	1.17	1.10	1.03	1.10
1980	2.16	1.72	1.70	1.26	1.26
1985	1.61	1.81	1.52	0.89	1.06
1986	1.49	1.74	1.43	0.86	1.04
1987	1.78	1.86	1.52	0.96	1.17
1988	1.59	1.92	1.70	0.77	0.94

Source: US Department of Agriculture (1990 and earlier issues); actual prices divided by 0.90 for cotton, 0.96 for rayon and polyester to estimate raw fiber equivalent for processing
a Delivered group B mill points (western North and South Carolina), net weight SLM $1\frac{1}{16}$ in b F.o.b. producing plants 1.5 and 3.0 denier regular staple c Reported average f.o.b. producing plants, 1.5 denier staple for cotton blending

oligopolistic market structure has resulted in relatively stable prices, especially in the short term, in comparison with natural fibers.

In the long run, prices received by cotton producers correspond closely to cost of production, while in the short run, prices may depend chiefly on other factors. Although statistically reliable data for determining the cost competitiveness of cotton with other fibers are not available, there is partial evidence to indicate that cotton is a viable competitor. For example, a worldwide survey of the costs of producing raw cotton indicated that the average cost of production in 14 countries was US\$1.79 kg^{-1} in 1986, compared with about US\$1.53 kg^{-1} in 1984 (International Cotton Advisory Committee 1990). Since this is an average cost that includes charges for all indirect costs such as land, overheads, unpaid family labor and management, as well as cash costs, many producers have the ability to produce at a substantially lower cost in the short run. Producers worldwide have increased efficiency by adopting improved varieties, using more fertilizer, irrigating more acreage, improving management of crop pests and adopting other cost reduction techniques. World cotton yields have increased markedly from 362 kg ha^{-1} in 1965 to 535 kg ha^{-1} in 1988. Another factor tending to hold down costs is the substantial reduction in numbers of small inefficient farms and the increase in size of remaining farms.

Comparable cost of production data are not available for manmade fibers, but the competitive prices of mill-delivered polyester and rayon staple in recent years indicate an ability to compete favorably from both a price and a cost standpoint. Fiber buyers indicate that, other factors being equal, the relative stability of manmade fiber prices, combined with dependable supply, tends to favor their use in many end products. However, cotton remains a viable competitor in the apparel and home furnishings market.

2. Jute

Jute is a soft fiber obtained from the inner bark of plants grown chiefly in Asia. It ranks second to cotton among vegetable fibers in total production and in world trade. Jute has historically been one of the least expensive natural fibers and has relatively good spinning properties. However, compared with other cellulosic fibers, it is relatively weak, difficult to bleach and not moisture-tolerant (Linton 1966). In recent years, the jute industries of India and Bangladesh have been faced with rising labor and raw material costs, variations in volume and quality of production and relatively low productivity. Jute also faces severe competition from noncellulosic fibers in all major end uses. However, it remains an important source of export earnings, especially in Bangladesh.

2.1 Production

World production of jute has varied substantially from year to year. While the overall trend in production since 1970 has been gently upward, weather problems, especially in Bangladesh and India, have restricted the crop in recent years. World production averaged about 3×10^6 t in 1987 and 1988, compared with about 5×10^6 t in 1985 and 1986 (Table 8). The cultivated area in most countries dropped in response to lower producer prices and profits in comparison with other crops, such as rice. Production is concentrated in Asia, with India by far the largest producer, followed by China and Bangladesh. India's share of world production, now about 40% of the total, has been maintained over the past 10 years.

2.2 Consumption

On a world basis, about 75% of all jute manufactured is used in commodity packaging, including sacks and bags (Food and Agriculture Organization of the United Nations 1986). Other uses include carpet backing, webbing, rope, cordage and twine.

Jute consumption has been growing in Asia but declining in the industrial countries of Europe and North America, though the USA remains an important consumer. In the late 1960s, jute accounted for a major share of the primary backing for tufted carpets. By the 1980s that market had been lost to manmade fibers. At the same time, jute has retained about one half of the secondary backing market. Jute has also experienced severe losses in its markets for bagging and sacks. Manufacture of jute is concentrated in producing countries. Sacks, hessian cloth and carpet backing are the three main categories of manufactured products.

2.3 Trade

Exports of jute fiber from producing countries have diminished since the mid-1960s (Table 9). However, Bangladesh remains a heavy exporter of its jute fiber and manufactured products and is the major exporter of raw jute. The value of exports of fiber from Bangladesh averaged US\$95 $\times 10^6$ in 1986–1988. Bangladesh accounted for about 75% of world jute fiber exports and about 50% of world total exports of raw and processed jute during 1986–1988.

In Bangladesh, most of the raw jute exports are handled by about 200 licensed merchant shippers, with the balance arranged by a government agency (United Nations 1982). Most of the raw jute is purchased from a small number of international traders, who sell to jute spinners and mills through other private firms or government agencies, depending on the purchasing country's form of government. India is also a major exporter of jute goods, and Thailand is the only other country having significant exports of manufactured jute products.

Major importers of raw jute have included Pakistan and the USSR. The USA imports very little raw fiber, but still imports jute products: carpet backing is the largest single import, followed by burlap. US imports

Table 8
Jute: area harvested, yield and production for selected countries and world

Year	Bangladesh	Brazil	China	India	Nepal	Thailand	World
			Area (1000 ha)				
1965	933	58	288	1117	34	381	2846
1970	910	44	139	1080	55	398	2764
1975	526	74	290	914	33	287	2253
1980	642	72	313	1300	52	185	2763
1985	1020	64	992	1496	27	209	4002
1986	782	64	345	1068	47	212	2688
1987	526	65	272	960	20	153	2165
1988	519	61	265	942	14	190	2152
			Yield (kg ha^{-1})				
1965	1354	1333	2185	931	1139	1413	1242
1970	1352	1200	2072	1033	964	981	1196
1975	1495	1029	2275	1165	1242	1104	1377
1980	1352	1081	3508	1130	1141	1300	1462
1985	1513	955	2328	1242	1058	922	1420
1986	1556	987	2058	1454	1298	1133	1510
1987	1634	1009	2092	1272	1150	1096	1443
1988	1702	1140	2208	1385	1143	1136	1527
			Production (1000 t)				
1965	1264	77	453	1040	39	538	3534
1970	1231	53	288	1115	53	390	3306
1975	787	76	660	1064	41	317	3101
1980	868	78	1098	1469	59	240	4039
1985	1557	62	2060	2291	33	274	6551
1986	1216	63	710	1553	61	240	4058
1987	859	66	569	1221	23	168	3123
1988	883	69	585	1305	16	215	3286

Source: Food and Agriculture Organization of the United Nations (1988a and earlier issues)

of raw jute fiber and manufactured products dropped from 373 000 t in 1971 to 111 000 t in 1988.

2.4 Competitive Factors

Noncellulosic fibers—chiefly polypropylene—had penetrated all major end-use markets of jute by the late 1960s. Prior to 1960, jute competed chiefly with cotton and hard fibers as materials for bags, sacks and cordage, and with paper for packaging materials. Technological innovations in packaging, transportation and bulk handling also accelerated the shift away from jute bagging, sacks and industrial cloth. The impacts of noncellulosic fibers, however, differed between countries. For example, jute consumption in the USSR continued to increase throughout the 1970s in response to a substantial increase in agricultural production as well as a substitution of jute for flax, another important natural fiber. Jute consumption in China continued to increase through the 1970s in response to a high degree of self-sufficiency in jute production and less conversion to other fibers.

A major concern of the jute industry is the price structure and variability in relation to competing fibers. The United Nations (1982) concluded that periodic shortages and price instability had been major factors causing shifts to manmade fibers. A related problem facing the jute industry is the reported lack of sufficient returns in manufacturing to cover reinvestment, market development and research needs.

Sharp fluctuations of jute prices in the early 1960s spurred efforts by producing countries to stabilize supplies and prices through various means, such as producer price supports, provision of buffer stocks and purchase quotas for mills and shippers (Food and Agriculture Organization of the United Nations 1982b). These actions by individual governments failed to provide the desired stability, leading to an attempt by the Food and Agriculture Organization's Intergovernmental Group on Jute, Kenaf and Allied Fibers to operate an informal price agreement for jute and kenaf. An "indicative" export price range was established with the objective of maintaining market prices within this range. Some improvement in price stability has occurred even though adherence to the agreed price level is voluntary. As with most commodity agreements, however, prices tend to fluctuate widely in response to extreme fluctuations in supply.

Table 9
Jute fiber trade for selected countries and world

	Exports (1000 t)						
Year	Bangladesh	Burma	China	India	Nepal	Thailand	World
1965	722	1	1	26	12	317	1151
1970	636	2	<0.5	27	14	257	996
1975	314	24	1	47	21	158	590
1980	342	44	36	18	26	30	526
1985	254	[a]	45	8	5	1	340
1986	387	[a]	77	20	5	4	518
1987	403	[a]	62	7	8	5	504
1988	250	[a]	50	4	14	5	349

	Imports (1000 t)						
Year	Belgium–Luxembourg	China	France	Pakistan	United Kingdom	USSR	World
1965	97	63	96		116	23	1038
1970	95	50	49	40	81	42	910
1975	32	21	29	48	32	23	569
1980	13	46	22	45	28	49	566
1985	6	20	3	80	17	28	361
1986	13	41	4	127	23	38	511
1987	7	22	4	130	21	53	500
1988	7	5	4	91	20	37	343

Source: Food and Agriculture Organization of the United Nations (1988b and earlier issues)
a No official report

3. Sisal

Sisal is the world's most important hard fiber—those fibers derived from leaves of perennial plants. It represents about 90% of all hard fiber production and ranks fourth among all vegetable fibers in world trade. Production is concentrated in very few countries, located in tropical and subtropical regions of the world. Sisal is an important export commodity in the developing countries of East Africa and Brazil. Often it is produced in poor areas where there are few alternative sources of employment and income. Like most natural fibers, it has lost market share to noncellulosic fibers since the 1960s. Sisal is used chiefly in the manufacture of cordage products.

3.1 Production

Brazil and Tanzania currently produce about 60% of world output, with lesser volumes being produced in Madagascar, Mozambique, Angola and Kenya (Table 10). Brazil has accounted for an increasing share of production in recent years, while Tanzania's share has dropped substantially. World production totalled more than 700 000 t in 1965, but had dropped to about 372 000 t by 1988.

Sisal had been a growth industry during the 1950s and early 1960s, but technological developments in materials handling and strong competition from noncellulosic fibers have led to a reduced market for sisal since the mid-1960s. Area harvested in sisal dropped from about 927 000 ha in 1965 to about 575 000 ha in 1988.

Historically, a high proportion of sisal production has been exported in raw form. However, a growing proportion of production is now processed in the producing countries. An estimated two-thirds of world output of sisal twine is spun in developing countries where it is produced (Food and Agriculture Organization of the United Nations 1982a).

3.2 Consumption

The major markets for sisal fibers and manufactured products are located in the developed countries of Western Europe and North America. The most important end use is the manufacture of cordage products, chiefly agricultural harvest twine. Other uses include woven goods such as sacks, bags, carpets, rugs, mats, padding, paper and handicraft products (Food and Agriculture Organization of the United Nations 1985). Projected world demand in 1990 for sisal, henequen and other minor hard fibers, in fiber equivalent by end uses, is shown in Table 11.

Cordage products consume about two-thirds of the

Table 10
Sisal: area harvested, yield and production for selected countries and world

Year	Angola	Brazil	Kenya	Tanzania	World
		Area (1000 ha)			
1965	70	326	115	216	927
1970	70	310	52	250	812
1975	50	327	48	163	718
1980	30	296	48	140	626
1985	2	333	38	63	630
1986	1	322	38	51	610
1987	1	296	35	50	579
1988	1	301	35	50	575
		Yield (kg ha^{-1})			
1965	994	598	557	1006	758
1970	881	661	845	808	762
1975	800	961	907	761	855
1980	667	794	988	614	724
1985	1500	875	1184	508	763
1986	1429	764	1316	587	717
1987	1429	646	1105	560	640
1988	1429	661	1143	560	648
		Production (1000 t)			
1965	70	195	64	218	703
1970	62	205	44	202	619
1975	40	314	44	124	614
1980	20	235	47	86	453
1985	3	291	45	32	480
1986	1	246	42	30	437
1987	1	191	40	28	371
1988	1	199	40	28	372

Source: Food and Agriculture Organization of the United Nations (1988a and earlier issues)

world output of sisal and henequen. These products are consumed largely in developed countries, where sisal has experienced substantial reductions in use since the mid-1960s. Although the trend in use of sisal is clearly downward, annual variations in use depend on such factors as the size of the hay and grain crops, carry-over stocks of manufactured products, prices and other factors.

3.3 Trade

The major producers of sisal have in the past exported much of their production in raw form. In the early and mid-1960s, over 600 000 t of sisal and other hard fibers were exported annually, compared with about 130 000 t in 1988 (Table 12). In 1965, about 88% of world production was exported as raw fiber. However, by 1988 only about 50% of total raw fiber production was exported in that form. In recent years, producing countries have expanded their manufacturing capacity and their exports of sisal manufactures—chiefly cordage (Food and Agriculture Organization of the United Nations 1989b).

Import destinations of raw fibers have changed

Table 11
Projected 1990 world demand for sisal, henequen and other minor fibers

Use	Demand[a] (1000 t)
Cordage	
Agricultural twine	200
Other twines	60
Ropes and cables	20
Total	280
Noncordage	
Sacks and bags	70
Padding	5
Carpets and mats	10
Other	55
Total	140
Grand total	420

Source: Food and Agriculture Organization of the United Nations (1985)
a Fiber equivalent

Table 12
Sisal fiber trade for selected countries and world

Year	Exports (1000 t)				
	Brazil	Kenya	Tanzania	Madagascar	World
1965	150	59	214	29	618
1970	149	45	217	21	586
1975	52	44	102	23	277
1980	100	40	47	13	229
1985	91	33	12	12	152
1986	76	32	12	9	141
1987	63	28	18	7	128
1988	65	31	18	9	140

Year	Imports (1000 t)						
	China	France	Italy	Poland	Portugal	USSR	World
1965	7	59	27	14	45	6	616
1970	8	61	34	14	47	7	541
1975	12	32	24	21	22	9	294
1980	4	22	18	16	34	30	225
1985	1	16	11	18	36	18	178
1986	2	17	7	13	30	15	158
1987	3	13	9	3	34	20	146
1988	3	16	5	5	33	13	129

Source: Food and Agriculture Organization of the United Nations (1988b and earlier issues)

substantially since the 1960s. Then, the USA was the single most important importer of raw sisal, but its imports were negligible in the early 1980s. There is currently very little harvest twine manufacturing in the USA. The UK, Belgium, the Federal Republic of Germany and the Netherlands were also important importers of raw fibers in the 1960s, but their relative importance has diminished. The most important fiber importing countries in the late 1980s were France, Portugal and the USSR.

The USA is by far the most important importer of manufactured products. Other important importers of manufactured goods, chiefly cordage products, are Canada, France and Germany.

3.4 Competitive Factors

The market for sisal is largely determined by the demand for agricultural twine. The demand for all agricultural twine, whether made from natural or manmade fibers, continues to diminish. As a result, sisal competes with manmade fibers for a larger share of diminishing market. Overcapacity of production in both sisal and polypropylene twine had led to relatively low prices of twine in recent years. There is some evidence to suggest that prices of both sisal and polypropylene twines were unprofitable in the early 1980s (Food and Agriculture Organization of the United Nations 1984). However, comparison of production costs is extremely difficult due to lack of appropriate data and variability of accounting procedures. In the short run, prices tend to reflect market factors rather than production costs. There is continuing debate concerning the price differential necessary to prevent further erosion of sisal's share of the market. However, there is general agreement that price levels and price stability are the chief determinants of market shares in agricultural twine. In markets other than harvest twine, including rope, cable and packaging twine, certain performance characteristics such as strength, durability and weight are often more important than price in determining market share (Grilli 1975).

As with jute, there have been organized attempts to stabilize international sisal production and prices. Overplanting and subsequent low producer prices in the late 1960s led to the establishment in 1968 of an international informal agreement under the auspices of the FAO Study Group on Hard Fibers. The goal of this agreement was to raise producer prices above the depressed levels of the late 1960s but below the level at which the use of manmade fibers would be encouraged (Grilli 1975). Two features of this agreement were (a) an informal export quota for sisal and henequen raw fibers and finished products and (b) an indicative or target price for sisal fiber. This agreement has had mixed results. The persistent oversupply of sisal was a major weakness of the informal agreement and prices acceptable to producers were seldom attained.

Governments of producing countries are involved in varying degrees in supporting their hard fiber industries. Most governments subsidize production, manufacturing and trade in one way or another. Perhaps as much as one half of the sisal twine manufactured in producing countries is spun in government-owned or controlled units (Food and Agriculture Organization of the United Nations 1982a). Government policies impact heavily on exports of both fiber and twine in most producing countries, although policies differ considerably among countries.

4. Flax

Flax ranks third among fiber crops in total production and world trade. Flax is an annual plant from which both oilseeds and fiber are derived. It is an ancient agricultural crop, with food and clothing applications predating written records (Eastman 1968). Flax was a major source of fiber for making linen cloth in Europe and the USA until the cotton gin was invented in the late eighteenth century, when cotton started to displace linen in clothing. Since the mid-1960s, manmade fibers have gradually entered the traditional markets of flax. Flax is now produced chiefly for its oilbearing seed (linseed). Fiber flax varieties represent only about 20% of all flax area in the world.

4.1 Production

The countries of the former USSR are the principal producers of fiber flax in the world, representing about 73% of the world harvested area and 46% of total world production of flax during 1986–1988 (Table 13). Other relatively important producers include Romania, Poland, France, Egypt and China.

World production area in flax has gradually diminished over the years as competing fibers have displaced flax. World area in 1965 totalled 1.9×10^6 ha, compared with 1.3×10^6 ha in 1988. Average fiber yields increased substantially during 1986–1988, chiefly in response to a growing proportion of world harvested area in China, where the highest fiber yields in the world are reported. Production in China has increased markedly since 1965, while that of the USSR has dropped.

Table 13
Flax: area harvested, yield and production for selected countries and world

Year	China	France	Poland	Romania	USSR	World
			Area (1000 ha)			
1965	6	51	114	23	1476	1853
1970	43	37	98	36	1284	1604
1975	52	43	78	60	1215	1555
1980	60	46	82	71	1120	1477
1985	65	59	43	76	1020	1346
1986	85	47	36	80	975	1295
1987	129	51	28	82	971	1332
1988	133	56	31	83	970	1345
			Yield (kg ha^{-1})			
1965	453	1172	498	567	325	396
1970	1773	1079	530	450	355	439
1975	1761	976	768	462	406	507
1980	1853	1478	602	464	260	427
1985	1092	1695	638	406	345	488
1986	2462	1383	638	400	345	562
1987	2490	1365	611	427	438	699
1988	1514	1312	577	434	333	531
			Production (1000 t)			
1965	2	59	57	13	480	734
1970	76	40	52	16	456	704
1975	92	42	61	28	493	789
1980	111	68	49	33	291	630
1985	71	100	27	31	352	657
1986	209	65	23	32	336	728
1987	321	70	17	35	425	932
1988	201	73	18	36	323	714

Source: Food and Agriculture Organization of the United Nations (1988a and earlier issues)

4.2 Consumption

Fiber flax is produced chiefly for its stalk, which is processed into fibers known as linen, and other fiber products. The better grades of flax fiber are used for manufacturing linen cloth for uses such as apparel, table linen, handkerchiefs, sheeting and decorative items and in making high quality thread (Dempsey 1975). Coarse grades are used in making twines, canvas and bags. The tow (short, twisted or broken fibers) is used in manufacturing such items as towelling and high-grade paper such as cigarette paper, bond and currency.

Flax fiber consumption in Eastern European countries is considerably higher than in most other countries of the world (Table 14). In 1980, the most recent year of comprehensive statistics, world per capita consumption of flax was 0.2 kg compared with 1.2 kg and 1.7 kg in the USSR and Poland respectively.

4.3 Trade

World trade in flax fiber has dropped substantially since the early 1960s. In 1965, world exports totalled about 450 000 t, compared with about 300 000 t in 1988 (Table 15). World exports dropped by about 3% per year during 1961–1981, followed by a partial recovery of earlier levels. Belgium–Luxembourg and France are major exporters and importers. Other leading exporters include China, Egypt, the Netherlands and the former USSR. Other leading importers include Italy, Japan and the UK. The total free-on-board value of world exports in 1988 was US\$306 $\times$ 10^6.

5. Wool

Wool accounts for about 90% of the total world production of animal fibers. It has been used to make apparel and other textile products from the beginning of recorded history and still enjoys a significant role in the world textile market. In the nineteenth century, wool probably accounted for about one fifth of all textile fibers consumed. During the four decades before World War II, when manmade fiber use was relatively insignificant, wool constituted about 10% of total fiber mill consumption in the USA. Because of the superior price and performance competition of manmade fibers, wool's share of the US textile market declined to about 1% in the 30 years following World War II. This trend of declining importance has also occurred in other industrial countries, but at a slower rate. In some of these countries, particularly in Europe, the preference for wool-containing textile products such as apparel and carpets has been relatively strong.

Table 14
Per capita consumption of flax in the world and selected countries, 1980

Area	Per capita consumption (kg)
World	0.2
USSR	1.2
Poland	1.7
Romania	1.4
France	0.4
United States	0.2
Netherlands	0.6
United Kingdom	0.1
China	0.1

Source: Food and Agriculture Organization of the United Nations (1983)

Table 15
World exports of flax and tow, 1965–1988

Year	Quantity (1000 t)	Total value (10^6 US\$)
1965	451	95
1970	328	86
1975	279	131
1980	244	192
1985	283	209
1986	298	241
1987	281	314
1988	299	306

Source: Food and Agriculture Organization of the United Nations (1988b and earlier issues)

5.1 Production

The world sheep population reached a record 1.2 $\times$ 10^9 head in 1988. Sheep numbers have grown at an average annual rate of about 1.0–1.2% since the end of the 1960s. Australia, the former USSR, China and New Zealand have the largest sheep populations, accounting for over 40% of the world total in 1988.

World wool production in 1988 was 3.2 $\times$ 10^6 t, grease basis (Table 16). In Australia, the world's largest wool producer, the 1988 clip reached a record 942 000 t, up almost 14% on 1985 output. The USSR reported 1988 wool production at 484 000 t, about the level of the previous three years.

Differences in yield, or raw wool weight per sheep, accounted for part of the variation in wool production by country. Australia and New Zealand averaged more than 5.2 kg per sheep and produced 39% of the world's wool from about 19% of the sheep. In contrast, China, with an average yield of 1.8 kg per sheep, produced 7% of the world's wool from 9.5% of the world's sheep. Breed of sheep, climatic conditions and type of feed are factors which influence both the fineness and the quantity of wool.

Most of the world's stocks of raw wool are held in the exporting producing countries of Australia, New

Table 16
World wool production and trade, grease basis

Country	1985	1986	1987	1988
		(1000 t)		
Wool production, total[a]	3039	3100	3160	3220
Australia	830	887	914	942
USSR	468	492	477	484
New Zealand	358	350	346	331
China	178	185	209	224
Argentina	152	150	157	167
Uruguay	87	90	89	87
South Africa	98	90	92	98
United States	41	39	39	41
Wool exports from main exporting countries[a]	1357	1385	1469	1401
Australia	718	714	807	732
New Zealand	307	306	272	304
Argentina	63	68	61	57
South Africa	53	38	33	40
France	37	33	41	37
Uruguay	26	39	40	33
Wool imports into the principal importing countries[a]	1261	1327	1412	1344
Japan	184	177	204	175
China	113	152	151	187
United Kingdom	128	118	139	127
USSR	109	115	134	128
Italy	120	109	122	114
France	132	132	118	115
West Germany	77	73	80	76
Belgium–Luxembourg	55	58	64	67
Taiwan	41	50	52	36
United States	35	44	48	44
South Korea	31	38	44	35
Yugoslavia	21	21	16	12

Source: International Wool Textile Organization and International Wool Study Group (1990)
a World totals

Zealand and South Africa. The available supply (stocks plus production) of raw wool at the beginning of 1988 totalled 1.94×10^6 t, slightly below the record supply of 1.97×10^6 t in 1986.

5.2 Consumption

The major types of wool consumed worldwide can be classified as merino, crossbred and "other." Merino wool is a fine grade of wool used in the better quality suiting fabrics. Crossbred wool is the medium grade of wool used to make shirts, sweaters, coats and blankets, for example. The "other" or coarser types of wool are mostly used in carpets. Since the mid-1980s, the merino types averaged 42% of world production, the crossbred types 28% and the carpet or "other" types 30% of production.

World consumption of wool totalled 1.77×10^6 t, clean basis, during 1988, approximately 8.5% above the 1985 level of use (Table 17). Four countries—the USSR, China, Italy and Japan—accounted for about 50% of global consumption in 1988. The countries of the former USSR are the world's largest consumers of wool, with use increasing slowly over a period of years. However, Chinese wool consumption has more than doubled since 1980, and China may soon replace the former USSR as the number one consuming nation. While part of this tremendous growth reflects increased domestic needs, China's emphasis on textile exports is the major factor. Wool consumption in most other countries has been static or has shown only slight growth in recent years, especially in the USA and EC countries.

Apparel and carpet face yarns represent the primary end-use markets for wool, accounting for about 90–92% of all wool used. Apparel products such as trousers, suiting and sweaters are the major wool-consuming items. Over the years, however, wool lost considerable market share largely to acrylic staple and

Table 17
World wool consumption, clean basis

Country	1985	1986	1987	1988
		(1000 t)		
USSR	308	309	321	318
China	190	243	248	299
Italy	130	125	140	140
Japan	123	114	127	126
United Kingdom	77	78	87	87
United States	52	62	62	58
West Germany	63	60	62	58
South Korea	28	39	46	54
Turkey	40	41	46	47
India	35	34	33	36
Belgium–Luxembourg	38	37	31	34
France	42	38	36	33
World total[a]	1634	1697	1743	1773

Source: International Wool Textile Organization and International Wool Study Group (1990)
a Total consumption by the wool spinning industry at the spinning stage

polyester staple in the manufacture of apparel products. Wool has lost most of the carpet market because of development of the tufting process, which produced a lower cost carpet and was better adapted to the use of noncellulosic fibers than the woven process. However, the popularity of wool as a natural fiber has helped to maintain a constant or slightly increasing share of noncarpet markets, despite the strong competition from manmade fibers.

5.3 Trade

The movement of raw wool in international trade has involved essentially a south to north direction. Most of the wool exported has been produced in countries in the southern hemisphere, while most of the imported wool has been taken by countries in the northern hemisphere. Six countries—Australia, New Zealand, Argentina, South Africa, France and Uruguay—accounted for about 86% of raw wool exports during the four years 1985–1988. Six countries—Japan, China, the USSR, France, the UK and Italy—took more than 62% of total raw wool imports.

The annual quantity of raw wool moving in international trade averaged 1.3×10^6 t, grease, during the four-year period 1985–1988. The volume of trade during that period increased about 5% annually.

During the 1980s, the major wool exporting countries shipped bales of higher density to achieve lower shipping costs. These dense bales (about 450 kg) were almost three times the weight of the regular or "farm" bales (about 160 kg). The finer grades of wool compact to a greater density than the coarser grades. It was this greater resistance to crushing that created a preference for coarse grades for use in carpets.

Australian wool producers sell their wool at auction using the services of wool brokers, or sell directly to wool merchants. In the 1980s about 80–85% of the Australian clip was sold by auction. The remaining amount was sold directly to raw wool merchants. The share of New Zealand wool sold at auction over the same period increased to about 75%.

The three major wool exporting countries found it necessary to organize wool stock schemes to minimize price fluctuations. New Zealand organized the first such plan in 1951. In 1958, South Africa began a similar scheme. Following the decline of merino wool prices in 1969–1970, the Australian Wool Corporation (AWC) was established to purchase wool offerings to maintain market prices. In September 1974, the AWC introduced a minimum floor price scheme and was given the power to control the supply of wool coming onto the market.

5.4 Competitive Factors

Following World War II, wool experienced intense competition from manmade fibers, particularly from the noncellulosic fibers. Historically, the other natural fibers—cotton, linen, silk, jute, sisal, hemp, ramie and others—competed very little with wool for the various textile markets. Wool and the other natural fibers were used in those products which most benefitted from each fiber's unique performance and economic characteristics. There were relatively few products which could be made by substituting one fiber for another while retaining equal or superior performance.

With the development of noncellulosic fibers and their commercial success by the early 1960s, wool began to lose its place in many traditional textile markets. Wool also experienced competition after World War II from nonfiber sources. Australian sheep growers, in times of relatively low financial returns from sheep production, switched to cattle raising or to wheat production.

In all sheep producing countries, the sale of sheep meat or live animals is important and these economic returns to the sheep grower must be combined with those from the sale of wool to determine ultimate profitability. The demand from the Middle East for live animals (supplied principally from Australia) and sheep meat (principally from New Zealand) became strong after the mid-1970s when the higher price of oil provided countries in that area with hard currency.

Costs and returns data for sheep wool producing operations are available only for the USA. During the four-year period 1984–1987, the return from sale of wool ranged from 27–34% of the sheep producers' gross returns. The sale of lambs and ewes comprised the remaining and major share of gross returns. Total variable costs ranged from 64–75% of total production costs. The most profitable sheep producing regions of the USA tended to have a lower ratio of variable feed cost to gross return. Such regions tended to have low concentrate feed costs and low pasture and range costs.

Bibliography

Dempsey J M 1975 *Fiber Crops*. University Presses of Florida, Gainesville, FL

Eastman W 1968 *The History of the Linseed Oil Industry in the United States*. Denison, Minneapolis, MN

Food and Agriculture Organization of the United Nations 1980 *Some Longer-Term Effects of Recent Developments in the Petrochemical Sector on the Competitive Position Between Sisal and Polypropylene*, CCP: HF 80/6. United Nations, Rome

Food and Agriculture Organization of the United Nations 1982a *Factors Influencing the Possibility for Setting Indicative Prices for Sisal and Henequen Baler Twine*, CCP: HF 82/7. United Nations, Rome

Food and Agriculture Organization of the United Nations 1982b *Recent Market Developments*, CCP: JU 82/2. United Nations, Rome

Food and Agriculture Organization of the United Nations 1983 *World Apparel Fiber Consumption Survey 1983*, M 62. United Nations, Rome

Food and Agriculture Organization of the United Nations 1984 *Short-Term Outlook On the Competitive Position Between Sisal and Polypropylene, Including Baler Twine*, CCP: HF 84/6. United Nations, Rome

Food and Agriculture Organization of the United Nations 1985 *Hard Fibers: Supply, Demand and Trade Projections to 1990*, CCP: HF 85/5. United Nations, Rome.

Food and Agriculture Organization of the United Nations 1986 *FAO Agricultural Commodity Projections to 1990*, FAO Economic and Social Development Paper 62. United Nations, Rome

Food and Agriculture Organization of the United Nations 1988a *FAO Production Yearbook*, Vol. 42. United Nations, Rome

Food and Agriculture Organization of the United Nations 1988b *FAO Trade Yearbook*, Vol. 42. United Nations, Rome

Food and Agriculture Organization of the United Nations 1988c *Jute Competition with Synthetics*, CCP: JU/SYN 88/1. United Nations, Rome

Food and Agriculture Organization of the United Nations 1989a *Techno-Economic Constraints on Jute Production and Policy Suggestions*, CCP: JU 89/4. United Nations, Rome

Food and Agriculture Organization of the United Nations 1989b *Statistics on Sisal and Henequen, 1984–89*, CCP: HF 89/2. United Nations, Rome

Food and Agriculture Organization of the United Nations 1989c *Jute, Kenaf and Allied Fibres: Quarterly Statistics*, CCP: JU/QS 89/4. United Nations, Rome

Food and Agriculture Organization of the United Nations 1989d *Comparative Cost and Price Developments in the Harvest Twine Market and Possible Indicative Price Recommendations*, CCP: HF 89/5. United Nations, Rome

Grilli E R 1975 *The Future for Hard Fibers and Competition from Synthetics*. Johns Hopkins University Press, Baltimore, MD

Hollen N R, Saddler J 1973 *Textiles*, 4th edn. Macmillan, New York

International Cotton Advisory Committee 1990 *Survey of Cost of Production of Raw Cotton*. Secretariat of the Committee, Washington, DC

International Wool Textile Organization and International Wool Study Group 1990 *Wool Statistics 1988–89*. International Wool Textile Organization, London

Linton G E 1966 *Natural and Manmade Textile Fibers: Raw Material to Finished Fabric*. Duell, Sloane and Pearce, New York

Ross P 1986 Materials of biological origin: an overview. In: Bever M B (ed.) 1986 *Encyclopedia of Materials Science and Engineering*. Pergamon, Oxford, pp. 2861–3

Starbird I R, Lawler J V 1986 Natural fibers in the world economy. In: Bever M B (ed.) 1986 *Encyclopedia of Materials Science and Engineering*. Pergamon, Oxford

Textile Organon (issued monthly) Textile Economics Bureau, Roseland, NJ

Tisdell C A, McDonald P W 1979 *Economics of Fiber Markets: A Global View of the Interdependence Between Man-Made Fibers, Wool and Cotton*. Pergamon, Oxford

United Nations 1982 The marketing of jute and jute products: areas of international cooperation, *Proc. UN Conference on Trade and Development, Geneva*. United Nations, Geneva

US Department of Agriculture 1977 *US Team Reports on Soviet Cotton Production and Trade*, FAS-M-277. USDA Foreign Agricultural Service, Washington, DC

US Department of Agriculture 1982 *China: Review of Agriculture in 1981 and Outlook for 1982*, Suppl. 6 to WAS-27. USDA International Economics Division, Washington, DC

US Department of Agriculture 1987 *The US Cotton Industry*. USDA Economic Research Service, Washington, DC

US Department of Agriculture 1989 *Foreign Agriculture Circular: Cotton*, FC-2-89, USDA Foreign Agricultural Service, Washington, DC

US Department of Agriculture 1990 *Cotton and Wool Outlook and Situation*, CWS-61, USDA Economic Research Service, Washington, DC

US Department of Agriculture *Statistics on Cotton and Related Data* (various issues). USDA Economic Research Service, Washington, DC

I. R. Starbird
[Manassas, Virginia, USA]

E. H. Glade, Jr
[US Department of Agriculture, Washington, DC, USA]

P

Paper and Paperboard: Environmental Considerations

Environmental considerations in the manufacture of paper and paperboard, a major world commodity, involve basic forestry, pulp manufacture, papermaking itself and the disposal and recycling of already used paper. Potentially affected environmental sectors include surface and ground waters, the air and the land itself (both by forestry operations and the disposal of solid wastes from manufacturing, consumer use and pollution control operations). The relative importance of these considerations is influenced by timberland and manufacturing site locations, the scale of operations and the type or level of technology employed in manufacturing and pollution control operations, and other cultural and societal factors.

1. Environmental Quality Considerations in Forestry Operations

Timberlands and their management and harvesting practices vary considerably. Their effects on environmental quality depend on the species being grown, forest terrain features and the principal uses of adjacent forest watersheds as well as the management practices followed. Environmental considerations in timberland management have increased greatly in number and complexity in recent years. Long-standing concerns include:

(a) sediment, organic matter and nutrient inputs to surface waters associated with road building, timber harvest and reforestation activities;

(b) contamination of surface waters by pesticides, herbicides and fertilizers;

(c) increases in stream temperatures due to harvest of trees that provided shade;

(d) changes in total, seasonal or episodic runoff due to timber harvest effects on evapotranspiration, infiltration and snow melt rates; and

(e) effects of prescribed burning on visibility and air quality.

Under most circumstances, these effects can be controlled effectively by following management guidelines developed and tested in recent decades. For example, water quality effects can be greatly reduced by limiting the amount of soil disturbance and timber removal in buffer strips along stream channels.

Newer environmental concerns related to timberlands management include:

(a) effects on endangered species and biodiversity;

(b) cumulative effects on multiple forestry operations within a watershed or region;

(c) effects on wetlands, old growth and other ecosystem components; and

(d) effects on carbon storage and atmospheric concentrations of carbon dioxide, methane and other gases that influence global climate through the "greenhouse effect."

Design of effective control measures to address the newer concerns is complicated by several factors. The effects of concern are usually defined only in general terms and are difficult to quantify. Moreover, the effects have many potential causes in addition to forest management, thus making it difficult to evaluate the efficacy of proposed control measures. Research is at the stage of better defining the effects of concern and developing techniques for their measurement, particularly those with global implications.

Concerns about the health of forests as affected by acidic deposition and other regional air quality problems have been the subject of an intensive research effort in recent years. Scientific studies have *not* supported claims of pervasive and irreversible forest decline in North America and Europe; the vast majority of forests are not affected by decline due to acidic deposition or any other stress factor. Sulfur dioxide causes damage to forests near uncontrolled point sources, but the extent of current damage is quite limited in most of North America and Western Europe. Tropospheric ozone pollution is a significant forest health problem in some areas (e.g., near Los Angeles and Mexico City) and is reducing the growth of sensitive tree genotypes in several forest regions.

Acidic deposition appears to be a relatively minor stress factor in most forests. Possible exceptions include certain montane and coastal forests that are frequently exposed to cloudwater or fog in which acidic pollutants are highly concentrated. There is experimental evidence that acidic cloudwater can reduce the ability of trees to tolerate weather-related stresses. Acidic deposition may also cause or exacerbate forest nutrition problems through effects on soil chemistry and nutrient cycling. The practical significance of such effects appears limited relative to effects of other natural and management variables that influence forest nutrition.

2. Water Quality Aspects of Paper and Paperboard Manufacture

Water quality concerns are related to potential changes in commonly regulated water quality parameters such as turbidity, suspended solids levels and biochemical oxygen demand (BOD); sensory quality indicators such as water color, taste and odor, or off-taste in fish taken from affected water bodies; and interference with use of waterways for propagation or harvesting of fishery resources of recreational or commercial importance. Recently, attention has been focused on the chlorinated organic materials that result from pulp bleaching and are discharged with bleachery wash waters. Intensive research continues into the possible environmental significance of these compounds. The regulation of these compounds has the goal of seeing that they are released only in quantities that pose no threat to human health and the environment.

Processing steps giving rise to potential discharges causing some of the above negative impacts include (a) debarking of wood; (b) mechanical and thermomechanical pulping; (c) chemical pulping by any of the principal sulfite- or sulfide-based pulping processes; (d) chemical recovery processes; (e) pulp bleaching to reduce the lignin content of the cellulosic fibers; (f) subsequent pulp cleaning; and (g) paper sheet or board formation, including use of additives for desired strength, appearance or printability properties, including surface coating.

The technologies used to control water pollution arising from these sources include:

(a) extensive water recycling in both papermaking and bleaching operations to reduce the effluent volume needing treatment and, in the case of papermaking, to reincorporate some lost material into the product;

(b) gravity and pressure-screen or filter-medium filtration to extract fibers and additives from in-process water or wastewaters;

(c) flotation and gravity sedimentation processes; and

(d) biological treatment of wastewaters.

Biological treatment means particularly the use of long- or short-term aerated or nonaerated stabilization basins, providing 5–100 d retention to achieve microbiological oxidation and stabilization of the oxygen-demanding organic effluent constituents before discharge. To an almost equal degree, the faster activated sludge process is employed in single-stage or occasionally multistage configuration, achieving comparable oxygen demand reductions of 85–95% in 4–24 h process retention. Compressed air or surface aeration supplies the oxygen needed by the microbial population with supplemental additions of nitrogen and phosphorus as needed to sustain microbial growth. Tonnage oxygen is gaining use in activated sludge systems. Excess sludges are generally separated by gravity sedimentation and concentrated further before disposal.

Effluent decolorization is infrequently practised, although several technologies involving lime or alum precipitation, bleaching and use of membrane separation or granular media have been developed or demonstrated to various degrees.

In-process load and loss control, particularly for spent pulping liquor losses, is gaining attention, along with stripping or specialized handling of evaporation condensates from spent liquor concentration, as a means of reducing the process load sent to external biological treatment. The recent interest in chlorinated organics has prompted extensive modifications to the bleaching process to minimize the formation and release of these compounds. Most modifications are aimed at minimizing the use of chlorine and related oxidants or reducing the amount of lignin that reaches the bleach plant. Engineered diffuser outfalls are employed as needed, to achieve the desired dispersion of effluent in receiving waters.

3. Air Quality Aspects of Pulp and Paper Manufacture

The manufacture of pulp and paper requires steam and power, which are typically generated at the mill site by the burning of fossil fuels (coal, oil, natural gas) or wood residues (e.g., bark) in boilers. Particulate emissions from solid fuel combustion are controlled with mechanical collectors, electrostatic precipitators, fabric filters, wet scrubbers or gravel bed filters. Sulfur dioxide emissions are limited by use of low-sulfur coal and oil, or flue gas desulfurization. Oxides of nitrogen emissions are limited by combustion modification.

Mills that produce chemical wood pulps have emission sources associated with the pulping and bleaching processes, and with the recovery of pulping chemicals. At mills using the kraft pulping process, emission sources include digesters, evaporators, pulp washers, kraft recovery furnaces, smelt dissolving tanks and lime kilns. Emissions of total reduced sulfur compounds are of most concern because of their objectionable and very low odor thresholds. A variety of practices have been instituted to severely curtail these emissions.

Kraft recovery furnace and lime kiln emissions have been reduced through modification of equipment design and improved operating practices. Off-gas collection and oxidation systems for digesters and evaporators have been installed at virtually all kraft mills. These measures have resulted in an estimated 90% reduction in TRS emissions from 1970 to 1990, despite an increase in kraft pulp production of 75%. Particulate emissions from kraft recovery furnaces are controlled with electrostatic precipitators, while lime kiln particulate emissions are controlled with wet scrubbers or electrostatic precipitators, and smelt tank particulate emissions are controlled by wet scrubbers.

Other chemical pulping processes include sulfite and semichemical processes, although the kraft process dominates. Sulfur dioxide from digesters and liquor-recovery operations at sulfite mills is well controlled by wet scrubbing systems, with the absorbed sulfur dioxide being used to make new cooking acid. There are a variety of semichemical pulping processes, many of which do not have significant atmospheric emissions.

In recent years, efforts have been made to reduce emissions of chlorine, chlorine dioxide and chloroform associated with the chlorine-based bleaching of pulps. Wet scrubbers have been widely employed to control chlorine and chlorine dioxide emissions from bleach plant washers, seal tank vents and bleaching towers. Emissions of chloroform have been reduced through modifications to pulp bleaching practices.

The objectives of emission controls at pulp and paper mills are to ensure mill emissions do not contribute to exceedances of ambient air quality standards or result in local nuisance situations. Most of the 600 mills in the USA are located in areas which have ambient air quality better than the National Ambient Air Quality Standards for particulates, SO_2, NO_2 and CO. Since ozone nonattainment areas are rather extensive, many mills are located in such regions. However, pulp and paper mills are not major sources of volatile organic compounds which contribute to ozone formation.

Within the workplace, exposure to airborne compounds is primarily of concern in mills with pulping and/or bleaching operations. In kraft pulp mills, concentrations of reduced sulfur compounds, such as hydrogen sulfide and methyl mercaptan, and turpentine vapors require attention. In sulfite pulp mills, SO_2 is of concern. Within chlorine-based bleach plants, chlorine, chlorine dioxide and chloroform are of interest. A number of monitoring techniques are available for measuring concentrations of these compounds in workplace atmospheres. Several mills have installed monitoring and warning systems for hydrogen sulfide and chlorine to protect employees.

4. *Solid Waste Disposal*

The solid wastes generated by the paper industry are primarily (a) wastewater treatment sludges, (b) ashes from the combustion of fossil fuels, wood waste and sludges, (c) inorganic wastes from chemical pulping processes, and (d) rejects from the processing of recycled paper. The amounts generated are highly variable from one mill to another depending on the production processes and wastewater treatment practices used. Wastewater treatment sludges are usually processed by mechanical dewatering before disposal. Belt filter presses are the equipment most commonly used for this purpose, but screw presses and vacuum filters are also in widespread use. Mechanical dewatering equipment is capable of drying paper-industry wastewater sludges to a solids content of 15–60% (wet weight basis).

About 70% of the industry's wastewater treatment sludges are disposed of in landfills or, less commonly, impoundments. Burning of sludges for energy accounts for another 20%, while about 8% is utilized in land application programs. Modern paper mill sludge landfills are being designed with features to protect groundwater quality. Leachate collection systems and liners (either in-place clay, imported clay or synthetic membrane) are common. Groundwater monitoring is also commonplace at newly constructed landfill sites.

5. *Used Paper Recycling*

Recovered paper and paperboard have long been an important source of fiber for the paper industry. In 1990, 29 million tons of paper and paperboard were collected for reuse, which translates into an overall recovery rate of 33.4% of paper and paperboard produced. The industry has adopted a goal of 40% used paper and paperboard recovery for reuse or export to be reached in 1995. If this goal is achieved, total recovery of recyclable paper and paperboard in the USA will increase from 29 million tons in 1990 to approximately 40 million tons in 1995. Areas of greatest recent and current growth in used paper recycling include (a) old corrugated carton fiber furnished to integrated linerboard and corrugated medium mills, and (b) production of newsprint, accelerated by growing regulatory requirements for recycled newsprint production. Attention is now being focused on recycled printing paper production and the possible reprocessing of used office papers and mixed used papers.

Wastes and emissions generated at recycled paper mills are highly variable in quantity and quality. Recycled paperboard mills produce little wastewater and solid waste. On the other hand, it is not uncommon for mills that deink recovered paper to generate 60 m^3 of wastewater and 200 kg of sludge solids per tonne of paper produced. Wastewaters from recycled paper mills are amenable to conventional wastewater treatment processes. Emissions to air include (a) emissions from power boilers and (b) chemicals used or generated in processing the recovered paper or making new products.

6. *Product Acceptability*

The discovery in the mid-1980s of the production of trace levels of dioxin in pulp bleaching raised a series of questions about product safety and acceptability. A series of risk assessments have largely allayed fears about human exposure to dioxin via (a) food contact and packaging materials, (b) dermal contact with printing and writing papers and personal care products, and (c) the inhalation of dust by workers in converting operations.

Pressures from overseas markets have given rise to new questions about the general chlorinated organics content of bleached pulps or products made from these pulps. The questions are driven by concerns over (a) chlorinated organics which may be released in wastewaters from nonintegrated mills which purchase these pulps, and (b) human exposure to materials which remain in the final product. Bleachery process changes that are part of the industry's response to the dioxin situation are also reducing the generation of other chlorinated organics. The nature of chlorinated organics remaining in finished products has been examined to some extent, and no showing of risk associated with these residues has been made.

See also: Recycling of Paper Products

Bibliography

Air and Waste Management Association 1991 *Air Pollution Engineering Manual*, AP-40. AWMA, Pittsburgh, PA

Barnard J E, Lucier A A 1990 Changes in forest health and productivity in the United States and Canada. National Acid Precipitation Assessment Program, State of Science and Technology Report No. 16. US Government Printing Office, Washington, DC

Megahan W F, King P N 1985 Identification of critical areas in forest lands for control of non-point sources of pollution. *Environ. Manage.* 9(1): 7–18

National Council of the Paper Industry for Air and Stream Improvement 1980 Emission trends for particulate matter and sulfur compounds from kraft pulping, 1968–1990. Special Report No. 80-08. NCASI, New York

National Council of the Paper Industry for Air and Stream Improvement 1987 First progress report on the assessment of potential health risks from use of bleached board and paper food packaging and food contact products. Special Report No. 87-11. NCASI, New York

National Council of the Paper Industry for Air and Stream Improvement 1988 Assessment of potential health risks to pulp and paper mill workers from dermal exposure to dioxin in bleached pulp, paper, and pulp-based products. Technical Bulletin No. 549. NCASI, New York

US Environmental Protection Agency 1982 Development document on pulp and paper effluent guidelines. US EPA, Washington, DC

I. Gellman
[National Council of the Paper Industry for Air and Stream Improvement, New York, USA]

Policy: Australia

During the last two centuries, Australia has been a major world supplier of pastoral and agricultural products, and of minerals. The fortunes of the mining industry have fluctuated, with present interest focused on energy-related developments. Here, the policies which influence Australia's ability and willingness to supply minerals and energy to world markets are examined.

1. Constitutional Framework

Australia is a federation made up of six states. It has a federal government which deals with specific matters, such as defence and foreign trade; the residual powers remain with the states. This division has resulted in conflict in a number of areas, of which the minerals and energy sector has been a major example. The states control mineral production, but federal government controls mineral exports. For offshore resources such as oil and gas fields, the situation has been more complex, with the states acting as designated authorities of the federal government.

2. Political Framework

Australia has three main political parties operating at the federal level—the National Country Party (NCP), whose support comes primarily from rural areas; the Australian Labor Party, which has a strong basis in the trade union movement; and the Liberal Party, which is the party most likely to receive support from the business sector and from the upper socioeconomic groups.

Both sides of federal politics believe that the Australian minerals and energy sector should produce for export as well as for domestic consumption. However, there are real differences in attitudes to the development of the sector. The Liberal Party tends to encourage foreign investment; the Labor Party prefers to limit such investment.

Politics at the state level tend to mirror federal politics, with a predominantly business–rural faction facing a trade union dominated Labor Party. One of the major characteristics of state governments of either political persuasion is that they are primarily interested in new economic developments without much concern about environmental devastation, the degree of foreign ownership, or input and output pricing. For example, with the recent spate of interest in siting aluminum smelters in Australia, both the Labor government in New South Wales and the Liberal government in Victoria were accused of selling the electricity input to smelters too cheaply—indeed, of having other consumers subsidize the proposed smelters.

3. Conservation of Resources

In the 1930s, the federal government imposed an embargo on the export of iron ore because it was thought that Australia had insufficient for its own needs. The lifting of that embargo, and the burgeoning of exploration, reserves, development and exports that followed, have been used as arguments for exports of other commodities which Australia may, in the future, wish it had retained. Thus there is not complete community agreement that the export of natural gas from the Northwest Shelf field makes sense for a nation whose own gas supplies are not ensured as far into the future as it might wish, and whose degree of oil self-sufficiency may dwindle in the foreseeable future. Those opposed to the gas exports may accept that the exports are essential to immediate development of the project, and would therefore argue that responsible resource management requires postponement of the project. The project developers (the Western Australian government) see immediate economic benefits to the state both from development expenditure and from the gas supplies thus made available within the state. The federal government has approved exports of crude oil only in the short term.

Although Australia is concerned about the availability of oil and gas for its own use, there has not been the same concern for supplies of coal, iron ore, nickel, bauxite, silver, lead and zinc, tungsten, and other minerals. This is not to say that such concern might not develop in the future (e.g., if coal liquefaction plants were established, Australians would hardly wish to see the output exported while they themselves were short of liquid fuel).

Further, although reserves of both coking and steaming coal seem adequate at present to support an expanding export market, it is possible that continuously expanding coal exports and greater use of other coals for liquefaction may ultimately lead to a review of export policy for coal and coal-based products. In the event, failure of the projected demand to materialize, and supply constraints such as absence of loading facilities and construction delays, have combined to render such a review unnecessary in the foreseeable future.

Provision of liquid fuels to other countries posed policy problems for Australia in two other areas. The first was oil shale. Significant deposits of shale are known, and were the subject of investigation by both Australian and overseas companies. Australia's main interest must be to serve its own liquid fuel needs first; if interest in shale quickens, it may be convenient to use overseas capital, and this may carry with it a commitment to export.

The second area of alternative liquid fuels in which Australia has been interested is the production of ethanol from plant material (including sugar and grains), and the production of middle distillate from oilseeds. This raises questions of economics, of energy efficiency and, on the part of some commentators, the moral question of whether land devoted to food production should be diverted to fuel production. It also raises the very practical issue of how far Australia's agricultural land can be extended, or more intensively cultivated, without causing further damage.

4. *Foreign Investment*

Economic nationalism in Australia seems to come in waves, with the strength of feeling not necessarily related to current capital inflow (e.g., an open door to foreign investment was a significant factor in the downfall of the federal government in 1972).

In the early 1970s, Australians were examining the great mineral boom of the 1960s and realizing how its benefits had flowed abroad. By 1980, as they contemplated the next phase of resource development they saw government anxious once more to encourage foreign investment.

At the federal level, the government was aiming for minimum Australian ownership at the development stage of 50%, with 75% for uranium. It did not insist on maximum Australian ownership, and experience showed that approval of a 50% foreign interest could be forthcoming even when further Australian equity was available. At federal level, there is a Foreign Investment Review Board whose operation remains important under the Labor government which took office in 1983.

At the state level, some state governments resent any federal government interference in project ownership. Some do not distinguish between foreign and Australian companies, and may award mineral areas to foreigners despite Australian interest. The New South Wales government, a Labor government, requires 51% Australian ownership before development.

5. *Taxation*

The federal government collects company income tax, withholding tax on dividends paid abroad, a coal export levy, a coal research levy, and a crude oil levy with a resource rent tax on new oil projects. State governments collect mineral royalties (with special arrangements for offshore petroleum). Some state governments also tax the mining industry by charging rail freight well in excess of cost, and using the proceeds to subsidize rail freight of other sectors. In addition, some states have accepted the concept of a lump sum payment as one of the conditions for granting a mineral area, and the federal government is examining the concept of lump sum bidding for relinquished petroleum permits.

6. *Structural Adjustment*

In the 1970s it was accepted that the growth of the mineral sector, through its pressure on the exchange rate, can cause problems for the rural sector, and for that part of the manufacturing sector which competes with imports. This caused some concern, as different sectors evaluated what resource developments meant for them.

The developments of the 1980s were much more attenuated than was at first thought; capital inflow for resource projects was matched to some extent by equipment imports; more importantly, capital inflow was for public consumption, not for private investment. By early 1985, with the Australian dollar in crisis, the fears of an overvalued exchange rate were no longer heard.

7. Uranium: A Special Case

Mining and export of uranium was in principle approved by the coalition federal government (1975–1983), subject to certain safeguards. Obtaining development approval was, however, a slow process, especially if foreign investment guidelines had to be met. If the deposit was on aboriginal land, as some uranium deposits are, then negotiations with the aboriginal people—for permission to mine and for agreed monetary compensation—had to be concluded. An environmental impact statement was to be prepared for government, and environmental approval was to be obtained. Sales contract negotiations were undertaken by prospective developers, but within constraints set by the Federal Minister for Trade and Resources on the advice of his Uranium Export Office. During both the development and construction phase, there was extensive monitoring of the operation by government.

The Australian Labor Party policy was to restrict uranium mining and export until certain safety aspects (e.g., waste disposal problems) could be regarded as having permanent solutions. The Australian Council of Trade Unions also had an anti-uranium policy. However, when Labor won federal office in 1983, uranium mining and export continued.

8. Value Added in Processing

Successive state and federal politicians have sought to increase the degree to which Australia processes its minerals before export. In the 1970s, several new aluminum smelters were planned, and the federal government approved the undertaking, by four Australian companies, of a feasibility study of a uranium enrichment plant.

Aluminum smelter proposals received public criticism on three points. Firstly, the Federal Treasury expressed a desire for all energy resources in Australia to be priced at world parity, adjusted for transport costs. It included in this low-quality coal supplied to power stations. Arriving at a world parity price for this coal may be difficult, but nevertheless the assumption seemed to be that electricity may be underpriced for one reason or another. The second category of public criticism was environmental. The Hunter Valley, an important producer of top-quality wines, was one area where smelter development was to occur, and there has been concern that vines may be damaged by fluoride emissions. Thirdly, the Portland smelter proposals faced conflict with aboriginal interests. In the event, most smelter proposals did not proceed, and the Labor government from 1983 would not contemplate uranium enrichment.

9. Pricing

Some minerals exported from Australia are subject to export control, so that the federal government can attempt to see that Australian exporters are not played off too much one against the other by unified purchasing. The degree of success depends on Australia's market strength at any time. In the falling markets of the 1980s, her ability to maintain prices was limited, and she attempted to diversify her markets.

See also: Policy: Canada; Policy: Europe; Policy: General; Policy: Japan; Policy: USA; Trade in Materials, International

Bibliography

Australian Foreign Investment Review Board *Annual Report*. Australian Government Publishing Service, Canberra

Bambrick S 1979 *Australian Minerals and Energy Policy*. Australian National University Press, Canberra

Cook L H, Porter M G 1984 *The Minerals Sector and the Australian Economy*. Allen and Unwin, Sydney

S. C. Bambrick
[University of New England, Coff's Harbour, New South Wales, Australia]

Policy: Canada

Canada is one of the world's most important producers of nonfuel mineral materials. Taking into account both the diversity and the quantity of production, Canada is ranked third after the former USSR and the USA. For most materials, there are large exportable surpluses: these accord Canada the first rank among world exporters of nonfuel mineral materials.

Canada possesses a highly developed industrial economy. However, because Canada is nearly self-sufficient in minerals, the predominant concerns of policy makers have not been closely allied with those of mineral-deficient industrialized countries. As a result, Canadian materials policy has until recently been virtually equivalent to mineral policy. Issues of exploration, production and marketing predominate. In earlier years the emphasis was on maximizing mineral development. In the 1970s it shifted towards nationalist economic policies designed to obtain the maximum benefit for Canadians from mineral development. In the 1980s, a worldwide recession having severe consequences for the primary industries underlined the realities of a global economy and Canadian economic policy once again emphasized competitiveness, productivity and the need to be governed by market forces. For the 1990s, a growing concern for the environment may bring increasing intervention and presages an uncertain future for Canadian mining.

1. Importance of Mineral Materials

The significance of Canada's production of mineral materials is shown in Table 1. These statistics include all primary nonfuel minerals and, in the case of metals,

Table 1
Value of production of Canada's leading minerals and rank in the world

	1988	1989[a]	
	($ million)[b]		Rank
Metals			
Copper	2393.6	2414.6	4
Gold	2332.0	2297.7	5
Iron ore	1323.2	1492.9	7
Lead	356.1	287.4	3
Molybdenum	121.1	122.4	3
Nickel	2790.4	3079.9	2
Silver	386.3	263.3	5
Uranium	1018.7	989.7	1
Zinc	2264.6	2843.6	1
Other	621.9	537.5	—
Total	13607.9	14329.0	
Nonmetals			
Asbestos	251.1	258.7	2
Potash	1167.7	947.0	2
Salt	246.7	270.2	5
Sulfur (elemental)	444.0	440.7	3
Other	607.7	624.3	—
Total	2717.2	2540.9	
Structural materials	2863.2	2890.8	
Total nonfuel minerals	19188.3	19760.7	

Source: Mineral Policy Sector, Energy, Mines and Resources Canada
a Preliminary b $1 Canadian = $0.85 US approximately

materials processed to the stage of smelting and refining.

In relation to total world production, Canada ranks first in uranium and zinc; second in asbestos, nickel and potash; third in lead, molybdenum and sulfur; fourth in copper; fifth in gold, silver and salt; and seventh in iron ore.

Over 80% of all minerals produced are exported, constituting around 15% of total Canadian merchandise exports. The mineral industries, including steel and aluminum, directly employed 183 000 persons in 1987. If the downstream employment in fabricating industries is included, about 381 000 jobs, or 3% of the total labor force, are accounted for. In 1987, the mineral sector accounted for about 5.6% of Canada's GDP. Imports of strategic materials include aluminum (as alumina or bauxite), chromium, manganese and phosphate rock. Canada is self-sufficient in virtually all other major mineral materials.

2. *Materials Policy and the Constitution*

Canada is a federation consisting of a central or federal government together with ten provinces and two federally administered territories. Ownership of natural resources, including minerals, is vested in the provincial governments. A survey of mining legislation in Canada is given by Barton et al. (1988). This proprietary interest confers on provinces the right to regulate access to, alienation of and production from natural resources. At the same time, the federal government has overriding responsibilities for interprovincial and international transport and trade and indirect taxation. In addition, the federal government is responsible for national economic policy. The division of responsibilities has been described by Burns (1976). The new Canadian constitution of 1982 has done little to change this division, other than to give the provinces some minor additional powers of indirect taxation. A firm grasp of this intergovernmental relationship is required for an understanding of the evolution of mineral policy in Canada.

For the greater part of the twentieth century, the objectives of federal and provincial governments have been largely parallel. At times, however, they have diverged, producing conflicting policies. This latter situation obtained during the early and middle part of the 1970s.

3. *Historical Overview*

Sporadic mineral extraction activities were carried on from the time of the earliest Canadian settlements. However, the first major impact was made on the national economy in the early years of the twentieth century. At that time, significant deposits of base metals were discovered in Canada. Improved methods for extraction and processing were also developed. Thus, at the beginning of World War I, Canada was well placed to take advantage of the wartime demand for metals. Although there was a hiatus at the end of the war, the booming North American economy of the 1920s caused a resurgence in demand. Even during the Great Depression of the 1930s, minerals provided a stabilizing influence in an otherwise moribund Canadian economy. Considerable assistance was offered by the effective doubling of the US dollar price for gold from 1933.

Before and during World War II, the demand for minerals was greatly accelerated. The expected postwar depression did not materialize and a period of stable economic growth for most western economies ensued during the 1950s and 1960s. During this period, the USA found itself a net importer of mineral materials. The prospects for continued deficiency of supply were identified in the Paley Report (US President's Materials Policy Commission 1952).

As the nearest supplier to a resource-deficient USA, Canada was an obvious place to seek increased production of mineral materials. Foreign capital was welcomed by Canada and a number of favorable policies were established to stimulate investment in new

Canadian mines. In such a policy environment, the minerals industries grew rapidly and steadily in Canada for 20 years. However, during the 1970s, other imperatives emerged. Before considering the specific policies operative at the present time, it is instructive to review those of earlier times.

4. Development of Mineral Materials Policies

Before World War II, Canadian economic and materials policy could be described as neutral. Whereas some common services, such as geological information, were provided by government, there was no major intervention in the market.

After the war, however, policy became more stimulative. An early example of a specifically mineral-directed policy was the Emergency Gold Mining Assistance Act (EGMA) of 1948. The Act provided a subsidy to assist mines to meet the fixed gold price and to protect dependent communities. It was not until 1973 that world prices rose sufficiently to eliminate payments under this Act (Canada 1976).

Among other federal measures designed to stimulate the mining industry during the 1950s was temporary relief from certain income taxes. In 1955 the tax incentives were made permanent. At the same time, provincial mining levies were at a moderate level. Since federal and provincial governments apparently agreed on the desirability of mining, the policies at both levels tended to be mutually reinforcing.

The first cracks appeared in the mid-1960s. The government of Quebec introduced policies to increase provincial control of resource development. Government departments were reorganized, alienation of mineral rights was tightened up, acreage taxes were increased, work requirements were stiffened and mining taxes were raised.

In 1967 the Carter Commission, established by the federal government to review all aspects of taxation, reported. The Commission recommended the abolition of most of the special privileges accorded the mining industry (Canada 1967). The federal government implemented a number of such changes in 1971 and 1972, some of which were to come into force after a transition period as long as four years.

At the provincial level, a wave of elections brought several activist governments to power, particularly in western Canada. An NDP (the New Democratic Party, a democratic socialist movement) government was elected in the province of Manitoba in 1968. In 1971, a new Alberta government promptly moved to increase revenues from oil and gas. In 1972 the Liberal government of Saskatchewan was defeated by an NDP administration and in 1973 an NDP government replaced the entrenched Social Credit (right-wing conservative) government of British Columbia.

Events of the early 1970s, including the OPEC embargo of 1973, the scarcity concerns of the Club of Rome's *Limits to Growth*, the formation of producer cartels in certain minerals and the forced renegotiation of mining concessions in less-developed countries, had the effect of persuading producers of nonrenewable resources that they possessed extensive market power. In these circumstances, governments determined to recover a greater share of the benefits from resource development for their citizens.

In the Canadian context, federal tax reform had just occurred. Several provinces moved quickly to increase royalties and taxes on a wide range of minerals, notably oil and gas. The federal government then responded to protect its revenue base by disallowing provincial levies as legitimate expenses in the calculation of federal income tax. A period of confrontation ensued, with the mineral industries caught in the middle.

Coincidentally, the federal Department of Energy, Mines and Resources, in cooperation with provincial mines departments, had launched a major review of mineral policy in Canada (Canada 1973, 1975). Unfortunately, the review bogged down largely because of the bitter federal–provincial dispute over energy revenues.

Throughout the 1970s, provinces moved more aggressively to gain control over the mineral industries. For example, the province of Saskatchewan greatly increased its tax level on the potash industry and later purchased over 40% of the productive capacity. In both Manitoba and British Columbia, mining taxes were increased to levels which the industry described as punitive. These measures were reversed only after a further change of government in both provinces. The provinces of Ontario, Quebec, Newfoundland and New Brunswick all increased their taxation levels and made other changes to policies affecting access, mineral alienation and mineral processing. The government of Quebec took over one major producer of asbestos, with the objective of pursuing further domestic processing and other socioeconomic goals.

It was not until 1978 that some measure of accord was reached on the federal–provincial taxation dispute and the policy environment stabilized (Canada 1978). In 1980, the first federal cabinet minister to be concerned exclusively with mines and mineral materials was appointed. Erola, this first Minister of State (Mines), left the post in 1983 and was replaced by Rompkey in early 1984. Later in 1984, with the election of the Progressive Conservatives, Layton took over to be replaced in 1986 by Merrithew, whose post combined Forestry and Mines. In 1988 the position was dropped, and a single minister again became responsible for Energy and Mines.

In December 1981, the Federal Department of Energy, Mines and Resources published *Mineral Policy: A Discussion Paper* (Canada 1981). This was one of four studies of resource-based industries commissioned by the federal cabinet in 1980. The paper dealt with a great number of current issues in the mining industry but did not really attempt to pick up the aborted attempt of the early 1970s to formulate an overall policy for minerals.

This discussion paper appeared at a very inopportune time, in that the industry was at the beginning of its worst depression ever. In 1982 nearly half of the mining work force was laid off, large losses were the order of the day, and many closures occurred.

The 1981–1982 recession was a world phenomenon. It was exacerbated by excess capacity in world mineral production and by an apparent structural slowing of demand growth. The causes of this "atrophy in metal demand" have been debated elsewhere (Tilton 1985, Vogely 1986, Nappi 1989, Tilton 1989). Although world consumption recovered to prerecession levels by 1984, excess supply and unfavorable exchange rate shifts restrained dollar-denominated metals prices until late 1987. To compensate, the industry maintained strict cost control throughout this period.

Fortunately, economic adversity was accompanied by a favorable policy shift. In 1983, the federal budget allowed "earned depletion," a tax incentive for mineral exploration, to be "flowed through" to investors in the shares of exploration companies. The so-called flow-through shares became a popular source of financing for mineral exploration, providing in 1987, the peak year, an estimated C$1200 million.

In 1984, the election of a Progressive Conservative government under Prime Minister Brian Mulroney initiated a renewed emphasis on economic performance and noninterventionist policies. Of importance to the mining industry were the abolition of the Foreign Investment Review Agency and the dismantling of the National Energy Program, a highly interventionist complex of resource taxes and exploration grants. Other aspects of the new government's economic policies involved the privatization of crown corporations (including Eldorado Resources, a government-owned uranium producer) and tax reform.

Tax reform proposals appeared in the 1985 federal budget. They involved a lowering of the income tax rate, combined with a broadening of the tax base. In the process, a number of special incentives were eliminated, including earned depletion. As a result, by 1988 flow-through shares were no longer an attractive investment. Under pressure from mining communities the government introduced the Canadian Exploration Incentive Program (CEIP), a system of exploration grants. In turn, CEIP was terminated in the cost-cutting federal budget of February 1990.

Another aspect of tax reform proposed in 1985 was the replacement of a Manufacturer's Sales Tax (MST) by a value added tax (called the Goods and Services Tax or GST). Due to have been introduced in 1991, it is expected to be of benefit to industries that produce internationally traded commodities, including metals and minerals.

In 1985, the Royal Commission on the Economic Union and Development Prospects for Canada (the Macdonald Commission, so named after its Chairman, Hon. Donald S. Macdonald, a former Liberal finance minister) submitted its report (Canada 1985). The Commission recommended several structural changes to make the Canadian economy more competitive and efficient. Among them were a free trade arrangement with the USA and a renewed reliance on market forces, implying a focus on privatization and deregulation. These recommendations reinforced the policy direction already taken by the new Conservative government.

Negotiations for a Canada–USA Free Trade Agreement (FTA) were initiated in the autumn of 1985. They culminated in the signing of an agreement on 4 October 1987. Throughout the period of negotiations, and until its final passage, the FTA was the subject of intense debate. A useful summary of the arguments for and against the FTA is given by Saunders (1987). It was bitterly opposed by economic nationalists, by most labor groups and by both opposition parties. Business was strongly in favor. The Conservative government was returned to power in the election of November 1988. Free trade legislation was passed immediately thereafter, and the Agreement became effective on January 1, 1989.

The Agreement provides for the gradual elimination of tariffs on virtually all goods originating in the two countries. In addition, the FTA liberalizes economic relations in areas such as investment, government procurement, border measures, energy, services, business travel and financial services. Special rules apply to agriculture, cultural industries, alcoholic beverages and automobiles. Institutions are to be established to oversee and facilitate the operation of the Agreement.

The Canadian minerals industry is assisted by the Agreement, as the USA is by far the largest market for its exports (representing about 65% of the total). US tariffs on metals and minerals had been moderate, except for a few items such as zinc alloys (19% duty). However, Canadian metals imports had been subject to a large number of antidumping and countervail actions, as well as other forms of US protectionism. The Free Trade Agreement (FTA) provides a dispute settlement mechanism that is less politicized than the one it replaces. Metals producers expected the FTA to provide more secure access to the US market, limiting the scope for unjustified protectionist actions.

Two other specific statements of mineral policy appeared in the 1980s. In December 1986, the Minister of Indian Affairs and Northern Development (responsible for mining in the Yukon and Northwest Territories) released a "Northern Mineral Policy" (Canada 1986). The policy had three stated objectives: providing certainty as to the federal government's intentions; increasing industry competitiveness; and encouraging dialogue and communication between industry, government and the public. The document held out the prospect of supportive policies for mining in the two northern territories.

In May 1987, the Minister of State for Forestry and

Mines published *The Mineral and Metal Policy of the Government of Canada* (Canada 1987). Together with an extensive review of the economic performance of the sector, the document identified six policy objectives. These were: a balanced fiscal and regulatory environment; regional development through minerals; technological advance; improved labor and community adjustment; enhanced commercial policy and market access; and economic and scientific information. The policy sketched out "a flexible and supportive role for the Government of Canada towards the minerals and metals sector."

Taken together, these two policy pronouncements show a concern for industry performance that was markedly absent from the 1981 document *Mineral Policy: A Discussion Paper*.

5. Current Policy Issues

For the present and future, several policy issues, discussed in the following sections, preoccupy Canadian policymakers and are of significance to the minerals and materials industries. Some have already been addressed in federal or provincial legislation; others have not. Nevertheless, the likely thrust of policy can be deduced from speeches, studies and discussion papers. Tentative assessments of the probable direction of policy are attempted in what follows.

5.1 The Environment

Popular concern for the environment has been growing for a number of years but until recently it has been subordinated to economic concerns. In common with other peoples, Canadians demonstrated a new level of environmental awareness in the 1980s, and policymakers responded eagerly.

While few new measures have been directed specifically at any sector, Canadian environmental legislation has been significantly strengthened and will affect mining. The Canadian Environmental Protection Act (CEPA) was passed in 1988. It provided new powers for the federal government and significantly increased penalties for wilful or persistent violation of standards. In 1989, detailed liquid effluent regulations were imposed, under CEPA, on the Canadian pulp and paper industry. It is probable that the mining industry will receive similar regulations in the near future.

Since the election of the Peterson (Liberal) government in Ontario in 1985, that province has moved aggressively on the environmental front. Its Municipal and Industrial Strategy for Abatement (MISA) calls for more extensive and costly monitoring, and it imposes new powers of regulation. It is expected to provide a model for other provinces.

Two 1989 judgements of the federal court found that the federal government was not sufficiently active in reviewing two dam proposals. These judgements have been interpreted liberally to put more onus on the federal government for effective review and assessment of any significant project remotely within federal jurisdiction. New legislation, the Canadian Environmental Assessment Act, is pending. The draft bill, tabled in June 1990, confers broad new powers. The federal government is also proposing a ten-year "Environmental Agenda," or action plan, to be released at the end of 1990. It will deal with a broad range of issues that could affect the minerals industry, including global warming, control of toxic substances, waste management and the use of novel policy mechanisms such as economic instruments.

Canada is a signatory to a growing number of international environmental protocols, conventions and the like. These are increasingly affecting the shape of the Canadian regulatory system and in many cases are driving domestic policies.

Several provinces are proposing new requirements under mining legislation for advance planning of mine reclamation and for prefunding of the reclamation plan. These provisions are intended to prevent the abandonment of mines that have not been placed in a stable and safe condition.

The Canadian minerals and metals industries are participating in the environmental policy revolution in several ways. The Mining Association of Canada has adopted an Environmental Policy and a Guide for Environmental Practice to guide the behavior of member firms. Research programs have been initiated to permit better understanding and resolution of outstanding environmental problems (notably the problem of acid rock drainage). Industry is also seeking to participate in the policy development process through constructive dialogue with government and other interested groups.

The industry has expressed support for environmental improvements. It promotes the concept of sustainable development, where environmental stewardship is accompanied by a concern for economic performance. It remains to be seen whether the current spate of new environmental policy initiatives will be consistent with the future growth and development of a viable mineral sector in Canada.

5.2 Land Use and Land Access

Mineral exploration and mining are prohibited in Canada's national parks and in many provincial parks. In addition, special restrictions apply on other categories of land such as wildlife sanctuaries.

The trend to restrictions on land access appears to be accelerating. For example, the federal government has announced its intention to complete the system of national parks by the year 2000. The World Commission on Environment and Development (the Brundtland Commission) recommended that a minimum of 12% of the world's landmass be protected, and an effective campaign has been mounted by environmental groups.

Moreover, public demands for protection of land

from development have led to the creation of a large number of provincial parks and other restricted areas. Furthermore, *ad hoc* responses to local controversies have resulted in closures of specific areas, such as the Carmanah Valley in British Columbia.

The Canadian mining industry supports the preservation of wilderness areas for species protection, for wildlife conservation and for recreation. However, trade-offs are involved whenever land is protected from development. Achieving a balance calls for clear, reasonable and consistent rules for access to public land. Industry has asked to be involved in consultations prior to a decision to protect land. Resource assessments are needed to provide some idea of the trade-offs involved.

5.3 Aboriginal Land Claims

While aboriginal land claims may be regarded as a special case of the problem of land access for mineral exploration and development, other issues are involved. Since the beginning of 1989, three major comprehensive aboriginal claims have been settled, with the conclusion of agreements or Agreements-in-Principle (AIPs) between the respective native groups and the federal government. The three groups are the Council of Yukon Indians (CYI), the Dene/Metis (D/M, Indians of the western arctic region) and the Tungavik Federation of Nunavut (TFN, "Inuit" or Eskimos of the eastern arctic). Each AIP provided that the natives would give up certain aboriginal rights in return for compensation. Compensation in each case consists of a cash award, the transfer of land ownership and the creation of new institutional arrangements that give the native community more control over economic development activities. Land ownership may carry only surface rights or, for a certain fraction of the land transferred, may include subsurface mineral rights.

Table 2 indicates the amount of the cash to be transferred in each AIP, as well as the total quantum of land and that proportion which carries with it subsurface rights.

The agreements will result in a new way of working throughout Canada's North. The mining industry has expressed satisfaction with the claims' settlement after nearly 20 years of negotiation. There may be a difficult transition ahead, but it is to be hoped that the industry and the natives will find a common purpose in developing the economic opportunities that exist in minerals in this region.

Table 2
Terms of aboriginal claims agreements

Claim	Cash transfer[a] ($ million)	Land transfer (km^2): Total	Land transfer (km^2): Subsurface rights
Yukon Indians	232 (1988$)	41440	25900
Dene/Metis	453 (1988$)	181300	10101
Tungavik Federation of Nunavut	580 (1989$)	350658	36260

Source: Various Agreements-in-Principle
a Canadian dollars, adjusted for inflation

5.4 Product Health Concerns

Many metals and minerals are known to have adverse health effects. Where exposure is excessive and prolonged, serious illness can result, as is demonstrated by several well-known occupational diseases of earlier times. However, these products can be used safely if appropriate exposure standards and procedures are observed.

In recent years, governments in many countries have reacted to public concerns about environment and health by drastically revising policies and legislation relating to metal and mineral exposure. In 1989 the US Environmental Protection Agency imposed a complete ban on the use of asbestos, to be phased in over several years. The International Agency for Research on Cancer has recently moved towards classifying nickel and lead (and their compounds) as possible or probable carcinogens. In 1989 a proposed amendment to the US Clean Air Bill contained provisions that would permit the banning of certain metals from any municipal refuse that might be incinerated. The 1989 Basel Convention on the Transboundary Shipment of Hazardous Wastes contains provisions that could well impede the legitimate import and export of metal scrap for recycling.

The long-term potential for the disruption of metal and mineral markets is obvious. Canada is not necessarily a prime mover in the international movement towards health-based regulation of metals and minerals. However, Canadian material-exporting industries could be severely affected by regulation that originates in customer countries or at the international level.

5.5 Economic Policy

A persistent problem for exporting industries in Canada is represented by large annual federal budgetary deficits since 1968 and a rapidly growing public debt. At the time of writing, the annual deficit stands at about C$30 000 million and the public debt over C$300 000 million, or 55% of GNP. The resulting economic stimulation promotes inflation, which is combated by the Bank of Canada through restrictive monetary policy. Tight money and high interest rates, together with artificially high exchange rates seriously affect Canadian exporting industries, including the metals and minerals sector.

The Mulroney government in 1984 and 1988 campaigned on a platform of fiscal responsibility, and successive federal budgets have contained expenditure-control measures. Nevertheless, these efforts have barely succeeded in stabilizing, but not significantly reducing, the annual deficit. The debt is growing correspondingly. Until this matter is brought in hand, the competitiveness and performance of the minerals and metals sector will be seriously and adversely affected.

5.6 International Trade

Canada is an open economy, with two-way trade typically accounting for about one-third of its gross national product. For the minerals sector, international trade is even more significant, as an estimated 80% of minerals production is exported. Moreover, the global economy is becoming ever more interdependent and integrated.

In these circumstances Canada, supported by the minerals industries, has been a leader in the quest for a liberalized trading environment. The Canada–USA Free Trade Agreement is considered only a step in the right direction. Canada also participates enthusiastically in Multilateral Trade Negotiations (MTN) under the General Agreement on Tariffs and Trade (GATT). The Uruguay Round of MTN was due to be completed in 1990 and Canadian mineral producers are hoping for significant progress in the reduction of tariffs and nontariff barriers, as well as the adoption of a more comprehensive subsidy code. Clarification of the provisions on technical barriers to trade is also sought, to prevent the use of health and environmental standards to create unnecessary barriers to trade.

A constant preoccupation of a trading nation, and a trading industry, is to remain cost-competitive. The mineral sector continues to keep a close watch on its controllable costs. Moreover, it is very concerned when actions of government in any policy area lead to cost increases. There is a real concern that an increased level of environmental regulation will combine with adverse fiscal and monetary policy to raise Canada's relative cost level in the 1990s.

6. Conclusion

Some issues which are of intense importance elsewhere are not perceived to be major problems in Canada. Among developed countries, Canada is probably the most nearly self-sufficient in the supply of basic raw materials. Thus, the strategic concerns of the USA, Western Europe and Japan for continued access to raw material supplies are not of first priority to Canadian decision makers. Naturally, there are a few deficiencies, notably chromium, manganese and phosphate rock. In addition, the aluminum smelting industry, a major industry in Canada because of cheap electric power, depends exclusively on imported bauxite and alumina. Canadians recognize that a major world conflict could jeopardize some of its supplies, particularly manganese and chromium. No overt policy response has emerged as yet, other than to investigate the technology of exploiting the low-grade chromium and manganese deposits which are available in Canada.

The 1970s and 1980s have been turbulent. The 1970s were marked in Canada by a pronounced shift from prodevelopment policies to a focus on social objectives and economic nationalism. In the early 1980s, a prolonged and difficult recession threatened the viability of the Canadian minerals sector. Towards the end of the decade, latent public concerns for environment and health found expression in a host of new government policies and initiatives, both in Canada and abroad. The prospect in the 1990s is for more of the same.

As an industry extremely involved in international trade, the Canadian minerals and metals sector views this outlook with some concern. For one thing, domestic regulation may result in increased production cost. Furthermore, initiatives related to health and environment, and taken by foreign governments and international agencies, may constrain the marketing of Canadian mineral products, either by inhibiting market access or by actually reducing or eliminating the use of the product.

Thus, the decade of the 1990s may well prove to be a turning point for the minerals and metals sector, not only in Canada, but globally. Noneconomic factors, motivated by health and environmental concerns, may come to dominate decision making. To the extent that producers develop strategies to influence these trends, or to profit from them, the 1990s may prove to be a decade of reasonable prosperity. One thing is certain: to operate successfully in the coming years will require an ability to work effectively with governments and with interest groups (both in Canada and abroad) whose objectives are not coincident with those of the traditional minerals and materials industries.

See also: Policy: Australia; Policy: Europe; Policy: General; Policy: Japan; Policy: USA; Trade in Materials, International

Bibliography

Barton B J, Roulston B, Strantz N 1988 *A Reference Guide to Mining Legislation in Canada*. Canadian Institute of Resources Law, Calgary, Canada

Burns R M 1976 *Conflict and its Resolution in the Administration of Mineral Resources in Canada*. Centre for Resource Studies, Kingston, Canada

Canada, Department of Energy, Mines and Resources 1976 *Final Report on the Administration of the Emergency Gold Mining Assistance Act*. Catalogue No. M21-1976. Minister of Supply and Services, Canada, Ottawa

Canada, Department of Energy, Mines and Resources 1981 *Mineral Policy: A Discussion Paper*. Report No. MP81-1E. Minister of Supply and Services, Canada, Ottawa

Canada, Department of Energy, Mines and Resources 1987 *The Mineral and Metal Policy of the Government of Canada*. Minister of Supply and Services, Canada, Ottawa

Canada, Department of Indian Affairs and Northern Development 1986 *Northern Mineral Policy*. Catalogue No. R72-198/1986. Minister of Supply and Services, Canada, Ottawa

Canada, Federal and Provincial Ministers Responsible for Mineral Policy 1973 *Mineral Policy Objectives for Canada.* Information Canada, Ottawa

Canada, Federal and Provincial Ministers Responsible for Mineral Policy 1975 *Towards a Mineral Policy for Canada: Opportunities for Choice: 1974.* Catalogue No. M37-21/1975. Information Canada, Ottawa

Canada, Federal and Provincial Officials 1978 *Joint Report on Federal–Provincial Taxation Review.* Document No. 800-9/018. Information Canada, Ottawa

Canada, Royal Commission on Taxation "Carter Commission" 1967 *Report.* Information Canada, Ottawa

Canada, Royal Commission on the Economic Union and Development Prospects for Canada 1985 *Report.* The Canadian Government Publishing Centre, Ottawa

Mining Association of Canada 1989 *Mining in Canada: Facts and Figures.* Mining Association of Canada, Ottawa

Nappi C 1989 *Metals Demand and the Canadian Metals Industry.* Centre for Resource Studies, Kingston, Canada

Saunders O J (ed.) 1987 *Trading Canada's Resources: Essays from the Third Banff Conference on Resources Law.* Conference sponsored by the Canadian Institute of Resources Law, Calgary. Carswell, Toronto, Canada

Tilton J E 1985 Atrophy in metal demand. *Earth and Mineral Sciences,* Winter 1985. Pennsylvania State University, University Park, PA

Tilton J E 1989 Changing trends in metal demand and the decline of mining and mineral processing in North America. *Resour. Policy* 15: 12–23

US President's Materials Policy Commission 1952 *Resources for Freedom: A Report to the President.* US Government Printing Office, Washington, DC

Vogely W (ed.) 1986 Proceedings of metals demand conference, May 19–22, 1986. *Mater. Soc.* 10: 237–539

Wojciechowski M J 1983 *Mineral Policy Update.* Centre for Resource Studies, Kingston, Canada

Yudelman D 1985 *Canadian Mineral Policy, Past and Present: The Ambiguous Legacy.* Centre for Resource Studies, Kingston, Canada

Yudelman D 1986 Maturity management: a new mineral policy for Canada. *Resour. Policy* 12: 29–39

C. G. Miller
[Mining Association of Canada, Ottawa, Canada]

D. Yudelman
[Ontario Hydro, Toronto, Canada]

Policy: Europe

The individual country write-ups in this article will become obsolete, probably within 10 years of 1992, as the European Community (EC) increasingly integrates. The only materials sector in the EC not to be handled solely by market forces will probably be iron and steel.

1. France

Unlike other European nations, the French have a highly centralized industrial system. It consists of a continuing series of 5-year plans as part of an ongoing system of indicative planning (i.e., sets goals and objectives only). Each plan provides guidance for specific policy measures to be administered by the various government ministries. With the exception of iron and steel, nonfuel mineral industry problems received little attention in the 5-year plans until the sixth plan (1971–1975) and seventh plan (1976–1980). In the sixth plan, the promotion of nonfuel minerals became a priority action program, the objective of which was to strengthen the position of the two large French metal companies that are world leaders in the production of aluminum, and nickel and lead, respectively. In the seventh plan, raw materials management was explicitly linked to the control of the balance of payments, and to the maintenance of economic and strategic security. The objectives and associated policies enunciated in this plan were:

(a) to reduce imports by encouraging recycling and efficient use in manufacturing, and by conservation;

(b) to double mine production within France over a 10-year period and to develop the natural resources of French Guiana;

(c) to strengthen the French raw materials industry;

(d) to undertake a geological survey of national resources;

(e) to establish an integrated line of supply for copper, with the aid of government subsidies for prospecting, for the start-up of exploitation and for processing of copper ores;

(f) to improve the position of France with regard to supplies from developing countries by providing aid to help these countries to prospect for, extract and process minerals; and

(g) to develop a national economic strategic stockpile of essential minerals.

The actual policies in pursuit of these objectives include tax incentives for domestic mining and recycling, financial assistance for domestic recycling, primarily through direct government programs, domestic exploration through direct government exploration and stockpiling for both economic and strategic purposes. The main instrument within France to accomplish these objectives is the Bureau des Recherches Géologiques et Minières (BRGM). The BRGM provides prospecting services and compiles inventories of mineral resources. It also participates in prospecting, developing and working ore deposits in the same manner as a private mining company. It is represented not only within France, but in more than 40 countries throughout the world, mainly former French colonies. Within the operations of this agency overseas, there is also backup of a political and military nature. The French take an active and supportive role in the nations which supply their resources; this involves government operations at all levels of the industry and explicit support for mineral producing activities in the developing countries. Essentially the French approach to overseas supply involves exchange of technical assistance, capital, and political and military support on a bilateral basis, in return for access to mineral resources.

In summary, the French approach is on an integrated basis, which allows the full weight of their technology, capital and political strength to be used to assure overseas supplies. These policies will not survive integration into the EC during the 1990s.

2. Germany

Germany places great weight on private decision making and private enterprise in the development of its industrial policy. Thus, there is no explicit or comprehensive statement of overall industrial policy, although the government may subsidize or direct private efforts when it feels it is in the national interest. Specifically in the nonfuels area, this involves subsidized freight rates and trade adjustment assistance to the iron ore industry, and a comprehensive package of government programs with respect to steel.

Until the 1974 oil crisis, the position of the government of the Federal Republic of Germany on materials was primarily one of watching. It followed the increasing dependence on imported raw materials, but no direct action was taken under the philosophy that materials supply policy should be the responsibility of private industry. After 1974, the government undertook some major studies concerning the flow of supplies to Germany and the importance of the raw-material-based industries in the German economy. In 1976, an interministerial committee stated some principles of mineral policy. The major statement was that securing supplies of raw materials is of utmost importance. The committee foresaw three types of supply problems: imbalances between supply and demand, disturbance of the market by political and other disruptions (e.g., transportation strikes), and natural disasters. In response to these findings, Germany has encouraged overseas investments by German firms, but like Japan does, they do not seek majority control. As part of this program, the government has provided mining companies with up to 50% of their overseas mineral exploration costs. In the area of overseas investment, policy measures involving capital investment guarantees include investment treaties, direct government investment in mineral development in developing countries, tax breaks for investment, long-term credits, and long-term credit guarantees.

The German government is opposed to international commodity agreements, and takes the position that international arrangements which tamper with the free working of markets are not acceptable. Consistent with this philosophy, Germany considers that stockpiling is principally a matter for private business, and therefore has no government run or sponsored nonfuel mineral stockpiles. They do, however, provide general geological survey services, both in Germany and overseas, as well as research and development programs for mineral extraction and conservation.

3. United Kingdom

The United Kingdom has no consistent body of policy or policy statements addressed to nonfuel minerals. Further, the instruments employed by nations such as France, Germany and Japan are not evident to any significant extent in the United Kingdom. This lack of direct government policy and action flows from the unique relationship of UK companies with the private mining sector in international mineral markets, through productive capacity in many of the former colonies and current members of the Commonwealth.

Nonfuel mineral problems which arise outside the traditional framework are treated on an *ad hoc* basis. The most significant *ad hoc* steps have concerned the supply of strategic minerals. Recognition of this problem is evident, but as yet no specific policy, philosophy or actions have been undertaken.

See also: Policy: Australia; Policy: Canada; Policy: General; Policy: Japan; Policy: USA; Stockpiling; Trade in Materials, International

Bibliography

Crowson P F 1980 The national mineral policies of Germany, France and Japan. *Min. Mag.* 142(6): 537–49

Economic Consulting Services 1981 *Nonfuel Mineral Policies of Six Industrial Countries.* Economic Consulting Services, Washington, DC

von Welk S F 1979 Non-fuels minerals policy in the Federal Republic of Germany. *Mater. Soc.* 3: 201–4

W. A. Vogely
[Pennsylvania State University, University Park, Pennsylvania, USA]

Policy: General

The materials policy of a nation is closely related to the economic system practised by that nation. In this Concise Encyclopedia the materials policies of several major industrial nations are discussed. This article explores the principles underlying materials policies under different economic systems, as well as the nature of governmental actions impacting upon materials.

1. Classification of National Economies

One significant way to characterize national economies is to classify them as primarily either market economies or centrally planned economies. Of course most national economies are mixed with free and planned sectors. In developing countries, there is a tendency to plan for those sectors leading to growth in exports. However, these sectors must sell in world markets and are thus tied to the world commodity markets and investment in productive facilities in them is dependent on world financial markets. As a developing nation grows it must choose to be either a planned or market economy. In the developed world, economic institutions are more static and stable, except in periods of national upheaval, such as that which occurred in Eastern Europe in 1990 and in the USSR in 1991. Thus it is not hard to classify the economy of a developed nation as market or centrally planned. It should be noted that the major centrally planned nations, the USSR and China, began undergoing changes in 1991 that may move their economies towards market organization, but it is too soon to evaluate the final outcomes. On the other hand, Eastern European nations are clearly moving to market economies.

2. *Materials Policy in Economic System Types*

In both market and centrally planned economies, materials are seldom important enough either economically or politically to merit special deviation from the norm for each nation. Thus, in market economies, government-dictated materials policies are not seen and, in planned economies, unplanned materials sectors are not found. An examination of the articles on the materials policies of various countries in this volume shows that materials are only singled out for special policies when they impact upon larger socioeconomic issues such as national security, economic development, trade, public land management and research and development.

3. *Materials Policy at Stages of National Development*

The stage of economic development can play a determining role in materials policy. A less developed country (LDC) is likely to look at the establishment of a raw materials extraction industry as an efficient way to increase national employment, provide infrastructure, raise GNP and earn foreign exchange through exports. A developed country, on the other hand, is concerned with a stable low-cost supply of raw materials to sustain its industrial system. An efficient international trading system allows these different goals to be achieved simultaneously.

The international organization UNCTAD (United Nations Conference on Trade and Development) was founded as a permanent organ of the UN General Assembly in 1964 and seven conferences had been held up to 1991. Each conference focused on a different aspect of mineral development. The World Bank plays a large role in financing programs in the LDCs (Gocht et al. 1988 Chaps. 8–10). The General Agreement on Tariffs and Trade (GATT) provides a means of maintaining a freer trade environment.

4. *Role of Market Structure*

Another factor militating against a special national materials policy is the worldwide integration into a single market of the major metals marketed through the London Metals Exchange or the New York commodities exchanges. These exchanges constitute a world market for traded metals. The existence of these exchanges means that the world markets are "contestable" and national markets are driven by world markets (see *Prices of Materials: Theory*).

5. *Effect of Public Policies*

The preceding discussion should not be interpreted as precluding by-product implications of many public policies upon materials production sectors and materials markets. Obvious examples are intentional government actions undertaken for economic development, environmental quality, health and safety, and international trade. Less obvious are national security research and development strategies, taxation methods and land tenure institutions. In fact, these by-product effects constitute the *de facto* materials policies of most developed countries.

The question is whether this situation has led to serious policy failures. The *ad hoc* nature of the materials policy of the USA was examined in 1981, when the National Materials Advisory Board (NMAB) was asked by the General Accounting Office to review how nonfuel materials issues were brought before the Executive and Legislative branches for attention under the Mining and Minerals Policy Act of 1970. The first finding of the NMAB committee (National Materials Advisory Board 1981 p. 2) was that

> "the non-fuel minerals supply system, while a subject of constant policy attention since the end of World War II, has sustained no failure of the policy process that caused significant damage to the national welfare"

In short, this sector in the USA did not suffer from the lack of an explicit materials policy.

See also: Policy: Australia; Policy: Canada; Policy: Europe; Policy: Japan; Policy: USA; Trade in Materials, International

Bibliography

Gocht W R, Zantop H, Eggert R G 1988 *International Mineral Economics.* Springer, New York

Gulley D A 1985 Environmental regulation and the minerals industry. In: Vogely W A (ed.) 1985 *Economics of the Minerals Industries*, 4th edn. American Institute of Mining, Metallurgical, and Petroleum Engineers, New York, Chap. 5.15, pp. 601–14

Hibbard W R Jr 1985 Non-fuel policy issues. In: Vogely W A (ed.) 1985 *Economics of the Minerals Industries*, 4th edn. American Institute of Mining, Metallurgical, and Petroleum Engineers, New York, Chap. 5.14, pp. 593–600
National Materials Advisory Board 1981 *National Non-Fuel Mineral Policy Planning Process*, NMAB-384. NMAB, Washington, DC
Vogely W A 1979 Why the United States has not adopted an explicit minerals policy. *Mater. Soc.* 3: 191–4

W. A. Vogely
[Pennsylvania State University, University Park, Pennsylvania, USA]

Policy: Japan

Japan has a highly coherent and explicit industrial policy, unlike most other industrialized countries. As part of the industrial policy, Japan has a clearly established materials policy. The basic facts on which this policy is based are that (a) Japan is highly import dependent, (b) raw materials are a major item of Japanese import trade and (c) the processing of raw materials is an important production sector within the Japanese economy and, because of the declining production of indigenous materials, the dependence on imports will substantially increase over the coming decades.

The structure of the Japanese raw materials industry is heavily weighted toward primary metals rather than mine production. Of Japanese total industrial production, less than 1% arises from mining, while iron and steel, aluminum, copper, zinc and other metal production in primary and fabricated forms provide 14% of the total. Japanese nonfuel minerals policy, then, is directed principally at providing the flow of raw materials for the highly important mineral processing sector in Japan's economy.

Japan depends primarily on privately owned concerns to conduct its economy. However, the relationship between the private sector and the government in Japan is unique, with the interaction aimed at building a consensus. Interactions occur at several levels involving three different types of institutions: a business association, a bureaucracy of a government agency and a deliberation council. The deliberation councils, which are the unique feature, are sponsored by the government and consist of representatives of business, government and other interests. They provide a mechanism whereby all views concerning an industry can be brought together so that a consensus can be built as to the direction of the industry, the role of the government and the content of policy.

Two of the most important deliberation councils are the Industrial Structure Deliberation Council and the Trade Council. Both of these councils have ministerial-level representation, as well as industrial and other leaders from Japanese society. The councils have under them numerous subcommittees, which deal with specific problems. There is, for example, an Aluminum Subcommittee to the Industrial Structure Deliberation Council. Further, there are lower-level deliberation councils corresponding to specific sectors of the economy, such as the Primary Commodities Committee and the Mining Industry Deliberation Council. This latter council, which includes the processing industries, as a rule convenes every five years.

These councils build a basis for industrial policies in Japan that are implemented with administrative guidance. Administrative guidance by government officials is informal, and the officials do not have statutory authority. Such a system can only work when there is a basis of trust between the business and the government. Of course, within the system firms vary as to their spirit and rate of compliance, but the system is generally accepted by industry and is taken as part of the price of performing within the quite favorable economic policies of Japan.

The overall element that guides Japanese raw material policy is the mandate arising from the fundamental need of Japan: that is, insuring adequate raw material flows to industry. The implementation of this objective, developed through the deliberation council arrangements, expresses the preferences of Japanese industry and government. The first such preference is reliance on long-term arrangements for import supplies, rather than spot markets. The second preference is to encourage direct Japanese investment in overseas resource production. The third preference is to maintain a relatively low profile, and thus limit exposure to risk from overseas investments. For example, Japanese corporations strive to limit their equity to 10–30% of the funds for a project in return for long-term contracts for the outputs from the projects. The fourth preference is avoidance of over-dependence on any one source for a material, thus reducing risks to the supply flow. Finally, Japan is willing to participate in producer–consumer commodity talks and arrangements. This involves arrangements to limit fluctuations in price, in return for guarantees of supplies of manufactured products.

In line with each of the named preferences above, there are specific programs under way within Japan. The Metal Mining Agency of Japan (MMAJ) is the mechanism whereby these governmental programs are administered. The objectives and the programs are as follows:

(a) To promote overseas exploration by means of subsidies for geological surveys by Japanese mining companies, subsidies for joint exploration projects with foreign partners and financing of overseas exploration. Subsidies for geological surveys are at rates of 50% of cost for exploratory drilling and 60% of other survey costs. Joint exploration projects are subsidized up to 50% of the Japanese share. In each case there are commodity limitations. For financing of overseas exploration, the MMAJ provides two types of financial assistance: loans and equity participation along with Japanese mining companies. The equity participation is only available for copper, lead, zinc and manganese exploration, whereas the loans are available for a broader group of commodities. The loans are normally available for up to 50% of the cost and carry a very low interest charge with repayment over 15 years.

(b) To promote overseas investment by means of loan guarantees, political risk insurance, coordination of bilateral aid, tax incentives, research and development incentives, technical assistance to suppliers and foreign policy support.

In addition to these direct aids to the exploration for and development of overseas supplies, Japan also adopts measures to provide security for these supplies. Specifically, the Japanese government helps to establish, and participates in, stockpiling efforts concerned with refined copper, zinc and lead, aluminum, and certain special metals (e.g., nickel, chromium, tungsten, cobalt and molybdenum). The actual stockpiling, however, has been limited, at least in the public record, to copper, zinc and aluminum. The price of the stockpiled materials is established as a refiner's average selling price for a specific month, with purchases and sales being made on the basis of inventory position and price movements up and down. These are then in essence economic stockpiles, although the special metal stockpile appears to be evolving into one with strategic implications.

Japanese policy is also directed toward protection of the domestic processing industries. For example, this involves a very low, often zero, tariff on unrefined nonfuel minerals with relatively high tariffs on the refined forms. The Japanese tariff system is very complex and in many cases is, in effect, a tariff quota system which sharply limits the actual volume of imports.

Protection of the environment is an integral part of the Japanese materials policy system. The legislative authority involves both air and discharge standards. Industry is given tax credits and other forms of incentives to help offset the cost of meeting pollution control standards.

In summary, Japanese materials policy is exerted within the context of private enterprise and a market economy, with a substantial accepted participation by the government bureaucracy through administrative and social mechanisms. The policy is influenced by the facts that the Japanese are almost completely dependent on world supplies for raw materials, and that the processing and fabrication industries are a very significant sector in their industrial economy.

See also: Policy: Australia; Policy: Canada; Policy: Europe; Policy: General; Policy: USA; Trade in Materials, International

Bibliography

Crowson P F 1980 The national mineral policies of Germany, France and Japan. *Min. Mag.* 142(6): 537–49

Economic Consulting Services 1981 Nonfuel mineral policy of Japan. *Nonfuel Mineral Policies of Six Industrial Countries*, Pt VI. Economic Consulting Services, Washington, DC, pp. 1–59

Nanao K 1979 Minerals policy in Japan. *Mater. Soc.* 3: 205–8

W. A. Vogely
[Pennsylvania State University, University Park, Pennsylvania, USA]

Policy: USA

The USA is endowed with a rich geological setting for materials extraction and has used these resources to provide the basis for a large and still growing industrial production sector. During the first century of the USA's existence, there was no question that the abundant raw materials should be used to foster the rapid growth of its manufacturing industries. Within the market economy, use of raw materials was favored by differential tax treatment and priority availability of public lands for development. Public lands were about 80% of land area in the USA at the time of the Civil War (1860–1865). Even in the early 1990s about 50% of the land area is public lands. During the second century of the USA's existence, the need to balance industrial growth with environmental quality has led to societal actions that place constraints on materials extraction, production and use. World War I and World War II, the rise of Japan and Korea as industrial nations, Western European recovery after World War II and the integration of European markets in the EC have changed the relative position of the USA among the world's industrialized nations, but the USA remains one of the largest producers and consumers of materials.

1. General Policies

Within the USA there is not a tradition of adopting sector economic policies as is the case in the French and Japanese economies. The USA tends to enact general tax, trade, environmental and health and safety laws that apply across the board to all industries. As a result, the problems related to nonfuel minerals industries must be considered at the policy level, within the context of dependence upon the private sector for production and consumption decisions and the tradition of operating without sector policies.

With respect to nonfuel minerals, the only major exception to the above statement relates to national security. Under the Defense Production Act of 1950, as extended, the US government does undertake an evaluation of the critical materials needs of the country during periods of emergency and provides for those needs through the maintenance of a large strategic stockpiling and other programs. However, it must be understood that criteria for stockpile management are based on security needs, and the stockpiles are not responsive to the nonfuel minerals industries themselves.

In short, the USA does not have an explicit industrial policy or sectorial policies dealing with specific industry groups. Thus, nonfuel mineral issues have been handled as they have arisen rather than by general policy measures of a continuing and consistent nature.

Although nonfuel minerals and materials issues have been at the forefront of discussion in the USA since the end of World War II, the USA has never adopted an explicit minerals or materials policy. Issues that have arisen on a commodity-by-commodity basis have been handled in an *ad hoc* way. Congress has twice enacted a policy statement with respect to nonfuel minerals, the first statement being contained in the Mining and Mineral Policy Act of 1970 (30 USC 21a) and the more recent one in Public Law 96-479 (30 USC 1601), cited as the National Materials and Minerals Policy, Research and Development Act of 1980, enacted in October 1980. That Act contains the following declaration of policy:

> "The Congress declares that it is the continuing policy of the United States to promote an adequate and stable supply of materials necessary to maintain national security, economic well-being and industrial production with appropriate attention to a long-term balance between resource production, energy use, a healthy environment, natural resources conservation, and social needs. The Congress further declares that implementation of this policy requires that the President shall, through the Executive Office of the President, coordinate the responsible departments and agencies to, among other measures—(1) identify materials needs and assist in the pursuit of measures that would assure the availability of materials critical to commerce, the economy, and national security; (2) establish a mechanism for coordination and evaluation of federal materials programs, including those involving research and development so as to complement related efforts by the private sector as well as other domestic and international agencies and organizations; (3) establish a long-range assessment capability concerning materials demands, supply and needs, and provide for the policies and programs necessary to meet these needs; (4) promote a vigorous, comprehensive, and coordinated program of materials research and development consistent with the policies and priorities set forth in the National Science and Technology Policy, Organization, and Priorities Act of 1976 (41 USC 6601 et seq.); (5) promote cooperative research and development programs with other nations for the equitable and frugal use of materials and energy; (6) promote and encourage private enterprise in the development of economically sound and stable domestic materials industries; and (7) encourage federal agencies to facilitate the availability and development of domestic resources to meet critical materials needs."

This Act covers the implementation of the policy, directing the President to achieve the objectives and to report to the Congress on program plans, organizational proposals and other matters required to fulfill the objectives of the Act. The Act also incorporated without change the provisions and requirements of the 1970 Act, and directs the President to have the Secretary of the Interior act immediately to obtain the goals contained in the 1970 Act. The 1970 Act's goals are stated as follows:

> "It is the continuing policy of the Federal Government in the national interest to foster and encourage private enterprise in (1) the development of economically sound and stable domestic mining, minerals, metal and metal-reclamation industries, (2) the orderly and economic development of domestic mineral resources, reserves, and reclamation of metals and minerals to help assure satisfaction of industrial, security and environmental needs, (3) mining, mineral, and metallurgical research, including the use and recycling of scrap to promote the wise and efficient use of our natural and reclaimable resources, and (4) the study of and development of methods of the disposal, control and reclamation of mineral waste products, and the reclamation of mined land, so as to lessen any adverse impact of mineral extraction and processing upon the physical environment that may result from mining or mineral activities...."

Thus, the official statement of the policy concerning nonfuel minerals adopted by the Congress and signed by the President identifies the private sector as the principal agent, and indicates the importance of an adequate and stable supply of minerals, the need to come to a balance between resource production, resource use and a healthy environment, and the necessity for a vigorous research and development program. However, these statements of goals and objectives have not been translated into legislation requiring specific actions dealing with this sector of the economy. Thus, the USA still does not have an explicit mineral policy in the way, for example, Japan does.

2. *Evaluation of the 1970 and 1980 Acts*

The *ad hoc* nature of materials policy in the USA was examined in 1981, when the National Materials Advisory Board (NMAB) was asked by the General Accounting Office to review how nonfuel materials issues were brought before the executive and legislative branches for attention under the Mining and Minerals Policy Act of 1970. The findings of the NMAB Committee were as follows.

> "1. The non-fuel minerals supply system, while a subject of constant policy attention since the end of World War II, has sustained no failure of the policy process that caused significant damage to the national welfare. On the contrary, events that have interrupted supplies have been handled by the public and private sectors in a manner that has imposed neither significant costs nor limitations on the economy. Thus past experience alone does not justify a major revision of the existing policy process. 2. There exists a continuing concern that the potential for major stress on the materials system is present, that strategies and mitigating actions designed to prevent serious disruptions to the economy are

not in place, and that the mechanisms in place are inadequate to foresee or develop contingency plans for such disruptions. 3. Concerns about materials supplies are focused in part on actions that have other societal objectives, but have long-term and poorly understood materials implications and may impose costs that are not considered in the decision process. 4. The reports submitted under the Mining and Minerals Policy Act of 1970 have not provided assurance that materials concerns are incorporated in federal decision making and properly presented to the President and the Congress. 5. Two important factors in the public perception that the materials policy process is inadequate are deficiencies in the data systems and analytical capabilities." (US National Materials Advisory Board 1981 pp. 2–4).

In short, the findings were that the materials sector in the USA did not suffer from the lack of an explicit materials policy.

3. Special Studies of Nonfuel Mineral Industry Issues

During the 1970s, there were three major studies of the problems of the nonfuel minerals industry. These studies were carried out by the National Commission on Materials Policy established by President Nixon, which reported in June 1973, the National Commission of Supplies and Shortages established by the Congress and reporting in December 1976 and a White House review of Materials Policy Issues conducted under the Carter administration, with draft reports issued in the fall of 1980.

The commissions and studies conducted during the 1970s, as well as those conducted earlier, focus on three nonsecurity minerals policy issues:

(a) the adequacy and effectiveness of the information and analytical system of the federal government to promptly identify and to provide information and analysis for the making of policy decisions;

(b) the implications for nonfuel minerals of federal land policies including, but not limited to, the withdrawal of large areas of land for wilderness purposes which in the long term precludes their use for mineral producing and consuming industries; and

(c) the environmental legislation adopted during the 1960s and 1970s.

It is noteworthy that these three issues are also mentioned in the 1980 Act as specific items to be considered by the Executive in administering that law.

4. Federal–State Responsibilities

The federal government has responsibility with respect to nonfuel mineral resources only for those resources that are produced from federally owned lands, which include public lands, acquired lands and the outer continental shelf. Public lands are lands that were reserved to the federal government at the time when the various states entered the union; they generally lie west of the Mississippi, and include Alaska. Federally owned lands in the other states are lands that had been acquired by the federal government for various purposes such as national forests, national parks and military bases. The disposal of mineral resources on privately owned lands is a decision of the private sector, and on state-owned lands is controlled by each state's laws. Of course, the federal government does have responsibility for aspects of mineral production when the products move into interstate commerce, under the various business, environmental, health and safety and other federal laws.

5. Federal Mineral Rights

In general, mineral deposits on public lands are locatable under the provisions of the 1872 Mining Law as modified by subsequent legislation and regulations. This law provides that discovery of a mineral deposit on public land enables the discoverer to develop that deposit without payment of royalty or a fee to the federal government. If the discoverer wishes to gain ownership of the mineral resources then he must apply to the federal government, and, upon payment of small fees and under certain legal conditions, title can be obtained for the mineral resources. However, the federal government has withdrawn at various times large areas of the public lands from the provisions of the mining law. These lands were generally withdrawn for classification purposes, and many of the withdrawals have been rescinded. However, if land is designated for a nonmineral use, mining may or may not be permitted under that designation. For example, with minor exceptions, mining is never permitted in national parks, and the wilderness areas will be eventually closed to prospecting and mining. Nonfuel minerals on acquired lands and on the outer continental shelf are leased, but common minerals, such as sand and gravel, are sold outright. Certain nonmetallic minerals (chiefly sulfur, phosphate and potassium) are leased under the Mineral Leasing Act of 1920.

6. Federal and State Taxation

There are, within the federal tax codes, special provisions relating to the production of nonfuel minerals. Producers of these minerals have the right to take percentage depletion at various rates throughout the productive life of a mine. Percentage depletion allows a deduction from income tax liability of a stated percentage of the total income arising from the property, with a maximum deduction of 50% of net income. Percentage depletion allows a reduction in tax liability not based on the actual cost of the project, and thus provides a lower tax rate in many cases, but the alternative of cost depletion is always available to the taxpayer. In addition, provisions of the tax code provide for write-off of exploration expenditure as expenses, and there are various accelerated depreciation provisions for development expenditure.

The most common special state taxes are severance taxes, that is, unit taxes on production. These taxes vary state-by-state, and can be used to discourage mineral production within a state. An example is the increase in state tax in Wisconsin that halted development of a large lead deposit in that state.

7. Federal Policy Interfaces

Many of the issues in nonfuel minerals and materials policy are those concerning interfaces with other policies. The primary ones are the interfaces between minerals and the environmental and worker health and safety legislation. Since minerals must be mined where they occur, and since the processing of minerals usually involves the emission of substantial particulates and chemical matter to the biosphere, environmental laws have unique implications for minerals. In the same way, since mining is one of the most hazardous of industrial activities, the worker health and safety laws also have a special importance for mining. Issues in these interfaces are handled on a case-by-case basis, and the criteria applied to the handling of these issues uniformly arise from those established for the environmental or health and safety acts rather than from the objectives involved in the mineral policy statements cited above. Most of the politically important questions concerning the domestic producing industry have arisen in the context of environmental regulations, particularly the question of sulfur emissions from smelting processes and worker exposure to toxic materials in the work place.

7.1 Environmental Legislation

The major environmental acts which impinge upon the nonfuel mineral industries are the Clean Air Amendments Act (1990), the Clean Water Restoration Act (1966), the Endangered Species Act (1973), the Solid Waste Disposal Act (1965), the Surface Mining and Reclamation Act (1977) and the Resource Conservation and Recovery Act (1976).

The Clean Air Amendments Act of 1990 remains the most important to materials, as it affects the smelting and refining industries. The 1990 amendments were largely directed towards tighter standards on particulate emissions, but also included some reforms and new approaches to regulation. The process of smelting ores generates substantial emissions to the atmosphere and substantial particulate waste. These emissions and wastes may contain sulfur and toxic materials, including heavy metals. Retrofitting of old smelters to meet current standards is often very expensive, and given other economic factors, such as age of plant, existence of ore supplies and markets, may be a trigger that leads to the closing of these industrial operations. Loss of smelter capacity in the USA has already occurred due to the 1970 amendments, but remains a possible issue facing society with respect to the assurance of supplies that may be at risk for political and economic reasons. Since mining involves the movement of large volumes of waste materials and their disposal, often requiring the use of a substantial amount of water, clean water standards and surface mining reclamation standards impact heavily upon the mining sector for nonfuel minerals.

The usual pattern of these environmental mineral conflicts is the closing or prohibiting of mineral development in areas which have been deemed environmentally fragile—as, for example, the expansion of phosphate mining in Florida and the development of molybdenum deposits in the Cascade Mountains. The control of the emissions, both to the air and to water, imposes substantial additional costs on the mining and smelting processes. In terms of total expenditure, the steel industry has been involved very heavily in capital expenditure to control emissions, while in the nonferrous smelting area there have been a number of closures for a variety of reasons, including environmental impact.

7.2 Health and Safety Legislation

Health and safety legislation impacts upon the nonfuel industry at two levels:

(a) mining regulations, and

(b) regulations on industrial exposure to toxic chemicals.

With respect to mining, nonfuel mineral industries are not nearly as hazardous as, for example, coal mining but, nevertheless, the provision for worker safety in mines imposes expenses and limitations upon mining methods. The major issue in this area, however, lies in exposure to toxic materials (e.g., in lead refining and the processing and use of asbestos). Exposure to lead is an inevitable consequence of lead refining and the question of what is an acceptable exposure is a critical one with respect to the cost and survival of the lead industry. The use of asbestos has also been heavily criticized for health reasons.

Health and safety regulations were not adopted to single out or to apply only to the nonfuel minerals and materials industries. It so happens, however, that the legislation does have a differential impact upon these industries, and the adjustment of the conflicting objectives of nonfuel mineral policy and environmental quality and health and safety issues make the problems of resolution in this area exceptionally difficult.

8. Emerging Policy Issues

The issue most talked about in federal government circles and in states with a heavy concentration of primary mineral industry sectors during the last half of the 1980s was "competitiveness," but concurrently, national policy concentrated on freer trade. President Bush signed a free trade treaty with Canada and is moving towards an agreement with Mexico. Within GATT (the General Agreement on Tariffs and Trade), which has been the mainstream treaty governing international trade among the industrialized nations, major efforts are under way to bring down trade barriers in Japan and in the EC. A new concept of "fair trade" is politically emerging. Proposed amendments to the Defense Production Act (1950) passed the Senate in the fall of 1990. They called for a "buy America" requirement for all critical weapons parts and components as well as broadening the definition of materials to "raw materials, articles, commodities, components, processes, products, supplies, and technical information." The amendments were not passed by the House and thus had not reached the White House as of May, 1991. In 1991, trade issues including "buy America" and "fair trade," as well as free trade under GATT were high on the list of important emerging materials issues.

See also: Critical and Strategic Materials; Policy: Australia; Policy: Canada; Policy: Europe; Policy: General; Policy: Japan; Trade in Materials, International

Bibliography

Gulley D A 1985 Environmental regulation and the minerals industry. In: Vogely W A (ed.) 1985 *Economics of the Mineral Industries.* American Institute of Mining, Metallurgical and Petroleum Engineers, New York, pp. 601–14

Hibbard W R 1985 Non-fuel mineral policy issues. In: Vogely W A (ed.) 1985 *Economics of the Mineral Industries.* American Institute of Mining, Metallurgical and Petroleum Engineers, New York, pp. 593–600

Materials and Society, special issues covering materials economics symposia
1979 *Mater. Soc.* 3(3)
1980 *Mater. Soc.* 4(1)
1981 *Mater. Soc.* 5(1)
1982 *Mater. Soc.* 6(2)
1983 *Mater. Soc.* 7(1)
1984 *Mater. Soc.* 8(3)
1985 *Mater. Soc.* 9(2)

US Comptroller General 1979 Learning to look ahead: the need for a national materials and planning process. Report EMD 79-30. General Accounting Office, Washington, DC

US National Materials Advisory Board 1981 National non-fuel mineral policy planning process. NMAB-384, Washington, DC

US National Research Council 1989 *Maintaining Competitiveness in the Age of Materials.* National Academy Press, Washington, DC

W. A. Vogely
[Pennsylvania State University, University Park, Pennsylvania, USA]

Prices of Industrial Minerals: History

The term "*industrial minerals*" refers to a broad category of nonmetallic mineral and rock materials that have been useful to mankind throughout its history. As such, these industrial minerals have economic value. Their price histories are a reflection of applications of technology, forces of supply and demand and the general economic conditions that affect the value or purchasing power of various currencies. In this article, price histories are considered in terms of US dollar equivalents as reported in the years of reference, generally understood as "current" dollar prices. Where possible, the measurement of price will be in terms of US dollars per metric ton unless otherwise indicated.

Commodities classified as industrial minerals cover such a wide range of materials and values that it is difficult to make general observations on their price history. Most of the industrial mineral commodities (listed in Table 1; note that zeolites, rare earths and a few other industrial minerals such as lithium minerals are not listed due to their great variations in price from one year to another) have been valued from US\$10 t^{-1} to US\$175 t^{-1} over the period 1950–1990. The more common materials, such as bauxite, crushed stone, gypsum, pumice, sand and gravel, have had relatively low values at their point of production, in the range of US\$1.00 t^{-1} to US\$15.00 t^{-1} over the period from 1950. More exotic industrial minerals, such as long-fiber asbestos, may be valued up to US\$2200 t^{-1} (nominal), and industrial diamonds at US\$1.00 to US\$8.00 carat^{-1} (equivalent to US\$5–40 $\times 10^6$ t^{-1}).

The value of industrial mineral materials depends largely on how they are used. More precise specifications of chemical purity, crystalline perfection or other physical properties such as color, hardness or particle size are generally associated with higher prices. Less exacting specifications and supply characteristics, such as common occurrence, are normally associated with lower-priced materials.

Prices shown in Table 1 give a 40-year (1950–1990) history of approximate values for most of the important industrial minerals in common and continuing use. In cases of significant price variations due to grading or specifications within a commodity type, there are some ranges presented. This is especially evident for asbestos, graphite, mica, nitrogen compounds and talc. Clays are subdivided into four commodity groups according to type and use: attapulgite (fuller's earth), used as an absorbent in cat litter and floor sweep and as a suspension agent in oil well drilling muds; ball clay, used principally as a ceramic raw material; bentonite, used as a bonding or suspension agent; and kaolin, which has a wide variety of uses (and prices) from ceramics and fillers to paper coating. Kaolin for paper coating may sell for prices up to ten times those shown for average grade kaolin, depending on brightness (whiteness and reflectivity) and crystal size. The increasing use of higher grades of kaolin has had a

Table 1
Industrial mineral prices[a]

Commodity	Price (US$ t^{-1})				
	1950	1960	1970	1980	1990
Asbestos					
average grade	56.00	72.00	85.00	356.00	460.00
lowest grade	30.00	40.00	60.00	120.00	140.00
highest grade	960.00	1800.00	1800.00	1800.00	2200.00
Barite or barytes	28.20	19.85	16.50	31.40	38.00
Bauxite	6.50	16.65	9.85	14.75	17.50
Borates	27.05	52.40	82.95	186.30	272.00
Bromine[b]	0.46	0.47	0.37	0.58	0.98
Cement	13.80	19.75	18.00	51.00	49.30
Clays					
attapulgite (fuller's earth)	18.10	24.75	35.30	49.60	85.15
ball clay	13.55	14.85	18.75	27.55	37.50
bentonite	9.70	13.05	16.30	27.60	30.50
kaolin (average grade)	18.75	18.75	26.50	38.60	82.00
Corundum (alumina)[d]	54.88	64.50	23.40	43.00	52.00
Diamond (industrial)[c]	3.23	4.06	3.67	4.95	0.84
Diatomite	44.00	54.60	60.25	172.40	219.00
Emery	13.25	19.20	22.00	83.00	96.50
Feldspar	6.90	10.50	14.65	35.25	43.75
Fluorspar	38.85	46.30	57.35	100.35	168.00
Garnet	94.00	103.00	113.30	143.30	170.00
Graphite					
flake	250.00	308.65	369.30	626.15	968.00
lump	260.00	330.70	405.65	535.75	1251.00
amorphous	12.50	20.95	39.70	153.25	105.00
Gypsum	3.05	4.00	4.10	8.45	7.50
Ilmenite	17.00	22.65	20.70	54.10	79.00
Iodine[b]	3.80	6.00	5.28	13.30	16.25
Kyanite	55.00	61.00	80.00	141.00	103.50
Lime	12.30	14.75	16.00	51.40	50.00
Mica					
scrap and flake	11.00	27.60	23.40	44.00	57.00
sheet[b]	6.00	13.20	1.75	6.00	80.00
Nitrogen compounds					
low content	32.00	32.00	45.00	120.00	115.00
high content	58.00	88.00	55.00	125.00	135.00
Perlite	7.05	10.55	12.90	28.75	30.66
Phosphate rock	6.00	7.20	5.80	24.25	23.00
Potash (average 60% K_2O)	22.00	28.00	22.00	86.00	130.00
Pumice	4.10	2.35	1.70	4.65	19.83
Rutile[d]	77.00	105.00	204.00	485.00	610.00
Salt	29.10	29.50	34.00	57.00	35.00
Sand and gravel	1.08	1.12	1.30	3.10	3.80
Soda ash (sodium carbonate)	28.65	38.00	23.10	79.60	83.00
Sodium sulfate	34.20	45.20	20.00	68.90	86.00
Stone (crushed)	1.35	1.55	1.90	3.30	4.60
Strontium minerals	59.55	67.70	24.25	60.00	82.00
Sulfur	17.75	24.60	22.75	87.85	81.00
Talc					
filler grade	10.00	10.00	5.00	10.00	20.00
cosmetic grade	27.00	40.00	250.00	280.00	325.00
Vermiculite	11.25	17.22	25.14	79.01	102.00
Wollastonite	22.00	43.00	38.50	86.00	114.50
Zircon	52.35	52.00	60.65	182.00	422.00
Consumer Price Index[e]	24.1	29.6	38.8	82.4	130.7
Producer Price Index[e]	28.2	33.4	39.3	88.0	119.2
Implicit price deflator (GNP price index)[e]	23.9	30.9	42.0	85.7	131.5

Source: *Economic Report of the President*, February, 1991
a FOB mine or processing plant prices in the USA, unless otherwise indicated b Price given in US$ kg^{-1} c Price given in US$ $carat^{-1}$ d Price is US$ t^{-1} CIF (cost, insurance and freight) or delivered price at East Coast (US) ports e 1982–84 = 100 for consumer price index, 1982 = 100 for producer price index, 1982 = 100 for implicit price deflator (GNP price index)

dramatic impact on the average prices shown in Table 1. Other clays show similar historical variations in prices.

Increased economic activity normally creates increased demand for industrial minerals that is not always possible to meet with new supplies or substitute materials. It may be noted that many prices of industrial mineral commodities shown in Table 1 had higher unit values in 1960 than in 1970. 1960 was a year of economic rebound from the 1958 recession, and many industrial mineral producers realized prices above the historic trend due to supply shortages in various geographic and end-use markets. 1980 was a year of robust demand and high inflation, creating upward pressures on some industrial mineral commodity prices that have since declined while other industrial mineral prices have continued to rise. Some of the commodities listed in Table 1 have increased in value in line with overall economic growth and the shrinking purchasing power of the US dollar. Other industrial minerals vary in both real and relative value from one point in time to another.

The long-term trend in real (constant purchasing power) US dollar prices was not evident in industrial mineral values until about the last decade of the 40-year time period in Table 1. Mineral prices from 1950 through to the mid-1970s displayed a trend that was below the rate of inflation. The real prices of industrial minerals often made them relative bargains when compared with the prices of other commodities during the 1950–1975 period. Since 1975, prices of most industrial minerals, with a few notable exceptions, have increased more or less in line with the general price level.

Most of the prices shown in Table 1 are free-on-board (FOB) or point-of-production prices. Since transportation costs can make a significant contribution to the delivered price of any bulk material, delivered prices (at the point of utilization) are commonly two to three times greater than the FOB values. This is especially true for commodities that must be taken long distances from the point of production to the point of use and/or whose unit values are relatively low in comparison to their bulk or volume. The value added by transportation to the more expensive industrial minerals is normally less significant.

Individual industrial mineral commodities have shown significant price variations due to swings in supply and demand over the 40-year period 1950–1990. Potash has experienced wide price swings due to supply variations over this period. When supplies were plentiful, such as upon the opening of several large Canadian mines in the late 1960s and early 1970s, potash prices were significantly depressed. At other times, growth in demand has outpaced supply, and prices have escalated.

While asbestos does not show significant price erosion over the 40-year period 1950–1990, the questions of continued use in many applications where human exposure is probable, combined with medical research on the carcinogenic potential of fibrous silicates, bring into question the continued upward trend in prices. This could be caused in the near-term by increased demand in developing nations where health impacts may not be as serious in threatening end-use markets or in the long-term by medical discoveries that may disprove or offset the carcinogenic potential of crystalline silicate fibers. It is becoming more widely recognized that only some forms of the family of minerals known as asbestos are potentially carcinogenic, and that different populations may have varying degrees of susceptibility to the carcinogenic effects of fibrous silicates. If the average life-span in a developing nation is less than 50 years, the health costs of known carcinogens may be more than offset by the use benefits of the suspect materials.

Abrasive materials such as corundum, industrial diamond, emery, garnet and dental grades of pumice have experienced substantial competition from synthetic abrasive materials such as silicon carbide and alumina. In the case of corundum, naturally occurring crystalline aluminum oxide, the availability of alumina produced from bauxite has virtually displaced all but a few selected markets for the natural material. The price erosion in corundum during the period 1960–1970 is mostly based on the impact of alumina on traditional corundum markets.

Prices for bauxite, used directly as an abrasive, are vastly overweighted by its major use as an ore for metallic aluminum. Abrasive bauxite is basically produced as a by-product of mining aluminum ore. Ilmenite and rutile prices vary in relation to the demand for paint and pigments. Demand for phosphate rock, potash and nitrogen compounds is tied to the demand for chemical fertilizers. Demand for gypsum, cement, crushed stone, sand and gravel is tied to the level of construction activity. Crushed stone demand is especially sensitive to new highway construction due to the large amounts used in base and pavement work.

Many of the industrial minerals in Table 1 are basic chemical raw materials. Borates and soda ash are used in glass and ceramic manufacturing, but are also important constituents in laundry detergents and chemicals. Sulfur prices largely reflect the market for sulfuric acid, and especially fertilizer demand, due to the significant consumption of sulfuric acid in phosphate processing. Barite (or barytes), bentonite and to some extent attapulgite, are especially important in oil well drilling muds, and their demand is heavily dependent on the level of drilling activity. Demand for ball clay, feldspar, kyanite, wollastonite and zircon tends to be tied to the demand for ceramic and refractory materials. Lime has become more closely associated with the level of activity of basic oxygen steelmaking. The demand for salt has little to do with its declining use as table salt, but is strongly tied to the demand for chloralkali chemicals.

Diatomite is largely consumed as a filter aid and specialized carrier or filler. Fluorspar is closely associated with aluminum, steel and chemical manufacturing. Graphite is used in electrodes for electric furnaces, in consumer goods such as flashlight batteries, in high-temperature ceramics and in lubricants. Mica is used in latex paints, for various electrical insulation purposes and in electronic applications. Perlite and vermiculite are closely associated with thermal and other insulation uses and increasingly in horticultural applications, such as soil conditioners and potting soil. Talc is used as a filler and in cosmetics such as face and body powder.

In all cases, the demand for industrial minerals is derived and dependent on their ultimate end use. These commodities have limited utility except after processing and as components of other products. Supplies of industrial minerals are largely controlled by the availability of deposits of suitable ores and the timing of mine development and the processing capacity of the producers. Prices vary according to growth in both supply and demand.

Over the period 1950–1990, markets for industrial minerals have tended to be more stable and orderly than markets for metals and energy raw materials. While prices of nonmetallics are unlikely to vary as greatly as prices of metals, fuel and energy materials, they are likely to continue to increase at an escalating rate into the next century as more developing nations expand their domestic economic capacity and their demand for all commodities, including minerals, becomes increasingly competitive with that of the industrialized nations.

See also: Prices of Materials: Theory; Prices of Metals: History and Current Problems

Bibliography

Industrial Minerals—Annual Review, published annually in *Mining Engineering*, 1971–1991

Lefond S J (ed.) 1983 *Industrial Minerals and Rocks*, 5th edn. American Institute of Mining, Metallurgical and Petroleum Engineers, New York

McCarl H N 1986 Prices of industrial minerals: history. In: Bever M B (ed.) 1986 *Encyclopedia of Materials Science and Engineering*. Pergamon, Oxford, pp. 3931–4

McCarl H N, Maclean J A, Wright D T 1980 *Non-Metallic Minerals*, Mahon's Guides for Accountants and Auditors, Guide 16. Warren, Gorham and Lamont, Boston, MA

Materials Prices, monthly feature in *Engineering News Record*

Mineral Commodity Summaries, annual editions 1978–1991. US Bureau of Mines, Washington, DC

Minerals Yearbook, annual editions 1950–1990, US Bureau of Mines, Washington, DC (since 1974: ASI Microfiche: 5604–)

Prices, monthly feature in *Industrial Minerals*

H. N. McCarl
[University of Alabama at Birmingham, Birmingham, Alabama, USA]

Prices of Materials: Stabilization

Materials prices may be valuable signals of relative scarcities. High prices tend to encourage limited use of a particular material and search for cost-reducing technological changes and cheaper substitutes, whereas low prices tend to encourage expanded use of a particular material instead of more expensive alternatives. In a world of imperfect information, data on prices that may reflect relative scarcities are often much more available and more timely than are data on quantities that directly reflect relative material scarcities. Some basic principles of how materials markets operate without price stabilization policies are discussed in Sect. 1.

Despite the possibly very useful roles of freely fluctuating prices, there long have been frequent attempts by governments and private entities to stabilize materials prices. Among the major policies that have been used for such efforts are buffer stocks; isolating a particular market and using exports from it to lessen downward price movements and imports into it to lessen upward price movements; announcing a price and rationing available quantities among users if demand at that price exceeds supplies; providing information; attempting to lessen the underlying causes of fluctuations; and intervening in futures markets for the material (Sect. 2).

The motivation for such efforts includes gains due to risk aversion (Sect. 3), macroeconomic gains regarding fuller employment and less inflation (Sect. 4), increases in consumers' and producers' surpluses (Sect. 5), and distributional gains (or losses) for particular groups often associated with changing the average level of material prices (Sect. 6).

Analysis with simple models suggests that some of these gains may indeed be attainable. But further analytical complications and empirical experience (see Behrman 1977, 1987) suggest that the costs of attaining such gains may frequently be considerable. Moreover, there is often confusion between the logically distinct efforts to stabilize materials prices and to change their expected (or average) values, usually to the benefit of some particular supplying or demanding group but with widespread efficiency costs (Sect. 7). Therefore, it is important that materials price stabilization policies should be analyzed carefully within a framework that incorporates the important features of a particular materials market, that considers the distribution of total costs and benefits, and that reflects the difficulties of predicting the appropriate expected prices—and that decisions should not be made on the basis of too simple analytical models.

1. Simple Materials Market without Price Stabilization Policies

Figure 1 is a graphical representation of a simple materials market. Underlying the market supply curve

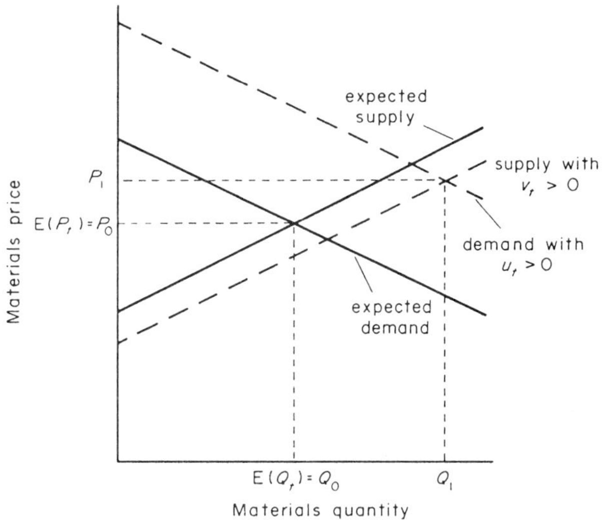

Figure 1
Simple linear materials market supply and demand curves with additive stochastic terms

is maximizing behavior by each of a large number of small suppliers so that no one supplier can affect the market price perceptibly, but each one responds instantaneously to the given price (from the point of view of the individual supplier) by supplying the profit-maximizing quantity (i.e., perfect competition). These quantities increase for higher prices since it is profitable to expand production at higher prices even if fixed inputs (e.g., mines and managerial capacities) have to be stretched further or higher prices paid for variable inputs (e.g., labor). By summing the quantities supplied at each price across all suppliers, the market supply curve is obtained. In Fig. 1 the market supply curve in period t (S_t) is assumed to be linear, with an additive stochastic term v_t due to fluctuations in, for example, the weather, with mean value of zero (as in the solid, upward sloping supply curve) and a finite variance σ_v^2:

$$S(P_t) = a + bP_t + v_t$$
$$b \geq 0, \qquad E(v_t) = 0, \qquad \sigma_v^2 > 0 \tag{1}$$

where a and b are constants, $E(v_t)$ is the expected value of v_t, and P_t is the short-run equilibrium price. Likewise, underlying the market demand curve is maximizing behavior by a large number of demanders, each of which is assumed to be unable to affect the market price perceptibly, but each of which responds to the given price (from the point of view of the individual demander) by purchasing the quantity that maximizes attainment of its objective (i.e., perfect competition again). At higher prices this quantity tends to be lower since other relatively cheaper goods and services are substituted for this material. By summing the quantities demanded across all demanders, the market demand curve is obtained.

In Fig. 1 the market demand curve in period D_t is assumed to be linear, with an additive stochastic term u_t due to fluctuations in income, for example, with mean value of zero (as in the solid, downward sloping demand curve) and a finite variance (σ_u^2) and zero covariance with the stochastic term in the supply relation ($\sigma_{uv} = 0$):

$$D(P_t) = c - dP_t + u_t$$
$$d > 0, \qquad E(u_t) = 0, \qquad \sigma_u^2 > 0 \tag{2}$$

where c and d are constants and $E(u_t)$ is the expected value of u_t. In the absence of any stabilization policy, in period t the short-run equilibrium price is that at which supply equals demand. (At a higher price the desired quantity supplied exceeds the desired quantity demanded, putting pressure for the price to fall, and *vice versa* for a lower price). This price P_t can be obtained by setting Eqn. (1) equal to Eqn. (2) and solving for P_t:

$$P_t = (c - a + u_t - v_t)/(b + d) \tag{3}$$

The short-run equilibrium quantity Q_t can be obtained by substituting Eqn. (3) into either Eqn. (1) or (2):

$$Q_t = S(P_t) = D(P_t)$$
$$= (cb + da + bu_t + dv_t)/(b + d) \tag{4}$$

Thus, the short-run equilibrium price and quantity both are stochastic since both depend on the stochastic terms v_t and u_t in the market supply and demand relations due to fluctuations in other relevant variables such as the weather and income. The dashed lines in Fig. 1 give an illustration in which both v_t and u_t are positive, with the resulting equilibrium price P_1 and quantity Q_1. From Eqns. (3) and (4) the expected values of the price $E(P_t)$ and quantity $E(Q_t)$ at which the stochastic terms take on their mean values of zero can be calculated:

$$E(P_t) = (c - a)/(b + d) \tag{5}$$

$$E(Q_t) = (cb + da)/(b + d) \tag{6}$$

In terms of Fig. 1, these expected values are just the equilibrium price (P_0) and quantity (Q_0) associated with the intersection of the solid supply and demand schedules (for which $v_t = u_t = 0$). However, Eqns. (3) and (4) indicate that the actual price and quantity from period to period fluctuate around these expected values, with variances that depend on the size of the variances of the additive stochastic terms in the supply and demand relations (σ_v^2 and σ_u^2) and on the slopes in the nonstochastic part of these relations (b and d):

$$\sigma_P^2 = (\sigma_u^2 + \sigma_v^2)/(b + d)^2 > 0 \tag{7}$$

$$\sigma_Q^2 = (b^2\sigma_u^2 + d^2\sigma_v^2)/(b + d)^2 > 0 \tag{8}$$

The reasons for these fluctuations are the underlying fluctuations in the supply and demand curves due to fluctuations in relevant factors such as weather and income. In response, the short-run equilibrium price varies so that the current quantity demanded and supplied respond to the current situation. If, for example, the weather is bad so $v_t < 0$ and income is high so $u_t > 0$, the short-run equilibrium price is relatively high, which induces a movement along the demand curve in that period to conserve use of the material in question and a movement along the supply curve to expand its supply. Thus, the relatively high price serves to improve the situation by reducing use and encouraging supply in a period of relative shortage (and *vice versa* if $v_t > 0$ and $u_t < 0$). If both suppliers and demanders are perfect competitors and there are not externalities (e.g., products like pollution which are distributed outside of markets), the short-run equilibrium materials price tends to reflect the true social opportunity cost of materials and using it to allocate the materials results in an efficient allocation. On the other hand, if there are market failures due to factors like externalities or imperfect competitors (e.g., monopolies), various policy interventions (perhaps including materials price stabilization) may make the allocation of resources more efficient. (For further considerations regarding modeling such markets and efficiency, see Varian 1990).

2. Major Price Stabilization Policies

Among the most used and most analyzed of major price stabilization policies is the institution of buffer stocks that try to stabilize material prices by buying the material when the price would otherwise be relatively low and selling it when the price would otherwise be relatively high. In the situation described in Fig. 1, a buffer stock authority could stabilize completely the price at its expected value (P_0 or $E(P_t)$ in Eqn. (5)) by trading in each period a quantity equal to $v_t - u_t$ (i.e., buying if $v_t > u_t$, selling if $v_t < u_t$). The variance in price would go from the positive value in Eqn. (7) to zero. The variance in quantity supplied would go from that in Eqn. (8) to σ_v^2 and the variance in quantity demanded would go from that in Eqn. (8) to σ_u^2 (which implies an increase in at least one of these two quantity variances). The expected trading of the buffer stock is zero ($= E(v_t) - E(u_t)$), so in the long run (assuming that the structure of the market is stable) the buffer stock tends to sell what it purchases at the same price. The buffer stock suffers an expected loss, therefore, equal to its expected transaction and storage costs. In a static market this cost could be covered by buying only when the price would otherwise go below some floor $\underline{P}$ below P_0 and selling only when the price would otherwise go above some ceiling ($\bar{P}$) above P_0, with the width of the price band ($\bar{P} - \underline{P}$) being a positive function of such buffer stock costs. The wider the price band, of course, the less the price stabilization and gains associated with price stabilization. In a dynamic market in which there is a secular trend in the expected price over time, the buffer stock costs may be covered by buying low and selling later at a higher price even without price bands if the secular trend in the expected price is positive enough, but the buffer stock may not be able to cover its direct costs even with very wide price bands if the secular trend in the expected price is negative enough.

A second widely used materials price stabilization policy is to isolate geographically a certain materials market (e.g., a nation or a group of nations like the European Community, EC) and regulate exports and imports of materials to offset fluctuations internal to the market. That is, if the objective is to stabilize the domestic materials price at the expected value P_0 in Fig. 1 (assuming that this figure refers only to the isolated market), let exports minus imports be equal to $v_t - u_t$. This strategy depends on there being some external rest-of-the-world market to which to sell exports or from which to buy imports when needed, and therefore is not applicable on a global or near-global level. This policy has the direct cost advantage over the buffer stock alternative of avoiding direct storage costs, though direct transaction costs remain (and probably are greater since direct administrative costs are incurred to isolate a particular market). The more important question regarding direct costs probably pertains to the comparison between the expected prices in the rest-of-the-world market and the market of interest. If the expected price in the rest-of-the-world market is less than that in the internal market of concern, as is apparently the case for many materials produced in the EC, direct subsidies are required to export surpluses and internal users pay more for the material than they would if they were able to buy on the world market. If the expected price in the rest-of-the-world market is greater than in the isolated market of concern (as, for example, for major material exports from the Ivory Coast and a number of other African nations), direct subsidy costs are not likely to be a problem, but internal production is discouraged and consumption encouraged relative to the alternative of trading with the rest of the world. For such reasons it would be best to stabilize around the world price (which generally implies expected net exports or imports with variations at the margin to offset internal stochastic fluctuations) unless there are good nonefficiency reasons to aim for a different target (i.e., a higher expected internal price to encourage development of internal "strategic" materials for national defense purposes). However, it is not clear that such advice implies anything much different than integration into the world market, which has the advantage of sharing one's own shocks with the rest of the world but the disadvantage of sharing shocks that originate in the rest of the world.

A third common procedure for stabilizing materials prices is to set a price and ration use if the quantity demanded at that price exceeds the quantity supplied. Such policies have been followed by a wide range of governments, both socialist and mixed capitalist, for strategic materials, particularly at times of war or other large shocks (e.g., oil price shocks in 1973–1974 and 1979–1980). Private entities have also sometimes used such a strategy, apparently because of the perception that a stable price with rationing at times of shortages leads to a larger longer-run demand (e.g., US copper producers in the 1960s and early 1970s). If the stabilized price is below the short-term equilibrium price and allocation by rationing is effective, one major cost of such a policy is that those whose demand is fulfilled pay less than the true scarcity cost of the material and are likely to use more of it than is socially efficient; others who are willing to pay an equal amount have their demands only partially (or not at all) fulfilled, and suppliers are discouraged from supplying as much of the material as they would at the equilibrium price. In addition to these inefficiencies, rationing is likely to encourage illegal activities and exploitation of "connections" for those who have them. Considerable costs are also incurred in administering any effective rationing system.

A fourth price stabilization policy is to provide information that can lead to integration of markets across space or time. If the stochastic terms are not perfectly correlated across such markets, integration resulting from such information may indeed reduce fluctuations by "pooling risks." Since information has the quality of being a "public good" (i.e., more of it for one person does not diminish what is available for others), private entities probably do not provide enough of it and public information policies possibly are efficient and are certainly undertaken. On the other hand, provision of information is not costless and it is not at all clear that governments have better information about materials markets than do private entities, who may also have strong incentives to integrate markets over space and time.

A fifth materials price stabilization policy that the discussion in Sect. 1 suggests, but which is more indirect, is to attempt to lessen the effects of the underlying stochastic terms. If supply fluctuations originate in weather or labor market conditions, for example, the effects may be lessened by explicit private or public measures. Many such measures are undertaken, in fact, although often with materials prices stabilization as only one objective (and, at times, a very minor one). Examples include private and public water control measures to lessen supply shocks, or public macroeconomic policies to lessen demand shocks.

Other policies could also be mentioned. For example, in the early 1980s a consortium of coffee producers attempted to affect current prices by trading in future markets in which their leverage is greater (since contracts can be purchased with margins much less than 100% of their value), but this can logically be considered a variant of buffer-stock type policies. As another variant of buffer stock policies, private entities could be subsidized to hold such stocks, with the subsidies related inversely to current price levels. However, these policies include the major materials price stabilization policies, with the possible exception of restrictions on supply and demand (see Sect. 6).

3. *Gains due to Risk Aversion*

If producers are risk averse in the sense that they are better off for a given level of expected profits (or surplus, see Sect. 4) if the variance in materials prices is lower, there may be a gain from price stabilization even if it leaves expected profits the same and has none of the other possible effects discussed in the following. If users are risk averse in a similar sense, there also may be such a gain, though it probably is less important for users since any particular material is probably a much smaller proportion of the users' expenditures than it is of the producers' revenues. If a nation is dependent on production, exports or use of a particular material for a significant share of its economic activity and risk aversion is widespread, there may be widespread gains since material price fluctuations may be amplified throughout the economy by government and private revenues and expenditures (see, for example, Adams and Behrman 1982 or Behrman et al. 1989).

Risk aversion is widely accepted as a reason for which price stabilization may be advantageous. However, at least two questions occur in this connection. First, is risk aversion likely to be with regard to prices or with regard to revenues, expenditures or surpluses (see Sect. 5)? If it is with regard to suppliers' revenues, for example, the observation in Sect. 2 that buffer-stock materials price stabilization may increase variance in quantities supplied is relevant. In fact this increase may be sufficient so that price stabilization actually increases the variance of suppliers' revenues. In such a situation price stabilization may make suppliers worse off, all other things being equal, if suppliers are risk averse with respect to revenues and not to prices. However, the trade-off between price and revenue stability is less likely if stabilization is partial rather than complete, the larger stochastic terms are in demand relative to those in supply and the less is the slope of the demand curve (Nguyen 1980). The available empirical evidence about materials markets suggests that, in general, there probably is not such a trade-off.

Second, is risk aversion empirically important? A large number of empirical studies have explored this question in recent years, albeit with measurement problems with regard to the representation of risk. The general conclusion seems to be that the effect of risk aversion as best it can be represented empirically often seems to be significant in a statistical sense, but not substantial in magnitude (see, for example, Adams and Behrman 1982). Thus, while risk aversion is a conceptually attractive reason for gains from materials price stabilization, it does not seem to be empirically important enough to warrant much economic or political cost.

4. Gains due to Macroinflation and Employment

Increases in materials prices may cause upward price pressures which are costly in terms of increased inflation or negative effects of anti-inflationary policies on output. For materials price stabilization to have a gain from lessening inflationary pressures, however, there has to be a reason why there is not a symmetrical offsetting downward anti-inflationary pressure when materials prices fall. In the period after World War II, in the industrial economies at least, there seems to be an asymmetry in price movements. Prices tend to move upwards much easier than downwards. The reasons for the asymmetry are not fully understood. Important factors probably include knowledge that governments have been committed to fairly full employment and are not likely to allow a repeat of the Great Depression, much better benefits for the unemployed than earlier and administrative pricing with an upwards bias by large (and not perfectly competitive in the sense of Sect. 1) companies. Although agreement is not universal on this matter, in this context there may be a considerable macro gain from materials price stabilization due to lessened inflationary pressure and lessened need to pursue anti-inflationary policies. One crude estimate calculates this gain for the USA alone to be three times as large as the gain to producers worldwide from international price stabilization of 10 major commodities: copper, tin and eight agricultural products (Behrman 1977).

5. Gains in the Form of Producers' and Consumers' Surpluses

The combination of an upward sloping supply curve and the assumption that all units are sold at the same price in period t (i.e., P_1 for the dashed lines in Fig. 1) has an important implication. For the last (or marginal) unit sold there is no producers' (or suppliers') surplus in that it would not be supplied if the price were lower, but for all other units sold there is a producers' surplus in that they would be supplied even if the price were somewhat lower. For the dashed lines in Fig. 1 this producer surplus is measured by the triangle bounded by the dashed supply line, the dashed horizontal price line at P_1 and the vertical price axis. In any period t the change in the producer surplus due to price stabilization at P_5 is the integral bounded by P_t and P_5 and by the supply curve and the vertical price axis. The expected value of this change in producers' surplus due to price stabilization under the assumptions of Sects. 1 and 2 is

$$\mathrm{E(SUR}_p) = [-b\sigma_u^2 + (2d + b)\sigma_v^2]/[2(d + b)^2] \quad (9)$$

This may be positive or negative, depending on the relative magnitudes of the terms in the numerator (Turnovsky 1978).

Likewise, the combination of a downward sloping demand curve and the assumption that all units are purchased at the same price in period t has a similar interpretation (with some technical qualifications that apparently are not critical empirically) for consumers' (or demanders') surplus, since demanders would be willing to purchase all but the last unit of the material even if the price were somewhat higher. For the dashed lines in Fig. 1 the consumers' surplus is measured by the triangle bounded by the dashed demand line, the dashed horizontal price line at P_1 and the vertical price axis. Under the assumptions of Sects. 1 and 2, the expected value of the change in consumers' surplus due to price stabilization is

$$\mathrm{E(SUR}_c) = [(2b + d)\sigma_u^2 - d\sigma_v^2]/[2(a + b)^2] \quad (10)$$

This, too, may be positive or negative depending on the relative signs of the terms in the numerator.

By summing Eqns. (9) and (10), the total expected surplus can be calculated

$$\mathrm{E(SUR}_p) + \mathrm{E(SUR}_c) = (\sigma_u^2 + \sigma_v^2)/[2(b + d)] > 0 \quad (11)$$

This total surplus or gain is unambiguously positive.

Equations (9)–(11) lead directly to the Waugh–Oi–Massell (WOM) conclusions about the impact of materials price stabilization.

(a) Demanders gain (lose) from materials price stabilization if the only source of price instability is random shifts in demand (supply).

(b) Suppliers gain (lose) from materials price stabilization if the only source of price instability is random shifts in supply (demand).

(c) If both supply and demand have stochastic terms, the gains to suppliers alone and to demanders alone may be positive or negative, depending on the variances of the stochastic terms (σ_u^2 and σ_v^2) and the magnitudes of the supply and demand slopes (b and d).

(d) However, the total gains to suppliers and demanders together are positive, with the gainers in principle able to compensate the losers.

Empirical estimates suggest that, for many materials, these types of gains could be considerable.

These conclusions from the WOM analyses have conditioned a widespread belief that there may be substantial gains from materials price stabilization.

However, in several important respects (which are discussed in more detail by Turnovsky (1978) and Newbery and Stiglitz (1981)) they are conditional on the special assumptions of the simple model of Sect. 1. If there are multiplicative (rather than additive) stochastic terms, if supply responds to expected prices generated by some common expectational mechanisms (instead of adjusting instantaneously to current price), or if some simple partial stabilization rules are followed (instead of there being a full stabilization at P_0), only the fourth conclusion may hold strictly. Of course, this is the most important conclusion since it states that total gains from materials price stabilization are positive.

However, even this conclusion is thrown into some doubt with further extensions of the analysis. First of all, the analysis ignores the transaction, administrative and storage costs of price stabilization. These may be considerable in some cases and subtracting them from expected producers' and consumers' surpluses may give a greatly reduced and even negative value. Second, the analysis presumes that the appropriate target price is known, but there may be uncertainty in regard to this price (Sect. 6). Third, there may be dynamics in regard to both supply and demand that complicate the analysis considerably (e.g., long gestations in expanding mine capacities or in shifting from use of one material to another). Fourth, the analysis assumes that there are no private holdings of stocks. However, at least for storable commodities, why would not suppliers or demanders (depending on who would gain) reap the expected benefits (up to the point at which they equalled the expected costs) by undertaking their own storage? Fifth, this and related analyses ignore the fact that there is a basic asymmetry in some materials price stabilization policies in that the desired price (or price ceiling) may not be defensible if the buffer stock runs out or if more supplies for imports are not available from the rest-of-the-world market.

These doubts are not conclusive. There may be good reasons why, for example, private inventory behavior for materials is not socially optimal: capital market imperfections, differences between private and social discount rates to value future events, externalities such as macro effects (Sect. 4), and distributional concerns (Sect. 6). For such reasons public buffer stock or other materials price stabilization government policies may be very sensible, although such possibilities should be evaluated carefully in light of the actual condition in a particular materials market, and the overly specialized WOM model results should not be used blindly to justify advocacy of ubiquitous materials price stabilization.

6. *Distribution of Gains*

Up to this point limited attention has been paid to the distribution of losses and gains from materials price stabilization policies among materials suppliers, materials demanders and the general public (through the government). Yet often such distributional effects are at the heart of the matter. If a particular group is thought to be more deserving or more needy (e.g., poorer producers of the materials), it might be socially desirable to promote price stabilization if this target group benefits enough even if the total costs outweigh the gains if special weight is not placed on the gains of the target group. Such distributional concerns seem very important in understanding the emphasis in many societies on price stabilization of basic mass consumption commodities used relatively often by the poor. It also is a factor in the call by many groups in developing countries for price stabilization of the materials important in their exports or imports. A related concern is to use price stabilization to redistribute gains from private inventory holders (i.e., middlemen or speculators) or traders with market power (i.e., sole buyers or one of a few buyers in a market) rather than perfect competition. In certain circumstances price stabilization may help in the desired redistribution, but the effects may be somewhat arbitrary (not all of the poor or even the poorest produce or use particular materials) and have negative side effects (e.g., discouraging socially useful private inventory-holding).

Alternatively, materials price stabilization may be perceived to benefit a powerful political group (e.g., well-organized producers) at a cost to a politically weaker one (e.g., diffused and unorganized consumers). In such a case, even if the total social costs of materials price stabilization outweigh the gains and there is nothing particularly socially deserving about the beneficiaries, the political system may bow to the pressures of the politically powerful pressure group and introduce the price stabilization program.

7. *Stabilization vs Changing Expected Materials Prices*

Recognition that materials price stabilization may be directed towards benefitting a particular group even if the total costs outweigh the total benefits is basic in understanding the widespread advocacy of materials price stabilization policies. A closely related fact is that materials price stabilization is often used to refer to a situation in which the real (often implicit) objective is to change the expected materials price, not just to lessen its variance. If special weight is placed on gains to suppliers (demanders), materials price stabilization may be a misleading characterization of an effort to raise (lower) the expected materials prices. Some stabilization policies, such as curtailing supplies (demands) by production or export (import) restrictions, are oriented not towards reducing materials price variances, but towards changing the expected materials prices.

The impact of changing the expected materials price

may be very different from the impact of stabilizing it, and logically should be separated from it. Unless the effort to change the expected materials price also offsets some market distortion (e.g., due to a monopoly that is restricting supplies), for example, it generally has inefficiency effects akin to those discussed in Sect. 2 with regard to isolating a particular market or rationing. The distributional impacts (as mentioned in Sect. 6) are also likely to be capricious in that the beneficiaries are not likely to be exactly those whom society considers to be particularly deserving, and these effects are likely to be larger the more the expected materials prices in fact are changed.

The frequent confusion between lessening the variances of materials prices and changing their expected values may partly reflect the efforts of those who wish to change the expected prices to obscure their intent, but they are possible only because there is considerable uncertainty about what is the expected price for a particular material at a point of time or, probably more importantly, how it changes over time. This same uncertainty creates an important problem in operating a pure materials price stabilization program and this problem is ignored in most of the analysis above: how can the appropriate expected materials price be identified? Simulations of price stabilization in materials markets suggest that such identification is difficult (Behrman 1977). If the expected materials price is wrongly identified, moreover, the results may include considerable unintended distortions and large operating costs (Behrman and Tinakorn 1979).

See also: Availability of Materials; Economic Theory of Mineral Resources and Markets; Markets for Materials: Modelling; Markets for Materials: Principles and Public Policy Implications of Competition; Prices of Materials: Theory

Bibliography

Adams F G, Behrman J R (eds.) 1978 *Econometric Modelling of World Commodity Policy.* Lexington, Lexington, MA

Adams F G, Behrman J R 1982 *Commodity Exports and Economic Development: The Commodity Problem and Policy in Developing Countries.* Lexington-Heath, Lexington, MA

Adams F G, Klein S A (eds.) 1978 *Stabilizing World Commodity Markets: Analysis, Practice and Policy.* Lexington, Lexington, MA

Behrman J R 1977 International commodity agreements: an evaluation of the UNCTAD integrated commodity programme. In: Cline W R (ed.) 1979 *Policy Alternatives for a New International Economic Order: An Economic Analysis.* Praeger, New York, pp. 63–153

Behrman J R 1987 Primary commodity instability and economic goal attainment in developing countries. *World Dev.* 15: 559–74

Behrman J R, Lewis J, Lofti S 1989 The impact of price instability: experiments with a general equilibrium model for Indonesia. In: Klein L R, Marquez J (eds.) 1989 *Economics in Theory and Practice: An Eclectic Approach.* Kluwer, Dordrecht, The Netherlands

Behrman J R, Tinakorn P 1979 Indexation of international commodity prices through international buffer stock operations. *J. Policy Model.* 1: 113–34

Materials and Society (5(3) 1981): issue devoted to market instability in the metal industries

Newbery D M G, Stiglitz J E 1981 *The Theory of Commodity Price Stabilization: A Study in the Economics of Risk.* Clarendon, Oxford

Nguyen D T 1980 Partial price stabilization and export earning instability. *Oxf. Econ. Pap.* 32: 340–52

Turnovsky S J 1978 The distribution of welfare gains from price stabilization: a survey of some theoretical issues. In: Adams F G, Klein S A (eds.) 1978 *Stabilizing World Commodity Markets: Analysis, Practice and Policy.* Lexington, Lexington, MA, pp. 119–48

Varian H 1990 *Intermediate Microeconomics: A Modern Approach*, 2nd edn. Norton, New York

J. R. Behrman
[University of Pennsylvania, Philadelphia, Pennsylvania, USA]

Prices of Materials: Theory

The prices of materials are determined by the characteristics of materials supply and demand, and by the structure of materials markets. Although the standard microeconomic theory of prices applies in general to materials, several factors distinguish materials price formation from that of general economic goods. These factors lie in the characteristics of materials demand and supply functions, and in the types of market structures encountered. As a result, market behavior of materials is markedly different from that of goods in general, being characterized by a high level of instability in both prices and production levels.

During the 1970s and 1980s, the market structure for the major nonferrous metals changed dramatically. On a world basis, they changed from a producer price or price leader system, to a competitive structure where producers are price takers, from prices established on the major commodity exchanges. The existence of these exchanges made metal markets "contestable" resulting in a competitive market structure. Even state-owned firms are impacted. This is discussed further in Sect. 3.

1. Materials Demand

The demand for a material is derived from the outputs of those goods which contain that raw material as an input. The derived nature of demand has implications for the shape of the demand function for materials, and for its short-run stability. Within the production function, the choice of materials will be determined by the demanding firm that will, with profit maximizing–cost minimizing behavior, choose the lowest cost combination of raw materials to produce the product. Thus the demand for materials arises from the production functions of the firms that use the materials to produce goods.

The theory of production choice (i.e., choice of technique/process) indicates that where there is a choice between material inputs, the cost-minimizing firm will choose the amount of each raw material input per unit of output that results in minimum costs given the prices of the material inputs (see *Substitution of Materials: Economics*). This situation leads to the condition that the raw-material inputs will be used in such a combination that the marginal rate of substitution between the alternative raw materials, given the production possibility function, is proportional to the prices of the materials. A firm will design or choose its production technology to minimize total cost perceived at the time, and usually is not able therefore to change instantaneously the proportions of raw materials used in its production processes. Given sufficient time for producers to change their production techniques in response to changing materials prices, there can be substantial substitution between raw materials (see *Substitution of Materials: Technology*).

The theory of production choice is illustrated in Fig. 1, where AA is the short-run equal product curve for the final product, and BB the long-run equal product curve for the same product. (Points along both curves represent situations in which the same quality of product is being produced—but with different levels of inputs of raw materials X and Y, reflecting the various technological possibilities.) Curve AA has its sharp shape because of the inability of a firm to change its production function in response to short-run conditions, whereas the shape of curve BB represents the substantial possibilities of substitution which exist with a redesigned production function. Line CC is a line whose slope represents the relative prices of the substitutable raw materials X and Y. The slope of this

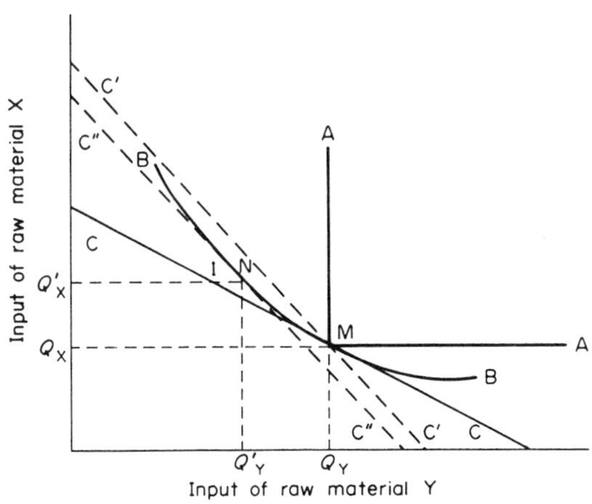

Figure 1
Choice of alternative raw-material inputs to produce the same quantity of final output

line is simply the ratio of the prices of Y and X. The minimum-cost combination of X and Y to produce that level of output represented by curves AA and BB is the point of tangency M between line AA or BB and the price line CC. At this point (i.e., the minimum cost point) the proportions of materials X and Y used, that is, Q_X and Q_Y, are such that the marginal rate of substitution of Y for X is proportional to the price ratio of X to Y.

Now consider a price change which alters the ratio of prices making material X cheaper relative to material Y, and where the new price ratio can be represented by the line C′C′. The short-run response to this price change is not to change the proportion of goods consumed. However, in the long run, there is a lower cost possibility, which is to substitute more of material X for material Y, moving to the new point of tangency at N between line C″C″ (parallel to C′C′) and the equal production curve BB. Figure 1 therefore illustrates the limited possibility of substitution in the short run, with a much higher possibility of substitution between raw materials in response to price changes in the long run.

The nature of the production choice as illustrated above means that the short-run elasticity of demand with respect to price is quite low. This inelasticity is reinforced by the fact that raw materials are normally a small part of the total cost of the final goods within which they are contained. Thus a change in the price of raw materials will not significantly change the price of the final goods, and the total production of these goods will not be affected by the change in price of the raw material alone. Therefore, there should be no direct reduction in the demand for the raw material from the change in outputs of the final goods. As a result of these two factors, the demand for a raw material in the short run tends to be very inelastic.

The long-run elasticity of demand is much higher than the short run. In the long run, substitution of one material for another will follow the long-run curve BB in Fig. 1. The long-run elasticities are also higher because a cost-minimizing producer will attempt to reduce cost through time. The producer will seek production functions that will use less raw material per unit of output, even if there is no substitution. Resource savings through time are a widely observed characteristic in the materials industries, arising from both technology changes and economizing on raw-material inputs by improving process efficiency.

A second effect of the derived nature of raw-materials demand lies in the high sensitivity of this demand to cyclical changes in the total (particularly industrial) output of the economy. Materials are largely used as inputs into the capital goods and consumer-durable industries. These industries show wide variation in output in response to the business cycles of the industrialized countries. Thus, the demand functions for raw materials will shift widely in response to changes in overall economic activity.

Over the longer term, the consumption of a material per unit of gross domestic product (GDP) tends to take an inverted U-shape function. That is, when a material is introduced, it enters the economy in replacement of other materials and creates new uses, and hence its consumption increases relative to total industrial activity. However, as the material becomes more widely used, substitution for other materials declines, and resource-saving technology causes its consumption per unit of output to decline. Some theorists argue that this behavior of consumption per unit of GDP reflects a functional relationship in which the independent variable is GDP per capita. Others argue that this behavior of a particular material represents the phenomenon of substitution of a material, followed by resource conservation, as indicated above. Regardless of the theoretical explanation, the record shows clearly that a material will reach a stage where its use per unit of total output will be declining (see *Intensity of Use of Materials*).

2. Materials Supply

The characteristic that distinguishes the supply of raw materials from that of other goods is that raw-materials supply functions are affected by nature. Mineral raw materials must be produced from deposits in the earth's crust and other materials must be produced from the productivity of the land. With respect to minerals, this creates a multistaged production function rather than the single-stage function discussed in the standard economic theory of supply. The stages of mineral production involve: discovery of a deposit (exploration); the determination of mining and metallurgical techniques to use the deposit discovered, since minerals occur in the earth's crust in many different physical and chemical states, and each deposit may require substantial research expenditure (feasibility stage); construction of the mine and processing facilities to win the mineral (development); and smelting and refining to produce the final material. Another characteristic of mineral supply is that there is no choice as to where the primary mineral is to be produced—it must be produced where it occurs. In many cases investment is required in new transportation facilities to service the mineral deposit and this forms part of the cost of producing and marketing the mineral.

In the short run, mineral capacity is fixed and there is little flexibility on the upward side for expanding production. This is because many mines, smelters and refineries are designed to operate continuously, so that increasing output beyond design capacity is very costly. There may also be bottlenecks in the transportation system that lead to short-run inability to increase output. Short-run mineral supply functions tend to be highly inelastic with respect to output expansion beyond the capacity level.

In the long run, however, exploration is constantly identifying new mineral deposits, which can be developed as long-term supply increases become appropriate. Exploration reveals new deposits that are lower cost, the same cost and higher cost than existing deposits. The long-term mineral functions tend to be quite elastic.

At any point in time, the deposits of minerals can be classified according to the degree of geological knowledge and the cost of producing the raw material from the deposit. This cross classification is at the basis of resource deposit characterization (see *Resource Appraisal*).

In the nonfuel area, reserves (i.e., deposits that are geologically identified and economically viable) typically represent a sufficient supply for a few decades at current production levels, while known resources and estimates of potential resources provide for a much longer period. The increments of minerals supply can come from: discovery of resources that are economic and therefore increase the reserve base from which future minerals supplies will come; technology that reduces the cost of sub-marginal deposits; and increases in real prices for the mineral.

Supplies of mineral raw materials have become available at decreasing real cost over the last century. Not only have the real costs been reduced because of discovery of new mineral resources, but also because of developing technology that allows the mining and concentration of the material at lower costs (see *Prices of Industrial Minerals: History*; *Prices of Metals: History and Current Problems*).

3. Market Structure

Economists argue that in a free market the structure will be endogenous; that is, the structure will be the most efficient allowed by the conditions of supply. Many producers with a homogeneous product will result in a competitive market. A few producers with either a homogeneous or differentiated product will organize as an oligopoly, with a producer-price system or a price-leader system. A single producer will act as a monopolist. However, even with as few as one existing producer, free entry for potential producers will make the market contestable, resulting in market outcomes the same as those for a competitive market. The word competitive will be used to refer to both competitive and contestable markets. During the 1970s and 1980s, the growth of the organized commodity exchanges made the world markets for traded materials contestable since any potential producer had access to a market for sales.

In materials markets, all kinds of market structures exist, although competitive markets have replaced oligopolistic markets for many commodities. Copper, lead, nickel, steel and aluminum are major examples. The markets for new advanced materials tend to be monopolies until the markets reach a sufficient size to attract potential producers making them competitive in nature.

Price and production behavior are different for producer price and competitive markets. Figure 2 illustrates the behavior of these two market structures during the periods of a business cycle. Note that because of the derived demand nature of the raw material, there is a very high income elasticity to the business cycle, represented by the shifting in the demand curves shown in Fig. 2. Competitive markets show high price variability relative to production variability, while producer markets show low price volatility relative to production volatility.

The behavior of state-owned firms cannot be explained within the context of standard economic production theory. These firms will, by their nature, have multiple objective functions rather than profit maximization as the overriding goal, and may tend to continue production in spite of falling prices in order to, for example, maintain export earnings or domestic employment. Unfortunately the state of economic theory is not such that the behavior of the state-owned firms can be integrated with the behavior of privately owned firms operating under different market structures. This is a serious problem in the explanation of mineral prices and production in the institutional setting of the 1980s. Note that state-owned firms must be price takers in competitive world markets, even though they maintain control of their production to achieve nonprofit goals.

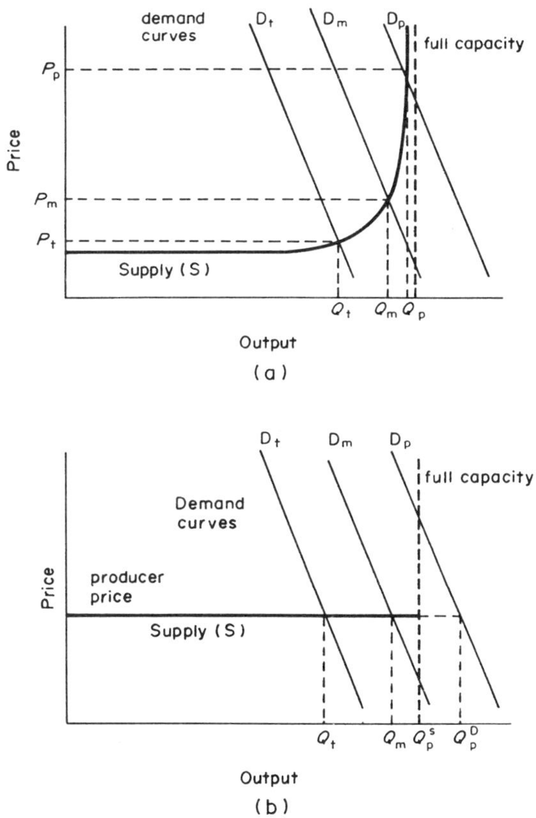

Figure 2
Nature of mineral supply and demand in (a) a competitive market, and (b) a producer market, over the business cycle. The subscripts t, m and p refer to the trough, median and peak of the business cycle, respectively (after Tilton and Vogely 1981)

4. Metal Market Behavior

The characteristics of the short-run metal demand and supply functions lead to instability in price, production and revenue from metal operations. Table 1 presents instability indexes for related metals for the period over the early 1950s to the mid-1970s covering mine production, primary production, prices and revenues. The instability index is the average absolute percentage deviation of price, output and total revenue for each year from the trend value for that year, measured by a five-year moving average. The indexes shown confirm in general the theoretical explanation of the behavior of competitive and producer markets. The competitive markets show a higher instability in price and revenue, while the producer markets show a relatively high instability in production.

Instability in mineral markets is of a different nature to instability in agricultural markets. Mineral market instability arises from the shifts in the short-run demand curves related to overall business activity, whereas the instability in agricultural markets comes from the fluctuating size of the annual harvest. Thus, while in agriculture low prices are associated with high production, in the minerals industries low prices are associated with reduced demand and therefore reduced output. Revenue instability, then, is very much higher for the minerals sector than for the agricultural sector.

Table 1
Instability indexes for selected metals by market type from the early 1950s to the mid-1970s

Market type	Copper	Aluminum	Steel
Ore production			
producer market	6.17	8.74	8.49
competitive market	4.06	6.30	6.81
Primary metal prices			
producer market	11.18	3.13	1.25
competitive market	15.21	4.07	12.53
Revenue			
producer market	9.69	6.40	6.64
competitive market	19.44	8.18	18.36

Source: Tilton and Vogely (1981)

See also: Prices of Materials: Stabilization; Prices of Metals: History and Current Problems

Bibliography

Baumol W J, Panzer J C, Willig R D 1982 *Contestable Markets and the Theory of Industry Structure*. Harcourt Brace Jovanovich, New York

Tilton J E 1977 *The Future of Non-Fuel Minerals*. Brookings Institution, Washington, DC

Tilton J E, Vogely W A (eds.) 1981 Market instability in the metal industries. *Mater. Soc.* 5: 243–354

Vogely W A (ed.) 1985 *Economics of the Mineral Industries*. American Institute of Mining, Metallurgical and Petroleum Engineers, New York, pp. 99–179; 383–415

Vogely W A 1987 Major metal market structure. *Mater. Soc.* 11: 408–9

W. A. Vogely
[Pennsylvania State University, University Park, Pennsylvania, USA]

Prices of Metals: History and Current Problems

The level and stability of metal prices are affected by traditional economic factors of supply and demand. Moreover, institutional factors such as the method of marketing and the structure of the market have come to play an increasingly important role. In this article, metal prices are considered from both a long- and a short-run point of view. After a discussion of historic trends, recent developments in the methods of metal-price determination are examined. In this examination, particular attention is paid to the causes and consequences of increased price instability.

1. Historic Metal Prices

This section takes a long-run view and analyzes a century of metal pricing. Long-run determinants are clearly different from the factors that influence month-to-month fluctuations—the subject of Sect. 3.

The behavior of metal prices often seems erratic and hard to predict. In the short run, prices fluctuate with random shocks in supply and demand, making it difficult to detect any regular behavior patterns. When examining the history of prices over a century, however, systematic tendencies can be seen and useful generalizations about price trends can be made, both for individual metals and for the metals taken as a group.

Current-dollar price paths for many metals are broadly similar to one another. In general, prices reached a minimum around the turn of the century. There were sharp price increases around World War I, dramatic declines during the depression and increases again after World War II. The 1950s and 1960s, which were periods of relatively stable prices, were followed by the commodity boom of the early 1970s, when the prices of many metals reached unprecedented highs only to fall precipitously. The 1980s have been characterized by two periods of rapid increases and declines. As with earlier booms, some of the activity was fuelled by speculation. The silver market played a particularly important role in the boom of 1980.

Current-dollar prices are highly correlated with the economy-wide rate of inflation. To detect patterns that are unique to the metals, therefore, it is necessary to correct for more general price fluctuations. That is, any trends that are purely inflationary or recessionary must be removed. This is achieved by dividing each price series by an overall deflator such as the US wholesale price index. The result is a set of real or constant-dollar prices—prices expressed in terms of a hypothetical dollar of constant value. The discussion in the remainder of the section will be in terms of constant-dollar prices.

Each metal has a unique price path. Certain general tendencies, however, emerge from the diverse behavior patterns. The most important factors that affect long-run trends in metal prices are growth of metal demand, changes in mining and metallurgical technology, discovery of new deposits, and depletion of high-grade ore bodies. The first three factors tend to predominate in the early phases of a metal's life cycle, whereas the fourth may be more important later on. Thus, it can be expected that real metal prices will initially fall, but, unless close substitutes can be found, they may eventually rise. Certainly the empirical evidence points to a decline in real metal prices up until the middle of the twentieth century (Barnett and Morse 1963, Slade 1982a). The reasons why this might be so are therefore summarized.

In the initial phase of a metal's commercial history, many factors can contribute to falling costs. Metal mining and processing is characterized by substantial economies of scale. As a consequence, it is possible for one mine to supply ten percent of current consumption. When a market is new, therefore, mines and smelters tend to be too small to be fully efficient. With significant scale economies, the growth in demand as additional metal uses are discovered allows mining companies to exploit cost-lowering economies of scale.

Learning-curve effects can also be important. As companies gain experience in mining, smelting and refining, they may discover new techniques that lower costs. Costs can therefore fall with cumulative metal production. In addition, there may be improvements in extractive, metallurgical and transport technology that are not specific to an industry but may be applicable across broad groups of industries. For example, a decrease in shipping costs can lower the delivered price of all metals.

Technical change can take place on the production or on the consumption side of a market. If it occurs on the production side, it generally lowers costs and therefore price. It if occurs on the consumption side, its effect is more ambiguous. It can, however, increase metal demand and allow mining companies to exploit economies of scale. When this is the case, the change will exert downward pressure on price in the long run.

The discovery of new deposits can increase metal supply and cause prices to fall. Even when new discoveries do not lower prices immediately, the expansion of known reserves can postpone the onset of higher costs due to the depletion of high-grade ore bodies.

There are thus many factors that can lead to falling real metal prices. Aluminum is the commodity which, in actuality, exhibits the most dramatic real-price declines.

A period when the trend is downward is often followed by one where the trend is flat or nonexistent. This period can be very long, as for lead and zinc, or fairly short lived. In this trendless period, production growth rates are often steady but not dramatic, and the average grade of ores mined falls only slowly.

Unless close substitutes can be found, however, a phase of rising prices can occur. As high-quality deposits are depleted, newly discovered deposits may tend to have lower metal content, be more deeply buried, of lower grade or in thinner veins. In addition, if prices are to rise, the rate of introduction of cost-lowering techniques must be insufficient to offset the rise in costs due to depletion. This has long been the case for tin and mercury. An increase in price will in general cause consumers to conserve on use and therefore consumption growth is expected to be small or negative for metals that experience rising costs due to depletion.

Figure 1 illustrates four possible constant-dollar metal-price paths. Superimposed on the paths are fitted linear and quadratic trends. The four metals illustrate falling, U-shaped, stable and rising real prices. These categories, however, should not be thought of as distinct patterns but perhaps should be considered to be different phases in the life cycles of the respective mineral commodities.

The real price of aluminum, a modern metal whose ores are very abundant, has been falling during most of its commercial history. Technological advances combined with economies of scale have made this decline possible. Copper, in contrast, displays a U-shaped price path, initially falling but eventually rising in real terms. Lead prices have been trendless for a century. Technical breakthroughs and grade declines

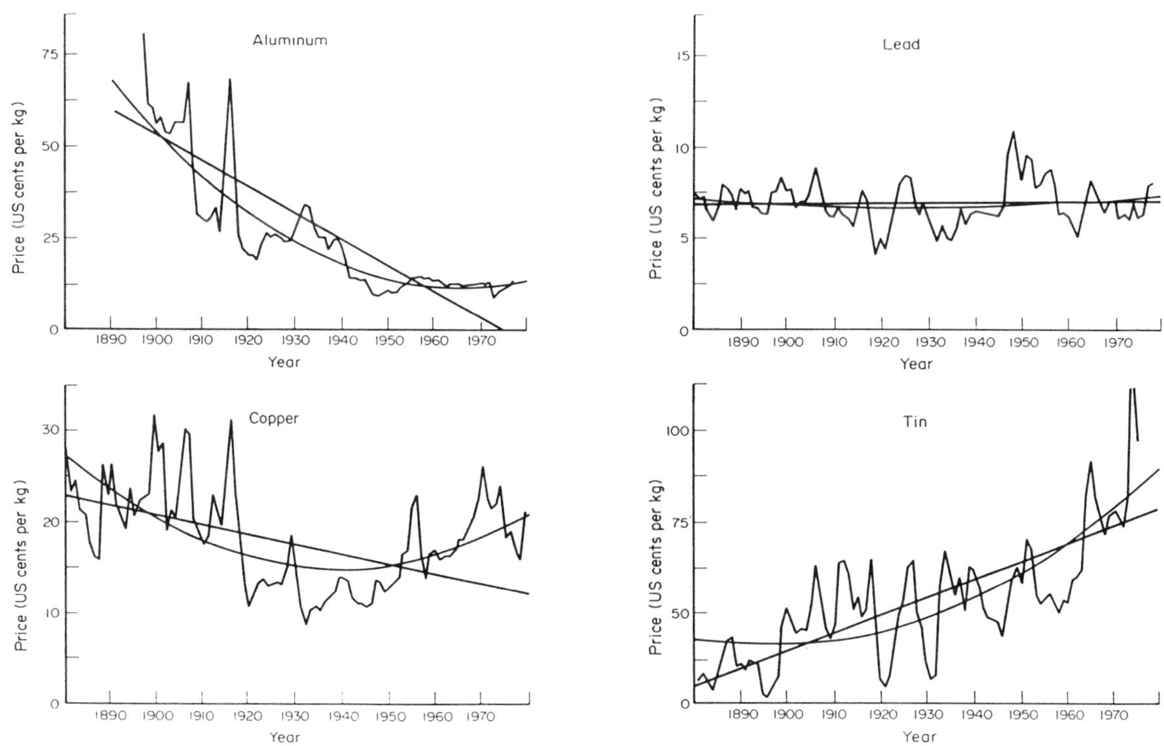

Figure 1
History of selected real metal prices

have both been modest and have almost exactly offset each other in determining lead-production cost. Finally, tin prices display a distinct upward trend. There have been few significant advances that have lowered tin-mining costs whereas the decline in the grade of tin ores mined has been both steady and sizeable.

The copper industry illustrates the historic counterbalancing influences of improvements in technology and deterioration in ore quality in determining production cost. In the first 75 years of the twentieth century, the average grade of copper ores mined in the USA declined from about 5% to 0.7%. In spite of this decline in grade, real copper price fell until about 1940. The fall in price was made possible by technological developments in the early part of the century, particularly the advent of large earth-moving equipment, which made possible the open-pit mining of extremely low-grade ore bodies, and the discovery of froth flotation, which made concentration of low-grade sulfide ores economical.

By 1940, however, the switch to this new technology had reached its natural limits. Since that time, in spite of new mining and processing techniques, the decline in grade has become an important factor in determining cost. In recent years, technological emphasis has been on energy-saving features and improved environmental acceptability. The former lowers costs but the latter tends to increase them. Both of these factors have contributed to post-World-War-II trends in real copper prices.

Petersen and Maxwell (1979) have documented the history of the relative impact of ore grade and technology on the cost of other metals (silver, tin, lead, zinc and iron) and assert that these factors have been much more important than the discovery of new deposits in the historic determination of metal prices.

Thus far we have examined long-run metal-price trends, a subject that is analyzed in greater detail by Slade (1982a, b). Long-run trends, however, are often swamped by short-run fluctuations. Within-decade fluctuations are surprisingly large relative to the 100-year price history. Section 3 examines these short-to-medium-run price fluctuations in greater depth. Prior to this examination, however, the principal methods of metal-price determination are described.

2. *Metal-Pricing Systems*

This section analyzes the changes that have occurred in nonferrous-metal pricing since the late 1970s. It is not a discussion of how price levels are determined. Instead, it attempts to explain why the North American producer-pricing system has weakened and to assess the consequences of increased reliance on metal exchanges. Particular attention is paid to the causes and consequences of the price volatility that has characterized nonferrous-metal markets. This issue is of concern to producers and consumers alike as it makes planning for orderly growth difficult and contributes to the gluts and shortages that have plagued these industries.

Nonferrous-metal sales in North America have traditionally been conducted under a system known as producer pricing. Producer prices are set by the major firms in the industry. These prices change only at discrete intervals and tend to have smaller variances than exchange prices. Exchange or terminal prices, in contrast, are related to price quotations on the major metal exchanges, principally the London Metal Exchange (LME). These prices have traditionally been more influential outside of North America.

In the late 1970s and early 1980s, radical changes occurred in the pricing of many nonferrous metals. This period saw the virtual demise of the producer price of copper and the introduction of aluminum and nickel contracts on the LME. These events were followed by the suspension of publication of producer prices for these commodities. Even those industries where producer prices remained relatively stronger were affected. The same period saw a tendency for the producer price of lead to follow exchange prices more closely and, more recently, the widespread practice of discounting the producer price of zinc.

2.1 Exchange Prices

Many commodity exchanges exist throughout the world. By far the most important for metal pricing, however, is the London Metal Exchange. This market dates back to the middle of the nineteenth century, when a need for forward dealing and hedging developed. It was at this time that the practice of trading in a ring began. It became the custom for some member to draw a large chalk circle on the floor of the coffee house where traders met and to call "ring, ring." Members then stood in their recognized places and cried their bids and offers across the floor.

Superficially, the LME today closely resembles its nineteenth-century ancestor. Membership in the Exchange is still limited to 36, the number of seats in the curved wooden benches that form the modern "Ring." Trading is still carried out by dealers calling out bids and offers across the floor. Bids and offers may not be directed at individual firms but must be to the entire Ring, a custom which is known as the open-outcry system. Dealings in each commodity are still limited to two five-minute intervals in the morning and two in the afternoon. Official prices are usually selected as the last best bid and offer prevailing at the close of the Ring. (The word Ring has two meanings: the physical circle of curved benches and the five-minute trading intervals for each commodity.)

In spite of superficial similarities with the past, the LME underwent a fundamental reorganization in 1987. This reorganization was due to general misgivings about the nature of regulation of investment business in the UK. It was further aggravated by the tin crisis of 1985, when LME tin trading was suspended indefinitely. For an analysis of the tin-market crisis, see Anderson and Gilbert (1988).

Three major changes were instituted. The first was the switch from a principals' market (a market where members act as principals for the transactions that they conclude across the Ring and with their clients) to a clearing-house system. A clearing house is an independent body that clears and guarantees business transacted between member brokers.

The second change was the introduction of margins. The LME system of margins is unusual, however, in that it applies only to the net position of brokers. The clearing house looks at a member's total exposure when calculating margins. In-credit positions can therefore compensate losses on risk positions. The third innovation was the introduction of formal options.

The only real competitor to the LME in terms of metal trading is the Commodity Exchange of New York (COMEX). With the exception of copper, however, it is of secondary importance. Nevertheless, a brief description of COMEX is in order.

Membership in COMEX is without formal limit and usually numbers in the hundreds. COMEX has always required margins, and its clearing-house system was established in the 1930s. The COMEX clearing system requires a daily cash settlement in contrast to the due-date settlement system on the LME. A final distinction is that the volume of speculative activity is greater on COMEX than on the LME, where "legitimate" trade dominates.

2.2 Producer Prices

Producer prices are those set by the major North American firms. These prices change only at discrete intervals and have smaller variances than exchange prices. Most often, the producer price is the price of refined ingot. Other products and shapes may sell at premiums and discounts, but their prices will usually be tied to this single price.

The exact method of producer-price determination varies by commodity. There are, however, broad similarities. The North American producer price of copper is therefore discussed in some detail to give a feel for this general class of pricing. Because the producer price of copper is no longer the predominant North American copper price, the following description is based on the period of the 1970s.

The producer price of copper was that set by the major primary producers and was the price at which most domestic refineries sold copper metal to affiliated and independent fabricators. In the mid-1970s, about 75% of refined copper in North America was sold at this price. The producer price represented a common set of price quotations for delivery of refined copper to any place in the nation. This price was based to a large extent on production cost and was moderately stable.

Some refinery shapes sold at slight discounts and others at premiums, but the prices of all shapes were tied to a single price—the producer price. For example, copper-product prices (the prices of sheet, tube, wire, castings and so forth) followed the price of refined copper closely. It therefore seems reasonable to speak about *the* producer price of copper.

It is important to note that the producer price is a price that is announced by the companies for current sales. It is not a long-term-contract price. Contracts can be based on this price, but they can also be based on exchange prices. The producer price guarantees neither security of supply nor stable future prices. Nevertheless, the practice of the leading firms in times of scarce supplies has been to serve existing contracts and regular customers first.

2.3 Relative Merits

Both exchange and producer-pricing systems have their virtues and their flaws. This subsection summarizes the salient pros and cons of each. The discussion, however, should not be interpreted as an attempt to pick a winner or to advise producers which system to use. It should become clear that the choice of system is determined to a large extent by structural conditions within the industry. It should also be clear that there is no single system that works best for all commodities.

A major difference between the two systems relates to the players in the game. Producer prices are determined by producers and, to a lesser extent, by consumers. On commodity exchanges, new actors are introduced—brokers, merchants and speculators. This introduction has both benefits and costs.

Hedging is one of the benefits. Hedging is defined as the establishment of an opposite position on a futures market from that held and priced in the physical commodity. The practice of hedging, by removing the risk of price fluctuations, provides insurance to both consumers and producers. In fact, the desire to hedge was one of the primary reasons for the founding of the LME and other exchanges.

Speculation can be both a cost and a benefit. In markets that are primarily speculative, prices may bear little relationship to underlying production cost or value in use. This, however, need not be the case. When commodity markets are dominated by legitimate trade, their prices, which are sensitive to supply-and-demand factors, possess the virtues common to all competitively determined prices.

A speculator enters a market to realize a profit. This in itself is not a bad thing and as a consequence the speculator provides services. In times of surplus, producers and consumers can always find a buyer for unwanted metal, and in times of shortage, they can find someone willing to sell. Rationing, therefore, never exists on commodity markets.

The most common complaint about exchanges is that prices are too volatile. Volatility creates problems if unstable prices provide the wrong signals. High prices, for example, may be taken as signals of long-run profitability and may therefore lead to excessive investment in new capacity. Low prices, in contrast, may signal bad times ahead and may discourage exploration more than is warranted by the underlying fundamentals.

Excessive volatility may also cause consumers to substitute away from metals sold on exchanges. If consumers prefer stable prices for planning purposes, and if substitute materials are sold under arrangements that facilitate price stability, metal-conserving technological developments can be triggered in the using sectors.

A significant advantage of exchange-determined prices is their transparency. LME and COMEX prices, which are published continuously, have worldwide visibility. In addition, unlike producer prices, they are always true transactions prices and are uniform across customers.

Other advantages include the fact that contracts are never broken. (An exception to this statement is when governments fail to honor their obligations, as may have happened with the tin crisis.) The LME, for example, does not have a *force majeure* clause in its contracts. An exchange therefore provides a dependable contractual arrangement in which everyone can participate at prices that are the same for all.

Another useful function of exchanges is their inventory-carrying role. With an exchange system, there is never a need for a potential seller to hold unwanted stocks. When excessive inventories cause the contango (the positive difference between the forward and spot prices) to widen, banks can step in. They do this because, by holding inventories, they can earn higher-than-normal returns. The size of the contango therefore does not normally exceed the cost of financing, insuring and storing metal.

Unfortunately, an exchange-pricing system does not insure against cartel manipulation and prices that are persistently above marginal costs. One need only consider the 1985 tin crisis to conclude that prices determined on exchanges are not synonymous with competitive prices. Nevertheless, it is generally the case that an exchange price more closely approximates the competitive norm than does its alternative.

In contrast to exchange prices, producer prices, which are normally based on fundamentals, are relatively stable. In addition, by doing away with the need for a broker, a producer-price system can foster a bond between buyer and seller. When the buyer is a vertically integrated subsidiary, the producer can, to some extent, fix the price of the basic raw material used by its downstream operation. In the extreme case of complete vertical integration, there is no useful role for an exchange to play.

Producer prices also facilitate control over market conditions. For example, by setting a price below the LME price, producers can clear warehouse inventories. Conversely, by setting a price above the LME price, they can provide stability in a falling market.

Control over price can, however, be costly if improperly exercised. For example, most specialists agree that the producer price of nickel was set too high in the late 1970s, a fact which contributed to its own downfall. Prices can also be set too low, as was the case with the Climax price of molybdenum in the late 1970s. By setting the price too low, AMAX exacerbated the already existing shortage.

When the producer price is the only price and it is suspended, the result is a free-for-all (a situation that occurred in the nickel market in 1977). Markets without well organized commodity exchanges are therefore more vulnerable to actions by a single firm or small group of firms.

At this point, it is perhaps helpful to list circumstances that tend to lead to greater reliance on exchange markets. These include

(a) an increase in the number of sellers,

(b) an increase in the number of buyers,

(c) a decrease in vertical integration,

(d) an increase in the geographic integration of markets,

(e) an increase in product standardization, and

(f) an increase in secondary and by-product production.

A glance at these factors should convince the reader that the greater reliance on exchange prices that has characterized the 1980s is an effect more than a cause. As metal markets have become more competitive and more international, the inevitable result has been a weakening of the traditional producer-pricing system.

3. *Factors Affecting Price Stability*

This section analyzes the principal determinants of price volatility in metal markets. Primary attention is given to the factors that vary across commodity markets. For example, all metals are industrial goods and therefore their prices are affected by fluctuations in industrial activity. Second, all are durable and therefore their prices are affected by inventory swings. Finally, all are capital intensive and subject to long delays in expanding capacity. These common factors, although very important, are not discussed in what follows.

3.1 Marketing Method

When commodity prices are set by producers, all purchasers of the good are consumers. When they are sold on exchanges, in contrast, new actors are introduced—brokers and speculators. Differences in motives across agents can lead to differences in price behavior.

Do speculators destabilize prices? Perhaps the first to ask this question was Adam Smith (1776). Since that time, many have written on the subject. The conventional wisdom says that speculation stabilizes prices. Speculators, if they are to realize a profit, must buy cheap and sell dear. Low prices should therefore be increased and high prices lowered. It would thus seem that speculation can destabilize prices only if speculators, on average, lose money.

The question, however, is not quite so simple. Speculators do not buy when price is low; they buy when the chance of price appreciation is high and sell when the chance is low. Recently, many economists have studied this question. For example, Hart (1977) notes that sophisticated speculators can earn a profit by exploiting the naive forecasting rules of less-sophisticated agents. And Newbery (1984) demonstrates that if a producer has market power, then even if all agents are equally rational and have the same information, it can pay the dominant producer to destabilize the spot market by engaging in destabilizing speculation.

Finally, although unlikely, it is possible for hedging to destabilize prices. By providing participants with insurance, hedging can encourage them to engage in more risky activities.

3.2 Horizontal Market Structure

Horizontal market structure refers to the number and size distribution of firms in a market. An industry with few firms or with a dominant firm that controls a large share of the market is said to be horizontally concentrated.

There are many reasons why prices might be more stable in concentrated industries. One of the most important is the availability of instruments to the firms in the market. In a competitive market, firms are price takers. As prices fluctuate, producers can alter the level of their output but they cannot control price directly. In a concentrated industry, in contrast, firms can choose to vary price, output, inventories or some combination of the three in response to fluctuations in demand. This higher degree of control usually results in more stable conditions.

When price is to some extent controlled by individual firms, managers must weigh the costs and benefits of changing prices. Economists have focused on two factors that might lead managers of firms in concentrated industries to choose not to adjust prices continuously and fully. Both are related to the costs associated with price changes. The types of costs are fixed and variable.

Fixed costs are incurred each time that a price is changed. Such costs are independent of the size of the change. When, for example, a new price schedule must be printed, a fixed cost is incurred. With fixed costs, there is an incentive to adjust prices infrequently but by large amounts.

Variable costs, in contrast, increase with the size of the price change. If it is costly for consumers to stay informed of relative prices of substitute materials, they may prefer frequent, small price changes. It is clear that behavior of producer prices in the nonferrous-metal industries more closely resembles the predictions of the fixed-cost models.

Inventories can also play an important role in stabilizing prices in concentrated industries. Dominant firms often hold much larger inventories than their market share would warrant. This behavior can be explained as follows. The benefits of price stabilization accrue to the industry as a whole whereas the costs are private. Price stabilization is therefore to some extent a public good. Only firms with a large share of the market can capture a sufficient fraction of the benefits to warrant engaging in private stabilizing activities.

3.3 Supply Factors

Factors affecting price stability on the supply side of the market include cost stability, recycling activity and by-product production.

Cost is clearly an important determinant of both the level and stability of price. For example, one of the factors that has contributed heavily to changes in relative metal prices since the early 1970s has been the cost of purchasing energy. In addition, the availability of cheap energy sources has been a major determinant of the location of new refineries and smelters that have been constructed in recent years.

Changes in wage costs across regions have also been very important in determining the competitive strengths of producing countries. For example, the USA has been more successful than Canada in controlling labor costs. This has enabled the USA to remain competitive in spite of a decline in the quality of many metal ores mined.

Secondary production, or recycling, is an important source of supply in many nonferrous-metal markets. Recycling costs are sensitive to the quality of material processed. For this reason, the recycler is often the marginal producer. If there is little secondary production when prices are low, coupled with heavy production when they are high, recycling will stabilize metal prices. If, on the other hand, there is a large recycling sector processing very low-cost material, prices will have to fall very far to clear the market in a downturn. The effect of increased recycling on price stability is therefore ambiguous.

Joint production, which is common in metal mining, is frequently associated with price instability. Metals occur in ores in roughly fixed proportions. This means that it is difficult to vary the proportions of the different metals that are recovered from a given ore. In addition, a by-product is (by definition) of secondary importance. This implies that by-product supply is apt to be more sensitive to the price of the primary commodity than to its own price. Markets where by-products are important can therefore be expected to have unstable prices.

3.4 Demand Factors

On the demand side of the market, there are two important factors that affect price volatility: the stability of end-use markets and the ease with which other materials can be substituted for metals in downstream production.

First, consider end-use stability. In recent years, two factors have adversely affected the consumption of all major metals. The first is a slowdown in the rate of growth of GDP in the West. Western economic activity grew at a rate of 5.2% per year in the 1960–73 period but slowed to a rate of 2.6% per year after 1973. In addition, the recessions that occurred in the mid-1970s and in the early 1980s were especially severe for the mining industries.

The second adverse factor is the recent decline in the intensity of metal use, where intensity is defined as the ratio of the dollar value of metal consumption to GDP per capita. This ratio grew by approximately one percent per year in the 1960–73 period. In contrast, the intensity ratio fell by about two percent per year after 1973 (Tilton 1989). As these trends were to a large extent unanticipated, they caused adjustment difficulties and metal-price instability.

Next, consider the ease of substitution. If metals were combined with other inputs in fixed proportions, end-use stability would be the sole determinant of demand stability. Fixed-proportion production, however, is rare. When substitution is easy, a high metal price will trigger a switch to an alternative material, which will facilitate price stability. On the other hand, variations in the price of the substitute material can also lead to changes in metal use which can have the opposite effect. For this reason, the net effect of flexibility in downstream-industry production processes is ambiguous.

3.5 Time-Period Factors

Two important determinants of price stability vary by time period but not by commodity. These are exchange rates and inflation rates.

Metals are traded in international markets. Both the dollar price of metals quoted in sterling and the sterling price of metals quoted in dollars will fluctuate with changes in the dollar/sterling exchange rate. In addition, mining costs are paid in many currencies. As sterling and dollar exchange rates fluctuate with respect to these currencies, changes in metal prices may be expected. Periods of exchange-rate volatility are therefore expected to be associated with metal-price instability.

Finally, high rates of inflation should also be associated with unstable prices. As the overall price level changes, firms must adjust their prices to reflect changes in input costs. As the nature of adjustment to inflation varies from industry to industry, both absolute and relative-price volatility can be associated with high inflation rates.

4. Concluding Remarks

This article examines both long-run trends and short-run fluctuations in metal prices. Price trends are very important. From a practical point of view, however, volatility is often of greater concern. For example, the US Nonfuel Minerals Policy Review (1979) found that on average nonferrous-metal prices were four times more volatile than prices in the economy as a whole. The gluts and shortages that result from severe price fluctuations have therefore been a traditional preoccupation of both producers and consumers.

Moreover, in the last decade, fundamental changes have occurred in metal pricing. Not only has the decade witnessed a tendency towards increased reliance on commodity exchanges, but these exchanges have themselves undergone radical reorganization. The switch to exchange-based pricing has been associated with increased price volatility, a cost that must be balanced against the considerable advantages of well-organized futures markets.

See also: Economic Theory of Mineral Resources and Markets; Prices of Industrial Minerals: History; Prices of Materials: Stabilization; Prices of Materials: Theory; Trade in Materials, International

Bibliography

Anderson R W, Gilbert C L 1988 Commodity agreements and commodity markets: lessons from tin. *Econ. J.* 98: 1–15

Barnett H J, Morse C 1963 *Scarcity and Growth.* Johns Hopkins University Press, Baltimore, MD

Hart O D 1977 On the profitability of speculation. *Q. J. Econ.* 91: 579–97

Newbery D M 1984 The manipulation of futures markets by a dominant producer. In: Anderson R W (ed.) 1984 *The Industrial Organization of Futures Markets.* Lexington, Lexington, MA

Petersen U, Maxwell R S 1979 Historical mineral production and price trends. *Min. Eng. (N.Y.)* January: 25–34

Slade M E 1982a Trends in natural-resource commodity prices: an analysis of the time domain. *J. Environ. Econ. Manage.* 9: 122–37

Slade M E 1982b Cycles in natural-resource commodity prices: an analysis of the frequency domain. *J. Environ. Econ. Manage.* 9: 138–48

Slade M E 1988 *Pricing of Metals.* Centre for Resource Studies, Kingston, Ontario, Canada

Slade M E 1989 The two pricing systems for nonferrous metals. *Resour. Policy* 209–20

Smith A 1776 *An Inquiry into the Nature and Causes of the Wealth of Nations.* Reprinted 1937, Modern Library, New York

Tilton J E 1989 Changing trends in metal demand and the decline in mining and mineral processing in North America. *Resour. Policy* 15: 12–23

US Nonfuel Minerals Policy Review 1979 *Background Papers.* US Department of the Interior, Washington, DC

Wolff's Guide to the London Metal Exchange, 3rd edn. 1987 Metal Bulletin Books, London

M. E. Slade
[University of British Columbia,
Vancouver, Canada]

Process Analysis

We use the phrase "process analysis" to mean the construction and use of systems models (Checkland 1981) to support the analysis of material and energy transformation processes and the socioeconomic structures that control and influence these processes. The socioeconomic resource systems of interest can be globally, regionally, nationally, subnationally, plant or unit operation specific in nature. The underlying processes can be either naturally occurring or man-made. In particular, process analysis aims at explicating the evolution of the patterns of physical flow–physical stock in socioeconomic resource systems. At the unit operation level, process analysis is used to support design and construction of components that are physically viable systems, that is, systems that are feasible because they are consistent with the physical laws of nature. These unit operations use some form of stock, appropriately controlled, to transform a set of flows of energy and material to another set. At the plant level, process analysis supports the design and construction of unit operations which can be coordinated for a specific purpose, that is, the production of some object or objects. At the corporate or company level, process analysis is used to support ongoing decision making concerning the operation of extant plant and additions to the corporate portfolio. At the subnational level, process analysis is used to support decisions concerning the coordination of regional development with all its concerns about zoning, tax base, employment and environmental impact. At the national level, process analysis is used to guide planning and policy as it partly represents the relationship of socioeconomic activity to the national resource base of a nation. At this level the concern with the resource base is both in its passive aspect as a source of materials and in its reactive aspect as a set of naturally occurring processes that must be used to ameliorate wastes. At the regional and global level, process analysis is used to examine potential trade patterns involving flows of materials, energy and service. This analysis supports decision making at the national level concerning trade agreements and industrial development strategy.

Process analysis is dynamic. It is dynamic in two senses: one is that it links levels, and the second is that the process-based systems models evolve in time as they are used. This dynamic time dependence of process analysis has two aspects: one is involutionary (i.e., how to exercise control) and the other is evolutionary (i.e., how to plan for development). The dynamic level linking of process analysis brings into relief an important distinction in the levels at which it is applied. That distinction can be cast as the ratio of involutionary to evolutionary concern of the decision making at each level. This ratio is low at the unit operation level, high at the plant and corporate level and decreasing through the remainder of the higher levels.

1. What is a Process?

The concept of a process (Jantsch 1981) is primordial. It can be defined as that which transforms one set of flows of material, energy and information into another. This implies that a process exists in an environment from which these flows arise and to which they go. What distinguishes a process from its environment is an arbitrarily chosen boundary. Although the choice of boundary is arbitrary, the wisdom of that choice will render analysis in a given context simple or difficult.

Networks of processes are again processes. This leads to concepts of process hierarchy, aggregation and embedability. For example, a unit operation can be considered to be a set of basic physical processes coordinated in some way. In turn, a plant can be seen as a set of unit processes, an economic sector as a set of plants, a nation as a set of economic sectors, and so on. In fact, societal decision-making processes are related to this process hierarchy, as is shown in Fig. 1.

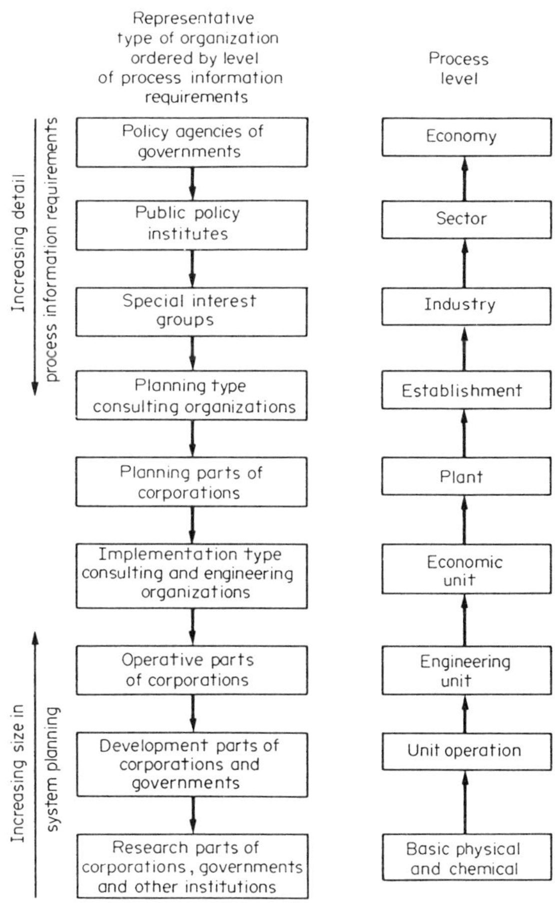

Figure 1
Hierarchy of processes

Depending on the time horizon over which an analysis is required, some processes operate so slowly that there appear to be objects, which, for our purposes, we shall refer to as funds. These funds, however, themselves undergo change and are simply networks (perhaps very complex) of processes. These networks or funds often provide a composite flow of information and energy to a process and such a flow is often referred to as a service. The importance of distinguishing these networks of slow-acting processes from the more active ones is that it simplifies analysis. The introduction of the concept of a fund is simply the addition of another boundary: thus the appropriate selection of boundaries for a given analysis becomes of prime importance in simplifying a system sufficiently so that tractable analysis can be carried out. The distinction of service (energy or information) flows from materials flows is important in process analysis because the principle of mass balance is an important consistency constraint on the correctness of any process description. The distinction of energy and information is also important since the second law of thermodynamics imposes an important consistency constraint on entropy generation. This leaves information flows identified, and that in itself is useful because of the relation of information to the notion of control and the notion of control to that of planning.

To make the above more explicit, consider a chemical reaction

$$A + B \rightarrow C + D$$

being carried out in a reactor vessel. The input flows of materials A and B are transformed by the process represented pictorially by the process node in Fig. 2 into output flows of materials C and D. The process receives an informational input flow from the reactor vessel expressed in the form of containment in space and the input flow of heat energy. The reactor vessel is shown as a fund in Fig. 2 by the triangle. Now the reactor vessel itself is changing slowly by corrosion and thus is essentially replaced. If the analysis in which this process was involved was of short enough duration that reactor vessel corrosion and the subsequent processes of replacement and/or repair were irrelevant, no flow lines of mass would be shown going to the fund. If not, then one would have input and output mass flows to the reactor vessel.

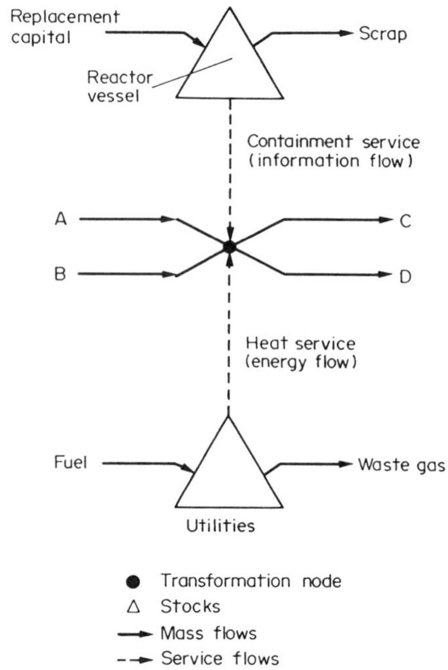

Figure 2
Example of a process topography

2. *Quantification of a Process*

Organized coordinated quantification of processes is an important prerequisite to process analysis. In order to carry out process analysis one requires a well-defined set of processes and an operational model of a connected set of processes. These are what may be called the systems components of the system under consideration in a particular process analysis and are the outcome of process quantification. Two largely independent structures are required for process quantification. These, their interrelation and relationship to process models (process analysis) are depicted schematically in Fig. 3. First is what may be called the topography of a process. This is simply a definition of the connective structure of the flows in a process. It may be thought of as a named diagraph or simply a conventional flowchart or blueprint of the process. It names the flows, denotes their composition and characteristics and indicates their origin and destination. In the construction of a topography an implicit boundary is formed which separates the process from its environment. This topography may be used as a framework to record observations of a process. However, a process description requires information on "how the process works" or a theory of operation of the process. The theory is in practice a model of the operation of the process. A model description consists of the specification of relationships, the definition and the quantification of parameters of the model.

Experience shows that many processes operate similarly from a generic perspective, thus the second structure required for process quantification is a generic model of the process. When a generic model of a process is brought into confluence with a topography of a specific process the result is a data structure appropriate for storing a measurement (calibration) of the process. The data take the form of values for parameters, the connection between the variables in the

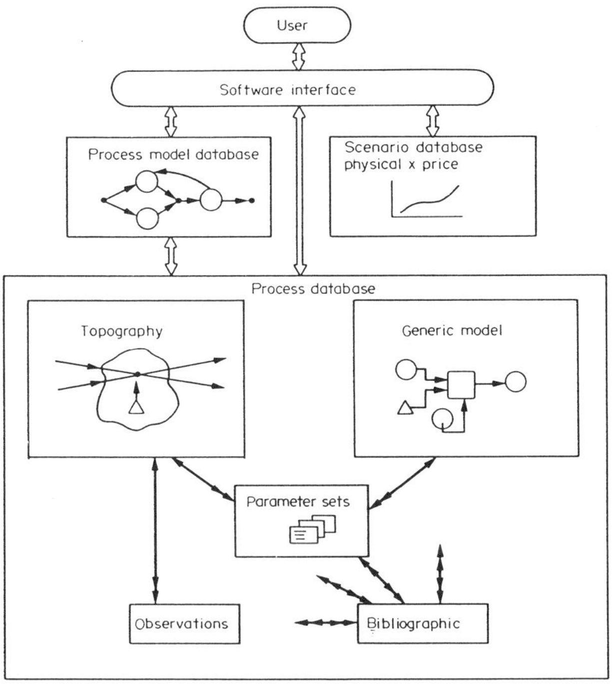

Figure 3
A database approach to process modelling

generic model and the flows in the process topography, and the names of variables and parameters.

3. Process-Based Systems Modelling

The data mentioned in Sect. 2 can be considered to constitute realized generic models, or simply models, of processes. These models can be used as the theory of the process in carrying out specific process analysis. Since generic models also represent a theory of operation for a process, the concepts of control and control variables are brought naturally into the process analysis. In most process analyses, realized generic models are coupled together through various linking rules and these resulting networks become the fundamental tools. These networks of realized generic models are referred to as process-based systems models. It should be noted that the linking rules themselves can be considered to be generic models and again have control variables associated with them. In this way two levels of control can be distinguished, the lower at the level of the process and represented in the generic model of the process, and the higher at the level of the linking rules. This allows process models, at one level of the hierarchy mentioned in Sect. 1, to become the "theory of the machine" at the next higher level and allows for the development of hierarchically co-ordinated systems models (Mesarovíc et al. 1970). It also allows the definition of levels of aggregation for representing processes that are appropriate at different levels in the organizational hierarchy.

Process models, like other model types, come in various subtypes, each suited to its own purpose. Four dimensions of these subtypes that are useful to distinguish in applying the art of choosing a particular subtype for a particular application are:

(a) theory versus data rich,

(b) degree and type of feedback,

(c) degree of buffering, and

(d) optimization versus simulation.

Theory-rich process models are models in which the behavior of the component is taken as *a priori* from some lower level "theory of the machine," whereas data-rich models are those whose component's behavior is estimated from observation. Theory-rich might also be called the white-box approach while data-rich could be termed the black-box approach.

The degree of feedback depends on the openness and complexity of the process model whereas the type of feedback can be user interactive or analytic at one level and integrative or differential at another level.

The degree of buffering is related to the amount of memoryless behavior of various process components in the model, that is, instantaneous communication via flows as opposed to communication via inventories.

Optimization process models are those that contain a large degree of decision making via internal programmed optimization routines as opposed to simulation process models whose decision making is a mix of heuristic and user-interactive.

At this point it is important to observe that a particular process model subtype is not equally useful at all levels of the organizational hierarchy. It is important to recognize this relation between process model subtype and organizational hierarchy as it can be a useful guide in allocating resources for process modelling efforts. As one moves up the hierarchy, one finds that increasing usefulness is obtained by moving to theory-rich, user-interactive, open, highly buffered simulation process models. In a word, one can describe this type of process model as loose compared with the tight models used in the lower reaches of the organizational hierarchy. Of course, this rule of thumb applies only generally and in any specific context may be overridden by other issues.

4. Examples of Process Analysis

There are five broad areas where process modelling has been extensively applied. These are:

(a) investment decision analysis,

(b) residuals management analysis,

(c) resource extraction analysis,

(d) process design, and

(e) process control.

In the first area, a detailed production function is derived from an engineering-based process model capable of articulating the investment decision in some prelimited space of decisions. In other words, questions as wide as portfolio decisions are not being addressed here. Extensive exposition of this area is given by Smith (1966), who addresses both input substitution and market penetration capability. Investment decisions can also be viewed from a broader policy perspective. In this area, process analysis has not been as widely used; however, Ayres (1978) provides an overview of it and a good reading list. The questions addressed here are materials substitution at the macroeconomic level and the degree to which certain technologies should be fostered.

In the area of residuals management, significant work was done in the early 1970s using process analysis. Russell (1973) and Russell and Spofford (1972) both give good overviews and specific examples. The former study deals with a specific technology while the latter deals with economic activity in a geographical region. The use of process analysis here is required by the need to relate economic activity to residual analysis. The importance of the common property character of the environment as it relates to its ability to ameliorate the waste streams of economic activity is at issue here.

In the area of resource extraction analysis, process analysis has long been used to address the question of the relation of economic activity and resource-base (in the sense of supplying materials) adequacy in a dynamic context. Examples of this are given by Kopp and Smith (1982) and Kydes and Rabinowitz (1981). There is an overlap here between investment decision analysis and resource extraction analysis in that models capable of contributing to one area also may contribute to the other.

In the more traditional area of process design, process analysis has long been used by the engineering profession to find physically feasible and least-cost solutions for the production of goods and services. Many works such as Peters and Timmerhaus (1968) may be referred to in this area. An important development since the 1960s in this area has been the advent of interactive computer-based and data-based tools to support this activity of process design. Other areas using process analysis can learn much from this development. The effect of this development is to significantly reduce the cost of process design, thereby allowing engineers to search a much larger portion of the physically feasible space of processes. This in turn contributes significantly to the overall productivity of an economy.

The last area is again an engineering area but is much newer than process design. Here a process model of an already designed process is used in a computer to aid in the control of that process. A discussion of this application is given by Rosenbrock (1974). There is again much that can be learned by social sciences in studying this area and transposing both the methods and certain results. Productivity rises very steeply when process control is aided by good process models encoded in control computers.

By looking at the commonality of these areas, one can discern a certain pattern or methodology of process analysis which is independent of the area of application. This methodology, once emancipated from its application-dependent manifestations, has significant implications for the mode of storage of, and mode of communication of, information in a modern society. It can be seen that the methodology forms a significant connecting link between the physical and social sciences. This link could, with the help of developing computer technology, enable the sciences in general to take a more responsible role in the guidance of the society in which they are embedded.

See also: Input–Output Analysis; Input–Output Analysis: Applications; Productivity and Technological Improvements: Managerial Evaluations; Residuals Management

Bibliography

Ayres R U 1978 *Resources Environment and Economics: Applications of the Materials/Energy Balance Principle.* Wiley, New York

Checkland P 1981 *Systems Thinking, Systems Practice.* Wiley, New York

Jantsch E 1981 *The Self-Organizing Universe: Scientific and Human Implications of the Emerging Paradigm of Evolution.* Pergamon, New York

Kopp R J, Smith V K 1982 The perceived role of materials in neoclassical models of production technology. In: Smith V K, Krutilla J V (eds.) 1982 *Exploration in Natural Resource Economics.* Johns Hopkins University Press, Baltimore, MD

Kydes A S, Rabinowitz J 1981 Overview and special features of the time-stepped energy system optimization model (TESOM). *Resour. Energy* 3: 65–92

Mesarovíc M D, Macko D, Takahara Y 1970 *Theory of Hierarchical, Multilevel Systems.* Academic Press, New York

Peters M S, Timmerhaus K D 1968 *Plant Design and Economics for Chemical Engineers*, 2nd edn. McGraw-Hill, New York

Rosenbrock H H 1974 *Computer Aided Control System Design.* Academic Press, New York

Russell C S 1973 *Residuals Management in Industry; A Case Study of Petroleum Refining.* Johns Hopkins University Press, Baltimore, MD

Russell C S, Spofford W O 1972 A quantitative framework for residuals management decisions. In: Kneese A V, Bower B T (eds.) 1972 *Environmental Quality Analysis; Theory and Method in the Social Sciences.* Johns Hopkins University Press, Baltimore, MD

Smith V L 1966 *Investment and Production: A Study in the Theory of Capital-Using Enterprise.* Harvard University Press, Cambridge, MA

R. B. Hoffman and B. C. McInnis
[Robbert Associates, Ottawa, Canada]

W. S. Page
[Kemptville, Ontario, Canada]

Product Liability and Materials: Europe

The liability exposure associated with the production and distribution of goods is less pronounced within the European Community (EC) than in the USA. The traditionally well-developed systems of health, accident and long-term disability insurance in the member states of the EC frequently obviate the need for product liability suits.

Several procedural factors also serve to deter litigation of this kind. Unlike in the USA, the plaintiff who loses at trial must reimburse the successful defendant for all costs, including attorney fees. This rule, coupled with the prohibition of contingent fee arrangements in all member states, poses a severe financial risk for potential litigants. Trials without juries and the tightly controlled fact-finding by the courts (a process that does not yield nearly as much evidence as does the US system of pretrial discovery) further discourage victims of product-related accidents from seeking redress through litigation in European courts. Finally, awards for pain and suffering of the plaintiff, a major financial strain for manufacturers and the insurance industry in the USA, are significantly lower in Europe than the corresponding US jury verdicts and punitive damages do not exist at all in any of the European systems.

Nevertheless, since the late 1960s, product liability law has been a rapidly developing area in most member states of the EC and it has been a subject of growing concern for both European consumers and the business community. This article will review an EC Council directive that was recently adopted in an effort to harmonize the traditionally diverse product liability rules of the member states. The article will also provide a brief outline of some of those traditional national rules which, unaffected by the EC directive, are of continuing importance for future litigation.

1. *Background of the EC Directive*

Until recently, product liability law in the member states of the EC was exclusively a matter of national legislation and case law. The approaches to tort liability questions involving defective products varied considerably, ranging from the application of pure negligence concepts requiring proof of the defendant's fault in some member states to the imposition of strict liability without a showing of individual fault in others. Since the costs of manufacturing goods tend to be higher in strict liability jurisdictions, this divergence of product liability rules was perceived as distorting competition within the EC. It was also felt that these discrepancies impeded the free movement of goods between the member states, a key prerequisite for the creation of a single market. Moreover, member states espousing lenient liability rules in favor of manufacturers afforded significantly lower levels of consumer protection than those states with more rigid liability regimes.

With these concerns in mind, in the late 1960s the Commission of the EC began to examine ways to coordinate the product liability laws of the member states. The attempts to harmonize this area of the law within the EC culminated in 1985, when the EC Council issued a *Directive on the Approximation of the Laws, Regulations and Administrative Provisions of the Member States Concerning Liability for Defective Products* (European Community Council 1985).

Although the legislative power of the EC Council is, of course, limited to reforming the product liability systems of the EC member states, the impact of the directive on nations that do not yet belong to the EC has also been considerable. All member nations of the European Free Trade Association (EFTA) felt compelled to draft national laws and regulations that comply with the directive's provisions. Indeed, Austria adopted the new rules long before many EC member states had transformed the directive into national law.

To date there has been little practical experience with the new rules because they do not apply to products that were marketed before the date the rules took effect. Most member states enacted legislation transforming the directive into national law only after the 1988 deadline prescribed by the EC Council. Since dangers associated with defective products frequently materialize years after their distribution, a significant number of future court decisions are likely to rest on the traditional, unharmonized product liability laws of the member states. Moreover, the new rules, even when they apply, do not provide solutions to every issue arising in product liability litigation. Indeed, important matters, such as the imposition of damages for pain and suffering, remain "unharmonized." The directive defers this question, as well as others, to the preexisting national provisions. Many proceedings will thus involve the application of both the harmonized rules implementing the directive and traditional laws of the member states.

1.1 Principles of Liability Under the EC Directive

The directive's basic liability rule is identical with the concept of strict liability in tort developed by US courts (see *Product Liability and Materials: General and USA*). In both systems producers are answerable for damage caused by defects in their products and recovery for such damage does not require a contractual relationship between the parties. The broad definitions of what constitutes a "product" and a "producer" under the directive leaves no doubt that the basic liability rule applies to materials producers. "Products" are defined as "all movables, with the exception of primary agricultural products and game, even though incorporated into another movable or into an immovable." The term "producer" specifically includes, aside from manufacturers of finished products and component parts, producers of any raw materials. Producers under the directive also include those who hold themselves out as producers by labelling products with their name, trademark or other distinguishing feature. The directive's liability rule further applies to persons who import goods into the EC for commercial distribution and, under certain circumstances, even to the suppliers of products.

In cases involving two or more liable parties, such as the original producer and the importer, liability is said to be "joint and several." This rule entitles the plaintiff to demand, at his choice, the entire amount of damages awarded at trial from any one of the defendants. A defendant subject to this rule may obtain contribution or recourse from the other liable parties in accordance with preexisting national law provided, of course, these codefendants are solvent.

In order to further one of its objectives, the improvement of consumer protection throughout the EC, the directive seeks to introduce liability without fault. Liability under this concept does not require the plaintiff to prove a wrongful act or omission on the part of the defendant. The plaintiff must prove, instead, that a defect in a product caused the damage. The definition of what constitutes a defect is thus of central importance for this (strict) liability concept. Article 6 of the directive enunciates that definition:

> "1. A product is defective when it does not provide the safety which a person is entitled to expect, taking all circumstances into account, including:
>
> (a) the presentation of the product;
> (b) the use to which it could reasonably be expected that the product would be put;
> (c) the time when the product was put into circulation.
>
> 2. A product shall not be considered defective for the sole reason that a better product is subsequently put into circulation."

The premise of this article, emphasizing safety expectations, resembles the consumer expectation test set forth in the US *Restatement (Second) of Torts*, Sect. 402A (see *Product Liability and Materials: General and USA*). Like its US counterpart, it does not pose problems of application as far as manufacturing (or production) defects are concerned. This type of defect exists when a product is not manufactured as contemplated. For example, a particular metal part used for car brakes and designed to resist pressure of up to 50 psi is defective in manufacture if it fractures under less stress while the other metal parts of the same production line perform as anticipated. Manufacturing defects can thus be said to manifest themselves in a product's deviation from the regular assembly line standard.

Article 6 of the directive also clearly applies to design defects and, since "the presentation of the product" is one factor of the safety expectation test, to cases involving failure to warn or instruct. In the age of industrial mass production, these types of defects involve far greater economic consequences and potential harm than manufacturing defects, which tend to occur as aberrations in single products only. A defect in design means that the entire production line does not meet reasonable safety expectations. Similarly, a failure to warn about the dangers of a certain product type is apt to affect all of its users and consumers. Like the US "unreasonably dangerous" formula and its manifold variants, however, the directive's safety expectation test fails to eliminate the traditional uncertainty about the existence of defects regarding design and failure to warn. It will remain difficult to determine, for example, whether a particular car model is "defective" in design because its frame is less sturdy than that of a competitor. Additional factors, such as increases in purchase price and gasoline consumption as a result of a heavier frame, must be considered. Since there are no objective criteria to establish the relative weighting of these factors, the outcome of this type of litigation is less predictable. Failure-to-warn cases involve even greater uncertainty. Decisions will hinge upon a determination of whether the presentation of the product was adequate in light of its dangers and the consumers' safety expectations influenced and conditioned by that presentation.

The directive does not prescribe liability for failure to warn of dangers that become apparent subsequent to marketing of the product. A failure to recall such products also has no consequence under the directive. However, duties to warn in the postmarketing phase and, where warnings are insufficient, to recall an entire product line may arise under preexisting national law.

1.2 Damages Under the EC Directive

Liability under the directive arises only for actual damage resulting from personal injury or death, such as medical expenses and loss of wages, or from damage to private property. The directive does not provide relief for pain and suffering. This and other types of nonmaterial damages may only be claimed in accordance with existing national laws.

These laws will also retain their importance in product liability litigation involving damage to commercially used property, since recovery for property damage under the directive requires that the damaged item be ordinarily intended and mainly used for private use or consumption. Moreover, existing national laws and not the rules of the directive control the recovery for damage to the defective thing itself and for minor damage that does not exceed 500 ECU (~US$650). Finally, on the insistence of Germany, the directive includes a provision that permits member states to limit a producer's liability to not less than 70 million ECU in cases of death or personal injury caused by identical items with the same defects. In addition to Germany, only Greece, Portugal and Spain have exercised this option. This global cap on liability will be reviewed and possibly eliminated by the Commission and the Council of the EC in 1995.

1.3 Defenses Under the EC Directive

There are several defenses available to producers to avoid liability under the directive. If the defendant producer proves that it did not put the product into circulation or that, if the product was circulated, the defect did not exist at the time of marketing, liability cannot arise. The producer will also prevail on proof that the product was not manufactured for sale or for any other form of distribution for economic purposes.

Liability under the directive is further barred if the defect is due to the product's compliance with mandatory regulations issued by the public authorities. This defense is limited to regulations that have legal force, and does not apply to privately issued industry rules and standards even if widely used. For example, technical norms and standards devised by the Deutsches Institut für Normung (DIN), a private organization, lack the requisite legal status despite their high public profile. However, since these self-imposed industry regulations influence and reflect safety expectations, they will have some relevance in product liability litigation under the directive's rules. Compliance with such standards may create the rebuttable presumption that the product is not defective. Future cases will have to determine the exact status of these private norms.

Producers of component parts are exonerated if they prove that the defect is attributable to the design of the final product containing the component part or to instructions provided by the manufacturer of the final product. This defence would apply, for example, to a producer of insulation material that is designed to resist heat of up to 250° but which, without the producer's knowledge, is incorporated by an assembler into finished products that are exposed to 300°.

The directive also contains the so-called development risk defense which precludes liability if the manufacturer proves "that the state of scientific and technical knowledge at the time when he put the product into circulation was not such as to enable the existence of the defect to be discovered." The directive's development risk defense is functionally equivalent to the "state-of-the-art defense" adopted in most US jurisdictions. Under both concepts, defendants would not be liable in cases like *Beshada vs Johns-Manville Products Corp.* involving asbestos-related dangers which were undiscoverable at the time of marketing and initial exposure of the victims (see *Product Liability and Materials: General and USA*). However, neither defense abrogates the duty of producers to monitor the performance of their products and to issue warnings or recall products when subsequent experience leads to the discovery of formerly unknown risks.

Manufacturers seeking to invoke the development risk defense must prove that, in light of all available scientific and technical knowledge, it was impossible to discover the defect at the time the product was placed on the market. Actual knowledge or ignorance of the individual manufacturer is immaterial. However, since the full scale of risks associated with newly developed materials and innovative substances is often objectively unknown at the time of their marketing, producers operating in these markets, notably manufacturers of pharmaceuticals, should be able to meet this burden of proof in many instances. As might be expected, the inclusion of this defense into the directive was highly controversial. On the one hand, it was argued that liability for unforeseeable risks would tend to discourage product innovation. It was difficult, on the other hand, to justify the exclusion of liability for development risks in light of the directive's stated goal to improve consumer protection. Development defects, particularly when they occur in pharmaceuticals, are apt to affect a great number of consumers as the thalidomide tragedy of the early 1960s and other drug-related cases of mass injuries exemplify. Indeed, the thalidomide disaster was a major force behind the original initiative to harmonize European product liability law on a high level of consumer protection.

In light of these strong policy concerns, it is no surprise that the decision to adopt the defense is neither final nor mandatory. Like the global cap on liability for death and personal injury, the development risk defense is not final because the EC Commission and the Council will evaluate its effects and decide whether to repeal it in 1995. The inclusion of the defense is not mandatory because the directive explicitly provides that national legislation implementing the directive may introduce liability for undiscoverable risks. Except for Luxembourg, however, all EC member states have chosen to incorporate the directive's development risk defense into their new legislation. This decision does not affect the few preexisting special liability systems under national law which do impose liability for undiscoverable risks. One example of such a statutory scheme is the German Drug Act, which was passed in the wake of the thalidomide catastrophe. Under the act, manufacturers of defective pharmaceuticals bear the consequences of development risks: within certain upper ceilings on the amount of recoverable damages, they are strictly liable for drug-related injuries and deaths, even if the defect was undiscoverable.

The victim's fault as contributing cause for the injuries sustained is a frequently asserted defense in product liability litigation. Under the directive, as under the traditional negligence rules of all member states, the plaintiff's own negligence results in a reduction of his recovery that is commensurate with the degree of his fault. The defendant can invoke this defense under the directive if the injury is caused both by a defect in the product and by the fault of the injured person or any person for whom the injured person is responsible. In contrast, the act or omission of a third party does not give rise to a reduction of the plaintiff's recovery. The directive leaves the exact implementation of this defense to the existing laws of the member states. It is safe to say, however, that national courts will disallow all recovery only in extreme cases, such as blatant product misuse or voluntary assumption of obvious risks.

In order to preserve the right of recovery, it is incumbent upon injured parties in all legal systems to advance claims of which they are aware within a fixed period of time. The directive recognizes this fundamental principle by subjecting claims arising under its rules to a three-year limitation period. This period begins to run when the plaintiff becomes aware or should have reasonably become aware of the damage, the defect and the identity of the liable party. The directive provides for an additional prescriptive period which was previously unknown in many European legal systems: rights under the directive expire ten years after the manufacturer has placed the product on the market. Like statutes of repose, enacted by many US state legislatures to shield producers from liability for injuries caused by older products, this time bar operates irrespective of the injured party's knowledge. The period only stops running if the victim institutes legal proceedings within ten years.

2. *Principles of Liability Under National Laws*

Although the directive establishes EC-wide rules for many vital aspects of product liability, it does not render obsolete the traditional laws of the member states. In many instances, plaintiffs will continue to rely on these "unharmonized" laws as coexisting sources for recovery. In others, such as business litigation involving damage to commercial property, traditional national laws will serve as an exclusive basis for the resolution of the dispute.

The legal systems of all member states recognize claims for product-related damage or injuries based on contractual as well as noncontractual liability. The former has not had much impact on the development of product liability law in the member states since, in most systems, the manufacturer owes contractual liability only to the immediate buyer, ordinarily the wholesaler or retailer. One notable exception exists under French law. The final purchaser of a defective product has a direct contractual claim against the manufacturer or any other professional seller further up the chain of distribution, provided the defect existed at the time the product left the control of the defendant seller.

Even this consumer-oriented concept, however, does not benefit victims who never purchased the product that caused their injuries. Unless their claims are covered by the harmonized rules under the directive, such "innocent bystanders" can recover damages only on grounds of noncontractual liability provided by traditional national tort law. These national solutions can be divided into roughly three different categories. Several member states, such as Italy and Portugal, require that the victim specifically allege and prove the defendant's fault. Others (such as the UK and Germany) operate on the presumption that the defendant manufacturer was at fault if a defect in an industrial product caused the injuries. In these systems, the defendant may nevertheless escape liability if he proves that he was not negligent. However, this burden is not easily overcome. For example, even if the production process was fully automatized on the basis of state-of-the-art equipment, the manufacturer would still have to document that competent and sufficient numbers of personnel supervised the production process at all times. In a third group of member states (France, Luxembourg), manufacturers will not prevail even if they present this kind of exonerating evidence; case law has created an irrebuttable presumption of their fault. Even in these member states, however, the victim ordinarily carries the burden of proving a defect, the extent of the damage and the causal link between the defect and the damage. In addition, this strict liability case law is limited to the manufacturer of the product. Those who merely supply defective products are not presumed to be at fault; the victim must prove that the supplier's negligence in handling, storing or treating the product caused the accident.

French law also includes a noncontractual remedy that does not fit any of the above categories. According to Article 1384 Sect. 1 of the French civil code, a person in control of a thing (*gardien* = keeper) is strictly liable for all damage caused by that thing. French courts applying this provision do not even require the victim to establish that the thing was defective. The *gardien* can avoid liability only if he establishes that the injury was a result of an act of God or was due to the victim's fault. Superficially, this rule would seem to apply only rarely to manufacturers, since it is ordinarily the ultimate purchaser who is in control of the thing at the time of the accident. For example, a patron in a bar who was injured by an exploding bottle of beer should have a cause of action against the bar owner under this rule, but not against the manufacturer of the bottle or the bottler. Several French courts nevertheless have held that the manufacturer is deemed to retain control over a product's structural characteristics, such as its pressure resistance, even though the manufacturer had already delivered the product to another. Although liability of the keeper of a thing does exist as a concept in other legal systems as well, this expansive consumer-oriented definition of the *gardien* is unique. Belgian and Italian law, for example, also recognize the concept of *gardien* liability, but limit its application to those defendants who are actually in physical control of the thing.

Increased industrial and technological activities have prompted the legislatures of all member states to enact a growing number of laws designed to guard against the risk of harm associated with specific activities. In England, for example, numerous regulations based on the English Consumer Protection Act and the Consumer Safety Act establish safety standards for the production and distribution of potentially dangerous goods, such as electrical equipment and asbestos products. Efforts to harmonize national safety regulations and technical standards are also taking place at the EC level. One example is the EC Council Directive on Machinery Safety, which seeks to create equivalent design and safety requirements for machinery throughout the EC. The application and enforcement of these national or EC-wide statutory standards is ordinarily a matter of public law, which regulates behavior through the imposition of sanctions, rather than through manufacturer liability to private parties. However, the relevance of these statutory standards in private product liability actions involving the application of traditional national laws must not be underestimated. Compliance with a pertinent legal standard, while not automatically absolving a manufacturer from liability, weighs heavily against a finding of a product defect or the manufacturer's fault.

Conversely, proof that the product did not meet a standard at least creates the presumption that the product was defective. In some jurisdictions, this proof may also give rise to a rebuttable presumption of the manufacturer's negligence. For example, German courts invoke this rule when the statute is intended, among other objectives, to protect the interests of the injured parties. English law goes even further. It treats the breach of certain statutory standards as a distinct cause of action that results in strict liability regardless of the defendant's fault, provided of course that the breach actually caused the injuries of the plaintiff. The breach of safety regulations under the English Consumer Protection Act and the Consumer Safety Act, for example, give rise to this independent tort action. This approach is very similar to the US concept of negligence *per se*, which also creates an independent basis of liability for damages that are the result of noncompliance with certain safety statutes, such as the Federal Consumer Product Safety Act.

In light of the limited scope of the directive and the diverging solutions under applicable national law, it is fair to conclude that product liability litigation in the EC will not follow a uniform pattern of dispute resolution. On the other hand, the new rules under the directive mark a first, albeit modest, step towards greater judicial harmony and improved consumer protection within the EC. They also reflect the ability of the EC to adopt a flexible approach in order to reach a compromise in highly controversial matters, as is illustrated by the tentative status of the risk development defense and the cap on liability.

In the USA, by contrast, all attempts to introduce a unified federal law of product liability have thus far failed. The expectation that the European experience could provide lessons for the US efforts, although tempting, would seem unwarranted in view of the differences between the interests at stake. As opposed to the EC directive, the proposed US bills involved fundamental departures from established state practice, such as restrictions on the availability of punitive damages and the elimination of strict liability causes of action. The influential lobbies of consumer organizations and, in particular, plaintiff's attorneys who have a vested interest in the legal status quo, have further narrowed the possibility for compromise. Even the most recent "product liability fairness bill," containing somewhat more moderate proposals than its predecessors, was rejected by the US Senate. It would appear unlikely, therefore, that Federal product liability law will be enacted in the years ahead.

See also: Product Liability and Materials: General and USA

Bibliography

Bouckaert C, Byrd R W II 1987 Product liability defense in France. *Def. Council J.* 54: 536–44

Corr M 1990 Problems with the EC approach to harmonization of product liability law. *Case W. Res. J. Int. L.* 22: 235–44

Dielmann H J 1986 The European Economic Community's Council directive on product liability. *Int. Lawyer* 20: 1391–1400

European Community Council 1985 Directive on the approximation of the laws, regulations and administrative provisions of the member states concerning liability for defective products. *Q. J. Eur. Comm.* (*No. L 210*) 28: 29

European Community Council 1989 Directive on the approximation of the laws of the member states relating to machinery. *Q. J. Eur. Comm.* (*No. L 183*) 32: 9

Miller C J (ed.) 1986 *Comparative Product Liability.* The British Institute of International and Comparative Law, London

Nilles K M 1985 Defining the limits of liability: a legal and political analysis of the European Community products liability directive. *Va. J. Int. L.* 25: 729–74

Sheinman L 1985 The EEC directive on product liability. *J. Bus. L.* 504–8

Tebbens H D 1980 *International Product Liability. A Study of Comparative and International Legal Aspects of Product Liability.* Sijthoff & Noordhoff, Germantown, MD

Thieffry P, Van Doorn P, Lowe S 1989 Strict product liability in the EEC: implementation, practice and impact on U.S. manufacturers of directive 85/374. *Tort & Ins. L. J.* 25: 65–91

United States Code 1988 *Federal Consumer Product Safety Act*, Vol. 15, Sect. 2072

van Wassenaer van Cattwijk O baron 1986 Products liability in Europe. *Am. J. Comp. L.* 34: 789–96

Whittaker S 1985 The EEC directive on products liability. *Yearb. Eur. L.* 5: 233–79

Zekoll J 1989 The German products liability act. *Am. J. Comp. L.* 37: 809–18

J. Zekoll
[Tulane Law School, New Orleans, Louisiana, USA]

Product Liability and Materials: General and USA

The threat and reality of product liability litigation are of continuing concern to materials producers and users alike. The increase in the number of cases filed and the size of awards, coupled with the substantial transaction costs involved in the litigation process, have taxed the resources of virtually every materials-related firm and have actually caused some to file for bankruptcy. This article reviews the history and policy origins of the product liability law, describes the current status and operation of the product liability system, and summarizes proposed changes to this system, particularly as they affect the decision-making process of materials producers and both industrial and consumer materials users.

1. History and Policy Origins of the US Product Liability System

Following the industrial revolution, damages caused by a defective product could be recovered if the seller of the defective product could be shown to have been negligent. Negligence has been defined as "conduct which involves great risk of causing harm" or conduct "which falls below the standard established by law for the protection of others against unreasonably great risk of harm." The focus here is not on the quality or performance of the product but rather on the conduct of the manufacturer or seller. Courts also required that there be "privity of contract," that is, a direct relationship between buyer and seller, to recover for product-caused losses stemming from negligent conduct. Hence most manufacturers were virtually immune from liability, since injuries almost always occurred to an ultimate user or consumer who did not purchase the product directly from the manufacturer. This barrier of privity was considered socially unacceptable and removed by a series of court decisions, starting with *MacPherson vs Buick Motor Company* (1916), where recovery was allowed directly from a remote manufacturer of a negligently produced defective wheel.

A parallel body of product liability law also developed, based on various forms of warranty which do not require the plaintiff to establish the presence of a specific defect in the injury-producing product. For example, Section 2-313 of the Uniform Commercial Code provides for recovery if the product fails to meet the seller's own express warranties or representations; Section 2-315 states that where "the buyer is relying on the seller's skill or judgement to select or furnish suitable goods, there is unless excluded or modified under the next section an implied warranty that the goods shall be fit for such purpose." If any of these warranties are breached, the buyer can recover from the seller.

Another form of warranty, the implied warranty of merchantability (Section 2-314 of the Uniform Commercial Code), does, however, focus on the quality of the product rather than the representations of the seller. This section requires that "Goods to be merchantable must be at least such as ... are fit for the ordinary purposes for which such goods are used ...." However, it was and is common practice for a manufacturer to disclaim this implied warranty of merchantability and hence avoid liability if this warranty is breached.

Today most product liability claims are based upon the theory of strict liability, which is a form of liability without fault. It focuses on the performance of the product rather than the conduct of the manufacturer or seller. Strict liability is described in the Restatement (Second) of Torts, Section 402A (1965) as follows.

> "1. One who sells any product in a defective condition unreasonably dangerous to the user or consumer or to his property is subject to liability for physical harm thereby caused to the ultimate user or consumer, or to his property if
> (a) the seller is engaged in the business of selling such a product, and
> (b) it is expected to and does reach the user or consumer without substantial change in the condition in which it is sold.
> 2. The rule stated in Subsection (1) applies although
> (a) the seller has exercised all possible care in the preparation and sale of his product, and
> (b) the user or consumer has not bought the product from or entered into any contractual relation with the seller."

The policy rationale articulated by the courts in adopting strict liability is that the costs of injuries stemming from defective products should be absorbed by manufacturers and spread among all users rather than being borne by injured individuals.

The advent of strict liability also marked the contraction of affirmative defenses available to defendants. Contributory negligence of any degree was an absolute bar to recovery under negligence law. Under strict liability a defendant must prove that the plaintiff voluntarily and unreasonably assumed a known risk to make out an affirmative defense.

2. Current Status and Operation of the US Product Liability System

To make out a case in product liability, regardless of the theory of the liability, the plaintiff must establish the following:

(a) the product was defective;

(b) the defect existed at the time the product left the defendant's hands;

(c) the defect caused the harm; and

(d) this harm is appropriately assignable to the identified defect, that is, the injury was not caused by a remotely probable or preposterous occurrence.

The defendant can prevail if he is able to attack the plaintiff's case successfully or prove an affirmative defense of his own: contributory negligence in negligence cases, voluntary assumption of a known risk in strict liability cases.

Estimates of the number of product liability cases filed annually in the USA vary considerably, but numbers around 80 000 for suits filed in both state and federal district courts are quite common. Between the years 1974 and 1981 the number of product liability suits filed in federal courts grew at an average annual rate of 28%, which is almost $3\frac{1}{2}$ times as fast as the average annual increase in civil suits filed in these same courts. This increase in number of suits filed, coupled with the sizes of awards, has increased the cost of product liability insurance. Testimony before Congress indicated that one small machine tool manufacturer reported that its cost of liability insurance rose from $200 per machine in 1970 to $11 000 per machine in 1982.

One reason offered for the increase in product liability litigation in the USA is the contingency fee system, whereby plaintiffs' attorneys accept cases under the condition that they will be paid for their services, usually a proportion of the award, only if the case is successful. Additions to the caseload are also caused by restrictions in the US Workers' Compensation System, whereby injured employees are compensated for their injuries according to a state-wide compensation schedule regardless of fault but lose their right to file a civil suit against their employers. Injured workers often turn for additional recovery to third parties, usually the manufacturers of the injury-causing product. Testimony before Congress also indicates that these third party suits are often actually initiated by the Workers' Compensation insurance carriers who are seeking to recover their awards to injured workers from manufacturers of injury-causing products. Many product liability cases have multiple defendants, many of whom are compelled to join by other defendants. Hence product liability suits are clearly not solely initiated by the plaintiff's bar but have become almost an everyday occurrence in the US commercial world as well.

Strict liability was first introduced in the context of production defect cases where the existence of a defect could be measured directly against a product of normal average quality. Strict liability is now commonly applied to design defect cases where there exist no comparable standards for determining whether or not a product is defective. Few products can be made absolutely safe. The original standard of "unreasonable danger" is not easily applied in design defect cases. The most common measure of a design defect is to weigh the risks and utility of the product.

Various other articulated measures of what constitutes a defective design are based upon either the perspective of consumer expectations or manufacturers' foreseeability. To complicate matters even further, different states have different criteria for what constitutes a defectively designed product. Among these are: if the product "fails to perform as safely as an ordinary consumer would expect it to when used in an intended or reasonably foreseeable manner"; and if it left the manufacturer's control "lacking any element to make it safe for its intended use or possessing any feature that renders it unsafe for its intended use."

One form of generic defect that is particularly difficult to address is failure to warn or instruct. Failure to warn is an appealing basis of liability for many courts which focus primarily on the notion that warnings are cheap. Seldom are the appropriateness and effectiveness of a warning explicitly evaluated. The question of whom to warn, if anyone, is particularly difficult for primary materials producers, whose products typically have multiple end uses. On the other hand, the "puffing" of salesmen may be interpreted as instructions by users and subsequently used as a basis of liability. The duty to warn has been carried to its extreme in *Beshada vs Johns–Manville Products Corp.* (1982) where the New Jersey Supreme Court held that Manville had a duty to warn of dangers which were *undiscoverable* at the time of manufacture.

Asbestos-related cases are prime examples of modern cumulative trauma litigation (DES, Agent Orange are other examples) which is characterized by significant delay between exposure and manifestation of disease. Many cases of asbestosis, mesothelioma and lung cancer are appearing today from exposures which occurred several decades ago. Prior to the Borel decision in 1973, victims of asbestos-related diseases were compensated only through the Workers' Compensation System, not the tort liability system. This situation has changed so dramatically that Johns–Manville, at the time of its filing for bankruptcy, was faced with 16 500 claims and new claims were being filed at a rate of 500 a month. Hence it appears that for the first time, injured plaintiffs with seemingly meritorious cases will not be compensated for their injuries through the tort liability system. To complicate matters further, there have been recent inroads into the traditional immunity shield of employers embodied in the Workers' Compensation System. These inroads threaten to expose manufacturers to tort actions from their own employees in addition to third-party users.

The tort litigation system has been characterized as creating an incentive to manufacturers not to produce and sell defective products. Various conferences and publications in both the public and private sector have been directed at preventing exposure to product liability litigation, principally by eliminating the manufacture and sale of defective products. ASTM Committee E-40 on Technical Aspects of Product Liability Litigation has concerned itself with developing recommended practices for (a) collecting, preserving, examining and evaluating potential technical evidence, and (b) reporting the results thereof for use in product liability litigation. Yet all these efforts will be fruitless if the product liability system becomes fundamentally irrational and ineffective. Feelings in the industrial community that this has already happened have engendered the tort reform efforts described below.

3. Adopted and Proposed Changes to the Product Liability System in the USA and Europe

Several studies in the mid-1970s resulted in draft legislation designed to address problems associated with the product liability system. In the USA, the Federal Interagency Task Force on Product Liability and later the Commerce Department's Task Force on Product Liability and Accident Compensation recommended the passage of legislation designed to make product liability insurance more readily available and establish balance and uniformity in product liability law. In Europe, the European Economic Community issued a Draft Directive on product liability and the Council of Europe produced a similar convention. Thus far, the only legislation enacted is the US Risk Retention Act, which essentially allows manufacturers from different states to form their own insurance company subject to the provisions of the Act.

Several recurrent themes are present in all these recent tort reform studies and draft legislation. No one is expressing serious concern about using strict liability in production defect cases, where a normal average product can serve as a standard for assessing defect. The difficulties arise with applying strict liability to design defect and failure-to-warn cases. These two concerns are paramount in the minds of materials producers and users alike. Other concerns are over limiting the time permitted between injury and filing of a lawsuit (statute of limitations) and possibly limiting the time between sale or manufacture and filing of a lawsuit (statute of repose).

In the USA, the most active efforts at passing a federal Product Liability Act have been spearheaded by Senator Kasten. One of Kasten's purposes in passing preemptive federal legislation is to remove the nonuniformities in tort law engendered by individual states in passing different tort reform bills. Three drafts of the Kasten bill have been produced, the last of which was dated November 30, 1982. Section 4 of this latest bill allows for liability if the product was unreasonably dangerous in construction or manufacture, or because the manufacturer failed to provide adequate warnings or instructions, or because the product did not conform to an express warranty made by the manufacturer with respect to the product. The tests for design defect as well as failure-to-warn claims essentially return to a negligence standard, that is, judging the manufacturer's conduct on the basis of what was known and practised by prudent members of the scientific, technical or medical community at the time of manufacture.

The bill also contains provisions for comparative responsibility, whereby damages would be reduced by an amount proportionate to the responsibility attributed to the claimant's conduct. Awards can also be proportionally reduced as a result of misuse or alteration of the product. Workers' Compensation benefits can also be subtracted from jury awards. Workers' Compensation insurance carriers would have no right of subrogation against product manufacturers, and Workers' Compensation would remain an employee's exclusive remedy against his employer. A statute of repose of 25 years is imposed on capital goods for design defect and failure-to-warn claims. Punitive damages would be awarded only after establishing through "clear and convincing evidence that the harm suffered was the result of the reckless disregard of the manufacturer or product seller for the safety of product users, consumers, or persons who might be harmed by the product."

The EEC Draft Directive on product liability applies strict liability for either a production or a design defect claim against a manufacturer, "whether or not he knew or could have known of the defect," which has a Beshada-like ring to it. The draft directive also states that "A product is defective when it does not provide for persons or property the safety which a person is entitled to expect." To limit defendants' liability, there would be a statute of repose of 10 years and a monetary limit to the total liability for all personal injuries caused by identical articles having the same defect. In addition, only noncommercial property damage is potentially compensable.

Hence substantial uncertainties concerning product liability will continue to face materials producers and users, as well as all other manufacturers and consumers, throughout the world. However, it is not unlikely that some form of relief will arise to prevent manufacturers from being indefinitely liable for risks which were unknowable at the time of sale. It is also not unlikely that Workers' Compensation will remain the exclusive remedy available to employees from their employers. It may be hoped that these and other potential adjustments to the product liability system can indeed serve to achieve the goal of balance and uniformity, not only in the USA but throughout the world.

See also: Product Liability and Materials: Europe

Bibliography

American Society for Testing and Materials 1977 *Standard Practice for Reporting Opinions of Technical Experts*, ASTM E620–77. American Society for Testing and Materials, Philadelphia, PA

American Society for Testing and Materials 1980 *Standard Practice for the Evaluation of Technical Data*, ASTM E678–80. American Society for Testing and Materials, Philadelphia, PA

American Society for Testing and Materials 1982 *Standard Practice for Examining and Testing of Items Which Are or Might Become Involved in Products Liability Litigation*,

ASTM E860–82. American Society for Testing and Materials, Philadelphia, PA
Birnbaum S L, Wrubel B 1982 Workers' compensation and the employer's immunity shield: recent exceptions to exclusivity. *J. Prod. Liab.* 5: 119–44
Bok D C 1983 A flawed system. *Harv. Mag.* 85(5): 38–45, 70–1
Court of Appeals of New York 1916 MacPherson v. Buick Motor Company. *Northeast. Rep.* 111: 1050–7
Epstein R A 1982 Manville: the bankruptcy of product liability law. *Regulation* September/October: 14–19, 43–6
Kasten R Product Liability Act March 1, 1982, June 8, 1982, November 30, 1982
Kolb J, Ross S S 1980 *Product Safety and Liability: A Desk Reference.* McGraw-Hill, New York
McGuire E P 1979 *The Product-Safety Function: Organization and Operations*, Conference Board Report No. 754. Conference Board, New York
Malott R H 1983 Let's restore balance to product liability law. *Harv. Bus. Rev.* May–June: 67–74
New Jersey Supreme Court 1982 Beshada v. Johns-Manville Products Corp. *Atl. Rep. 2nd* 447: 539–49
Noel D W, Philips J J 1981 *Products Liability in a Nutshell*, 2nd edn. West Publishing Company, St Paul, MN
Orban F A III 1978 Product liability: a comparative legal restatement—foreign national law and the EEC direction. *Ga. J. In. & Comp. L.* 8: 342–406
Prosser W L 1971 *Handbook of the Law of Torts*, 4th edn. West Publishing Company, St Paul, MN
United States Consumer Product Safety Commission 1975 *Handbook & Standard for Manufacturing Safer Consumer Products.* US Government Printing Office, Washington, DC
United States Court of Appeals for the Fifth Circuit 1973 Borel v. Fiberboard Paper Products Corporation. *Fed. Rep. 2nd* 493: 1076–1109
Weinstein A S, Twerski A D, Piehler H R, Donaher W A 1978 *Products Liability and the Reasonably Safe Product.* Wiley–Interscience, New York

H. R. Piehler
[Carnegie Mellon University, Pittsburgh, Pennsylvania, USA]

Productivity and Materials

Productivity and materials are closely related. Three types of relations can be distinguished:

(a) the relations between the output of a material-producing industry and inputs such as capital and labor required in the production of the material—the productivity of materials-producing industries;

(b) the relations between the output and input of a given material in a production or manufacturing process—the productivity of materials utilization; and

(c) the relations between the input of a material and the productivity of other inputs such as capital and labor required in a production or manufacturing process—the productivity effect of materials.

1. Materials in the Literature on Productivity

Productivity is defined as the ratio of outputs to inputs. This ratio is widely interpreted as a measure of the efficiency with which inputs are utilized in production. The input most often considered is labor and "productivity" usually refers to labor productivity. Other inputs are being increasingly considered in the definition of productivity: "Productivity indices describe the efficiency with which inputs of capital, labor, technology and managerial skills are converted into goods and services" (Atkinson in US House of Representatives 1979). It may be noted that this statement does not mention materials or other intermediate inputs.

The tendency to neglect materials in production theory and productivity analysis was criticized by Christ (1961). Others have also recommended that all factors of production be included in the definition of productivity (Fabricant 1959, Kendrick and Creamer 1965, Salter 1966). A more inclusive approach to productivity has been developed in industrial economics, resource economics and input–output analysis.

Industrial economics has focused on the role of intermediate inputs in the productivity of firms. Gold (1955, 1975, 1979) questioned the usefulness of any single measure of economic efficiency. He introduced the concept of a "network of productivity relationships" which includes the interaction between materials inputs and other inputs (see *Factor Dominance in Industry*; *Productivity and Technological Improvements: Managerial Evaluations*). Eilon et al. (1976), Boylan (1975), Gold and Soesan (1976) and Rosegger (1980) applied network analysis to examples in the chemical and steel industries on the levels of plants and production units within plants.

Resource economics has considered the relation of materials and productivity. The productivity of materials-producing industries and its bearing on resource scarcity was analyzed by Barnett and Morse (1963). The concept of the "intensity of use" (Malenbaum 1972) is defined as the ratio of a material consumed to the GNP of a country or region.

Input–output analysis is concerned with the relations among primary and intermediate inputs at the level of industries simultaneously for all sectors of an economy (Leontief 1966, 1986) (see *Input–Output Analysis*). Carter (1970) justified the emphasis of input–output analysis on changes in intermediate input requirements:

> "Explicit introduction of intermediate inputs ... contributes to the understanding of the conventional capital and labor-output relationships.
> Superficially it is clear that long-run changes in labor productivity are rooted in changes in the organization of production, that new materials, components, communications, as well as new types of capital goods, have been prerequisites to continued rises in primary factor efficiency ... (p. 5)."

Several empirical studies of electric utilities examined the role of fuels in productivity change, for example Nerlove (1965) and Fuss and McFadden (1978). Griliches and Ringstad (1971) considered materials inputs in a study of Norwegian industry. Some other economy-wide empirical studies have also explicitly included intermediate inputs (Gollop and Jorgenson 1980, Gollop and Roberts 1981). There is now a trend to include intermediate inputs in productivity analysis as "economists become conscious of the constraints to growth from nonprimary inputs" (Sato and Ramachandran 1980).

Public concerns, especially of the 1970s, contributed to the recognition of the role of materials in productivity. They included concerns over the decline in labor productivity, the slowdown of technological advance, environmental degradation and perceived scarcity of critical materials. Evidence that the approach to defining and measuring productivity is changing is the recommendation by the Panel to Review Productivity Statistics of the National Research Council (1979) to collect statistics on intermediate inputs, including materials, for use in productivity indices.

2. *The Productivity of Materials-Producing Industries*

In the production of a material, various inputs are used more or less effectively; the effectiveness with which these inputs—other than the natural resource or main raw material—are used will be designated as "the productivity of materials-producing industries." This type of productivity is of interest at the levels of firms, individual plants and production units, where it affects profitability, and at the level of entire industries, where it affects viability and international competitiveness as in the case of the steel industries of various countries.

2.1 Nonferrous Metals

The productivity of primary and intermediate inputs in materials production can increase through gradual improvements in a given technology or through the replacement of one process technology by another. The history of various nonferrous metals industries provides illustrations. For example, gradual improvements in the Bayer process for producing aluminum oxide from bauxite and in the Hall–Héroult process for the reduction of the oxide to aluminum have accounted for productivity increases since the early 1960s.

The zinc industry since the 1930s has experienced process replacement. An obsolescent batch process using batteries of small, manually charged horizontal retorts was replaced by several novel continuous processes using mechanically charged vertical retorts of large capacity. These processes use labor and energy more efficiently. At the same time, capital requirements increased, especially for those zinc smelting processes which substitute electrothermic for traditional heating by combustion gases. The aluminum industry also provided an example of process innovation: the chloride smelting process.

2.2 Iron and Steel

The technology of iron and steel production since the 1940s has undergone changes which have had major effects on productivity. The iron blast furnace has experienced a large increase in size, the introduction of top pressure, the use of oxygen-enriched blast and the injection of fuel; the blast furnace charge has changed to sinter and pelletized ore. In competition with the blast furnace, processes for the direct reduction of iron have been introduced. Developments in steelmaking were the replacement of the open hearth by the basic oxygen process, the growth of electric furnace capacity and the adoption of continuous casting.

These changes in iron and steel technology have had marked effects on the industry's demands for raw materials, fuels, capital and labor. For example, pelletizing iron ores resulted in substantial increases in the productivity of labor and energy in steelmaking and permitted the substitution of taconite ores for richer ores, the US deposits of which were being depleted rapidly. As another example, the emergence of the basic oxygen process increased labor productivity, materials yield and thermal efficiency.

2.3 Nonmetallic Materials

The production of nonmetallic materials involves a great variety of processes characterized by different input mixes and productivity relations. Some nonmetallics are extracted from natural resources (e.g., building stone) or produced by synthesis (e.g., nylon), while others are grown (e.g., hevea rubber). The production processes may be labor intensive (e.g., hemp production) or capital intensive (e.g., chemical processing). These processes may permit attainment of economies of scale.

The general economic and productivity literature has paid even less attention to nonmetallic materials industries than to metal mining and processing. Forest products—in particular saw logs—are an exception (Manthy 1978). Labor productivity in the chemical processing industry has also received attention (American Chemical Society 1973, Wei et al. 1979).

2.4 Recycled Materials

The recycling of materials may be considered from the viewpoint of productivity. Recycling, if conducted efficiently, accomplishes, in addition to materials conservation and waste disposal, several kinds of savings: it generally requires less processing and therefore less energy and equipment than primary production; it also usually incurs smaller pollution abatement costs (Bever 1976a, b, 1977). Expressed differently, the inputs of energy and capital into secondary production are, as a rule, appreciably lower than those of primary production; this is not necessarily true of labor requirements since the collection and physical preparation of metal scrap and other secondary materials tend to be labor intensive.

Some scrap metal enters the primary production circuit and becomes indistinguishable from primary metal. In other cases, secondary materials are produced in a separate circuit. This may not produce metal of the same quality as primary metals; some recycled aluminum alloys are examples of this.

The production of some nonmetallic materials from secondary sources raises other problems. The energy efficiency of paper recycling can be compared with the energy efficiency of plants using fuels derived from by-products in the production of paper from wood (Hunt and Franklin 1973). The recycling of cullet (scrap glass), although not yielding marked benefits with respect to raw materials costs, promotes melting of charge materials in glassmaking and thus productivity of energy, labor and equipment.

2.5 Materials Availability

In the long run, the productivity of the materials-producing industries is a leading factor in the continuing availability of materials. Barnett and Morse's classic studies of materials scarcity in the USA showed that despite decreasing ore grade, real unit costs and relative prices of most mineral materials declined or remained constant from the Civil War to 1970 (Barnett and Morse 1963, Barnett 1979). They attributed this finding to technological advances in the extractive industries which increased the productivity of labor, capital and other inputs and permitted the substitution of lower-grade, more plentiful ores for higher-grade, scarcer ones. The continuation of this trend has become a matter of controversy. It has been argued that the trend will be reversed as returns on mechanization in mining decline (Lovering 1969, Dasgupta and Heal 1979) and costs increase owing to energy, environmental and other factors.

3. *The Productivity of Materials Utilization*

Production processes convert raw materials into bulk materials and engineering materials; manufacturing converts the latter into products. Materials are recovered in production or utilized in manufacturing with varying degrees of effectiveness which will be designated "the productivity of materials utilization." In extractive industries this productivity is measured by the ratio of material output to material input ("resource productivity" or yield). In the production of iron in the blast furnace, for example, this is the ratio of the iron content of the hot metal to the iron content of the charge. In manufacturing operations the corresponding ratio is the amount of a given material contained in the product to the amount of the material used in its manufacture; it will be designated as the "conversion ratio."

3.1 Yield

Only a fraction of the metal content of an ore is recovered, because of losses in the various extractive operations. The following average recoveries of typical metals in the different processing stages have been suggested (Hassialis 1975):

Mining 75%
Milling 85%
Smelting 90%

Based on these figures, the average overall recovery from deposit to bulk metal is approximately 60% and from mine run ore to bulk metal approximately 77%. However, some metals, and for a given metal, some ores, differ appreciably from these averages. The recovery of copper from the mine run ore to the final wire bar is approximately 85%. The recovery of the aluminum content in producing alumina from bauxite is 85%; this stage accounts for nearly the entire processing loss in the production of aluminum.

As in the materials-producing industries, yields can be increased by gradual process improvements or by innovation. At some point, efforts to increase the yield become uneconomical as the marginal costs of recovery increments become too large. For the productivity of extractive processes, the optimal combination of recovery and grade of the concentrate produced by mineral dressing is the decisive consideration.

An example of process innovation which has increased yields is the argon–oxygen decarburization (AOD) process. This process, which substituted duplex processing under controlled atmospheres for traditional arc furnace practice, has improved the yield of chromium.

In another example, continuous casting has improved the yield of finished from liquid steel, which with ingot-making practice averages about 80%. Improvements in rolling technology have also increased product yields. Changes in product mix affect the yield; for example, switching from heavier to lighter gauges adversely affects the yield of sheet steel production. (Most of the losses incurred in finishing operations are recovered by recycling home scrap; this, however, requires additional inputs of labor, capital and energy.)

Yield considerations also apply to nonmetallic materials. In the production of synthetic materials, yield is a function of the degree of reaction and separation of product. Recycling of unreacted reagents, however, improves the overall yield of synthesis. By contrast, unrecovered metal contents entering mill tailings or furnace slags have generally been lost, though some of these are now being reworked.

3.2 Conversion Efficiency

Another factor determining the productivity of materials utilization is the conversion ratio in manufacturing. The processes involved are largely mechanical operations for the shaping of metals and other materials. These operations tend to produce waste, as is obvious for machining, forging and stamping. Scrap generation in successive manufacturing stages is compounded at each stage (Becker and Pick 1975). Conversion ratios are improved by reducing the amount of scrap and other wastes generated.

The overall rate of scrap generation for iron and steel, aluminum and copper in the manufacture of different products was found to be about 20% for each of these metals (US Department of Commerce 1957). Specific cases can reach extreme values, as in the generation of 85% scrap in the machining of certain aircraft parts from aluminum alloy shapes.

In contrast to machining, chipless metal forming processes generate little or no scrap. The outstanding example is powder metallurgy, which produces final shapes by pressing powders with little waste. Metal powders, however, are expensive and, for economic and technical reasons, powder metallurgy is used mostly for small parts. Another materials conserving process is net shape forging.

As a valid generalization, some types of manufacturing processes are to be preferred over others because they minimize materials waste: additive over subtractive processes; solventless over solvent processes; closed loop over open loop processes; and feedback controlled over sample controlled processes (see *Conservation of Materials: Technology*). A general ranking of metalworking processes in terms of decreasing materials losses is: machining; stamping; hot forging; cold or warm pressing; and die casting (van den Kroonenberg 1979).

4. The Productivity Effect of Materials

The materials used as inputs in a production or manufacturing process affect the productivities of other inputs required by the process. Pick (1972) stated:

> "The manufacturing properties of materials determine the manpower and the level of investment in plant and machines ... required for the manufacture of a given product .... A decision to select a given material and to specify it in a particular way will have a primary effect on productivity; productivity ... is critically 'materials dependent.' "

Frequently, the material is the determining factor in the choice of the process (e.g., tungsten and powder metallurgy). Conversely, the choice of process can determine the material used (e.g., injection molding and plastics).

4.1 Materials Production

The substitution of lower-grade for higher-grade raw materials strongly affects other inputs. The substitution of nonbauxite ores for bauxite as a source of aluminum is likely to require more complicated processing and larger inputs of energy and capital. The adoption of the AOD process involved tradeoffs in stainless steel production between higher chromium yields, the use of a lower grade of ferrochromium and an increase in the ratio of scrap to primary metal input on the one hand and larger inputs of capital, energy and indirect materials on the other.

4.2 Manufacturing

The automobile industry provides examples of the effects of materials substitution on productivity in manufacturing. One such example is the use of steel in place of wood in automobile bodies. The greater temperature tolerance of steel allowed paint to be baked on, which accelerated the finishing process (Abernathy 1978); this reduced the required labor, inventories and space. As another example, the introduction in the 1930s of wide cold-rolled sheets combined with suitable welding techniques altered the design and manufacture of automobile bodies with attendant effects on productivity (Abernathy 1978).

The front end panel of some automobiles is now made by forming a sheet molding compound in one piece. This panel replaces the traditional panel made from ten or more parts requiring separate manufacture and a comparable number of joining operations. The elimination of manufacturing and assembling operations resulting from the substitution of plastics for stamped steel sheet and zinc diecastings has reduced other required inputs. Increased labor productivity has been a factor in the substitution of plastics for wood in the furniture industry. In contrast to molded plastics, however, the use of a plastics matrix reinforced by fibers requires a large input. In particular, filamentary winding and hand layup techniques are intrinsically labor intensive, but this labor intensity can be reduced by automation.

The use of plywood and other sheet materials in building construction reduces labor requirements and increases labor productivity. Plywood is also an example of improved materials productivity (discussed in Sect. 3) since it utilizes a larger fraction of the harvested wood than traditional wood products intended for building use.

The efficiency with which machining operations can be carried out depends on the material of the cutting tool and the material being machined. The use of high-speed steels, special ceramics, polycrystalline diamonds and cemented carbides has permitted large increases in machining speed. This in turn has reduced labor inputs per unit of output (US Department of Labor 1982). The effect on capital productivity, however, is complex. While output per hour increases, machine tools operating at great speeds must be more rigid and thus may increase capital requirements. The changes in construction of machine tools necessitated by greater speed and higher levels of performance have been studied (Boucher 1981). The use of free machining steels where acceptable has an effect on labor productivity similar to that of using high-speed steels or cemented carbides.

Indirect materials also affect productivity. This can be seen for lubricating oils, cutting liquids and heat treating oils.

The substitution of materials and the introduction of new processes have transient effects on other inputs and their productivities. In particular, the productivity of labor is reduced when training and experience are needed for work with a new material or process. This effect was analyzed for the substitution of aluminum for copper in automobile radiators (National Research Council Materials Advisory Board 1972). A temporary decline in materials productivity is likely to result from lost heats and other forms of spoilage that occur during the learning period.

Some developments involving materials broadly affect the productivity of entire industries. An example is the introduction of optical fibers for light transmission, which affects the productivity of the telecommunications industry. The effect of packaging materials on the productivity of labor and materials in the distribution of food is another illustration of the close connection between materials and productivity.

See also: Productivity and Technological Improvements: Managerial Evaluations

Bibliography

Abernathy W J 1978 *The Productivity Dilemma: Roadblock to Innovation in the Automobile Industry.* Johns Hopkins University Press, Baltimore, MD

American Chemical Society 1973 *Chemistry in the Economy: An American Chemical Society Study.* American Chemical Society, Washington, DC

Barnett H J 1979 Scarcity and growth revisited. In: Smith V K (ed.) 1979 *Scarcity and Growth Reconsidered.* Johns Hopkins University Press, Baltimore, MD, Chap. 8, pp. 163–217

Barnett H J, Morse C 1963 *Scarcity and Growth: The Economics of Natural Resource Availability.* Johns Hopkins University Press, Baltimore, MD

Becker P E, Pick H J 1975 Resource implications of materials waste in engineering manufacture. *Resour. Policy* 1: 142–53

Bever M B 1976a The recycling of metals— I. Ferrous metals. *Conserv. Recycl.* 1: 55–69

Bever M B 1976b The recycling of metals—II. Nonferrous metals. *Conserv. Recycl.* 1: 137–47

Bever M B 1977 Systems aspects of materials recycling. *Conserv. Recycl.* 2: 1–17

Boucher T 1981 Technical change, capital investment, and productivity in U.S. metalworking industries. In: Dogramaci A, Adam N R (eds.) 1981 *Aggregate and Industry-Level Productivity Analyses.* Nijhoff, Boston, MA

Boylan M G Jr 1975 *Economic Effects of Scale Increases in the Steel Industry: The Case of U.S. Blast Furnaces.* Praeger, New York

Carter A P 1970 *Structural Change in the American Economy.* Harvard University Press, Cambridge, MA

Christ C F 1961 Comment on the design of consistent indices. *Output, Input and Productivity Measurement: Studies in Income and Wealth.* Princeton University Press, Princeton, NJ

Dasgupta P S, Heal G M 1979 *Economic Theory and Exhaustible Resources.* Nisbet, Welwyn, UK

Fabricant S 1959 *Basic Facts on Productivity Change.* National Bureau of Economic Research Occasional Paper 63. NBER, New York

Fuss M, McFadden D (eds.) 1978 *Production Economics: A Dual Approach to Theory and Applications*, Vol. 2. North-Holland, Amsterdam

Gold B 1955 *Foundations of Productivity Analysis: Guides to Economic Theory and Managerial Control.* University of Pittsburgh Press, Pittsburgh, PA

Gold B (ed.) 1975 *Technological Change: Economics, Management, and Environment.* Pergamon, Oxford

Gold B 1979 Long run prospects for the world steel industry: some internal potentials for improvement. *The Steel Industry in the Eighties.* Metals Society, London, pp. 1–6

Gold B, Soesan J 1976 Complex steel processes. In: Eilon S, Gold B, Soesan J (eds.) 1976 *Applied Productivity Analysis for Industry.* Pergamon, Oxford, Pt. 2, Chap. 5, pp. 75–99

Gollop F M, Jorgenson D W 1980 U.S. productivity growth by industry, 1947–73. In: Kendrick J W, Vaccara B N (eds.) 1980 *New Developments in Productivity Measurement and Analysis.* University of Chicago Press, Chicago, IL, Pt. 1, Chap. 1, pp. 17–124

Gollop F M, Roberts M J 1981 Imported intermediate input: the impact of sectoral productivity in U.S. manufacturing. In: Dogramaci A, Adam N R (eds.) 1981 *Aggregate and Industry-Level Productivity Analyses.* Nijhoff, Boston, MA

Griliches Z, Ringstad V 1971 *Economies of Scale and the Form of the Production Function; An Econometric Study of Norwegian Manufacturing Establishment Data*, Contributions to Economic Analysis Vol. 72. North-Holland, Amsterdam

Hassialis M D 1975 Improvements in minerals recovery. *Proc. Joint Meeting of the National Academy of Sciences—National Academy of Engineering National Materials Policy.* National Academy of Sciences, Washington, DC

Hunt R G, Franklin W E 1973 Environmental effects of recycling paper. *Forest Products and the Environment*, American Institute of Chemical Engineers Symposium Series Vol. 69, No. 133. AIChE, New York, pp. 67–78

Kendrick J W, Creamer D 1965 *Measuring Company Productivity: Handbook with Case Studies*, National Industrial Conference Board Studies in Business Economics No. 89. National Industrial Conference Board, New York

Leontief W W 1966 *The Structure of American Economy, 1919–1939: An Empirical Application of Equilibrium Analysis*, 2nd edn. International Arts and Sciences Press, White Plains, NY

Leontief W W 1986 *Input–Output Economics*, 2nd edn. Oxford University Press, New York

Lovering T S 1969 Mineral resources from the land. *Resources and Man: A Study and Recommendations.* National Academy of Sciences, Washington, DC, Chap. 6, pp. 109–34

Malenbaum W 1972 *Materials Requirements in the United States and Abroad in the Year 2000.* National Commission on Materials Policy, Washington, DC

Manthy R S 1978 *National Resource Commodities—A Century of Statistics: Prices, Output, Consumption, Foreign Trade, and Employment in the United States, 1870–1973.* Johns Hopkins University Press, Baltimore, MD

National Research Council Materials Advisory Board 1972 *Mutual Substitutability of Aluminum and Copper.* National Academy of Sciences, Washington, DC

National Research Council, Panel to Review Productivity Statistics 1979 *Measurement and Interpretation of Productivity*. National Academy of Sciences, Washington, DC

Nerlove M 1965 *Estimation and Identification of Cobb–Douglas Production Functions*. Rand McNally, Chicago, IL

Pick H J 1972 Some economic consequences of technical decisions. *Mater. Sci. Eng.* 10: 301–23

Roberts M C 1991 Advanced materials and trends in the material composition of product. *Mining Engineering Transactions*, 1991 Annual Volume

Rosegger G 1980 Basic oxygen furnace vs. open hearth: empirical findings. In: Gold B, Rosegger G, Boylan M G Jr (eds.) 1980 *Evaluating Technological Innovations: Methods, Expectations, and Findings*. Heath-Lexington, Lexington, MA, Pt. 2, Chap. 8, pp. 155–74

Salter W E G 1966 *Productivity and Technical Change*, 2nd edn. Department of Applied Economics Monographs 6. Cambridge University Press, Cambridge

Sato R, Ramachandran R 1980 Measuring the impact of technical progress on the demand for intermediate goods: a survey. *J. Econ. Lit.* 18: 1003–24

US Department of Commerce, Business and Defense Services Administration 1957 *Industrial Scrap Generation*. US Government Printing Office, Washington, DC

US Department of Labor, Labor Statistics Bureau 1982 *Technology and Labor in Four Industries*: *Meat Products, Foundries, Metalworking Machinery, Electrical and Electronic Equipment*. US Government Printing Office, Washington, DC

US House of Representatives, Ninety-Sixth Congress, 1st Session, Committee on Science and Technology 1979 *National Science and Technology Policy Issues, 1979*. US Government Printing Office, Washington, DC

van den Kroonenberg H H 1979 Solutions to future resources problems by engineering design. *Mater. Soc.* 3: 271–9

Wei J, Russell T W F, Swartzlander M W 1979 *The Structure of the Chemical Processing Industries: Function and Economics*. McGraw-Hill, New York

M. B. Bever†
[Massachusetts Institute of Technology, Cambridge, Massachusetts, USA]

S. Nasar
[Tarrytown, New York, USA]

Productivity and Technological Improvements: Managerial Evaluations

One of the extraordinary anomalies of past efforts to appraise the effectiveness of industrial operations has been the continuing divergence among the criteria employed by the engineers directly involved in improvement efforts, by economists seeking to measure resulting progress and by the managements whose evaluations decide which among available innovations are to be adopted. The resulting disparity is generated partly by the ambiguity of the concept of operational effectiveness or "productivity," for it encompasses multiple dimensions and levels of aggregation, and partly by the tendency of each group of analysts to concentrate on the particular dimensions and hierarchical levels of greatest relevance to their own purposes.

Thus, engineers tend to concentrate on the relationships between the physical magnitudes of the specific inputs for which they are responsible and resulting changes in the physical output levels of the operations affected by them. Economists tend to focus rather on the more aggregative relationships between total inputs and total outputs—sometimes at the level of firms, but more often at the level of industries and entire national economies—using financial values as the basis for combining the physical quantities both of various inputs and of various products (Baird 1977, Erdilek 1977). Managements, however, because of their responsibilities for evaluating and improving all operations under their control, have to be concerned with a structure of successively more aggregative physical as well as financial measures reaching from individual production units to total plant or firm operations.

In order to analyze changes in the effectiveness or "productivity" of industrial operations, therefore, the basic need is not to choose among these different criteria, but to develop and utilize those which are most relevant to the purpose of the inquiry. Because applications of technological innovations and of other sources of productivity improvements tend to depend on managerial approval, it is of critical importance for an understanding of such progress in industry to clarify the evaluative criteria and related considerations which underlie adoption decisions.

1. Requirements for Managerial Evaluations of Operating Performance

1.1 Dimensions of Effectiveness in Production

Contrary to widespread beliefs, it is inherently impossible to measure the "physical efficiency" of manufacturing processes, for the concept is based on a false analogy. Such a measure would require comparison of a combined measure of all relevant inputs with a combined measure of all relevant outputs in terms which reflect the primary purposes of the undertaking. Thus, the physical efficiency of an engine with respect to energy conversion can be calculated by comparing the energy content of its fuel consumption with the energy equivalent of the useful power delivered by it. But it is not possible to measure the physical efficiency of the engine as a whole, or of the process of producing it.

This inability is attributable to the absence of any important physical common denominators for combining the contributions to production of labor efforts and skills, many kinds of materials and supplies, even greater varieties of facilities and equipment, and a wide array of technical and managerial services, in view of the obvious irrelevance of the available measures of numbers, volume or weight. Nor are there important physical common denominators for combining the wide range of quality and service characteristics which differentiate the many available types of engines from one another. Hence, instead of seeking to measure the efficiency of any production process, such studies must be designed to appraise the effects on a production system's performance objectives of the interacting effects of changes in the array of input–output relationships which it encompasses.

This conception also clarifies the serious shortcomings both of "partial" productivity measures, each comparing the quantity of a given category of inputs with total output, and also of overly aggregative "production function" and "total productivity" measures. The former, which are commonly illustrated by such measures as output per man-hour or per unit of energy consumed, implicitly attribute to changes in the selected input factor the entire accompanying change in output, although the latter actually represents the contributions and interactions of changes in each of the inputs utilized as well as of changes in the product composition of output. At the other extreme, "total factor productivity," which compares changes in total product value at fixed product prices with concomitant changes in total costs at fixed factor prices, combines the effects of so many changes in relevant variables as to make interpretations of the resulting amalgam highly vulnerable (Gold 1982). Moreover, merely determining that output per unit of any specified input, or that "total factor productivity," had increased or decreased by 5% last year reveals nothing about what caused the change, or how rewarding or burdensome it was, or what might be done to improve future results.

In order to fulfill management's need to evaluate and improve operating performance, productivity analysis must reach beyond customary simplified approaches to cover:

(a) changes in the level of each category of input requirements per unit of output, including materials and capital goods as well as labor;

(b) changes in the proportions in which inputs are combined, in order to take account of substitutions (for example, buying more highly fabricated components instead of making them) and also in order to differentiate between changes in the effectiveness with which major as against minor inputs are used;

(c) differences between the productive contributions of inputs when they are fully utilized and when they are partly idle (as in the case of underutilized equipment);

(d) changes in input prices attributable to alterations in their quality as differentiated from changes in the prices of identical inputs; and

(e) changes in product-mix (to take account of differences among products in their input factor proportions as well as in their relative prices) as differentiated from changes in the volume of output of an unchanged mix of products.

1.2 The Network of Productivity Relationships

One means of meeting these needs is offered by the model in Fig. 1 covering direct inputs. It identifies six components of the "network of productivity relationships." Three cover the input requirements per unit of output of materials as well as of labor and fixed capital. The latter input is compared with productive capacity rather than with output in order to take full account of the contribution that it could make if fully utilized. The remaining three links cover the proportions in which these inputs are combined, covering the possibilities, for example, that more highly processed materials or additional equipment may be substituted for labor. Because decreased market demand may lead to the underutilization of capital facilities, the factor proportions relate labor and materials to "actively utilized" net fixed investment.

By presenting productivity relationships as a network of interactions, this approach emphasizes that changes can be initiated within any of the six components and also that the effects of any change must then be traced through successive interactions around the network until the production process has been effectively reintegrated. Thus, a shift to more highly processed material inputs would involve an increase in materials volume per unit of output, along with increases in the ratio of materials to man-hours and to actively utilized net fixed investment, as the more highly processed materials eliminate the need for the labor and equipment inputs which had been required to process the former cruder materials up to the levels now provided by suppliers. These interactions would lead in turn to an increase in output per man-hour and to an increase or no change in the ratio of capacity to net fixed investment, depending on whether the capital goods made redundant by the more highly processed materials could be utilized for other purposes, or could be sold, or would be left idle.

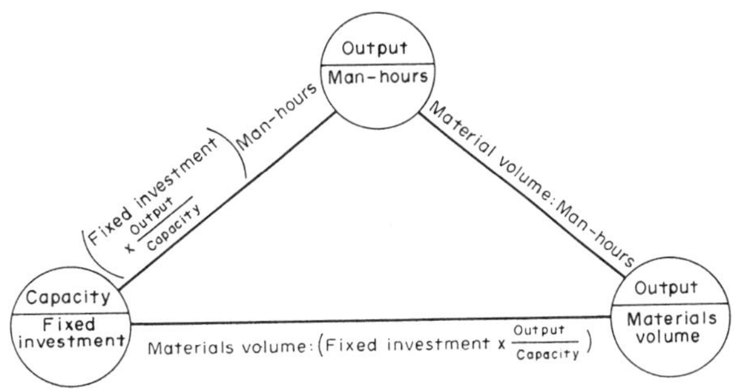

Figure 1
The network of productivity relationships among direct input factors (after Gold 1982 p. 48)

The preceding example illustrates two additional important principles in appraising changes in the network of productivity relationships. First, a decrease in the ratio of output to one of the inputs may be entirely favorable if it engenders more than offsetting gains in the output per unit of requirements of the other inputs. Second, a change in any of the six components need not originate at that point—as in the case of an increase in output per man-hour actually attributable either to an increase in the productive capabilities of the capital facilities and equipment manned by an unchanged labor force, or to the increasing mechanization of formerly manual tasks. Thus, Fig. 2 shows that increases in output per man-hour in the basic steel industry of the USA were almost entirely attributable during 1904–1965 to increases in the productivity (i.e., capacity) of net fixed investment, due to advances in technology and increases in the scale of operations. Thereafter, however, further gains in output per man-hour were due rather to increases in the volume of actively utilized facilities per man-hour because sharp increases in the price of capital facilities increased net fixed investment more rapidly than capacity (Boylan 1975, p. 162).

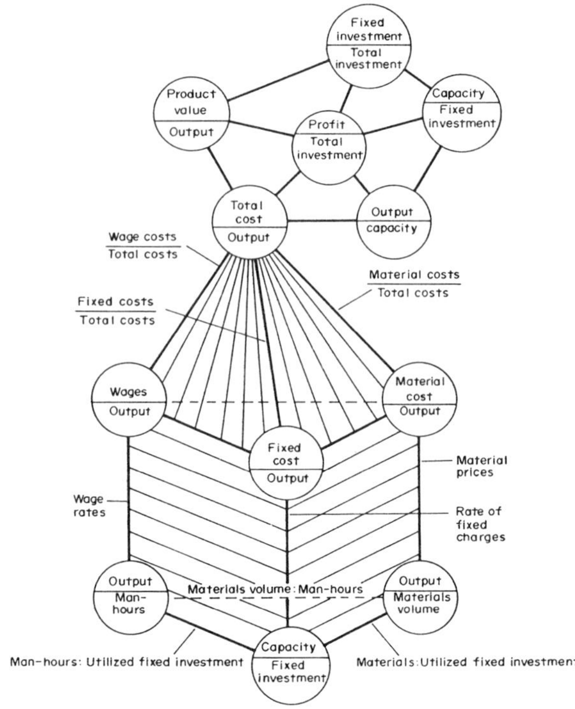

Figure 3
Productivity network, cost structure and managerial-control ratios (after Gold 1982 p. 50)

2. *Integrating Productivity Network Analysis into Cost and Profit Analysis*

In order for management to evaluate the economic effects of past or prospective changes in the network of productivity relationships, however, the analysis must also be extended to cover resulting changes in costs as well as in profitability. These may be examined systematically by superimposing "the structure of costs" and "managerial control ratios" on the productivity network (Fig. 3).

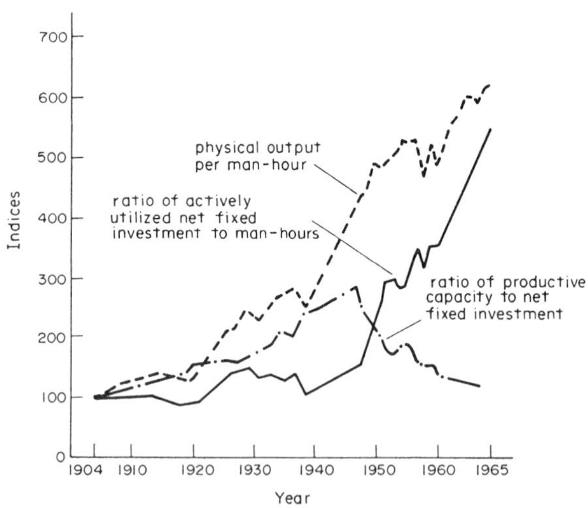

Figure 2
Basic steel: physical output per man-hour, productive capacity relative to net fixed investment and actively utilized net fixed investment relative to man-hours for basic steel industry, 1904–1965 (1904 = 100) (after Gold 1982 p. 179)

2.1 Productivity Changes and the Structure of Costs

The effect of changes in output per man-hour on unit wage costs depends, of course, on concomitant changes in wage rates. Similarly, the effect of changes in the volume of unit material requirements on unit material costs depends on accompanying changes in the prices of such purchased inputs. The effects of changes in the productivity of net fixed investment on capital charges per unit of output depend on the annual charges on such investment, on the ratio of capacity to fixed investment and on the rate of utilization of capacity. The point being emphasized is the necessity of considering possible interactions between productivity adjustments and factor prices instead of assuming that the latter necessarily remain unchanged.

Specifically, increases in output per man-hour commonly engender largely or totally offsetting increases in average hourly earnings as a result of various factors including piece rates, or wage incentives, or trade union pressures for proportionate gains in wage rates (Fig. 4). But increases in output relative to material requirements are more likely to yield comparable and even enhanced benefits. For example,

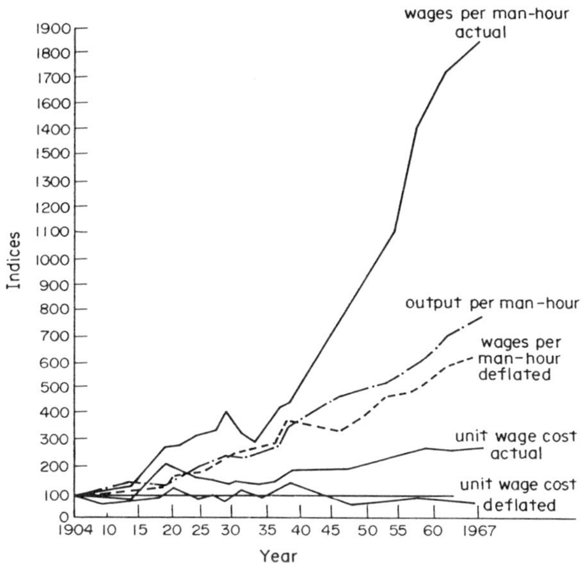

Figure 4
Basic steel: unit wage costs, output per man-hour and wage rates, 1904–1967 (1904 = 100) (after Gold 1982 p. 178)

reducing such unit input requirements by decreasing material wastage and spoilage need not involve any offsetting increase in materials prices. When such unit input requirements are reduced through establishing tighter dimensional or higher quality specifications, however, the short-term effect may be to increase their prices to some extent. But the increased diffusion of material-saving practices, whether through waste reduction or through improved specifications, will tend to reduce the demand for such materials relative to the output level of users, and hence to encourage lower prices over the longer run than would otherwise have prevailed, thereby further augmenting the benefits of the productivity improvement.

Resulting effects on total unit costs depend in turn on the relative changes in each unit cost multiplied by its share of total costs. For example, if wages account for 20% of total costs, a decrease in unit wage costs of 5% would tend to reduce total unit costs by only 1%, whereas an equal relative decrease in unit material costs would tend to have three times as great an effect on total unit costs when materials account for 60% of total costs (both cost proportions representing reasonable approximations to the average in all manufacturing industries combined in the USA).

2.2 *Productivity, Costs and Profitability*

Managerial decisions in private industry, however, are much less concerned with minimizing total unit costs than with increasing the rate of profit on investment. Accordingly, the productivity network and the structure of costs must be further integrated with some model of the determinants of changes in profitability. According to the managerial control ratios model shown in Fig. 3, variations in the rate of profits on total investment are traceable to changes in, and interactions among, average product prices, total unit costs, capacity utilization, the productivity of net fixed investment and the internal allocation of investment between fixed assets and working capital:

$$\frac{\text{Profit}}{\text{Total investment}} \equiv \left(\frac{\text{Product value}}{\text{Output}} - \frac{\text{Total costs}}{\text{Output}}\right) \times \frac{\text{Output}}{\text{Capacity}} \times \frac{\text{Capacity}}{\text{Net fixed investment}} \times \frac{\text{Net fixed investment}}{\text{Total investment}}$$

Within this integrated productivity–cost–profitability system, it should be emphasized that changes in the productivity network may affect profitability not only through total unit costs and capacity relative to net fixed investment. In addition, any associated changes in the quality of products could also affect the average prices charged and the rate of capacity utilization attained.

In short, the preceding discussion suggests that:

(a) changes in the productivity of material inputs, although widely ignored except for recently developing concern about the energy component of such inputs, offer potentially greater impacts on the total unit costs of most manufacturing industries than changes in any other cost category;

(b) increases in the productivity of materials tend to yield greater net benefits than comparable increases in the productivity of labor, because the latter are far more likely to engender largely or wholly offsetting gains in the price of such inputs; but

(c) the ratio of output to the volume of material inputs measures only the "apparent productivity of materials," because resulting changes may actually be due to a variety of possible changes in the quantity, quality, prices and mix of products as well as of all other inputs. Hence, some decreases in the ratio of output to materials volume may be as beneficial as some increases in such "apparent productivity," depending on the combination of adjustments underlying the changes in this measure.

3. *Major Sources of Changes in the "Productivity" or Effectiveness of Industrial Operations*

3.1 *General Sources of Change in Productivity*

Among the multiplicity of changes which affect the productivity of industrial operations, there are eight categories which tend to have significant effects. These are: changes in product design, changes in the product composition of output, changes in the nature of production processes, changes in the scope of processing and fabrication operations, changes in the effectiveness with which operations are integrated, changes in the productive capacity of operations, changes in the degree of utilization of available capacity, and changes in the quality and composition of inputs utilized (Gold 1955).

In exploring the effects of the foregoing sources of potential change, it is important to recognize not only that each tends to interact with one or more of the others, but also that they tend to differ in the frequency with which each is likely to undergo significant changes. Those which are most likely to account for short-term variations in productivity levels are changes in: product mix, product designs, capacity utilization rates, the quality of labor and materials, and the effectiveness with which operations are integrated. Over the longer run, productivity levels are most likely to be affected by: advances in production processes; improvements in the integration of operations; advances in the quality of materials and capital goods; and changes in the scope of production (involving forward or backward integration or changes in the degree of diversification).

3.2 *Sources of Change in the "Apparent Productivity" of Individual Inputs*

As was indicated earlier, changes in output relative to any given input are affected not only by changes in the contributions by the given factor but also by changes in the contributions of all other factors entering into the output. Hence, such measures may be more accurately described as measuring only the "apparent" rather than the actual productivity of that factor.

Adjustments in the contributions derived from any category of inputs may be attributed to four types of change: in the average quality of this input, which determines its potential contributions in each task; in the effectiveness with which the tasks are designed to harness such potentials, thereby affecting the actual contributions derived from the input while it is employed; in the level of idleness, underutilization or wastage of the factor during the period of operations; and in the combination of higher and lower productivity tasks assigned to this input factor, which tends to alter the average level of its contributions even if its contribution in any given task remains unchanged.

Thus, changes in the productive contributions derived per unit of labor may be rooted in each of the above factors. For example, changes in labor's performance potentials might reflect its physical capabilities; its training, skills and experience; and the effectiveness of its work incentives, including wage arrangements, job security and the satisfactoriness of labor–management relations. Actual productivity in given tasks would be affected by changes in performance specifications, work tempo, working conditions and quality specifications, as well as by changes in the degree of idleness during working hours and in the volume of paid absences from work. In addition, the plant-level productivity of labor would obviously be affected by changes in the combination of relatively high- and low-productivity tasks performed by wage earners as a result of changes in the product composition of output, in the level of fabrication of purchased materials, in the production technologies and equipment employed, and in the scope of the production processes.

Similarly, changes in the volume of material inputs per unit of output can be affected by changes in the level of total output, in the design and quality

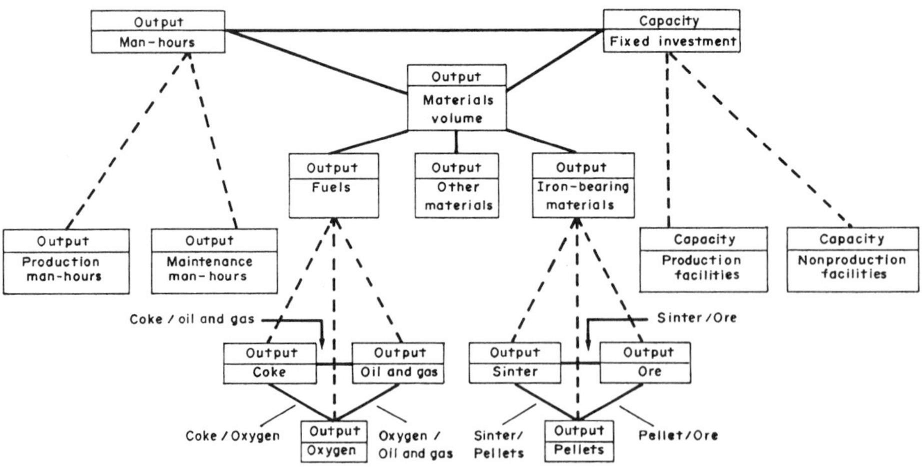

Figure 5
Elaborated productivity network for blast furnaces (after Gold 1982 p. 54)

specifications of individual products, and changes in the product composition of output. In addition, however, changes in the "apparent productivity" of materials may also be engendered by alterations: in the proportions of different material inputs purchased; in the level of processing as well as in the quality grades of each category of such inputs; and in the ratio of waste and rejects as well as in offsetting gains in the use or sale of salvaged materials and by-products. Such more detailed probing of the sources and effects of changes in materials inputs per unit of output may be facilitated by the extension of Fig. 3 represented by Fig. 5, which illustrates the further decomposition of unit material costs for blast furnace operations in the steel industry.

See also: Factor Dominance in Industry; Input–Output Analysis; Input–Output Analysis: Applications; Productivity and Materials

Bibliography

Baird R N 1977 Production functions, productivity and technological change. In: Gold B (ed.) 1977, pp. 11–41

Boylan M G 1975 *Economic Effects of Scale Increases in the Steel Industry: The Case of U.S. Blast Furnaces.* Praeger, New York

Boylan M G 1977 Reported economic effects of technological changes. In: Gold B (ed.) 1977, pp. 131–65

Chew W B 1988 No nonsense guide to measuring productivity. *Harv. Bus. Rev.* 66: 110–18

Dogramaci A (ed.) 1981 *Productivity Analysis: A Range of Perspectives.* Nijhoff, Boston, MA

Erdilek A 1977 Productivity, technological change and input–output analysis. In: Gold B (ed.) 1977, pp. 43–85

Freeman C 1989 Technical change and productivity. *Finance Dev.* September

Gold B 1955 *Foundations of Productivity Analysis: Guides to Economic Theory and Managerial Control.* University of Pittsburgh Press, Pittsburgh, PA

Gold B 1971 *Explorations in Managerial Economics: Productivity, Costs, Technology and Growth.* Macmillan, London

Gold B (ed.) 1975 *Technological Change: Economics, Management and Environment.* Pergamon, Oxford

Gold B (ed.) 1977 *Research, Technological Change and Economic Analysis.* Heath–Lexington, Lexington, MA

Gold B 1982 *Productivity, Technology and Capital: Economic Analysis, Managerial Strategies and Governmental Policies.* Heath–Lexington, Lexington, MA

Gold B, Peirce W S, Rosegger G 1975 Diffusion of major technological innovations. In: Gold B (ed.) 1975, pp. 129–48

Gold B, Rosegger G, Boylan M G 1980 *Evaluating Technological Innovations: Methods, Expectations and Findings.* Heath–Lexington, Lexington, MA

Rosegger G 1975 Technological change, managerial incentives and environmental effects. In: Gold B (ed.) 1975, pp. 43–64

B. Gold
[Claremont Graduate School, Claremont, California, USA]

R

Reclaiming of Elastomers

The terms "reclaiming" and "recycling" are generally used with different connotations when applied to the rubber and elastomer fields. Recycling is a relatively new term that is being used in a broader context than reclaiming. Recycling did not actually begin to appear commonly in dictionaries until the early 1970s. Reclaiming, on the other hand, was defined in Webster's Seventh New Collegiate Dictionary (1965) as "to obtain from a waste product" or "to obtain the return of."

Reclaiming of some types of scrap materials is easy to accomplish and is driven by economics. However, many of the materials currently targeted for recycling and facing various obstacles are driven by environmental awareness and may not be economically recyclable.

Normal requirements for when economics are driving the reclaiming/recycling effort are as follows.

(a) The scrap must be easily collectable in sufficient quantities to justify the effort required to recycle and to yield a reasonable return on the required capital investment.

(b) The scrap should be relatively uniform in composition with only minimal contamination.

(c) Scrap must be easily processed without appreciable change in the physical properties of the materials.

(d) The recycled material must be reusable at some level in the original application without appreciable degradation in properties and/or impart other desirable properties that will foster its use.

(e) The scrap collection system must yield a reasonable financial return for each participant in the cycle (i.e., generator, collector, hauler and final user).

The economic and social value of recycling elastomers for reuse was recognized even before vulcanization was discovered. Goodyear's invention was not only the start of the rubber industry but also of the rubber reclaiming industry.

Recycled crumb rubber and reclaimed rubber make up the bulk of recovered elastomeric materials at present available to the materials engineer. Unvulcanized and thermoplastic elastomers present no special reclaiming problems since they can be simply chopped, milled and blended with new materials for recycling directly back into the same or equivalent products as those from which they came.

Many elastomers have only limited utility in the virgin state and must be compounded with reinforcing materials and vulcanized to develop the desired physical properties. Vulcanization of an elastomer serves to crosslink chemically the very long chains of a polymer so that their movement is restricted. This changes the character of the polymer from a millable, plasticlike material to a dry, nontacky thermoset which, in most cases, cannot be replasticized by simple milling.

Most vulcanized elastomers can easily be mechanically ground to a fine particle size with minimal agglomeration. Mechanical grinding, accompanied by the removal of contaminants such as wire and fiber, was one of the first methods developed for recycling elastomers. It is still widely used. None of the compounding ingredients contained in the scrap elastomer are removed during mechanical grinding so the physical properties remain essentially unchanged. The quantity and gradation of the crumb to be used in a new compound are dictated by the properties of the desired finished product.

Reprocessed rubber involves taking used rubber products, such as tires, trim, scrap or flash, and processing them in a manner that produces a form of rubber filler or ingredients that can be reincorporated into virgin rubber compounds.

There are two general methods for producing reprocessed rubbers:

(a) the severing of crosslinks by chemical or steam digestion to produce a product known as reclaim; and

(b) the grinding of rubber compounds by ambient grinding, cryogenic grinding or solution grinding in water.

This article reviews the various methods used to produce recycled rubbers and compares their characteristics and applications.

1. Current Usage

All the methods of recycling mentioned previously are currently in use in the rubber industry. Chemical digestion was extensively practised for many years with the products being used mainly in tire compounding. However, the advent of radial tires requiring high green strength for processing soon reduced or eliminated reclaim from tires. It is estimated that during and after World War II reclaim rubber usage reached a volume of 2.7–2.9 $\times$ 10^8 kg year^{-1}. This has fallen drastically to a current level of 0.4–0.5 $\times$ 10^8 kg year^{-1} of reclaim with about half of this used in tires. The major single component is reclaimed butyl. It represents perhaps 40% of all chemical reclaim used.

There are three methods of grinding rubber used in the USA. These are ambient grinding, cryogenic

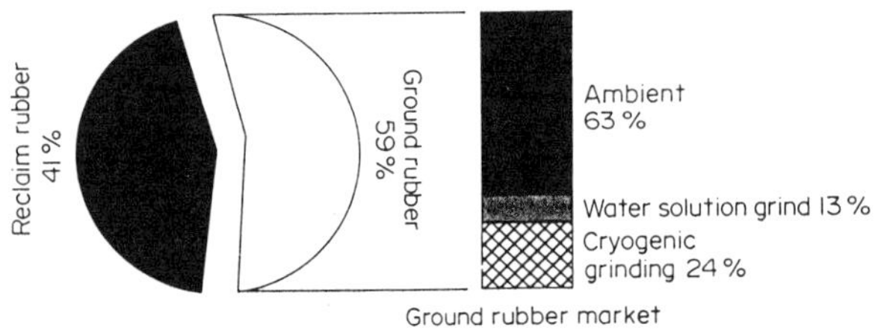

Figure 1
US recycled rubber market in 1990 (total market 110×10^6 kg)

grinding and solution or wet grinding. It is estimated that 0.5–0.6 $\times$ 10^8 kg of rubber are ground in the USA using these methods. About 70–75% of this is reincorporated into rubber compounds with the remainder being added to other products such as asphalt (see Fig. 1).

2. *Reclaim Rubber*

The process of reclaiming rubber by chemical digestion was discovered over 100 years ago and, with minor modifications and improvements, has remained in use. It involves the use of chemical reagents, such as mercaptans, that react with the carbon–carbon bonds. This allows control of the devulcanization process. In addition, the use of refining mills provides a method of producing the reclaim more homogeneously and uniformly.

The major benefit of using reclaim is that it lowers cost. Reclaim typically sells for 20–30% of the cost of virgin rubbers. In addition to cost reduction, reclaimed rubber also imparts some very desirable improvements in processing. Since reclaim has much lower nerve than virgin polymers, compounds containing reclaim have a much lower die swell. This results in uniform die swell and extrusion rates. It also increases calender rates and, in general, improves flow and mold filling.

One of the shortcomings of reclaim is that it lowers the green strength and tensile strength of the compounds in which it is used. Reclaim is in fact a mixture of rubber, carbon black, oil, zinc oxide, stearic acid and other compounding ingredients used in the original compounds. The levels of reclaim used are generally 15–45 parts per hundred (phr), about half of which is rubber and the other half compounding ingredients. In some low-performance-requirement rubber products, such as mats, much higher levels can be used.

Steam digestion of silicone rubber produces a useful silicone reclaim. The product is used to reduce the cost of silicone compounds. It has been widely used in silicone automotive ignition systems. The resulting compounds have excellent aging characteristics and retain their outstanding electrical properties.

Reclaim was extensively used in bias tire components for many years, but the advent of the radial tire and its need for increased properties has all but eliminated this use. However, large volumes of some high-quality rubbers, such as natural rubber and butyl, are still being reclaimed. These are used in some solid tires and other mechanical goods where they extend the compound without having too adverse an effect on the product. Butyl reclaim made from tubes is still used in the inner liner of both bias and radial compounds while some reclaim still finds application in low-performance bias tires.

3. *Ground Rubber*

The first step in the manufacturing of reclaim is grinding of the rubber part to be reclaimed. The rubber or rubber fiber mixture is ground to increase the surface area. This increases the rate of the chemical reaction in reclaiming and produces a more uniform product. As mechanical grinding became more effective, it was found that small-particle-size vulcanized ground rubber could be added to the rubber compound to reduce cost. Ground rubber used in compounding varies from 10 mesh to 200 mesh. Figure 2 shows particle sizes for comparison.

3.1 *Ambient Grind*

Using effective types of mechanical grinding techniques such as serrated grinders, recyclers grind various types of scrap and used rubber articles. Articles with steel and fiber have the reinforcing materials removed by various mechanical methods. The rubber is then ground to a particle size in the range 10–30 mesh.

Ambient ground rubber is widely used in tires and mechanical goods. It is generally used at levels of 5–20 phr. The larger the particles (lower mesh size), the rougher the processing compounds will be that contain

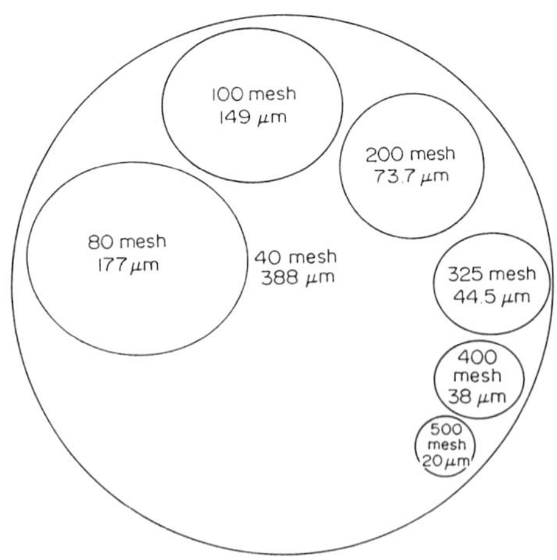

Figure 2
Particle size comparison

Table 1
US recycled rubber usage in 1990

Reclaim	Ground rubber		
	Ambient ground	Cryogenically ground (CGR)	Wet or solution grinding
40–45 × 10^6 kg 14–16 × 10^6 kg butyl Others include NR, PBD, SBR and specialty polymers	36–40 × 10^6 kg 18–20 × 10^6 kg in asphalt modification	14–16 × 10^6 kg	6–8 × 10^6 kg

ambient (room-temperature ground). The name "ambient grind" is something of a misnomer since the grinding generates some heat during processing. With high-modulus or high-durometer compounds the amount of heat generated can be high, resulting in heating of the rubber and perhaps some degradation.

3.2 Cryogenically Ground Rubber

In the mid-1960s, the technique of grinding scrap rubber and trim in liquid nitrogen was developed. The process involves placing small pieces (1 in × 1 in × $\frac{1}{2}$ in) into liquid nitrogen and grinding them into a fine powder. The particle size varies from 30 mesh to 100 mesh for most products. The particle size is controlled by the time spent in the liquid nitrogen and by screens placed in the grinding chamber controlling the discharge of the rubber.

Cryogenically ground rubber (CGR) is used in tires, hose, belts and mechanical goods, wire and cable, and numerous other applications. It is especially useful in producing a product for tire inner liners. The particle size chosen is controlled by cost and the fineness needed to achieve the desired processing. The finer the particle size, the smoother the calendered sheets and the finer the edge that can be produced on extrusions. Generally, the cost increases as the particle size decreases.

The inclusion of CGR can improve physical properties, as in the following.

(a) During molding, CGR in all mesh sizes aids in removing trapped air. The cured rubber particles provide a path for the air to escape by bleeding air from the part. Once cured, the compound behaves like a composite with the impermeability being determined by the composition of the polymer blend.

(b) Mold flow has been observed as generally improving with CGR addition.

(c) Shrinkage is usually less for compounds containing CGR. The shrinkage reduction is proportional to the amount of CGR in the compound. If CGR is used in concentrations sufficient to reduce shrinkage (5–10%), less mold hang-up occurs with intricate parts or undercut parts.

It has been found that certain particle sizes are more suitable for specific applications. For optimum cost and performance benefits, use of a finer mesh size (60–100 mesh) and somewhat lower percentages (5%) are recommended for extrusions and calendered products where feathered edges and smoothness are critical. For less critical conditions, 40–60 mesh at 10–20% usage is practical. Molding will generally benefit in terms of mold flow and reduction of blisters when particle sizes are around 40 mesh.

3.3 Wet or Solution Process Grinding

The Gould Company and GenStar both developed an ambient grinding process that reduces the particle size of rubber by grinding in a liquid medium. The process involves putting a coarse ground rubber crumb (10–20 mesh) into a liquid medium, usually water, and grinding between two closely spaced grinding wheels. The particle size is controlled by the time spent in the grinding process. By placing a screen of the desired size at an exit port, the ultimate particle size can be controlled. Particle sizes as small as 400–500 mesh (20 μm) have been reported, while the commercial sizes are commonly reported in the 80–200 mesh range.

Wet-ground rubber is used in many of the same applications as ambient and cryogenically ground rubber. This includes tires, retreads, solid tires, and molded and extruded mechanical goods. It is usually used at levels of 5–20 phr. The advantage of the fine-particle ground rubber is that it allows good processing, producing fairly smooth extrudates and calendered sheets.

3.4 Applications

Use of the different types of ground rubber has been growing steadily. Ground rubber has replaced reclaim in many tire applications including treads, carcasses, inner liners, subtreads and bead components. It is also used in mechanical goods, footwear, solid tires, mats and retreads.

In cryogenically and wet-ground rubbers, smaller particle size allows recycled rubber to be used at moderately high levels while still allowing compounds to be processable. The use of these products and ambient ground rubbers will continue to grow. Table 1 summarizes the types and amounts of various recycled rubbers used in the USA in 1990.

See also: Recycling of Plastics

Bibliography

Ball J M 1956 *Manual of Reclaimed Rubber.* Rubber Reclaimers Association, Akron, OH

Bowers B, Barber D, Allinger R 1986 A profile in silicone reclaim. *ACS Rubber Division Meet.* October

Eckart D 1980 Cryogenics advances ground rubber technology. *Modern Tire Dealer* June

Eggleton R, Fesus E 1986 Reprocessing uncompounded elastomers. *ACS Rubber Division Meet.* October

Lynch J, LaGrone B 1986 Ultrafine crumb rubber. *ACS Rubber Division Meet.* October

Midwest Elastomers Inc. 1990 *Fine Particle Rubber Technology: Properties and Processing.* Midwest Elastomers, Wapakoneta, OH

Morton M (ed.) 1987 *Rubber Technology,* 3rd edn. Van Nostrand-Reinhold, New York

Nakajima M, Harrell E R, Seil D A 1986 Rubber processing. In: Bever M B (ed.) 1986 *Encyclopedia of Materials Science and Engineering.* Pergamon, Oxford, pp. 4276–8

Novotny D, Marsh R, Masters F, Tally P 1976 Microwave devulcanization of rubber. US Patent No. 4,104,205

1978 Scrap tire disposal procedures. *Rubber Chem. Technol.* 51(3)

Smith F 1978 Reclaim rubber. *R T Vanderbilt Handbook.* Vanderbilt, Norwalk, CN

Smith F 1986 Cryogenically ground rubber. *ACS Rubber Division Meet.* October

Smith F, Klingensmith W 1990 Tires as a source and consumer for recycled rubber. *ACS Rubber Division Meet.* October

Snyder R 1986 The shape and size of the scrap tire problem and some potential solutions. *Clemson University Tire Technology Conf.* October 29

Stark J III 1987 The tire cycle solution: Minnesota's answer to the scrap tire problem. *Clemson University Tire Technology Conf.* October 28–29

Sudekum J H, Varga R S 1986 Elastomers: uses. In: Bever M B (ed.) 1986 *Encyclopedia of Materials Science and Engineering.* Pergamon, Oxford, pp. 1346–8

Swor R, Jensen L, Budzol M 1980 Ultrafine recycled rubber. *ACS Rubber Division Meet.* May

Winters R 1986 Ambient ground rubber crumb. *ACS Rubber Division Meet.* October

B. D. Lagrone
[US Rubber Reclaiming Company,
Vicksburg, Mississippi, USA]

F. G. Smith
[Environmental Technology Alternatives,
Lima, Ohio, USA]

Recycling of Demolition Wastes

The costs of land disposal of demolition wastes—and the closely related construction wastes—are continually growing and the practice is increasingly subject to limiting regulations because of concerns over, principally, ground-water contamination. A new and significant incentive towards recycling has therefore been added. In other respects, the large volume and low specific value of demolition and construction materials lead to a level of recycling that depends not only on where and in what condition and quantity waste materials are found, but also on the materials themselves.

In this article the approximate flows of different demolition materials from different sources are given, and the manner in which the principal materials are recycled is reviewed. The data presented come from US and European studies, but the proportions of the materials involved should be broadly applicable to the demolition of similar structures elsewhere.

The origins of demolition materials can be classified into two groups: pavements and structures. Pavements include highways, sidewalks, parking lots, malls, airport runways, quays and stockyards; these are all cases where a hard, strong surface has been laid over the ground to withstand the passage of motor vehicles, people or animals. Structures include buildings, bridges, tunnels, ramps and walls. The distinction is significant because pavements are usually demolished only for repair, and not for a complete change of form and use, as is often the case for structures. As energy, materials and landfill space become relatively more expensive, there is an increasing tendency for pavement materials to be recycled *in situ* and be incorporated into the new pavement. The proportion of pavement materials being recycled in this way is, however, still small; it is almost certainly under 10% of available wastes. In particular areas (e.g., Berlin, Germany), the publication of guidelines and encouragement by local government have increased pavement recycling from 15% to 36% (Dudenhöfer and Herper 1989).

In situ recycling is more usual with asphalt pavements, which can be lifted by travelling machines, crushed, screened, mixed with new binders and pressed or rolled back into place. Concrete, whether from pavements or structures, can also be crushed and screened, and that fraction falling within certain size limits can be incorporated into new concrete in place of aggregate (Nixon 1976, Frondistou-Yannas and Ng 1977, Mather 1980, Sierra 1980). The strength of concrete made in this way is typically reduced to 80% of that made with crushed stone, but it is normally adequate for use in pavements. However, machinery capable of crushing large pieces of concrete is not yet portable enough to be made part of pavement resurfacing equipment. The presence of steel reinforcing mesh in pavement concrete, or heavier reinforcing bars in most structural concrete, is a complicating factor that only a few plants are equipped to handle, and most demolition concrete is used as coarse-fill or base material. At present in the USA and in the European Community (EC), the demand for this material is such that it can be dumped on site free of charge (Wilson et al. 1979, Environmental Resources 1980). Crushed and graded concrete can be sold. Contractors wishing to dispose of mixed demolition wastes, particularly mixtures including wood and other degradable or combustible material, must often transport them considerable distances (typically 50 km in the neighborhood of large cities) and pay an ever-increasing tipping ("dumping") charge.

Demolition materials from structures can be classified into those from residential and those from nonresidential buildings. In North America, residences are usually constructed either mostly of wood or mostly of masonry. The resulting demolition flow has slightly more wood than bricks by mass for the USA (see Fig. 1). It is expected that the proportions for Canada would be similar. In the EC, the proportion of wood in demolition materials is almost negligible compared with brick (Environmental Resources 1980). The ready availability of wood in the USA results in very little recycling of wood as lumber, fuel or chips for building board, despite the considerable proportion of wood in residential demolition materials. Companies formed in the period of energy uncertainty in the 1970s to process demolition wood in the USA have ceased operations. In contrast, Mexican demolition firms operating near the US border may remove all lumber into Mexico (Wilson and Malarkey 1974).

In most countries, a high proportion of the unbroken bricks in demolition wastes is recycled. It is still profitable in the USA for contractors to employ people to remove mortar from bricks by hand; some machines that were developed to clean bricks mechanically were found to be uneconomic (Wiesman and Wilson 1976).

Most of the "heavy" (i.e., thick-section) metals in demolition wastes, both ferrous and nonferrous, are recycled. This category includes copper pipe, but not insulated copper wiring, although machinery has been developed to strip insulation from wiring. However, while there is always a market for recycled copper, brass, lead and aluminum which will repay the contractor's work in separating these metals from the overall waste stream, scrap prices can fall low enough in times of recession to make the recycling of steel and all but the heaviest nonferrous sections uneconomic. At these times the larger contractors often stockpile steel beams, for example, until market conditions improve.

Other demolition materials such as glass, plaster and plastics are recycled only to a very small extent.

The quantities of materials which have been used in the USA in building construction (excluding pavement, bridges and tunnels) are shown in Fig. 1. The average age of demolished buildings has been approximately 60–80 years for the older cities and 40–45 years for the newer cities of the west and south. The resulting quantities of materials produced in the USA in the 1970s from building demolition are estimated in Table 1 and in the countries of the EC in Table 2. Because of the frequent immediate reuse as base or fill, there are no good data for the quantities of pavement (especially highway) demolition materials. An excellent general reference is by Kreijger (1980).

An area related to the recycling of demolition materials is the use of secondary materials from any source in new building construction. This received a setback, at least in the USA, because cement blocks

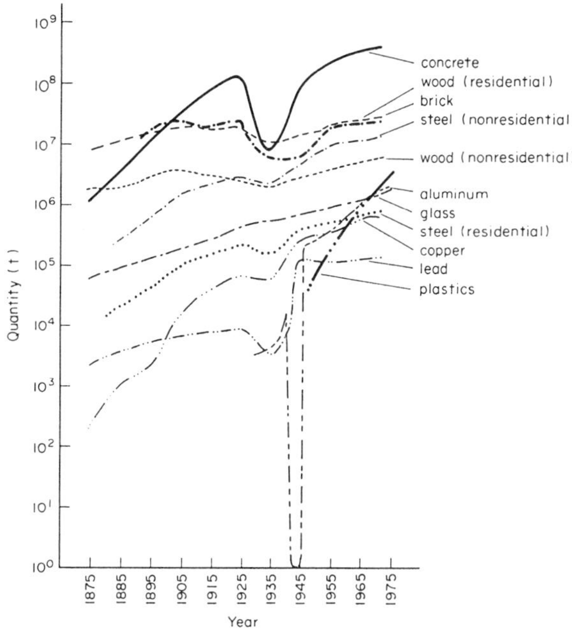

Figure 1
Quantities of materials used in building construction in the USA

Table 1
Estimated quantities of materials from the demolition of buildings in the USA (mean values for the 1970s)

Material	Quantity (kt year^{-1})	Percentage share
Steel and iron	477.4	1.57
Aluminum	3.0	0.01
Copper	15.2	0.05
Lead	18.2	0.06
Glass	100.3	0.33
Brick and clay	4564	15.01
Concrete	19255	66.33
Wood	5972	19.64
Plastics	0.9	<0.1
Total	30405	100.0

Table 2
Estimated quantities of materials from the demolition of buildings in the EC (mean values for the 1970s)

Material	Quantity (Mt year^{-1})	Percentage share
Masonry	40.85	57
Concrete	26.52	37
Wood	1.43	2
Steel	0.22	0.3
Other	2.65	3.7
Total	71.67	100.0

incorporating pulverized fly ash were found to be a source of radon. However, The Netherlands, through its national environmental policy plan, has taken the initiative in a new thrust to use secondary materials, where it is safe to do so, in construction: this is the theme of the 1991 WASCON international conference.

See also: Recycling of Metals; Resource Recovery: Economic and Institutional Aspects

Bibliography

Campbell J A, Russell S H 1986 Resource recovery: technology. In: Bever M B (ed.) 1986 *Encyclopedia of Materials Science and Engineering.* Pergamon, Oxford, pp. 4231–8

Davidson T A, Wilson D G 1982 US building-demolition wastes: quantities and potential for resource recovery. *Conserv. Recycling* 5: 113–32

Dudenhöfer B, Herper M 1989 Recycling von Baustoffen—Verwendung in ungebundenen Schichten des Strassen-, Wege und Sportplatzbaus. *AbfallwirtschaftsJournal* (7/8): 3–6

Environmental Resources 1980 *Demolition Waste.* Construction Press, New York

Frondistou-Yannas S A, Ng H T S 1977 Use of concrete demolition wastes as aggregates in areas that have suffered destruction: a feasibility study, NSF Report AEN 75-23221. National Science Foundation, Washington, DC

Kreijger P C 1980 *Adhesion Problems in the Recycling of Concrete,* NATO Conference Series VI, Vol. 4. Plenum, New York

Mather B 1980 Crushed concrete as concrete aggregate. In: Kreijger P C 1980 *Adhesion Problems in the Recycling of Concrete,* NATO Conference Series VI, Vol. 4. Plenum, New York, pp. 293–302

Nixon P J 1976 The use of materials from demolition in construction. *Resour. Policy* 2: 276–83

Saylak D, Gallaway B M, Epps J A 1976 Recycling old asphalt–concrete pavements. *Proc. 5th Mineral Waste Utilization Symp.* Illinois Institute of Technology Research Institute, Chicago, IL, pp. 16–26

Sierra R 1980 Example of re-use of concrete in France. In: Kreijger P C 1980 *Adhesion Problems in the Recycling of Concrete,* NATO Conference Series VI, Vol. 4. Plenum, New York, p. 347

WASCON 91. International environmental conference on the application of secondary raw materials in the construction industry. MECC, Maastricht, The Netherlands, March 24–29, 1991

Wiesman R M, Wilson D G 1976 An investigation of the potential for resource recovery from demolition wastes, NSF Report AEN 75-14197. National Science Foundation, Washington, DC

Wilson D G, Davidson T A, Ng H T S 1979 Demolition wastes: data collection and separation studies, NSF Report RA-790407. National Science Foundation, Washington, DC

Wilson D G, Malarkey J M 1974 The resource potential of demolition debris, NSF Report 74-SP-0889. National Science Foundation, Washington, DC

D. G. Wilson
[Massachusetts Institute of Technology, Cambridge, Massachusetts, USA]

Recycling of Materials: Economics

Many materials, such as glass, aluminum or steel, may be recovered from industrial scrap or consumer discards again and again before being relegated to municipal landfills or other waste-disposal facilities. Metals may be returned for reprocessing from industrial plants, from the demolitions of buildings or from discards of obsolete consumer goods, such as automobiles or home appliances. Glass bottles may be recovered from commercial facilities, such as eating and drinking establishments or from households, and remelted. Newspapers are readily collected to be deinked and reprocessed. This reuse of materials is commonly referred to as recycling.

Recycling has occurred through all recorded history. In modern, industrial societies, markets exist for scrap metals, waste paper, glass and even some plastics. These markets have been expanded in recent years by environmental concerns, particularly in the more developed economies. Problems in disposing municipal waste—the solid waste generated by households and small commercial establishments—has stimulated a variety of public policies designed to encourage or even to mandate recycling of the more valuable components of this waste stream. In some areas, industrial consumers of primary materials have even been required to use a specified percentage of secondary or recycled material in their production processes.

1. Recycling of Municipal Waste

The recycling of high-value industrial wastes poses few interesting issues for economic analysis. Industrial enterprises must either recycle the home scrap generated by their production processes or remove it from their premises. If this scrap is valuable, with or without further processing, it may be collected and shipped to processors or other users. Scrap dealers or brokers may provide the intermediary services and secondary processors may refine or otherwise convert the scrap to useful materials for further use. The extent of this recycling will depend on the value of the secondary material, the cost of collecting and reprocessing it, and the cost of the alternative, virgin material.

The interesting economic issues in recycling occur in the choices among the possible techniques for disposing of obsolete consumer products, refuse from commercial establishments and a variety of packaging and reading materials. This refuse is called "municipal" waste because municipalities typically assume the responsibility for disposing of it. The practical choices for such disposal are dumping in a landfill, combustion in an incinerator or recycling. The choice among these options depends on a number of factors, including the value of the material, the cost of transporting and reprocessing it, the availability (value) of land for landfills, and the price of energy.

2. Degree of Waste Recycling

The degree to which wastes are recycled varies substantially across societies. Table 1 shows the distribution of disposal, incineration and recycling of municipal waste for the three largest industrial countries—the USA, Japan and Germany. The USA has the lowest population density and the greatest availability of land for landfills; therefore, it uses landfills far more extensively than do Japan or Germany. Because Japan has far fewer virgin resources, such as wood pulp or metal ores, it recycles the most heavily. Both Germany and Japan utilize incineration far more extensively than does the USA, undoubtedly because their energy prices are much higher than US energy prices.

All materials do not have the same recycling potential. Traditionally, plastics in their numerous formulations and forms have been difficult to recycle, although this is changing. Many paper products may not be easily recycled because they are combined with other materials or are otherwise contaminated by foodstuffs, coatings or ink. In addition, paper fibers degrade with repeated recycling and further processing, reducing the strength of the paper and limiting its further use. Iron and steel scrap is readily recycled if it can be separated from other materials, particularly from nonferrous metals. Aluminum is among the easiest materials to recycle because it enters the refuse stream as beverage cans, siding, storefronts and other products from which it can be easily extracted.

The relative importance of various materials in the municipal waste stream and the degree to which they are recycled in the USA are shown in Table 2. Paper products constitute the largest share by weight of municipal waste and are the largest contributors to total recycling. Aluminum is more intensively recycled, but it constitutes only 2% of the refuse. Ferrous metals are not heavily recycled from municipal waste because they enter this waste stream mainly as discarded food containers, other packaging and parts of small obsolete consumer durables and implements. The major sources of ferrous scrap—motor vehicles, railroad equipment and large household durables—do not generally enter the municipal waste stream.

Table 1
Disposal methods for municipal solid waste (%)

Method	United States	Japan	Germany
Recycling	11	50	15
Incineration	6	23	30
Landfill	83	27	55

Source: Menell (1990)

Table 2
Projection of materials recovery from municipal waste in the USA, 1995 (10^6 t)

Material	Total waste	Total recovery[a]
Paper	85.5	29.5
Glass	11.1	2.6
Metals		
ferrous	11.7	2.0
aluminum	3.1	1.2
other	1.4	0.8
Plastics	18.6	0.8
Textiles	4.1	
Wood	7.4	
Food wastes	13.2	0.5
Yard wastes	33.0	8.8
Other	5.7	1.7
Total	194.8	47.9

Source: US Environmental Protection Agency (1990)
a Average of high and low estimate

3. Optimal Level of Recycling

Economists have shown that private market prices and public handling of municipal waste may not induce the optimal use of materials and the appropriate degree of recycling. If discarded materials create social costs that are not reflected in market prices, consumers will consume too much of these materials and discard them at too great a rate. These external social costs include the disamenities of landfills, the leaching of wastes into the groundwater and the smoke created in burning waste, as well as the undesirable effects of littering. Moreover, if virgin materials are subsidized through tax policy or the failure to control harmful pollution from extracting and processing the materials, there will be excess consumption of these materials.

If disposal charges are levied against all materials to reflect the marginal costs of disposing them through dumping or incineration, the demand for products requiring the highest-cost disposal will decline or the degree of recycling will increase or both. Obviously, raising the price of primary materials to cover all of the social costs of extraction and processing reduces the demand for these materials and increases the demand for substitutes, including recycled or "secondary" materials. If, however, the collection and processing of recycled materials also cause environmental harm, these external costs should also be reflected in charges against the production of these secondary materials. It is therefore possible that environmental considerations could require a reduction in the degree of recycling if secondary production processes are particularly damaging to the environment.

A diagrammatic exposition of these effects in a market for secondary materials is shown in Fig. 1. The demand for a given type of scrap or secondary material

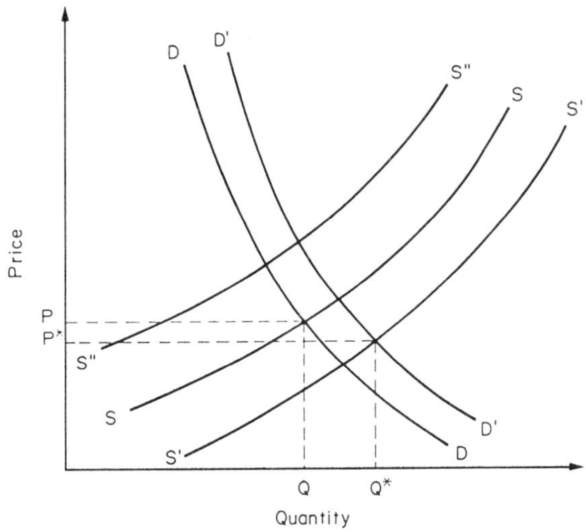

Figure 1
The market for recycled material

is derived from the demand for final products, but it also depends on the prices of other inputs. This demand is shown as the schedule DD in Fig. 1. If a tax is levied on the primary material, the demand schedule shifts to D′D′ as producers shift somewhat from the primary material to the secondary (recycled) input.

The supply of secondary material in Fig. 1 is shown as the upward-sloping schedule SS. The location of SS depends on the availability of scrap and the ease with which it is recycled. At higher prices, more scrap is recovered and reprocessed. The imposition of disposal charges on discarded obsolete materials increases the degree of recycling and, therefore, the supply of secondary material to S′S′.

At equilibrium, demand equals supply. Without disposal charges or taxes on primary materials, the price of the secondary material settles at P, and the quantity of the recycled material supplied is Q. With the imposition of a primary-product tax and a disposal charge on discards, the price falls slightly to P* and the quantity recycled rises to Q*. The ultimate effect on price and quantity depends on the price sensitivity, or "elasticity," of the supply and demand relationships. If the environmental costs from processing the recycled material are large, pollution taxes on this process could actually shift the supply curve upward and to the left to S″S″, reducing the equilibrium level of recycling.

The proportion of a material that is recycled is a function of the price of the material, the degree to which recycled material can be substituted for primary materials and the cost of collecting and reprocessing the material. Valuable materials are more likely to be recycled than low-value materials—all other things equal. If primary and recycled materials are close substitutes, small changes in the price of the recycled material may induce substantial changes in the quantity recycled, a phenomenon reflected in a more price-sensitive (or "price-elastic") demand curve in Fig. 1.

Finally, low costs of collection and processing of recycled materials are obviously conducive to recycling. Moreover, if these costs rise slowly with the amount of recycling, the supply curve in Fig. 1 will be relatively price-elastic. If both the demand and supply curves are price-elastic, small changes in the overall economy or other external conditions that affect the position of both schedules may generate large swings in the quantity recycled, but not in the price of recycled materials. If, on the other hand, both schedules are relatively price-inelastic, these external forces are more likely to lead to large swings in prices but relatively small changes in the degree of recycling.

4. Policies to Induce Recycling

Public policies to induce recycling may be directed at either the supply of recycling or the demand for recycled material.

4.1 Supply Incentives

Many countries have introduced mandatory recycling programs to reduce the burden on waste-disposal facilities and to address environmental demands. These mandatory programs are likely to increase the amount of material collected for reuse, but they are unlikely to provide the proper incentives for an efficient degree of recycling because they do not allow for differing costs of disposal or differences in the cost to consumers of sorting or otherwise preparing various wastes for recycling. For instance, a requirement that all glass be separated by consumers and deposited in separate bins for collection may induce a substitution of paper or plastic for glass containers, and these paper or plastic products may be more harmful or costly to bury or burn than is the disposal of the glass. Moreover, mandatory recycling policies might increase unlawful dumping or littering.

An alternative approach is to use some form of disposal charge to reflect the full social cost of disposing of the material. This charge may be levied against the material when it is used by the manufacturer or packager, or it may be levied against the consumer when he attempts to dispose of the material. The charge should reflect the marginal social cost of disposal and should therefore be lower if the material is separated by the household for recycling than if it is discarded with other refuse. Deposit–refund systems for beverage containers are one form of this incentive system.

Optimal disposal charges are likely to be difficult to estimate and even more difficult to enforce, particularly when they are imposed on millions of households. Taxes may be levied on primary materials to reflect the discounted future social costs of disposing them, but such taxes will not necessarily induce consumers to recycle the discarded products. If the cost of separating products from a municipal waste stream is high once all wastes are combined by the household or commercial establishment, it is essential that the household or commercial establishment be provided the appropriate price incentive for performing the separation. The price to these entities for collection and disposal of the mixed wastes should be greater than the price charged for collecting separated refuse.

4.2 Demand Incentives

Relying on incentives to increase the supply of recycled materials may not be successful if the demand for these materials is insensitive to changes in price. The increase in recycling will depress the market price of the secondary material, inducing some substitution of secondary for primary materials or for other materials. However, if the elasticity of substitution is low, there may be little increase in total recycling, inducing governments to attempt to stimulate the demand for recycled or secondary materials.

Among the most obvious demand incentives are fees on primary materials designed to offset the underpriced social costs of extracting and processing them. Pollution taxes on sulfur emissions increase the price of virgin copper, for example, and thereby induce a substitution of secondary copper for primary copper.

In addition, subsidies may be employed to reward producers' use of recycled materials, such as newsprint, to overcome the higher marginal costs of using them. Or subsidies may be applied to the consumption of products that contain a specified proportion of recycled content. Finally, use of recycled materials may simply be mandated by government, but such mandates are unlikely to be efficient given the variability in the trade-off between virgin and secondary materials across producers.

5. Limits to Recycling

There is an upward limit to the degree to which recycling can be induced efficiently, that is, without reducing economic welfare. This limit depends in large part on the value of consumers' time, the social cost of primary resources and the nature of production processes in a modern economy. Clean, safe disposal technologies exist and if the marginal costs of such disposal are charged to those who would dispose of household, industrial or commercial waste, the efficient degree of recycling should emerge as households attempt to minimize their total disposal costs (including the value of their own time spent sorting their trash).

Alternatively, the volume and weight of municipal waste can be reduced substantially by burning it and potentially recapturing the energy that is generated from such combustion. In societies with high energy costs, incineration may be superior to recycling of combustible wastes.

6. Conclusion

The degree of recycling varies across countries due to differences in the cost of disposal, the price of energy and government policies. In advanced societies, the cost of separating municipal trash at the source is high, reducing the voluntary level of recycling. Mandatory recycling programs, deposit fees, disposal charges levied on material inputs in manufacturing industries and mandated consumption of recycled materials may be used to induce increases in the degree of recycling. Mandatory programs are likely to be inefficient, requiring recycling from high-cost and low-cost sources alike.

Given the price of handling and processing recycled materials and the limit to efficient use of some recycled materials, an efficient policy may involve greater incineration of the solid-waste stream as well as increases in recycling. The efficient degree of recycling is in practice unknowable. The best that government can achieve is the establishment of the proper price incentives to induce a relatively efficient mix of disposal, incineration and recycling.

See also: Reclaiming of Elastomers; Recycling of Demolition Wastes; Recycling of Metals; Recycling of Paper Products; Recycling of Plastics; Resource Recovery: Economic and Institutional Aspects

Bibliography

Baumol W J 1977 On recycling as a moot environmental issue. *J. Environ. Econ. Manage.* 4: 83–7

Menell P 1990 An incentive approach to regulating municipal solid waste. John M. Olin Law and Economics Working Paper No. 9004. Georgetown University Law Center, Washington, DC

Smith V L 1972 Dynamics of waste accumulation: disposal versus recycling. *Q. J. Econ.* 86: 500–16

US Environmental Protection Agency 1990 *Characterization of Municipal Solid Waste in the United States: 1990 Update.* US EPA, Washington, DC

R. W. Crandall
[Brookings Institution, Washington, DC, USA]

Recycling of Metals

Collection and processing of scrap are the first two steps in recycling. Scrap can be classified into three categories: (a) home, revert or run around; (b) prompt industrial; and (c) postconsumer or obsolete. Purchased scrap consists of the last two categories. Since home scrap is recycled within the plant where it is generated and requires no additional processing prior to melting, only purchased scrap will be discussed. Prompt industrial scrap is usually new, clean material generated during an industrial operation (e.g., trimmings which result from the stamping of automobile fenders) or reject products (e.g., defective cans). Prompt industrial scrap is usually sold by the generator and may be purchased by either scrap processors or final consumers. Obsolete scrap can come from a variety of sources (e.g., discarded automobiles, outdated electronic gear, old buildings, railroad cars and ships).

Since scrap comes from a variety of sources in many different forms, it must be processed to facilitate efficient use. The primary roles of the scrap processor are to collect, accumulate, sort, grade, prepare, market and distribute scrap. Sorting is done following identification of the scrap. Preparation may involve any of a number of activities, such as shearing, baling or briquetting.

1. Ferrous Scrap

Ferrous scrap has supplied a slowly increasing proportion of the iron consumed in the USA (Table 1). Almost all of this scrap was used to produce steel and cast iron. Small amounts have been sold for copper cementation (about 450 kt per year) and for production of ferroalloys.

Ferrous scrap is usually identified and sorted according to source or prior use. As ferrous scrap is low in value compared with nonferrous scrap, stainless steel and specialty alloy scrap, only simple identification techniques are economic. When mixed scrap is received, it is sold as such and commands an appropriately lower price.

Processing of ferrous scrap depends on the requirements of the ultimate consumer and the form in which it is received by the scrap processor. Oxyacetylene torches are used to cut up obsolete railroad cars and ships, structural steel from buildings, and other very large objects. Shears reduce large pieces of scrap (e.g., structural members) to a dimension suitable for furnace charging, usually 1–1.5 m long. Automobiles pressed into logs also are sheared. Two types of shears are used: guillotine and alligator shears. The guillotine shear is the larger of the two, consisting of a movable head which is hydraulically forced past a stationary anvil. The scrap to be cut rests on and protrudes beyond the anvil. Alligator shears function like giant scissors, except only the upper part is movable.

Loose, light-gauge scrap is frequently baled or briquetted. Examples of this type of material include sheet remaining after a stamping or punching operation, and turnings (metal removed in a machining operation). In addition, entire automobiles and appliances are sometimes baled. Balers consist of hydraulic cylinders that compress the scrap by pushing it against fixed walls of a cube. Steel briquetters have a pair of rotating drums with a series of cavities around the periphery of each drum into which the scrap is compressed. The scrap (e.g., turnings) assumes the shape and dimensions of the cavities.

Shredders are used mostly for preparing light scrap (e.g., automobiles and large home appliances). They consist of a series of steel hammers mounted on a rotating drum. The hammers pass by a series of fixed anvils. Feeding the car, appliance or other scrap between the hammers causes it to be quickly ripped into fist-sized chunks. Magnetic separation removes the ferrous metal from nonferrous metals and nonmetallic materials such as glass, rubber, plastics and dirt. The resulting ferrous product has the advantages of increased purity and improved consistency, handling, transportation and melting characteristics. In 1990 there were an estimated 200 automobile shredders operating in the USA with a capacity of more than 13 Mt per year.

The ferrous fraction of municipal solid waste presents a unique case. This material may be sold directly to steel mills for blending with other scrap materials, to a detinner, or to a scrap processor. The cans must be cleaned of paper and food waste by the seller or scrap processor prior to detinning. The detinner will leach (dissolve) the tin away from the steel using an $NaOH–NaNO_3$ solution. This process also removes any aluminum present. The resultant ferrous scrap has considerably increased value, and the tin can be recovered from solution by electrolysis.

Lower-purity scrap usually is restricted to applications with less stringent chemical and mechanical property specifications. These include the manufacture of reinforcing bars and manhole covers. High-purity scrap can be used to make any grade of steel or cast iron.

Table 1
US consumption of ferrous scrap

Year	Consumption (kt)	Percentage of total ferrous consumption
1977	41040	35
1979	44970	37
1981	39640	36
1983	32230	39
1985	37270	39
1987	40290	42
1989	43070	40

2. Copper

Copper-bearing scrap materials are among the oldest used by man, and indications are that these materials were recycled by the earliest civilizations. Extensive copper recycling has continued up to the present time, when copper-bearing scrap accounts for over 40% of all copper raw materials (Table 2). Depending on the purity of the scrap and the melting technique employed, using scrap to produce copper requires only 3–40% of

Table 2
US consumption of copper scrap

Year	Consumption (kt)	Percentage of total copper consumption
1977	1085	40
1979	1553	47
1981	1418	45
1983	1084	41
1985	1139	41
1987	1214	41
1989	1299	44

the energy needed to produce pure copper from ore.

Copper-bearing scrap is divided into a number of grades. No. 1 copper scrap must be unalloyed, as it is used directly to make wirebars. No. 2 copper scrap is refined electrolytically and therefore may contain some impurities, though a copper content of at least 94–96% is normally required. Light copper also passes through an electrolytic refinery; its copper content must exceed 92%. Refinery-grade brass must contain at least 61.3% copper, while copper-bearing material generally contains less than 60% copper. These last two grades normally are melted in a blast furnace and blown in a converter prior to electrolytic refining.

Copper-bearing scrap can be used by tube mills, refiners, brass mills, ingot makers and specialty companies. The end-use determines the preparation of the scrap by the processor. As tube mills do not refine electrolytically, metallic contamination of scrap to be used by tube mills is unacceptable. However, small amounts of nonmetallic contaminants may be acceptable, depending on the user's ability to fire refine. Refiners can produce wirebar directly from No. 1 copper scrap, but No. 2 and light copper scrap must first be cast into anodes and refined electrolytically before it can be used for this purpose. Brass mills usually use No. 1 copper and brass scrap in their melts in combination with pure copper.

The composition of brass scrap must be known and kept within narrow limits so as to keep the product made from the scrap within composition specifications. Brass and bronze ingot makers supply alloys of known, fixed composition to foundries. Hence, they can and do melt a wide variety of copper-bearing scrap. Specialty uses for copper-bearing scrap include chemicals, electroplating, castings (both copper and gray-iron), copper–aluminum alloying and steelmaking.

Prompt industrial copper-bearing scrap, which accounts for 60–70% of recycled copper, is generated by a variety of industrial operations. For example, scrap turnings from brass screw machines amount to up to 65% of the brass feed. Other sources include stamping plants, wire-drawing rejects, die shavings and short rod wire lengths. Postconsumer scrap includes metal in surplus and obsolete items from communication, rail, electronics, power utility and defense operations; metal in disused automobiles and ships; and metal in spent shell casings.

Identification may be made using the material's color, spectroscopy, spot testing with acids, chemical analysis or, in some cases, magnetism. Highly trained and experienced sorters are required for the first three methods. Together with the shape of scrap and its past use, the color of a clean surface may be all that is needed to determine the alloy type. Spectroscopy offers only a semiquantitative analysis of an alloy, but this information is often sufficient for sorting scrap into alloy groups. Spot testing involves observing the reaction, or lack thereof, between various reagents and the metal.

A variety of techniques is used to prepare copper-bearing scrap materials by the scrap processor. Traditional processes include shearing to the size required by the customer, briquetting or crushing of thin and lightweight materials (e.g., turnings and borings), magnetic separation to remove iron, cleaning and degreasing.

One new technique is chopping of copper wire scrap. Large, stiff pieces of wire are presheared prior to chopping. Bales of wire must be broken down and are usually cut up with an alligator shear. The major steps then consist of chopping in three stages, with magnetic separation to remove ferrous contaminants following the first stage and sizing following the second stage. The third stage handles unliberated wire from the air separation process which follows sizing. Both the coarse and intermediate size fractions are processed by air separation to produce four fractions: (a) clean metal, (b) liberated metal with some insulation contamination, (c) unliberated wire, and (d) clean insulation.

Shredded automobiles also yield copper-bearing scrap. Following magnetic recovery of the ferrous metals, the nonferrous rejects can be processed by a number of techniques. The nonmetallics can be removed by air separation and water elutriation, and/or heavy media separation. The copper-bearing metals can then be hand picked after the aluminum and zinc are floated in a heavy medium.

Cryogenics (i.e., cooling scrap to temperatures as low as $-185\,^{\circ}C$) can be used to recover copper from items such as small motors, transformers, generators, and their components. When exposed to these very low temperatures, certain materials, including steel and plastics, become brittle, whereas copper remains malleable. Therefore, crushing of an item immediately after cooling will allow relatively large pieces of copper to be separated from the other, finer materials by screening.

3. *Aluminum*

The increase in aluminum recycling has been dramatic. In 1960, recovery of secondary aluminum was about

Table 3
US consumption of aluminum scrap

Year	Consumption (kt)	Percentage of total aluminum consumption
1977	1456	26
1979	1612	26
1981	1789	28
1983	1773	35
1985	1762	33
1987	1986	37
1989	2054	41

300 kt but by 1977 that figure had increased to over 1.4 Mt, accounting for 26% of total aluminum production (Table 3). By 1989 consumption of aluminum scrap had increased to 41% of consumption.

Prompt industrial aluminum scrap is produced by a variety of industrial operations, such as foil rolling, canmaking, airplane manufacturing and wire drawing. It is usually kept sorted and moves in bulk to the scrap processor or recycler. Prompt industrial scrap is, therefore, easy to handle.

Major suppliers and potential sources of obsolete scrap include industry, government and municipal solid waste. Products from which aluminum scrap is obtained are automobiles, household appliances and furniture, used cans and containers, trucks, aircraft, boats and buildings.

Scrap sorted by alloy class brings the highest prices, so the first step in scrap processing is identification. The first step in the identification process is to consider the nature (e.g., sheet, plate, forging, casting) and source of the scrap. If more information is needed, a spectroscope may be used. In the case of castings, chemical spot tests are used to determine whether the alloy is based on aluminum, magnesium or zinc.

Once the scrap has been sorted, one or more processing techniques may be employed to prepare it for shipment. Borings and shredded material may be shipped loose for bulk handling and unloading. Other types of scrap, such as clippings and sheets, usually require baling or briquetting. Large pieces of aluminum are often sheared to a size specified by the customer.

In addition to the technology mentioned above, there are five other processes for recovering aluminum: wire chopping, shredding, aluminum can recycling, sweating and dross processing. Depending on the type of wire metal, wire chopping recoveries are usually in the range 55–65%. The technology is similar to that described for copper. However, air classification of the coarse and intermediate aluminum wire size fractions gives three products, not four fractions. These products are clean metal, plastic and middlings (a mixture of some metal with insulation). The middlings are sent to a tertiary chopper or granulator to separate the metal from the insulation. The chopped middlings are then recycled to join the product of the second-stage chopper before sizing. Wire chopping produces a well-accepted product containing about 99.5% aluminum.

Most shredded aluminum scrap is obtained from three sources; shredded automobiles, scrap cans and aluminum shredded for handling in the scrapyard. For every tonne of shredded steel scrap produced from disused automobiles, about 13.5 kg of aluminum is recovered. With the potential recovery in 1979 being about 135 kt of aluminum from the 10 Mt of shredded automobile scrap, actual recovery was approximately 100 kt. Given projections for increased use of aluminum in cars, these numbers should increase.

Following automobile shredding, the material undergoes magnetic separation. A number of techniques exist to recover the aluminum from the nonmagnetic shredded material. Initially, the nonmetallic materials can be recovered by air classification, water elutriation, eddy current, or a friction sliding board to produce a heavy fraction. Alternatively, a sink–float technique, in which the density of a heavy medium is controlled between that of the metallic and nonmetallic materials, can be used. Finally, a second sink–float tank, using a medium of higher density, can be used to float the aluminum and zinc from the other metals. Final separation is made by hand picking or in a sweat furnace.

Scrap aluminum cans come from both prompt industrial and postconsumer sources. Since 1987, collections of cans from postconsumer sources have exceeded 50% of the aluminum used for cans that year. There are three levels of aluminum can recycling: (a) minimal, with a scale and sorting table; (b) intermediate, with scales, magnetic sorters and balers or flatteners; and (c) full-scale, with shredders, material handling and truck or rail car loading equipment in addition to (b). New separation processes (e.g., the eddy current linear-induction motor) are the key to recovering aluminum from municipal solid waste.

Sweating is the process of separating materials based on differences in their melting points. In this process the scrap is heated to a temperature between the melting points of the higher and lower melting point materials. Sweating should be used only to recover aluminum from contaminated scrap. This scrap should: (a) be free of contamination with other low-melting metals (e.g., zinc and magnesium, which should have been sweated off first); (b) be at least 6 mm thick so that the particles will melt into droplets large enough to flow in the furnace; and (c) contain at least 30% (preferably 50–60%) aluminum. Ingots produced from the sweated aluminum may not contain more than 1% iron, 1% magnesium and 1–3% zinc depending upon requirements.

There are three kinds of sweating furnaces: (a) direct-fired sloping hearth, (b) direct-fired rotary, and (c) indirectly fired rotary. The direct-fired sloping hearth is the original style of sweating furnace. It consists of a refractory brick lined box with a sloping floor and a tap hole at its lowest point. The aluminum in the scrap charge is melted and then runs down the sloping hearth to the tap hole and into an ingot mold or collection chamber. The scrap on the hearth must be manually raked and rabbled to free any trapped aluminum which cannot flow off the hearth. Manual raking is unnecessary if a rotary hearth furnace is employed. These may be heated directly, with a burner firing into a kiln from the discharge end, or indirectly with a ceramic-lined, stainless steel tube rotating in a temperature-controlled furnace. Both styles of rotary furnaces are inclined, with weep holes at their lower (discharge) end. The aluminum runs down the hearth through the weep holes and is collected, while the remaining solid material passes over the weep holes and is discharged.

Dross recovery is the separation of metallic aluminum from the nonmetallic residues produced whenever aluminum of any type is melted. The dross is a mixture of aluminum oxide, dirt, spent fluxes and entrained aluminum. The aluminum content, which depends on the care that was taken to prevent oxidation in and after removal from the furnace, can range from zero up to 70%. When the aluminum content is sufficiently high, usually in excess of 40–45%, the dross may be sent directly to a rotary furnace where it is heated along with a flux consisting of equal proportions of NaCl and KCl, and up to 5% cryolite. However, the majority of drosses are milled and screened at about 20 mesh (0.84 mm) with the oversize processed in the rotary furnace and the undersize discarded or sold to the steel industry for use as hot-topping compound.

Wrought aluminum scrap of sufficient purity can be sold directly to a primary smelter where it is used to make new alloy. Economics are the primary incentive for the dramatic increase in this practice since the mid-1960s. For example, construction of an aluminum reduction plant with a capacity of 73 kt per year would cost about US$250 million, whereas construction of the same size plant producing can stock from can scrap would cost about US$16–20 million (1983). In addition, recycling aluminum scrap requires only about 5% of the energy needed to produce new metal.

Most secondary aluminum smelters produce recycled secondary ingot from both wrought and cast scrap (i.e., an ingot with a known composition). This material is used primarily for aluminum castings. In some cases notch bar or shot is produced for use as a deoxidizer or alloy addition. Scrap (e.g., from wire chopping) can also be used in these applications. Other possible uses for aluminum from wire chopping include production of powder, pharmaceuticals, and paints and coatings.

4. Zinc

The USA is the world's largest consumer of zinc. From 1977 to 1989 the fraction of domestic consumption met from scrap ranged from 24% to 29% (Table 4). One reason for the relatively low level of zinc recycling is that about 45% of zinc consumption is used dissipatively (e.g., for galvanizing).

Table 4
US zinc recovered from zinc scrap

Year	Recovery (kt)	Percentage of total zinc consumption
1977	330	24
1979	370	27
1981	305	29
1983	349	28
1985	347	27
1987	383	28
1989	347	26

Zinc is recovered primarily as alloy scrap, which may contain zinc, copper, aluminum or magnesium as the major constituent. Zinc is also recovered from skimmings, drosses and fumes. The major sources and uses of zinc scrap materials are presented in Table 5. Galvanizing provides the largest single source of scrap zinc, in the form of skimmings, drosses, fumes and spent fluxes. These zinc-containing wastes are sold directly to recyclers. Zinc die-casters also produce a dross, as well as rejected castings. The availability of fragmentized die-cast scrap has grown markedly since the early 1970s as a result of the tremendous expansion of automobile and appliance shredding. However, owing to the smaller amount of zinc used in the automobile during the last few years as a result of increased use of plastics, future zinc recovery from automobile shredding is expected to diminish considerably.

The major role of the scrap processor is in the recovery of zinc from shredded automobiles and appliances. If this scrap zinc is recovered by mechanical means (e.g., flotation and/or screening) rather than by sweating, a number of problems are created for the consumer. If the zinc content of this material is too low, operation of the recycling furnaces is inefficient, resulting in higher production costs. Excessive aluminum and finely divided zinc in the scrap increases the amount of skimmings and therefore the loss of zinc. The recovery of zinc is decreased by the presence of glass, rubber, rocks and trash. The degree of purity required is determined by agreement between the scrap processor and the consumer.

5. Magnesium

Consumption of magnesium-based scrap for 1977–1989 is shown in Table 6. Only a small fraction of magnesium is consumed as magnesium-based alloys that are later identifiable as magnesium scrap.

Table 5
Major sources and uses of zinc scrap materials

Type of scrap	Sources	Uses
Fragmentized die-cast	nonmagnetic portion of crushed automobiles and appliances	distillation
New industrial die-cast	rejected castings from die-cast operations	die-cast alloy distillation
Old die-cast	old, oxidized and contaminated accumulators, mostly from automobile wrecking	distillation
Die-cast skimmings	ashes, drosses from melting die-cast scrap	distillation chemicals
Aircraft forming dies	forming dies from aircraft and automobile industry	die-cast alloy distillation
Zinc skimmings	hot-dip galvanizing kettles	distillation chemicals (fluxes), agricultural compounds
Sal-ammoniac skimmings	hot-dip galvanizing kettles	chemicals (fluxes)
Zinc dross (top)	top of continuous galvanizing kettles	distillation
Zinc dross (bottom)	bottom of continuous and hot-dip galvanizing kettles	distillation
Brass ingot makers fume	baghouse collectors	chemicals (agricultural compounds)
New zinc, painted, oxidized, corroded	printing plates, battery casings, addressograph plates, punchings from rolled zinc	remelt zinc brass ingot makers galvanizing
Old zinc, painted, oxidized, corroded	roofing, boat anodes, jar tops, spent plating anodes	remelt zinc distillation
Prime die casters dross	die-casters kettle (high metallic skim)	die-cast alloy distillation
Remelt die-cast (blocks)	sweating iron-containing and fragmentized die-cast	distillation

Magnesium, scrap or primary, has both nonstructural and structural applications. Nonstructural uses are as alloying agents, as a reducing agent for the production of other metals, as an addition to nodular cast iron, as anodes for the cathodic protection of steel structures and as a scavenger for sulfur in blast-furnace iron destined for primary steel production. Except when used as an alloying agent, the composition of magnesium scrap does not need to be closely controlled. However, when magnesium scrap is used for structural purposes, composition of the scrap is critical. For example, high-grade casting alloys may contain no more than 0.05% copper and 0.05% nickel, with some alloys limited to 0.002% nickel. Therefore, scrap processors must take great care to remove these metals or their alloys from the magnesium scrap. Removal is usually done manually.

Because magnesium and aluminum are very similar in appearance, sorting by visual identification of these two metals, or their alloys, is extremely difficult. A reliable method of distinguishing between magnesium and aluminum is to apply a 5% silver nitrate solution to a scratched surface of the metal in question. The presence of magnesium is indicated by the surface turning black. Aluminum is not affected.

Magnesium-bearing wastes, such as skimmings, generally have not been recycled owing to the low prices offered by magnesium smelters for them.

Table 6
US consumption of magnesium scrap[a]

Year	Consumption (t)	Percentage of total magnesium consumption
1977	7818	7
1979	8893	7
1981	7644	6
1983	7424	7
1985	6140	5
1987	4703	5
1989	8220	8

a Not including magnesium in aluminum-based alloy scrap or magnesium used to make aluminum-based alloys

6. *Lead*

More than half the lead consumed in the USA is recycled from scrap (Table 7). In 1980, 86% of lead scrap was obsolete and 83% of the obsolete scrap was obtained from recycled batteries.

Lead scrap is of three types: whole battery scrap (both prompt and depleted batteries); prompt industrial scrap (such as drosses, skimmings and lugs); and other scrap (such as cable sheathings, bearings, babbits and solders). Of these types, only batteries require significant processing to recover the lead.

Batteries are processed to drain the contained sulfuric acid and to liberate the lead. Four techniques are used to accomplish this, of which two, sawing and

Table 7
US consumption of lead scrap

Year	Consumption (kt)	Percentage of total lead consumption
1977	758	53
1979	801	59
1981	641	55
1983	504	44
1985	616	54
1987	710	58
1989	809	63

guillotining, are similar. Both remove the top of the battery, including the posts and connectors of the battery section, from the rest of the case and the plates. The remainder of the battery is hit against a striker bar to allow plates, separators and acid to fall into a conveyor. Acid drains off the conveyor into a sump where it is neutralized with lime or ammonia. The battery top is crushed and the metallic posts and connectors are separated from the crushed casing material by air classification. A more recent concept in processing scrap batteries is to break the whole battery and separate the battery plates from the casing, followed by heavy media and hydrometallurgical techniques to separate the grid metal from the paste. An alternative method involves breaking the battery to drain the acid, then charging the entire battery to the smelting furnace.

Although about 60–65% of the lead used in batteries is recycled, the same is not true of lead used in other products. For example, only about 10% of the lead used in solder, 25% of the lead consumed by the cable industry and 30% of the lead in bearings is recycled. The reasons for this lower recycling rate are diverse. Most solder applications are dissipative (i.e., the solder is used in very small quantities bonded to other metals). Cable sheathing, while easy to recover, has an average life of 40 years. Recovery of lead in bearings frequently requires disassembly of a large piece of machinery, which is uneconomic.

In some applications, which by themselves would be dissipative, lead is recovered in the process of recovering another material with which the lead is associated. Examples are the recovery of lead from solder and bearings associated with recycled copper-base alloys. This mode of recycling has been called corecovery.

Virtually all lead scrap is sold to a secondary smelter and/or refiner. The resulting product finds its principal use in the production of new batteries. However, the resulting lead also may be used to produce other alloys for type metals, solders, shot and bearing metals.

Increasingly stringent environmental controls have forced the closure of some secondary smelters. Some new secondary plants have used hydrometallurgical/electrowinning processes because of the difficulty of controlling airborne emissions of smelters.

7. Nickel Alloy and Stainless Steel Scraps

Although treated together because of the importance of nickel, nickel alloys and stainless steels are fundamentally different. Scrap nickel alloys can be used in stainless steel production. However, because of the high iron content, stainless steel scrap cannot be used as a feed material for making nickel alloys. Table 8 presents data on the rate of nickel recycling.

Nickel alloys are usually classified as follows.

(a) Superalloys possess high strength at high temperatures. A typical application of these alloys, such as the Inconels, Hastelloys and Waspaloys, is in gas turbine engines. Nickel content can vary from 20% to 80%. Most of the prompt industrial scrap is generated by the aerospace industry in the form of turnings, rejected turbine components, and casting stubs and sprues. A major source of obsolete scrap is jet aircraft engines.

(b) High-nickel alloys and heating alloys are used in corrosion- or oxidation-resistant applications. Nickel contents vary from 20% to 100%. As usual, prompt industrial scrap results from metal forming and fabrication. Obsolete scrap is recovered from discarded industrial equipment.

(c) Pure nickel scrap is recovered as depleted anodes from electroplating. This represents only a small portion of the nickel available for recycling.

(d) Other nickel-bearing wastes include Monels, small amounts of special alloys, grinding swarfs, sludges and spent catalysts. Their importance is not sufficiently significant to warrant discussion here.

Stainless steel alloys are classified into three broad categories: 200, 300 and 400 series. As the 400 series contains only chromium it will not be discussed. The 300 series contains about 18% chromium and about 8% nickel. Other elements such as niobium and

Table 8
US consumption of nickel scrap

Year	Consumption (kt)	Percentage of total nickel consumption
1977	45.8	22
1979	52.1	23
1981	41.4	27
1983	45.2	25
1985	48.7	25
1987	42.3	23
1989	39.8	24

molybdenum may also be present. Applications of 300 series stainless steels are found in kitchen utensils, hospital equipment, chemical plants and the aerospace industry. In the 200 series, 5–10% manganese is added to the alloy to promote the formation of austenite. Because nickel serves the same function, the nickel content can be reduced to under 4%.

The users of superalloys and stainless steels have rigid specifications for these alloys, and so reliable segregation and identification of this type of alloy scrap is essential to its recycling and composition. The source or form of the scrap is often a good indication of its identity. In addition, four relatively simple procedures (magnetic testing, spark testing, chemical spot testing and x-ray fluorescence) can be used for initial identification. If more reliable information or a guaranteed composition is required, a sample is taken for chemical analysis.

The magnet test is conducted with a small permanent magnet. Unless it is used to separate two alloys of known composition, it can provide only an approximate or initial classification of alloys. Spark testing can be used to differentiate among iron- and nickel-based alloys. However, the reliability of this procedure depends strongly on the skill of the operator. Spot tests include susceptibility to acid attack and reaction of test reagents to form specific colors.

Following identification of the scrap, it is processed for sale. Processing steps may include cleaning and/or sizing. Turnings are usually crushed in a ring-type or hammer mill crusher and then sampled for chemical analysis. Superalloy scrap is degreased with trichloroethylene, screened to remove fines and processed through a magnetic separator. The screened fines are sold to a nickel refinery, while the prepared turnings are sold to a superalloy manufacturer or a stainless steel mill or are exported, depending on their composition. Stainless steel scrap may be crushed and, if series 200 or 300, passed through a magnetic separator, but degreasing is rarely economic and screening is never performed.

Superalloy solids are cut with a plasma torch, and the surface is cleaned by sand blasting. With the exception of cleaning the surface by a combination of steel and grit blasting, solid, stainless steel scrap is handled in an identical manner to solid, plain, carbon-steel scrap.

Stainless steel dusts and grinding swarfs can be pelletized or briquetted prior to charging back to the steelmaking furnace. Alternatively, service companies convert this material to alloy ingots of known composition suitable for charging to the steelmaking furnace.

8. Titanium

Titanium alloys are classified as alpha, beta and alpha–beta. Alpha alloys include unalloyed metal or low-alloyed titanium grades. Typical alpha alloys contain aluminum and tin or palladium. Of the beta alloys, the alloy Ti–13V–11Cr–3Al was widely used during the 1960s, but has since been displaced by Ti–8V–4Mo–6Cr–4Zn–3Al. The alpha–beta alloys are the most important, with a production volume comparable to that of the alpha and beta alloys combined. Typical alpha–beta alloys are Ti–6Al–4V, Ti–4Al–3Mo–1V and Ti–8Mn.

Physically, titanium scrap is classified as chips (light) and solids (heavy). This material is recycled to the smelter in the ratio 1:2. The processing method employed by scrap processors is determined by the physical properties of the scrap.

Table 9 presents data on the tonnage and degree of recycling of titanium from 1977–1989. Note that 5% of total ingot weight becomes home scrap in the form of croppings, collars and turnings. About 86% of total ingot weight ends up as prompt industrial scrap. Because of restrictions on contamination, only about 35% of total ingot weight is recycled to the titanium-melting furnace. The remainder is used for alloying in the steel and aluminum industry, exported, or is lost as grinding and other wastes. A small amount of obsolete scrap, mostly from aircraft maintenance, is also recycled.

Titanium scrap must meet very strict standards for purity, especially for aerospace applications, because there is very little refining action during melting. Contaminants that must be handled by processing include surface oxides resulting from exposure to high temperatures during use, hydrocarbons from metalworking lubricants and pieces of cutting tool bits (usually tungsten carbide).

Chips are crushed in a hammermill to improve their handling and their packing density for shipping and subsequent briquetting. The crushed scrap is then solvent-vapor degreased in trichloroethylene. This also removes tramp impurities and any foreign metal pieces entrained in the grease. Next, magnetic separation is used to extract any ferrous or ferromagnetic impurities

Table 9
US consumption of titanium scrap[a]

Year	Consumption (t)	Percentage of total titanium consumption[b]
1977	9880	40
1979	12690	37
1981	13420	32
1983	9500	39
1985	13350	41
1987	16360	48
1989	17960	42

a Excludes titanium scrap used as an alloying addition b Total consumption includes only scrap and sponge; pigment is excluded

Table 10
US consumption of gold scrap

Year	Consumption (kg)	Percentage of total gold consumption
1977	32340	21
1979	52100	35
1981	50070	49
1983	55480	58
1985	49820	52
1987	63840	63
1989	36895	37

that are present. A representative sample is then taken for chemical analysis. Care should be taken in handling and storing turnings as they are easily flammable, particularly if they are light and oily.

Solids are subjected to minimal processing by the scrap processor. Following magnetic separation each piece is analyzed by x-ray spectroscopy, segregated by grade and shipped to a melter. The melter cuts the heavy scrap with a plasma torch, descales the titanium by abrasion, further cleans the surface with caustic soda and pickles the surface with nitric acid containing some hydrofluoric acid.

9. Precious Metals

This section covers gold, silver and platinum-group metals recycling. With the exception of periods, such as early 1980, when the price of precious metals was exceptionally high, most secondary precious metal bearing materials have been obtained from industry. Once the secondary precious metals bearing scrap is put into a concentrated or segregated form, it is sold to an appropriate refinery where it is converted into pure metal or alloys identical to those obtained from primary sources. Consumption of gold, silver and platinum-group metals scrap is presented in Tables 10, 11 and 12, respectively.

Table 11
US consumption of old and new silver scrap

Year	Consumption (t)	Percentage of total silver consumption
1977	3098	65
1979	2378	49
1981	2607	72
1983	2102	58
1985	2254	61
1987	2126	58
1989	2230	44

Table 12
US consumption of old platinum-group metals scrap

Year	Consumption (t)	Percentage of total platinum consumption
1977	6077	12
1979	9620	11
1981	12182	12
1983	9429	16
1985	8043	11
1987	5177	8
1989	3954	5

Precious metals are obtained from a variety of secondary sources. Film users, such as hospitals and printers, generate scrap film and hyposolutions from which silver is recovered. Electrical and electronic components may contain gold, silver and/or platinum-group metals (e.g., as contacts or solders). Catalytic converters from automobiles contain platinum and palladium, although converters on newer models also contain rhodium. Precious-metals scrap also is generated during plating and jewelry manufacture, and by scientific and dental laboratories. Aircraft and aerospace equipment contains precious metals in spark plugs and brazing alloys.

Precious metals are sometimes found in conjunction with base metals, most frequently in an assembled part (e.g., electrical equipment and automobile catalytic converters). The function of the scrap processor is to remove as much of the base metals and nonmetals as possible.

Complicated devices may be chopped or shredded to produce a fairly uniform material. This material may be subjected to magnetic separation, screening, air classification, and/or incineration to produce precious metal concentrates. When appropriate, hand dismantling may be practised.

The material is sampled and analyzed prior to shipment. Preparation for sampling would be by grinding to fine powder or by melting and either taking a dip sample or pouring into an ingot from which saw cuttings or drill turnings may be obtained.

Automobile catalytic converters are removed from junked cars and then sold for further processing. The processor opens the stainless steel containers and removes the platinum–palladium covered alumina balls or honeycomb that constitute the catalyst. The stainless steel is sold to a stainless steel producer and the catalyst is sold to a precious-metals refiner specializing in platinum-group metals.

See also: Conservation of Materials: Technology; Recycling of Materials: Economics; Resource Recovery: Economic and Institutional Aspects

Bibliography

Alter H 1982 Toward a national policy for secondary materials. Symposium on resource recovery and environmental issues of industrial solid wastes. *Resour. Conserv.* 9: 17–27

Arthur D Little Inc. 1979 *Proposed Industrial Recovered Materials Utilization Targets for the Metals and Metals Products Industry.* US Department of Energy, Washington, DC

Battelle Columbus Laboratories 1972 *Identification of Opportunities for Increased Recycling of Ferrous Solid Waste.* National Technical Information Service, Springfield, VA

Battelle Columbus Laboratories 1977 *A Study to Identify Opportunities for Increased Solid Waste Utilization: I. General; II. Metals.* National Technical Information Service, Springfield, VA

Campbell J A, Russell S H 1986 Resource recovery: technology. In: Bever M B (ed.) 1986 *Encyclopedia of Materials Science and Engineering.* Pergamon, Oxford, pp. 1560–4

Carrillo F V, Hibpshman M H, Rosenkranz R D 1974 *Recovery of Secondary Copper and Zinc in the United States.* US Bureau of Mines, Washington, DC

Fine P, Rasher H W, Wakesberg S (eds.) 1973 *Operations in the Nonferrous Scrap Metal Industry Today.* National Association of Recycling Industries, New York

Institute of Mining and Metallurgy 1990 *Recycling of Metalliferous Materials.* IMM, London

Institute of Scrap Recycling Industries 1990a *Recycling Nonferrous Scrap Metals.* ISRI, Washington, DC

Institute of Scrap Recycling Industries 1990b *Recycling Scrap Iron and Steel.* ISRI, Washington, DC

Kusik C L, Kenehan C B 1978 *Energy Use Patterns for Metal Recycling.* US Bureau of Mines, Washington, DC

Rheimers G W, Rholl S A, Dahlby A B 1976 Density separations of nonferrous scrap metals with magnetic fluids. *Proc. 5th Mineral Waste Utilization Symp.* US Bureau of Mines, Washington, DC

Siebert D L 1970 *Impact of Technology on the Commercial Secondary Aluminum Industry.* US Bureau of Mines, Washington, DC

U S Bureau of Mines 1989 *Minerals Yearbook 1989.* US Bureau of Mines, Washington, DC

US Bureau of Mines 1990 *Mineral Commodity Summaries 1990.* US Bureau of Mines, Washington, DC

R. S. Kaplan and F. J. Schottman
[US Bureau of Mines, Washington, DC, USA]

Recycling of Paper Products

Wastepaper can be classified into five major categories: corrugated boxes, newspapers, mixed papers, pulp substitutes and high-grade stock suitable for deinking. Of these, corrugated is the easiest to segregate, mixed papers are the lowest grade, and high-grade stock suitable for deinking and pulp substitutes are the highest grades of waste. The latter include such materials as clippings and office waste (e.g., computer printouts and files). Figure 1 shows the growth in consumption of recyclable paper since 1970. Since the mid-1980s the rise in consumption has increased significantly. This increased rate of consumption is expected to continue, with almost 25 000 000 short tons per year being consumed in 1992.

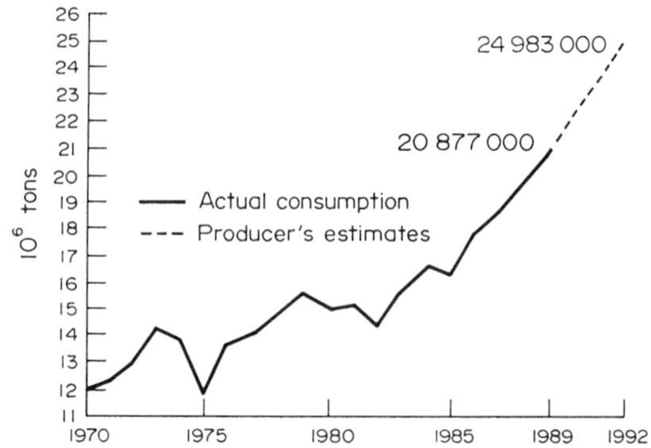

Figure 1
Growth in the consumption of recyclable paper, 1970–1992 (courtesy of the American Pulp and Paper Institute)

In 1985, W. E. Franklin of Franklin Associates Ltd. estimated waste paper consumption based on data from the American Paper Institute (API) and his own sources of information (see Table 1). His estimates for 1990 (made in 1985) agree very closely with the 1990 consumption estimates made by the API in 1989. Franklin estimates that usage by the year 2000 will be 30 590 000 short tons but he presents an alternative forecast for the year 2000 of 36 860 000 short tons based on a markedly increased emphasis on recycling.

Paper properties are altered by recycling, with some characteristics being improved and others reduced. Contaminants such as adhesives, hot melts and plastic film are potentially troublesome to all recycling operations, and removal methods, although improved greatly in recent years, are still not completely efficient.

1. Grades of Wastepaper

1.1 Corrugated Boxes

Used corrugated boxes are mainly accumulated by commercial establishments such as supermarkets and department stores. After being repulped, they are recycled into several different grades of paper.

Corrugated boxes made from virgin fiber are composed of linerboard containing unbleached kraft softwood pulps (long fiber, 3–5 mm) and corrugated medium (fluted element) containing mostly high-yield semichemical hardwood pulps (short fiber, 0.5–1 mm). The linerboard must be reasonably puncture resistant whereas the fluted medium, which maintains the structure of the box, must be resistant to crushing. During recycling, the hardwood and softwood fibers are mixed together and the resultant mixture does

Table 1
Waste paper recovery and utilization trends 1970 to 2000 (in short tons and percent)

Year	New supply (consumption)	Other waste paper use in paper and board mills	Utilization rate	Uses of waste paper	Exports	Total recovery	Percent recovered
1970	55969	11803	22.8	418	408	12562	22.4
1979	69961	15361	23.8	509	2233	18025	25.8
1980	67233	14922	23.4	473	2664	17972	26.7
1981	68101	15037	23.3	488	2213	17659	25.9
1982	64897	14433	23.6	489	2127	16975	26.2
1983	71491	15504	23.1	490	2573	18467	25.8
1984	NA	16948	NA	500	3235	20683	27.5
1990	85000	20160	24.4	420	3480	23960	28.2
1990	85000	21950	26.5	450	5000	27400[a]	32.2
2000	104600	25820	24.9	420	4450	30590	29.2
2000	104600	29940	28.9	420	6500	36860[a]	35.2

Source: American Paper Institute and Franklin Associates, Ltd
a Alternate forecast of recovery

not have the individual characteristics of either original component.

Recycled products include bogus linerboard and fluting medium made completely from used corrugated boxes. Although the quality in terms of bursting strength of the linerboard and crush strength of the medium are generally not as good as the virgin products, the economic advantages nevertheless create a market for the recycled products. Recycled corrugated boxes are also used in the filler plies of solid folding boxboard to provide good bending and strength properties, in absorbent grades of paper such as industrial towelling, and for industrial wrapping paper.

The closure for a corrugated box might be staples, asphalt tape (often with glass-fiber reinforcing) or hot-melt adhesive. These closure materials result in contaminants in the defibered wastepaper. Mills processing waste corrugated boxes use pulpers equipped with raggers, junk removers, rough screening systems and so forth to remove as many of the contaminants as early in the process as possible. Residual asphalt particles in the reused fiber can cause objectionable dark spots in the final product being manufactured.

Residual asphalt can be made essentially trouble-free by processing the fiber through an asphalt dispersion system in which pulped stock is thickened to 35% consistency (percent fiber on slurry basis) and steamed in a pressure vessel at 150 °C to melt asphalt present. Generally the hot fiber and softened asphalt is refined to complete the dispersion of the asphalt onto the surface of the fibers. The use of asphalt for tape and to provide barrier protection for moisture and so on is not as common today as it was during the 1970s and 1980s.

1.2 Newspapers

Newspapers for recycling are obtained by the collection of postconsumer material or from overissue, the cleanest source. When not deinked, old newspapers are used mostly for the production of inner plies or backing of solid paperboard, such as folding boxboard, and also construction board (e.g., gypsum and insulation board). When deinked, newspapers are used for recycled newsprint and to a lesser extent for tissue and towelling paper.

Deinking of newspaper is accomplished either with froth flotation cells, as used for example in mine ore processing, or by washing, using repetitive dilution and thickening techniques. Flotation is more common in Europe using a combination of newspapers and magazines, whereas washing systems are more common in the USA. A typical flotation deinking system is illustrated in Fig. 2.

New types of inks both for oil-base high-speed offset printing and water-base flexographic printing have put greater demands on the deinking operation. Some mills and suppliers are now advocating a combination of flotation and washing (displacement) deinking.

Machinery builders worldwide have made much progress in developing screening and cleaning equipment for improved contaminant removal. As a result, the pressroom runnability of recycled newsprint can be competitive to virgin newsprint. Even though recycled newsprint mills have sophisticated cleaning and screening systems, they still prefer clean, sorted waste newspapers.

Old newspapers are used for purposes other than recycling back into paper grades. Material free of coated inserts makes an acceptable blown fiber insulation if the proper type and amount of fire

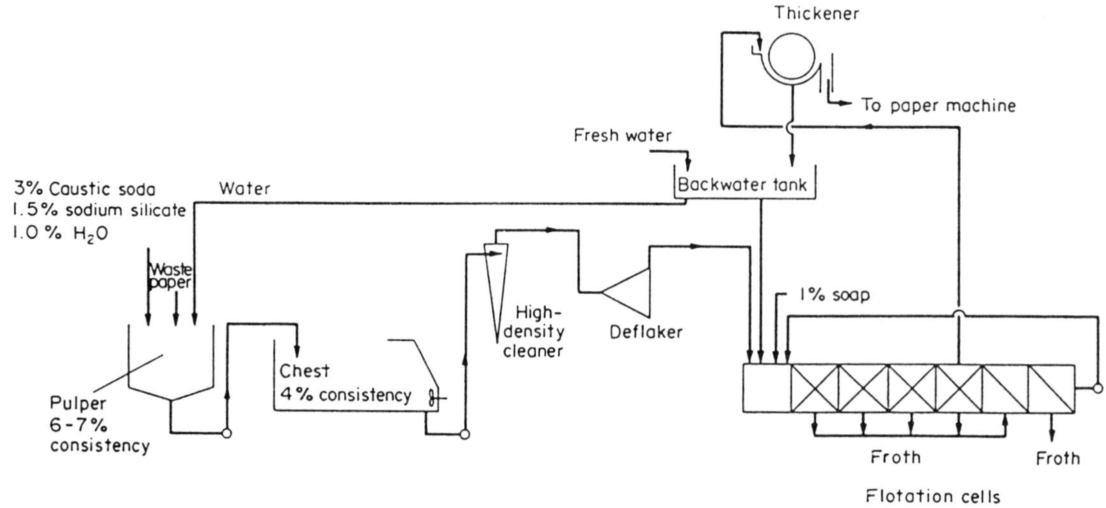

Figure 2
Flotation deinking system

retardant has been added. Extensive studies have been made using shredded newspapers for bedding material for cows, horses, chickens and so on. It is more absorbent than straw and decomposes readily when applied to fields in normal farm manure application.

1.3 Pulp Substitutes and High-Grade Stock for Deinking

Pulp substitutes represent premium material usually in the form of shavings, clippings and sheet waste from converting operations, such as the production of envelopes and paper cups. These residuals are used directly, without being deinked, in paper mills which make the original grade. Conventionally, a mill reuses its own waste, known as broke, and the pulp substitute that it buys from converting operations is similar to the mill's own broke. The pulp substitute ranges from envelope cuttings to various publication blanks, and also includes kraft-bag waste. These recycled materials are also widely used in the manufacture of tissue and multi-ply boxboard (especially in the liner plies). Such grades of waste have a high market value and a very high recycling rate.

High-grade stock suitable for deinking includes printed materials such as business papers, tab cards and computer printouts. The fiber components of these materials are bleached chemical pulps having good strength and aging properties. Generally these high grades contain very little or no mechanical pulp. Deinking of high grades is usually achieved by washing rather than flotation, with the basic chemical being caustic soda which is added in the pulping stage. Deflaking, screening and cleaning are similar to those used in other waste paper recycling operations. High-grade deinked pulps are used for printing and writing grades, tissue (where the softness and whiteness of chemical pulps are important) and to a lesser extent for liner plies of boxboard.

1.4 Mixed Paper

Mixed papers contain grades of paper of vastly different characteristics and are not readily subject to sorting. They are therefore the lowest grade and are the most economical source of waste paper. They have little use in papermaking, but wide use in multi-ply boxboard and construction board. Mixed papers are usually heavily contaminated, and a series of screening and cyclone-cleaning steps are necessary for the separation of plastic, sisal, fabric and metal objects. Adhesives are also a problem, and are mostly in the form of hot-melt materials.

There is increasing interest in the use of waste products as a fuel for generating electricity. Processes have been developed for removing usable fiber grades of paper before burning the residual waste. With the concern over global warming, the common preference is to have all recyclable paper and paperboard products removed before the solid waste is burned or landfilled.

1.5 Other Waste Materials

Another source of high-grade recycled chemical pulp (primarily sold to manufacturers of sanitary tissue) is polyethylene-coated food board. This material is processed in pulpers with very fine (0.3175 cm) extraction plates and with careful control of temperature (28 °C) to maintain the plastic film at its original size without either melting (at high temperatures) or fracturing (at low temperatures). The recovered pulp material is screened, cleaned and then

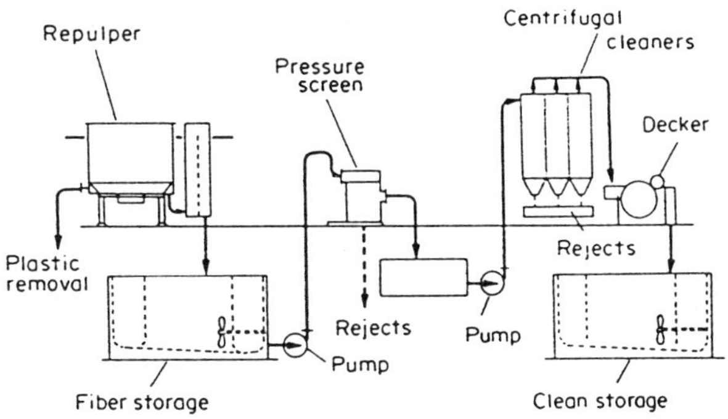

Figure 3
Commercial system for processing polyethylene-coated waste

dewatered by pressing. A commercial system for cleaning polycoated waste is shown in Fig. 3.

2. *Effect of Recycling on Paper Properties*

The surfaces of fibers are composed of windings of fibrils which on refining are raised from the surface, thereby providing greater surface area for hydrogen bonding. These fibrils collapse somewhat on drying and on rewetting during recycling the refibrillation is less extensive. Thus, strength properties that rely on bonding such as burst, tensile and fold resistance are decreased with recycling. With less bonding, the density of the paper tends to be lower, resulting in higher caliper, higher stiffness and greater resistance to tear. Recycled paper also has greater permeability to air and is more opaque.

The effects of the number of repulping steps on paper properties are shown in Figs. 4–6, indicating the effect on swelling (i.e., change in dimension of wood fibers, due to the absorption of water), tension characteristics and strength, respectively. Among these properties, tear strength, which relates to improved printing-press runnability, is improved by repulping. The changes in characteristics are greater with the first recycling, diminishing with each subsequent recycling.

3. *Control of Contaminants*

Contaminants are a common problem in all paper recycling processes, regardless of whether the end product is tissue, printing paper or paperboard. Heavy and large contaminants are easy to remove. Those which are small in size and similar in density to wood fibers are more difficult to remove. From a final recycled fiber quality standpoint, it is better to try to keep contaminants out of the waste paper being processed than it is to try to remove the contaminants in the recycling process.

Perhaps the most troublesome contaminants are

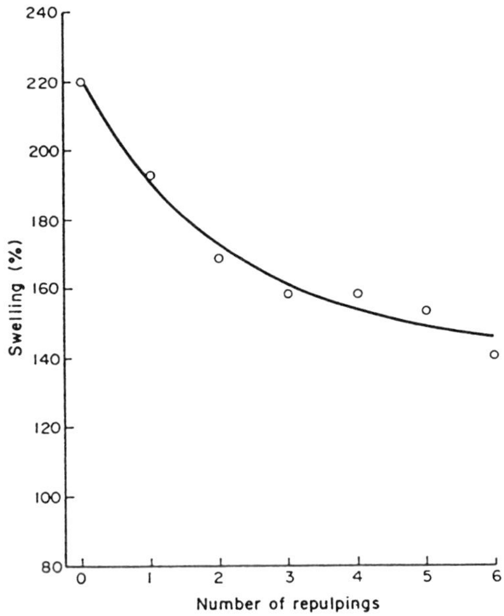

Figure 4
Effect of repulping on swelling of paper

those which are sticky. They can either cause breaks in the paper machine operation or breaks when rolls of paper are continuously unwound (e.g., on a roll-fed printing press). These sticky contaminants are derived from adhesive materials, typically with pressure-sensitive adhesives used in splicing paper rolls, or hot-melt adhesives used for closures with paperboard.

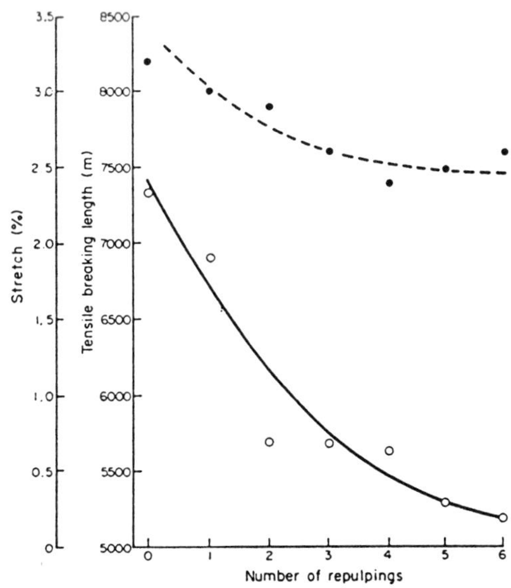

Figure 5
Effect of repulping on tension characteristics of paper: ○, tensile breaking length; ●, stretch

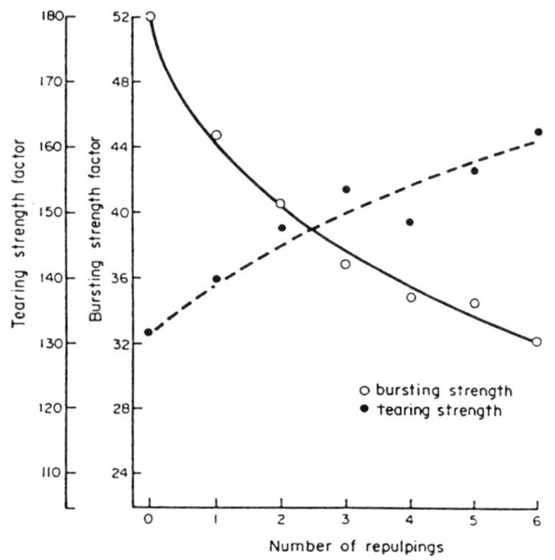

Figure 6
Effect of repulping on paper strength

In recent years, paper machinery manufacturers have made excellent progress in the design of more efficient screening and cleaning equipment for the removal of contaminants—including the sticky types. A typical paper recycling process will have coarse screens containing either holes or slots early in the process followed by screens later in the process that have slots as narrow as 0.2 mm (0.008 in.). A hole size for a typical coarse screen would be 1.57 mm (0.062 in.). Much progress has been made in the design of these screens which allow effective contaminant removal to take place at consistencies as high as 2–4% (two to four parts fiber to 100 parts of slurry).

Forward-flow (conventional) centrifugal cleaners will remove contaminants that have a specific gravity greater than that of the fiber. Reverse-flow cleaners are designed to remove contaminants that have a specific gravity less than that of the fibers.

Even when using fine slotted screens as well as both forward- and reverse-flow cleaners, there may be enough small ink or sticky specks remaining to be visually objectionable in the final product. Some recycling mills use dispersers following the screening and cleaning equipment. The fibrous slurry is dewatered to a consistency of 12–20% and heated before going through the disperser. The heat will soften most stickies and many types of inks, and distribute them on the surface of the fibers in such a thin layer that they are less visually objectionable.

Thermosetting inks such as those used in Xerox-type copy machines or the so-called laser printers cause a unique problem to the waste paper recycler. These thermosetting inks are not dissolved by normal deinking chemicals. They fracture into smaller particles some of which are thermally bonded to the fiber. Flotation systems have some limited success in removing them. A new approach is to add chemicals that cause the thermosetting ink particles to agglomerate in large enough particles to be removed by normal screening and cleaning.

With the advent of computerized image analysis, it is possible to more accurately determine the amount of residual ink, stickies and other visual contaminants in the fiber after deinking, screening and cleaning. By using image analysis on the final product containing reused fibers, it is possible to have a day-to-day audit of the quality of the paper being made as determined by the amount of residual contaminants.

See also: Recycling of Materials: Economics; Resource Recovery: Economic and Institutional Aspects

Bibliography

Bennett R M, Koffinke R A 1985 Mill experience with a total reject system. *TAPPI Pulping Conference*, Book 1. TAPPI, Atlanta, GA, pp. 53–5

Darlington W B 1989 A new process for deinking electrostatic-printed secondary fiber. *TAPPI J.* 72(1): 35–8

Estes T K 1986 Important properties of recycled fibers relating to paperboard. *TAPPI Pulping Conference*, Book 3. TAPPI, Atlanta, GA, pp. 621–4

Franklin W E 1985 Trends in recovery and utilization of waste paper in recycling mills, and other uses of waste paper 1970 to 2000. *TAPPI Pulping Conference*, Book 1. TAPPI, Atlanta, GA, pp. 135–9

Guest D, Weston J 1986 Fiber–water interactions: modified by recycling. *TAPPI Pulping Conference*, Book 3. TAPPI, Atlanta, GA, pp. 589–95

Heise O 1990 Slotted headbox screening for fine, publication, and newsprint grades. *TAPPI J.* 73(4): 165–9

Hisey A B 1986 Quality control of recycled newspaper. *TAPPI Pulping Conference*, Book 1. TAPPI, Atlanta, GA, pp. 197–8

Holusha J 1989 Old newspapers hit a log jam. *N. Y. Times* 10 September 1989

Klungness J H, Fernandez L E, Plantinga P L 1989 Image analysis for measuring adhesive contaminants in pulp. *TAPPI J.* 72(1): 89–93

Lapointi M, Marchildon L, Bonnelly B 1988 The deinking of Xerographic paper by flotation. *TAPPI Pulping Conference*, Book 1. TAPPI, Atlanta, GA, pp. 81–94

LeBlanc P E 1986 A breakthrough in pressure screening. *TAPPI Pulping Conference*, Book 1. TAPPI, Atlanta, GA, pp. 75–8

McCready M 1990 API capacity survey forecasts rising waste paper use. *Pap. Age* 106(1): 17–24

McKee R C 1971 Effect of repulping on sheet properties and fiber characteristics. *Pap. Trade J.* 155(21): 34–40

Matzke W and Selder H 1986 Additional benefits from pressureless high consistency disperging of secondary fiber stock. *TAPPI Pulping Conference*, Book 3. TAPPI, Atlanta, GA, pp. 597–605

Pekkarinen T 1985 Fractionation of OCC waste paper with a pressure screen. *TAPPI Pulping Conference*, Book 1. TAPPI, Atlanta, GA, pp. 37–9

Schall H E 1986 Asphalt dispersion systems, past and present. *TAPPI Pulping Conference*, Book 2. TAPPI, Atlanta, GA, pp. 285–9

Seldin I 1985 Xerographic copy recycling. *TAPPI Pulping Conference*, Book 2. TAPPI, Atlanta, GA, pp. 303–11

W. C. Martin
[Wilmart Associates, Allendale, New Jersey, USA]

F. W. Lorey
[Garden State Paper Company, Saddle Brook, New Jersey, USA]

Recycling of Plastics

Plastics accounted for about 14.4 million tons or 8% of the total 179.6 million tons of municipal solid waste generated in the USA in 1988, an increase from 5.8% in 1980. If volume is the measure, plastics required about 19.9% of the municipal landfill space utilized in 1988 (US Environmental Protection Agency 1990a). While more than 75% of plastic waste from the industrial sector is recycled (Curlee and Das 1991a), only about 1.1% of all postconsumer plastics was recycled in 1988, compared to 25.6% for paper and paperboard, 12% for glass and 14.6% for metals (US Environmental Protection Agency 1990a). The low recycling rate for plastics, in combination with growing quantities of plastic waste and public concern about the environmental effects of plastics disposal, has put plastic waste in the forefront of the debate over the management of nonhazardous solid waste.

Compared to the recycling of other components in the solid waste stream, plastics recycling faces greater uncertainties from environmental, technical, regulatory, institutional and economic perspectives. An understanding of why plastics recycling succeeds in some cases and fails in others is based on an understanding of these uncertainties and the underlying incentives and barriers that face the groups that jointly determine the viability of plastics recycling. A review of the technologies that exist to recycle plastics and the composition of the plastic waste stream sets the stage for a discussion of these uncertainties.

1. Technical Approaches

Figure 1 presents a simple flow diagram of the steps involved in recycling or disposing of plastic waste. Landfill is the primary means of disposing of plastics, with incineration growing in popularity in regions where landfills are more expensive and difficult to site. Plastics recycling involves four main steps: collection (either as part of or outside of the municipal waste stream), cleaning and separation of plastic resins from other materials, processing of the segregated plastics into new products, and disposal of waste from the recycling process.

Recycling technologies are typically divided into four main types: primary, secondary, tertiary and quaternary. Primary recycling refers to processes that manufacture new plastic products with physical and chemical properties equivalent to the properties of the products from which the plastic waste is derived. Primary recycling requires very clean plastics and is usually limited to clean manufacturing waste that can be collected before the waste plastics are contaminated with other materials. In the typical operation, manufacturing scrap is collected immediately following its generation, remelted, and extruded or otherwise made into new plastic products. With very few exceptions, postconsumer plastics are not recycled in a primary sense because of unacceptable contamination levels.

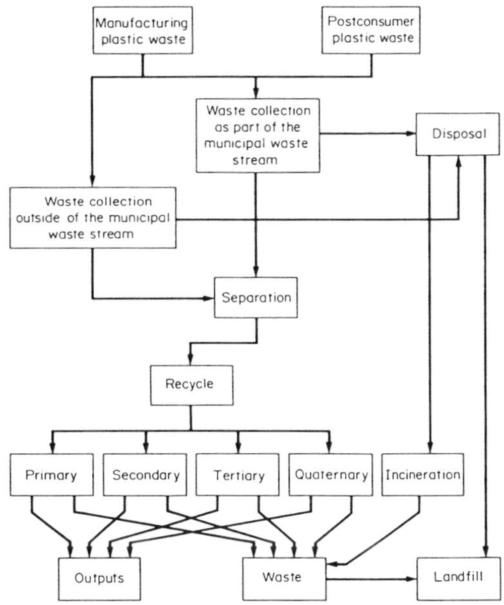

Figure 1
Flow diagram of plastic waste disposal or recycle

Secondary recycling refers to processes that manufacture new plastic products with less-demanding physical and chemical properties than those required of the products from which the waste is generated. Typically, manufacturing and postconsumer plastics are cleaned before being remelted and formed into new products. Numerous processes have been developed in Europe, Japan and the USA to recycle commingled or multiple-resin plastics, as well as single-resin wastes. Products such as plastic lumber and bulky items usually made from metal or concrete (e.g., parking curbs and roadway markers) can be made from a waste stream that contains commingled plastics and as much as 50% nonplastic materials. Higher-valued products can be made from waste containing only one resin. For example, recycled high-density polyethylene (HDPE) from containers such as milk jugs and bottles for household detergents is used to make drainage pipe, flower pots, flooring materials and containers for motor oil. Reclaimed polyethylene terephthalate (PET) from beverage bottles is cleaned and used to make a variety of products, from fiberfill insulation for jackets to industrial strapping.

Tertiary recycling refers to processes that chemically or physically break down plastic polymers into their basic building blocks or, alternatively, into fuels. Pyrolysis, a process in which plastics are heated in the absence of oxygen, can produce basic chemicals and fuels. Another tertiary process is hydrolysis, which involves the chemical decomposition of plastics. In general, higher valued products can be produced from cleaner waste. Tertiary recycling of dirty commingled plastics is limited to the production of fuels, including combustible gas and substitutes for #2 and #4 fuel oils. Alternatively, clean PET from beverage containers can be depolymerized into the basic acid and alcohol monomers from which new PET can be produced for use in new beverage bottles.

Quaternary recycling is another term for incineration with heat recovery. Although about 14% of municipal waste (inclusive of plastics) was incinerated with heat recovery in 1988, quaternary recycling usually refers to the combustion of segregated plastic wastes. Technologies have been customized to burn individual resins with high energy efficiencies and emissions that are within regulatory limits. Continuous rotary kilns to burn PET, which is collected outside of the municipal waste stream in large quantities, are an example. Although the heat value of plastics varies from about 15–45 MJ kg^{-1}, 28 MJ kg^{-1} is commonly assumed for a mixture of widely used resins. In comparison, distillate fuel oil contains 44.6 MJ kg^{-1}, bituminous coal about 30.5 MJ kg^{-1} and unprocessed municipal waste about 12 MJ kg^{-1} (Curlee 1986). If all plastic wastes were utilized to produce heat energy, the amount produced would equal about 1% of total US energy consumption (Curlee and Das 1991b). When compared to secondary technologies, quaternary technologies dedicated to the incineration of individual resins have met with little commercial success.

2. Sources and Composition of Plastic Waste

The viability of plastics recycling largely depends on the types of resins available and the extent to which plastic waste is contaminated with other materials. This, in turn, depends on the sources and composition of plastic wastes.

In response to the rapid growth of plastics usage during the 1960s, 1970s and 1980s, the quantity of plastics entering the waste stream grew quickly. While plastics accounted for only 0.4 million tons or 0.5% of the 87.8 million tons of municipal waste generated in 1960, plastics made up 3.5% of municipal waste by 1975 and 8.0% by 1988 (US Environmental Protection Agency 1990a). If waste categories not considered part of the municipal waste stream are included (e.g., scrap automobiles, building and construction debris and industrial machinery), total postconsumer plastic waste totalled about 14.8 million tons in 1988 and increased to 17.9 million tons in 1990. The quantity of manufacturing waste entering the waste stream (i.e., manufacturing waste that is not recycled as a normal part of the manufacturing process) has decreased over time due to technological developments and is small compared to postconsumer waste. Manufacturing waste entering the waste stream totalled only 0.8 million tons in 1990 (Curlee and Das 1991a).

Figure 2 disaggregates total postconsumer plastic waste by product area for 1990. Packaging represents more than 46% of the total, with consumer and institutional products making up more than 11%. The ease with which plastic waste is collected outside of the general waste stream depends largely on the products in which plastics are used. Cleaning and separation of waste plastics into individual resins depends on the level of contamination of incoming waste, which in turn depends on the source of the waste and method of collection. Collection of plastic waste as a clean, segregated waste stream is obviously easier for plastic beverage bottles in states with bottle deposit laws than it is for miscellaneous household goods (e.g., brushes) that are only partially made from plastic and are generally collected with other municipal wastes.

Figure 3 disaggregates total postconsumer plastic waste by resin type for 1990. Resins differ in their ease of recycling because of technical and other reasons. The distinction between the two main types of plastics (thermosets and thermoplastics) is particularly important for plastics recycling. Thermoplastics (e.g., polyethylene, polypropylene and polyvinyl chloride) differ from thermosets (e.g., epoxy, phenolic and urea) in that they can be melted repeatedly and reformed into new products with only limited damage to their molecular structures. Because of their interlinking molecular bonds, thermosets cannot be melted and reformed, which eliminates thermosets from consideration as a source material for primary and secondary recycling. Although some specialized thermosets may be candidates for tertiary recycling, the only possibilities for recycling the vast majority of thermosets are

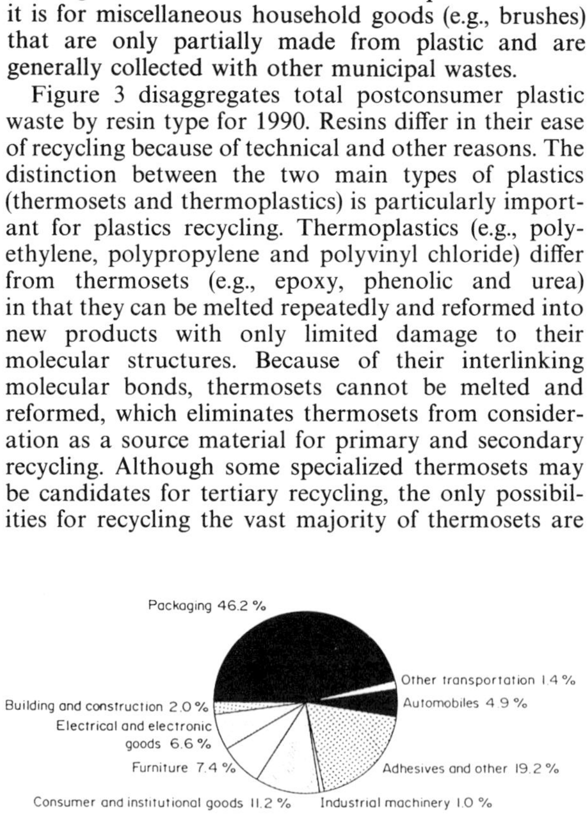

Figure 2
Composition of postconsumer plastic waste by product area (1990)

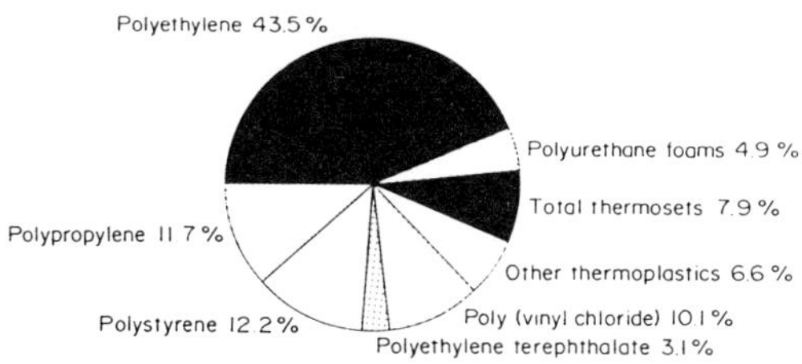

Figure 3
Composition of postconsumer plastic waste by resin type (1990)

incineration with heat recovery and the use of ground thermosets as filler materials with thermoplastics. Thermosets represented 7.9% of all plastics entering the waste stream in 1990.

3. *Incentives and Barriers to Recycling Plastics*

3.1 Environmental Issues

A primary consideration in recycling plastics is the cost and environmental uncertainty associated with plastics disposal. Stricter environmental regulations and difficulty in siting new landfills and incinerators have contributed to rapid escalations in disposal costs. Further, a lack of credible and/or complete information on the environmental effects of plastics disposal has led to widely differing views about the acceptability of plastics disposal. Other things being equal, the more costly and environmentally damaging plastic disposal is perceived to be, the more attractive the alternative of recycling becomes.

There are two main views of the environmental effects of plastics in landfills. One view suggests that plastics are not acceptable in landfills because most plastics do not degrade quickly. It is argued that because plastics do not degrade and do not compact easily they contribute to leaching of potentially toxic substances in landfills. It has also been suggested that when plastics are not distributed evenly in landfills, they result in "spongy" areas, thus limiting the potential uses of the landfill site after closure. Another argument suggests that plastics are not acceptable in landfills simply because they are bulky and require a large volume of space compared to their weight contribution.

The other main view is that plastics are of little concern when landfilled because plastics do not degrade and thus do not contribute to either liquid or gaseous toxic substances. In addition, when plastics are distributed evenly with other waste materials, plastics are claimed to provide structural stability to the landfill site after closing. As other wastes degrade, plastics are claimed to support the soil above the waste level, and therefore postclosure use is enhanced. The body of scientific evidence leans toward the second main view, that is, plastics are not of great environmental concern when landfilled properly.

Degradable plastics have been developed that degrade by light or microorganisms and have been touted as a major achievement in overcoming the environmental effects of plastics when landfilled or littered. There are two basic approaches to degradability: photodegradability and biodegradability. Photodegradable plastics require direct sunlight to initiate the degradation process. Biodegradables consist of two basic types: polymers that are completely digested by microorganisms and plastics that contain additives which are digested, leaving the remaining plastic in the form of dust. While degradables present obvious advantages when plastics are disposed of improperly (e.g., littered on land or water), degradables also pose two potential problems. The first is the effect that mixing degradable plastics with conventional plastics will have on the secondary recycling of commingled plastics. Recycled products that contain both conventional and degradable plastics may have unsatisfactory properties. The second problem concerns potential environmental damage from the products of degradation. The products of degradation are in some cases not well understood, nor are the potential impacts on animals that may ingest the degraded materials.

The environmental effects of incinerating plastics are more controversial than those of landfills. Some plastics, especially polyvinyl chloride (PVC), produce hydrogen chloride when burned, which can react with water to form hydrochloric acid. About one half of the weight of PVC is composed of hydrogen and chlorine. One position argues, therefore, that plastics are not acceptable for incineration on the basis of their contribution to acid rain and potential acidic damage to incineration equipment.

Greater concern has been raised over the potential link between the combustion of plastics and emissions of furans and dioxins, which are widely considered to be carcinogenic. Some studies have suggested that furan and dioxin emissions may be linked to plastics. Other studies have found that when incinerators are operated at the correct temperature, plastics do not contribute to emissions of furans and dioxins (Visalli 1987).

Concerns have also been raised about plastics incineration and heavy metal emissions. Heavy metals, such as lead, cadmium and mercury, are contained in some additives in commonly used plastics. Plastics may thus contribute to concentrations of heavy metals in incinerator ash and gas emissions. Suspicions have also been raised that plastics may play a role in metal emissions by forming particulates to which metals attach and escape from the stack. Moves are generally under way to eliminate or substitute for the heavy metals used in plastics, especially in plastics used for packaging. In many cases, substitutes are available that provide similar performance characteristics.

Many environmental experts argue that plastics present no significant environmental concerns if available emission-control technologies are used correctly. Virtually all acidic, furan and dioxin emissions can be removed with current technology. Others suggest that the potential emissions from plastics incineration are, at worst, trivial when compared to large-scale polluters, such as coal-fired electricity generation plants. Incineration of plastics may, in fact, be advantageous because the high thermal energy content of plastics helps burn other materials which are difficult to burn.

3.2 Technical Issues

A variety of secondary products can be produced from commingled and single-resin wastes with properties that meet or exceed those of similar products made from alternative virgin materials. For example, plastic lumber (a relatively low-technology secondary product utilizing commingled waste that is at least 50% thermoplastic) is suitable for many applications normally fulfilled by wood, such as deck flooring, fencing materials, park benches and boat docks. While plastic lumber cannot be used for structural purposes because its physical properties may degrade at higher temperatures and over time, plastic lumber is superior to wood in applications where decay is a problem. Further, the technology utilized in most secondary processes is not extremely sophisticated by modern industrial standards. Unlike the recycling of, for example, aluminum, steel, paper or glass, technical constraints do not necessarily require that secondary recycling facilities be large scale. If recycling is defined to include low-technology secondary, tertiary and quaternary processes, all plastic wastes can be recycled from a technical perspective.

If recycling is limited to the production of higher-valued and more demanding products that require a clean plastic waste stream, then the majority of plastic waste is not suitable for recycling from a technical perspective. Two main technical obstacles arise: (a) incompatible mixtures of resins and (b) difficulties in separating plastics from other materials and in separating individual resins. The properties of different resins vary in some cases as much as the properties of glass differ from those of metals. These varying properties severely limit the capabilities of some mixtures of resins.

While compatibilizers can help to bind mixtures of plastics and thus make those mixtures more useful, the more direct solution is to perfect separation technologies. Systems based on shredding followed by air classification, magnetic separation and/or screening can separate plastics and paper from heavier materials. Wet and dry systems can be used to separate plastics from paper. In some dry systems, differences in the reaction of plastics and paper to heat and electrical charges are utilized. Most wet systems depend on the different water absorption rates for plastics and paper.

Systems to separate individual resins depend on a variety of approaches. The simpler approaches include sink/float methods that make use of the different densities and wetting characteristics of different resins. Solvent extraction and electrostatic methods are also used. Solvent extraction is based on the tendency of some resins when dissolved in a solvent to form different phases, which facilitates separation. Electrostatic methods depend on the abilities of different resins to hold varying electrical charges. The most sophisticated methods use, for example, cryogenic grinding with magnetic separation, separation assisted by microorganisms and the identification and subsequent separation of resins using x rays.

Separation utilizing x rays is representative of the problem of resin incompatibility and the extent to which technology must go to separate some plastics. For example, PVC and PET have similar densities, which limits the effectiveness of simple separation approaches. Further, PVC and PET are highly incompatible when mixed together, especially in the processes that utilize PET from beverage bottles. X rays are used to excite the chlorine atoms in PVC. Once a PVC bottle is identified on a conveyor system, it is separated using sophisticated air-blow techniques.

Although technologies exist to separate or compatibilize different resins, there are practical limits to which technology can be expected to overcome the problems of plastics recycling. The cost of separating resins can quickly exceed the value of the segregated waste streams. The general opinion among experts is that once plastics are mixed with other materials in the municipal waste stream, the separation of plastics from other materials is a costly but feasible process. The separation of individual resins can be achieved in most cases. However, in a practical sense, the separation of individual resins is limited to commingled waste streams that contain plastics with significantly different physical and/or chemical properties.

Polymer degradation is another practical technical limit to recycling plastics. Most primary and secondary processes involve the melting of thermoplastics, with subsequent damage to the plastic's molecular structure. While there is disagreement over the extent to which damage occurs, there is agreement that recycled plastics have a limited number of lives. Eventually, the plastic will become unacceptable for further recycling because of molecular degradation or contamination with other materials. At this point, tertiary or quaternary methods must be used or the plastic must be disposed of. From an environmental perspective, this technical constraint is important because plastics cannot be recycled in an "infinite loop," with the exception of tertiary processes that produce monomers from which new polymers are constructed. Secondary processes delay, but do not eliminate the eventual need for or the environmental implications of, the incineration or other disposal of plastic waste.

3.3 Regulatory Issues

Legislative and regulatory actions at the federal, state and local levels have in most cases promoted, but at times discouraged, the recycling of plastics. Included are measures which directly mandate that communities collect plastics for recycling, measures to make the option of recycling more attractive and measures to promote alternative responses to the "plastics problem," such as degradability standards, product and material bans, and source reduction. Measures that promote additional plastics recycling have focused on reducing technical and market uncertainties and on utilizing moral suasion and government leadership to establish and promote recycling goals.

Technical constraints call for clean plastics and make the collection of segregated plastic waste very important; it is in this area that legislative and regulatory actions have been most effective. Bottle deposit legislation in at least nine US states has led to the segregated collection of PET beverage bottles. The extension of deposit legislation to other plastic containers and the passage of a national bottle deposit law continue to be debated within legislative bodies. About 78 kt of PET were recycled in 1988, or about 43% of the total 180 kt of plastic recycled in that year. PET bottle recycling increased by 8.6% in 1989 to some 85 kt, which equalled 28% of all PET bottles produced. Further, approximately 95% of all recycled PET comes from states with bottle deposit laws. Without the PET collection system that has resulted from bottle deposit legislation, the recycling rate for plastics that normally enter the municipal waste stream would be reduced by 40%.

Other government measures have directly or indirectly promoted the collection of clean plastic wastes. Recycling goals for municipal solid waste established by the US Environmental Protection Agency (EPA) (25% by 1992 and 40% by 1996) have encouraged state and local government to adopt curbside collection of segregated wastes, including PET and HDPE containers. Several states have adopted mandatory plastic labelling that identifies the type of resin used in containers and assists consumers and waste processors in segregating individual resins.

Some measures specifically address technical and market uncertainties. For example, some states have passed legislation to define what is meant by terms such as "recyclable" and "recycled" in an attempt to bring uniformity to markets. Rules and regulations on the procurement of recycled products by federal, state and local governments have fostered market demand for some recycled plastic products. Some governments allow a premium to be paid for recycled goods, as compared with similar goods made from virgin resins. Various tax incentives are available in some states to promote the construction and operation of recycling facilities.

Other government actions serve as a direct or indirect barrier to additional plastics recycling. For example, zoning ordinances may prohibit the construction of a recycling collection point or facility in areas where large quantities of plastics and other materials are available. Local actions to mandate the use of degradables may lead to a plastics waste stream contaminated with plastics that are unacceptable for many secondary recycling technologies. Plastic bans may reduce the quantity of plastics available for recycling to levels below which recycling programs are not justified in economic terms. Even bottle deposit laws may indirectly limit recycling of non-PET resins. Once PET and HDPE bottles are removed from the postconsumer stream, the remaining plastics are of such low value that the collection and processing of those remaining plastics may not be justified economically. The hundreds of federal, state and local laws addressing plastics recycling create such a complicated and rapidly changing regulatory environment that firms entering recycling markets devote significant resources to monitoring and responding to changing "rules of the game." Finally, government actions that reduce the potential environmental effects of plastics disposal indirectly decrease the incentive to recycle. For example, bans on the use of heavy metals in plastic packaging reduce the concern about the incineration of plastics. Further, stricter regulations on emissions from landfills and incinerators make disposal a more viable option and indirectly discourage recycling.

3.4 Institutional Issues

Plastics recycling requires at least limited cooperation and coordination among plastic and product manufacturers, consumers, waste processors and government. Each of these groups can take actions that promote or hinder recycling, and each group faces specific incentives and disincentives to take particular actions. If any one group faces a net disincentive to recycle, that group can, in effect, block a recycling effort that is attractive from the perspectives of the other groups and which would result in positive net benefits from a social perspective. Institutional structures can help "smooth out" the incentives and disincentives among the groups so that barriers to recycling can be overcome. Institutions that facilitate market development and foster the flow of information are particularly important.

Industry sponsored groups have greatly facilitated the flow of information about various aspects of recycling and provided financial support for the development of new recycling technologies and for market assessments and environmental studies. Organizations such as the Plastics Recycling Foundation, the Council for Solid Waste Solutions and the National Association for Plastic Container Recovery have been instrumental in promoting the secondary recycling of PET, HDPE and commingled plastics. These organizations act as an information clearing house to collect and disseminate information to consumers, municipalities, waste processors and plastic manufacturers. The Society for the Plastics Industry developed and helped implement a digital and letter code to identify the particular resin used in a product. Industry has also promoted recycling by developing industry standards for recycling, by designing products and resins that are more easily recycled, and by using resins for which recycling networks exist. For example, the bases and labels on PET beverage bottles have been designed with recycling ease in mind. Multilayered, multiresin ketchup bottles have been replaced by bottles made from PET to take advantage of the well-established PET recycling industry. Some automobile manufacturers consider the compatibility of resins used in automobile parts and the ease with which plastic parts can be removed for recycling. Concerns raised by consumers, government and environmental groups about the environmental effects of plastics disposal have prompted the plastics industry to promote recycling as an alternative to initiatives to curb the use of virgin resins.

Consumer organizations and environmental groups have been instrumental in addressing environmental concerns and acting as a watchdog over claims of "recyclability" and "recycled content." For example, the Green Cross Certification Company has given its so called "Green Cross" seals to products made from recycled plastics to verify to consumers that recycled plastics are, in fact, used. The Environmental Defense Fund has worked with fast food restaurants to promote recycling of plastics, in addition to minimizing plastic waste. Recycling organizations within communities play key roles in collecting waste plastics and educating the public about recycling possibilities.

Agreements between waste collectors and large companies to process plastic wastes have been instrumental in expanding the plastics resource base beyond PET collected from bottle deposits. Agreements between Waste Management Inc. and DuPont Corporation and between Browning Ferris Inc. and Wellman Inc. have promoted the curbside collection and recycling of PET and HDPE bottles and, to a more limited extent, commingled plastics. In these agreements, waste processors agree to purchase the plastics collected by waste management firms. In so doing, the waste collector ensures a demand for the plastics collected, and the waste processor ensures a supply. Market uncertainties are thus reduced.

3.5 Economic Issues

The economic viability of plastics recycling depends on the perspective. Three perspectives are particularly important: (a) that of the private firm involved in the recycling activity, (b) that of the community attempting to minimize "out-of-pocket" cost of waste management, and (c) that of the community attempting to minimize the cost of waste management in the larger context (e.g., inclusive of potential external environmental costs).

From the perspective of the private recycling firm, recycling is economically viable in a limited number of cases. The most successful recycling operations involve PET beverage bottles and HDPE milk jugs, which are collected as a result of bottle deposit legislation or as part of curbside collection programs. Market observations suggest that the recycling of PET and HDPE bottles is economically viable when the cost of collection is not borne by the recycler. In other words, if curbside collection of plastics is paid for or subsidized by households or through general tax revenues or, alternatively, if households and the beverage industry are forced to pay for the collection of PET bottles in a bottle deposit program, private firms can profit from recycling the resulting plastic waste streams. With few exceptions, any clean thermoplastic waste, which is made available to recyclers at no cost or which is simply sold to the highest bidder, can be recycled economically. Less, but nonetheless significant, success is evident in the recycling of commingled and contaminated plastics that are collected at public expense or which can be otherwise collected at very low cost (e.g., plastic waste that results as a by-product of recycling the metal components of automobile batteries or telephone equipment).

Studies suggest that the potential markets for recycled PET, HDPE and commingled plastics are large compared to the quantities of those resins collected for recycling. By some estimates, small penetrations into potential markets of only 2% to 10% account for all plastics recycled currently. Recycled PET and HDPE enjoy significant price advantages over virgin resins, which has spurred their demand.

The case for commingled resins is somewhat less clear. While the potential markets (e.g., plastic lumber) are huge in comparison to the quantities of commingled plastics collected outside of the general waste stream (Curlee and Das 1991b), the economic viability of commingled recycling is uncertain. For very clean commingled waste, which can substitute for virgin resins in products with less demanding properties, market evidence suggests economic viability. However, for products such as plastic lumber, which may cost two to four times the cost of alternative products made from wood, the evidence is less clear. Whether consumers are willing to pay a premium price for the different and, in some cases, superior properties of plastics is an issue being determined in the market place.

Turning to the perspective of the community attempting to minimize "out-of-pocket" disposal cost, there is even greater uncertainty. The average US cost of landfill increased from US\$13.43 to US\$20.36 per ton (not adjusted for inflation) between 1986 and 1987. The average cost of incineration increased from US\$30.42 to US\$33.64 per ton over the same time frame. Costs vary tremendously from location to location, with disposal costs approaching US\$100 per ton not uncommon. Obviously the economic advantage of plastics recycling over disposal is enhanced by escalations in disposal costs. The collection and recycling of plastics displaces the use of landfill or incineration capacity and should be credited with the value of that displacement. Further, recall that the volume contribution of plastics to the municipal waste stream is 2.5 times their weight contribution; and no landfill has ever closed because it weighed too much. Thus, the exact cost of disposal at which plastic recycling becomes the preferred alternative is complicated to determine and dependent on local conditions.

A key question concerns the marginal cost that plastic waste collection imposes on a community's curbside recycling program. Evidence from a Milwaukee curbside collection program suggests that plastics (PET and HDPE bottles) contribute 4% of the weight but 25% of the volume of all waste collected. One study concluded that the net cost of collecting plastics in a curbside program (i.e., the cost of collection less the revenues received from the sale of the collected materials) is US$800 per ton, while the net cost of collecting glass was only US$52 per ton. Although many experts disagree that the marginal burden imposed by plastics on a curbside collection program is of this magnitude, there is general agreement that plastics are relatively expensive to collect.

If the perspective is that of the community attempting to minimize disposal costs inclusive of external environmental cost, the viability of recycling plastics is even more uncertain. In addition to considering the "out-of-pocket" savings that recycling provides in terms of avoided disposal costs, such an assessment must also consider the environmental costs that may be avoided by recycling, as compared to disposing of, plastics. Recall from Sect. 3.1 that there is a lack of agreement on the environmental consequences of plastic disposal. Experts also disagree about the environmental consequences of some recycling operations. Thus, from this most inclusive perspective of net economic benefit, there is insufficient evidence to suggest that expanded plastics recycling is or is not economically justified.

4. Future Recycling: Crucial Trends

The degree to which postconsumer plastics recycling will increase in future years above its 1988 level of 1.1% will depend on several crucial trends: (a) the extent to which a consensus can be reached about the environmental consequences of plastics disposal, especially the incineration of plastics; (b) the degree to which new technically and economically viable processes can be developed to utilize mixtures of plastic resins and/or separate individual resins from a commingled stream; (c) the legislative and regulatory incentives and barriers to recycling enacted by federal, state and local governments and the degree to which a consensus and uniformity of goals can be achieved across governmental and regional boundaries; (d) the degree to which institutional structures can be put in place to generate and disseminate credible technical, market and environmental information among industry, consumers, environmental groups and government; (e) the extent to which communities adopt curbside or other "household-friendly" recycling collection systems and whether those systems are designed to collect commingled plastics, only single resins (e.g., PET and HDPE bottles) or no plastics at all; (f) trends in the costs of landfill and incineration and the abilities of communities to site waste disposal facilities at any cost; and (g) the extent to which consumers are willing to pay a premium for the special properties of products made from recycled plastics as compared to similar products made from lower-cost materials.

An expansion of the recycling rate to between 7% and 14% is thought to be possible if technical, regulatory, institutional and economic conditions are such that recycling is limited primarily to PET and HDPE packaging. If conditions allow the recycling of commingled postconsumer plastics, recycling percentages in the 50% range may be achieved (Curlee and Das 1991b). If incineration with heat recovery is found to be acceptable economically and environmentally, virtually all plastic waste becomes a candidate for recycling. Plastics recycling is likely to be the subject of significant controversy in the foreseeable future as firms test the technical and economic viability of new approaches, as consumers and households argue the merits of recycling, and as government debates its appropriate role in the management of plastic waste.

See also: Reclaiming of Elastomers

Bibliography

Curlee T R 1986 *The Economic Feasibility of Recycling: A Case Study of Plastic Wastes.* Praeger, New York

Curlee T R, Das S 1991a Plastics recycling: quantity projections and cost estimates. *Mater. Soc.* 15(1): 41–73

Curlee T R, Das S 1991b Identifying and assessing targets of opportunity for plastics recycling. *Resour. Conserv. Recycling* 5: 343–63

Derry R 1990 The recycling of post consumer plastics in Europe. *Plast. Rubber Process. Appl.* 13(4): 221–7

Hawkins W L 1987 Recycling of polymers. *Conserv. Recycling* 10(1): 15–19

Leidner J 1981 *Plastic Waste: Recovery of Economic Value.* Dekker, New York

Plastics Institute of America 1990 *Plastics Recycling as a Future Business Opportunity.* Technomic, Lancaster, PA

Spaak A 1985 Recycling a mixture of plastics: a challenge in today's environment. *Conserv. Recycling* 8: 419–28

US Congress, Office of Technology Assessment 1989 *Facing America's Trash: What Next for Municipal Solid Waste?* OTA-O-424. US Government Printing Office, Washington, DC

US Environmental Protection Agency 1989 *The Solid Waste Dilemma: An Agenda for Action.* EPA/530-SW-89-019. US Government Printing Office, Washington, DC

US Environmental Protection Agency 1990a *Characterization of Municipal Solid Waste in the United States: 1990 Update.*

EPA/530-SW-90-042. US Government Printing Office, Washington, DC
US Environmental Protection Agency 1990b *Methods to Manage and Control Plastic Wastes: Report to Congress.* EPA/530-SW-89-051. US Government Printing Office, Washington, DC
Visalli J R 1987 A comparison of dioxin, furan, and combustion gas data from test programs at three MSW incinerators. *J. Air Pollut. Control Assoc.* 37 (12): 1451–64

T. R. Curlee
[Oak Ridge National Laboratory, Oak Ridge, Tennessee, USA]

Residuals Management

A decent quality environment is an important element in the economic and social development of all societies, rich and poor alike. Environmental degradation can affect the well-being of people in two ways: directly through exposure to pollutants and contaminants which can result in both acute and chronic health effects and in reduced recreational and other opportunities; and indirectly through damages to life-support systems, to water and food supplies, and to marketed goods and services. An important public policy issue for all nations is to strike an appropriate balance between levels of environmental quality and levels of other public and private goods and services.

Protection of the environment requires the same resources—human, land and capital—that could be put to productive use in other sectors of the economy. Tradeoffs between levels of environmental quality and levels of material well-being must be made by all nations regardless of the political–economic systems used to produce and allocate goods and services. Within the environmental sector, further tradeoffs must be made between the efficient use of resources employed to protect the environment and equity. That is, are the desired levels of environmental quality being purchased at least cost in terms of other opportunities foregone? Who benefits from improvements in environmental quality, and by how much? What are the costs of these improvements, and who pays for them? Are the costs and benefits of pollution control distributed equitably?

In this article, environmental pollution and its control are viewed from the perspective of a public management problem. In designing government policies for pollution control, three primary functions are distinguished. First, an institutional and legal framework within which pollution control is to be conducted must be established. Second, environmental goals and targets—usually expressed in the form of ambient standards—must be adopted. Finally, a management strategy for achieving these goals and targets within the institutional and legal framework that has been established must be identified, adopted and implemented.

The intent of this article is to provide a brief overview of residuals management in the context of environmental pollution and its control. First, the nature of residuals is discussed. Second, the major elements of a residuals management system are delineated. Third, alternative procedures for establishing environmental targets are presented. In Sects. 4 and 5, the steps to be taken in preparing a regional management plan for pollution control, steps to be taken in implementing the plan, and assessment of the plan in terms of achieving the environmental goals and targets established by government are discussed.

1. The Nature of Residuals

Residuals—sometimes referred to as wastes—are generated at virtually every stage in the production and consumption of goods and services. An example for iron and steel production is shown in Fig. 1. Residuals are of two basic types, material and energy. Most material residuals are readily transformed from one form to another in production, consumption and treatment activities. Examples include: solids to gases as in the combustion of municipal refuse and sewage sludge; gases to liquids as in the wet scrubbing of gaseous effluent streams containing SO_2; and gases to solids as in the electrostatic precipitation of gas-stack particulates. Changes in the form of residuals, however, often involve inputs of energy and the generation of additional residuals.

In the context of residuals management, the principle of mass conservation is important. Application of this principle in the analysis of environmental problems is referred to by some environmental economists as the materials-balance approach. Residuals can be changed from one form to another, but the total weight to be handled and disposed of does not change. (In some cases, it increases.) Consequently, solids are sometimes referred to as the irreducible limiting form of residuals. By the application of appropriate equipment and energy, all undesirable substances can be removed from water and air streams (except CO_2 which may be harmful in the long term). As restrictions on the discharge of gaseous and liquid residuals to the atmosphere, water courses and land become more stringent, problems of handling greater and greater quantities of bottom ash, sludge and solid wastes can be expected to gain in importance. This has already been demonstrated in countries with well-established environmental protection programs. Both air and water pollution can be reduced, and in some cases eliminated entirely, but the solid residuals problem cannot be eliminated without total materials recovery and reuse.

The quantities of material residuals disposed of in the environment may be reduced by: (a) decreasing the quantities of residuals generated in production and consumption activities, and (b) increasing levels of

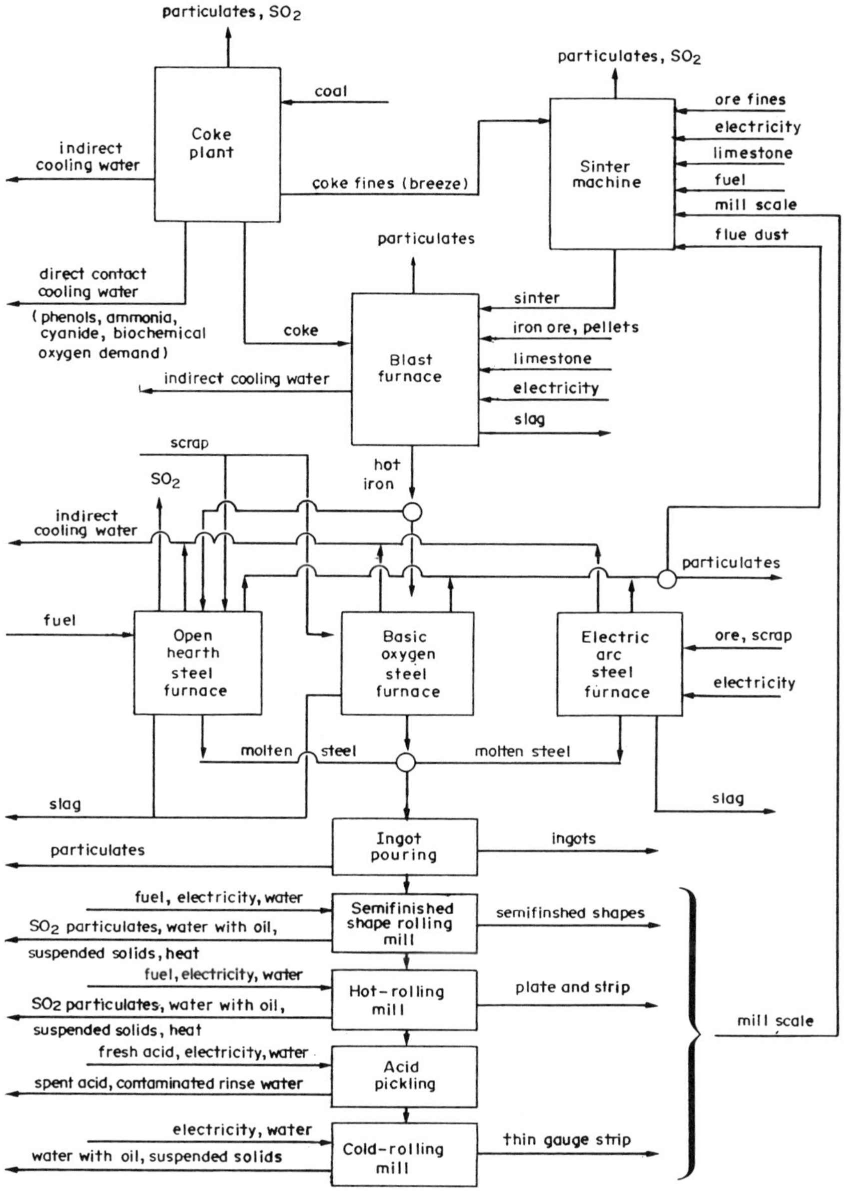

Figure 1
Schematic representation of iron and steel production indicating the generation and discharge of liquid, gaseous and solid residuals (after Russell and Vaughan 1976. © Johns Hopkins University Press, Baltimore, MD. Reproduced with permission)

materials recovery and reuse through recycling, by-product production and reclamation.

The generation of residuals in production and consumption activities can be reduced by: changing the nature of intermediate and final products; reducing levels of production of a particular product; using substitute raw material inputs; employing different technologies of production and consumption; and increasing the useful lifetimes of consumer goods, such as cars and household appliances.

The extent of materials recovery and reuse is a function of economic factors and public and private restrictions and incentives. The economic feasibility of materials recovery and reuse depends on: the demand for final products, the relative prices of virgin and secondary materials, the availability of a consistent supply of secondary materials of a specified quality and quantity, processing and reprocessing technologies, and product specifications. The demand for secondary materials is highly sensitive to changes in the demand for final products. The output of the secondary materials industry typically expands during periods of shortages of competitive virgin raw materials. The most notable example of this occurred during World War II.

Of the nonproduct outputs generated in production and consumption activities, those that are not recovered and reused must be disposed of, usually in the natural environment. The ability of the environment to assimilate residuals is a valuable resource. Complete elimination of all discharges of residuals to the environment would impose unnecessary cost burdens on society. However, if no restrictions were placed on the use of the environment to assimilate residuals, it would become overused, as most nations have already experienced. Controlling discharges of residuals to the environment in the most cost effective way is the essence of residuals management.

2. Residuals Management System

A conceptual framework for residuals management that has found considerable use in the study of pollution problems is shown in Fig. 2. This framework is applicable to residuals management in all countries. It is useful for delineating (a) the range of structural and nonstructural alternatives available to society for reducing the damaging effects of residuals in the environment, and (b) the range of policy instruments available to government for reducing discharges of residuals from individual production and consumption activities.

The management system is conceived as consisting of flows of materials and energy from production and consumption activities (reflecting decisions about production processes, raw-material input mixes, recycling and reclamation opportunities, and residuals treatment and disposal alternatives), through the environment where residuals are stored, diluted, transported, and physically, chemically and biologically transformed, to receptors—humans, animals, plants and inanimate objects—where damages occur. Some damages are experienced by humans directly and some indirectly. In some cases, final protective measures, such as air filtration, water treatment and noise barriers are used to reduce the impacts on receptors of pollutants and contaminants in the environment.

The flows of residuals and the damages which result from them give rise to corresponding flows of information from people who are affected by pollution to response organizations, such as conservation groups and other defenders of the environment, and local and national governments. The concerns of people ultimately reach the environmental quality management agencies that have jurisdiction. These agencies, in turn,

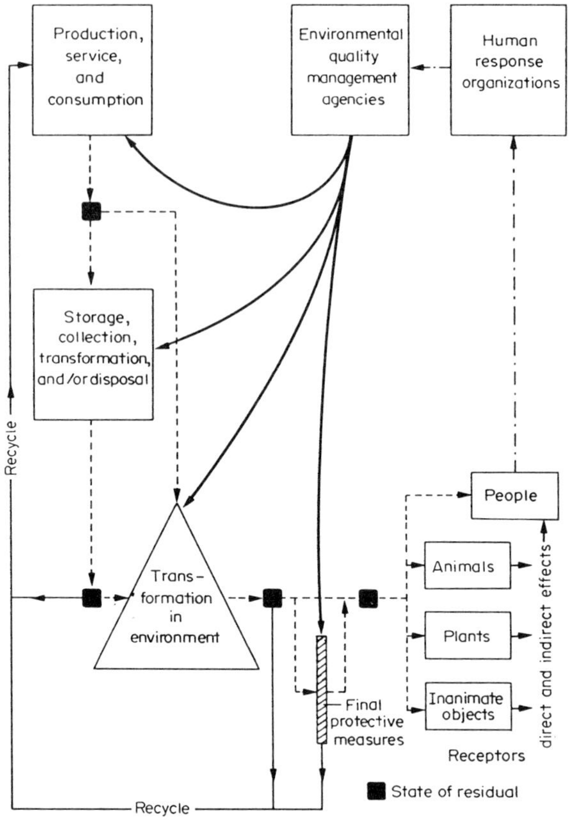

Figure 2
Schematic representation of an environmental quality management system: —— government action, -- flow of residuals, —·— information flow (from Resources for the Future 1968 *Annual Report*. © Resources for the Future, Washington, DC. Reproduced with permission)

take whatever actions seem desirable or are required by law. These actions may include requiring certain levels of treatment, issuing discharge permits, imposing effluent charges, subsidizing recycling operations, modifying the assimilative capacity of the environment, or installing measures to protect people from environmental pollution, such as water treatment. Whatever actions such agencies take will modify the damages caused by pollutants and contaminants in the environment.

3. Establishing Environmental Targets

Various procedures can be used to establish target levels of environmental quality for a region or nation. The procedure used in a given management context depends on the institutional and legal framework that has been adopted for pollution control. The three most commonly used procedures include: (a) cost–benefit analysis, where ambient standards derive directly from a comparison of the costs and benefits of increments of pollution control; (b) cost–effectiveness analysis and public deliberation, where ambient standards derive indirectly from a consensus on the willingness to pay for different levels of environmental quality; and (c) human health, ecological, or esthetic considerations where ambient standards (and in some cases effluent standards) are based on ethical, legal, or political concerns rather than on costs.

Ideally, the objective of an environmental quality management program for a given region or river basin should be to maximize the value of expenditures in pollution control. This requires, in principle, that the benefits and costs of alternative pollution control strategies be estimated, and that the benefits and costs of increments of pollution control be compared. The most efficient pollution control strategy for the region as a whole is the one in which the net benefits (the benefits minus the costs) are maximized, or alternatively the one in which the sum of pollution control costs and environmental damages is minimized. This analysis simultaneously defines the optimal levels of the ambient standards and the optimal levels of investment in pollution control.

Although widely used in assessing other public investment projects, cost–benefit analysis has not found extensive use in environmental quality management because of problems in valuing certain kinds of benefits. Not only is it difficult to assess the physical, chemical and biological impacts of alternative pollution control strategies, but even where these impacts can be assessed with a reasonable degree of accuracy, it is often not possible to assign unambiguous values to reductions in them. This is particularly true of human health and esthetic impacts and damages.

In cases where the benefits of environmental improvements cannot be estimated with a reasonable degree of accuracy, the next most useful information for purposes of environmental decision-making is the cost of achieving alternative levels of environmental quality. Given this information, a region or nation can determine the level of environmental quality it is willing to pay for in terms of other opportunities foregone. This process involves setting ambient standards in light of the resources that will be required for pollution control. The damages associated with pollution and the benefits associated with pollution control are dealt with subjectively, in a qualitative sense, rather than objectively as is required in cost–benefit analysis. This approach to establishing ambient standards is often used, especially for those situations where the damages are primarily esthetic.

The third procedure that has been used to establish acceptable levels of environmental quality for a region or nation is based on health, ecological, or esthetic considerations independent of estimates of either the costs or the benefits of pollution control. Costs, to the extent that they are reflected at all in decisions, are not considered until after the pollution control program has been implemented and they are about to be incurred. At this point, some adjustments to the ambient standards based on the actual costs of achieving them may be made. For countries with well established environmental protection programs, such as those in Europe, North America, Australia and Japan, this is the approach that has been used to select target levels of environmental quality.

4. Regional Management Plans

4.1 Preparation

After agreement on levels of the ambient standards for a region or nation has been reached, the next step is to identify alternative residuals management strategies that can meet these standards. A management strategy for a region consists of specifying: (a) the structural and nonstructural measures that are both capable of meeting the ambient standards and available for use, and (b) regulations and/or economic incentives that will ensure that these measures will be adopted. The results of this analysis provide the basis of a management plan for pollution control in the region. A management plan consists of a management strategy, technical guidelines and administrative procedures for implementing the management strategy, and a compliance schedule for pollution control.

A large number of alternatives is available to any region for meeting the ambient standards. However, they will differ in total cost to the region, in the distribution of costs among dischargers in the region, and in other pertinent characteristics. An important consideration in residuals management is the cost of pollution control.

The potential cost savings to a region of the least-cost strategy can be substantial. Experience has shown, for example, that the total cost of meeting ambient standards by requiring a uniform percent reduction in discharges at individual sources are typically two to five times more than the cost of the regional least-cost strategy. Alternatively, in situations where there are severe limitations on the availability of capital for pollution control, as is the case in most developing nations, substantially more progress can be made in protecting the environment if the available resources are used as productively as possible. The cost savings associated with the least-cost strategy are due partly to differences in the costs of pollution control at individual discharge activities and partly to the relative locations of dischargers and of areas of low environmental quality in the region.

A residuals management strategy which may be highly desirable from the perspective of least cost to the region, however, may not be feasible because of the distribution of pollution control costs or because of concerns with equity more generally. With the least-cost strategy, dischargers who can treat their discharges most cheaply are asked to reduce them proportionally more than those whose treatment costs are greater. In a number of countries, particularly those with market economies, the least-cost strategy has not been popular because of the inability to arrange an appropriate institutional mechanism for dividing the potential cost savings among the dischargers in the region in an acceptable manner.

Because the regional least-cost strategy is often considered inequitable and also because it is generally difficult to administer, more equitable and more easily administered management strategies should also be identified and their costs assessed. Uses of uniform effluent standards, uniform effluent charges and effluent standards based on best available technology are examples of strategies that are generally considered to be more equitable and more easily administered than the least-cost strategy. Management strategies based on the use of marketable discharge permits are particularly promising because of their desirable features with respect to both the total costs of pollution control and the distribution of these costs among the dischargers in the region.

After the regional analyses have been completed, and a residuals management strategy for the region adopted, there are two final steps in the preparation of a regional management plan. Guidelines for establishing discharge standards and for issuing discharge permits, if they are to be used, and guidelines for using economic incentives, such as subsidies for construction, effluent charges, and marketable discharge permits, if they are to be employed, must be prepared. Finally, a compliance schedule for the completion of treatment facilities, for meeting permitted discharges at individual activities, or for implementing an incentive system for discharge reduction must be established. The management plan should provide for both population and economic growth, and for the additional pollution associated with this growth, for a planning period of at least 15–20 years into the future.

4.2 Implementation

Having developed a management plan for pollution control in the region that prescribes the allowable discharge levels at all sources, a pollution control compliance schedule for individual dischargers, and the administrative steps to be taken to attain the desired levels of control, the next step is implementation of the plan.

Implementation involves three major activities: (a) construction of pollution control facilities, (b) operation of pollution control facilities, and (c) enforcement of pollution control laws by government. For those countries with well-established pollution control programs, implementation has tended to focus more on the design and construction of pollution control facilities than on the operation and maintenance of the facilities once constructed. Examples exist in all countries of pollution control facilities not operating properly, or indeed at all, in order to reduce operating costs. Without adequate enforcement with a focus on emissions rather than on installed equipment, the management plan will not fulfill its purpose.

Enforcement requires well-trained inspectors, ambient and effluent monitoring equipment, and laboratory support, backed up with sufficiently severe penalties for offenders. The success of a pollution control program depends on frequent inspections of emissions.

5. Assessment of the Pollution Control Program

To ensure that the pollution control program achieves the environmental goals and targets established by government, it will be necessary to monitor the quality of the air, water and land throughout the region, including the toxic substance contents of plants, fish and human tissues. This information is used to ascertain whether or not the pollution control program is achieving its goals, and if not, what changes are needed in the regional management plan to ensure that these goals will be achieved.

See also: Materials and the Environment

Bibliography

Armstrong E L (ed.) 1976 *History of Public Works in the United States, 1776–1976.* American Public Works Association, Chicago, IL

Ausubel J H, Sladovich H E (eds.) 1989 *Technology and Environment.* National Academy Press, Washington, DC

Basta D J, Lounsbury J L, Bower B T 1978 *Analysis for Residuals-Environmental Quality Management: A Case Study of the Ljubljana Area of Yugoslavia.* Resources for the Future, Washington, DC

Bower B T (ed.) 1977 *Regional Residuals Environmental Quality Management Modeling.* Resources for the Future, Washington, DC

Bower B T, Löf G O G, Hearon W M 1971 Residuals management in the pulp and paper industry. *Nat. Resour. J.* 11: 605–23

Brown F L, Lebeck A O 1976 *Cars, Cans, and Dumps: Solutions for Rural Residuals.* Johns Hopkins University Press, Baltimore, MD

Freeman A M III 1979 *The Benefits of Environmental Improvement: Theory and Practice.* Johns Hopkins University Press, Baltimore, MD

Gill C B 1986 Environmental control in metals production. In: Bever M B (ed.) 1986 *Encyclopedia of Materials Science and Engineering.* Pergamon, Oxford, pp. 1560–4

Kneese A V, Ayers R U, d'Arge R C 1970 *Economics and the Environment: A Materials Balance Approach.* Johns Hopkins University Press, Baltimore, MD

Kneese A V, Bower B T 1968 *Managing Water Quality: Economics, Technology, Institutions.* Johns Hopkins University Press, Baltimore, MD

Kneese A V, Bower B T 1979 *Environmental Quality and Residuals Management: Report of a Research Program on Economic, Technological and Institutional Aspects.* Johns Hopkins University Press, Baltimore, MD

Mills E S (ed.) 1975 *Economic Analysis of Environmental Problems*. Columbia University Press, New York
Page T 1977 *Conservation and Economic Efficiency: An Approach to Materials Policy*. Johns Hopkins University Press, Baltimore, MD
Quimby T H E 1975 *Recycling, The Alternative to Disposal: A Case Study of the Potential for Increased Recycling of Newspapers and Corrugated Containers in the Washington Metropolitan Area*. Johns Hopkins University Press, Baltimore, MD
Russell C S 1973 *Residuals Management in Industry: A Case Study of Petroleum Refining*. Johns Hopkins University Press, Baltimore, MD
Russell C S, Vaughan W J 1976 *Steel Production: Processes, Products, and Residuals*. Johns Hopkins University Press, Baltimore, MD
Sawyer J W Jr 1974 *Automotive Scrap Recycling: Processes, Prices, and Prospects*. Johns Hopkins University Press, Baltimore, MD
Spofford W O Jr 1971 Solid residuals management: some economic considerations. *Nat. Resour. J.* 11: 561–89
Spofford W O Jr 1973 Total environmental quality management models. In: Deininger R A (ed.) 1973 *Models for Environmental Pollution Control*. Ann Arbor Sciences Publishers, Ann Arbor, MI
Spofford W O Jr, Russell C S, Kelly R A 1976 *Environmental Quality Management: An Application to the Lower Delaware Valley*. Resources for the Future, Washington, DC

W. O. Spofford Jr
[Resources for the Future,
Washington, DC, USA]

Resource Appraisal

Resource appraisal refers to the evaluation of mineral materials in the ground, both discovered and undiscovered. Attention is centered on materials in such form, concentration and location that they might be extractable under foreseeable economic and technological conditions.

Among the users of such appraisals are mineral exploration planners, economic analysts, land-use planners and policy makers. Each group looks for aspects that are most pertinent to its own field and time frame of interest. There is, however, no such thing as an all-purpose resource appraisal. Any type of appraisal has its conceptual limitations.

1. Terminology

Discussions of resources have long been bedevilled by a vague and inconsistent use of terms. For more than a century, efforts have been made to establish definitions of the chief concepts, and of their subdivisions, that would be widely accepted and applied. Unfortunately this has resulted in a multitude of definitions, of which none has won general acceptance, largely because of differences in purpose and emphasis. Even the commonly used expressions *resource base*, *resources* and *reserves* convey different things to different people. Their most usual meanings are described below.

Resource base refers to the total amount of a mineral material presumed to exist in the earth's crust, regardless of technological feasibility and costs of extraction. Limits may be specified, such as geographic area or depth.

Resources are the part of the resource base that is considered extractable under current and foreseeable technological and economic conditions. (In this context, the foreseeable future usually does not extend further than 25 years or so; beyond that, the effects of changes in technology and economics become totally conjectural.)

In practical terms, resources may be thought of as mineral concentrations—the known "deposits" as well as the merely speculative—that represent realistic opportunities for supply during the next few decades.

Reserves are restricted to those resources that, at a given time, have been fairly well measured and are expected to be profitably minable in the immediate future, given current technology. The term reserves is applied to tonnages of economically minable rock (ore reserves) as well as to quantities of any particular mineral commodity contained in such rock (e.g., copper reserves).

2. Resource Base—Some Indications of Size

The concept of resource base provides a useful backdrop to resource appraisal. It affords a perspective that helps in envisioning the immense mineral supply possibilities in the remote future through exploration and technology, as well as some of nature's hurdles that may have to be cleared.

Column 1 of Table 1 shows that a considerable number of useful metals collectively make up $<0.77\%$ of the continental crust. Nonetheless, on a human scale the absolute amounts are still vast. For most of these metals, the elemental abundance as a fraction of the total mass of the earth's crust (taken as 24×10^{18} t) is in the range 10^8–10^{11} times the amount of current annual world production.

Conveniently the constituent elements are unevenly distributed in the crust. For most metals, however, concentrations have to be very high to be of interest to miners (see Table 1, column 2). Society is willing to pay the cost of extraction only for those deposits that are large, of preferred mineralogical compositions and in accessible locations. Such deposits are geological rarities.

3. Possible Link Between Ore Concentrations and Crustal Abundance

Many geologists believe that the size and number of a metal's largest ore deposits are related to its crustal

Table 1
Crustal composition and minable grades

Element	Composition of continental crust[a] (wt %)	Lowest grade economically minable in 1975, as multiple of average crustal grade[b]
Oxygen	45.20	
Silicon	27.20	
Calcium, magnesium, sodium, potassium, hydrogen	11.97	
Aluminum	8.00	2
Iron	5.80	3
Titanium	0.86	16
Phosphorus	0.10	70
Manganese	0.10	190
Subtotal	99.23	
Copper	most of balance	56
Cobalt		80
Nickel		100
Lithium		240
Carbon		310
Uranium		350
Zinc		370
Molybdenum		770
Gold		1000
Silver		1330
Tin		2000
Chromium		2100
Lead		3300
Tungsten		4000
Mercury		11200

a Skinner (1976, p. 260) b Cook (1976, p. 678)

abundance (Brobst 1979). It has even been suggested (McKelvey 1973) that a rough proportionality exists between the reserves and many well-explored nonfuel commodities and their crustal abundance, the tonnages of reserves being equal to 10^9–10^{10} times the crustal abundance (expressed in percent). However, this remains difficult to explain, given the many changing economic factors that affect the size of reserves over time.

4. Grade Distributions in the Earth's Crust

The analytical sampling of the crust carried out to date is insufficient for a decisive test of any of the hypotheses that have been developed concerning the relation between the tonnage of rock and its content of any individual metallic element. This content or "grade" is usually expressed as weight percent.

The total low-to-high spectrum of grades at which a metal occurs in the earth's crust may be divided into small equal-sized ranges. A plot of the total amounts of metal in each of these ranges would, in the case of geochemically abundant metals such as aluminum, iron and titanium (see Table 1), follow a pattern that many geologists envisage as approximating a symmetrical bell-shaped curve. In the case of the scarcer metals, it has long been assumed that a closer approximation would be given by a log normal distribution, that is, by an asymmetrical bell-shaped curve with a peak at low grade and a long drawn-out tail towards the high grades. However, Skinner (1976) has challenged this assumption by proposing that, for the relatively scarce metals, the grade distribution might have two distinct peaks.

Skinner built his argument on the fact that metallic elements generally occur combined with other elements in crystalline compounds. The production of metals involves their liberation from such compounds by physical or chemical processes. Separation is easiest in the case of sulfides, oxides, hydroxides and carbonates. Silicate minerals, which make up the bulk of the earth's crust, are more difficult to break down. Furthermore, in common silicate rock, the scarce elements are present chiefly as randomly distributed atoms trapped by isomorphous substitution within minerals of the geochemically abundant elements; for example, a zinc atom may replace a magnesium atom. To release such zinc atoms would involve a complicated and energy-hungry process, well beyond the realm of current practicality.

These basic differences in mineralogical associations led Skinner to suggest that the amounts of scarce metals occurring at different grades would show bimodal distributions as illustrated in Fig. 1. The large peak represents the distribution of a scarce element in common silicates, the small peak its distribution in nonsilicate minerals. Skinner suggested that a point would be reached—for some metals, soon—where a scarce metal was no longer available in adequate quantities in the kinds of deposits from which the metal

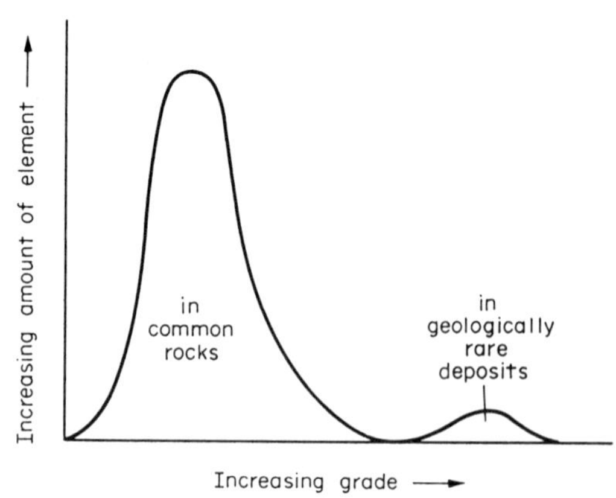

Figure 1
Possible distribution, by grade, of geochemically scarce metals (Skinner 1976)

could be recovered relatively easily. If Skinner is right, mankind may eventually (although the time scale is by no means clear) be forced to rely on low-grade silicate minerals for supplies of certain metals. These minerals may be physically inexhaustible in human terms, but they would yield their metals only at high costs, unless progress is made to the use of cheaper, cleaner and more plentiful forms of energy and acceptable ways of processing immense volumes of rock are developed.

The question of whether progressively larger quantities of the scarce elements exist at lower and lower grades or whether gaps occur is of vital importance to long-range metal supply. However, data available for studies on this subject (see Sects. 6, 7) come chiefly from the high end of the grade spectrum, which for economic reasons has attracted the greatest interest. This represents not only a very small but also a biased sample of the crust. Radical extrapolation from such samples toward lower grades in the crust is difficult to justify.

5. *Deposit Size Versus Deposit Grade*

The smaller peak in Fig. 1 includes the amount of a geochemically scarce metal contained in all nonsilicate deposits collectively. Focusing on individual deposits within this group, studies have been carried out to determine the relationship between deposit tonnage (total mineralized material) and average grade. This has been done by deposit "type," a designation encompassing all deposits found in the same general geological environment and characterized by similar mineral assemblages, such as porphyry copper deposits.

The results of such studies by the US Geological Survey in the 1970s have been mixed. Whereas it is commonly believed among mineral scientists and engineers that, for a given deposit type, larger deposits may be expected to occur at lower average grades, such an inverse relationship between size and grade has actually been found only in some deposit types. For most, grades and tonnages of the deposits covered by the studies were statistically independent (DeYoung and Singer 1981).

For any deposit type in which deposit tonnage and grade are in fact independent, the total low-grade tonnage of mineralized rock is likely to contain less metal than the total high-grade tonnage. If more metal exists at lower grade, which may well be the case, it would have to be in other deposit types, some of which may have remained unrecognized to date.

Singer and DeYoung (1980) observed that typically more than half of the total known metal in any one deposit type is contained in the largest 10% of the deposits. This means that there can be no great expectations of large numbers of small undiscovered deposits, as they could contribute relatively little to total metal supply.

The implication of the above findings is that the high-grade end of the curve in Fig. 1 may well have several peaks for some metals, each peak reflecting the metal content of a deposit type. Thus, although such analyses do not allow projection very far below observed grades, they have provided some grounds for questioning the comfortable assumption of continuity in grade–tonnage relations from higher to lower grade.

6. *Other Tonnage–Grade Relationships*

The studies referred to in Sect. 5 sought a measure of statistical predictability of average deposit grade, given the size of a deposit, or vice versa, and found none for many deposit types. Other researchers, with a different purpose in mind, have developed mathematical equations linking tonnage and grade in a general way, pertaining alike to parts of deposits, entire deposits and groups of deposits within deposit types. For example, Lasky (1950), starting with the highest-grade material and successively adding tonnages of mineralized rock of lower and lower grade, observed that as the cumulative tonnage of mineralized rock grew at a constant *geometrical* rate, the average grade of this cumulative tonnage decreased at an *arithmetical* rate. This relationship, which has since become known as Lasky's law, also applies only to certain high-grade sections within the smaller peak of Fig. 1. It cannot be validly extrapolated far enough toward lower grades to shed light on Skinner's idea of a grade gap.

DeYoung (1981) has warned that the mathematical and geological limitations, assumptions and complexities underlying mathematically expressed relationships of this type are not widely appreciated. Moreover, he has noted that cumulative tonnage in Lasky's law has, by several later authors, been taken to refer to metal instead of to mineralized rock, a misinterpretation that may well have contributed to the common notion that the lower the grade, the more there is in any deposit type. DeYoung stressed, however, that this notion is implied neither by Lasky's law nor by the mathematical expressions of tonnage–grade relationships formulated by later researchers.

7. *Approaches to Resource Appraisal*

Resource appraisals may be divided into two main types.

(a) Primarily geoscientific appraisals have the chief purpose of providing long-range land-use planners, exploration planners and mineral supply analysts with the best possible geoscientific judgements on the likely distribution and character of undiscovered mineral resources by region. Mostly applied to regions in which there has been little or no mineral development, such appraisals may be only qualitative. Economic constraints imposed tend to be loose and implicit.

(b) Geoscientific–economic appraisals have the aim of providing policy analysts with estimates of the magnitudes (as now perceived) of the sources of short-range and longer-range mineral supply, so that appropriate efforts may be directed toward exploration and development and toward technology to aid in recovery. Sources of supply that can be quantified reliably are confined mostly to those from which supplies may come only within the next 15–20 years. This type of appraisal tends to be conducted on the scale of entire countries. Economic considerations are stated explicitly.

Harris (1984) is the only single source providing a complete state-of-the-art coverage of all aspects of mineral resource appraisal. The quarterly *Materials and Society* dedicated a special issue (1984) to the history, state-of-the-art and applications of resource appraisal; this issue provides a convenient overview of the subject. McLaren and Skinner (1987) brought together up-to-date resource assessments in the context of relevance to future world development.

7.1 Primarily Geoscientific Appraisals

Worldwide observations have shown that mineral deposits typically occur in particular geological environments and have characteristic relationships to their host rocks. This is explained by the assumption that all are the end result of the same series of earth processes, although we are far from knowing all the processes that have taken place and from understanding their effects. Thus, recognition of favorable geological environments provides the key: evaluating a region's potential for undiscovered mineral resources involves determining the presence or absence of geological features believed favorable for the existence of particular deposit types.

The result of such evaluations may be shown on maps indicating regions favorable for the occurrence of deposits of specified metals or other mineral commodities. Refinements may be introduced to show degrees of favorability of occurrence, such as high, moderate, low or nil, and a distinction may be made between certain deposit types. It is usually implied that the favorability refers to occurrence of deposits discoverable with current exploration tools, but the questions of how much may actually be discovered, and when, are not generally addressed in these resource appraisals.

Regional resource appraisals in such form are increasingly in demand for the classification of land according to the future uses that may be made of it. Singer and Ovenshine (1979) have described the variety of approaches taken by the US Geological Survey in assessing metal resources in Alaska, where large regions had not even been mapped on a reconnaissance level. Tracts were delineated—144 for the whole of Alaska—in which the occurrence of various types of mineral deposits was considered conceivable. For each tract, a tabulation was provided giving succinct descriptions of: deposit types; contained metals; geological settings; geological, geochemical and geophysical indications of favorability; extent and adequacy of exploration and geological knowledge; and, for most of the large-tonnage deposit types, estimates of the number of deposits and indications of tonnages and grades extrapolated from models based on well-explored prototype mineral districts elsewhere in the world. Where the level of available geoscientific information warranted it, experienced geologists were asked to make subjective judgements, in terms of probabilities, of the number of deposits by type likely to occur in each tract.

Shawe (1981) prepared a compilation of mineral resource appraisal procedures that may be applied to wilderness areas and to quadrangles of the Conterminous United States Mineral Assessment Program (CUSMAP), patterned after the Alaskan Mineral Resource Assessment Program. His report also contains examples of how results may be portrayed. Appraisals of this sort call for integrated efforts of team members such as field geologists, geophysicists, geochemists and geostatisticians, who may be aided by specialists in, for example, isotope geology, remote sensing and paleontology. Thus, a wide range of geoscientific thinking is brought to bear on the description of the probabilities of existence and distribution of many types of mineral deposits and on the estimation of amounts. In such regional appraisals, estimates may be made of tonnage and grade ranges and of the chemistry and mineralogy of deposits, but explicit economic considerations tend to be avoided.

Among all mineral commodities, uranium is one for which the potential for as yet undiscovered resources has been subjected to exceptional scrutiny on an international scale. A report prepared jointly by the Organization for Economic Cooperation and Development (OECD) Nuclear Energy Agency and the International Atomic Energy Agency (1978) reviewed the possible extent and location of "speculative" resources of six different deposit type categories for 185 countries. Speculative resources were defined as resources that, mostly on the basis of indirect indications and geological extrapolation only, were thought to exist in deposits discoverable with existing techniques. They were restricted further to resources that, if discovered in 1978, would have been considered exploitable at costs of less than US\$130 kg^{-1} of uranium. Areas were identified that were believed to be favorable for the existence of such uranium resources, and the potential for occurrences was ranked as low, moderate, moderate to high, high, and high to very high, or very high. For the purpose of obtaining an aggregate measure of this potential, tonnage ranges were then applied to these rankings. The report took care to emphasize that judgements of speculative resources should be viewed only as a measure of the then current state of geological knowledge, with all its inherent uncertainties. The tonnage ranges given in this

1978 report were critically reassessed and updated in 1983 (OECD Nuclear Energy Agency and the International Atomic Energy Agency 1983 p. 25).

7.2 Geoscientific–Economic Appraisals

In multidisciplinary resource appraisals, two aspects have generally been considered separately, namely (a) the degree of assurance about an estimated quantity actually being present in the earth's crust, and (b) the degree of economic attractiveness. Given these two dimensions, the simplest subdivision of resources is into four main categories as shown in Fig. 2. This should be viewed as a dynamic picture, the arrows portraying the directions of possible movement between the various categories.

The clean separation of discovered from undiscovered resources along the horizontal axis of Fig. 2 is a simplification. Inferred extensions of known ore bodies and nearby mineralization surmised on the basis of analogous geology are examples of material that lies in a transition zone between the definitely discovered and the entirely speculative.

The level of assurance about the physical existence of certain tonnages of given grades (regardless of economic considerations) is determined by the degree of confidence in geological continuity between observation points and in assumed geological analogies. The assurance level may cover a continuous spectrum from very high to nil, but in practice it is generally expressed in discrete gradations described by terms such as proved, measured, probable, indicated, possible, inferred, speculative and many others. The use of a series of such terms expressing diminishing confidence levels is an improvement on the simplistic division discovered–undiscovered, but the dividing lines between them are difficult to define satisfactorily, so that what is "proved" to one person may be "probable" to another.

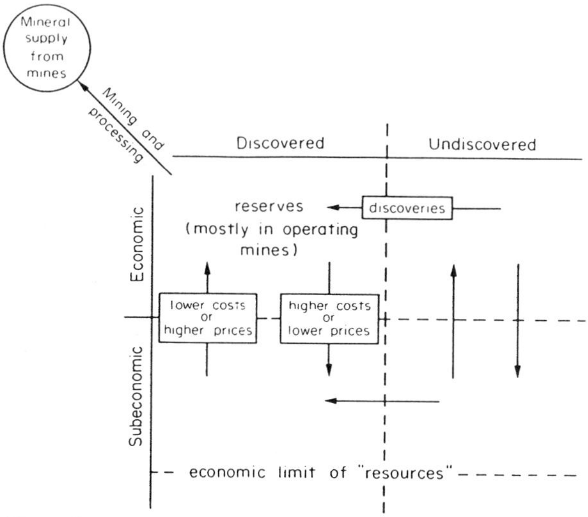

Figure 2
The essentials of a resource appraisal

The distinction between economic and subeconomic, along the vertical axis of Fig. 2, is also vague. This distinction can be made properly only on the basis of mining feasibility studies that consider all the details of mining methods, costs and revenues. Subdivisions in the economic dimension are therefore highly judgemental as well, largely because of uncertainty about future costs and prices. The explicit criterion used for subdivision in the economic dimension is generally one of the following: (a) price, (b) cost, (c) cost–price ratio or (d) probability of becoming economically minable within a given time period. For some metals, economic subdivision is complicated by association with other metals in the same deposits, which may enhance the economics of exploitation.

Detailed knowledge about reserves, the discovered-and-economic category in Fig. 2, together with past production, is invaluable as a basis for extrapolation in resource appraisal. However, the cost of upgrading knowledge about the exact tonnage of new discoveries to the "reasonable assurance" level of reserves is high. So also is the cost of economic analysis. Therefore, the necessary drilling and analysis to establish reserves are normally undertaken only to the extent needed for production planning. As to methodologies for computing reserves, which cannot be discussed here, the reader may consult Popoff (1966) for conventional methods and Clark (1979) for geostatistical methods.

Especially imprecise is the outer limit of resources (bottom of Fig. 2) determined by foreseeable technology and economics, twins that respond closely to one another. Among technological improvements that would extend the economic limit of resources are techniques for locating deposits at greater depths and methods for better extraction of ores and recovery of the mineral commodities from them. These would represent changes in the technology of raw material *production*. The economic limit of the resources of a commodity can also be affected by the technology of material *use*, for example, by lowering the demand for it, and thus its price, through the use of another material in its place.

Therefore, the economic limit of resources reflects an analyst's perceptions about the future evolution of technology and other factors influencing prices. It is worth noting that this dependence on perceptions has generally made resource estimates grow over time; while some deposits were being exhausted, the growth in possibilities for the future envisaged by analysts has tended to more than offset the losses.

A well-known practical framework using the same two dimensions as in Fig. 2 is shown in Fig. 3, the classification used by the US Bureau of Mines and the US Geological Survey for reporting resources of individual mineral commodities. Variations on this principle have been devised in many countries, maintaining the same two dimensions but applying

	Identified resources			Undiscovered resources	
	Demonstrated		Inferred	Probability range	
	Measured	Indicated		Hypothetical	or Speculative
Economic	reserves		inferred reserves		
Marginally economic	marginal reserves		inferred marginal reserves		
Subeconomic	demonstrated subeconomic resources		inferred subeconomic resources		

Figure 3
A scheme for classifying resources of a mineral commodity (US Geological Survey 1980)

slightly different criteria for separating the categories, in efforts to make the divisions more practical or more informative.

The proliferation of resource classification schemes has worsened the terminological confusion that has always plagued resource appraisals. Difficulties in interpretation of descriptive terms such as those shown in Fig. 3 are compounded if some important details and assumptions remain unstated (Zwartendyk 1976). For example, in the case of undiscovered or subeconomic resources, appraisals usually describe the mineral content of rock in place, but, in the case of reserves, they tend to report only the mineral content of the ore that would be recovered in the mining process.

On an international basis, a common classification scheme with a common set of clearly defined terms has been applied only in appraising uranium resources, specifically those whose existence is assumed with some confidence. To date, quantified estimates have been provided biennially at two or more levels of assurance about existence, and in two or more cost categories (OECD Nuclear Energy Agency and the International Atomic Energy Agency 1983). Within the USA, the potential for undiscovered resources has been studied more thoroughly for uranium than for any other mineral commodity. Within the US Department of Energy, assessments of geological favorability have been subjected to economic analysis to describe "uranium potential" by forward cost (US Department of Energy 1980).

Given the lack of international consistency and comparability of resource appraisals in general, the United Nations has endeavored to standardize mineral resource classification by proposing a common set of definitions and terminology that would be suitable for reporting to the UN about mineral resources (Schanz 1980). To sidestep a choice from the multilingual profusion of existing terms, a letter–number system was designed to label the resource categories (Fig. 4). The term *reserves* (corresponding to the recoverable part of R-1-E) was deliberately avoided entirely to minimize possible confusion, in several languages, between resources and reserves.

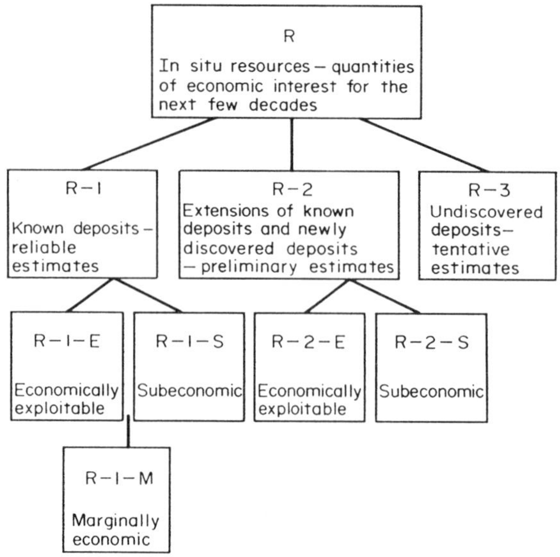

Figure 4
The United Nations scheme for classifying resources

8. Geostatistics

Geostatistics refer to the application of statistical methods to problems in mining and geology. Applied to the mining problem of reliably estimating ore in known deposits, geostatistical methods have proved their worth, for instance in estimation of grade distributions between control points through kriging, a sophisticated interpolation technique (Clark 1979). A basic assumption in such methods is that the difference in value between two positions in a deposit depends only on the distance between them and their relative orientation.

Apart from applications in ore estimation, geostatistics has been used also to assess links between regional geological variables and mineral concentrations. However, as statistical analysis of geological variables distributed in three dimensions has evolved only since the early 1960s, it is too soon for estimates of undiscovered resources obtained through this technique to have been adequately tested by actual production. Many geologists believe that, on the basis of only indirect geological evidence, credible quantitative estimates cannot be made of totally speculative resources well away from known deposits. However, others are convinced that it is valid to evaluate undiscovered resources geostatistically to produce quantitative values with probability limits, provided that sufficient objective data are available for such an approach. Brinck has even estimated resources ignoring geology and using the statistical (log normal) distribution of element concentration in the earth's crust (Harris 1984).

Given our imperfect knowledge about the associations between mineral deposits and certain other geological features presumed to have been caused by the same earth processes, geostatisticians believe that the most appropriate way to view the observed associations is in terms of probabilities. For regional resource appraisals, the geostatistical approach has consisted essentially of observing statistically nonrandom features in the geology of mineral deposits and of inferring the probability of occurrence of undiscovered deposits on the basis of similar observed geological features.

Deposit models developed in this way employ human experience with economic deposits and observations of the anomalies of metal concentration as a basis for inferring the existence of undiscovered economic and subeconomic deposits. In contrast, models based on the geochemical distribution of elements (referred to in Sect. 2) draw their inferences from all or large parts of the earth's crust. Harris (1984) has described both types of models in detail. Applications of geostatistics are no longer unusual; such techniques were employed in most of the regional mineral resource appraisal methods analyzed by Singer and Mosier (1981) in a review of more than 100 publications describing a great variety of such methods.

A useful twofold division was followed by Harris and Agterberg (1981) in a comprehensive perspective of approaches to appraisal methods:

(a) those that estimate or describe one or more of the components by a mathematical relationship; and

(b) those that rely on subjective judgements of a geologist but require them to be expressed quantitatively, usually as probabilities or percentile values.

The subjective (judgemental) approach has been used more commonly, typically for large regions, as it takes less time and can usually integrate more miscellaneous information than can be captured formally and in uniform quality for a mathematical model. However, subjective appraisals done through unconstrained judgements may lack credibility because they are not reproducible. To lessen this weakness, much experimentation has taken place to construct explicit appraisal systems in which a geologist is asked questions in a step-by-step logical sequence. This provides a record of the individual judgements made and of how these were synthesized by subsequent computations. It has also shown the limitations of the mind's capabilities when faced with the complexities of evidence, geoscience and inference, and has pointed to the deficiencies in geoscience when it comes to explaining the relations between earth processes and mineral concentration.

Harris and Agterberg (1981) concluded that improving the reliability of resource appraisals will require significant advances not only in the accuracy with which we can relate geology to mineral concentrations but also in geological and resource data required to achieve that accuracy. Current resource data primarily describe discovered economic resources, an unrepresentative sample of what occurs in nature. Anything too small or low-grade to be of economic interest tends to go unreported, but knowledge about it is vital for the development of geoscientific theory.

9. Existence vs Availability of Resources

Resource appraisals presented in the form of aggregated resource categories are snapshots without a time dimension. They represent some analysts' current perceptions of the relative magnitudes of various resource categories. As such, they illustrate or imply possible targets for exploration and for applying technology.

Appraisals that confine themselves to estimating the likelihood of existence of resources—as distinct from their availability—are a necessary foundation for judgements pertaining to future supply, but they are not sufficient for that purpose. When it comes to finding answers about actual mineral supply flows, which have a time dimension, rates of discovery and production must also be considered. Estimates of resources in the ground should not be (but often are) mistaken for amounts that will be available at acceptable prices when and where needed.

Much useful information gathered on individual deposits or deposit types for a resource appraisal is not visible in the aggregated totals displayed in classification schemes such as those in Figs. 3 and 4. To progress towards judgements about rates of future production, analysts cannot use these aggregated totals; they must go back and continue the analysis by individual deposits and deposit types, because production rates are constrained by deposit characteristics and place of occurrence.

For discovered resources, estimates can be made, with an indication of the degree of reliability, of the timing and future rates of production on a deposit-by-deposit basis. Consequently, for discovered resources, existence and the likely rate of future availability can be combined in a portrayal such as Fig. 5 (Zwartendyk 1981). This shows future supply from reliable sources (category I), from likely sources that can be reasonably counted on (category II) and from additional sources (category III) for which, at this stage, the timing and rate of supply cannot be estimated.

For undiscovered resources, size, quality characteristics and exact location of individual deposits are matters of pure speculation. Even more so are the timing and rate of future production from them. Long-term supply from undiscovered resources can be forecast only on a generalized statistical basis, extrapolating from past experience in geologically comparable settings. In any case, because the entire process of exploration–discovery–deposit delineation–

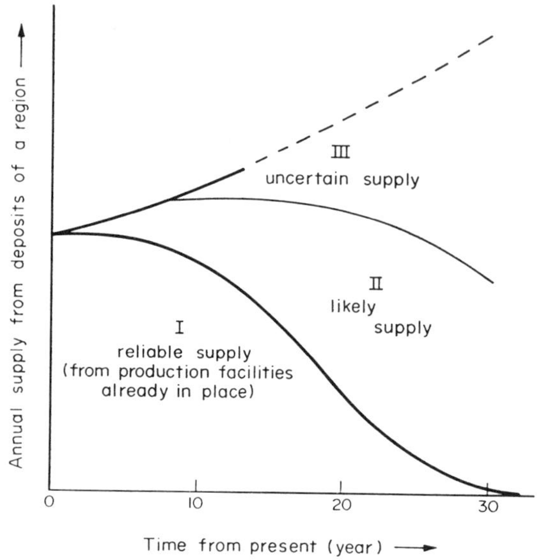

Figure 5
Mine supply of a mineral commodity

economic assessment–mining development usually takes at least 10 years, undiscovered resources are unlikely to contribute much to supply in the immediate future. For some 10–15 years, most of the mineral supply will have to come from already identified deposits and their likely extensions.

Resource appraisal is no longer an exercise based solely on geological judgements. The determination of the amounts of minerals likely to be available for those living in the distant future has always been shrouded in uncertainty and this has not changed of late. However, with regard to possible supply sources for the coming decades alone, evaluation techniques have evolved markedly, having received growing attention in the 1970s from geoscientists, engineers and economists. The resulting flow of ideas has contributed to setting resource appraisal on its way from a tentative art to a respectable science.

See also: Availability of Materials; Mineral Resources: Definitions, Uses, Classifications and Future Availability

Bibliography

Brobst D A 1979 Fundamental concepts for the analysis of resource availability. In: Smith V K (ed.) 1979 *Scarcity and Growth Reconsidered*. Johns Hopkins University Press, Baltimore, MD, pp. 106–42

Clark I 1979 *Practical Geostatistics*. Applied Science, London

Cook E 1976 Limits of exploitation of nonrenewable resources. *Science* 191: 677–82

DeYoung J H 1981 The Lasky cumulative tonnage–ore relationship—a re-examination. *Econ. Geol.* 76: 1067–80

DeYoung J H, Singer D A 1981 Physical factors that could restrict mineral supply. *Econ. Geol.* 75th Anniversary Volume: 939–54

Harris D P 1984 *Mineral Resources Appraisal. Mineral Endowment, Resources, and Potential Supply: Concepts, Methods and Cases*. Oxford University Press, New York

Harris D P, Agterberg F P 1981 The appraisal of mineral resources. *Econ. Geol.* 75th Anniversary Volume: 897–938

Lasky S G 1950 How tonnage and grade relations help predict ore reserves. *Eng. Min. J.* 151(4): 81–5

McKelvey V E 1973 Mineral resource estimates and public policy. In: Brobst D A and Pratt W P (eds.) 1973 *United States Mineral Resources*, US Geological Survey Professional Paper 820. US Government Printing Office, Washington, DC, pp. 9–19

McLaren D J, Skinner B J (eds.) 1987 *Resources and World Development: Energy and Minerals, Water and Land*, Dahlem Konferenzen. Wiley, Chichester, UK

Materials and Society 8 1983 Issue devoted to Resources Assessment: History, State of the Art and Applications

OECD Nuclear Energy Agency and the International Atomic Energy Agency 1978 *World Uranium Potential*. Organization for Economic Cooperation and Development, Paris

OECD Nuclear Energy Agency and the International Atomic Energy Agency 1983 *Uranium—Resources, Production and Demand*. Organization for Economic Cooperation and Development, Paris

Popoff C C 1966 *Computing Reserves of Mineral Deposits: Principles and Conventional Methods*, US Bureau of Mines Information Circular 8283. USBM, Washington, DC

Schanz J J 1980 The United Nations' endeavour to standardize mineral resource classification. *Nat. Resour. Forum* 4: 307–13

Shawe D R 1981 *US Geological Survey Workshop on Nonfuel Mineral-Resource Appraisal of Wilderness and CUSMAP Areas*, US Geological Survey Circular 845. USGS, Arlington, VA

Singer D A, DeYoung J H 1980 What can grade–tonnage relations really tell us? In: Guillemin C, Lagny P (eds.) 1980 *Ressources Minérales*. Bureau de recherches géologiques et minières (BRGM), Orléans, pp. 91–101

Singer D A, Mosier D L 1981 A review of regional mineral resource assessment methods. *Econ. Geol.* 76: 1006–15

Singer D A, Ovenshine A T 1979 Assessing metallic resources in Alaska. *Am. Sci.* 67: 582–9

Skinner B J 1976 A second iron age ahead? *Am. Sci.* 64: 258–69

US Department of Energy 1980 *An Assessment Report on Uranium in the United States of America*, GJO-111(80). US DOE, Grand Junction, CO

US Geological Survey 1980 *Principles of a Resource/Reserve Classification for Minerals*, US Geological Survey Circular 831. USGS, Arlington, VA

Zwartendyk J 1976 Problems in interpretation of data on mineral resources, production and consumption. *Nat. Resour. Forum* 1: 7–15

Zwartendyk J 1981 Economic issues in mineral resource adequacy and in the long-term supply of minerals. *Econ. Geol.* 76: 995–1005

J. Zwartendyk
[Energy, Mines and Resources Canada,
Ottawa, Canada]

Resource Recovery: Economic and Institutional Aspects

Resource recovery concerns both resource utilization and waste management. Viewed from the first perspective, it is the means by which waste resources are expected to become valued inputs to a nation's economy, fulfilling both producer and consumer needs. Resource recovery includes both waste-to-energy applications and materials recovery. Resource recovery is also seen as a means of reducing demands on the solid waste management system. In fact, currently, this second perspective is the predominant one underlying the decision to incorporate recovery as a component of a community's solid waste management system. There are a number of concerns that affect this resource recovery decision. Economic and institutional aspects are of particular importance, including household behavior, as a result of a movement towards reliance on separation at the source of the discards as a primary means of attaining recycling goals. Hence, the extent to which a community responds to the conservation ethic must be considered along with more traditional concerns such as the availability of markets for the recovered materials, costs and revenues and financing.

The current ordering of waste management priorities in the USA reflects hierarchical goals promulgated by the Environmental Protection Agency:

(a) to *reduce* the amount of waste generated,

(b) to *divert* from the waste stream through recycling,

(c) to *combust* for volume reduction if feasible and attainable, and

(d) to *landfill* the diminished amount that remains.

Reacting to the difficulty of access to new landfill capacity and the environmental problems and costs associated with combusting discards, public officials and citizens alike are currently focusing on reducing waste generation, called "source reduction," and diverting selected discards from the waste stream by recycling.

Post-consumer waste, what households throw away, consists of consumer durables (such as refrigerators, water heaters and automobiles, which traditionally have been recycled, as well as cans, bottles and newspapers, which for the most part have been discarded) and also, unfortunately, materials with little recycling possibility (such as mixed paper, food wastes, rocks, stones, ash and grass clippings). Now, even this latter group, particularly grass clippings, has become targeted for diversion through separate collection and composting. The situation in the USA is reflected in waste management activity throughout the industrialized world. The examples used in this article mainly concern the USA. However, they are generally applicable elsewhere with little variation as they illustrate the basic economic and institutional concerns associated with implementing recycling as a waste management option.

1. Historical Perspective

During the early years of the recycling movement, most communities experiencing increasing amounts of household discards attempted to preserve the efficiencies of commingled collection by looking to some sort of "back-end" sorting, in contrast to the current emphasis on separation at the source. Collecting source-separated waste involves either an entirely separate collection or new compartmentalized trucks. With collection costs already amounting to about 75% of the waste management budget, communities were reluctant to increase this further.

However, back-end separation proved to be more difficult than initially supposed. The materials extracted from the mixed waste were of poor quality and generally failed to find applications in competition with virgin materials and, perhaps of even more importance, in competition with off-grade, newly produced materials and in-plant scrap, both of which are of known specification and generally free of contamination. This situation was not readily duplicated by the sorting machinery of these waste processing plants, at least not at investment and operating costs competitive with simply disposing of the collected waste by traditional means, principally by landfill. With few exceptions worldwide, these plants were closed.

Waste-to-energy plants suffered less. Community dependence on these plants generally is greater as, given their high initial cost, they would not have been built had the availability of landfill not already been an acute problem. Because their energy product is electricity, in the USA they were favored by federal legislation requiring the electricity utilities to purchase their output, although the determination of the selling price remains a matter of contention. It must be said, however, that those waste-to-energy plants whose technical approach involved processing of the feedstock, ostensibly to improve its fuel characteristics, along with some materials recovery by-products, suffered the same type of processing-related and materials handling difficulties as the materials recovery facilities.

The lessons learned from the back-end processing era related to technological feasibility, cost and "staying power." In the last case there was no institutional or public constituency committed sufficiently to stay the course in order to overcome the technological problems, assuming that they could be overcome. Individual communities have no research and development charter. The market is rather small and fragmented from the standpoint of the private sector. Municipal purchasing practices are designed to procure public works, few of which involve an uncertain revenue stream. Public decision makers, by the nature of their responsibilities, are risk-averse and, as has been demonstrated, recycling involves risk. Overall, the circumstances have not proved recycling to be attractive as a place for investment by well-financed, risk-taking innovators. There have been a number of marginal operators and their demise has further depreciated the image of back-end sorting. Firms that have done well in the waste management industry have focused on "picking up" the discards and "putting them down" using traditional methods. They have taken little risk as far as advancing the state of the art is concerned.

During this earlier period, there was some movement toward separation at source, particularly of packaging materials. In the USA, major buy-back recapture programs were instituted, mainly under the auspices of the aluminum manufacturers. A number of factors provided the motivation. During periods of high levels of facility utilization in the industry, recycling was seen as a replacement for construction of new capacity. Perhaps of even more importance were marketing factors such as the impact of environmental concerns on consumer purchasing behavior. From a recycling standpoint, aluminum was seen as having a competitive advantage over other packaging materials. Across the beverage industry, there was the threat of deposit legislation, which negatively affects the sales of beverages in convenience packaging. This is because deposits greatly diminish the utility of the "one-way" container. For the consumer, preserving the utility means forfeiture of the deposit. Recycling is seen by the industry as alleviating the need for deposits.

Some smaller communities asked, or required, householders to separate recyclables by taking cans, bottles and paper to designated drop-off points. A few abandoned commingled collection in the same container and instituted separate collections for recyclable materials. This mainly was for paper segregated by the householder for either *same day* (as the remainder of the waste) or *different day* pickup.

Recently, there have been a number of changes, emanating mainly from an environmentally concerned public. This has been reflected in political decisions that accommodate this concern, unknowing or unmindful of its costs. These concerns are led by the search for alternatives to landfill, space for which is becoming increasingly more difficult to obtain. These difficulties are related both to scarcity of land suitable for such use and to political opposition by residents. In the first regard, the potential for ground-water contamination is a principal concern. In the second, the recently coined acronym NIMBY ("not in my back yard") tells the story.

The urge to reduce has not yet resulted in any significant reduction in waste generation as a result of changing industry or consumer purchasing habits, although it has resulted in a great deal of environmental posturing. However, it has drawn much more attention to enhancing the recyclability of the discards once an item has performed its designated function. Recycling has taken a number of forms, including programs sponsored by industries to take back the waste generated by the use of their products, such as disposable diapers, additional deposit legislation requiring the consumer to return convenience packaging in order to reclaim the deposit and government-sponsored buy-back programs, paid for by special taxes on the packaging and beverage industry. However, most frequently, the urge to recycle has led to the community-run, mainly "curbside" pickup programs in which special collections divert the potentially recyclable discards from the regular waste stream. There are advantages and disadvantages associated with each of the methods proposed to convert the problem of waste to something approaching a societal advantage, including energy from waste. The evaluation of whether the benefits exceed the cost depends on an individual's viewpoint on a number of issues that are discussed in the following sections.

2. The Conservation Ethic

Although the amount of solid waste generated in the USA has grown at only about 2% annually since the beginning of the 1960s, which is considerably less than gross national product (GNP), many people believe that greater utilization of waste resources should be attained even if it cannot be justified on economic grounds. This school argues that greater attention to recycling will stimulate the populace to a higher level of environmental consciousness and therefore to less wasteful behavior. In many instances, the efforts of this group focus more on reshaping consumption patterns than on improving recovery practices. Another school calls this designing "good waste" instead of "good products," an imposition regarded as incompatible with basic consumer utility. Those who give waste reduction a higher priority than recycling often oppose recovery attempts as a form of second best, which would not be required if citizens only would pay "proper" attention to waste reduction possibilities.

3. Market Imperfections

Another argument for taking steps to increase the use of waste resources relates to the belief that the economy improperly discounts the future in determining whether to use virgin rather than secondary materials. The controversy centers on the appropriate level for the social rate of discount. At lower rates, the current price of depletable resources would be higher. If so, the rate of depletion would be less and future generations would have more nonrenewable resources available at lower prices. This would also stimulate recycling. The operative recommendation is for a price concession in favor of products made from recycled materials or for products that are recyclable. Government procurement is the usual target. This implies a willingness to pay more for government and the services it provides.

Concern is expressed, particularly in the USA, that taxation policies applied to the extractive industries stimulate production and depress prices of virgin materials. Examples of the tax concessions available are depletion allowances, special capital gains considerations and foreign tax credits. Utility regulators are seen as favoring electricity generation at large central plants rather than as a product of cogeneration at industrial facilities. The call is for a redress of these presumed inequities through government actions which would stimulate the demand for recycled materials.

4. *Cost Avoidance*

Landfill costs in most countries that stress environmental protection have risen dramatically and are expected to continue to increase. Avoiding landfill costs is the usual reason advanced to justify the expense of the separate collections required by the curbside approach. Cost avoidance is not as simple as assuming that for each ton of waste diverted, the average per ton cost of disposal is avoided. However, this is usually the assumption made, because many people seem to confuse average cost with marginal cost and assume that if, say, 10% is diverted from the waste stream, 10% of the cost of waste management will be saved. Unfortunately, waste collection and disposal involves large fixed costs and capacity is usually sized for peak periods, such as the first collection after a weekend. As a result, meaningful savings from incremental diversion are seldom seen.

In fact, as tonnage decreases, per ton charges often have to be increased in order to realize the revenue necessary to cover fixed costs such as the bonds floated to construct plants or landfills. This is usually evident to private operators and regional waste authorities. In contrast, many units of government expense capital outlays and therefore do not adequately provide for replacement. This can lead to an illusion of cost savings. The fact is that most curbside collection programs increase current costs, rather than decreasing them. Curbside programs across the USA typically cost far more than landfilling the collected materials, frequently twice as much even when sales revenues and avoided waste disposal costs are included in the calculation.

Cost avoidance also involves the estimated cost of replacing current capacity when it has been exhausted. Seldom is the time factor taken into account. In deciding whether to spend today to delay a cost at some future date, the discounted present value of the future expenditure must be taken into account. Even at extremely high anticipated future land prices, coupled with the expected costs of making an otherwise acceptable site environmentally safe, one would not spend much today to postpone the need for a cubic yard of landfill capacity to some time into the future. The need has to be very immediate to warrant the outlays on curbside programs.

Outlays are being made, however, because they are denominated in political as well as financial terms and the former probably are more significant. In some areas, the available land is not suitable for use without undue risk of ground-water contamination, although at some level of cost, most land could be made safe for use as a disposal location. But this presumes agreement on what level of risk should be regarded as safe. This debate has its political cost. Therefore, there are incentives on all sides—government officials and the general public—to pay to postpone it. Combine this more generalized concern with the more particular one represented by NIMBY and this shows the main elements underlying the cost avoidance argument of the curbside movement. The arguments made to support investment in waste-to-energy facilities are a variant of this basic theme.

5. *Uncertain Revenues*

A critical barrier faced by the back-end plants of earlier times was control of the waste feedstock. To be viable, any plant must have a reasonable degree of certainty about the supply of its raw material. Waste-to-energy plants have the same concerns for the calorific value of the waste they receive. Diversion of large portions of the paper and plastics discards can greatly affect the economics of such facilities. Even though the curbside approach leaves less of a gap between discard and collection, it is also susceptible to some of the same diversion possibilities.

When prices for secondary materials rise, householders seek to realize the potential financial benefits themselves. They reject their prior practice of simply throwing the materials away or separating the recyclables and placing them by the curbside for municipal or franchised private hauler pickup. Instead, they sell the materials themselves. In other words, waste is not waste when householders see it as having sufficient value to warrant a more purposeful and financially motivated decision to recycle it for their own benefit. There is an intermediate point, when the value of the material in postconsumer waste is such that it is not high enough to motivate individual household recycling, but when the materials are put out for pickup by the designated waste collector, they are instead taken by other people whose entrepreneurial instincts impose themselves between the householder and the designated collector (this is known as "scalping"). There is a high probability of this happening if the program design calls for aluminum cans to be put out separately. Because of their value, they are often the target for nondesignated collectors, particularly in urban areas. For this reason, beverage containers are generally put out mixed. Policing against such activity, as well as enforcing compliance, is difficult at best. Communities have instituted the "trash cop," which some citizens see as an invasion of privacy.

The important point is that the current and, of course, the future character of the recycling feedstock are of concern, particularly when planning a major investment in new and perhaps specially configured vehicles, processing plant and equipment. A change in the recyclable content of the waste, whether brought about by scalping, changes in the pattern of consumer demand for recyclable products or changes in producer technology making products less recyclable, will cause revenue projections to be miscalculated.

6. Participation

Legislatively mandated programs are said to achieve participation rates of 90–95% of the community. The best voluntary programs may obtain 75–80%. Participation usually means placing recyclables at the curb at least once during a program month even though collections may have been more frequent. Others suggest that the participation rate, particularly for major cities, is more like 40%. Surveys conducted before the implementation of such programs show almost universal support. However, there is a considerable gap between what householders say they plan to do and how often and to what degree of accuracy they make the separations when a program actually begins. Even after a program is under way, there are indications that there is a significant difference between householders' perceptions of how often and how effectively they separate and the data acquired from analysis of their actual throw-aways. Participation is a behavioral phenomenon and one that has not been studied to any great extent. Intuitively, the fewer separations that are required, the greater will be the participation, and probably the more frequent the pickup, the more likely the household will be to participate. These and other policy considerations warrant investigation. Included are the linkages between motivation and actual participation, measurement of the disutility of participation by recyclers and nonrecyclers, identification of the most productive information sources and analysis of the impact on program goal attainment of community involvement in decision making along with the traditional correlations of performance with social, education and income characteristics. These questions will, it is hoped, attract increased interest among the social science research community.

7. Availability of Markets

The implementation of numerous curbside recycling programs, some voluntary, but many legislatively mandated, has caused renewed concern about the sufficiency of markets for the collected material, both in the short term and the long term. In the short term, this concern has proved justified. For example, within the span of a year, municipalities which were receiving US\$25 t^{-1} for collected newsprint found themselves in a situation where they were forced to pay paper brokers US\$5–25 t^{-1} to haul the separately collected material away.

Secondary materials markets have traditionally been volatile. Prices are high when business conditions are good. On the low side of the business cycle, materials users retreat to the lower-risk purchasing of virgin inputs or, if not virgin, known seconds from primary materials suppliers rather than run the risk of unknown secondary inputs. "Known," as used here, refers to the specifications of the material. The major problem of postconsumer materials reuse is the unknown quality. Aluminum, steel, plastic, paper and even glass suffer from this. Households throw away a variety of materials. Even such a seemingly simple package as an aluminum beverage can is made up of more than one piece and more than one alloy: the sides and bottom are of different material from the top. Plastic containers are made of different types of plastics and sometimes they are composites, with one type of plastic layered upon another. Hence, even if plastics are separated from aluminum and the aluminum pots, pans and window parts separated from the cans, the purity of the separated materials does not match that of virgin aluminum ingot or in-plant plastic scrap left over from a production run.

8. Source Reduction

Environmentalists have long voiced concern about how products are manufactured, distributed, purchased and used. They point out that products are not manufactured to last, that distribution and sales involve too much packaging and that too often they are purchased for single service use. Products that are serviceable but are out of style are discarded and those that are broken but could be repaired are discarded and replaced. In a market-driven economy, change must largely originate with the consumer. There is evidence that change is taking place in some areas: reuse of shopping bags; buying in bulk, facilitated by the introduction of warehouse outlets; purchasing liquid or powdered concentrates as in the case of detergents; and keeping automobiles longer. In a sense, this last point is linked to technological advances, as is the advent of longer-lasting automobile tires. Variable waste disposal charges (discussed in Sect. 15) are seen as influencing buying habits with the direction of their impact favorable to source reduction. If they have to pay for disposal as a function of the amount discarded, it is less likely that consumers will buy bulky, heavy products if less bulky, lighter ones are available. How less likely is a matter for speculation. The same line of reasoning supports the argument that the consumer will be less likely to discard an item, but will fix it or otherwise make the product last longer before replacement. Some observers feel that the end result will be a simple diversion to illegal dumping, manifested by an increase in roadside litter. While there may be changes taking place in lifestyles, the prominent ethic is still based on convenience, time saving and newness. It is unlikely that major gains through source reduction will materialize in the short run, if ever.

Design for recycling is another matter. There is a heightened awareness among consumers and industry that some products, especially certain types of packaging, recycle better than others or are less harmful to the environment in terms of manufacturing, distribution, use and disposal. However, which is best is sometimes a matter of contention. The controversy over fast-food packaging is an example. Both paper and plastic container manufacturers have conducted studies that indicate that their product is more environmentally friendly. It depends on perspective. The popular press seldom examines the assumptions of the analyses and often exaggerates the issues. An example is the case of polystyrene packaging which has been labelled as a large part of the solid waste problem when in reality its discard takes up less than 0.25% of landfill volume. The question of the relative environmental friendliness of products is not easily answered.

9. Goals, Guidelines and Governmentally Mandated Requirements

Democratic societies are particularly fond of setting voluntary goals, guidelines and what are sometimes called targets. Perhaps such societies are at their best when functioning in this manner, although it sometimes raises constitutional concerns such as when a ban is legislated against a particular type of packaging. Put simply, government actions without a firm footing in public health or other valid government concern may be illegally discriminatory. Likewise, industry activity to enhance recycling may, on the surface, appear to be in the public interest, but may be seen by the courts as anticompetitive, involving collusion.

There have been several goals and guidelines established in recent years in the USA that are of more general interest. In its "Agenda for Action," the US Environmental Protection Agency (EPA) (Office of Solid Waste 1989b) set a national goal of achieving by 1992 a 25% reduction in net discards through a combination of source reduction and recycling, the first two steps in their integrated solid waste management planning concept. Incineration is to handle 20% on a gross discard basis with the remaining (55%) being landfilled. In 1991, it was stated that it was unlikely that this goal would be achieved on a national level (Alter 1991). Adjusting for participation and for the actual yield of usable materials, but assuming all is recycled, a 16% reduction appears to be a high estimate. About half the yard and garden waste would have to be recycled to reach the 25% target. Nevertheless, the EPA pronouncement offers support to local community efforts to increase recycling. It is not clear what it does in terms of the elusive concept of source reduction.

Methods that central government has to influence local government behavior short of direct program financing are limited. A proposal of the EPA is of interest and illustrates the difficulties encountered. As a means of requiring communities to progress toward the national recycling goal, the EPA proposed to invoke its authority under the Clean Air Act which mandates the use of "best available control technology" (BACT) to reduce air emissions. The EPA proposed to make an explicit link between recycling and the operation of an incinerator by including a 25% precombustion recycling requirement as necessary in order to satisfy BACT. The environmental groups favored the idea but claimed that the 25% figure was too low. The incinerator manufacturers claimed discrimination and pointed out that, in fairness, disposal by landfill should be held to the same standards. Municipalities objected to the additional costs. The EPA shelved the concept. Another "trial balloon" from the EPA called for a special US\$20 t^{-1} tax on paper, glass, plastic and metals made from virgin materials with the proceeds used as rebates for bottlers, publishers and other users of recycled products, in recognition that using recycled materials often costs more. This proposal was also withdrawn.

Other levels of government also set goals. To encourage newspaper recycling, one US state intends to require publishers to certify annually the tonnage and recycled content of newsprint in order to determine if the industry has met the recycled content of 10% for 1993, 40% by 1996 and 50% by the year 2000. Fines for noncompliance are anticipated.

A final example of goals comes from the US paper industry, which in 1990 announced its commitment to a 40% recovery rate for all grades of paper by 1995. In terms of tonnage recycled this is a 50% increase over the measured 1988 level. This commitment involves the sponsorship of a national public education program to promote recycling, the development of instructional material explaining how to implement a paper recycling program, the creation of a national "matchmaker" database to help communities identify recycling mills or waste paper dealer intermediaries and, most importantly, a multimillion-dollar investment by the industry in paper recycling capacity.

10. Role of Government

As described earlier, for the most part central government has taken only an indirect role in the municipal waste issue. Many people believe that the solid waste disposal issue is a local and not a national concern and that it is wrong to redistribute funds collected through the national tax system to subsidize local waste management activities. Even if no redistribution were involved (i.e., if the exact amount collected from each locality by the federal tax mechanism were returned to the same locality), the argument is that the decision should be made locally, letting the local authorities decide on how much is needed, raise the funds and determine how the funds are to be spent.

In the USA, there have been relatively small amounts of money spent by the federal government on research and development and some on demonstration projects. The focus of much of this effort was on energy from waste, which resulted from the oil crises of the 1970s. Some spending spilled over to materials recovery research and development and demonstration, but very little. In the main, government support has been confined to financing planning activities at the state, and sometimes the local, level and to paying for collecting and providing information to aid local decision making. For example, there is a Solid Waste Information Clearinghouse described as the "one-stop source for solid waste information." There is also a Resource Conservation and Recovery Act/Superfund hotline.

11. The Green Movement

According to a recent poll, eight out of ten Americans are environmentalists. Seven out of ten say the best way to deal with the growing waste management problem is to require manufacturers to use less packaging and consumers to do more recycling. Rightly or wrongly, packaging is seen as the chief villain even though there is evidence that packaging as a percentage of the weight of household waste has been falling since the early 1970s (Alter 1991). It is further argued (Testin and Vergano 1990) that consumer packaging reduces the volume of the waste originating in the household by about a third because there is less food waste than there would be otherwise. This waste is left at the food processing plant. Less spoilage during transit and storage also plays a part in this comparison. Nevertheless, industry has moved to provide "greener" alternatives such as concentrates, larger sizes, packages that are made of biodegradable materials and refillable containers for products such as detergents. As a form of source reduction, glass, can and plastic beverage container manufacturers have carried out major lightweighting programs averaging about 40% each.

Companies increasingly see a good environmental reputation as an important aspect of corporate good will. This extends to the industries to which these companies belong. For example, the National Soft Drink Association now has a Solid Waste Programs Division to highlight the beverage industry's concern for waste management problems. Industries have funded new single-purpose organizations such as the plastic industry's Council for Solid Waste Solutions, its Council on Plastic and Packaging in the Environment, the Degradable Plastics Council and National Association for Plastic Container Recovery.

The environmental reputation of products has garnered increased attention, particularly the relative merits of different types of packaging. Many environmental life-cycle studies have been undertaken, with each industry seemingly able to produce or cite one favoring its products. But controversy remains. In 1990, McDonald's shifted from its traditional polystyrene "clamshell" to paper as a means of indicating its ecological conversion only to find the National Audubon Society and other environmental groups questioning the replacement decision as some of the plastic containers were being recycled while it was not clear that the new, according to the Audubon Society "forest-depleting," coated cardboard would be.

Many feel it would be helpful if there were standards that differentiate products from an environmental standpoint. Several national governments, such as those of Japan and Canada, as well as private groups in the USA are attempting this. Canada is implementing its Environmental Choice labelling. Germany has put its Blue Angel seal on more than 3000 products. In the USA, a group known as the Alliance for Social Responsibility seeks to implement a Green Seal labelling program.

12. Environmental Labelling

Environmental labelling is an emerging issue. There are two types of claims: "made of recycled material" and "recyclable." In the first instance, the concern is the origin of the recycled content. There is a linkage between the designation "made of recycled material" and consumer purchasing behavior and therefore the creation of markets for recycled materials, but this is broken if the recycled material is not postconsumer, but originates in plant. Several industries, including the paper industry, oppose this distinction. The amount of recycled material required to make the claim is also a point of controversy. Does "made of recycled material" signify that the total content is recycled? If not, how much recycled material should be required?

Turning to the recyclable claim, the issue here is whether the product is currently being recycled or is it simply that it could be recycled? And, does it have to be actually recycled in the community where the product is being marketed or simply recycled somewhere? Some claims appear to rely on the fact that in theory almost everything can be recycled if one wants to spend enough on the processing, even though there is no recycling capacity currently in place.

In the USA, a task force of the State Attorneys General recommended that the federal government adopt national standards for environmental marketing claims used in the labelling, packaging and promotion of consumer products. The report also called for the adoption of uniform definitions—highlighting the terms "degradable," "compostible," "recycled" and "recyclable" specifically. The position of the report is that only postconsumer material should be referred to as "recycled," with preconsumer waste identified as "reprocessed (or recovered) industrial" material. The report also advocated that preconsumer waste include only "factory waste that is routinely discarded and not that which is routinely fed back into the industrial process." Further, a package or product that claims to be recyclable must indicate the maturity of the recycling structure for that material, the number of locations where the product is being recycled, the nature of the collection sites utilized if there is no curbside pickup available and the states where collection and recycling facilities are located.

The report illustrates the complexity of what at first seems to be a relatively simple subject. Clearly, standardization is important to consumers and industry alike. In the latter case, differing rules and regulations impede the flow of products across state lines. This is a major concern of manufacturers whose goods are distributed nationally. However, if made to follow the recommendations of the report, it is not likely that manufacturers will make many claims of materials having been recycled or of recyclability.

13. The Materials Recycling Facility

The need for some kind of materials processing/storage facility accompanies the decision to effect recycling based on the source separation concept. Programs vary in terms of the types of materials collected and to what extent the individual household is responsible for separation. Household separation is not perfect; in fact, it is often far from it. The materials recycling facility (MRF) is the interface between the material collected and what prospective buyers will accept from the quality, quantity and form perspectives. Most MRFs utilize some hand sorting either off the floor or from picking belts. There usually will be a magnet for the extraction of ferrous metal. As currently configured, MRFs are labor intensive as a result of the manual separation. Separations include paper, magazines, cardboard, various colors of glass, aluminum and perhaps plastic by type. Glass is crushed and paper, plastics and cans baled or otherwise compacted to reduce volume. MRFs will probably become relatively more capital intensive in response to several forces: shortage of labor willing to do the unpleasant picking task, availability of more efficient equipment and growth of throughput which makes investment in the equipment worthwhile.

14. Siting

Siting of solid waste management facilities currently ranks as perhaps the most inhibiting barrier to resolving the discard situation. As mentioned earlier, it is the problem of siting of landfills that is a driving force behind the desire to recycle. Yet the siting of a recovery facility, whether it is a waste-to-energy plant or a materials recycling facility, is as difficult as that of a landfill even though less land is required. There is an even greater desire than in the case of the landfill to locate the resource recovery facility close to the center of waste generation, because the recyclables cannot be compacted to the same extent as refuse destined for landfill.

"Not in my back yard" describes the public attitude toward the siting of all waste management facilities. Attempts have been made to overcome this resistance by instituting extensive public participation procedures, establishing state-level siting boards with the power to overrule local decision makers and requiring compensation. The issue of compensation arises because the disbenefits of the facility are concentrated in the immediate vicinity of the plant while the benefits are distributed to the community at large. The disbenefits are risks associated with ground-water contamination and air pollution along with noise from plant operations and increased truck traffic. The perceived risks to the local inhabitants may be higher than the actual risks, because of a general mistrust of available technical information and of the people who communicate it. Technical information is often seen as a tool that is manipulated by vested interests. The credibility of public officials appears to be just as suspect as that of private waste management companies.

Putting the level of perceived or actual risk aside, there have been various proposals to effect compensation schemes as a means of overcoming the NIMBY problem. These include guarantees against property value declines, reduced taxes, independent monitoring of emissions, accident insurance, donation of land for use as parks and in-kind services such as free waste disposal for nearby residents and businesses. In Japan, new waste-to-energy plants are often constructed concurrently with recreational sports and community activities centers, usually with a swimming pool heated by energy from the waste-burning facility. The cost, including the compensation, is considerably above US standards, recognizing a broader definition of waste management which merges waste management costs with other community activities.

State siting boards generally have the power to preempt local governments' zoning and permitting procedures which are often the methods used to block the siting of waste processing facilities. The concurrent establishment of schemes for compensation represents a movement in the opposite direction (i.e., away from permitism). The reader is guided to the innovative siting law of Massachusetts which contains both features but leans heavily toward compensation (O'Hare et al. 1983).

Citizens in opposition to the siting of facilities have often resorted to the courts to forestall or defeat the proposition. Often this has been combined with protests such as sit-ins, letters to the media and petition campaigns. Interestingly, a new form of litigation has come about aimed at the latter kind of opposition. Called SLAPP, which stands for **s**trategic **l**awsuits **a**gainst **p**ublic **p**articipation, the plaintiffs—public bodies or private organizations—allege libel, defamation or interference with business. The suits have a chilling effect on any process that requires public participation. Seeing this as an effort to stop protesters from voicing their concerns and a violation of their First Amendment rights, Ralph Nader, the American consumer activist, has offered free defense counsel to SLAPP suit targets and is encouraging them to file countersuits.

15. Ownership, Fees and Financing Issues

There is a general movement to clearly identify solid waste management costs. In several jurisdictions this has led to the establishment of solid waste enterprise accounts where the full costs of discard management are collected in fees charged to the residents of an area rather than, as has generally been the practice, subsuming them in property taxes.

"Pay as you throw" or volume-based disposal charges have become quite popular in recent years. In addition to internalizing the overall cost, or a portion thereof, at the household level, they are used as a means of widening the gap between gross discards and net discards by offering an incentive to householders to participate in source separation programs. The householder is not charged for the pickup of the separated recyclables, only for the regular waste. Communities report success with this approach. Seattle, Washington, reports a decrease among participants in its curbside recycling program from 3.5 cans of nonrecyclable disposables per family per week to slightly more than one.

While the main proponents of deposit legislation and of other initiatives to increase recycling have been the environmental groups, some of whom have operated drop-off and buy-back centers, the current shift to more governmentally supported recycling efforts has placed their operations at a disadvantage. For all practical purposes, deposit legislation makes the free-standing recycling center redundant. Consumers prefer to take the containers back to the market on a shopping trip. Even when there is legislated financing for stand-alone redemption centers, larger, better capitalized firms generally supersede the smaller operators. The story is the same when curbside programs are implemented. Typically, community drop-off centers that might have expected to become the MRFs for the new curbside programs are superseded by firms that can both collect and process and often have better access to markets for the recycled materials. This has been a point of contention and community groups have taken their local government to task for issuing requests for proposals that required that the bidders offer both collection and separation, which the community groups are not able to do.

The integration of recycling into a municipal solid waste management system has also raised problems of financing. Tax-exempt revenue bonds have been the main method of financing for waste-to-energy facilities. Tax-exempt here refers to the interest earned on the bonds by the lenders. To establish the economic versatility of these facilities, the community has had to underwrite a great many risks. These include guaranteed delivery of waste through "put or pay" agreements and rate covenants where the municipality pledges to set disposal rates (tipping fees, household user fees, assessment fees, etc.) at a level high enough to cover all operating costs as well as debt service coverage. Also required are flow control covenants which obligate the municipality to direct all waste generated in the municipality to the plant, and finally a service covenant which protects the plant in the case of "force majeure." The result is to transfer the economic risks associated with the private aspects of the recovery activity to the public sector. In effect, these risks no longer exist as far as the facility itself is concerned since they have been assumed by the community. The rationale is the same as justifies other "public works"; that is, that the citizens enjoy the benefits associated with the presumed "better" management of their waste and therefore the public should cover the risks.

System financing extends the risk insurance of the covenants as described above to the system as a whole and not simply to the recovery facility. It thus encompasses activities that in their own right could not be financed by the issuance of revenue bonds. A community would be forced to use general obligation financing which usually requires voter approval. As the siting issue described in Sec. 14 exemplifies, waste management issues coalesce opposition. Those affected are easily mobilized. The remainder of the community is much less concerned and is difficult to induce to vote for what is generally a significant public expenditure. As a result, municipal officials favor revenue-based financing since it does not involve voter approval. How to oblige this desire has created new challenges for the purveyors of tax-exempt financing.

See also: Recycling of Demolition Wastes; Recycling of Materials: Economics; Recycling of Metals; Recycling of Paper Products; Recycling of Plastics

Bibliography

Alter H 1991 The future course of solid waste management in the U.S. *Waste Manage. Res.* 9: 3–20

Center for Economics Research 1989 *The Effects of Weight- or Volume-Based Pricing on Solid Waste Management.* Research Triangle Institute, Research Triangle Park, NC

Chertow M 1989 *Garbage Solutions: A Public Official's Guide to Recycling and Alternative Solid Waste Management Technologies.* US Conference of Mayors, Washington, DC
Franklin W 1989 *The Role of Beverage Containers in Recycling and Solid Waste Management.* Franklin Associates, Prairie Village, KS
Jacobs H, Bailey J 1982 Evaluating participation in a residential recycling program. *J. Environ. Syst.* 12(2): 141–52
Mielke G 1988 *A Planning Guide for Residential Recycling Programs in Illinois.* Illinois Department of Energy and Natural Resources, Springfield, IL
Office of Solid Waste 1989a *Decision-Makers Guide to Solid Waste Management*, EPA/530-SW-89-072. US Environmental Protection Agency, Washington, DC
Office of Solid Waste 1989b *The Solid Waste Dilemma: An Agenda for Action.* EPA/530-SW-89-019. US Environmental Protection Agency, Washington, DC
Office of Solid Waste 1990 *Sites for Our Solid Waste: A Guidebook for Effective Public Involvement.* US Environmental Protection Agency, Washington, DC
Office of Technology Assessment 1989 *Facing America's Trash: What Next for Municipal Solid Waste?* OTA-O-424. US Government Printing Office, Washington, DC
O'Hare M, Bacow L, Sanderson D 1983 *Facility Siting and Public Opposition.* Van Nostrand Reinhold, New York
Sellers J 1987 *Feasibility of Tax Incentives for Purchases of Recycling Equipment or Recycled Products.* Franklin Associates, Prairie Village, KS
Skumatz L 1990 *Variable Rates in Solid Waste: Handbook for Solid Waste Officials*, Vol. 1, EPA/530-SW-90-084A. US Environmental Protection Agency, Seattle, WA
Testin K F, Vergano P I 1990 *Packaging in America in the 1990s.* American Paper Institute, Washington, DC

J. G. Abert
[Georgetown University, Washington, DC, USA]

Risk Management

In a world where international economic interdependence is increasingly the norm but stability and predictability in contracts and traditional commercial relationships are not, industrial materials users have been giving attention to ways to manage and reduce their materials risks. This article discusses a few of the industrial approaches to risk management and describes one specific case history.

1. Background

Historically, industry in the USA could depend on domestic sources for the bulk of its raw materials and minerals needs. For a variety of reasons, however, it has increasingly had to rely on imports—to the point that they now exceed 50% of consumption for about half of the 36 most essential industrial materials and 90% for some of the most critical ones. Some of this trend results from depletion or exhaustion of the most economically viable domestic deposits and from higher labor and environmental costs in the USA, and some reflects the growing needs of mineral producers, especially in the third world, to earn foreign exchange income and their increasing ability to produce and export, often from richer and more accessible deposits, on a cost-competitive basis.

Concurrent with these developments has been a rise in political and economic instability in many resource-rich countries—Africa offers several examples, the countries of the former USSR have shown inclinations to involve themselves in overseas minerals supplies and markets for politicostrategic as well as commercial reasons, and OPEC's geopolitical success in cartelizing world oil supplies aroused hopes of emulation by other raw materials producers. In addition, there has been a decline in the respect for sanctity of contracts and in what may be considered due process in international commercial law. Business materials planners have also had to contend with more than normal volatility in the prices of metals and minerals, with peak increases or "bubbles" during shortages of commodities.

In the USA, the federal government has assumed responsibility for the national security aspects of strategic materials through its National Defense Stockpile and Defense Production Act agencies and through other programs. However, the stockpile program in particular has long been plagued by problems, and during the Reagan administration the proposed goals were dramatically reduced and responsibility for the stockpile was transferred to the Defense Department. Recent appropriations for new acquisitions have been relatively small. Although critical materials can be released from the stockpile by presidential determination for purposes of national defense only, under less than national emergency conditions many US manufacturers have become concerned over their own peacetime vulnerability under the global conditions described above.

2. Types of Approach

TRW's internal risk analysis efforts showed that one-third of the company's purchasing volume of supplies and materials contained critical materials resulting in a dependency ratio of sales revenues to procurement costs of nearly 7 to 1 (US Department of Commerce 1982). Cobalt, tantalum, chromium, niobium and platinum group metals are essential in a number of important defense and civilian products manufactured by the company; the USA as a whole is over 90% import-dependent for supplies. Looking ahead, the company made assumptions that this situation would continue as foreign suppliers continued to seek increased value added in their own countries and improve their processing technology and capacity, leading to a concomitant reduction in US-based production and processing capacity. Moreover, industrial competitors of the USA would continue to seek to secure their own supplies of critical minerals, possibly reducing the quantities available in normal markets.

TRW has developed planning strategies including a long-term (five-year) strategic marketing forecast to determine the amounts of strategic materials needed to meet production goals, track forecasting trends, analyze the risks and impact of supply disruptions and establish vigorous materials management policies and programs including research and development (R & D), substitution and reclamation. Associated procurement and technical strategies have also been developed by the company.

Westinghouse Corporation followed a similar general approach and developed a two-axis matrix for internal and external criticality (see Sect. 3) with risk management strategy focused on product redesign and specification change as well as R & D for development of substitutes (US Department of Commerce 1982).

Cabot Corporation developed a fine-tuned management approach involving defining, priority allocation and acting through various means, including exploring and investing in producing operations, R & D and recycling and, in some cases, maintaining contingency stockpiles (US Department of Commerce 1982).

Union Carbide Corporation formerly monitored raw materials vulnerabilities with high sales and profit ratios and looked to new technologies to permit use of lower-grade and more readily available ores, for example, chemical instead of metallurgical grade chromite. The effort has, however, been suspended owing to changes in the company's product lines.

In addition to such individual company efforts, many of those concerned have also sought more active US public policies of materials risk management for the nation as a whole by such approaches as improved strategic stockpiling and domestic production and standby capability, increased access to minerals on public lands, adjustments to environmental standards, conservation and recycling, more government R & D on substitutes, trade actions where imports threaten national security, and diversification in US overseas sources through aid and foreign policy initiatives. Many such proposals were also included in recommendations of the Interior Department's National Strategic Materials and Minerals Program and Advisory Committee (US Department of the Interior 1986). In addition, proposals have been made for a trial system of private stockpiling on the basis of government sharing of additional costs and benefits. The objective is to create additional stocks of the most critical materials over and above those likely to be obtained in the near future by appropriations for the defense stockpile or by individual company programs.

Various approaches to risk management have also been followed in other industrial countries (see *Stockpiling*).

3. The Corning Experience

The complexity of the problem of specialty glass supply is illustrated by the fact that Corning Incorporated, a leader in specialty glass and ceramics manufacture, consumes in any given year some 250 different raw materials in more than 6000 glass and ceramic compositions for more than 60 000 products that are sold to 47 000 customer firms. In all, the list of approved raw materials contains over 400 entries. For these materials, the company spends more than US$100 million annually supplying its domestic operations.

As a consumer of minerals and chemicals, Corning is distinctive because of its need for exacting purity in its specifications and its relatively small tonnage requirements. These factors suggest the possibility of supply interruptions and indicate the size of the potential risk involved. The management of these risks, both to price and availability, constitutes a major challenge to consumers of raw materials in the 1990s.

The risks and associated anxiety are expressed in publications by Szuprowicz (1981) and Hargreaves and Fromson (1983). A Congressional Handbook on import dependency and vulnerability has been published (US Congressional Research Service 1981). Generally, the problems are caused by world economic conditions, the takeover phenomenon as business after business changes ownership and direction, and scarcity of the materials.

Following a period of relative abundance in the early 1980s, interest in critical materials management has renewed as the economy of the late 1980s continued to be robust. The specialty glass business has been confronted with shortages and escalating prices of alumina, copper, nickel, titania and zircon. Over a two-year period beginning in 1986 prices have at least doubled, and in the case of nickel a fivefold increase was seen. Strong markets and limited production capacity of these commodities, especially zircon, have increased the glassmaker's costs directly, both as raw material and through the increased costs and lead time in obtaining refractories used to build furnaces.

The industrialized nations are dependent to a large degree on less developed nations for continued supplies of industrial minerals and metals. Corning, like the rest of US industry, depends on southern Africa for chromium, cobalt, diamonds and platinum group metals. While substantial import tonnages originate in South Africa and Mexico, most of the dollars for imported goods are spent in Europe.

During the 1970s and 1980s Corning formalized a strategic materials program in response to the pressures mentioned above, as well as to four patterns in supply that were seen to be emerging.

(a) Environmental and economic factors were forcing suppliers to examine the profitability of the lines they produced. In several cases, production of needed commodities has ceased.

(b) Political action such as economic sanctions,

nationalization of industries and political insurrections had disrupted markets.

(c) Richer deposits were being depleted and substitute deposits were not meeting the quality standards of previous ore bodies.

(d) Energy-dependent industries were passing along cost increases rather than trying to find more efficient processes.

Corning's approach was to take a number of specific actions which included:

(a) organizing a corporate strategic raw materials function;

(b) identifying all raw materials that were the ingredients in its most important products;

(c) deciding on the definitions for criticality, and classifying materials by means of an analytical matrix;

(d) examining the economics of protective action;

(e) stockpiling certain critical materials for use in an emergency; and

(f) placing long-range contracts with producers.

3.1 The Criticality Matrix

With its consultants, Corning sought to design a hierarchical ranking of critical materials that communicated the urgency of the situation clearly to managers authorized to commit resources to raw material programs. The result was a two-dimensional matrix ranking materials by their relative importance to the corporate objectives and by their relative vulnerability to supply disruption (Fig. 1). Materials listed towards the right-hand side of the matrix are the ingredients of products that are important to the present and future objectives of the corporation. Those listed towards the left have less importance. Materials listed at the top of the matrix are highly vulnerable to price escalation or supply interruptions. Materials ranked further down are deemed to have a greater degree of supply safety.

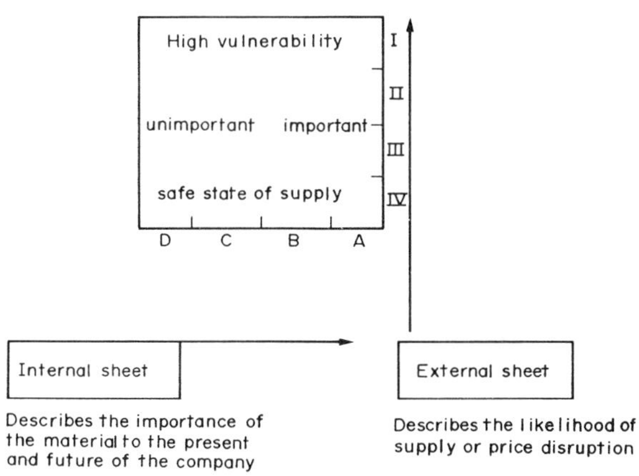

Figure 1
Criticality matrix for materials supply

Before any material is assigned a location in the criticality matrix, it is given two dimensions called internal and external criticality. The internal dimension is measured on the horizontal scale and states the company's estimate of the consequence of a supply disruption. On the vertical scale, the probability that a supply interruptions or price escalation will occur is judged.

The supporting documentation for the analytical process is essential to the ranking. The external documentation summarizes data concerning vendors and the marketplace into five criteria that determine the external rank.

3.2 The Analytical Process

To illustrate the principle, the example of a particular material used in Corning's manufacturing processes is considered. The name and data are hypothetical, to protect proprietary interest. The material in this example will be called "unobtanium trioxide." It is a derivative of the ore mineral, bewilderite, an accessory mineral in many enigmatite bodies. It is a by-product of enigmatite mining in central Erehwon and from the new deep-sea mining venture in the Republic of Atlantis. It is a compound of rare occurrence. The basic inputs to the analysis consist of "internal" and "external" data as follows.

Unobtanium trioxide is consumed to make several products that net the company over US\$60 million in operating margin. The consequences of a six-month supply allocation of one-half of Corning's normal consumption would force customers to seek substitutes for the Corning product if the company failed to meet deliveries. Corning sells to a growing market and the company's future position as a supplier depends on its maintaining its reputation of reliability. Purchases of unobtanium trioxide total more than US\$8 million.

External data are derived from the analysis of the present market. The company buys from one supplier who has wisely chosen to use ore from two sources. There is a potential supplier but it is only marginally interested in producing unobtanium trioxide. Both suppliers prefer to manufacture unobtanium nitrate instead of trioxide, but could convert some of their product into the trioxide. It is estimated that there is a 20% chance that the deep-sea mine will go bankrupt in the near future, causing market allocations by the supplier. The present supplier is stable, but about 20% of the time his shipments arrive late or are out of specification, requiring replacement. Because of the potential for increased labor and energy costs in mining and processing, the analysts expect that a price increase more than 3% above the inflation rate will occur during the year.

These data are condensed into the nine indices necessary to specify the coordinates and place the material in the matrix, and the analysts proceed to ask

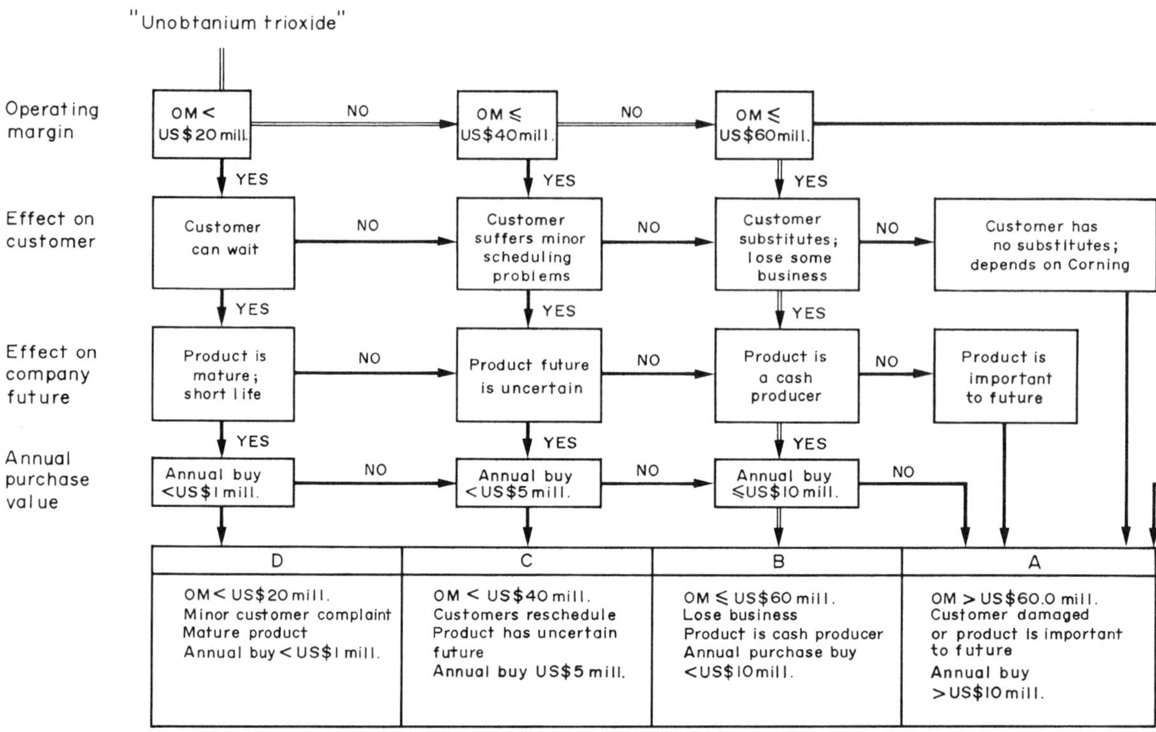

Figure 2
Internal screening logic

questions as illustrated in Figs. 2 and 3. The screening process gives unobtanium trioxide an internal rank of B and an external criticality of I, according to the tests. It is placed in the matrix location BI as indicated in Fig. 4.

In the Corning system, this material has sufficient criticality to warrant the assignment of a project manager who, with a team of technological and commercial specialists, evaluates the options open to relieve the problem. He considers such strategies as stockpiling, multiple sourcing, substitution or strengthening the relationship between the company and its supplier. The project manager defines the goals of the project and leads the work until the state of criticality is reduced to an acceptable level. These project managers can be assigned either from corporate staff or the line division, depending on the requirements of the strategy. The project managers must seek funding from the appropriate line division, which means the funding organization must understand and accept the challenge of mitigating risks.

The final buying decisions are reserved for the Vice President Corporate Purchases. In the hypothetical example, the company might decide to invest in a stockpile of unobtanium trioxide while the team has time to work out a more permanent solution to risk management.

3.3 Conclusion

The method of risk analysis outlined has been applied to all the raw materials that contribute to 90% of the profits of Corning Incorporated. Despite the very large dimensions of this task, it has been accomplished with a minimum of staff: four full-time professionals and four part-time technologists.

The results have been gratifying for the company. Corning has been successful in communicating the urgency of its risk positions to top management. The incidence of crises in the raw material supply is distinctly lower than before the analytical matrix was applied. A reduced incidence of crises helps to reduce costs, improve productivity and enhance Corning's position as a reliable supplier to its customers. Unlike the period of the early 1970s, when the company was reacting rapidly to the fast-moving changes in world conditions, it is now able to anticipate many changes. Consequently, it can act in advance to solve supply and price problems free of the tensions and time pressures inherent in crises.

As noted, this is simply one illustrative example of a company that has pioneered the application of the materials risk management strategy described. Other firms which judge themselves particularly vulnerable to supply risks use similar or analogous approaches. Still more companies are only in the early analytical stage

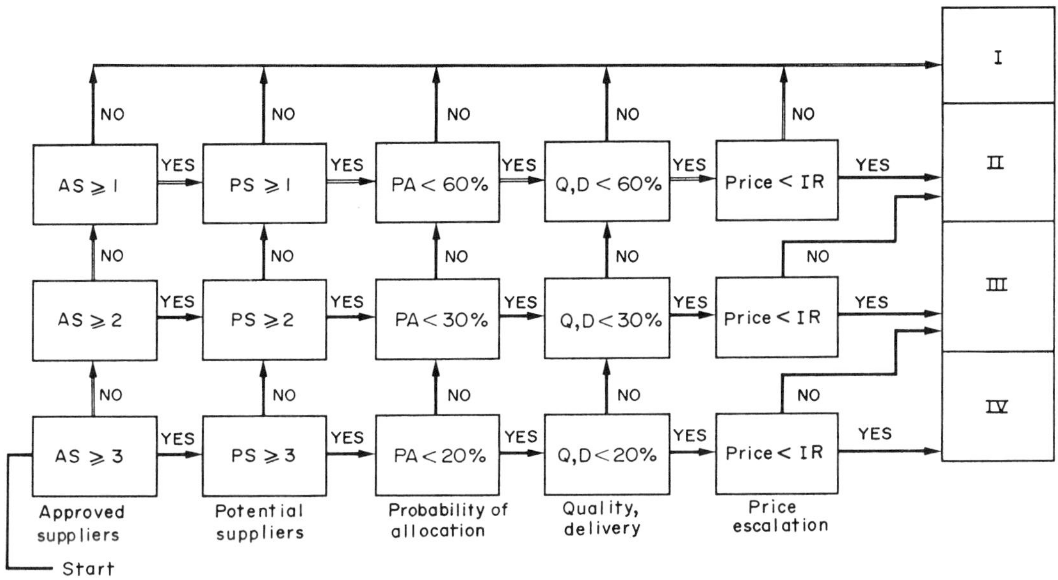

Figure 3
External screening logic: AP, approved supplier; PS, potential supplier; PA, probability of allocation; Q,D probability of a quality or delivery problem; Price < IR, increase is less than rate of inflation

of assessing their own vulnerability, while others, apparently, regard the risks as acceptable—that is, they feel that they can safely rely on the market (or the defense stockpile in emergency situations) for critical materials, or else they assess the costs of remedial or risk reduction measures as too high in relation to those risks. However, there are possibly many other US companies which have yet to address the problem in a serious way, even though the supply problems described are likely to be with industry for many years to come.

Acknowledgement

Section 3 of this article draws heavily on the work by Edwards and Anderson cited in the Bibliography. The authors are grateful to the Editor of *Industrial Minerals*, New York, for permission to use this material.

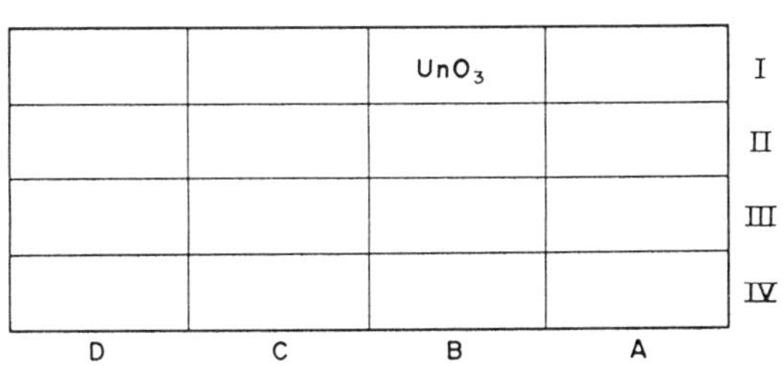

Figure 4
Position of "unobtanium trioxide" in the criticality matrix

See also: Critical and Strategic Materials; Materials Requirements Planning; Trade in Materials, International

Bibliography

Anderson D W 1990 Strategic materials in glass manufacturing. *Mater. Soc.* 14(2): 145–58

Edwards G H, Anderson D W 1983 Assessing risks in raw materials supplies for the specialty glass industry. In: Coope B M, Clarke G M (eds.) 1983 *Proc. 5th Industrial Minerals Int. Congr.* Metal Bulletin, London, pp. 243–8

Edwards G H, Copley J G Jr 1977 Trends in raw material needs for the specialty glass industry. In: Harben P W (ed.) 1977 *Raw Materials for the Glass Industry*. Metal Bulletin, London, pp. 87–91

Hargreaves D, Fromson S 1983 *World Index of Strategic Materials.* Facts On File, New York

Holt J L, Stanley T W 1983 *Nonfuel Mineral Stockpiling Policies: The Private Stockpiling Alternative*, Contemporary Issues No. 6. International Economic Studies Institute, Washington, DC

International Economic Studies Institute 1987 *Raw Materials and Foreign Policy*. Westview, Boulder, CO

Piepgras J J, Metz H J 1980 Managing critical materials in the 1980s. *Met. Prog.* 118: 39–46

Sinclair J E, Parker R 1983 *The Strategic Metals War: The Current Crisis and Your Investment Opportunities.* Arlington House, New York

Szuprowicz B 1981 *How to Avoid Strategic Materials Shortages.* Wiley, New York
US Congressional Research Service 1981 *A Congressional Handbook on US Materials Import Dependency/Vulnerability,* Committee Print 97–6. US Government Printing Office, Washington, DC
US Department of Commerce 1982 *Materials Risk Management, Proc. Government/Industry Seminar.* Office of the Under Secretary for Economic Affairs, Washington, DC
US Department of the Interior 1986 *Status of Strategic Materials and Minerals Relating to US Vital Interests.* National Strategic Materials and Minerals Program Advisory Committee (NSMMPAC), Office of the Secretary, Washington, DC

D. W. Anderson
[Corning Glass Works, Corning, New York, USA]

T. W. Stanley
[International Economic Studies Institute, Washington, DC, USA]

S

Stockpiling

Events in the world political and security fields have removed the issue of stockpiling for security objectives from an important public policy issue to a nonproblem in the industrialized nations of the world. The return to sovereignty of the Eastern European nations; the decline of the USSR as a military threat to the world; a new conception of war, as illustrated by the United Nations effort in Operation Desert Storm to counter the Iraqi invasion of Kuwait; the integration of Europe into a true common market; and the establishment of a free trade union of Canada and the USA, with the probable addition of Mexico are the major events involved.

For these reasons, it seemed advisable not to update this article, but to reprint it for historical reasons as it appeared in the *Encyclopedia of Materials Science and Engineering* in 1986.

Webster's Unabridged Dictionary had no listing for "stockpiling" in its 1922 edition, but in 1951 it defined a stockpile as "a reserve supply of an essential raw material—processed food, or the like, accumulated within a country for use during a war-induced shortage." By 1973, "war-induced shortage" had evolved to "a shortage caused by emergency conditions (as war)."

Private firms, especially those involved in mineral-intensive industries, maintain stocks of raw materials both as normal business inventories and as a hedge against shortages, price escalations or other contingencies (see *Risk Management*). Governments also stockpile and the USA maintains an elaborate national defense stockpile for use in an "emergency with respect to the national defense" as well as a strategic petroleum reserve. Stockpiling in the USA is detailed elsewhere (see *Stockpiling in the USA*).

Broadly speaking, stockpiling is only one, albeit important, component of national materials policy (see articles on policy in various countries). This article deals primarily with the stockpiling of nonfuel strategic minerals in six industrialized countries that are mainly raw-materials importers: France, the FRG, Japan, Sweden, Switzerland and the UK. Other countries have also considered various approaches to the problem. Spain, for example, planned in 1979 to develop a US$350 million industry-held, government-financed stockpile of critical nonfuel minerals, but has not yet acted because of budget limitations. Other smaller countries have elected to rely on world markets or on industrial suppliers in their larger neighbors, particularly within the European Economic Community (EEC). There does not, however, seem to be much synchronizing of stockpile-type programs even within the Community. There are sharing provisions for oil under the International Energy Agency accords, but it is unclear whether and to what extent any one country's stocks of nonfuel minerals might be available to an ally under emergency conditions.

Most of the Communist nonmarket economies also have significant stockpiles. Some are incidental to normal and supplemental inventories at state enterprises; some are stored at production sites, some held pending favorable export marketing opportunities, and some maintained as mobilization or emergency reserves. Soviet mineral policy, for example, continues to pursue the goal of maximum self-sufficiency even at uneconomic costs. Although there is an extensive literature on the mineral industries of China and the USSR, there is insufficient publicly available information on the types, purposes and holdings of stockpiles in the major Communist countries to warrant coverage in this article.

1. Overview

The USA maintains by far the world's largest strategic stockpile (if the state holdings of Communist countries are disregarded), largely because it inherited one at the end of World War II, but other countries have outstripped the USA in other minerals vulnerabilities risk management techniques. France and the UK have developed relationships with their former colonies as material suppliers through the Lomé Conventions and their related Stabex and Sysmin programs. Japan likewise has a network of supplier relationships tied up through investments, long-term contracts, aid programs, price stabilization schemes and the like. Access to resources in the postcolonial era has generally played a much larger role in the foreign policies of these countries than has been the case for the USA. Like Switzerland and the Netherlands, for example, the FRG has dozens of bilateral investment treaties with developing countries providing for national treatment and other standards, while the USA is still struggling to initiate a few such treaties. However, the FRG's plan for a cooperative government–industry stockpile has been shelved for the time being because of budget limitations and disagreements over control.

On the other hand, France and the UK maintain modest governmental (or quasi-governmental) stockpiles of critical nonfuel minerals and Japan, Sweden and Switzerland have what might be considered "private" stockpiling policies. Japan operates through associations of private firms which maintain inventory levels under the guidance of the Ministry of International

Trade and Industry (MITI) and receive government-subsidized financing in which stocks are pledged to the Metal Mining Agency. These serve both as a hedge against shortage and a cushion to maintain processing capacity during slack times. The Swedish government maintains an elaborate national economic defense mechanism, including extensive stocks of food, supplies, machines, minerals and fuels. In addition, its income-tax treatment of business inventories (partly designed to avoid taxing inflationary increases in value) allows accelerated tax write-offs for reduced inventory valuation, encouraging higher levels of stockholding for private firms.

Switzerland has both mandatory stocks financed by the government and voluntary stockpiles which are facilitated by government credit arrangements, of which half can be retained by the holder and half are subject to emergency allocation. Their combined value is about two-fifths that of the current US national defense stockpile—in a country that has only 3% of the US population!

Both the Organization for Economic Cooperation and Development (OECD) and the EEC have sought to coordinate members' efforts in this field. The former is in the process of gathering and comparing data, but its eventual report is unlikely to be made public. The Community has also performed useful analyses but is blocked because some countries, notably France, consider policies in this field a purely national responsibility. France, however, did play a key role in the negotiations of the first and second Lomé Conventions between the Community and former colonies and dependencies of its member countries.

2. *France*

Faced with a dearth of domestically produced nonfuel minerals, France's mineral-intensive industries (which contributed 5% of its gross domestic product in 1980) rely primarily on foreign producers for an adequate supply (55%) of their principal manufacturing inputs. In the light of the recently acquired independence of many of the mineral-rich countries (including its own former African colonies), the influence of the Organization of Petroleum Exporting Countries (OPEC) on energy supplies and prices, and several contemporary commodity shortages, the government of France undertook a major analysis of that country's vulnerability to sudden supply disruptions of raw materials. The strategic analysis, geological surveys and domestic mineral inventories were published in 1978 (Bureau des Recherches Geologiques et Minières 1978).

Consistent with the general recommendations put forth in that report, industrial policymaking as outlined in the government's seventh five-year plan (from 1976 to 1980) was revised to pay more attention to the economic benefits attainable through better management of imported raw materials. This "calculable risk" study of mineral supply shortages illustrates the combination of foreign-policy considerations and economic and political forces that drive international minerals markets.

To this end, France has pursued several options, including: (a) improving conservation and recycling through technology developed in government and business research programs; (b) increasing domestic production through incentives to mining companies for long-term financial and commercial exploration; (c) promoting economic, political and scientific (transfer) agreements with France's major suppliers of raw materials, as demonstrated by the Lomé Convention; and (d) to a lesser extent, stockpiling. All these measures serve to reduce dependence, ensure availability (judged more important than price) and offset foreign-exchange costs.

Of the major Western European countries, France is one of the few to devote substantial financial resources to the creation and maintenance of a strategic stockpile. Hampered by budget constraints in the last few years, however, expansion plans devised in 1975 have not lived up to expectations, so that what was originally to be a US$681–851 million materials stockpile has reached only US$340 million. Controlled by the GIRM (a centralized purchasing agent of the Ministry of Industry), stockpiled minerals are purchased at favorable market prices and sold to domestic buyers when market supplies dry up.

Significantly, the French assign degrees of vulnerability. The "very great" category includes silver, platinum, diamonds, phosphates, zirconium, titanium, cobalt and vanadium. The "great" class includes antimony, copper, manganese, molybdenum and tungsten. "Medium" includes bauxite, chromium and tin, while France considers itself merely "weak" in iron, nickel, lead, zinc and potash.

Determined not to establish a special budget for the stockpile or a specialized purchasing agent, and opposed to the financing of commodity stockpiles through bank loans, the French government addressed its stockpile shortfalls with the creation of the CFMP (Caisse Française de Matières Premières) under the jurisdiction of the Ministry of Industry for issuing and administering bonds. The intention was to use funds raised in this manner to increase holdings in copper, lead, tungsten and chromium—none, curiously, in the "very great" vulnerability category. So far CFMP has issued 500 million French francs in 14% ten year bonds, which were fully subscribed. Long-term goals, revised in 1980, envisioned spending up to three billion French francs for the stockpile, but overall progress remains inconclusive. Only 169 million were earmarked for purchases in the 1980 budget, which figure has undoubtedly been further reduced under the current government. Therefore, the same tensions between planners and budget officials are evident as have characterized the US nonfuel minerals policy in the past three decades.

3. Federal Republic of Germany

The FRG consumes over 10% of the world's annual mineral production but produces less than 1%; thus its dependence level is somewhat above that of the UK and France. Moreover, because of common EEC trade policy, FRG refined-metal exports are shipped primarily to other EEC countries, thus distorting the dependency of Western European nations on imported raw materials.

The economic policy of the FRG is to rely entirely on market forces in order to obtain the socially optimum supply and demand equilibrium. Eschewing large-scale, statist-type industrial policymaking, the German government prefers to treat different sectors of industry as separate branches, which merit special attention only if they affect the country's security or international competitiveness (either directly or through forward linkages). While focusing on industries of the future such as aerospace, electronic data processing and nuclear energy, the FRG also follows a national tradition of high standards in science and technology, including metallurgical technology.

Although worldwide leaders in minerals processing and refining, German firms have not been able to transfer this dominance into overseas mining operations, where they have been overshadowed by US, British, French and Japanese firms. At the crux of German concern over raw materials are long-term imbalances in the development of supply and demand (owing to insufficient investment for exploration and development) and the growing appetite of the FRG's mineral-intensive industry.

The German government can more easily deal with the problem of overseas investment. Indeed, it has been a leader, along with the Netherlands and Switzerland, in negotiating bilateral investment treaties with developing countries, and well ahead of the USA in this respect. However, a second supply problem that deeply troubles Germans is the politically motivated supply shortfalls that might be caused by "other OPECs" and general Third World intervention in resource markets, for these factors are essentially out of their hands.

Despite their declared position of nonintervention in commodity markets, concern over such nonmarket forces has stirred considerable debate in the FRG over whether to stockpile strategic minerals. The 1979 proposals for both governmental and private-sector stockpiling efforts have been shelved for budgetary reasons, but may be reconsidered as the economy recovers. The government and industry never agreed on who should have operating control, but the essence of the scheme proposed was for voluntary participation of companies in a consortium with the government. Each company would receive incentives in the form of interest-free loan and tax credits to add 3–12 months to its normal inventory of materials such as chromium, cobalt, manganese and vanadium and store it until an emergency, when the extra stocks would be split 50–50 between the companies and the government. Costs would have run to about US$100 million a year for five years.

4. Japan

No other country has such an absolute dependence on the importation of raw materials as Japan. In contrast to the early periods of Japan's modern history, when minerals such as base metals, gold and even coal were major export earners, the burgeoning demands of Japanese "miracle" industries, coupled with depletion of the scattered home reserves, have made it necessary to import over 90% of the nation's crude nonfuel minerals requirements.

The Japanese acceptance of state involvement in industrial affairs, shared by government and corporate executives alike, has ensured that formulation and periodic reevaluation of raw materials management policies, objectives and philosophies are prime elements of industrial policy. Furthermore, cooperation between the two principal actors in the procurement of overseas raw materials permits Japan to exercise a considerable degree of monopsonist power when negotiating with foreign producer nations. Indeed, the strength and diversity of equity investments, loans and grants, long-term production contracts, price-stabilization agreements and development assistance projects that Japan has established with a variety of developed and developing country mineral suppliers have led its industrial policymakers to feel relatively secure in the area of raw materials despite their overall dependence.

In essence, Japanese industry has accepted the high costs of capital investments in offshore exploration and mining facilities, and of stabilized prices through long-term contracts, in order to obtain reliable supplies to fuel its sustained economic growth.

Although it has invested management efforts and funds in offshore suppliers of strategic minerals, the Japanese government has also participated in three nonferrous-metal stockpiling associations: (a) the Metallic Minerals Stockpiling Association, for various forms of refined copper, zinc and lead; (b) the Light Metal Stockpiling Association, for primary aluminum; and (c) the Special Metals Stockpiling Association, for nickel, chromium, tungsten, cobalt and molybdenum (including ore and intermediates). Strategically, these stockpiling associations provide the Japanese with a cushion to fall back on in their negotiations with raw-material suppliers. However, the more explicit objectives of the stockpiles are less strategic in nature.

First of all, by protecting the economies of minerals-supplying countries when demand from Japanese industries falls below projected levels, stockpiling of surplus contracted materials enables Japan to ensure continued flows, avoid international friction and preserve the integrity of long-term contracts. Conversely, in the event of a severe supply shortage, timely releases of stockpiled material to the consuming industries would permit production to continue with reduced or no loss of export earnings. Finally, the stockpiles have provided relief to the domestic smelting and refining industry, insofar as excess production is purchased from them, stocked during surplus periods and returned to them when they must meet increased demand.

Japan has made an explicit decision to use its stockpiles to ensure price and supply stability, maintain assured GDP and sustain relationships with key suppliers, and it is willing to act to limit vulnerability to disruptions in overseas supply. The Japanese have also considered more direct government stockpiling of certain critical materials beyond those covered by the associations, utilizing facilities of MITI and the Metal Mining Agency of Japan; but at the time of writing, details on what has been done are not available. This is one of the areas in which Japan provides a textbook example of business, government and labor cooperation to meet a major national problem.

5. Sweden

Given the traditional neutrality of Sweden, as demonstrated in World War II, economic defense is considered a part of the country's national security posture. In what are called "storage villages," the government stores, maintains, rotates and upgrades supplies of food, fertilizers, clothing, medical supplies, chemicals and solid and liquid fuels as well as metals, machines and tools. Sweden spends 300–500 million kronor (about US$65 million) annually on these activities, not counting petroleum. Total capital expenditures to date run to several billion kronor.

The National Board of Economic Defense maintains stockpiles of a large number of mineral raw materials to be used in the event of war or a blockade. In addition, there are critical peacetime stockpiles of chromium ore, manganese ore, cobalt and vanadium, to be used if supply lines are disrupted in peacetime. A recent government study proposed that molybdenum, tungsten, nickel, titanium, niobium, tantalum and borates be added to the peacetime stockpiles, and that critical peacetime stockpiles of manganese, chromium and cobalt be expanded. Peacetime stockpiles equivalent to three-months' imports were recommended.

Strategic stockpiling in Sweden was initiated in the 1930s to strengthen national security in the event of war. After World War II, the Korean crisis led to a further buildup of such stockpiles. Present strategic stocks, administered by the National Board of Economic Defense, are calculated to enable the country to cope with a wartime cutoff for about one year.

Minerals stocked under this system (strategic stockpiles in a traditional sense) include chromium, cobalt, nickel, manganese, vanadium, titanium, molybdenum and tungsten (ore, alloys and metal) as well as tin, zinc, aluminum and a number of industrial minerals.

In addition, Sweden is in the process of building up stocks for needs that may arise in a peacetime crisis. The purpose is to maintain, as far as possible, normal levels of production, employment and exports. The new program includes certain alloy metals such as chromium, cobalt, manganese and vanadium. The stocks will correspond to normal use of chromium and cobalt for two months and one month for manganese and vanadium.

Recent Royal Commissions on defense and minerals policies have suggested that the peacetime program be developed further. For instance, stocks of chromium, vanadium, manganese, cobalt, molybdenum, nickel and titanium should be increased to three-months' normal use. It has also been suggested that private industry might be induced to carry a larger share of stockpiling for peacetime purposes.

In addition to these economic defense activities, Sweden has a *de facto* private stockpiling system in the way in which materials inventories are treated under Swedish tax laws. Inventories are valued at purchase (FIFO) or replacement value, whichever is lower, and can be written down by as much as 60% annually after adjustment for obsolete or unmarketable items under various procedural rules. Taxable income is also reduced by the amount of this write-off in inventory valuation; given Sweden's high tax rates, some corporations use the tax savings to purchase larger stocks than they would otherwise hold. Others prefer to pass the tax savings on to shareholders through higher dividends.

6. Switzerland

Switzerland, with another long and fiercely maintained tradition of neutrality and a dearth of most natural resources, stockpiles imported foodstuffs to last one year and fuel for at least six months. In addition, the Swiss have established a system of voluntary stockpiling programs administered by the Federal Office for Economic Defense Measures. Importers of strategic raw materials and industrial items are encouraged to hold stocks corresponding approximately to a normal twelve-month supply. As an incentive to maintain these voluntary levels, under the government stockpiling program importers are offered substantially lower interest rates, tax rebates and preferential treatment in times of war or other emergencies. Items of primary interest to Switzerland are iron and steel, nonferrous metals, chemical raw materials, plastics and textiles. Reportedly, defense stocks of these commodities are high.

Expenditure for national defense in Switzerland currently amounts to 2.2% of its gross national product, of which military items represent 1.9% and civil defense takes 0.3%. This must be regarded as low, even if allowance is made for a certain cost reduction stemming from unpaid voluntary services under the militia system. Procurement outlays have increased tremendously and are therefore being staggered.

The system of stockpiling strategic reserves has been maintained and strengthened throughout the years following World War II. Its basic features are as follows. For essential goods on which Switzerland has a critical import dependence, compulsory stocks are maintained by private firms, grouped into "consortia" according to economic sectors, such as food, fodder, fertilizer, seeds, fuel and detergents. The Swiss maintain safety stocks in decentralized warehouses for normal nine-months' requirements of the total population. The cost of financing and handling is covered by a special import levy, based on consumption level prices.

In addition, industry is encouraged to form voluntary stockpiles in essential raw materials and intermediates, such as iron and steel, nonferrous metals, cotton, antibiotics, pulp and fuel rods. These stocks can be financed by bank credits at favorable rates, with federal guarantees, up to 90% of purchase value. For a long time, the rule was to value them at prewar price levels (i.e., constant prices). Now this level is usually negotiated, both for federal and cantonal corporation taxes. Strategic inventories are thus treated preferentially. The capital risk rests with the individual firm. The value of these safety stocks, compulsory and voluntary inventories taken together, amounts to 10–12 billion Swiss francs (roughly US$5 billion). This is equivalent to three times the Swiss current expenditure on national defense, showing the importance the Swiss assign to their stockpiling. (For comparison, the US strategic stockpile was valued at US$10.9 billion in 1982.)

Through careful economic management, the social utility of such reserves in goods can potentially be much higher than might be revealed by current consumption and import patterns. If an international crisis creates the monetary reserves of the banking system alone cannot hedge against material distortions in supply. The total safety stocks amount to approximately a quarter of official monetary reserves. The real needs in an emergency may have to be reduced drastically and rapidly to survival levels.

Through planning and preparing basic responses, the Swiss hope to gain valuable time for adjustment, including ensuring their continued autonomy through indigenous production via the Swiss food plan, largely based on the success of Plan Wahlen during World War II. All these emergency preparedness measures are linked and include provisions for fuel, which is critically important to a modernized economy, especially under wartime or siege conditions.

The Swiss have pragmatically solved one of the problems that plague stockpile planning: in an emergency, who gets what? The mandatory stocks are available under special conditions for disposition on government order, but 50% of the voluntary stocks remains with the holder, while the other 50% is subject to emergency allocation by the government. This could serve as a useful example for the USA.

7. United Kingdom

The UK, the founder of the industrial revolution, is without a significant raw materials base and has long been a highly dependent consumer of strategic minerals, this being a major factor in its long colonial history. Confirmation of the British government's intention to set up a strategic minerals stockpile came on February 14, 1983, in the form of a House of Commons statement. It was initially clouded in secrecy, apparently to take advantage of low prevailing prices and avoid disruption of markets, and there are still only speculative press reports as to the scope of the program.

The British divide strategic materials according to criticality, which refers to the role which a specific mineral commodity plays in the country's reliance on imported raw materials. The House of Lords Select Committee Report on Stockpile Minerals (1982) identified four minerals of prime strategic importance: chromium, manganese, phosphate rock and platinum. It also listed eight minerals less important strategically (because of substitutability) but still supplied by sources outside the EEC: antimony, cobalt, molybdenum, nickel, niobium, tantalum, titanium and vanadium. Costs are not known, but one earlier British plan would have involved £100 million in subsidies to private-sector companies.

Focusing primarily on southern African imports, the UK stockpiling scheme is designed to mitigate their short-term disruption by targeting levels equivalent to a three-months' supply. The possibility of disruption of these imports has long been recognized but never seriously acted on, owing to severe budget restrictions. Apparently, given the increasing ability of the USSR to intervene both in source countries and on the sea lanes, past strategies (such as development of strategic minerals within the EEC, aid to Third World countries as in the Lomé Convention's Sysmin scheme and promoting recycling and substitution) have been deemed insufficient to meet real danger.

Because until recently there was no British government-run strategic stockpile or privately operated reserve system, it is unclear why both the government and (to a lesser extent) mineral-intensive industries did not perceive the need to prepare for a war or peacetime crisis. Unlike the French, whose dependence most resembles their own, the British viewed threats to overseas mineral supplies primarily as a foreign-policy problem. As long as the large British mining houses continued exploration and development in the supplier countries, government efforts other than maintaining relatively friendly relations were not deemed necessary.

This view accorded with Prime Minister Thatcher's

effort to cut back on government involvement in the national economy, but the outlook may be changing as Britain's dependence and vulnerability increase. Because the mining houses are mainly living on past investments and not looking to the future, supply interruptions of "strategic minerals" are of mounting concern in the UK.

8. Summary

The shortages and price escalations during the early 1970s, especially the oil exporters' ability to manipulate petroleum prices to unprecedented levels, gave the major impetus to stockpiling policies in the post-World-War-II period. However, the outbreak of minor crises in the Middle East, Africa, the Falklands and Central America, plus instabilities in the domestic situation of many key supplier countries, uncertainties over the future of southern Africa and a propensity on the part of the USSR to be involved in resource issues and markets for strategic and political as well as commercial reasons, will continue to make stockpiling options of interest to policymakers in the key industrial nations.

Acknowledgements

The author is indebted to a translation of an article *Emergency Stockpiling in Sweden* by K.-G. Bartoll of the National Board of Economic Defense; and also thanks to Dr Carl Eugster of Basle, Switzerland, for his unpublished analysis of strategic preparedness and stockpiling in that country. The Publisher would like to thank Professor W. A. Vogely for reviewing this article following the changes in Eastern Europe.

See also: Critical and Strategic Materials; Stockpiling: Economics; Stockpiling in the USA

Bibliography

Bureau des Recherches Géologiques et Minières 1978 *Annual Report.* Bureau des Recherches Géologiques et Minières, Paris

Holt J L, Stanley T W 1983 *Nonfuel Mineral Stockpiling Policies: The Private Stockpiling Alternative*, Contemporary Issues No. 6. International Economic Studies Institute, Washington, DC

House of Lords 1982 *20th Report of the Select Committee on the European Communities, Strategic Minerals*, House of Lords Papers and Bills, Session 1981–82. HMSO, London

International Economic Studies Institute 1977 *Raw Materials and Foreign Policy.* Westview, Boulder, CO

Kaplan J J, Stanley T W 1977 *Dependence and Vulnerability: The United States and Imported Raw Materials*, Contemporary Issues No. 3. International Economic Studies Institute, Washington, DC

Minerals Yearbook, Vol. 2, *Area Reports International.* US Government Printing Office, Washington, DC

Stanley T W 1982 *A National Risk Management Approach to American Raw Materials Vulnerabilities*, Contemporary Issues No. 5. International Economic Studies Institute, Washington, DC

US Congress, Office of Technology Assessment 1985 *Strategic Materials: Technologies to Reduce U.S. Import Vulnerability.* US Government Printing Office, Washington, DC

US Congressional Budget Office 1983 *Strategic and Critical Nonfuel Minerals: Problems and Policy Alternatives.* US Government Printing Office, Washington, DC

US Federal Emergency Management Agency (semiannual) *Stockpile Report to the Congress.* US Government Printing Office, Washington, DC

US General Accounting Office 1982 *Foreign Governments' Stockpile Policies*, Report to the Chairman, Preparedness Subcommittee of Senate Armed Services Committee (GAO/ID83-16). US Government Printing Office, Washington, DC

T. W. Stanley
[International Economic Studies Institute, Washington, DC, USA]

Stockpiling: Economics

This article presents the economic considerations commonly used to evaluate the effectiveness of materials stockpiles held for different public policy purposes. The emphasis is placed on (a) stockpiling to protect against curtailed imports of certain key materials, (b) stockpiling to protect against substantial and sudden increases in the prices of key imported materials even if unaccompanied by disruptions in supply, and (c) stockpiling to stabilize the prices of key raw materials.

The considerations of interest derive from aspects of the microeconomic theory of demand, supply, inventory and price adjustments. Stockpiles have been well analyzed in a wide literature, some general and theoretical, some commodity-related. These are well reviewed with respect to public interests, for example, in *Studies on Economic Stockpiling* (US National Commission on Supplies and Shortages 1976), *Considerations in Choice of Form for Materials for the National Stockpile* (US National Materials Advisory Board 1982) and works such as *Contingency Planning for Materials Resources* (Berman 1977). This focus avoids extensive review of inventory theory and private stockpiling as well as alternatives to stockpiling, among them substitution (Labys 1973, Labys and Lord 1992). Finally, it avoids consideration of hedging and speculative motives, production surges and involuntary inventory buildup.

Successive sections of this article deal with a framework for analysis; public stockpiling; the interaction between public and private stocks; and stockpile decision making.

1. A Framework for Evaluating Public Stockpiling

Economic evaluation of the public interests in materials stockpiling and of its market effects is a complex task. Like all inventories, public stockpiles can be held for three objectives: commodity transactions, precautionary purposes and price stabilization.

The transactions motive arises from the time required for production processes and the need for sufficient inputs and outputs at the various production and sales stations to keep processes running efficiently (US National Commission on Supplies and Shortages 1976 p. 9). The stabilization motive can arise from concerns over inflation or expectations about future prices and market conditions; a large expected price increase (or increase in delivery times) will, other things being equal, induce larger inventories. Finally, the buffer or precautionary motive arises out of the uncertainties about the quantities required at any time. As a result of uncertainty, inventory costs are judged to be worth incurring to secure a buffer against the possibility of unexpected increases in prices, purchases or desired production levels. A common solution to these problems is practising materials requirements planning (MRP) to manage inventories and to control replenishment (see *Materials Requirements Planning*).

1.1 Stockpiling and Supply Restriction

The most frequent requirement for government stockpiles is to protect against a materials supply crisis (see *Critical and Strategic Materials*). While all crises are not the same, public stockpiling is generally seen as reducing the costs of potential damage from the crisis. These costs have been defined by the US National Commission on Supplies and Shortages (1976) to include: (a) losses to users from restricted consumption and the adoption of more costly alternatives; (b) increased resource costs in the production of the material in short supply and its substitutes; and (c) increased payments for the short material. While the stockpile would prevent the price of the material from rising too sharply, the extent of the price rise depends on the cost to users of restricting their consumption, on user expectation as to future supplies and prices, and on the responsiveness of alternative suppliers to a price increase.

In addition, a government could incur costs above those of a private firm. These could arise where the timing and sources of stock purchases as well as of stock sales are not efficiently arranged. Costs could also be incurred because of the impact of the uncertainty involved in the reaction of private firms to the government's decisions.

A final consideration is whether the benefits and costs of materials stockpiling can be maximized from the point of view of social optimality. In the context of a possible supply crisis, the most socially efficient level of stocks is that which when properly released and allocated results in the largest reduction of expected crisis costs, net of stockpiling costs.

One measure of social optimality frequently employed is the level of economic welfare measured as the sum of consumers' and producers' surplus (Burrows and Beggs 1977 pp. 116–21). The conventional approach for explaining how stockpiling adjustments affects consumers' and producers' surplus is to employ a set of material supply and demand schedules, such as that shown in Fig. 1. Domestic consumption is shown to be much greater than domestic production, requiring that imports be made to achieve domestic market equilibrium.

1.2 Public Stockpiling and Supply Restriction

An example of a supply crisis or market disruption occurring abroad can be illustrated through its impact on world prices. The latter are shown to increase severalfold from their normal level to the illustrated disrupted level. This increase may lessen or become more severe as the disruption continues, depending on a number of factors. These include estimates of the likelihood of the restriction, its potential severity and price impacts, the response of prices during the restriction to stockpile releases, and the costs of accumulating, holding and releasing various amounts of stocks over the relevant time period.

Domestic demand and supply are assumed to be fairly unresponsive to price in the short run, as illustrated by the more nearly vertical short-run curves in Fig. 1. Thus, at the beginning of the crisis imports drop only from the normal quantity indicated to the smaller quantity shown as applying during disruption in the short run. In the long run, the short-run domestic demand and supply curves move along their respective long-run curves. Imports then decrease further to the quantity shown as applying during disruption in the long run.

The economic cost to the domestic economy of the foreign supply disruption can be illustrated geometrically by defining the various areas between the disrupted market price and the normal market price on

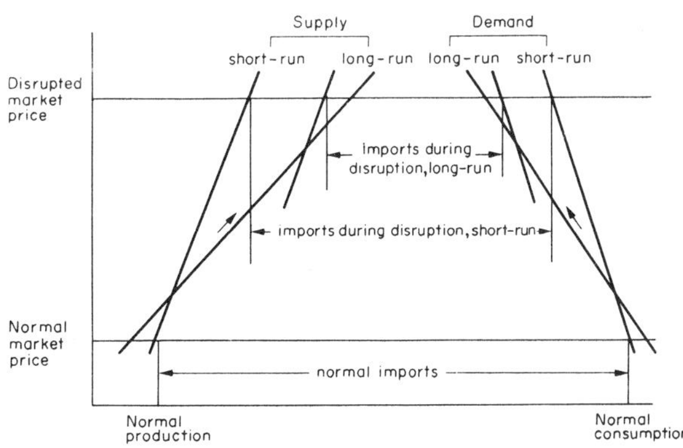

Figure 1
Schematic representation of the supply crisis market (adapted from Burrows and Beggs 1977)

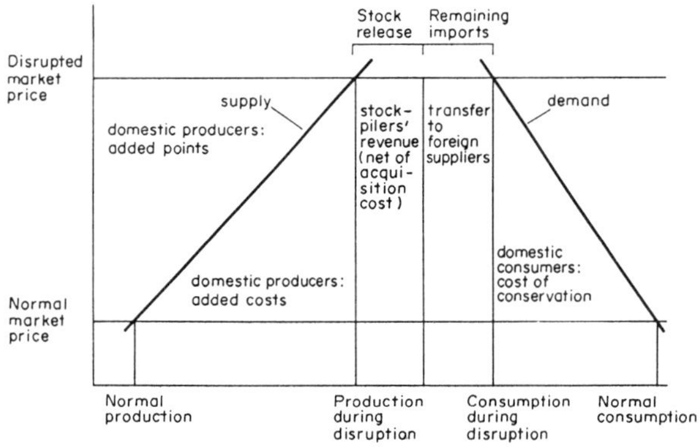

Figure 2
Supply crisis impact (adapted from Burrows and Beggs 1977)

the vertical axis, as shown in Fig. 2, representing monetary or dollar values. In considering the cost impact on consumers, the cost of conservation to domestic materials users is represented by the triangle under the demand curve on the right-hand side of Fig. 2. This triangle conceptually includes the cost of domestic industry utilizing substitute materials and the cost to ultimate users of switching to other finished goods. Other input prices are assumed to be constant. The entire rectangle to its left represents the extra payment made by domestic consumers for their remaining consumption during the disruption. The parties gaining at the expense of domestic users are indicated by the subdivisions of this rectangle, which will be described in the following, from left to right.

Turning to the cost impact on producers, the latter increase their production in response to the higher price during the market disruption. Part of the additional revenue they earn is expended for added costs of production, represented by the triangle under the supply curve, but the remainder is added profits. It is generally agreed that it is desirable for domestic suppliers to produce at the high point on their supply curve which corresponds to the disrupted world market price. It is very difficult to design practical policies which reduce domestic producers' windfall profits but leave unimpaired the high price incentive inducing this desirable expansion of production. Here, it is assumed that the price system is allowed to operate unimpeded during disruption. If a government policy, such as a dual pricing system for domestic crude oil, is imposed, then the magnitudes of losses in various categories will change. In the conventional treatment of losses inflicted in the domestic economy by an increase in the price of imports, the transfer of income from domestic consumers to domestic producers is a neutral consideration, so the net loss to a nation is given by the area bounded by the supply and demand curves and by the normal and disrupted market price. However, different evaluation of shifts in the distribution of income can be readily incorporated in the analysis. Broader measures of macroeconomic impact can also be evaluated (Davis and Cherniack 1987).

Domestic consumption which is not satisfied by domestic production must be met from stock releases or by imports from foreign suppliers. Stockpilers' revenues, net of acquisition costs, are represented by the small rectangle which has been labelled in the middle of Fig. 2. Acquisition costs, assumed to be at the normal market price, are represented by the smaller rectangle just below "stockpilers' revenue (net of acquisition cost);" gross stockpilers' revenue is the "stock release" times the "disrupted market price." It is convenient here to speak of stockpilers as though they were separate economic entities with their own accounting records, though in practice domestic consuming and producing firms often do most of the stockpiling. Out of their revenues, stockpilers must pay the interest charges in their acquisition costs and the costs for other resources required to maintain the stockpile (e.g., land, buildings, security and supervision). Averaging out years in which prices are normal with those in which releases can be made at high prices, stockpilers may make either an overall profit or an overall loss.

The "remaining imports" indicated in Fig. 2 must be purchased during the disruption at a much higher market price than during a normal period. The "transfer to foreign suppliers" of revenues beyond what this quantity would cost in normal times is the last of the major categories of economic losses that are indicated in the figure.

The situation on the demand side of the domestic market is symmetrical with that just described on the supply side; it is thus not necessary to repeat the above argument. In brief, when domestic consumers maintain their capability to use substitutes or to forego consumption altogether when prices are high, they incur additional costs during periods of normal prices which can be measured by the area under the demand curve.

Another type of economic loss not represented in Fig. 2 is indirectly suggested by the diagonal arrows in Fig. 1. This loss reflects the fact that markets adjust dynamically as well as statically. These arrows indicate movement in the short-run domestic supply and demand curves during disruptions. If capacity is expanded very gradually, capital can be acquired and installed at normal costs which are captured by the long-run supply and demand curves. Once a serious disruption begins, it is usually efficient to expand capacity rapidly even if costs incurred are greater than normal.

2. *Public Stockpiling*

Public stockpiling has often been viewed as the only viable approach for meeting a supply crisis. The rationale for this approach appears in a number of studies reviewed by Labys (1980). However, most of these studies deal with international buffer stocks which serve more to stabilize international materials prices than to provide contingency stockpiles to meet domestic supply crises (see *Prices of Materials: Stabilization*). This section will concentrate more appropriately on the rationale for the latter.

A major goal in establishing a contingency stockpile should be to achieve a socially optimal level of stocks that would maximize the sum of consumers' and producers' surplus expected from the acquisition and sale of stock. This concept is strictly an economic one that omits other benefits and costs of stockpiling which are difficult to measure. For example, it excludes any income distribution or transfer effects as well as noneconomic benefits such as greater freedom in political action on an international scale. There are a number of reasons private stockpiling may tend to diverge from the market equilibrium level which is analogous to the socially optimal level. First, if there are tariff or trade barriers, then the amount private firms must pay to stockpile may exceed the social cost of these stocks (i.e., since the tariff causes the private costs to differ from the social costs, the private stockpile is likely to differ from the socially optimal amount). Second, the presence of public stockpiles might act as a deterrent to private stockpiles. This results mainly from the likelihood that existing stockpiles might cause the expected future price to be less than it might otherwise be, largely because such stocks would prevent any agent from attempting to control or to maintain prices. Third, firms may feel that they would not be allowed to profit from their stockpiles should a supply crisis occur.

Thus, it would appear that an argument could be made for public stockpiling which might move material stockpiles closer to a socially optimal level than what could be held by private firms alone.

3. Interactions Between Public and Private Stockpiling

Given a raw material market that is perfectly competitive and government interaction that is perceived as no different from private interaction, an increase in government stocks of one unit is likely to cause private stocks to decrease by the same amount. This surplus proposition implies a stock trade-off that is essentially one for one. It also assumes that the optimal social level of stocks is the same as the amount private firms want to stockpile. Where the former is greater than the latter, the government may have to stockpile all the amount required.

There are several other underlying factors that could also affect the proposition of a one-for-one trade-off. To begin with, if the government is expected to impose price controls on expected profit gains during a crisis, then the private sector will anticipate lower profits from stock sales and will reduce the amounts held. Where imperfect information exists regarding stockpiles held by other firms, then the level of stocks held would be higher than the equilibrium level market. Deviations from competition in the form of a monopolist supplier can also affect stocking. In particular, a monopolist supplier can influence prices, and hence profits, even when government stockpiling occurs such that his stocks will not be decreased by one unit when the government increases its stockpiles by one unit. Finally, when the government stockpile is recognized as a deterrent against future supply crises, then private firms will perceive lower expected profits from stock sales and hence will decrease their purchases by more than one unit for a full unit government purchase.

This exposition is no more than one simple theoretical interpretation of stockpile interaction behavior. A more comprehensive analysis would involve considerations of varied and more complex modelling. Some additional factors that would receive consideration include the actual nature of government stockpile release rules as well as of stockpile and market conditions. Concerning the first of these, three kinds of release rules would appear to be most likely. First, government stocks would be released only when the price of the commodity threatens to rise above a certain prespecified limit or ceiling. Such a role is applicable to several different forms of economic stockpiles, as mentioned earlier. Second, government stocks would be released only when the government declares a state of economic emergency (i.e, independent of prices). Third, government stocks would be released only when a specific political action occurs (e.g., an embargo by foreign producers). In this last case, private firms are likely to have greater uncertainty about stockpile releases than in the case of economic emergencies.

Other factors that add to the complexity of the modelling experiment have been analyzed by Labys (1980). Most important are the price elasticities of supply and demand adopted. These elasticities not only vary over periods of structural change but their impact will vary, depending on whether linear compared to nonlinear supply and demand relationships are assumed. Outcomes will also vary with the actual stockpile size as well as the time period of analysis. Expectations and risk response are also crucial in predicting the current price response (Newbery and Stiglitz 1981). Definitions of consumers' and producers' surplus must also be formulated to accurately measure the optimal social value of the stockpile. Where stockpiles for several materials might be employed simultaneously, it cannot immediately be assumed that sales from one stockpile could be used to purchase additions for a second stockpile (Labys 1977).

4. Stockpile Decision Making

This analysis of economic stockpiling has revealed several results which can be of further use for decision making in this area. First, private firms accumulate stocks which under certain circumstances will satisfy economic efficiency measured in terms of benefits and costs. Second, if the resulting level of stocks is not the same as the socially optimal level of stocks needed to satisfy a materials supply crisis, then the government must also accumulate stockpiles. Since public stockpiling will have an effect on private stockpiling behavior, a careful analysis should be made of the resulting economic efficiency. Third, if public stockpiling is undertaken, its performance will be difficult to evaluate even when overall social welfare is measured in terms of producers' and consumers' surplus. A number of underlying assumptions and related factors must be considered, including a range of foreign actions, each of which must be weighted by the probability or its likelihood of occurrence. A possible domestic factor is the actual form in which the material is held in the stockpile. This form should be as permanent as possible so as to avoid technological obsolescence (US National Materials Advisory Board 1982).

Finally, the analysis of the performance of stockpiling is obviously complex. Only by careful consideration of the many factors involved and using varied theoretical applications can effective decision making take place.

See also: Materials and Innovation; Policy: USA; Stockpiling; Stockpiling in the USA; Substitution of Materials: Economics; Substitution of Materials: Technology; Trade in Materials, International

Bibliography

Berman E G (ed.) 1977 *Contingency Planning for Materials Resources.* National Science Foundation, Washington, DC

Burrows J C, Beggs S D 1977 Policy implications of producer country supply restrictions: approach and conclusions. In: Berman 1977, pp. 105–30

Davis H C, Cherniack H 1987 Interindustry approaches to the analysis of a supply disruption of a critical resource. *Resour. Policy* 3: 47–54

Labys W C 1973 *Dynamic Commodity Models: Specification, Estimation and Simulation.* Heath-Lexington, Lexington, MA

Labys W C 1977 Optimal portfolio analysis of multicommodity stocking arrangements. *Oxf. Bull. Econ. Stat.* 39: 219–29

Labys W C 1980 Commodity price stabilization models: a review and appraisal. *J. Policy Model.* 2: 121–36

Labys W C, Lord M 1992 Inventory and equilibrium adjustments in international commodity markets: a multi-cointegration approach. *Appl. Econ.* 24: 77–84

Newbery D M G, Stiglitz J E 1981 *The Theory of Commodity Price Stabilization.* Clarendon, Oxford

US National Commission on Supplies and Shortages 1976 *Studies on Economic Stockpiling.* US Government Printing Office, Washington, DC

US National Materials Advisory Board 1982 *Considerations in Choice of Form for Materials for the National Stockpile.* US National Academy Press, Washington, DC

W. C. Labys
[West Virginia University, Morgantown, West Virginia, USA]

Stockpiling in the USA

Stockpiling food in times of plenty for use in future times of need antedates humanoid origins. Formal guidance for food stockpiling was set forth by the highest authorities in the "2-year stockpile rule," as divined by Joseph four millenia ago: one-fifth of the produce of the land set aside in each of the seven good years (Genesis Chap. 41). In later centuries stockpiles not only of food but also weapons, including the ingredients for chemical warfare, were accumulated in secure places. *Histoire d'une Forteresse*, written shortly after the Franco-Prussian War by the noted architect and military engineer Colonel E Viollet-le-Duc, pointed out the elaborate precautions taken in earlier times to have stocks of food, water and weapons on hand in major fortified areas, to which the peasantry in time of war could retreat to be armed to repel invaders.

Isaiah prophesied the beating of swords and spears into plowshares, but the reverse process often also took place. After a major war, bronze cannon were melted down and recast as heavy fences. Later such bronze objects were remelted for cannon again, thus serving useful purposes in any season of man's affairs. In World War I many materials and fuels were in short supply. A variety of controls and coordinated allied efforts were needed to channel supplies to priority programs. Supplies of materials were expanded and new industries created. For example, synthetic nitrates were produced for explosives, forming the basis of the peacetime nitrate fertilizer industry.

Again, in World War II, materials and fuels were in short supply. The Axis nations accumulated supplies of strategic materials before the war, which made up in part for their natural resource shortages. In his book *Wartime Economy of the USSR*, Voznesensky, Deputy Premier and Chief of the State Planning Commission during World War II, stated that a program for the accumulation of state and mobilization reserves—especially oil products, ferrous metals and foodstuffs—was planned and implemented before the start of World War II under direct instructions from Stalin, and that such stockpiles played a significant role in the Soviet war effort. He pointed out that in modern warfare victory depends upon reserves of production capacity, manpower, raw materials and other commodities.

1. Strategic Considerations

Because liquid fuels are among the most essential supplies, not only as fuels *per se* but also as basic raw materials for a variety of plastics and other hard materials, the US President, on 26 April, 1989, expressed "strong concerns about this country becoming further dependent on foreign oil." In 1990 imports supplied 45% of US demand and in 1990 the invasion of Kuwait by Iraq led to the 1991 Middle East war, fought in part over free-world concern about continuity of petroleum supplies and control of petroleum reserves. US net import reliance of selected nonfuel minerals and metals is shown in Table 1. Western European nations and Japan are even more dependent on imports. In contrast, the Comecon bloc has been largely self sufficient—the result of a long-pursued policy of autarky on the part of the USSR, which, with an area of 22×10^6 km^2, is ideally situated for such a policy. Further, not just import dependence but also many other factors must be considered. Materials on hand in strategic stockpiles, other government stocks and industry stocks are often significant. The use of flywheels to deliver power to offset sudden increases in load is accepted practice in machinery design, and the use of surge bins is standard practice in materials handling. Expanding such analogies to national economies points to the desirability of maintaining larger than normal working inventories to deal with unexpected interferences with normal supplies which may be occasioned by wars, politicoeconomic disruptions or such natural disasters as earthquakes, landslides, hurricanes and floods.

Escalating costs and higher interest rates in recent years have tended to discourage the private sector from maintaining large inventories. Tax incentives for the maintenance of larger than normal working inventories were in effect in Sweden for a number of years. In the Swedish system a large fraction of the acquisition costs of raw materials could be written off in computing taxable income for the year in which materials were acquired. Such incentives permit informed judgments by numerous buyers and sellers, and lessen the need for direct market intervention by government entities. Switzerland, too, is among countries that provided some tax incentives to encourage holding of inventories.

Several Western European nations have considered stockpiling imported materials, but the costs have been a major deterrent to most. In Japan, the Ministry of International Trade and Industry (MITI) sometimes helps to finance major metal stockpiles, but these are more in the nature of assistance to domestic smelters to carry stocks in periods of economic downturn rather than strategic stockpiles *per se*. Stockpiles in nations with centrally planned economies are considered state secrets. As a consequence of the dislocation of oil markets in the 1970s, many nations adopted a variety of programs to increase oil stocks.

2. Early Stockpiles in the USA

The USA was early to recognize the importance of stockpiles in the form of known mineral deposits in the ground. Commodore Perry's Far East naval expedition of 1852–1854 had specific instructions to look for deposits of high-quality steam coal in the far Pacific. In later years, the Army and Navy were granted preference purchase rights to coal in Alaskan deposits and before World War I the USA established naval petroleum reserves in the ground in California and Wyoming, and naval oil shale reserves in Colorado and Utah.

As early as 1921 the US War and Navy Departments were interested in a program for the stockpiling of strategic and critical materials. During the period 1933–1936 these departments and the National Resources Board made recommendations concerning defense stockpiles, but it was not until the Naval Appropriation Act of 1938 that US$3.5 million was made available for stockpiling, followed by US$0.5 million in fiscal year 1939 and in 1940. With these funds the Navy purchased small stocks of tin, ferromanganese, tungsten ore, chromite, optical glass and manila fiber. In late 1938 and early 1939 the Army and Navy Munitions Board, aided by government departments, made studies leading to Public Law 76-117, June 7, 1939—the first legislation for the primary purpose of establishing stockpiles of strategic and critical materials for national defense, and also authorizing exploration and development of domestic resources by the Department of the Interior. Prewar appropriations under this Act were US$10 million in August 1939, US$12.5 million in March 1940 and US$47.5 million in June 1940. Also in August 1939 the Commodity Credit Corporation of the Department of Agriculture was authorized to exchange surplus US agricultural products for foreign strategic and critical materials.

However, by mid-1940 it became evident that the 1939 Stock Piling Act, designed primarily for acquisitions in a period of relative peace, was inadequate. Instead, stimulation of the flow of domestic and foreign materials directly into munitions and essential industrial production was required. Accordingly, on June 25, 1940, the Reconstruction Finance Corporation (RFC, which dated from 1932) was given broad powers and vastly greater financial authority. The RFC then proceeded to charter the Metals Reserve Company, the Defense Plant Corporation and the Defense Supplies Corporation to acquire, store, produce and sell strategic and critical materials. Stockpiles were totally inadequate at the start of US active participation in World War II. Some of the larger holdings were: 290 000 t of chromite, 815 752 carats of industrial diamonds, 523 900 t of manganese ore, 818 500 kg of quartz crystals, 50 842 t of tin, 6574 t of tungsten ore and 80 466 t of zinc concentrates.

During World War II, shortages of materials required major priorities and allocations programs,

Table 1
US net import reliance for selected nonfuel mineral materials

	Percentage imported 1990	Major foreign sources 1986–1989
Arsenic	100	France, Sweden, Chile, Mexico
Cesium (pollucite)	100	Canada
Columbium (niobium)	100	Brazil, Canada, Germany, Thailand
Graphite	100	Mexico, China, Brazil, Madagascar
Manganese	100	Republic of South Africa, France, Gabon
Mica (sheet)	100	India, Belgium, France, Brazil
Rubidium	100	Canada
Strontium (celestite)	100	Mexico, Germany, Spain
Thallium	100	Belgium, UK, Germany, France
Bauxite and alumina	98	Australia, Guinea, Jamaica, Surinam
Gem stones (natural and synthetic)	98	Belgium, Israel, India, Republic of South Africa
Diamond (industrial stones)	92	Ireland, UK, Republic of South Africa, Zaire
Asbestos	90	Canada, Republic of South Africa
Fluorspar	90	Mexico, Republic of South Africa, China, Spain
Platinum-group metals	88	Republic of South Africa, UK, USSR
Tantalum	86	Germany, Thailand, Brazil, Australia
Cobalt	85	Zaire, Zambia, Canada, Norway
Nickel	83	Canada, Norway, Australia, Dominican Republic
Chromium	79	Republic of South Africa, Turkey, Zimbabwe, Yugoslavia
Tin	76	Brazil, China, Indonesia, Malaysia
Tungsten	73	China, Bolivia, Germany, Peru
Stone (dimension)	70	Italy, Spain, Canada, Taiwan
Barite	69	China, India, Mexico, Morocco
Potash	68	Canada, Israel, USSR, Germany
Antimony	64	China, Republic of South Africa, Mexico, Hong Kong
Cadmium	54	Canada, Mexico, Australia, Germany
Selenium	54	Canada, UK, Japan, Belgium–Luxembourg
Peat	47	Canada
Pumice and pumicite	40	Greece, Mexico, Turkey, Ecuador
Zinc	37	Canada, Mexico, Spain, Peru
Iodine	35	Japan, Chile
Gypsum	30	Canada, Mexico, Spain
Silicon	30	Brazil, Canada, Venezuela, Norway
Iron ore	26	Canada, Brazil, Venezuela, Liberia
Quartz crystal (industrial)	17	Brazil, Namibia
Nitrogen	14	Canada, USSR, Trinidad & Tobago, Mexico
Cement	13	Mexico, Canada, Spain, Greece
Sodium sulfate	13	Canada, Mexico
Iron and steel	12	EC, Japan, Canada, Republic of Korea
Sulfur	11	Mexico, Canada
Salt	10	Canada, Mexico, Bahamas, Chile
Vermiculite	10	Republic of South Africa, China
Magnesium compounds	9	Greece, China, Canada, Ireland
Mica (scrap and flake)	8	Canada, India
Copper	5	Canada, Chile, Mexico, Peru, Zaire
Lead	4	Canada, Mexico, Australia, Peru
The following also have significant import dependency, but data are withheld		
Andalusite		Republic of South Africa
Bismuth		Belgium, Mexico, Peru, UK
Ilmenite		Australia, Canada, Republic of South Africa
Mercury		China, Spain, Turkey, Japan
Pyrophyllite (wonder stone)		Republic of South Africa
Rhenium		Chile, Germany
Rutile		Australia, Republic of South Africa, Sierra Leone
Tellurium		UK, Canada, Japan, Philippines
Vanadium		Republic of South Africa, EC, South America, Mexico
Zirconium		Australia, Republic of South Africa

Source: BuMines

conservation and use limitation directives, domestic and foreign supply expansion programs and allocations of ships to priority cargoes. Mindful of the problems encountered in the war, the Army and Navy Munitions Board in March 1944 adopted the following definition: "Strategic and critical materials are those materials required for essential uses in a war emergency, the procurement of which, in adequate quantities, quality and time, is sufficiently uncertain for any reason to require prior provision for the supply thereof." Also during World War II, extensive cooperative measures with allied nations were undertaken to expand supplies in strategically accessible nations and to move them to high-priority production programs.

3. *Post-World-War-II Planning*

After the end of World War II, increased attention was paid to defense planning and industrial mobilization readiness. In 1946 the US Congress reviewed and strengthened the Stock Piling Act and materials were acquired through direct purchases, transfers of war surpluses, procurement through use of foreign counterpart funds generated by US foreign aid programs and barter of US surplus agricultural commodities. The legislative intent of the 1946 Stock Piling Act was much broader than just acquiring strategic and critical materials for a national stockpile. A second major purpose of the Act was to encourage the conservation and development of sources of these materials within the USA.

In 1947 the National Security Act established the National Security Council (NSC), the Central Intelligence Agency (CIA) and the National Security Resources Board (NSRB). The NSRB was charged specifically "to advise the President concerning the coordination of military, industrial, and civilian mobilization, including ... programs for the effective use in time of war of the Nation's natural and industrial resources ...; policies for unifying, in time of war, the activities of Federal agencies and departments engaged in or concerned with ... materials; the relationship between potential supplies of, and potential requirements, for, ... resources ...; (and) policies for establishing adequate reserves of strategic and critical material, and for the conservation of these reserves ...." The importance of appropriate planning with US allies led in 1949 to the formation of the Joint United States–Canada Industrial Mobilization Committee to coordinate plans for industrial mobilization "in order that the most effective use may be made of the productive facilities of the two countries." These agreements have been extended over the years resulting in the US/Canada North American Defense Industrial Base Organization.

4. *Korean War Activities*

The start of the Korean War on 25 June, 1950, found US strategic and critical materials stockpiles valued at only US$1.6 billion, compared with stockpile goals valued at US$4.1 billion. It soon became apparent that the government needed a much broader range of authorities to deal with the emergency. Consequently, Title I of the Defense Production (DP) Act of 1950 provided specific authority for priorities and allocations and Title III provided broad authority for expanding supplies of materials, making specific provisions for:

(a) loans or guarantees of loans for exploration, development and mining of strategic and critical metals and minerals;

(b) purchases of or commitments to purchase metals, minerals and other materials, for government use or resale;

(c) the encouragement of exploration, development and mining of strategic and critical minerals and metals; and

(d) the development of substitutes for strategic and critical materials.

Coordination with allied nations took place through the International Materials Conference, which, through a number of committees, stimulated needed production and assisted in implementing allocation programs.

The Defense Production Act of 1950, as amended, was long the basis for defense mobilization efforts, but it lapsed on 20 October, 1990. However, the Selective Service Act of 1967 continued to give the US President authority to require priority treatment of defense orders. The creation or enlargement of domestic productive capacity, if it can be done at reasonable cost, should in most cases provide greater flexibility than stockpiles. In fact, 1 t of domestic productive capacity is considered equal to 3 t of stockpiled material under current stockpile planning. DP Act supply expansion programs as of 30 June, 1956, reached US$8.4 billion in gross transactions with probable ultimate net cost of US$0.9 billion. Major materials programs in US$ billion included:

aluminum	1.5	tin	0.2
titanium	0.9	molybdenum	0.2
rubber	0.9	magnesium	0.1
nickel	0.9	cobalt	0.1
copper	0.7	niobium–tantalum	0.1
manganese	0.4	other materials	0.4
tungsten	0.4	(more than 50)	

In a few years these programs doubled US aluminum production, increased US copper mine capacity by a quarter, initiated US nickel mining, created the titanium industry, quadrupled US tungsten mining and greatly expanded the world niobium–tantalum mining and processing industries as well as expanding supplies of many other materials for production needs and stockpiles. The latest major nonfuel use of the DP Act was a US$83 million loan in 1967 during the Vietnam War period to facilitate the Duval Sierrita copper mine development.

Also in the Korean War period, accelerated amortization ("rapid writeoffs" of capital expenditure), authorized by Sect. 168 of the Internal Revenue Code, was a powerful device to encourage capital investment in areas of national concern. This incentive alone was sufficient to add nearly one quarter to US steel capacity. In the period from October 1950 through December 1956, industrial expansion valued at US$37 billion was certified, of which 61%, or US$22 billion, was eligible for rapid amortization. The total amounts in US$ billion then certified in materials-related industries were as follows:

mining	2.2
primary metal industries	5.7
chemical and allied products	2.9
products of petroleum and coal	3.0
pipeline transportation	1.4

Use of the US tax laws to encourage domestic production of strategic minerals is not limited to periods of war or national emergency. As a general rule, minerals considered more strategic are allowed higher percentage depletion rates. Exploration for minerals is also specifically encouraged by the tax laws.

5. Post-Korean-War Activities

In 1954 the Congress created a "supplemental stockpile" of strategic materials to be acquired by barter of surplus agricultural commodities. Several billion US dollars of materials were acquired by this route and were later merged into the total government stockpiles.

Under the 1946 Stock Piling Act, administration policy in the period from 1946 to 1958 based stockpile planning on a five-year emergency, from 1958 to 1973 on a three-year emergency, from 1973 to 1976 on a one-year emergency and from 1976 to the present again on a three-year emergency, following a major interagency review. The Department of Defense currently oversees stockpile management while its Defense Logistics Agency buys, sells, rotates and stores stockpile materials.

In 1979 the new Strategic and Critical Materials Stock Piling Revision Act of 1979, Public Law 96-41, reaffirmed the need for stockpiling, conservation and development of domestic sources. The Stock Piling Act states "... that the natural resources of the United States in certain strategic and critical materials are deficient or insufficiently developed to supply the military, industrial and essential civilian needs of the United States for national defense ..." and that the purpose of the Act is "... to provide for the acquisition and retention of stocks of certain strategic and critical materials and to encourage the conservation and development of sources of such materials within the United States and thereby to decrease and to preclude, when possible, a dangerous and costly dependence by the United States upon foreign sources for supplies of such materials in times of national emergency." The Stock Piling Act further provides that "the purpose of the stockpile is to serve the interest of national defense only ... (and) is not to be used for economic or budgetary purposes" and that "the quantities of materials to be stockpiled under this Act should be sufficient to sustain the United States for a period of not less than three years during a national emergency that would necessitate total mobilization of the economy of the United States for a sustained conventional global war of indefinite duration."

The Act provided definitions as follows: "For the purposes of this Act: (1) the term 'strategic and critical materials' means materials that (A) would be needed to supply the military, industrial, and essential civilian needs of the United States during a national emergency, and (B) are not found or produced in the United States in sufficient quantities to meet such need. (2) The term 'national emergency' means a general declaration of emergency with respect to the national defense made by the President or by the Congress." Section 8 of the new Act also reaffirmed the injunction to develop domestic resources, assigning that responsibility directly to the President (a trend in more modern legislation), who then by Executive Orders 12155 and later 12626 reassigned responsibility for minerals to the Secretary of the Interior and for agricultural materials to the Secretary of Agriculture. Additional reenforcing policy statements dealing with materials and the national security are provided by the National Critical Materials Act of 1984 (98 Stat. 1248; 30 USC 1801), the National Materials and Minerals Policy, Research and Development Act of 1980 (94 Stat. 2305; 30 USC 1601 ff.), the Mining and Minerals Policy Act of 1970 (84 Stat. 1876; 30 USC 21a) and many other more narrowly focused laws.

6. Planning in the 1980s

On April 5, 1982, the President transmitted to the Congress his "National Materials and Minerals Program Plan and Report to the Congress." His letter of transmittal stated "This national minerals policy recognizes: the critical role of minerals to our economy, national defense, and standard of living; the vast, unknown and untapped mineral wealth of America and the need to keep the public's land open to appropriate mineral exploration and development; the critical role of government in alerting the Nation to minerals issues and in ensuring that national decision-makers take into account the impact of their decisions on minerals policy; and the need for long-term, high potential pay-off research activity of wide generic application to improve and augment domestically available materials," and "This policy is responsive to America's need for measures to diminish minerals vulnerability by allowing private enterprise to preserve and expand our minerals and materials economy."

The report stated "It is the policy of this Administration to decrease America's minerals vulnerability by taking positive action that will promote our national security, help ensure a healthy and vigorous economy, create American jobs, and protect America's natural resources and environment." It enumerated several major areas of action including:

(a) increased availability of lands for exploration, development and multiple use;

(b) reform of excessively burdensome or unnecessary regulations and statutes;

(c) increased stockpiling, including use of exchanges and barter and ensuring the quality of the stockpile;

(d) encouraging private sector materials research and development, stimulating coordination between industry and government in technical innovation and focusing government-financed research and development on long-term, high-risk technology with the best chance for wide generic application to materials problems; and

(e) improved minerals and materials information collection, analysis and dissemination.

On July 22, 1982, the President issued National Security Decision Directive 47 which states "... It is the policy of the United States to have an emergency mobilization preparedness capability that will ensure that government at all levels, in partnership with the private sector and the American people, can respond decisively and effectively to any major national emergency with defense of the United States as the first priority."

The section on industrial mobilization of NSDD 47 states "It is the policy of the United States to have a capability to mobilize industry in order to achieve timely and sufficient production of military and essential civilian material needed to prosecute successfully a major military conflict, to lend credibility to national strategic policy, and to respond to national security emergencies." The industrial mobilization program will:

"improve the capability of United States industry to meet current and mobilization requirements by identifying production and supply deficiencies and initiating actions to overcome them;

"increase the capability of industry and infrastructure systems, including transportation and energy, to meet national security needs through use of improved guidance on resource-claimancy, continued use of import and export controls, and appropriate use of Defense Production Act authorities in cases where the free market cannot be reasonably expected to provide the required national security capability in a timely manner;

"provide for assessment of the impact on the industrial base resulting from existing and proposed agreements for coproduction of defense material, related offset arrangements with our allies, and other reciprocal trade agreements; and

"ensure the availability of strategic and critical materials by primary reliance on the National Defense Stockpile; the President may authorize the use of DPA Title III in those instances where the free market cannot be reasonably expected to provide the required national security capability in a timely manner."

7. Stockpile Status in the USA 1990

In 1983, the US President ordered a major review of strategic stockpiling led by the National Security Council (NSC) and involving a dozen major federal agencies. The NSC analyses included assessment of US military, industrial and essential civilian needs as well as foreign demand, trade patterns and shipping losses in a three-year conventional global military conflict—commonly considered to require the largest supplies of materials. After two years of interagency study and review, the US President, on July 8, 1985, proposed significant modernization, concluding that significantly lower goals were appropriate but that an additional supplemental reserve of materials already owned would be held to provide assurance against shortages during a period of military conflict. The proposed new goals were valued at about US$0.7 billion and the supplemental reserve at US$6.0 billion, compared to previously existing goals valued at US$16.6 billion and total inventories valued at US$10 billion. The US Congress resisted such changes and the US President's proposals were not adopted. However, in more recent years, some goal changes have been authorized by Congress. The US President also emphasized the need to determine requirements for new high-technology materials for future weapons systems. These could be items already on the stockpile list or new candidates for some form of government program not necessarily limited to stockpiling. Currently under Title III of the DP Act the Department of Defense is actively implementing several such projects.

As listed in Table 2, 98 materials are designated by the US Government as "basic stockpile materials," largely on the basis of import vulnerability, but it would be fallacious to conclude that only these are strategic. For example, neither steel nor iron ore, coke, coking coal and limestone are on the stockpile list, yet steel is considered a highly strategic material in every modern economy. Many materials have been recognized as stockpile candidates ever since World War I, while newer ones have been added to the list only after research and development have proved their utility, such as titanium. Of the materials listed, 84 are of

Table 2
Department of Defense stockpile goals and inventories (as of 30 June, 1991)

	Goal		Inventory
Aluminum metal group			
alumina	0		0
aluminum	700000	ST	2082
bauxite, metal grade, Jamaica type	21000000	LDT	12457740
bauxite, metal grade, Surinam type	6100000	LDT	5299597
Aluminum oxide, abrasive grain group			
aluminum oxide, abrasive grain	374000	ST	50904
aluminum oxide, fused, crude	0	ST	249867
bauxite, abrasive grade	0	LCT	0
Antimony	88500	ST	36011
Asbestos, amosite	0	ST	34006
Asbestos, chrysotile	3000	ST	10703
Bauxite, refractory	1240000	LCT	276067
Beryllium metal group			
beryl ore 11% BeO	18000	ST	17729
beryllium copper master alloy	7900	ST	7387
beryllium metal	400	ST	320
Bismuth	1060000	lb	1888655
Cadmium	11700000	lb	6328570
Chromium, chemical and metallurgical group			
chromite, chemical grade ore	675000	SDT	242414
chromite, metallurgical grade ore	3200000	SDT	1582133
chromite, ferro, high carbon	185000	ST	692927
chromite, ferro, low carbon	75000	ST	317168
chromite, ferro, silicon	90000	ST	58357
chromium, metal	20000	ST	3763
Chromite, refractory grade ore	695000	SDT	391414
Cobalt	85400000	lb Co	52778142
Columbium group			
columbium carbide powder	100000	lb Cb	21372
columbium concentrates	13270000	lb Cb	2019217
columbium, ferro	0	lb Cb	930911
columbium, metal	0	lb Cb	44851
Copper	1000000	ST	29047
Cordage fibers, abaca	155000000	lb	0
Cordage fibers, sisal	60000000	lb	0
Diamond, industrial group			
diamond dies, small	60000	PC	25473
diamond, industrial, crushing bort	0	KT	14820961
diamond, industrial, stones	7700000	KT	7777225
Fluorspar, acid grade	900000	SDT	892856
Fluorspar, metallurgical grade	310000	SDT	410822
Germanium	146000	kg	65072
Graphite, natural—Ceylon, amorphous lump	6300	ST	5492
Graphite, natural—Malagasy, crystalline	14200	ST	17266
Graphite, natural—other than above	1930	ST	2803
Indium	1350000[a]	TrOz	0
Iodine	5800000	lb	6055906
Jewel bearings	120000000	PC	78748518
Lead	1100000	ST	601053
Manganese dioxide, battery grade group			
manganese, battery grade, natural ore	25000	SDT	202936
manganese, battery grade, synthetic dioxide	25000	SDT	3011
Manganese, chemical and metallurgical group			
manganese ore, chemical grade	170000	SDT	172744
manganese ore, metallurgical grade	2700000	SDT	2702878
manganese, ferro, high carbon	439000	ST	868540
manganese, ferro, low carbon	0	ST	0
manganese, ferro, medium carbon	0	ST	29072
manganese, ferro, silicon	0	ST	23574

***Table 2**—continued*

	Goal		Inventory
manganese, metal, electrolytic	0	ST	14172
Mercury	10500	flasks	158356
Mica, muscovite, block, stained and better	2500000	lb	5207431
Mica, muscovite film, first and second qualities	90000	lb	1176338
Mica, muscovite splittings	12630000	lb	14335260
Mica, phlogopite block	210000	lb	130745
Mica, phlogopite splittings	930000	lb	1486596
Molybdenum	0		0
Morphine sulphate and related analgesics			
crude	0	AMA lb	29195
refined	130000	AMA lb	39508
Natural insulation fibers	0	lb	0
Nickel	200000	ST	37214
Platinum group metals, iridium	86000	TrOz	29590
Platinum group metals, palladium	2150000	TrOz	1264601
Platinum group metals, platinum	1310000	TrOz	452641
Platinum group metals, rhodium	30000[a]	TrOz	0
Platinum group metals, ruthenium	65000	TrOz	0
Pyrethrum	500000	lb	0
Quartz crystals	240000	lb	1476028
Quinidine	10100000	AvOz	2473050
Quinine	4500000	AvOz	3246065
Rayon (aerospace grade fiber)	3000000[a]	lb	0
Ricinoleic/sebacic acid products	8800000	lb	5009697
Rubber	850000	LT	127227
Rutile	106000	SDT	39186
Sapphire and ruby	0	KT	16305502
Silicon carbide, crude	29000	ST	47691
Silver (fine)	0	TrOz	84839492
Talc, steatite block and lump	0	ST	1081
Tantalum group			
tantalum carbide powder	0	lb Ta	28688
tantalum metal	0	lb Ta	201133
tantalum minerals	8400000	lb Ta	2837943
Thorium nitrate	600000	lb	7097687
Tin	42674	MT	167341
Titanium sponge	195000	ST	36831
Tungsten group			
tungsten carbide powder	2000000	lb W	2032942
tungsten, ferro	0	lb W	2024143
tungsten, metal powder	1600000	lb W	1898831
tungsten ores and concentrates	79230000	lb W	76356308
Vanadium group			
vanadium, ferro	1000	ST V	0
vanadium pentoxide	7700	ST V	721
Vegetable tannin extract, chestnut	5000	LT	11692
Vegetable tannin extract, quebracho	28000	LT	121652
Vegetable tannin extract, wattle	15000	LT	14998
Zinc	1425000	ST	378768

Source: BuMines 24 July, 1991 a Added 1 June, 1989 with interim requirements
Key: ST, short ton; LDT, long dry ton; LCT, long calcined ton; SDT, short dry ton; TrOz, Troy ounce; 1 flask = 76 lb; AMA lb, anhydrous morphine alkaloid (pounds); AvOz, avoirdupois ounce; LT, long ton

mineral origin and 14 of agricultural origin. Despite the vast synthetic rubber industry, natural rubber is still preferred for tires for aircraft and off-the-road vehicles. Quebracho is used not only in tanning leather but also in oil well drilling muds. Abaca is preferred for some hawsers. Materials have been removed from the stockpile list because substitutes or alternatives were found. For example, hog bristles for paint brushes have been replaced by tapered nylon and Egyptian extra-long-staple cotton by domestic sea island cotton or synthetic fibers.

Serious consideration must be given to the forms in

which materials are stockpiled, particularly in view of the need for "surge capability" in the defense industrial mobilization base. Upgraded forms (e.g., aluminum metal instead of bauxite, ferroalloys or pure metals instead of ores) would increase stockpile costs but would also stockpile energy, transportation, production facilities, manpower, and time—all likely bottlenecks in time of mobilization. In recent years the government has been upgrading stockpiled chromium and manganese ores to ferrochromium and ferromanganese, thus also assisting in maintaining the defense industrial base.

Other government stockpiles (with holdings as of the end of 1990) are the US Department of Energy's strategic petroleum reserve (590 million barrels), the US Bureau of Mines' helium stockpile (one billion cubic meters) and the US Treasury Department's gold stocks (260 million troy ounces).

8. Related Programs

In addition to strategic stockpiles, various price support programs, especially for agricultural commodities, have been significant at various times in the last several decades. These price support programs sometimes result in the acquisition by the US Government of substantial quantities of certain commodities. Further, from time to time suggestions have been advanced that there should be peacetime US "economic stockpiles" which would attempt to moderate the peaks and valleys of cyclic demand for materials. As of mid-1991, however, no such deliberate schemes have been put into operation other than for certain agricultural commodities. In addition, there have been a variety of international schemes proposing "buffer stocks" which would act as economic stockpiles. The USA did participate in the 5th International Tin Agreement and in the International Natural Rubber Agreement and it has also participated in international discussions involving other materials.

See also: Critical and Strategic Materials; Policy: USA; Risk Management; Stockpiling: Economics

Bibliography

Department of Defense, FEMA, FPA, OEP, OCDM, ODM, Munitions Board 1947–1990 *Stockpile Report to the Congress* (published semiannually). DOD, Washington, DC

Morgan J D 1958–1991 Mining annual review: the United States. *Mining Journal (London)*

US Bureau of Mines 1985 *Mineral Facts and Problems.* USBM, Washington, DC

US Bureau of Mines 1936–1989 *Minerals Yearbook* (annual editions). USBM, Washington, DC

US Congress Joint Committee on Defense Production 1951–1976 *Annual Report of the Joint Committee on Defense Production.* US Government Printing Office, Washington, DC

Stockpile items are regularly reported in various trade publications, including:

Mining Journal, Mining Magazine, Metal Bulletin, Metals Week, Engineering and Mining Journal, American Metal Market, Mining Engineering, Wall Street Journal

In recent years the National Materials Advisory Board of the National Research Council/National Academy of Sciences/National Academy of Engineering has prepared a number of studies dealing with stockpiling and/or major strategic materials. Published by the National Academy Press, Washington, DC, these include:

NMAB-335	*Contingency Plans for Chromium Utilization*
NMAB-378	*Considerations in Choice of Form for Materials for the National Stockpile*
NMAB-381	*Assessment of Selected Materials Issues*
NMAB-385	*Analytical Techniques for Studying Substitution Among Materials*
NMAB-391	*Tantalum and Columbium Supply and Demand Outlook*
NMAB-392	*Titanium: Past, Present, and Future*
NMAB-403	*Priorities for Detailed Quality Assessments of the National Defense Stockpile*
NMAB-406	*Cobalt Conservation Through Technological Alternatives*
NMAB-424	*Quartz for the National Defense Stockpile*
NMAB-426	*Magnetic Materials*
NMAB-452	*Beryllium Metal Supply Options*
ISBN 0-309-03928-2	*Materials Science and Engineering for the 1990s*

J. D. Morgan
[US Bureau of Mines, Washington, DC, USA]

Structure of Materials Industries

Industry structure refers to the ways in which firms organize to conduct and coordinate their activities and also to the degree of competition that prevails in markets. Industry structures are influenced both by the technologies involved and by the differences among the administrative costs of adopting alternative structures.

Another key distinction is between vertical and horizontal integration of a firm. Horizontal integration is doing more of the same thing, such as expanding capacity in raw steel production. Vertical integration is doing more things. A celebrated example is when the steel mill undertakes pig iron, coke, raw steel and finished steel production and its owner engages in coal and iron ore mining and owning steel warehouses.

Study of optimal organization stresses the combined influence on choice of the direct cost of production and the additional transaction costs involved in organizing firms, markets and transactions. A central theme stressed (Coase 1937, 1991, Williamson 1975, 1985a) is that making firms larger can, up to a point, decrease transaction costs sufficiently to outweigh the costs of running a larger firm.

These arguments are more critical to vertical than horizontal integration. Vertical integration always involves major changes in the nature of transactions. Inputs to production are bought instead of the output, for example, labor, mining equipment and supplies instead of iron ore. Horizontal integration can occur with the same number of input purchases and output sales as if several smaller firms provided the output.

As is true for most economic phenomena, the optimal degrees of vertical and horizontal integration differ with circumstances in an unknown way. Beliefs in the area have changed radically among economists and the business community. Enthusiasm for increased integration and diversification particularly in minerals was widespread into the 1970s. This enthusiasm has subsequently diminished.

An even more controversial aspect of structural analysis has been that of attempting to use information on market structure to guide the design and enforcement of antimonopoly policies, particularly in the USA. This article emphasizes the descriptive aspects of materials industry structure and then notes the severe problems of using structural analysis in policymaking.

1. Descriptive Aspects of Materials Industry Structures

At least four key issues arise in describing structures. First is the optimal degree of horizontal integration for a firm and each of its plants. Second is the choice among vertical integration, contracting and other procurement methods for securing inputs. Third are questions of the connections among minerals. Fourth is the geographic extent of a given firm's activity.

1.1 The Size of Firms and Plants

A major issue concerning mineral production is the efficient size of operation at different stages of the operation. In extraction, geology, technology and economics interact to determine the optimal pattern of extraction. The size of the occurrence sets an upper bound on output, but technical and economic forces keep outputs well below the geologic limits.

Production is likely to be extended over long periods of time. More rapid extraction tends to generate various types of increases in costs. Higher investments must be made in immobile facilities to support the production at a given site. The productivity of labor and capital may decline as the annual rate of output is increased. A more rapid rate of extraction can greatly raise the cost of recovering the remaining resources.

Extraction, therefore, necessarily occurs in discrete limited units. For most widely used materials and fuels, the number of efficient units that operate is large enough to allow considerable competition. The main barriers to competition consist of concentration of ownership of many operations, predominantly by government-owned organizations.

The organization of mineral processing industries is determined by economics and technology. Within various limits, scaling up has considerable advantages grounded in basic geometric principles. Costs are a function of the surface area of facilities while capacities depend on volumes. The percentage increase in area needed to produce a given percentage increase in volume is less than the volume rise. This contributes to lower costs per unit of output for larger facilities.

For example, to double the volume (s^3) of a cube, the surface area ($6s^2$) need only increase about 59%. (If $V' = s'^3 = 2s^3$, $s' = 1.26\,s$, and the new surface area $A' = 6s'^2 = 6(1.26\,s)^2 = 1.59\,A$.) If cost only rose in proportion to the increase in area, unit costs fall to 79% of their prior level. (Total costs are 1.59 times the old level but output is double so that costs per unit are $1.59/2 = 0.79$.)

These advantages of scaling up have various offsets. Rises that occur in the costs of fabricating and transporting larger components can cause costs to rise more rapidly than does area. Reliability may become a problem. In practice, numerous competing facilities are found in such critical areas as smelting, refining and fabricating metals. Thus, the limits to scaling up seem to take hold at outputs that comprise a small portion of industry sales.

Another force that has become increasingly important is the radical differences in the minimum efficient size for different technologies. Electric furnace steel making, for example, is efficient at much smaller scales than is basic oxygen furnace steel making.

Circumstances can and have arisen in which the smaller-scale technologies look more attractive. In steel, the availability of scrap, rising construction costs and changes in the location of consumption have encouraged development of smaller-scale steel mills, generally called minimills. These mills make steel from scrap using electric furnaces.

1.2 Degrees of Vertical Integration in Processing

The number of activities a firm undertakes is heavily influenced by the nature of the markets for different goods. In particular, firms are more likely to produce goods used in their main business or further process the output of their principal products when the markets for these additional products are inadequately competitive or expensive to use.

Classic ways to avoid exploitation by a monopolist are to undertake self-production or negotiate favorable contracts (see Stigler 1968). Firms can also undertake investments that deliberately limit their ability to change suppliers or customers (see Williamson 1975). If a plant is located next to a coal mine, the boiler may be tailored to best use that mines coal and neither the mine nor the coal user may be linked to rail lines. In this case, changing suppliers or customers may be costly. Integration or other devices may be used to prevent one participant from taking advantage of the other.

Vertical integration is one polar extreme among possible methods of coordinating different steps in a production chain; open market purchases are the antipodes. In the latter, producers at one stage sell to independent firms at the next stage to fulfill orders placed to meet the needs of limited periods of much less than one year.

A wide spectrum of alternatives lies between the extremes. The arrangements with independent suppliers can be formally or informally extended over considerably longer periods. Long-term contracts can be written for various durations. Alternatively, an informal agreement for continued purchases can be made.

In principle, formal contracts provide greater assurance, but this is not necessarily true in practice. Contracts that poorly anticipate future market conditions may prove less satisfactory methods for continued dealing than reliance on the commercial record of the supplier. In fact, the evidence suggests that less turmoil has occurred in the metals area in which informal relations are more prevalent than in fuels where at least the electric utility sector became enamored of heavy reliance on contracts.

Contract renegotiation became widespread in energy markets during the 1980s. Electric utilities quietly renegotiated a shift from US to foreign suppliers of uranium. Many coal contracts were terminated or adjusted. In some cases, difficulties in resolving disputes led to extended litigation or arbitration.

Parallelling the increase of concerns among contracting parties has been an explosion of analyses of theory and practice of contracting. The theoretical work has explored the basic characteristics of markets and of contracts that will ensure that contracts produce outcomes satisfactory to the participants. Case studies have ranged over coal, natural gas and petroleum coke.

Theory and practice have converged to recognizing that contracting can often produce the results desired by the contracting parties and others affected by the actions of these contracting entities. The critical requirement is the ability of contract terms to allow response to market changes whether or not they have been anticipated.

Such adaptation is facilitated by definition of mechanisms for adjusting to changing circumstances. Contract clauses of this type are deliberately phrased loosely to allow the desired flexibility.

This can be associated with widely different methods of assigning liabilities. At one pole, the "take or pay" contract, the buyer agrees to compensate the seller for any lost investment. The other pole is assumption of the costs and benefits of changing market conditions by the sellers. Seller moves during periods of euphoria to ensure adjustments to rising prices may lead to unexpected price cuts when markets weaken.

Take or pay contracts are widely used by US gas and coal buyers. The most striking known example of seller assumption of the liabilities is in sales of coking coals to Japan, particularly by Australian producers. Prices and quantities are annually negotiated with all Australian suppliers, several of the Canadian producers and others.

Alternatively, various stages of partial integration can be employed. A firm more expert in processing can choose to conduct mining in partnership with a firm that is a mining specialist; the latter firm may be the manager of the project. Alternatively, the processing firm may totally own the mining venture but hire a mining firm to manage the property. Still another alternative is for the processor to own the mineral rights while the mining and beneficiation plants are owned by the mining company.

Firms need not depend solely on one method but may use a mixture. A firm at any given stage may choose to produce more or less than it can absorb in its own facilities at the next stage. It may mine more than it smelts or smelt more than it mines. The second case necessitates that some ore be purchased from others. Another widely employed arrangement is tolling—an accord between a mining company and a processor to secure processing for a fee. The mining firm then undertakes the sales.

Moreover, overall balance need not imply avoidance of purchase or sale, because the locations of mines and smelters may be such that it is not the best strategy to avoid transactions with other firms. It may be better for a company to sell the ores from its own mine to another company's smelter nearer the mine, and simultaneously to supply its own smelter with ores from a third company whose mine is closer by.

The ultimate example of integration undoubtedly was Henry Ford's venture into manufacturing steel and glass for his automobile plants. The steel industry historically involved a considerable degree of integration of all stages from mining of iron ore to fabrication of steel shapes. Steel companies also tended to produce coal for their own use. An integrated steel company simultaneously produced pig iron, raw steel and shapes, and was heavily involved in iron ore and coal mining.

However, independent iron ore producers have always been a significant factor in the market. Firms independent of the steel producers dominate the production of ferrochrome, nickel, molybdenum and other alloying metals.

The Japanese steel industry grew by relying on imported ore and coal purchased from independent suppliers. South Korea and Taiwan are pursuing a similar approach. Some companies have long concentrated on producing a limited number of products—often the more specialized ones. More recently, as noted already, minimills producing and fabricating steel predominantly from scrap have arisen. While these mills have thus far been limited to producing simpler products such as bars and rods, one leading company has invested in a pioneering plant to roll the thin sheets previously only producible in traditional pig-iron-based integrated steel mills.

Participation in coal and iron ore similarly has declined. Radical changes have occurred in the structure of both markets. Supplies have increased sharply and demand has declined. Major new sources of coal supply have arisen in Australia and Canada; important iron ore developments have arisen in Brazil and Australia. While the US remains a major supplier of coking coal, supplies now appear sufficiently ample that integration no longer appears as attractive as was previously the case.

An even more radical change occurred in the FRG. Its coking coal mines are kept alive by massive injections of subsidies. As part of this effort, Ruhrkohle AG was founded to merge the interests of the previously independent separate mining companies. In the mid-1980s, in response to the changed nature of consumption, several steel companies sold their shares in Ruhrkohle to electric utilities. (In 1960 59.8 Mt of coal were used for coking and 25.4 Mt were used to generate electricity; the respective 1989 figures were 23.9 Mt and 40.4 Mt.)

Integration has been even more extensive in aluminum. Companies are typically involved in mining as well as processing and fabricating. Moreover, many also own the electric power facilities to provide this critical input to the reduction process. The situation in other metals is more complex. While full vertical integration is widespread, numerous smelters operate to process ores from independent mines. As noted, such mines often choose to toll—buy processing services and market the processed metal.

1.3 Variety in Materials Output Mix

Both the number of materials involved in production and the number of stages that are conducted in the processing of a given material are important in determining industry structure.

In both cases, with a few key exceptions, the technology leads to diversity of various sorts. For example, both the steel industry and most nonferrous metal industries produce a wide variety of products. However, the basis of diversity is quite different between the two industries. The dominant products of the steel industry have always been carbon steels. However, they come in a wide variety of forms; many different sizes of bars, rods, plates, pipes, wires, sheets and strips. In addition, the industry provides pig iron for final use (which is typically much smaller than the use of pig iron in steel production) and a variety of stainless and other alloy steels. Steel companies, in short, attain diversity through the enormous variety of ways that one element—iron—can be processed, combined and formed. (Steel is, moreover, by far the most important metal with world production in the late 1980s of around 800 Mt compared to 15–16 Mt for aluminum and 8 Mt for copper. The most important steel products each account for more tonnage than the total output of any other single metal.)

Diversity in nonferrous metals consists of involvement in various elements from which many products are manufactured. The classic example is the tendency of lead and zinc to occur together, and thus for companies to be involved simultaneously with both metals. Copper deposits generally contain various amounts of other materials such as gold and silver, and some of the leading nickel deposits also have large amounts of copper (see *By-Products and Coproducts*).

A major exception is aluminum in which the highly specialized reduction processes lead to comparable specialization of activity. The pioneering aluminum firms only produced aluminum (but were involved in all stages from extraction to fabrication of shapes). Later entrants included established producers of other metals, but aluminum activities were still separate operations.

1.4 Geographical Extent of a Firm's Production Facilities

The geographical extent of a given firm's activities can differ considerably. At one extreme, simple construction materials such as sand, gravel, stone and cement are often produced by firms serving a small region. The raw materials are amply available in many regions, and the mining cost differences among regions are often too small to justify long-distance hauling. The steel industry tended to limit its international ventures to the mining stage. Iron- and steelmaking itself was for long predominantly conducted by companies operating in a single country. The main exceptions were Western European companies with interests in several countries.

The general increase in international interactions has altered these steel industry characteristics. International trade in steel products has expanded. The resulting competition caused steel producers in the USA and Western Europe to seek and secure protection from imports. Technology transfers among countries have become widespread. Through joint ventures and stock purchases, foreign steel producers have begun to participate in the US market.

In contrast, many nonferrous metal producers have long operated in many countries. The leading US copper companies, for example, had numerous foreign ventures, notably those in Chile taken over by the government. Many other multicommodity, multinational companies exist based in the USA, the UK, South Africa, Australia and Canada.

2. Evolution in Structures

Structures are continuously evolving in response to changing market conditions, public policies and investor appraisals of opportunities. Steel again is a prime example of the range of possibilities. In the USA, the merging of numerous independent entities into the United States Steel Corporation was clearly designed to create what was expected to be a more efficient organization. However, advocates of strong governmental antimonopoly policy have long argued that another motive of the merger was to lessen competition. The US government lost a case in 1920 when trying to prove that the creation of US Steel had led to further monopolization. Critics of US Steel felt that the decision was based on defective economics. The weight of this criticism was sufficiently strong that changes in laws and their interpretation prevented a merger of Bethlehem Steel and Youngstown Steel in the late 1950s.

Whatever may have been true previously, the situation since the 1970s is that vigorous foreign competition has had to be faced by US steel companies. Youngstown Steel merged with Jones and Laughlin and then went into reorganization. Other US companies have gone out of business or abandoned plants, sometimes offering to sell them to their workers.

Europe has also experienced upheaval in steel industry structure. After World War II, pressures to restructure German industry led to the breakup of Vereinigte Stahlwerke into several companies. In the process, each of the resultant individual steel companies was allowed to retain coal mines to meet direct requirements. The remaining mines were placed in the hands of a separate company. Difficulties in the European steel industry led to numerous mergers in the 1970s.

Another major development has been the influence of rising nationalism, particularly, but not exclusively, in less developed countries. In many cases, this has led to nationalization (e.g., Chilean copper and Bolivian tin). Another approach, used in Australia and Canada as well as in some less developed countries such as Mexico, is insistence on domestic participation in ventures.

A more recent effort was the move by oil companies and others to enter metals mining by merger or creating new companies. Atlantic Richfield acquired Anaconda; what was then Standard of Ohio (now BP America) purchased Kennecott. Standard of California after acquiring about one-fifth of Amax was rebuffed in efforts at a total takeover.

The efforts were justified by claims of natural linkages among the businesses and belief that metals companies were unsatisfactorily managed and could benefit from new direction. However, these expectations were not realized. The revision of expectations produced by the weak markets starting in the late 1970s produced disenchantment with the diversification. Shutdowns of operations, sales and spin-offs to stockholders became widespread.

These experiences have inspired recognition that drawbacks to the diversification of a firm are considerably greater than many industry executives and academic economists believed. The fully vertically integrated, multiple-commodity, minerals company clearly no longer seems an attractive model. More critically, theory and practice now involve more serious consideration of reducing the activities of firms.

These developments affirm the insights of such pioneering analysts of firm structure as Stigler. Such observers suggested that the extent to which a firm undertook additional activities depended critically on the vigor of competition in supplying these goods. When open-market trading is easy, it is less profitable to add an activity to those already undertaken.

Market expansion due to demand growth and transportation improvements that increase the linkages among buyers and sellers in different regions has made open-market trading more feasible. Brokers and commodity exchanges have reacted by increasing their roles. Conversely, producers have moved to reduce the extent of their activities.

3. Structural Analysis in Policymaking—Horizontal Concentration

In the analysis of antimonopoly policy, the attempt has been made to extend structural analysis beyond the description of how production is organized. In particular, efforts have been made to develop and use measures of structure to determine the vigor of competition. Basically, the approach is to define a "market" and measure the importance of given firms in that market. The simplest measure is the concentration ratio (the percentage of sales accounted for by a firm or group of firms), but more complex measures have been devised which also adjust for differences in sizes of firms within a group. The same total share of the five, eight or twenty largest firms might be divided in many ways. It is presumed to be more troublesome if the shares differ so that a few firms have very large shares. (One method that has become widely used is an index summing the squares of shares. Squaring increases the weight given large shares and thus better accounts for differences in shares.)

These efforts date back to the pioneering work in the 1930s to produce concentration ratio data. However, it was later work, notably by Bain (1968) and Kaysen and Turner (1959), that more fully developed the approach and led to proposals for using the ratios in public policy analysis. Eventually the US Department of Justice adopted concentration ratio criteria in establishing 1962 guidelines on whether mergers were acceptable. (These guidelines were modified in 1984 to use the alternative measure of concentration based on summing the squares of market share and provide more flexibility. The rules are less mechanistic than their predecessors but retain the problems of reliance on overly simple numerical tests.)

Actually, the structural approach has severe drawbacks. Markets are difficult, if not impossible, to define conclusively. Even if they were not, no clear guidelines exist about what constitutes a concentration ratio high enough to produce serious monopoly problems. Economic theory can only reach broad qualitative conclusions on the relevant issues. First, although it is agreed that the fewer firms there are in a market, the less vigorous competition is likely to be, neither the theory nor observation of practice has determined even a rough relationship between number of firms and vigor of competition. Second, the theory suggests the number of firms is not the sole determinant of competitive vigor; another major question is the attitude of firms towards each other. If other firms are viewed as reliable supporters of restricted competition, restriction is more likely than if other firms are viewed as undependable.

A more fundamental problem is that modern economic theory has increasingly stressed the invalidity of the concept of a distinct market. The central modern concept is general equilibrium—the vision of economies as a collection of interdependent components. All producers are potential rivals of all others for both the patronage of consumers and for the services of land, labor and capital. Certainly, though, the degree of rivalry will differ considerably: for example, aluminum is more a rival of steel than is balsa wood.

The concept that distinct markets exist is, however, a denial of differences in the degree of rivalry. The market is supposed to be limited to those participants with a high degree of rivalry. This adds to the design of structural measures another problem for which no good solution exists. The situation has not been improved by use of Bain's suggestions, particularly in his pioneering text (Bain 1968), that conduct and performance measures be employed to supplement the structural ones. The result can be a rigid formulation that never quite fits any actual case. A preferable approach is simply to examine all the critical economic characteristics of an industry, organize their review in the form most appropriate for the case, and then deduce what can and cannot be said about the vigor of competition.

Another issue is the validity of the conclusions Bain and others reached about the optimal size of firms. Advocates of breaking up large firms contend that such firms are larger than necessary to minimize costs. Their critics counter that the evidence is inadequate to support this claim. (See Goldschmid et al. (1974) for papers presenting both sides of this debate and transcripts of the discussions indicating that the advocates of deconcentration are aware of the inadequacy of the proof that large firms can readily be broken up.)

Again, theory and practice in the 1980s moved strongly in the direction of developing more flexible approaches to monopoly problems. The merger guidelines, as noted, were modified to employ the more complex measures of firm roles and were applied less restrictively. The new generation of texts on industry analysis (e.g., Tirole 1988, Schmalensee and Willig 1989, Carlton and Perloff 1990) stresses approaches involving elaboration of basic economic models of market behavior under various combinations of market characteristics.

Such writers as Coase (1988), Stigler (1968), Demsetz (1968), McGee (1979) and Peltzman (1976) have long argued that the strength of competitive forces and the inherent weaknesses of actual government intervention make regulation a less necessary and less desirable activity than policymakers often believe. These criticisms of intervention have become increasingly influential.

The key defects of regulation are that costs may often exceed benefits and that regulators may follow inappropriate goals. Regulatory costs necessarily include far more than these expenses of the regulatory agencies. Those regulated must incur considerable expenses in simply responding to regulators' requests for information. Regulation is often distorted by the continuing struggle for influence. Those regulated exert pressures to be protected from competition. Powerful customers seek favoritism over others.

4. Structural Analysis in Policymaking—The Vertical Integration Issue

A perennial issue in analysis of market structure is the role of vertical integration. It is agreed that vertical integration can be harmful only if monopoly problems arise in one or more levels of production. Disputes arise about what role, if any, integration can play in furthering monopoly power. McGee and Bassett (1976) point out that the fundamental issue concerns whether and how the maximum profits can be extracted out of the whole chain of stages leading to a final product.

It has long been well-known that when an earlier stage is monopolized, vertical integration can raise profits over the level occurring on an unintegrated basis. Without integration, price-taking firms at a later stage may have higher costs and lower output than would maximize profits of the whole chain. In particular, the monopoly price makes the input more costly to buy than its cost of production. This can lead to processors substituting other inputs for the monopolized input. Such inputs may cost society more to produce than those produced by the monopolist. The buyers make the shift to lessen the transfer of profits to the monopoly. Actual production costs are higher and thus total chain profits lower than they could have been under integration.

A further reduction may result from lower output than the integrated firm would choose. Thus, integration can be a way to produce higher profits than separate operations when an earlier stage is monopolized.

As shown by various writers (as summarized by Tirole 1988 pp. 174–81), an alternative pricing system could prevent this outcome. The monopolists could shift to a system in which buyers paid a cost-based price plus a fixed "franchise" fee for the right to buy at the cost-based price.

McGee and Bassett (1976) begin by recognizing the implicit supplementary conclusion that the most profitable stage at which to monopolize is in final production. If the last step could be readily monopolized, integration would be unnecessary. They add that another way to offset the distortions from monopoly of one input is to institute a monopoly of all inputs. The issue thus broadens into which of the logically possible ways to establish monopoly is the most effective.

The implicit assumption in the vertical integration literature is that more easily monopolized earlier stages supply more competitive later stages. This may or may not be correct. Mineral extraction, for example, is not necessarily more readily monopolized than processing (as McGee and Bassett (1976) explicitly note for the illustrative case they selected of iron ore and steelmaking).

To complicate matters further, should monopoly arise at the earlier stage, or worse, at both stages, the wisdom of prohibiting integration becomes ambiguous. As long as the monopolies can be expected to persist, integration lessens the damage. The case against integration centers on whether restricting integration decreases the immunity of the industry to increased competition. Again, the standard anti-integration case implicitly considers one effect of integration control more likely than another. The vision is that integration will lead to an industry of a smaller number of firms that will be difficult to challenge. However, restrictions may actually reduce the ability to create more strong firms. The policy could prevent those best able to benefit from integration from doing so.

See also: Economic Theory of Mineral Resources and Markets; Economics of the Semiconductor Industry; Markets for Materials: Principles and Public Policy Implications of Competition; Structure of Metals Industries; Wood-Based Industries in the USA: Economic Structure in a Worldwide Context

Bibliography

Areeda P 1981 *Antitrust Analysis, Problems, Texts, Cases*, 3rd edn. Little, Brown, Boston, MA

Bain J S 1968 *Industrial Organization*, 2nd edn. Wiley, New York

Barnett D F, Crandall R W 1986 *Up from the Ashes.* Brookings Institution, Washington, DC

Bull C 1987 The existence of self-enforcing implicit contracts. *Q. J. Econ.* 102(1): 147–159

Carlton D W, Perloff J M 1990 *Modern Industrial Organization.* Scott, Foresman/Little, Brown, Glenview, IL

Coase R H 1937 The nature of the firm. *Economica.* 4(4): 386–405

Coase R H 1988 *The Firm, the Market and the Law.* University of Chicago Press, Chicago, IL

Coase R H 1991 The nature of the firm: origin, influence, meaning. In: Williamson O E, Winter S G (eds.) 1991 *The Nature of the Firm: Origins, Evolution and Development.* Oxford University Press, New York

Demsetz H 1968 Why regulate utilities? *J. Law Econ.* 11: 55–65

Dye R A 1985 Costly contract contingencies. *Int. Econ. Rev.* 26(1): 233–50

Elzinger K G, Hogarty T F 1973 The problem of geographic market delineation in anti-merger suits. *Antitrust Bull.* 18(1): 45–81

Elzinger K G, Hogarty T F 1978 The problem of geographic market delineation revisited: the case of coal. *Antitrust Bull.* 23(1): 1–18

Furobotn E G, Pejovich S (eds.) 1974 *The Economics of Property Rights.* Ballinger, Cambridge, MA

Goldberg V P, Erickson J R 1987 Quantity and price adjustment in long-term contracts: a case study in petroleum coke. *J. Law Econ.* 30: 369–98

Goldschmid H J, Mann H M, Weston J F (eds.) 1974 *Industrial Concentration: The New Learning.* Little, Brown, Boston, MA

Grossman S J, Hart O D 1986 The cost and benefits of ownership: a theory of vertical and lateral integration. *J. Polit. Econ.* 94: 691–719

Hart O D, Moore J 1988 Incomplete contracts and renegotiation. *Econometrica* 56(4): 755–85

Hicks J R 1946 *Value and Capital*, 2nd edn. Oxford University Press, Oxford

Joskow P L 1977 Commercial impossibility, the uranium market and the Westinghouse case. *J. Leg. Stud.* 6: 119–76

Joskow P L 1985 Vertical integration and long term contracts: the case of coal-burning electric generating plants. *J. Law, Econ. Org.* 1: 33–88

Joskow P L 1987 Contract duration and relationship-specific investments: empirical evidence from coal markets. *Am. Econ. Rev.* 77(1): 168–85

Joskow P L 1988 Price adjustment in long-term contracts: the case of coal. *J. Law Econ.* 31(1): 47–83

Joskow P L 1990 The performance of long-term contracts: further evidence from coal markets. *Rand J. Econ.* 21(2): 251–74

Kaysen C, Turner D F 1959 *Antitrust Policy: An Economic and Legal Analysis.* Harvard University Press, Cambridge, MA

Klein B 1980 Transaction cost determinants of unfair contractual relationships. *Am. Econ. Rev.* 70: 356–62

Klein B R, Crawford G, Alchian A A 1979 Vertical integration, appropriable rents, and the competitive contracting process. *J. Law Econ.* 22: 297–326

Klein B, Leffler K B 1981 The role of market forces in assuring contractual performance. *J. Polit. Econ.* 89(4): 615–41

Kreps D M 1990 *A Course in Microeconomic Theory.* Princeton University Press, Princeton, NJ

Krouse C G 1990 *Theory of Industrial Competition.* Basil Blackwell, Cambridge, MA

Lima J G H *Restructuring the U.S. Steel Industry: Semi-Finished Steel-Imports, International Integration, and U.S. Adaption.* Westview, Boulder, CO

McGee J S 1958 Predatory price cutting: the Standard Oil (N.J.) case. *J. Law Econ.* 1: 137–169

McGee J S 1979 Predatory pricing revisited. *J. Law Econ.* 22: 289–329

McGee J S, Bassett L R 1976 Vertical integration revisited. *J. Law Econ.* 19(1): 17–38

Masden S E, Crocker K J 1985 Efficient adaptation in long-term contracts: take-or-pay provisions for natural gas. *Am. Econ. Rev.* 75(5): 1083–92

Mény Y, Wright V (eds.) 1986 *The Politics of Steel: Western Europe and the Steel Industry in the Crisis Years (1974–1984).* de Gruyter, Berlin

Peltzman S 1976 Towards a more general theory of regulation. *J. Law Econ.* 19(3): 211–40
Posner R A 1976 *Antitrust Law, An Economic Perspective.* University of Chicago Press, Chicago, IL
Posner R A 1986 *Economic Analysis of Law*, 3rd edn. Little, Brown, Boston, MA
Posner R A, Easterbrook F H 1981 *Antitrust: Cases, Economic Notes, and Other Materials*, 2nd edn. West, St Paul, MN
Rogerson W P 1984 Efficient reliance and damage measures for breach of contract. *Rand J. Econ.* 15(1): 39–53
Scherer F M, Ross D 1990 *Industrial Market Structure and Economic Performance*, 3rd edn. Houghton-Mifflin, Boston, MA
Schmalensee R, Willig R D (eds.) 1989 *Handbook of Industrial Organization.* North-Holland, Amsterdam
Shavell S 1980 Damage measures for breach of contract. *Bell J. Econ.* 11(2): 466–90
Shavell S 1984 The design of contracts and remedies for breach. *Q. J. Econ.* 99(1): 121–48
Spulber D F 1989 *Regulation and Markets.* MIT Press, Cambridge, MA
Stigler G J 1968 The division of labor is limited by the extent of the market. *The Organization of Industry.* Irwin, Homewood, IL
Stiglitz J E, Mathewson G (eds.) 1986 *New Developments in the Analysis of Market Structure.* MIT Press, Cambridge, MA
Tirole J 1986 Procurement and renegotiation. *J. Polit. Econ.* 94(22): 235–59
Tirole J 1988 *The Theory of Industrial Organization.* MIT Press, Cambridge, MA
Walker G, Weber D 1984 A transactions cost approach to make or buy decisions. *Adm. Sci. Q.* 29: 373–91
Williamson O E 1975 *Markets and Hierarchies: Analysis and Antitrust Implications.* Free Press, New York
Williamson O E 1979 Transaction cost economics: the governance of contractual relationship. *J. Law Econ.* 22: 233–61
Williamson O E 1983 Credible commitments: using hostages to support exchange. *Am. Econ. Rev.* 73: 519–40
Williamson O E 1985a *The Economic Institutions of Capitalism: Firms, Markets, Relational Contracting.* Free Press, New York
Williamson O E 1985b Assessing contract. *J. Law, Econ. Org.* 1(1): 177–85
Williamson O E, Winter S G (eds.) 1991 *The Nature of the Firm: Origins, Evolution and Development.* Oxford University Press, New York

R. L. Gordon
[Pennsylvania State University, University Park
Pennsylvania, USA]

Structure of Metals Industries

This article looks at the international location of minerals mining and metals processing industries, and at the relative costs in different countries of some of the main factors of production, such as capital and skilled and unskilled manpower. It also discusses the gradual shift which has been taking place towards a higher proportion of world processing of metals being located in developing countries.

1. Determinants of International Structure

The structure of the minerals industries depends upon the location of the basic mineral deposits, the costs of transportation, the distribution of capital and labor resources and technology. Each of these factors influences the location, size, distribution and integration of the industries that transform the minerals into useful materials. Obviously, these objective determinants of industry structure vary with different materials; moreover, they change over time.

Minerals may be processed or transformed into useful materials near the location at which they are mined or they may be transported considerable distances before further processing. Iron ore is often transported thousands of miles before it is sintered, smelted and refined. Copper, on the other hand, is usually beneficiated near the mine site and is generally smelted at a nearby location. The relationship between bauxite mining and alumina refining is changing, with an increasing share of refining occurring near the bauxite mine.

The international location of production depends very much on political and economic institutions as well as on comparative advantage. The shift of steel production from developed to developing countries reflects, in part, conscious attempts by developing countries to industrialize in directions consistent with their comparative advantage (Crandall 1981). Such industrialization requires a large inflow of capital and, therefore, economic and political conditions under which lenders may be assured of receiving a return on their investment. In 1982–1984, many developing countries began to experience difficulties in servicing these loans and were thus forced to slow their rate of investment in energy and minerals production.

1.1 Mining

For most of the more important metallic minerals, new deposits have increasingly been discovered and developed in the less developed countries (Table 1). Iron ore mines have been developed in Brazil, Liberia, Venezuela and India and, as a result, the relative importance of the USA, Canada and Western Europe in mining ore has fallen substantially. Similarly, copper, nickel and zinc have increasingly been mined in the less developed countries. Of the major metal ores, only bauxite evidences a relative increase in developed-country production—solely because of the sharp growth in Australian output. The development of the mineral industries in the less developed countries has increased with intensified exploration, intense competition among multinational companies and the steep relative decline in shipping rates since World War II.

Table 1
Share of world mine output of selected metal ores in developed and developing countries

	Developed countries[a]		Less developed countries[b]	
	1962–1964	1986–1988	1962–1964	1986–1988
Iron ore	0.730	0.487	0.270	0.513
Bauxite	0.432	0.532	0.568	0.468
Copper	0.495	0.394	0.505	0.606
Nickel	0.840	0.574	0.160	0.426
Zinc	0.728	0.691	0.272	0.309

Source: *Minerals Yearbook*
a Including the USA, Canada, Japan, Western Europe, Australia, New Zealand and South Africa b Including the remainder of the noncommunist bloc countries

1.2 Processing

The structure of the minerals production industries—embracing various combinations of beneficiation, smelting, refining or other processing—depends very much upon government policies, factor endowments and technology. In industries such as steel, few countries depend upon private market decisions alone (Hogan 1982). Since integrated steel production requires very large scale operations and investments of several billion dollars per efficient plant, most countries' integrated steel industries are very concentrated. Nonferrous metals, on the other hand, generally require substantially smaller investments per new plant. Most nonferrous metals, such as lead, copper, nickel and zinc, are produced in smaller-scale facilities than those required in steel production. As a result, most of these industries have attracted substantially more entry since the 1950s than the integrated iron and steel industry. Moreover, because nonferrous metals are more valuable per unit weight than pig iron or carbon steel, they move more freely in world trade in response to small differences in price. For this reason, domestic concentration of most nonferrous metals industries is much less important than concentration in other materials industries.

Virtually all major metals are transformed from ore by means of capital-intensive technology. Large smelters or refining operations typically use relatively little labor except for maintenance operations. Labor intensity increases only when the metal is shaped into semifinished products. Despite this capital intensity, most metals industries have shifted slowly towards the less developed countries. In some cases, this shift is towards countries with large mineral deposits, for example, Zambia, Mexico, Peru and Chile for copper. In other cases, however, developing countries without major mineral deposits are increasing their share of world production, for example, Korea, Taiwan, Turkey and Argentina in iron and steel production.

The shift of minerals production towards developing countries is especially noticeable in the case of aluminum, steel and copper. Why these countries should expand their output of these materials with capital-intensive production facilities has been the source of considerable controversy in applied economics.

2. *Specialization and the "Leontief Paradox"*

Classical economic theory predicts that economic activity will be distributed across societies according to comparative advantage in production, which in turn derives from countries' factor endowments (Travis 1964). Rich countries with relatively large pools of accumulated capital should specialize in capital-intensive industry, while poorer, less developed countries concentrate on labor-intensive production. Despite this well-reasoned theory, several researchers have discovered that developed countries (especially the USA) often import more capital-intensive manufactures than they export. This finding has become known as the "Leontief paradox," named after its original discoverer.

Explanations for the Leontief paradox are numerous, but the most common is that developing countries often specialize in adding value to minerals because minerals industries require relatively little skilled labor, which is scarce in these societies, but considerable capital, which can be borrowed from developed countries. In many cases, rich local deposits of nonfuel minerals and energy provide the comparative advantage for these less developed countries. Developed countries, on the other hand, specialize in the production of goods and services that require relatively skilled labor (Bowen 1983, Maskus 1983). Since these more advanced societies have a large share of their capital invested in the education and training of their labor forces, they do not thrive on relatively simple, capital-intensive production.

Less-developed countries are likely to find that they enjoy a comparative advantage in relatively capital-intensive industries that require a high ratio of

Table 2
Hourly compensation of production workers in selected developed and developing countries, 1988 (US dollars per hour)

Country	All manufacturing	Iron and steel	Nonferrous metals
Developed countries			
USA	13.85	23.58	16.90
West Germany	18.00	20.60	19.29
France	12.96	15.57	14.90
UK	10.46	13.41	11.41
Japan	12.80	22.80	16.65
Developing countries			
Korea	2.50	4.40	3.14[a]
Taiwan	2.82	4.22[a]	3.69[a]

Source: US Bureau of Labor Statistics
a Includes foundries

unskilled to skilled labor. Capital and technology can be obtained at competitive prices on a world market, but unskilled labor and major mineral deposits—in countries such as Mexico, India and Brazil—are likely to provide a substantial advantage to these countries in producing such basic materials as copper, steel or aluminum from locally mined ores.

The labor cost advantage in developing countries is readily seen in Table 2. Developed countries, such as the USA, Japan or Germany, pay workers in basic metals industries far more than countries such as Taiwan or Korea. These differentials for relatively unskilled labor do not reflect differences in the inherent productivity of labor. Rather, they reflect social policies, the premium demanded by workers for working conditions, relative collective bargaining power and the other opportunities for employment in developed countries.

3. *Shift of Production Towards Developing Countries*

As wage rates for unskilled and semiskilled workers rise in developed countries and technology becomes more freely available to developing economies, the production of many basic materials shifts towards the newly developing economies. This shift is gradual because the sunk capital in the developed economies has a relatively long lifetime and world demand for these materials has not been growing rapidly.

Production of most major metals has been shifting to the newly industrializing countries over the past two decades. Table 3 shows that this shift is most pronounced for aluminum, copper and steel. Of the metals, only lead has failed to shift measurably to the less developed countries, in large part because of the role of secondary production in a slow-growth industry.

With declining economic growth in most developed countries after the 1973–1974 increase in oil prices, the rate of increase in consumption of many materials has slowed dramatically. Consumption growth for most metals averaged less than 2% per year in developed countries since 1974 after averaging more than 4% per year in the 1950s and 1960s (Considine 1990). This slow growth in consumption in developed countries has slowed the inevitable shift of materials production to the developing countries.

Recycling of metals has increased modestly since the 1960s. Slow economic growth in the 1970s and early 1980s created a favorable climate for recycling. As economic growth slows, the ratio of recycled or secondary production to primary production generally increases, since total demand for the material declines relative to the supply of obsolete, discarded goods from which secondary materials are extracted. The more rapid economic growth in developed countries after 1982 in turn reduced the relative growth of recycling.

Table 3
Share of noncommunist world primary metals production in developed and developing countries

	Developed countries[a]		Less developed countries[b]	
	1968	1988	1968	1988
Steel	0.934	0.779	0.066	0.221
Aluminum	0.937	0.787	0.063	0.213
Copper	0.866	0.653	0.134	0.347
Lead	0.829	0.823	0.171	0.177
Zinc	0.900	0.811	0.100	0.189
Tin	0.296	0.158	0.704	0.841

Source: *Minerals Yearbook*
a Including the USA, Canada, Japan, Western Europe, Australia, New Zealand and South Africa b Including the remainder of the noncommunist bloc countries

Table 4
Relationship between scrap recovery and metals consumption (Mt) in developed Western countries

Metal	1968 Consumption	1968 Scrap recovery	1968 Ratio of scrap recovery to consumption	1988 Consumption	1988 Scrap recovery	1988 Ratio of scrap recovery to consumption
Aluminum	8441.2	1890.1	0.224	17202.6	5090.8	0.296
Copper	4940.1	2998.6	0.607	5665.3	4021.1	0.710
Zinc	3398.1	601.4	0.177	4235.8	455.8	0.108
Tin	155.6	54.1	0.348	137.9	35.5	0.257

Source: *Metal Statistics*

As Table 4 demonstrates, the ratio of recycling to total consumption of aluminum and copper in the Western world has risen substantially since 1968. Zinc and tin, on the other hand, evidence modest declines. In the steel industry, growth in the supply of obsolete scrap has been offset by an increase in product yields in steelmaking, with the result that scrap consumption has remained between 45% and 50% of crude steel output since the 1960s. Since reuse of home scrap at the steel mill does not constitute recycling of finished steel products, this suggests that the rate of recycling has actually risen in steel as well.

The slow growth of metals consumption since 1973 and the growth of internal recycling has ameliorated the shift towards the developing countries. Table 5 shows the different patterns in the growth of import reliance for each material in the developed Western countries. The developed countries have modestly reduced their self-sufficiency in aluminum and tin, but have increased their self-sufficiency in lead, zinc and copper. Finally, the developed world's steel producers have reduced their share of Western world steel consumption, but not by much, because of the sharp decline in steel consumption in the 1970s and early 1980s.

Table 5
Self-sufficiency in selected metals in the developed Western world

	Ratio of production to consumption	
	1968	1988
Aluminum	0.95	0.93
Lead	0.88	1.01
Zinc	1.00	1.05
Copper	0.84	0.93
Tin	0.36	0.20

Source: *Metal Statistics*

4. Outlook

There is little doubt that the mining and processing of metals have shifted towards the less developed countries since the 1960s. Slow consumption growth and increased recycling have reduced this trend somewhat, but it seems likely that metals production will continue to shift away from the developed countries. The mining and processing of most metals is rather capital-intensive but requires relatively little skilled labor. Unskilled and semiskilled labor are simply too expensive in developed societies to permit metals production to grow with world demand.

See also: Structure of Materials Industries

Bibliography

Bowen H P 1983 Change in the international distribution of resources and their impact on U. S. comparative advantage. *Rev. Econ. Statist.* 15(3): 402–14

Considine T J 1990 Recent trends in material consumption: the role of technology and economics. *Mater. Soc.* 61(1): 167–82

Crandall R 1981 *The United States Steel Industry in Recurrent Crisis: Policy Options in a Competitive World.* Brookings Institution, Washington, DC

Hogan W T 1982 *World Steel in the 1980s: A Case for Survival.* Lexington Books, Lexington, MA

Leontief W 1954 Domestic production and foreign trade: the American capital position re-examined. *Econ. Int.* 7: 9–38

Maskus K E 1983 Evidence on shifts in the determinants of the structure of U. S. manufacturing foreign trade, 1958–76. *Rev. Econ. Statist.* 15(3): 415–22

Metal Statistics 1976, 1986, 1988. Metallgesellschaft, Frankfurt

Mikesell R F 1986 Materials industries: financing. In: Bever M B 1986 *Encyclopedia of Materials Science and Engineering.* Pergamon, Oxford. pp. 2844–9

Minerals Yearbook 1966, 1969, 1984, 1988, 1989. US Bureau of Mines, Washington, DC

Travis W P 1964 *The Theory of Trade and Protection.* Harvard University Press, Cambridge, MA

R. W. Crandall
[The Brookings Institution, Washington, DC, USA]

Substitution of Materials: Economics

Material substitution comprises several different types of events that may substantially alter material usage.

Material-for-material substitution entails the replacement of one material for another in the production of a particular good. Aluminum and plastic, for example, have been increasingly used in place of wood siding in residential construction in recent years.

Other-factors-for-material substitution diminishes material use by increasing other inputs into the production process, such as labor, capital and energy. The hand soldering of many electronic components, for example, requires less solder but more labor than more automated procedures which make use of printed circuit boards.

Quality-for-material substitution occurs when material savings are produced by a reduction in the quality of the final product. For example, the thinner glass beverage containers produced during World War II consumed less glass but also had a higher breakage rate. Similarly, the amount of nickel, molybdenum and other alloying elements used in alloy steels can be reduced if lower levels of performance are acceptable.

Interproduct substitution produces a change in the mix of goods used to achieve a particular function or to satisfy a given need. Public transportation can be substituted for private automobiles, a fax for a letter, frozen vegetables for fresh or canned produce. Unlike the other types of substitution, interproduct substitution does not alter the manufacturing processes and the use of materials in the production of individual goods, but rather affects material use by changing the composition of goods and services. This type of substitution includes what Chynoweth (1976) and others call functional substitution. This occurs when a product (along with the materials it contains) is replaced by an entirely different means of achieving a desired function, such as the use of satellites instead of underground cables to transmit long-distance telephone messages.

Finally, if material substitution is defined broadly, it includes technological substitution, whereby an advance in technology reduces the amount of material needed to produce a particular product. The development of electrolytic tinplating, which greatly decreased the tin required per tinplate can, is a classic example of this type of substitution. New technology can also lead to the introduction of new products, such as the aluminum can, and thus increase the possibilities for interproduct substitution.

In short, material substitution may result from the introduction of new technology, from shifts in the composition or quality of final goods and from changes in the mix of factor inputs used in producing these goods. The technological aspects of substitution are examined in the article *Substitution of Materials: Technology*; this article concentrates on the economic aspects.

1. Importance of Material Substitution

For economists, material substitution is of interest for several reasons. First, by allowing society to replace expensive and scarce resources with cheaper and more abundant resources, it has historically helped offset the upward pressure on resource costs resulting from the depletion of high-grade, readily accessible and easily processed mineral deposits (Barnett and Morse 1963, Barnett 1979). As for the future, there are those who believe that material substitution will have to play an even greater role if the adverse effects of resource depletion are to be held at bay (Goeller and Weinberg 1976, Skinner 1976).

Second, material substitution may in some instances help alleviate shortages caused by war, embargoes, labor unrest, natural disasters and cartels. Such shortages rarely last for more than a few years. However, they often arise quickly with little warning and as a result can be quite disruptive while they last.

Third, the threat of material substitution forces producers to behave in a more competitive manner. Even in highly concentrated industries where dominant firms charge a uniform producer price, material substitution may prevent high prices and monopoly profits. In a similar manner, substitution reduces the likelihood that major producing countries will join together to form a cartel, and in the event they do, that it will be successful.

Finally, economists are often called upon to forecast future material requirements. Since substitution may greatly alter the demand for materials, it must be considered in making such predictions.

2. The Economic View of Substitution

In analyzing material substitution, or more generally the substitution of any factor of production, economists rely on the branch of microeconomics called production theory. An important conceptual relationship found in this theory is the production function, which shows how much output of any particular good can be derived from given quantities of labor, capital, energy, materials and other inputs. This relationship is governed by the existing state of technology, and so changes as new innovations are introduced. For this reason, the production function shown below includes not only the output Q of the good being produced and the inputs $(A, B, \ldots)$ of all factors of production but also a variable T reflecting current technology:

$$Q = f(A, B, \ldots ; T)$$

The extent to which any two inputs, such as A and B, can be substituted for each other in producing Q is portrayed by isoquant curves. Three such curves, identified as Q_1, Q_2 and Q_3, are shown in Fig. 1. Each of these curves indicates the various quantities of A and B that are just sufficient to produce the specified output,

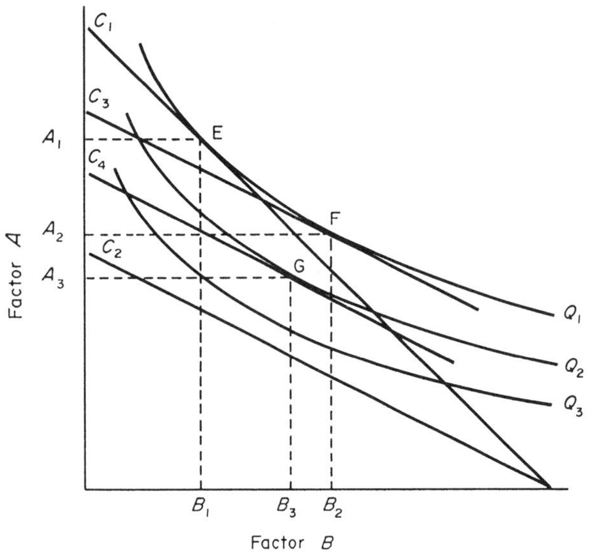

Figure 1
Factor demand determined by isoquant and isocost curves

assuming all other inputs and technology remain fixed at some prescribed level. The further an isoquant is from the origin, the larger the output level it reflects and the greater the inputs of A and B required to produce it.

The isoquant curves shown in Fig. 1 are downward sloping, indicating that a specified level of output can be maintained while one factor input is reduced if the other is increased. That is, some substitution between A and B is presumed possible. The curves are also drawn as continuous functions convex to the origin. Convexity implies that as more of one factor is substituted for another, greater quantities of the factor being added are required for each unit of the factor being replaced. The presumption here is that the most economical substitutions, those requiring the least amount of the factor being added, will be made first.

While isoquant curves are commonly drawn as shown in Fig. 1, the implicit assumptions in such constructions can be relaxed. For example, if two inputs must be used in fixed proportions and no substitution between them is possible, the curve collapses to a point representing their requisite quantities. Alternatively, if three different production processes are feasible each using A and B in fixed but different proportions, the isoquant curve is composed of two line segments connecting the three points determined by the fixed quantities of A and B required by each process. (This assumes that all three processes are economic at some input price ratio; otherwise, the resulting isoquant is simply one line segment connecting the points determined by the two processes than can be economic.) As the number of possible processes grows, each using fixed but different proportions of A and B, the isoquant curve is defined by an even larger number of line segments and so increasingly approximates a continuous curve, such as those shown in Fig. 1.

The determination of just where on an isoquant curve a firm will operate, and hence how much of A and B it will use, requires the introduction of the isocost curve. This curve shows the various quantities of A and B a firm can purchase for a given expenditure. Its slope is determined by the ratio of the price of B to the price of A, and its distance from the origin by the level of expenditure. Figure 1 depicts four isocost curves—C_1, C_2, C_3 and C_4.

If a firm produces the output represented by the isoquant Q_1 and if the slope of the isocost curve C_1 reflects the existing ratio of input prices, the firm minimizes its costs by purchasing A_1 of factor A and B_1 of factor B. Any other combination of A and B on isoquant Q_1 lies on an isocost curve further from the origin and hence involves a greater level of expenditure for factors A and B.

Now let the price of factor A increase, causing the isocost curve to rotate from C_1 to C_2. This shift reflects the reduction in the amount of factor A the firm can purchase for a given expenditure. If the firm still desires to produce the same level of output, it minimizes its costs, which have increased as a result of the rise in price of factor A, by moving from point E to point F on the isoquant Q_1. In the process, it reduces its consumption of factor A to A_2 and increases its consumption of factor B to B_2.

Since the rise in the price of factor A increases the firm's costs, it may decide to reduce its output. If other firms do likewise, this would produce a rise in prices to cover all or part of the increase in costs. This implies a shift to a lower isoquant curve, such as Q_2. In this case, the firm would end up producing at point G, consuming A_3 of factor A and B_3 of factor B.

The decline in the consumption of factor A can be separated into two parts: the reduction from A_1 to A_2 which reflects the substitution of factor B for factor A motivated by the rise in the price of the latter; and the further reduction from A_2 to A_3 caused by the fall in output. In the case of factor B, the substitution effect increases consumption from B_1 to B_2, which is to be expected since its price has fallen relative to that for factor A. The output effect, however, reduces consumption from B_2 to B_3. The net result as shown in Fig. 1 is an increase in consumption of factor B, though this need not be the case.

This conceptual view of material substitution reflects the economists' concern with the role of prices in influencing material demand, and has little to say about other factors affecting substitution. It is most appropriate for examining material-for-material and other-factors-for-material substitution. It can accommodate technological substitution when the latter reflects a reduction in material requirements used to produce existing goods, for such innovations simply result in a shift towards the origin of the isoquant curves. It is of little use in examining quality-for-material substitution, interproduct substitution, or technological substitution resulting from the introduction of new goods.

3. Economic Research on Substitution

Economic research on substitution falls into three general categories. The first covers studies that estimate aggregate production functions (or the corresponding information in the form of cost functions) for entire industries, economic sectors or even the economy as a whole. These efforts, which Slade (1981) and Field and Berndt (1981) have reviewed, generally treat materials as a single factor of production and focus on the extent to which they can be replaced by other factors, such as labor, capital and energy. Not surprisingly, their findings vary greatly depending on how they specify the aggregate production function, how they take account of technological change and how they resolve other issues. Many of these studies also suffer from the use of aggregated input data and from their static nature (Holmes 1988, Kopp and Smith 1980, Slade 1981).

The second category of research consists of more micro investigations that differentiate individual materials and often focus on specific end uses. These studies involve the estimation of formal models, the most common of which are econometrically estimated supply-and-demand models, such as the model for copper by Fisher et al. (1972) and the model for aluminum by Woods and Burrows (1980).

The demand equations in these models typically take account of material substitution by including price variables for close substitutes as well as for the material whose demand is being estimated. Thus, they assume that substitution is primarily driven by material prices. In addition, the model specification generally assumes that the relationship between demand and material prices is continuous and reversible. The latter implies that the demand lost when the price of a material increases can be completely recaptured if price returns to its original level. These assumptions are all consistent with the economic view of substitution described earlier.

The short-run and long-run effects of material prices can be differentiated by introducing a lag structure, such as lagged price variables, into the demand equations. When this is done, material demand is almost always found to be more responsive or elastic to prices in the long run. This, as may be expected, implies that a change in material prices will effect more substitution the longer the response time.

The third category of research encompasses case studies of material substitution in particular end uses, such as beverage containers (Demler 1983, Nappi 1990), automobiles (Eggert 1990) and electric batteries (Holmes 1990). Such studies do not normally entail formal modelling, but rather attempt to determine more about the nature of material substitution by analyzing how and why the mix of materials used in certain applications has changed over time.

While the findings vary depending on the end uses examined, several conclusions are fairly common (Tilton 1983). First, material substitution may, and often does, substantially alter material requirements in specific uses, particularly over the long run. Second, a major determinant, perhaps *the* major determinant, of substitution is technological change. Third, changes in material prices often have little effect on the material mix in the short run as producers are constrained by existing technology and equipment. Over the long run, both technology and equipment can change, and prices have more of an influence. Indeed, the major effect of material prices may occur indirectly over the long run by altering the incentives to conduct research and development on new material-saving technologies. Finally, material substitution is often motivated by changes in government activities and regulations, such as building codes or worker health and safety standards.

These findings have a number of implications. The possibility that prices may influence material substitution most by inducing technological change suggests that the relationship between price and material demand may be discrete and irreversible, rather than being continuous and reversible. This, coupled with the finding that other factors, such as government regulation, are often important determinants of substitution, raises reservations about the traditional economic view of substitution as described earlier.

Forecasts of future mineral requirements, particularly over the long run, need to take account of substitution, for the latter can greatly alter material consumption. This, however, is not easy to do, for substitution is often the result of technological change, which by its nature is uncertain and difficult to predict.

The findings also suggest that substitution can play an important role in alleviating material shortages caused by the depletion of high-quality mineral deposits. Such shortages tend to arise slowly and persist for extended periods, providing ample time for rising material prices to induce material substitution. For other types of shortages, those caused by war, cartels, embargoes and strikes, material substitution is likely to provide much less relief. Such shortages tend to arise abruptly and unexpectedly, and the resulting disruption is likely to be greatest in the short run. By the time sharply higher prices stimulate material substitution by inducing changes in technology and equipment, these shortages are likely to have passed.

See also: Conservation of Materials: Economics; Conservation of Materials: Technology; Prices of Materials: Theory; Substitution of Materials: Technology

Bibliography

Barnett H J 1979 Scarcity and growth revisited. In: Smith V K (ed.) 1979 *Scarcity and Growth Reconsidered.* Johns Hopkins University Press, Baltimore, MD, pp. 163–217

Barnett H J, Morse C 1963 *Scarcity and Growth.* Johns Hopkins University Press, Baltimore, MD

Chynoweth A G 1976 Electronic materials: functional substitutions. *Science* 191: 725–32

Demler F R 1983 Beverage containers. In: Tilton 1983, pp. 15–35

Eggert R G 1990 The passenger car industry: faithful to steel. In: Tilton 1990

Field B C, Berndt E R 1981 An introductory review of research on the economics of natural resource substitution. In: Berndt E R, Field B C (eds.) 1981 *Modeling and Measuring Natural Resource Substitution.* MIT Press, Cambridge, MA, pp. 1–14

Fisher F M, Cootner P H, Baily M N 1972 An econometric model of the world copper industry. *Bell J. Econ.* 3: 568–609

Goeller H E, Weinberg A M 1976 The age of substitutability. *Science* 191: 683–9

Holmes M J 1988 The role of input prices in the process of material substitution: the case for a more disaggregated approach. *Mater. Soc.* 12: 207–21

Holmes M J 1990 Material substitution in battery electrodes. *Resour. Policy* 16: 22–34

Kopp R J, Smith V K 1980 Measuring factor substitution with neoclassical models: an experimental evaluation. *Bell J. Econ.* 11: 631–55

Nappi C 1990 The beverage and food container industry: change and diversity. In: Tilton 1990

Skinner B J 1976 A second iron age ahead? *Am. Sci.* 63: 258–69

Slade M E 1981 Recent advances in econometric estimation of materials substitution. *Resour. Policy* 7: 103–9

Tilton J E (ed.) 1983 *Material Substitution: Lessons from Tin-Using Industries.* Resources for the Future, Washington, DC

Tilton J E (ed.) 1990 *World Metal Demand: Trends and Prospects.* Resources for the Future, Washington, DC

Woods D W, Burrows J C 1980 *The World Aluminum–Bauxite Market.* Praeger, New York, pp. 184–230

J. E. Tilton
[Colorado School of Mines,
Golden, Colorado, USA]

Substitution of Materials: Technology

Substitution as a technological process is the replacement of one material, process, design or technology for another. These types of substitution occur at varying levels of complexity and under the principal responsibility of a design engineer.

In general, substitution is an evolutionary process and one motivated by cost reduction, functional advantage or supply considerations. To these historical factors are now added regulatory constraints, strategic concerns, environmental issues and public attitudes. These, together with a quickened pace of innovation, have led to more abrupt substitution and a growing emphasis on the technological planning for substitution.

Substitution is one facet of a broad conservation effort aimed at conserving scarce materials, reducing energy use and eliminating waste. By its very nature, substitution can alter materials use and recycling patterns, manufacturing processes and facilities and the industry in which it occurs.

1. Types of Substitution

There are four basic types of substitution: material, process, design and functional substitution. They are closely related in that a substitution of one type may well cause others. Thus, for example, a material substitution often requires a modification of the process used and design changes almost invariably lead to material and process changes. Similarly, substitution stimulates competitive development and innovation in the material, process or system that it seeks to displace.

Material substitution is probably the most familiar type and that of greatest concern because of its relation to the supply of critical and strategic resources. Under this heading, three categories of materials substitution may be distinguished: physical, quantitative and invisible.

Physical substitution is the replacement of one material for another and countless examples of this could be cited: aluminum for copper in electrical conductors, glass for steel in containers and synthetic detergents for soap, to name but a few. Such substitutions only occur if the substitute material has properties which permit the function to be achieved at lower costs under the prevailing circumstances.

Quantitative substitution involves a quantitative reduction in the amount of material used per unit of output. Examples would include using a thinner tin coating on tin-plated steel, less gold on an electrical contact or the use of a composite to replace a thicker, monolithic material.

Invisible substitution is less obvious and occurs when a new product entering the market uses material(s) other than that traditionally used. This type is more difficult to trace, although some examples are clear, such as when a new car model or a new piece of sporting equipment is introduced.

Substitution in the area of processing and fabrication is a major activity in product manufacture and an important source of cost reduction. It can range from eliminating a manufacturing step to adopting a wholly new manufacturing process as in going from a machined to a powder-metallurgy part. This type of substitution is closely correlated with technical development and material property improvement through processing.

Design substitution involves alteration of the

product. Miniaturization is one example. Others include redesigning a product for easier maintenance or to operate on a different mechanical or electrical principle.

In functional substitution, a completely different approach to performing a needed function is found. A classic example is the substitution of transistors for vacuum tubes. Jet engines replacing piston engines and propellers in aircraft and nuclear reactors replacing boilers fired by fossil fuels are other examples. Functional substitution can lead to the profound revision of consumption patterns for materials and energy and can inspire the creation of entirely new industries (Chynoweth 1976).

2. *Levels of Substitution*

It is possible to identify several levels of substitution, depending on the complexity of the substitution process involved, and to attach typical development times for their implementation. These levels are adapted from Jantsch (1976) and are listed in Table 1. A related ranking is given by Altenpohl (1980).

Noninteractive materials substitution, or drop-in substitution, can be implemented easily while retaining product performance. It could be as simple as changing suppliers for a given material, but typically involves a different material. The development of a new material or chemical process is more common and interactive materials substitutions are so named because they often involve associated processing and design modifications.

The majority of substitution activity takes place in the first four levels listed, which correspond to the time frame for much of present corporate planning. Higher levels of substitution induce significant lower-level changes.

3. *The Substitution Process*

Once a substitution need or opportunity has been identified and defined in terms of cost and performance objectives, the steps in its implementation are:

(a) to draw on the available research and development stockpile,

(b) to begin design and development,

(c) to conduct screening tests of viable alternatives,

(d) to develop hard design and reliability data, and

(e) to evaluate on a pilot scale in the manufacturing environment.

These steps are the same as those for an original design process differing only in being more narrowly focused for low-level substitution.

The design engineer has the principal responsibility for the substitution process and the critical nature of this role can be seen in Fig. 1. At each stage, alternatives exist for the material and the part, showing that there is no unique path to a material for use or to obtaining the behavior of a part to meet its intended use. This is illustrated by Cochran's (1988) study of basic materials substitution dynamics in manufacture. Cochran found that most long-term shifts in their consumption could be explained by users choosing these on the basis of lowest cost per unit volume among technically suited materials. Thus, aluminum and plastics replaced iron and zinc in many applications where volume was a reasonable proxy for the engineering properties needed.

The choices made, then, at these various stages determine the kinds and amounts of materials (and energy) used, attendant pollution, reuse and recycling possibilities and the related technologies to be impacted. Design guides are seldom available to help

Table 1
Levels of substitution and associated development times

Level		Development time (years)
1	Noninteractive, or drop-in, material substitutions Assembly or component change	1–3
2	Development of new material or chemical process	3–4
3	Interactive materials substitution Electronic technology change Subsystem or small system development	4–5
4	Systems of reasonable complexity	~7
5	Complex weapon systems	~10
6	Telephone exchange technology	~13
7	Time for scientific discoveries to find large-scale technical application	~15

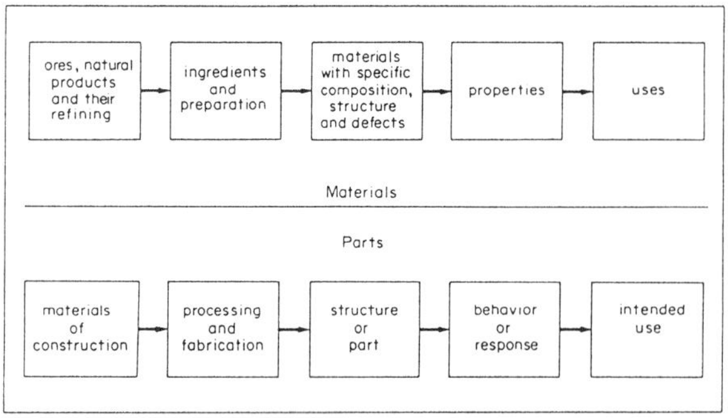

Figure 1
Schematic relationship between starting materials and their final uses

make these choices wise ones for our resources and environment. It is useful to remember also that design engineers are conservative in their technical judgements and are usually rewarded on the basis of the initial effectiveness of their designs and not on end-of-life considerations such as ease of recycling.

4. *Substitution and Technology Planning*

The critical element in facilitating future substitution is to ensure that an adequate technology stockpile exists on which to draw. This first step in the substitution process allows for the widest range of options to be considered. On a narrower basis, technological forecasting is used to identify likely future directions and products and the substitution opportunities they present. A substantial methodology exists for doing this (Lindstone and Sahal 1976). On yet a narrower level, the materials and components most critical to the product(s) are identified and specific substitution strategies developed to deal with their potential unavailability. The economics of substitution is dealt with in the article *Substitution of Materials: Economics.*

5. *Substitution Constraints*

Many factors act to inhibit substitution. At the design level, there may be a lack of property or reliability data. In manufacturing, there are retooling and start-up costs to consider. Within the organization, there is a tendency to continue with past methods and delay converting to the new. Government, society and individuals can also act to block or delay substitution.

A particular constraint arises when a design depends on a unique property or material. For example, certain superconducting systems uniquely require liquid helium. In this, and even in less demanding applications, design lock-in can occur that prevents substitution. Lock-in can also occur with dedicated manufacturing facilities which are inflexible to change. For these reasons, substitution offers little hope for mitigating short-term, unanticipated shortages. These are constrained by the technology in place and must be a direct fit in order to work.

Substitution can also be inhibited because of its impact on other materials, facilities or related areas of technology. For example, demand for a substitute material could exceed known supply or the substitution may require developments in several related areas before it itself could be implemented.

6. *Aluminum and Copper: Cost-Driven Substitution*

Sections 6–9 illustrate some of the principles discussed in the earlier sections and identify certain important driving forces for substitution.

The mutual substitutability of aluminum and copper in electrical applications has received wide study and close monitoring. It is a substitution driven principally by cost considerations (Mathur and Clark 1983) that begins with the relative cost required for equivalent conductivity. As seen in Table 2, aluminum was more attractive than copper on that basis in 1990. The choice, however, is dictated by many other factors including relative availability, ease of manufacture, size, connector technology, performance in the field and customer acceptance.

In the specific instance of substituting aluminum for copper in communications cables, it required some five years to accomplish because new alloys, connectors, cable constructions and manufacturing methods had to be developed and tested. This is a case where now either could be used depending on overall economics, but there are other cases where only copper or only aluminum can be used for technical reasons.

7. *Cobalt: Availability-Driven Substitution*

Long-term materials availability and the potential for supply disruption can be important factors driving substitution efforts and cobalt offers an excellent example of this. In 1978, due to production difficulties in Zaire, the producer price for cobalt quadrupled and a user allocation of 70% was imposed temporarily. Since Zaire is the world's major producer of cobalt, and

Table 2
Electrical conductor materials

	Copper	Aluminum
Resistivity ($\mu\Omega$ cm)	1.71	2.83
Specific gravity	8.89	2.70
Price (US\$ kg^{-1})[a]	2.86	1.76
Relative cost per mho	1.00	0.31

a *American Metal Market*, April 1990

Table 3
US cobalt consumption (%)

	1977	1980
Superalloys	26	45
Magnetic alloys	25	15
Cutting and wear tools	16	15
Catalysts and driers	24	20
Other	9	5

Source: US Bureau of Mines, *Mineral Commodity Summaries*

since most countries are heavily dependent on imports for cobalt, these events sparked a wide effort to find substitutes. As seen in Table 3 and discussed by Chin et al. (1979), magnetic alloys were one area where several commercially viable substitutes were found. These and other substitutes served to markedly alter the pattern of cobalt consumption between 1977 and 1980 in the USA. By contrast, the continued high use of cobalt in superalloys reflected the difficulty of substitution, where long and expensive testing is required to qualify new jet engine alloys.

8. Cars: Regulatory-Driven Substitution

Government-mandated fuel economy requirements in the USA promulgated in the 1970s have acted to substantially reduce the weight of cars and to alter their pattern of material use. This weight reduction required smaller cars and greater use of lightweight, high-strength materials. This has meant increased use of high-strength steels, aluminum and plastics, as seen in Table 4. By 1988, it was estimated that some 20% of a typical car's weight was in materials selected according to cost per unit fuel saved (Stark 1988).

Since the automotive industry is a large user of materials, this changing pattern has had a major impact on materials suppliers and on recycling. In addition, these newer materials require significant changes in manufacturing and assembly.

Table 4
Estimated material content in a typical US car

	1978		1988	
Material	(kg)	(%)	(kg)	(%)
Cast iron	233	14	208	14
Steel:				
high strength	60	4	105	7
other	907	56	689	48
Aluminum	51	3	68	5
Plastics	82	5	101	7
Other	289	18	268	19
Total	1622		1439	

Source: Stark (1988)

Table 5
Effect of technology on copper consumption in telecommunications

	Annual consumption decrease (10^6 kg)		
Technology factor	1985	1990	1995
Fiber optics	16	51	63
Other electronics	33	63	62
Total	49	114	125

9. Light-Wave Communication: Technology-Driven Substitution

In the 1970s, optical fibers having very low optical signal attenuation and good mechanical strength were developed and offered commercially. Reliable, long-lived, light-emitting and laser diode sources also became available at about the same time. Combining this transmission medium with reliable sources and detectors led to light-wave communication systems of high information carrying capacity or bandwidth. Their introduction, especially in telecommunications where they offered major advantages, decreased the use of traditional copper cables for this purpose as is reflected in Table 5 (Key and Schlabach 1986). Over this same period, changes in other electronics used in telecommunications had a comparable effect in lowering copper consumption. The combined impact of these technology changes on copper consumption is significant.

Light-wave communication systems, joined with current integrated chip technology, are creating a communications–computer environment with major commercial and social impacts. The availability and low cost of these systems facilitate electronic mail, teleconferencing and electronic meetings, electronic journals and books, robotics and computer-aided design and worldwide computer networks. Functional substitution of this breadth and depth profoundly alters our use of materials and energy.

See also: Conservation of Materials: Economics; Conservation of Materials: Technology; Critical and Strategic Materials; Multiattribute Utility Analysis; Substitution of Materials: Economics

Bibliography

Altenpohl D G 1980 *Materials in World Perspective*. Springer, Berlin

Chin G Y, Sibley S, Betts J C, Schlabach T D, Werner F E, Martin D L 1979 Impact of recent cobalt supply situation on magnetic materials and applications. *IEEE Trans. Magn.* MAG-15: 1685–91

Chynoweth A G 1976 Electronic materials: functional substitution. In: Abelson P H, Hammond A L (eds.) 1976 *Materials: Renewable and Nonrenewable Resources.* American Association for the Advancement of Science, Washington, DC, pp. 123–30

Cochran C N 1988 Long-term substitution dynamics of basic materials in manufacture. *Mater. Soc.* 12: 125–50

Jantsch E 1976 *Technological Forecasting in Perspective.* Elsevier, New York

Key P L, Schlabach T D 1986 Metal demand in telecommunications. *Mater. Soc.* 10: 433–51

Lindstone H A, Sahal D (eds.) 1976 *Technological Substitution.* Elsevier, New York

Mathur S C, Clark J P 1983 An econometric analysis of substitution between copper and aluminum in the electrical conductor industry. *Mater. Soc.* 7: 115–24

Stark H A (ed.) 1988 *Wards Automotive Yearbook.* Wards Communications, Detroit, MI, pp. 33–7

Tilton J E 1986 Substitution: economics. In: Bever M B (ed.) 1986 *Encyclopedia of Materials Science and Engineering.* Pergamon, Oxford, pp. 4723–6

T. D. Schlabach

[AT&T Bell Laboratories, Murray Hill, New Jersey, USA]

T

Taxation: Effect on Materials Policy

As a result of increasing interdependencies between materials and economic policies, national and subnational government tax policy is becoming an increasingly important economic management tool. Taxes and taxlike devices (including royalties, and equity and profit sharing) are used to finance public expenditures and to implement aspects of materials policy. The resource and materials industries have historically been subject to differential tax treatment and rapidly changing markets and technology will reinforce these trends. This article discusses the most important taxes, reviews how they affect decisions of materials firms and consumers and summarizes how economists evaluate tax policy.

1. Tax Bases and Rates

A tax base is the thing, person, income, commodity or transaction on which taxes are levied. Materials industries are subject to all standard business taxes such as payroll and sales taxes at the virgin, processed and recycled stages for natural, synthetic and manmade materials. Other taxes pertain specifically to extractive industries (virgin material) and these are applied at differential rates depending on the material or the extractive technology and on inputs to the extractive processes. No matter how complex the tax base and rate structure, a tax can be traced as falling on industry output, inputs which include purchases from other firms, capital investment and labor, and on the raw material itself.

Severance taxes and royalties paid to government are applied to output at differential rates based on the physical material, the quality of output and the value of output. Furthermore, various royalty arrangements include one-time and extraction-based, negotiated or competitively determined payments for the mineral (lease and bonus bids), profit sharing and required equity participation by government, and a large number of other provisions exist which partition the profits and risks for individual projects between private owners and government. Taxes on inputs used by the materials industries are as variable and complex as those on output. The property tax is applied (sometimes at rates different from those for other classes of property) to plant and equipment, to the imputed market value of the natural resource *in situ* and to royalty payments which accrue to private owners. It may also fall on possessory interest, the difference between the market value of a lease and lease payments to the resource owner. In certain cases, corporate income taxes are applied to materials industries at differential rates or net proceed taxes are invoked; the more frequent case is for special provisions to compensate for the depletion of exhaustible resources (cost and percentage depletion allowances) and for exploration and development expenses which are deducted ("written off") more quickly than normal investment amortization.

2. Tax Effects on Materials Industries

The entity which remits tax payments is but one among many which are affected. Because tax policy alters production costs and gross sales price in the taxed industries, depending on market conditions, the true tax burden may be passed in part or *in toto* to others. Tax-induced price changes in turn affect production and market prices of downstream products and alter materials selection. In the economist's short run, plant and equipment are assumed to be fixed in quantity and quality, and the firm manager decides what and how much to produce given a fixed technology. The possibility of shutting down and varying the number of employees, shifts and raw and semifinished materials used allows for variation in output. Taxes falling on fixed inputs, such as property taxes on plant and equipment, are unavoidable short of bankruptcy, whereas taxes on variable inputs (labor and purchased materials) increase production costs so that tax liabilities then depend on output levels.

In the economist's long run, *all* inputs and outputs are variable, and consequently all taxes on the firm or on the product can be viewed as increasing production costs, reducing the price net of tax received by the firm or increasing costs to the using industry. Taxes can be minimized by changing location, technology, plant size and inputs, especially materials.

The effects of taxes on the materials industries can be analyzed by means of an economic model of the firm. Management is depicted as being an optimizer and virtually all models specify profit maximization (return on equity investment) as the sole objective. Assuming that management acts as if its actions have no effect on prices (a price taker or pure competitor), the conventional firm maximizes profits by increasing production to the point where the incremental cost (marginal cost) of the last unit produced (which rises with output beyond a certain stage of production) equals the price at which they believe it can be sold. A manager in a materials industry extracting an exhaustible resource realizes that the more that is processed today the less will be left for the future because of finite reserves of various quality grades, which are inversely related to extraction costs. Consequently, this manager must look forward and evaluate how much today's extraction reduces future production and profits. This is called "user cost," and the literature shows that when the rate of extraction is increased to the point where the incremental costs of extraction (marginal production costs) plus the incremental user cost (marginal scarcity cost) equals the prevailing price, profits are maximized (Hotelling 1931). This implies that lower-cost (higher-grade) reserves should be extracted first.

When taxes affect production costs, user costs or market prices, they provide a reason for the manager to alter decisions. Markets are then destabilized, later to converge to a post-tax equilibrium which the economist attempts to forecast. For the materials-using firm, taxes may cause a change in the level of output resulting in a reduced demand for the material. In addition, when tax differentials prevail, there is an incentive to substitute a more lightly taxed for a more heavily taxed material. The fixed plant and equipment and technology of the short run constrains materials substitution, but in the long run it becomes more dominant. Economic models can be developed which reveal the effects of taxes on prices, manufacturing costs and tax effects, combined with physical properties and technology on materials selection and substitution.

In the short run, managers of nonrenewable-resource firms (materials-producing firms) select the rate of extraction and the ore cutoff grade which maximize profits. The latter decision is important because lower-grade resources left in the ground may never be economically recoverable once a mine is abandoned, for example, when ceilings collapse or when the overburden in strip mining operations is placed on top of low-quality reserves. On the other hand, tailings from milling operations are reserves which can and may be recovered in the future. In the short and long run, taxes which affect recovery and processing costs will engender changes in these critical decision variables (ore cutoff grade and extraction rate), which jointly determine total recovery from each geologic reserve over time.

In the long run, not only are the rate of extraction and ore cutoff grade decision variables, but abandonment, incremental investment in existing projects and entirely new projects are possible. For the natural-resource firm, a major portion of investment goes for exploration and the development of sites once discovery has been made and feasibility analyses have been completed. International, national and local government tax policy is included in investment feasibility and profitability studies, so tax policy is a determinant of the location and technology of extraction. (For example, after the enactment of Montana's 30% coal severance tax in 1975, all new Northern Great Plains mine openings occurred in Wyoming and North Dakota.) Thus, taxes affect prices, output, materials selection, exploration and development and location of investment.

The economist seeks to model the materials firm and the industry, to forecast how taxes affect the decisions of firm managers and how these decisions affect supply, demand and market equilibrium. These changes result in the tax burden being shifted from the taxpayer to others. The usual consequence of taxes on output is for firms to slow the rate of extraction and this is true of any tax that increases marginal extractive cost such as the corporate profits tax which falls both on pure profits and on the cost of equity capital (or accelerate it in response to a depletion allowance deduction from the corporate income tax). The response to a property tax on the estimated future profit and the resource *in situ* is to "mine out" from under the tax by accelerating the extraction rate. These responses reveal themselves in industry supply curves. Materials demand is derived from the demand for final goods and services produced and consumed which incorporate the resource. Market equilibrium occurs at the point where the quantity supplied equals demand and a market clearing price is determined.

3. Evaluation of Tax Policy

The likely effects of taxes on prices, production, economic reserves and investment allow the economist to evaluate tax policy. Judging what makes a good or bad tax requires ethical judgment. Two criteria are employed.

(a) How does the tax affect economic efficiency?

(b) Who bears the final burden of the tax?

Competitive markets ensure that economic efficiency is achieved as long as there are no "market failures." Such "failures" associated with the materials industries are attributable to extraction activity (e.g., strip mining effects on land) that produces harmful by-products which are transmitted in the air and by water. When firms are not properly regulated or "charged" for these socially costly activities, private and social objectives diverge and inefficiencies occur. The second source of divergence between private decisions and social optimality occurs when firms possess sufficient power to affect market prices, or firm managers are myopic and discount the future too heavily compared with the "proper" rate of time discount. Government policy creates its share of inefficient materials use. Taxes, other than those correcting a market failure, drive a wedge between what sellers receive and what buyers pay, and this can be shown to introduce allocative distortions, as do inappropriate regulations. One goal of tax policy is to minimize tax-induced or other distortions. The second widely recognized tax goal is that of equity; namely, that those in equivalent situations should be treated equally (horizontal equity) and those in different positions should be treated consistently according to criteria which are determined by the political process (vertical equity and income distribution objectives).

Economists have devised methods to make monetary

measurements of tax-induced distortions. Whenever a tax can be shown to change the quantity of a good or service produced, a distortion occurs, although the presence of market "failures" may mean that it can be a beneficial distortion. The difference between what people are voluntarily willing to pay for the quantity not produced and consumed because of the tax and what it would have cost to produce it measures the "efficiency loss." Taxes also cause prices to change, and these changes can be estimated as well.

The change in relative prices from the pre-tax situation, which affects incomes to persons (e.g., profits, wages and interest rates) and the goods and services on which they spend their incomes (prices), measures who bears the burden of a tax. Society's income distribution criteria are invoked to determine which is a good tax and which is a poor one from a standpoint of tax burdens.

One source which can be taxed *in toto* or in part and is devoid of efficiency loss is economic rent. Expropriating economic rent causes no changes in production and consumption, but only changes the price of the taxed item, thereby creating a nonshifted tax burden. There are four sources of economic rent in the materials industries:

(a) the constant portion (over time) of "user cost";

(b) additional profits created by lower-cost reserves being recovered simultaneously with higher-cost reserves (differential-quality rents);

(c) windfalls resulting from an unanticipated large-scale discovery of resources or technology; and

(d) windfalls occurring because of a change in government tax, regulatory or expenditure policy and unforeseen increases in market prices.

Economic rents cannot be equated to profits as defined by accountants. However, the two are associated and are often confused. Economic profits equal economic rent and exceed the necessary return to attract capital investments and other resources necessary to find and extract the material.

Although most economists advocate a tax policy goal of neutrality (no tax-induced distortions), pragmatic considerations undoubtedly dominate in actual tax policy making. Elected officials strive to satisfy various special interests and the general public simultaneously. The result of attempting to satisfy conflicting interests is highly complex tax codes and tax administration. One way these conflicts have been characterized is by analyzing the inferred trade-off in actual tax policy between maximizing economic development and maximizing tax revenues. Others have suggested that an alternative to maximizing tax revenues is maximizing the exported share of tax burdens, whereby tax burdens are shifted to those living outside of the taxing jurisdiction and residents receive the benefits.

See also: Policy: General; Policy: USA

Bibliography

Burness S 1976 On the taxation of nonreplenishable resources. *J. Environ. Econ. Manage.* 3: 27–38

Church A M 1981 *Taxation of Nonrenewable Resources.* Lexington Books, Lexington, MA

Church A M 1982 *Conflicts Over Resource Ownership.* Lexington Books, Lexington, MA

Church A M, Zimmerman M B, Foley P T, Clark J P 1981 The effects of taxation on the United States copper and coal industries. *Mater. Soc.* 5(1): 81–7

Conrad R F, Hool R B 1980 *Taxation and the Theory of the Mine.* Lexington Books, Lexington, MA

Gaffney M (ed.) 1967 *Extractive Resources and Taxation.* University of Wisconsin Press, Madison, WI

Gillis M S (ed.) 1978 *Taxation and Mining: Non-Fuel Minerals in Bolivia and Other Countries.* Ballinger, Cambridge, MA

Hotelling H 1931 The economics of exhaustible resources. *J. Polit. Econ.* 39: 80–100

Nellor D C L 1987 Sovereignty and natural resource taxation in developing countries. *Econ. Dev. Cult. Change* 35(2): 36–45

Villamor G, Mendelsohn R 1985 The taxation of exhaustible resources. *Q. J. Econ.* 100(1): 165–81

Yucel M 1989 Severance taxes and market structure in an exhaustible resource industry. *J. Environ. Econ. Manage.* 16(2): 134–48

A. M. Church
[University of New Mexico, Albuquerque,
New Mexico, USA]

Timber: Trade, Prices and Consumption in the USA

The USA is a major consumer of timber, consuming 583×10^6 m^3 in 1986. Through the mid-1980s, physical scarcity of wood fiber was not, with limited exception, of major concern. During that period, the consumption of wood fiber was outpaced by the net annual growth of the nation's forests and removals from growing stock were less than net annual growth. However, future consumption of timber is expected to be greater than the net annual growth of US forests and greater than the expected removals from growing stock inventories. Imports will then play a significant role in meeting expected levels of consumption. Net annual growth of US forests will exceed removal rates for the foreseeable future. Economic scarcity as measured by changes in resource and product prices is apparent for a number of timber species and product categories.

1. Prices, Production and Consumption: General Trends

For decades, wood has played an important role in the development of the economic and social fabric of the USA. As a material, it has found its way into a variety of uses. Since the turn of the century, however, notions

Table 1
US roundwood consumption, timber removals, net annual growth and growing stock inventory, 1952, 1962, 1970, 1976, 1986 and projections to 2040

	Amount of wood (10^6 m^3)									
Item	1952	1962	1970	1976	1986	2000	2010	2020	2030	2040
Roundwood consumption	308	314	348	382	583	637	713	777	804	821
Growth stock inventory	15761	17150	18116	19096	20196	19578	19973	20322	20752	21312
Net annual growth	390	469	556	616	632	555	575	594	619	625
Removal from growing stock	287	265	306	312	383	440	456	486	505	519

of wood scarcity have surfaced at regular intervals. In fact, scarcity concepts applied to wood fiber supplies often serve as foundations for many of the current public and private forestry programs. These concepts are basically twofold, namely, physical scarcity and economic scarcity. The distinction between the two is critical to the development of policies focused on wood supplies and has been recognized by numerous authors (Potter and Christy 1962, Barnett and Morse 1963, Manthy 1978, Smith 1979). Physical concepts of scarcity are embodied in Malthusian doctrine which implies that wood supplies are insufficient to meet

Table 2
US lumber consumption and producer prices by hardwood and softwood, 1960–1988

	Consumption (10^6 m^3)			Price (1982 US dollars)		
Year	Total	Softwood	Hardwood	All lumber	Softwood lumber	Hardwood lumber
1960	85.0	69.9	15.1	29.6	28.8	34.6
1961	83.8	69.6	14.2	28.1	27.3	32.8
1962	88.0	72.7	15.3	28.6	28.0	32.8
1963	92.5	75.0	17.5	29.3	28.6	33.8
1964	96.3	78.8	17.5	29.9	29.0	35.1
1965	97.0	78.8	18.2	30.2	28.9	37.1
1966	96.3	77.4	18.9	32.2	30.4	41.4
1967	91.5	73.5	18.0	32.2	31.1	38.1
1968	97.9	80.2	17.7	37.8	37.5	39.8
1969	96.8	78.4	18.4	42.3	41.8	45.8
1970	93.2	76.0	17.2	36.6	35.2	43.7
1971	102.7	85.7	17.0	43.8	44.0	43.2
1972	108.1	91.6	16.5	51.3	52.1	48.1
1973	109.0	91.8	17.2	66.0	66.6	64.4
1974	94.6	77.9	16.7	66.6	65.7	72.2
1975	87.3	73.4	13.9	61.9	62.4	61.1
1976	102.4	87.1	15.3	75.0	77.1	67.1
1977	114.2	98.2	16.0	89.0	92.5	76.3
1978	120.3	103.8	16.5	103.7	107.6	89.9
1979	118.0	100.2	17.2	114.0	118.1	99.1
1980	100.7	84.5	16.3	104.9	107.3	96.0
1981	91.6	77.2	14.4	104.6	106.6	97.3
1982	88.0	76.7	11.6	100.0	100.0	100.0
1983	105.0	92.5	12.7	113.5	115.0	108.1
1984	114.4	100.3	14.4	112.5	110.0	121.8
1985	116.8	102.9	13.9	109.6	107.4	117.1
1986	127.9	111.4	16.5	110.5	108.4	118.2
1987	134.7	117.8	16.9	118.2	116.1	126.8
1988	128.6	112.3	16.3	122.1	120.0	131.0

consumption requirements and needs and that social standards will decline as a result. Major concern is focused on the adequacy of supply of wood fiber. In contrast, economic concepts of scarcity imply price as a device which signals changes in the condition of supply and demand for a resource. Prices rise because of supply and/or demand factors. Supply conditions leading to long-term price increases include reduced physical stock of wood as a raw material, declining quality of timber resources, rapid rises in the cost of labor or capital required to process timber and the advantages of monopoly power. Demand changes resulting in long-term price increases include increases in expansion of timber consumption rates faster than the market can supply (Manthy 1977).

Prior to the mid-1980s, physical scarcity of timber growing stock was not readily apparent in the USA, at least nationwide and for all raw material categories combined (see Table 1). Growing stock volumes increased by nearly 28% between 1952 and 1986 and are expected to increase by 6% over 1986 levels by 2040 (0.6% per year). Fuelwood and pulpwood consumption account for the largest portion of this increase in expected volume. Consumption levels in 2040 are expected to be 5.13×10^8 m^3 greater than those of 1952—an increase that is substantially behind the 5.6×10^9 m^3 increase in growing stock that will occur during the same period.

2. *Price and Consumption Levels, 1960–1988: Wood Product Classes*

Wood as a material is consumed in a variety of manufactured forms. When categorized according to product class, trends in price and consumption of wood become more apparent and meaningful to policy-setting organizations. In 1988, domestic wood fiber consumption in the USA (estimated at 2.235×10^8 t air-dry weight) was distributed as follows: lumber, 26%; plywood and veneer, 6%; panel products, 5%; woodpulp, 30%; fuelwood, 29%; and other industrial products, 4%.

2.1 Lumber

Approximately one-quarter of all wood fiber consumed in the USA is in the form of lumber (see Table 2). Since

Table 3
US plywood consumption and producer prices by hardwood and softwood, 1960–1988

	Consumption (10^6 m^2)			Price (1982 US dollars)		
Year	Total	Softwood	Hardwood	Softwood plywood	Hardwood plywood	Plywood
1960	889.1	720.6	168.5	40.1	58.1	47.2
1961	977.6	789.2	188.4	39.0	57.3	46.2
1962	1088.3	865.0	223.3	37.7	55.3	44.6
1963	1206.2	963.1	243.1	38.6	55.0	45.1
1964	1336.0	1062.0	274.0	37.4	55.7	44.6
1965	1439.3	1152.2	298.1	37.5	55.5	44.6
1966	1498.0	1189.5	308.5	37.6	56.0	44.8
1967	1478.1	1185.3	292.8	35.4	55.2	43.1
1968	1692.2	1331.5	360.7	45.8	55.5	49.8
1969	1608.5	1240.6	367.9	49.3	57.5	52.8
1970	1656.0	1304.2	351.8	40.3	56.6	46.7
1971	1923.8	1510.4	413.4	45.0	55.6	49.4
1972	2117.1	1637.8	479.3	54.9	57.6	56.3
1973	2027.1	1628.3	398.8	68.8	62.3	66.9
1974	1648.7	1372.1	276.6	66.2	71.9	69.4
1975	1655.9	1386.3	269.6	71.1	66.0	69.4
1976	1910.1	1598.1	312.1	87.8	67.7	80.6
1977	2042.0	1728.8	313.3	104.9	70.5	91.4
1978	2127.7	1789.0	338.7	115.7	77.4	101.5
1979	1952.8	1656.3	296.5	114.2	93.4	107.9
1980	1611.2	1407.2	204.1	109.5	97.5	106.2
1981	1605.0	1659.8	223.9	108.7	99.3	105.9
1982	1550.7	1365.9	184.9	100.0	100.0	100.0
1983	1890.6	1649.3	241.3	110.0	99.3	105.2
1984	1945.2	1724.9	220.4	107.6	99.6	104.1
1985	2006.8	1775.0	231.8	107.4	89.9	99.6
1986	2097.5	1848.6	248.9	109.4	91.0	101.4
1987	2280.1	2012.2	267.9	109.8	92.9	102.6
1988	2200.0	1967.7	232.3	109.1	94.2	103.4

1960, total consumption has increased significantly, while real prices for lumber have increased dramatically. Prices for all types of lumber in 1988 were over four times those in 1960, with the bulk of the increase occurring since the early 1970s. The most dramatic increases in price have occurred in the softwood lumber market, although hardwood lumber prices have also experienced substantial increases.

2.2 Plywood

Consumption of plywood in the USA more than doubled between 1960 and 1988 (see Table 3). The bulk of this increase is attributed to rising demand for softwood plywood. Plywood prices were fairly stable between 1960 and the early 1970s, rose dramatically in the period to 1979 and since then have remained stable. Hardwood plywood prices have followed trends similar to those occurring in plywood markets generally. The same cannot be said for softwood plywood prices, which are almost three times their early 1960 levels. This can be attributed in large measure to growing physical and economic scarcity of the raw material required to manufacture softwood plywood.

2.3 Pulpwood

Domestic pulpwood consumption rose dramatically between 1960 and 1988, from 1.78×10^8 m^3 to 3.80×10^8 m^3 (see Table 4). In general, this is consistent with slow but positive increases in per capita consumption of pulp products nationwide. For selected regions and species, pulpwood prices have risen consistently but not in an unusually dramatic fashion. Over the 28-year period 1960–1988, southern pine pulp increased by a factor of 5.9, Midsouth hardwoods by 3.4, Wisconsin pine by 2.2 and Wisconsin aspen by 1.7. A major structural change in the pulpwood market has been the advent of chips produced from roundwood and by-products of primary processing plants. In 1950,

Table 4
US pulpwood consumption and prices by selected regions and species, 1960–1988

		Price (1982 US dollars m^{-3})			
		Midsouth[a]		Wisconsin[b]	
Year	Consumption (10^6 m^3)	Southern pine	Hardwoods	Pine	Aspen
1960	178.1	4.43	3.64	1.40	0.82
1961	178.7	4.40	3.62	1.53	0.55
1962	190.4	4.39	3.67	1.43	0.68
1963	195.7	4.38	3.64	1.46	0.62
1964	209.8	4.42	3.65		
1965	226.1	4.53	3.83		
1966	242.8	4.68	4.18		
1967	242.4	4.76	4.22	1.35	0.62
1968	253.5	4.93	4.40	1.32	0.76
1969	272.2	5.15	4.68	1.51	0.83
1970	274.3	5.22	4.71	1.61	0.83
1971	274.5	5.31	4.82	1.53	0.77
1972	285.5	5.78	5.15	1.53	0.78
1973	308.5	6.62	5.92	1.50	0.88
1974	316.2	7.85	7.03		
1975	259.3	7.98	7.14		
1976	297.4	8.26	7.36	2.24	1.00
1977	308.4	8.72	7.80	2.22	1.21
1978	320.2	9.21	8.40	2.57	0.85
1979	337.5	11.14	9.42	3.17	1.35
1980	331.1	12.03	10.19	3.89	1.78
1981	322.9	14.10	10.97	4.36	2.01
1982	311.7	14.69	11.06		
1983	339.3	17.51	11.14		
1984	368.0	15.35	11.42	3.65	2.00
1985	350.8	15.13	11.58	3.33	1.95
1986	368.9	14.03	11.47	3.10	1.88
1987	376.4	14.17	11.58	2.90	1.78
1988	380.1	14.30	12.42	3.06	1.39

a Weighted average for all stumpage points b Delivered to mill

6% of US pulpwood consumption was in the form of chips (i.e., 4.5×10^6 m^3). By 1985, this had risen to 41% (1.269×10^8 m^3).

2.4 Particleboard, Hardboard and Insulation Board

Hardboard and particleboard have seen major increases in consumption since 1960 (see Table 5). Consumption increases have been especially sharp for particleboard, which between 1960 and 1978 increased by a factor of 17, only to decline in the early 1980s, followed by a significant recovery between 1982 and 1988. Real prices for particleboard have increased significantly in the seven-year period 1982–1988. Hardboard over the 28-year period beginning in 1960 has also experienced significant increases in real prices.

2.5 Paper, Paperboard, Building Board and Woodpulp

Between 1960 and 1988, consumption increased 127% for paper, 128% for paperboard and 137% for woodpulp (see Table 6). Per capita consumption of these products has also increased, often markedly, during the same period. For example, US paper consumption per capita in 1960 was 110.2 kg, rising to 184.2 kg by 1988. In constant US dollars (1982 = 100), prices have been relatively stable through the mid-1970s. Woodpulp and paperboard have experienced especially sharp increases since 1986.

2.6 Sawtimber Stumpage

Some of the most dramatic shifts in US timber product prices have occurred at the stumpage level, notably sawtimber stumpage (see Table 7). Increases were especially dramatic for Western sawtimber species and most apparent in the 1960s and 1970s. For example, between 1960 and the early 1970s, real prices for Douglas fir sawtimber varied generally in the range of US\$0.032–0.035 m^{-3}. Between 1973 and 1978, prices increased substantially—often more than 25–50% per year. Since 1981, a dramatic decline in the demand for

Table 5
US particleboard, hardboard and insulation board consumption and producer prices, 1960–1988

Year	Consumption[a]			Price (1982 US dollars)	
	Particle-board	Hardboard[b]	Insulation board	Particle-board	Hardboard
1960	25	201	265		44.4
1961	30	225	263		44.6
1962	38	259	268		45.0
1963	46	291	284		45.4
1964	59	309	309		44.7
1965	75	321	315		44.7
1966	93	325	288		44.6
1967	103	317	300		43.8
1968	132	396	327		43.0
1969	159	447	340		43.7
1970	163	438	302		44.7
1971	221	527	361		44.2
1972	287	616	369		44.7
1973	319	647	369		46.0
1974	278	595	302		51.6
1975	250	483	271		51.5
1976	322	586	314		57.5
1977	384	651	324		62.5
1978	422	762	326		68.7
1979	391	721	315		72.1
1980	344	588	262		82.0
1981	334	554	197		95.4
1982	290	535	171	100.0	100.0
1983	368	581	218	106.3	102.6
1984	416	546	255	116.6	102.3
1985	428	526	267	110.3	103.0
1986	456	454	224	112.5	103.7
1987	478	443	222	121.1	101.9
1988	492	429	220	125.4	103.5

a Particleboard, 1.91 cm basis; hardboard, 0.32 cm basis; insulation board, 1.27 cm basis b Estimated for 1982–1988

Table 6
US paper, paperboard, building board and woodpulp consumption and producer prices, 1960–1988

	Consumption (Mt)				Price (1982 US dollars)			
Year	Paper	Paper-board	Building board	Wood-pulp	Paper	Paper-board	Building board	Wood-pulp
1960	21.983	15.365	1.869	26.563	32.4	41.0	46.1	27.0
1961	22.403	16.053	1.933	27.812	32.4	38.2	45.8	25.6
1962	23.246	17.058	2.066	29.511	32.6	38.4	44.2	25.1
1963	23.927	17.782	2.255	31.474	32.5	39.1	43.6	24.7
1964	25.369	18.740	2.454	33.777	32.9	39.8	42.7	25.9
1965	26.769	19.885	2.565	35.721	33.0	39.8	42.1	26.4
1966	28.846	21.541	2.396	38.388	34.1	40.1	42.1	26.4
1967	28.801	20.833	2.407	38.126	34.9	39.2	41.8	26.4
1968	30.157	22.783	2.831	42.522	35.6	37.6	42.1	26.4
1969	31.794	24.212	3.000	44.751	36.8	39.0	44.1	26.4
1970	31.699	23.530	2.828	43.969	38.8	39.7	42.2	28.9
1971	32.347	23.916	3.366	45.243	39.9	40.2	42.9	29.6
1972	34.351	26.378	3.780	48.243	40.6	41.4	44.4	29.4
1973	35.704	27.307	3.875	49.986	42.4	45.2	47.1	33.8
1974	35.498	25.718	3.452	49.670	51.9	59.7	51.6	57.5
1975	30.137	22.765	3.052	43.380	60.4	66.8	53.1	74.8
1976	34.466	25.850	3.634	48.930	63.7	69.1	57.9	75.5
1977	36.490	27.039	3.800	50.363	67.8	69.1	65.6	74.2
1978	38.452	28.137	4.142	51.443	72.0	70.5	78.2	70.3
1979	39.703	28.942	3.910	52.559	80.2	79.3	76.2	82.9
1980	39.142	27.764		53.204	89.7	92.0	86.1	100.3
1981	39.034	28.918		53.199	97.7	101.2	96.7	104.8
1982	37.942	26.508		51.247	100.0	100.0	100.0	100.0
1983	41.364	29.301		54.505	98.5	98.4	104.4	91.5
1984	44.831	31.443		58.644	105.8	110.4	108.2	104.8
1985	44.842	30.493		58.364	106.0	107.7	107.4	91.4
1986	46.338	32.568		60.697	107.0	106.6	108.8	94.7
1987	48.485	34.359		62.371	111.5	118.1	111.2	111.5
1988	49.970	34.916		63.208	123.2	133.2	113.3	136.7

Douglas fir sawtimber has led to an equally dramatic decline in price. Similar, although less dramatic, price changes have occurred for sawtimber from other species.

2.7 Price Outlook for Products and Species Groups

The real price of wood fiber and its products is almost certain to be higher in the future (see Table 8). By the year 2040, softwood lumber prices are likely to have increased by 65% over 1986 levels, while hardwood lumber prices are expected to increase by 81% over the same period (averaging 1.1% per year). This is consistent with the historical trend in lumber prices. Since 1900, the average price of lumber measured in constant US dollars has been rising at an average of 1.4% per year. Typically, however, there have been periods (a decade or two) when prices showed little change (e.g., in the 1950s). These periods have been followed by periods, such as the 1940s and 1970s, when prices rose rapidly. Similar price movements are expected in the future.

Price increases for structural panels are also expected to increase, especially plywood which is expected to rise by a factor of 1.7 by the year 2040 (averaging 1% per year). Most of this increase occurs early in the projection period. As with lumber, the price increases for plywood reflect changes in raw material (stumpage) costs. For example, the softwood stumpage cost component rises nearly four times faster than processing costs in the western Douglas fir region and nearly five times faster in the South Central region.

In contrast to structural panels, there is little change expected in projected prices for nonstructural panels. Prices for plywood are expected to decline over the projection period, while the price of other panel products is expected to remain constant for the next several decades.

Paper and board prices are expected to increase by nearly half the 1986 levels by the year 2040. Such increases reflect changing raw timber material prices and are similar in magnitude to price increases occurring since 1962.

Table 7
Average stumpage prices for sawtimber sold from National Forests in the USA by species group, 1960–1988

	Price (1982 US dollars m^{-3})			
Year	Douglas fir	Ponderosa pine	Southern pine	Eastern hardwoods
1960	42.04	25.12	45.33	29.96
1961	36.38	15.96	35.33	23.75
1962	32.58	21.17	34.17	24.46
1963	36.79	20.83	33.08	28.33
1964	50.25	25.04	36.67	29.67
1965	54.87	25.54	40.88	32.25
1966	62.58	24.79	48.29	36.67
1967	52.04	27.71	47.79	33.67
1968	74.54	36.79	51.42	28.75
1969	96.20	83.08	60.50	35.33
1970	47.33	36.25	49.79	30.38
1971	53.58	41.12	57.08	26.92
1972	75.08	68.88	68.67	35.92
1973	127.88	21.68	86.50	42.58
1974	157.62	78.33	59.33	35.75
1975	120.92	50.79	40.67	24.17
1976	120.17	69.42	59.33	23.79
1977	145.04	84.38	64.38	24.33
1978	149.21	98.17	80.17	24.50
1979	208.79	126.54	82.17	24.79
1980	200.54	95.62	72.12	24.33
1981	148.88	83.00	73.12	21.62
1982	49.25	27.88	53.00	23.50
1983	66.46	42.79	57.83	24.71
1984	53.42	49.29	56.00	36.21
1985	50.96	40.96	36.62	26.42
1986	66.83	65.12	43.08	29.08
1987	77.08	84.83	55.00	35.71
1988	99.79	70.96	55.29	58.96

Table 8
Price indexes for selected timber products in the USA by softwoods and hardwoods, 1962, 1970, 1976 and 1986 with projections to 2040

Product and species group	Index of price per unit[a]								
	1962	1970	1976	1986	2000	2010	2020	2030	2040
Lumber (m^3)									
Softwood	32.2	34.8	46.0	41.7	55.3	58.6	64.4	64.7	68.8
Hardwood	33.4	38.2	35.5	41.7	46.5	52.2	59.3	67.5	75.6
Structural panels (100 m^2, 0.95 cm basis)									
Plywood	116.0	106.5	140.0	107.6	154.8	155.8	161.4	163.5	180.4
OSB and waferboard			190.8	107.6	102.9	108.3	110.1	106.4	135.0
Nonstructural panels (100 m^2, 0.95 cm basis)									
Plywood	207.0	181.9	131.1	100.0	104.8	102.8	100.9	98.8	96.9
Other panels[b]	152.1	115.6	92.8	100.0	100.0	100.0	100.0	100.0	100.0
Paper and board (t)	90.9	87.6	87.6	100.0	109.7	117.2	125.1	133.7	142.8

a Prices are in constant (1986) US dollars and are net of inflation or deflation; prices measure changes relative to the general price level and most competing materials b Hardboard, particleboard and fiberboard

Table 9
US trade balance in wood products by commodity, 1990

Commodity	Exports ($US\$10^6$)	Imports ($US\$10^6$)	Net balance ($US\$10^6$)
Softwood logs	2136.678	5.246	2131.432
Hardwood logs	252.243	10.447	241.796
Poles, piles and posts	58.144	28.661	29.483
Wood chips	261.709	22.362	239.347
Softwood lumber	150.309	0.188	150.121
Hardwood lumber	1325.664	2518.725	(1193.061)
Softwood flooring	802.231	140.656	661.575
Hardwood flooring	3.424	1.169	2.255
Siding	35.842	43.997	(8.155)
Molding	1.568	121.056	(119.488)
Treated lumber, flooring,	50.379	156.137	(105.758)
siding and molding	10.063	11.329	(1.266)
Railroad ties	14.201	4.677	9.524
Softwood veneer	15.331	22.670	(7.339)
Hardwood veneer	195.911	189.278	6.633
Softwood plywood	303.453	12.567	290.886
Hardwood plywood	34.020	524.630	(490.610)
Hardboard	44.271	47.837	(3.566)
Particleboard	82.204	190.659	(108.455)
Cellular wood panels	48.168	8.374	39.794
Gypsum and plaster board	97.704	43.729	53.975
Other panel products	46.625	23.219	23.406
Miscellaneous wood products[a]	534.970	1175.966	(640.996)
Total[b]	6505.112	5303.579	1201.533

a Includes such products as wood waste, pulpwood, building components, containers and woodenware b Note that totals may not add up due to rounding

3. *Trade in Timber Products*

The USA has been a major net importer of wood products since the early 1900s. In 1988, the net volume imported was $3.52 \times 10^7\ m^3$, roundwood equivalent. In that year, $9.06 \times 10^7\ m^3$ of roundwood equivalent was exported (30% of roundwood produced in the USA), while $1.258 \times 10^8\ m^3$ roundwood equivalent was imported (23% of total US timber product consumption).

In 1990, softwood logs represented nearly 33% of the value of wood products exported from the USA (see Table 9). Well over 65% of the softwood log export volume in 1990 was shipped to Japan. Hardwood lumber was the second most valuable US wood product export in 1990. As for imports, nearly 47% of US expenditure on imported wood products was incurred for purchase of hardwood lumber. Almost all (US$$2.5 \times 10^9$) of this hardwood lumber import total was paid to Indonesia and Canada. In 1990, the US experienced over a US$$1.2 \times 10^9$ trade surplus in wood products. This figure contrasts with 1986, when the USA had a trade deficit of more than US$$2.1 \times 10^9$.

The US$$6.5 \times 10^9$ of wood products exported by the USA in 1990 was sold to markets in the following countries: Japan (43%), Canada (15%), the Republic of Korea (6%), Mexico (4%), the UK (4%), Germany (4%), Italy (3%), China (Mainland) (3%), Belgium–Luxembourg (2%), Taiwan (2%), Spain (2%) and other countries (12%). Countries leading in terms of contribution to the US$$5.3 \times 10^9$ imports of wood products into the USA in 1990 were Canada (66%), Indonesia (7%), Taiwan (6%), Mexico (4%) and Brazil (3%).

Future US trade in timber products will depend largely on the economic supply of timber in major forested regions of the world, and on wood product demands originating in major markets. Being a significant source of US wood imports, the timber situation in Canada will have a major influence on future US timber trade prospects.

See also: Wood-Based Industries in the USA: Economic Structure in a Worldwide Context; Wood: Future Availability

Bibliography

Barnett H J, Morse C 1963 *Scarcity and Growth: The Economics of Natural Resource Availability*. Johns Hopkins University Press, Baltimore, MD

Bureau of the Census 1990 *Lumber Production and Mill Stocks*, Current Industrial Report Series MA-24T. US Department of Commerce, Washington, DC
Bureau of the Census 1990 *Particleboard*, Current Industrial Report Series MA-24L. US Department of Commerce, Washington, DC
Bureau of the Census 1990 *Pulp, Paper, and Board*, Current Industrial Report Series MA-26A. US Department of Commerce, Washington, DC
Bureau of the Census 1990 *Softwood Plywood*, Current Industrial Report Series MA-24H. US Department of Commerce, Washington, DC
Foreign Agricultural Service 1991 *Wood Products: International Trade and Foreign Markets*, Circular Series 1–91. US Department of Agriculture, Washington, DC
Forest Service 1982 *An Analysis of the Timber Situation in the United States: 1952–2030*, Forest Resource report No. 23. US Department of Agriculture, Washington, DC
Forest Service 1990 *An Analysis of the Timber Situation in the United States: 1989–2040*, Pts. I and II. US Department of Agriculture, Washington, DC
Luppold W G, Araman P A 1988 *Hardwood Trade Trends: U. S. Exports*, Research Paper NE-611. US Department of Agriculture, Broomall, PA
McKeever D B 1984 *Trends in the Production and Consumption of Major Forest Products in the United States*, Resource Bulletin FPL 14. US Department of Agriculture, Madison, WI
Manthy R S 1977 Scarcity, renewability, and forest policy. *J For.* 75: 276–78
Manthy R S 1978 *Natural Resource Commodities: A Century of Statistics.* Johns Hopkins University Press, Baltimore, MD
Potter N, Christy F T 1962 *Trends in Natural Resource Commodities: Statistics of Prices, Outputs, Consumption, Foreign Trade and Employment in the United States 1870–1957.* Johns Hopkins University Press, Baltimore, MD
Skog K, Risbrudt C 1982 *Trends in Economic Scarcity of U.S. Timber Commodities*, Resource Bulletin FPL 11. US Department of Agriculture, Madison, WI
Smith V K (ed.) 1979 *Scarcity and Growth Reconsidered.* Johns Hopkins University Press, Baltimore, MD
Ulrich A H 1990 *U. S. Timber Production, Trade, Consumption, and Price Statistics: 1960–88*, Miscellaneous Publication No. 1486. US Department of Agriculture, Washington, DC

P. V. Ellefson
[University of Minnesota, St Paul, Minnesota, USA]

Timber Use: History

The tree and its wood have played a prominent role throughout history. The *International Book of Wood* points out that "man has no older or deeper debt" than that which he owes to trees and their wood (Bramwell 1976). "A culture is no better than its woods" writes W. H. Auden in *Woods.* The Bible records that Noah built the ark from gopherwood, the first ark of the covenant was made from acacia wood, and the cedars of Lebanon framed Solomon's temple. The best known of the early cultures developed along the Tigris–Euphrates, Nile, Indus and Yellow Rivers where forests were quickly depleted. Imports were soon heavily depended on to supply people's needs for wood.

Man's long relationship to the forest as a source of timber has been primarily a destructive one, as the resource has been exploited for fuel, for other products and for land clearance, as well as to supply wood. Consumption of wood has been a major force behind forest depletion from earliest recorded history to the present. However, records show that as early as the fourth century BC in China there was reaction to this societal and economic disruption by specific attention to sustaining timber supplies. Sustainable management of forests has been practiced since at least the thirteenth century in Europe. However, it was not until the eighteenth and nineteenth centuries that forest management for wood and other benefits was really established on a scientific basis. More recent experience in several parts of the world has shown that forests can be restored for their value in supplying timber as well as for a wide variety of environmental values. In some areas, such as the Mediterranean region, heavy use of wood over thousands of years without forest management has left restoration of the earlier forest virtually an unattainable objective.

Developments in the technology of timber use have, until rather recently, been essentially refinements of older technologies. The new developments in polymer chemistry and in some areas of engineering made possible new concepts in wood products, such as composite boards and new approaches to processing and conversion efficiency. Developments in the technology have come about as the result of three major influences: improved opportunities to use new science and engineering; the changing nature of the timber resource from natural forests to plantations; and technological, economic and market factors resulting from consumer demand and expectations.

Utilization of forest products provides the economic incentive for management of forest resources. This is an important element both in sustaining the economies of industrialized nations and in economic development in the developing world. The role timber plays in this is to a large degree basically the same role it has played over thousands of years. There is a close and dynamic relationship between economic development and the demand for forest products and services. Special attention has been given to more complete and efficient use of underutilized timber species to reduce the demand for removals from the forest, reduce the areas degraded by logging and strengthen long-term social and economic development. Over the years, the interaction of timber supply, cost, characteristics, technology, economics, social situation and market demand has been a key, but very complex, basis for timber utilization. Much of the history recounted here is related to the recognized, well established industries, because they were the ones best recorded. In the developing world, however, much of this is being replayed on a smaller scale in forest-based processing enterprises that are driven by economic incentives in the presence of a wide variety of wood materials and for a wide variety of local and export markets.

1. The Early Years

Wood has been one of the most important raw materials from early Paleolithic times, both for building and for the manufacture of tools, weapons and furniture.

From the earliest times, the use of wood involved consideration of quality, cost and availability, as well as the intended use of the product. Early woodwork in Egypt used native trees of acacia, zizyphus and tamarisk. These were not suitable for high-quality craftsmanship and were used only beneath plaster, paint or gold leaf. Higher-quality, more expensive material, such as cedar, cypress, juniper, pine and yew, was imported into Egypt from Syria and Lebanon by sea. The most expensive wood ebony was carried overland from Sudan.

Scarcity of valuable timber led to careful and economic use. Boards were carefully matched and fitted. Blemishes were removed and filled. Practices begun early in the recorded history of wood use were carried forward to the present and still appear in economical use of lumber and veneer for high-quality applications.

Wood was worked early in human existence because no elaborate tools were needed. At that time, however, the quality of wood products depended more on the characteristics of the wood than on the tools available for woodworking. The availability of copper tools by about 5000 BC made possible the higher degree of craftsmanship evident in some few surviving relics of the time. This high degree of craftsmanship led to the use of wood in carefully worked form for coffers and chests for the storage of precious possessions by about 2600 BC. Closely related to this was the first coopering to make barrels of various types by about 2800 BC.

2. The Last Millenium

From the tenth to the eighteenth centuries in Europe, wood was the material primarily used for buildings, tools, machines, mills, carts, buckets, shoes, furniture and beer barrels, to name a few of the myriad articles of wood of the time. The first printing press was made of wood and such presses continued to be made of wood for a hundred years. Most of the essential machines and inventions to establish the machine age were developed in wood during this period. Wood played a dominant role in all industrial operations and much of the art and culture of the times. In Europe, wood use reached a high plateau around the sixteenth century. About that time, however, the availability of timber began to diminish owing to the many demands for both fuel and materials and the expansion of agriculture.

North America's wood use reached a plateau during the middle to late nineteenth century, 150–200 years after that phase had peaked in Europe. The seemingly inexhaustible forests of colonial days were exploited, along with other natural resources, to feed a rapidly growing economy. Railroads, telegraph lines, steel mills and other industries were consuming wood at a dramatically increasing rate. During the latter half of the nineteenth century, the volume of sawn wood produced annually increased from 10 million to 85 million cubic meters, a level that has been maintained up to the present. Traditional uses of wood—for fuel, shipbuilding, construction of large buildings and bridges—were taken over by petroleum, coal, iron and steel, stone and brick. However, new uses for plywood, paper, poles, sleepers and chemicals, as well as the continuing demand for lumber, maintained a high level of timber production.

3. Wood for Transportation

Wood has played a key role in the transportation of people and their possessions, both as fuel and as a material, for thousands of years. Sledges made of wood were used for transport in northern Europe from 7000 BC. These were used for heavy loads such as stones and archaeologists believe that the massive stones of Stonehenge must have been moved on wood sledges placed on rollers. From this, the cart or wagon was created by putting the sledge on wheels. There are pictures of wheels that date from 3500 BC and actual wheeled vehicles found in tombs from 3000–2000 BC. In medieval times the spoked wheel became a great achievement of the joiner's art and in classical Greece both the spoked wooden wheel and the three-part solid wooden wheel were in common use. Solid wheels of planks were used on farm carts. Spoked wheels with 10–14 spokes were found in Roman forts in northern England. Rods of wood inserted into grooves and turning between the hub and the axle formed the first roller bearings.

In the nineteenth century in North America, railroads used wood for fuel, as well as for tracks, crossties, cars, bridges, trestles, tunnel linings, sheds and stations. In the cities public transportation was mostly of wood, including horsecars, electric trolleys, cable-cars, carriages and buggies. Roads made of planks laid across parallel rows of timbers embedded in the earth had come from Russia to London in the 1820s and spread to the USA during the period 1850–1857. When they were in good condition, these were the best roads in the country and more than 3000 km of such road were constructed in the mid-1800s. However, their demise soon came about as the result of excessive cost of maintenance.

Wood bridges employing advanced design concepts, laminated structural timbers and trusses were common in the nineteenth century. Covered bridges were common in North America and parts of Europe by the late nineteenth century, with the wood covering designed to protect the wood framework of the bridge itself. These were gradually replaced as the technology of wood preservation provided more economical means of wood protection and as iron, steel and concrete became the common materials for most bridge construction.

One of the first uses of wood for water transport was probably a raft or hollowed-out log. The earliest wooden ships copied the hull form of the reed boats that had been made by the Egyptians in about 4000 BC. Larger ships were built in Egypt using cedar imported from Lebanon, sometimes on a grand scale. The barge built for Queen Hatshepsut in 1500 BC to transport granite obelisks from Aswan to Thebes had a displacement of some 6800 t and required 30 oar-powered tugs to tow it. Theophrastus, a pupil of Aristotle, recorded that the shipbuilding woods in ancient Greece were silver fir, fir and cedar— silver fir for lightness, fir for decay resistance, and cedar in Syria and Phoenicia because of lack of fir. The Phoenicians dominated trade in the Mediterranean for a thousand years up to the time of Christ with their biremes, large galleys with two banks of oars. By 1000 AD, Vikings from Scandinavia were travelling mostly in wooden ships that were at least 25 m long. The wooden ship evolved somewhat in design up to the late nineteenth century. By this time, paddlewheel steamboats were cruising the major rivers of the USA and consuming large quantities of wood fuel in the process. American clipper ships were the fastest seagoing ships of the late nineteenth century, their sleek shape and large expanse of sail taking them to 20 miles (32 km) per hour. However, by the 1880s iron steamships dominated the fleets of most naval powers, with wooden ships passing even in North America as the economics became unfavorable.

The discovery in 1982 of a 15 m ship sunk at some time around 1400 BC at Ulu Burun, off the coast of Turkey, with valuable cargo from around the Mediterranean, added substantially to our knowledge of the shipping, ships and wood use of the time. While most of the wood of the ship had deteriorated, enough remained of fragments, along with distribution of remains of the cargo on the ocean floor, to tell much about the size and nature of the ship. Included in the cargo were wooden tablets that served as a base for the wax used in writing and logs of wood known in Egypt as "hbny" and generally considered to be ebony. However, the latter was identified as African blackwood (*Dalbergia melanoxylon*) rather than true ebony (*Diospyros* sp.).

4. *Wood for Weapons and Tools*

Wood was also a key material for building the war devices of the ancient world. The battering ram, the scaling ladder, the tortoise and the siege tower are examples. Another example is the catapult, used to attack enemies from a safe distance. The properties of wood made it particularly suitable for such uses. High strength-to-weight ratio was a valuable and desired characteristic of wood then, just as it is now. Instruments of war such as those mentioned were essential to the expansion of Greek and Roman civilizations and of the science and technology that developed under the tutelage and guidance of the great thinkers and teachers of the time.

One of the oldest weapons, and the oldest surviving wood relic, is the pointed end of a spear made of yew from the Lower Paleolithic era found waterlogged in an English bog. Spear handles of wood were common in the Upper Paleolithic. The bow was invented during that period and there is evidence of use of the lever concept in tools and weapons as well as of hafting stones in wood to make axes and adzes. In the Mesolithic and Neolithic eras there is much evidence of new techniques for working wood, better axes and adzes, and chisels. Woodworking improved greatly with the advent of efficient carpenter tools during the Neolithic era, about 2000 BC, and still more during the Bronze Age. The lathe spread from the Mediterranean to northern Europe in about 1500 BC, but was not in common use until iron cutting tools were available, by about 700 BC. The plow was in common use at this time, beginning with devices made from a single piece of wood and evolving to the use of various woods according to their characteristics.

5. *Wood in Construction*

Wood has been the most versatile and useful building material in its many forms and adaptations. Furthermore, until the relatively recent development of metallic and plastic structural materials, it was the only material from which complete structural frameworks could be fabricated readily. The type and durability of structures built at various times and places has depended on the type and quality of timber available and the conditions of use, as well as on the culture and way of life of the people concerned. Early pole structures were built in Paleolithic times from small trees growing along the rivers. The techniques used have been carried down over the ages and are very similar to those used by nomadic peoples today. However, the most significant developments in the use of wood as a building material have taken place where the culture and living conditions favored the erection of permanent structures.

In forested zones, where timber was plentiful, solid walls of tree trunks or heavy planks were built. Walls of timber houses in Neolithic Europe were frequently made of split trunks set vertically in or on the ground or a bottom sill plate, as indicated by examples excavated in Moorish settlements in Germany. One of these also shows a beam with mortise holes. This method of wall construction continued through the Iron Age up to Norman times. A Viking fortress built in Denmark in about 1000 AD had this type of construction, as did stave churches built during that century. Later construction used similar principles, but employed vertical squared timbers or sawn planks. This "palisade" type of house construction was brought to North America by early French settlers along the Mississippi River.

Another common style of heavy timber construction, "log cabin" style with wall timbers placed horizontally rather than vertically, can be traced back nearly three thousand years. It has been most frequently used in the northern, central and mountainous areas of Europe and North America where there have been plentiful supplies of relatively large, straight trees. In this type of construction, round, squared or sawn timbers are laid horizontally and interlocked at the corners of the building. Some large farmhouses built in Central Europe about 900 BC were apparently of this form. Vitruvius, a Roman architect writing at the time of Christ, referred to houses of this type built in a district of Asia Minor bordering the Black Sea. Records of the Roman period of a few centuries before and after the birth of Christ indicated considerable use of wood for roof construction, interior storage areas and lofts, and other parts of buildings.

Stone construction had been developed in Egypt by at least 2000 BC for both strength and durability, which were of special concern for temples and pyramids. From Egypt, it diffused through the eastern Mediterranean region to Greece, to northern Italy and to Rome. It was in Rome that the natural cement from pozzolana was discovered, then its mixture with broken stone and brick to form concrete. Using wood forms, structural members of such material were made and buildings erected at least by the first century BC. The Romans also made substantial advances in timber construction and are credited with inventing the truss for efficient use of material in structures. This concept for bridges, roof support and other structural applications was in place by the third century AD.

Gothic construction in wood was introduced during the eleventh to fourteenth centuries in Europe. In this construction the roof timbers form an integral part of the frame and the principal framing members were typically large and heavy. This type of construction persisted until the seventeenth century in Europe, but variations of it were used in the USA even into the nineteenth century. Popularity of Gothic construction peaked during the last part of that century.

Wood construction has had an interesting evolution in North America because of the relatively abundant timber resource and the scattered development of much of the country. Native Americans in the forested areas of the East and Northwest built homes and community houses from indigenous woods, the style depending on the culture and nature of wood available. In the East, structures were commonly made of poles covered with bark of birch or elm and palisades of vertical logs were used for protection. Along the northern Pacific coast, Native Americans built houses from planks of split cedar or redwood and even had gabled roofs and decorative carvings. Even in some of the sparsely forested areas of the Great Plains and the western mountains, Native Americans built frames of timber and covered them with earth to make strong permanent dwellings.

Architecture of the early colonists evolved from that of their homelands, adapted to the climatic and cultural conditions of the times and materials availability. The log cabin was introduced into North America by Scandinavian immigrants in about 1638 and was adopted in the eighteenth century by Scottish–Irish immigrants. In the far Northwest, explorers andsettlers from Russia moved south from Alaska and built houses, forts and churches of log-cabin-type construction.

The dwellings built in New England by immigrants from England during the seventeenth and eighteenth centuries followed the pattern of the English timber frame house with wattle and daub or brick between the framing members. The classical "saltbox" type of house evolved from adaptations to that pattern to provide more protection against the severe weather—covering with oak or pine clapboards, improvement of the timber frame, and exclusion of the brick. Houses in the Middle Atlantic states and the South were more likely to have exteriors of brick or stone, but elaborately decorated interior woodwork was not uncommon. The ready availability of building materials, especially wood, led to the detached or freestanding home becoming the standard dwelling in that part of the world.

Public buildings of the eighteenth century in North America followed similar trends. Churches were frequently made of wood, evolving from the simple to the ornate. Balconies were supported by wood pillars, often elaborately carved. A wooden canopy was frequently placed above the carved wood pulpit as a sounding board. Public buildings in the Georgian style frequently combined wood and brick in ways that took advantage of good features of both materials. An outstanding example is Independence Hall in Philadelphia which combined with its basic brick construction a wood superstructure for the tower clock, a wood balustrade, cedar shingles and elaborate interior woodwork.

Wood remained the principal construction material in North America well into the nineteenth century and remains so for housing today, as it does in some parts of Scandinavia and other regions where timber supplies are plentiful and the tradition of wood construction remains strong. A critical element in this continued use was the development of "balloon frame" construction in the 1830s in Chicago. This was developed to increase flexibility in building construction and to overcome the problem of the lack of trained artisans needed to make the intricate connections necessary for the heavy timber and half-timber then in use. Rather than the heavy beams, posts and diagonals connected by precise and intricate joinery, the balloon frame concept employs relatively light studs held together by panels and nailed joints. Lumber and wood or other panel material can supply the necessary components inexpensively and flexibly. Following the disastrous Chicago fire of 1871, what had been done in wood was recreated in iron. Problems of fire and the need for high-rise buildings have contributed to the demise of wood-frame construction in the centers of large cities, but it is still the major form of house construction in most of the country.

6. *Wood for Furniture*

Furniture originated as basic utilitarian pieces to meet everyday human needs. Wood was the predominant material from the beginning and is to this day. Decoration and style in furniture have developed as part of the artistic, cultural and technical development of society. They have advanced most rapidly where nobility or others with great wealth have invested in and supported such development. Availability and preferences in woods and economical processing possibilities have influenced use of wood, as well as other materials, for thousands of years.

Wooden furniture appears to have been developed first in Egypt, beginning in the early Dynastic times (c. 3000 BC). It was rare until copper tools became available, though considerable refinement in woodworking had been made possible by the development of polished stone tools. Reeds and rushes had earlier been used for furniture and early wood furniture designs copied the light style of that material. Shortages of furniture wood in that part of the world led to early developments in economical wood use. Planks cut from the same log were laid side by side and carefully fitted together. Defects were cut out and replaced with patches or plugs. Short pieces were joined with scarf joints to make longer ones. A toilet casket of Amenhotep II (1447–1420 BC) also shows the use of precious wood as a face veneer over common wood. Mortising, tenoning and dovetailing to make strong attractive furniture joints appeared by 2600 BC. Parts of furniture were originally lashed together with linen cords or rawhide thongs. Later, copper bands were used, especially on large coffins. Bronze pins or nails entered the picture in about 1440 BC.

Carvers of wood, who had evolved from carpenters or joiners, were found frequently in Europe during the Middle Ages. By the early thirteenth century, painting and gilding of wood had evolved into a separate craft. The joiner gradually became a complete furniture maker, while turners concentrated on turned furniture, which had been developed during Roman times and moved into northern Europe. By the sixteenth century, however, the two crafts came back together as turners worked with joiners to decorate higher-quality chairs and bed frames.

At that time furniture- and cabinet-making reached a very high level in several parts of Europe, with new designs and styles developed and executed by outstanding designers and furniture makers. This was evident as Renaissance furniture came into prominence then spread to England as well as to France, Spain and other parts of Europe. This began early in the fifteenth century and remained in vogue until the end of the sixteenth century. Regional variations were increasingly evident as furniture styles evolved in various parts of Europe.

During this same period, furniture styles and craftsmanship were developing in China, Japan and India. They reflected both the social role and wealth of the nobility and the culture of the times. Chinese furniture was especially noted for the high quality of workmanship, with joints being fitted so well that they remained tight without pegs or nails and often without glue.

Furnishings used in colonial America were at first quite simple and usually home made. However, a rich array of woods and highly skilled craftsmen soon led to more ornate styles of furniture. Some of those copied European styles; others were original. By the latter part of the nineteenth century, the earlier small furniture shops in the USA had been largely superseded by large furniture factories that could take advantage of the new planers, mortisers, carving machines, saws, veneer lathes and veneer slicers that were becoming available.

7. *Plywood and Veneer*

The material we now know as plywood can be traced back to ancient times, even though the term "plywood" did not appear in an English language dictionary until 1927. The tomb of King Tutankhamun from 1325 BC, discovered in 1922, revealed some of the world's oldest examples of plywood manufacture. However, gluing of veneers to make plywood was an old art even at that time. Historians generally place the art of veneering at about 3000 BC. The purpose of decorative veneering was then, as it is largely now, to extend as far as possible the utilization of the valuable, attractive woods. Such woods had to be imported into the centers of artistic wood products manufacture in Egypt and other countries and were undoubtedly expensive, and supplies were uncertain. A tablet from Amenhotep III to King Harundaradu of Arzawa, found at Tell-el-Amarna, states "one hundred pieces of ebony I have dispatched." Ornamental woods found in Egypt in that time were limited to palms, sycamore, tamarisk and acacia. More exotic woods, such as ebony, walnut, rosewood and teak, were imported from India.

The ancient Greeks also used veneer for ornamentation, but not to the same extent as the Egyptians. The highest degree of craftsmanship and artistry was first reserved for the furnishings of temples and public buildings, but later the furnishings used by private citizens also became more luxurious.

The Romans were fond of using figured veneers on their furniture, particularly on their tables. The most expensive Roman tables were those made with the veneer of citrus trees grown in Africa. While the Romans desired comfort and utility, first consideration was given to style and quality of workmanship, with the decorative value of fine veneer being highly prized.

As compared with decorative plywood, manufactured primarily from hardwoods, softwood plywood is of relatively recent origin. It began in the early 1900s in the USA and the industry is still concentrated primarily in that country. It was developed as an alternative, and in some ways an improvement, to lumber by gluing together thin layers of knife-cut wood (cut usually on a rotary lathe) with the grain of alternate layers at right angles. Early development of the industry was in response to the need for wood panels for doors, but the industry has now expanded to produce a wide array of building panels. An exhibit at the Lewis and Clark Exposition in Portland, Oregon, in 1905, provided the interest and awareness on which was based the growth of the industry over the next decades.

The increased demands for structural wood during World War I provided the first basis for rapid growth of the softwood plywood industry. The new design and application potential of this panel product led to its being adopted for many structural applications. Exterior applications were limited, however, until the availability of synthetic resin adhesives in 1935. Those superior adhesives, coupled with improvements in manufacturing and quality control, set the stage for a further rapid expansion of the industry as a means of overcoming the declining size and quality of the timber resource. This led to further development in response to the needs of World War II and to subsequent additional expansion of the industry. Softwood plywood was made in the northwestern USA originally from Douglas fir, but in 1911 a plant was established in California to make it from ponderosa pine. Plywood made from southern pine entered the market in 1963. Several other softwoods are now also used in plywood manufacture.

Production of plywood panels has increased substantially during the past few decades in most parts of the world other than Europe. From 1964 to 1985, world production doubled from 22 to 44 million cubic meters per year. Currently, something over 40% of that is produced in the USA.

8. *Wood-Based Composites*

The development of wood-based composite materials, mostly within the twentieth century, has had a significant effect on wood use and opened new opportunities for creative products. The capability to make engineered structural products and a variety of forms and combinations with resins and other materials and the opportunity to use wood residues from other types of production provide incentives for application of this concept.

Wet-process fiberboard was developed during the latter part of the nineteenth century and commercially produced as insulating board early in the twentieth. It was based on papermaking technology. The board ranges in specific gravity from 0.16 to 0.40 and is widely used for sheathing, interior panelling, roof insulation and siding.

Wet-process hardboard was the next type of composite board to enter the picture. William H. Mason discovered the process in 1924, leading to the process and board type marketed as masonite. This is a hot-pressed board with a specific gravity of about 1.00.

Platen-pressed particleboard evolved from a series of developments in the late nineteenth and early twentieth centuries using shavings, sawdust or thin wood particles. The concept was further refined through the early part of the twentieth century, leading to commercial production after World War II. Featured in development since that time has been the concept of board made from flakes or strands of wood and suitable for structural use as well as for decorative use. This type of board, called flakeboard, waferboard or oriented strandboard has grown into competition with softwood plywood during the period since 1960. It has typically been manufactured from either softwood or medium-density hardwood, such as aspen (*Populus* sp.).

This is a field that is still evolving rapidly. Medium-density fiberboard came along in the 1960s and is used primarily as core stock in the furniture industry. Mineral-bonded building products made of excelsior (wood wool) and various mineral products, commonly cement, are suitable for a wide variety of structural products. Products molded from wood particles are relatively common in Europe and may become so in other parts of the world if the economics are favorable. World production of particleboard has increased spectacularly in recent decades, from 7.6 million cubic meters in 1964 to 43.5 million cubic meters—essentially equal to production of plywood—in 1984. Expansion of the particleboard industry has been strongest in Europe, which accounts for half of the world's production. Fiberboard production has increased relatively little over the same period, amounting to 16 million cubic meters.

9. *Wood for Paper*

The term "paper" comes from the papyrus that was pressed together by the ancient Egyptians for use as a writing material. Paper in the form we now know can be traced back to China in about 105 AD. It spread to

Central Asia and to the Arab world by the eighth century and to Europe by the fifteenth.

Up to the middle of the nineteenth century, paper had been made primarily from rags. Their increasing cost and decreasing availability, together with a rapidly increasing demand for paper, had led to studies of other possible raw materials. The naturally fibrous structure of wood and its ready availability made wood a prime candidate and by the middle of the century three commercially feasible processes for making paper from wood had been developed. Hugh Burgess in the UK and Morris Kean in the USA developed the soda process, which could yield pulp suitable to mix with that from rags or straw in papermaking, and established a paper company in Philadelphia to use the process. Heinrich Voelter of Germany perfected the groundwood process, which became the principal method for producing woodpulp. In this process which is still the primary method of producing pulp for newsprint, the yield is high but the strength of the paper is low. The sulfite process was perfected in Sweden and became the basis for development in the USA by Benjamin Tilghman, who had previously developed a way to grind rags mechanically as an improvement in industrial papermaking.

Paper from woodpulp greatly reduced the cost of books and papers and encouraged literacy. It also led to manufacture of paperboard, wallpaper and paper collars and bonnets. Wood is the source of pulp for most of the paper made and used in the world today. Creative new approaches to pulping and papermaking during recent years are leading to increased efficiency in use of the raw material, paper products with new use capabilities, including structural applications, and reduced environmental impacts. Deterioration of the cellulose fiber in paper over time due to the acidic sizing agents used to improve printability is now being overcome by the use of alkaline sizing agents.

See also: Timber: Trade, Prices and Consumption in the USA; Wood-Based Industries in the USA: Economic Structure in a Worldwide Context; Wood: Future Availability

Bibliography

American Academic Encyclopedia, Vol. 8 1990 Grolier, Danbury, CT, pp. 373–6

American Academic Encyclopedia, Vol. 15 1990 Grolier, Danbury, CT, pp. 69–70

Bass G F 1987 Oldest known shipwreck reveals wonders of the Bronze Age. *Natnl. Geogr.* 172(6): 693–733

Bramwell M (ed.) 1976 *The International Book of Wood.* Simon and Schuster, New York

Daumas M 1969–79 *A History of Technology & Invention*, Vols. 1–3. Crown, New York

Davey N 1963 *A History of Building Materials.* Phoenix House, London, pp. 32–48

Derry T K, Williams T I 1961 *A Short History of Technology.* Oxford University Press, Oxford

Encyclopaedia Britannica, 15th edn., Vol. 9 1990 Encyclopaedia Britannica, Chicago, IL, p. 126

Encyclopaedia Britannica, 15th edn, Vol. 15 1990 Encyclopaedia Britannica, Chicago, IL, pp. 324–8

Encyclopedia Americana, Vol. 12 1991 Grolier, Danbury, CT, pp. 187–8

Freas A D 1989 Building with wood. In: Schniewind 1989a, pp. 31–7

Hindle B (ed.) 1975 *America's Wooden Age: Aspects of Its Early Technology.* Sleepy Hollow Restorations, Tarrytown, NY

Kauman W G 1987 Technological development in mechanical wood processing. In: Kallio M, Dykstra D P, Binkley C S (eds.) 1987 *The Global Forest Sector: An Analytical Perspective.* Wiley, New York, pp. 198–223

Mumford L 1963 *Technics and Civilization.* Brace and World, New York

Peters T F 1987 The rise of the skyscraper from the ashes of Chicago. *Am. Heritage Invent. Technol.* 3(2): 14–23

Schniewind A P (ed.) 1989a *Concise Encyclopedia of Wood and Wood-Based Materials.* Pergamon, Oxford

Schniewind A P 1989b Archaeological wood. In: Schniewind 1989a, pp. 14–18

Sedjo R A 1987 Forest resources of the world: forests in transition. In: Kallio M, Dykstra D P, Binkley C S (eds.) 1987 *The Global Forest Sector: An Analytical Perspective.* Wiley, New York, pp. 7–31

Singer C, Holmyard E J, Hall A R, Williams T I (eds.) 1954–59 *History of Technology*, Vols. 1–5. Clarendon, Oxford

Steiner P R 1989 Adhesives and adhesion. In: Schniewind 1989a, pp. 8–14

Tillman D A 1989 Energy generation from wood. In: Schniewind 1989a, pp. 102–5

Tsoumis G 1988 The depletion of forests in the Mediterranean Region: a historical review from ancient times to the present. *Scientific Annals of the Department of Forestry and Natural Environment*, Vol. KH, No. 11. Aristotelian University of Thessaloniki, Greece, pp. 283–300

World Book, Vol. 7 1990 World Book, Chicago, IL, pp. 569–73

World Book, Vol. 15 1990 World Book, Chicago, IL, pp. 136–7

Youngquist W G, Fleischer H O 1977 *Wood in American Life: 1776–2076.* Forest Products Research Society, Madison, WI

Youngs R L 1982 Every age, the age of wood. *Interdisciplin. Sci. Rev.* 7(3): 211–19

Zivnuska J A 1989 Resources of timber worldwide. In: Schniewind 1989a, pp. 238–42

R. L. Youngs
[Virginia Polytechnic Institute and State University, Blacksburg, Virginia, USA]

Trade in Materials, International

International trade in materials has reached a new level of importance in the economic and political relations between nations. Factors that have long affected materials trade patterns have also received new prominence. Resource depletion and rising production costs in the industrialized countries have increased their materials imports from the rest of the world, particularly from the developing countries. The unequal geographic distribution of these resources has made geopolitics an important factor in national materials decision making. Multinational materials companies have also expanded their activities, as have their counterparts in the form of state trading organizations. Institutional factors also can influence the competitive or free trade characteristic of materials flows. Examples at the international level include producer organizations such as the Intergovernmental Council on Copper (CIPEC), the Organization of Petroleum Exporting Countries (OPEC) or the European Community (EC).

1. Structure of Trade

The starting point for analyzing the structure of materials trade is the commodity composition of that trade, that is, the relative importance of the major materials that are traded internationally. Table 1 provides this information for the major commodity groups traded. Although the fuels are shown to dominate total trade, the main concern here is with the agricultural materials and mineral commodities. Among these, the agricultural raw materials and iron and steel are the most important. Exports of both of these materials groups have more than doubled since 1973, iron and steel even tripling from US$28.5 billion in 1973 to US$81.7 billion in 1987. Next in importance are the ores and minerals with an export total of US$45.0 billion in 1987. Nonferrous metals exports have also increased, reaching some US$43.0 billion in 1987, but represent a smaller fraction of minerals exports because of their high transport costs relative to those for metals that have been further processed. Nonmetals surprisingly do not represent a large share of world minerals trade. Since the magnitude of their transport costs is similar to that of ores, the low unit values of the nonmetals would not result in significant returns after shipping them over long distances.

Moving from trade composition to trading areas, it is obvious that these areas can be grouped only in a very general way. Different groupings are possible, but the one considered here is based on United Nations data: (a) industrial or developed market economy countries, (b) developing market economy countries, and (c) former or partially socialist economy countries. Table 2 provides the geographic patterns of materials trade among these areas. The materials data available for this purpose divide total materials trade into agricultural raw materials and into ores and metals. The percentage distributions reported show the destination of exports and sources of imports for each area.

The higher value materials trade flows both in terms of exports and imports occur among the industrial or developed economies, that is, basically those belonging to the Organization for Economic Cooperation and Development (OECD). The developing countries export materials mostly to the developed countries. They also import materials from them, most of which have been processed in some way. The centrally planned economies of Eastern Europe in the past have exported and imported mostly to and from each other, with the exception of agricultural raw materials exported to the industrial countries.

What Table 2 fails to indicate in describing materials trade is the degree of processing that is embodied in that trade. It also fails to indicate trading patterns between individual countries. Although Table 2 cannot be disaggregated completely to show more detailed trade flows, an attempt has been made in Table 3 to perform a disaggregation for the more important materials and countries. For each material the five most important countries have been selected and ranked according to percentage market share.

Examining the agricultural raw materials first, one finds a similar trade pattern for cotton, timber, hard fibers, jute and rubber. These materials are mostly exported from developing countries and are imported by the industrial countries. Some exceptions are the USA as a major cotton and timber exporter and Korea

Table 1
Value of world export trade in major materials

Materials	Export trade (billion US$)			
	1973	1978	1980	1987
Agricultural raw materials	34.6	52.0	75.0	88.5
Ores and minerals	14.9	24.5	39.0	45.0
Nonferrous metals	17.3	27.8	52.0	43.0
Nonmetals (crude)	5.9	7.7	10.0	—
Iron and steel	28.5	57.2	76.5	81.7
Fuels	63.5	233.6	468.0	284.0
Total merchandise exports	574.3	1303.0	1990.0	2882.0

Source: *International Trade*. General Agreement on Tariffs and Trade (GATT), Geneva, various issues

Table 2
Gross exports and imports of agricultural and minerals materials 1986

	Total value (billion US$)	Distribution (%)			
		World total	Developed market countries	Developing countries	Socialist countries[c]
Exports (fob)					
Developed market countries					
Agricultural materials	35.2	100.0	78.2	16.2	5.6
Ores and metals	47.3	100.0	71.4	18.2	10.4
All products[a]	1471.4	100.0	76.6	18.6	4.8
Developing countries					
Agricultural materials	13.4	100.0	55.6	32.0	12.4
Ores and metals	20.2	100.0	62.4	26.3	11.3
All products[a]	421.7	100.0	76.6	15.0	8.4
Socialist countries[b]					
Agricultural materials	4.8	100.0	37.2	19.4	43.4
Ores and metals	6.2	100.0	21.7	23.7	54.6
All products[a]	190.0	100.0	20.1	17.0	62.9
Imports (cif)					
Developed market countries	45.9	100.0	74.6	18.1	7.3
Agricultural materials	63.5	100.0	79.0	17.2	3.8
Ores and metals	1447.3	100.0	77.8	18.6	3.6
All products[a]					
Developing countries					
Agricultural materials	11.5	100.0	51.9	35.0	13.1
Ores and metals	13.6	100.0	64.4	23.3	12.3
All products[a]	439.0	100.0	62.3	26.6	11.1
Socialist countries[b]					
Agricultural materials	4.2	100.0	24.2	22.5	53.3
Ores and metals	5.0	100.0	36.9	9.9	53.2
All products[a]	170.0	100.0	21.0	11.8	67.2

Source: *Handbook of International Trade and Development Statistics*, 1988. UNCTAD, United Nations, Geneva
a Total trade including fuels b Socialist countries of Eastern Europe. This classification has been retained until it is replaced in international data sources c Socialist countries of Eastern Europe and Asia

as a major rubber importer. A similar, though more mixed trade pattern emerges for the metals. Australia, for example, is a major exporter of bauxite, iron ore and manganese and Canada is a major exporter of iron ore and copper. All of the major importers of these metals are industrialized countries. This structure of trade has led to the so-called "North–South debate" between materials-producing and consuming countries. This important materials trade issue will be returned to later.

2. *Determinants of Trade*

A number of different determinants underlie the trade patterns described in Sect. 1. Geographic determinants relate to the geological distribution of resources, political determinants relate to particular national strategies, and social determinants relate to the attempts to accelerate economic development in different parts of the world. However, economic determinants are the primary concern here.

These determinants can be understood by organizing the trade flows between the above countries into a more abstract or general trade table or matrix (Table 4). Its elements T_{ij} indicate the quantity of the material shipped from country i to country j during a given period. Total exports of the ith country ($T^{X}j$) can be calculated from this matrix simply by summing the ith row:

$$\sum_{i=1}^{n} T_{ij}$$

and the total imports of the jth country (T^{M}) by summing the jth column:

$$\sum_{j=1}^{m} T_{ij}$$

Table 3
Geographical distribution of major materials exporters and importers (and percentage market share) 1987

Materials	Major exporting nations[a]	Major importing nations[a]
Rubber	Malaysia (40.9), Indonesia (25.3), Thailand (21.0), Sri Lanka (2.6), Liberia (2.2)	USA (18.8), Japan (13.4), China (8.3), FRG (5.3), Korea (4.9)
Timber[b]	Malaysia (21.6), Indonesia (20.4), USA (8.0), France (4.4), Singapore (3.9)	Japan (23.2), USA (7.9), FRG (7.8), Italy (6.5), UK (6.2)
Bauxite	Guinea (40.9), Jamaica (13.3), Guyana (10.0), Brazil (9.3), Australia (5.8)	USA (28.9), USSR (14.1), FRG (10.1), Venezuela (7.4), Spain (6.8)
Copper	Chile (24.5), Canada (11.1), Zambia (9.0), Zaire (6.8), Peru (6.0)	Japan (19.9), FRG (12.3), USA (9.6), Belgium–Luxembourg (8.8), Italy (6.4)
Iron ore	Brazil (23.7), USSR (17.9), Australia (17.4), Canada (10.7), India (7.2)	Japan (31.7), FRG (14.6), USA (6.3), UK (3.9), Korea (3.4)
Tin	Malaysia (25.9), Indonesia (12.4), Brazil (11.2), UK (10.3), China (10.3)	USA (18.8), Japan (14.7), FRG (8.9), Singapore (7.1), Malaysia (6.9)
Cotton	USA (13.8), USSR (11.6), Egypt (10.9), China (10.7), Pakistan (8.4)	Japan (13.5), FRG (8.4), Italy (8.2), Hong Kong (7.1), Korea (5.3)
Hard fibers	Brazil (20.6), Philippines (12.2), Portugal (10.6), India (8.2), Sri Lanka (6.0)	USA (22.4), UK (8.9), FRG (8.5), France (8.5), Canada (6.1)
Jute	Bangladesh (46.1), India (21.1), Thailand (8.5), China (7.1), Belgium–Luxembourg (5.2)	USA (10.6), USSR (10.5), Saudi Arabia (6.6), Belgium–Luxembourg (5.9), UK (5.3)
Manganese ore	Gabon (34.1), South Africa (23.0), Australia (9.6), Brazil (8.0), USSR (7.3)	Japan (19.6), France (10.8), Norway (8.2), Italy (5.2), FRG (5.2)

Source: *UNCTAD Commodity Yearbook*. United Nations, Geneva, 1989
a A single country may be classified as both importer and exporter because it may import a raw material and then re-export the same material after processing b Nonconiferous

Total world trade in the material T is the sum of all elements:

$$\sum_{i=1}^{n} T_{ij} \sum_{j=1}^{m} T_{ij}$$

Economic analysis of this trade is normally divided in three ways: (a) imports into a country $T^{M}i$, (b) exports from a country $T^{X}i$, and (c) flows between any pair of countries T_{ij}.

Materials import demands depend on a number of factors which range from a country's general economic level and its military goals to its prices relative to those of other countries and the cost of producing that material domestically. Steel imports may thus depend on the size of the domestic automobile industry, the extent of the armaments industry and the costs of producing steel domestically. Economic demand theory would refine this explanation by stating that the quantity of material imports purchased is based on real domestic income or product, the price level of that import relative to the price level of an equivalent domestic material, and other factors, such as foreign exchange rates, credit availability and so-called dummy variables, which reflect unusual import-adjusting events or institutions. The substitutability of the material imports against other imported or domestic goods is also important (see *Substitution of Materials: Economics*).

Table 4
Generalized trade matrix

$$\begin{array}{cccccccc|c} T_{11} & T_{12} & \cdots & T_{1j} & \cdots & T_{1m} & T_1^X \\ T_{21} & T_{22} & \cdots & T_{2j} & \cdots & T_{2m} & T_2^X \\ \vdots & \vdots & & \vdots & & \vdots & \\ T_{i1} & T_{i2} & \cdots & T_{ij} & \cdots & T_{im} & T_i^X \\ \vdots & \vdots & & \vdots & & \vdots & \\ T_{n1} & T_{n2} & \cdots & T_{nj} & \cdots & T_{nm} & T_n^X \\ \hline T_1^M & T_2^M & & T_j^M & & T_n^M & T \end{array}$$

Materials exports depend on the demand for a material in other countries as well as the domestic capacity to produce that material. Economic trade theory normally considers exports in the demand context; they are primarily the demand for imports in other countries or areas of the world. A country's material exports are thus dependent on world income or product (or national income in the respective importing countries) and its export prices relative to world prices or to domestic prices of other countries.

Whether or not a country performs well in providing exports is often considered in terms of its ability to increase its market share. Since this depends on its relative export price level, export analysis must inevitably deal also with the conditions underlying the domestic supply of that material. These often include differential growth rates of available productive input factors and the responsiveness of export supply to the domestic supply of these factors, differential rates of productivity increases and the extent to which the country is concentrated in exports to very rapidly growing markets. Behind these factors is the quality of the materials resource base, whether it be the richness of mineral deposits or the high yields from agricultural production.

Instead of dealing with just total imports or exports, economic analysis also attempts to explain the individual trade flows between countries, as given in the matrix in Table 4. In this context, a country's choice of trading partners is seen as depending on a number of complex factors in addition to prices and incomes. The material trade flow between countries *i* and *j* is thus seen as depending on: (a) the potential of country *i* to export the material; (b) the potential of country *j* to import the material; and (c) the resistance to trade between country *i* and country *j*. The distance between the two countries is seen as a major factor determining resistance to trade, since transportation costs rise with distance. However, other factors can be equally if not more important.

Membership in the same political bloc such as the European Community could be expected to stimulate trade. Also in this category is proximity or being a neighboring country. Similarly, when multinational companies based in the importing country own production facilities in the exporting country, ownership ties may promote trade. Product differentiation such as that arising when mineral concentrates contain particular impurities that only certain smelters can remove may also influence trade. Finally, the nature of purchase agreements such as long-term mineral contracts can also be significant.

An analysis of the relative importance of these determinants of trade flows in certain materials has been made by Demler and Tilton (1980); materials studied included bauxite, alumina, aluminum copper concentrates, blister copper, refined copper, iron ore and refined nickel. The results of the analysis generally support ownership ties and the multinational companies that create them as greatly influencing trade, at least at one stage of production. For aluminum, copper and steel they are very important at early stages of processing but they are not so important for copper at the refined stage. Political blocs also are significant for trade in refined products but not for trade in ores and concentrates. Sharing a common border has also been found to stimulate trade in refined products such as aluminum, nickel and copper, but not in ores other than iron ore. Finally, the nature of long-term contractual arrangements has been found to be an important determinant.

An alternative view for explaining trade flows or why countries are net material importers or exporters concentrates on the concept of comparative advantage. This concept assumes international markets are freely competitive and that trade improves in those countries which have lower relative production costs due to efficient factors of production. If the factors of production can be extended to include materials resources as well as labor and capital, then countries with a higher quality (or lower cost) materials resource base will finally become major exporters. However, a number of imperfections in the world materials markets can complicate this simple concept. Included are the stage of process of materials and the growth in demand for it, transport problems and costs, integrated and oligopolistic market conditions, and tariff and nontariff barriers levied on processed or semiprocessed materials as they move to other markets.

Tariff and nontariff barriers are of particular importance here. In theory, they are seen as altering levels of economic welfare in particular countries. In practice, they are seen as influencing the benefit that can be derived from producing materials or having access to them. Thus, tariffs on imports provide a form of domestic subsidy that protects and fosters growth of an infant materials industry. Domestic production of such materials might additionally be encouraged for security reasons. Tariffs could also preserve an inefficient yet active domestic mining industry to sustain exploration and to stimulate employment. Both arguments also seem to be influenced by the industrialized countries having higher trade barriers for processed as compared to unprocessed materials. These tariffs are seen as being harmful for a number of reasons: materials cannot be obtained from lowest cost sources; rapid exhaustion of resources may occur, sometimes under premature technology; and trade patterns can emerge that will lead to less than optimal growth in the long run.

3. Important Trade Issues

Several issues have tended to reappear frequently in recent discussion of international materials trade (Labys 1989a). In this section, they are briefly reviewed and their significance is assessed.

3.1 International Commodity Arrangements

Materials prices are known to fluctuate more frequently and widely than prices of semiprocessed or processed goods. The explanation normally given for this phenomenon is that materials markets experience variations in demand caused by changing industrial and economic activity in various nations. Given that materials producers cannot change their output rapidly to meet such demand variations in the short run, materials prices will then have to adjust or fluctuate considerably to move a particular market towards equilibrium. This instability has thus been seen as harmful to materials producers, particularly those located in developing countries where materials earnings dominate total export revenues or foreign exchange earnings.

As a consequence, considerable international deliberations and negotiations arose during the 1970s. The focus of these negotiations was the United Nations Conference on Trade and Development (UNCTAD) and in particular the establishment of the Integrated Program for Commodities (IPC) and the Common Fund. The set of policy initiatives contained therein relates to a set of corrective commodity arrangements as well as a so-called "common fund." These commodity arrangements would attempt to maintain commodity prices at levels that in real values provide producing countries with favorable terms of trade (i.e., maintain material export purchasing power in relation to manufactured goods import requirements). In addition they would also attempt to reduce price and volume fluctuations that cause fluctuations in export earnings and balance of payments. The issue of raising prices has tended to be met with disagreement (Brown 1980) because of the lack of international institutional structures that could accomplish this task efficiently.

Price stabilization would seem to be easier to accomplish and could benefit consumers as well as producers. However, the possibilities of a favorable outcome from enacting stabilization arrangements are difficult to assess and forecast (see *Prices of Materials: Stabilization*). Impacts of price instability on consuming countries can be in the form of inflation and recession (Labys 1989b) and on producing countries in the form of the mentioned export revenue fluctuations (Labys 1980a). In reality, the tin buffer stock stabilization operation associated with the International Tin Agreement is the only materials arrangement of this type which has been actually enacted. However, it has not been notably successful in its attempt to stabilize international tin prices and buffer stock operations have been terminated. Further negotiations involving the IPC also have come to a near halt.

3.2 International Market Competitiveness

Materials trade and in particular the ability of a country to export can depend on the competitive conditions in the markets in which it operates. The competitiveness of materials producers thus depends on underlying production conditions, the influence of surrounding economic and monetary factors, and the overall market structure. Regarding production conditions several factors are important. Producers' ore grades which are better or worse than those of the marginal mine will provide cost advantages or disadvantages. The availability of less expensive energy benefits those producers whose mining and processing activities are energy intensive. The relative international level of wage costs can benefit or disbenefit producers, depending on the country of location, and strict environmental legislation can raise production costs. The adoption of modern technology improves productive efficiency. Those producers experiencing more favorable conditions will thus be able to compete more effectively.

Market competitiveness can also be influenced by differences in economic and monetary conditions among countries. One of the most important of these conditions is the exchange rate regime. Exchange rate fluctuations can cause the export position of a country to deteriorate when other foreign countries devalue their currency against that country's currency (Adams and Vial 1988). Likewise, cost competitiveness can deteriorate if that country's costs rise faster than the costs of competitors. As an example, Table 5 summarizes relative consumer price index movements between the USA and its principal copper competitors, Canada, Chile and Peru. Although the Canadian consumer price index rose faster since 1980 than the US index, there had been very little difference in US competitiveness with respect to Canada, after adjusting for exchange rate changes through 1984. In 1985, however, a 5% appreciation of the US dollar against the Canadian dollar more than offset slightly larger price increases in Canada, worsening US competitiveness in relation to Canada.

Consumer prices in Chile have risen considerably

Table 5
Exchange rates and domestic price differences in the USA, Canada, Chile and Peru (1980 = 100)

	Consumer price index in domestic currencies			Consumer price index in US dollars			
Year	Canada	Chile	Peru	USA	Canada	Chile	Peru
1980	100.0	100.0	100.0		100.0	100.0	100.0
1981	112.4	119.7	175.4	110.4	109.7	119.7	120.1
1982	124.6	131.6	288.4	117.1	118.1	100.8	119.4
1983	131.8	167.4	609.0	120.9	125.0	82.8	108.0
1984	137.5	200.7	1280.2	126.1	124.1	79.3	106.7
1985	143.0	262.3	3372.0	130.5	122.4	63.5	88.8
1986	148.9	313.4	5999.5	133.1	125.3	63.3	124.7
1987	155.4	375.7	1115.0	137.9	138.7	64.2	203.4

Source: *International Financial Statistics*. International Monetary Fund, Washington, DC, various issues

faster than in the USA, but when adjusted for exchange rate changes, US competitiveness with respect to Chile deteriorated considerably. Compared with the US consumer price index of 137.9 in 1985, the exchange-rate-adjusted index in Chile was 64.2. In other words, the Chilean peso appeared to be devalued far more than was necessary to offset the rise in relative inflation rates. In fact, of the three countries, the competitiveness of US copper producers relative to Chile has worsened the most since 1980.

In Peru prices have risen dramatically since 1980; by 1985 prices were more than thirty-three times their level in 1980. After adjusting for exchange rate changes, however, the consumer price index in Peru was 88.8, compared with 130.5 in the USA. Again, as in the case of Chile, currency devaluations in Peru had more than exceeded the difference in the relative inflation rate between Peru and the USA. Although US copper-producing cost competitiveness in relation to Peru worsened, the deterioration was less severe than in relation to Chile.

A final issue influencing market competitiveness is whether producers in individual materials markets can develop sufficient market power to control international materials supplies and prices. To a large extent, this issue has been raised because of the success of the Organization of Petroleum Exporting Countries (OPEC) in raising the levels of crude oil prices. The possibility of market control can be seen in the context of a study showing the relatively high degree of concentration in materials markets (Labys 1980b). A small number of countries or multinational firms have the capacity to dominate the supplies of materials in a number of individual materials markets. This has given rise to two institutions in particular. First, materials producers have organized into producer associations such as CIPEC for copper, BAUPEC for bauxite and PHOPEC for phosphate (see *Cartels and Producer Associations*). Second, an increasing number of countries have concentrated their materials selling and buying power into state trading organizations (STOs). However, recent studies have shown that the latter organizations do not presently represent any form of market threat (Radetzki 1985).

3.3 Import Reliance and Supply Security

In the above explanation of materials trade patterns, attention was drawn to the fact that many of the developed or OECD countries import a considerable portion of their materials from other countries. Although international political tensions are now lessening, this import dependence has raised issues concerning the availability of strategic and critical materials and the need for stockpiling them (see *Critical and Strategic Materials*; *Stockpiling: Economics*). Also relevant is the possibility of increased recycling and increased substitution.

The extent of materials import reliance varies from nation to nation. In the case of the USA, Japan and Western Europe, this dependence is relatively high, particularly in the ferroalloying minerals such as manganese, chromium and nickel. The extent to which this reliance might involve risks depends on the following four factors in particular (Hargreaves and Fromson 1983).

(a) *Production risks* are composed of the adequacy of existing capacity to meet current demand, the possibility of labor disputes, a violent conflict such as war or revolution, the range of supply sources (the concentration and location of reserves and existing mine and refinery capacity) and the time required to obtain alternative sources of supply.
(b) *Transportation risks* involve the likelihood of disruption of moving primary raw materials from the production to the fabrication sites, and of moving the processed and manufactured products to the final consumer.
(c) *Consumption risks* consist of economic impacts which involve the extent of the world market effects on key growth industries, the availability of substitutes and longer-term substitutability.
(d) *Trade risks* comprise possible collusive price or quantity agreements and embargoes.

Possibilities for offsetting these risks in addition to the maintenance of government stockpiles include:

(a) the subsidization of the domestic production of minerals;
(b) diversification of foreign supply;
(c) the evaluation of the technical performance and cost-effective aspects of substituting engineering materials for strategic minerals in a number of end uses;
(d) the development of improved processing practices through conservation of the losses of materials at all stages of the materials cycle, from extraction of ore and conversion into intermediate forms, through fabrication of the finished product, and to the eventual disposal or recycling;
(e) private inventory holding under certain circumstances; and
(f) trade policies along protectionist lines.

At present, the possibility of an import supply crisis is not seen as particularly serious (Mikesell 1987). New mineral investments are taking place in different parts of the world which tend to diversify sources of supply. In addition, the maintenance of a competitive or free trade situation in raw materials can assure ready access to low-cost supplies, at least under normal conditions.

3.4 Commodity Tariffs and Processing

Materials trade between nations has shown to be affected by a variety of tariff and nontariff barriers. Among these are specific and *ad valorem* tariffs, fixed and variable quotas, price and export controls and voluntary restraint arrangements. Import tariffs or duties tend to be lower per unit of material contained in the commodity to be imported. For example, duties are lower per unit of metal contained in ores and concentrates than on smelter or refinery products. While such a differential allows in part for metallurgical losses, it also tends to encourage the importation of crude rather than refined materials. This would give a nation the benefit of the economic activity resulting from the processing of foreign crude materials for domestic consumption.

In recent years, developing countries have suggested that existing world tariff structures have not worked to their benefit. Nations are ready to purchase the materials the latter countries export, but impose relatively high tariffs on semiprocessed or processed materials. This issue has received attention in the international tariff negotiations, particularly in the context of recent tariff liberalization proposed in the Uruguay Round of the General Agreement on Tariffs and Trade (GATT). Some tariff concessions have taken place which would permit an expansion of materials processing in developing countries. However, the negotiation of tariff and nontariff barriers is an arduous and difficult process. In many cases the developed countries have comparative advantage (greater efficiency) in processing these materials. In other cases, the developed nations have large and long-established materials processing industries which cannot easily be displaced without serious employment and economic impacts.

4. Materials Trade and Development

Primary commodity or materials exports have long been recognized as a means for importing goods and technology which can then serve to accelerate domestic economic growth. The argument for exported growth stemming from a materials base applies principally to developing countries where materials are abundant and the manufacturing sector is not very strong (Bosson and Varon 1977). In this case materials exported are often those which reflect the natural resource base of a country and those in which it should be a reasonably efficient producer. By concentrating production in one or two commodities, the country should also be able to take advantage of productive economies of scale.

An even stronger rationale is that the export industry would feature certain linkages which can stimulate growth in the related national economy. Backward linkages exist because the materials export industry purchases other materials, energy and other inputs from the rest of the economy. Forward linkages exist because other export industries or domestic industries can process the material produced for export, thus increasing added value and employment. Final demand linkages exist because expenditures by the materials export industry in the form of salaries or wages create further demands for domestic goods. Fiscal linkages exist because the export industry through taxation provides a major source of government revenues, and technological linkages can exist in the form of knowledge, skills and other externalities passed to the other sectors of the economy.

As a consequence of this rationale, a number of developing countries have seized on their materials base as a way of accelerating their economic growth. For example, Bolivia has attempted to exploit its tin base, Chile its copper base, Liberia its iron ore base, Malaysia its rubber base, and Zaire and Zambia their copper base. For a number of reasons, however, this opportunity base has not always led concomitantly to rapid growth. First, materials prices which fluctuate considerably have often led to similar fluctuations in export earnings with a subsequent stop–go effect on development plans. This effect is particularly severe where one or only a few materials dominate the total export earnings of a country.

Second, dependence on one or two export materials is often coincident with a situation where the high costs of investment require that the country depends on foreign capital sources, such as that provided by a multinational company. The host country may then view the multinational company as a sole and major source of fiscal revenues. This unfortunately has led to an "obsolescing bargain" whereby the initial dependence lessens and the host country increases taxation and capitalization until nationalization or expropriation occurs. In this case, the host country may lose the required operational skills as well as possible sources of further investment.

Third, the economic linkages obtained have not always been as strong as hoped for. For example, mining activity does not always produce strong backward or final demand linkages since its technology is generally highly capital and skill intensive, and hence it has its input requirements that diverge from domestic factor supplies. The fiscal linkage, though stronger, has often not been combined with an ability of the host government to invest productively.

The developing countries with a strong materials base thus view themselves as being in a dilemma. They possess materials representing a potential source of wealth, yet they cannot develop their materials export industries sufficiently rapidly to capture this wealth. Moreover, attempts to do so have at times led to overinvestment, sometimes at lower than market rates of interest, and hence overcapacity in markets. The obtaining of low interest rates has been attributed to (a) financing of government-owned operations by public funds that do not require a market rate of return, and (b) developing countries' access to loans by the multilateral development agencies or international financial institutions, whose terms are perceived to be more favorable than those available to the major OECD countries

These problems, however, are not insoluble. Multinational materials companies have in many cases provided an economic stimulus to governments of developing countries and will continue to do so. The former can provide the stimulus for materials-based industrialization by helping to overcome the uncertainty of exploration and related investment decisions, by providing access to markets together with increased materials processing, and by organizing consortia to provide the large amounts of capital necessary for materials investment projects.

See also: Economic Theory of Mineral Resources and Markets; Prices of Materials: Theory; Structure of Materials Industries

Bibliography

Adams F G, Vial J 1988 Explaining recent metals price swings. *Resour. Policy* 14: 85–97

Bosson R, Varon B 1977 *The Mining Industry and the Developing Countries.* Oxford University Press, London

Brown C P 1980 *The Political and Social Economy of Commodity Control.* Macmillan, London

Demler F R, Tilton J E 1980 Modeling international trade flows in mineral markets. In: Labys W C, Nadiri I M, Núñez del Arco J (eds.) 1980 *Commodity Markets and Latin American Development: A Modeling Approach.* Ballinger, Cambridge, MA

Hargreaves J, Fromson A 1983 *The World Index of Strategic Minerals.* Croom-Helm, London

Labys W C 1980a Commodity price stabilization models: a review and appraisal. *J. Policy Model.* 2: 121–36

Labys W C 1980b *Market Structure, Bargaining Power and Resource Price Formation.* Heath Lexington, Lexington, MA

Labys W C 1989a United States mineral trade: where do we go from here? Contract Report No. J0189025. US Bureau of Mines, Washington, DC

Labys W C 1989b *Impact of Commodity Market Instability on the Developed Economies.* World Institute for Development Economics Research, United Nations University, Helsinki

Mikesell R F 1987 *Nonfuel Minerals: Foreign Dependence and National Security.* University of Michigan Press, Ann Arbor, MI

Radetzki M 1985 *State Mineral Enterprises.* Resources for the Future, Washington, DC

W. C. Labys
[West Virginia University, Morgantown, West Virginia, USA]

Trading of Materials on Commodity Exchanges

Many important industrial materials are traded on commodity exchanges—sometimes referred to as futures markets. These institutions trade in commodities on a daily basis in a manner that resembles the trading of securities on stock exchanges. On a commodity exchange a seller may sell or a buyer may buy a contract involving a stated quantity of a commodity of standard specifications for delivery at a stipulated time. The time of future delivery may be as much as two years after the date of the transaction or delivery may be due in the same month as the date of the transaction. In certain markets outside the USA the time may be more constrained and in some it is possible to stipulate delivery on the day of the transaction.

Among the major industrial materials actively traded on US exchanges in early 1990 were copper, gold, silver, platinum, palladium, lumber, cotton, crude oil, heating oil and gasoline. Hides, rubber, silk, mercury, aluminum and zinc have been listed for trading but these markets in the USA have been inactive. On the London Metal Exchange there is active trading in aluminum, nickel, lead, zinc and tin in addition to copper.

Trading on exchanges exerts an important influence on worldwide markets for these industrial commodities. Other commodities of equal importance from an industrial standpoint are not deemed suitable for exchange trading. Most major foodstuffs—including livestock—are actively traded on commodity exchanges but lie outside the scope of this article.

1. Trading Conditions

Each exchange organization establishes the trading rules for the commodities in which it trades. There are significant differences among the individual exchanges as to the applicable rules, but certain general conditions apply to all.

(a) Trading is conducted by open outcry on the floor of the exchange by individuals who are members of the exchange or representatives of members of the exchange. Members may trade for their own account or on behalf of clients, but members accept full responsibility for each transaction in which they are involved.

(b) Exchange contracts are irrevocable. *Force majeure* may not be declared by either party to a particular trade, regardless of circumstance. On occasions, if necessary to maintain orderly conditions in the market, the governors of an exchange may suspend or limit trading in a specific delivery month to the liquidation of existing contracts. This is a decision made by the governors and not by the parties to the contract. In times of national emergency, such as the outbreak of war, government may act to suspend trading.

(c) The hours of trading for each specific commodity are set by the exchange authorities. Trading in all positions (i.e., in all active trading months) is conducted on each business day.

(d) In the USA, trading is based on the month of agreed delivery—usually the current month and specified months as far forward as two years. The seller is obligated to deliver (or to repurchase his contract) and the buyer is obligated to accept delivery (or to resell his contract) prior to the end of the specified month. On the London Exchange, trading is based on specified dates of delivery, with the transaction to be completed prior to that date.

(e) The exchange contract defines for each listed commodity the specific amount involved (in weight or number of units). As an example, the copper contract on the New York Commodity Exchange is based on 25 000 lb (11 363 kg) of metal, whereas the copper contract on the London Metal Exchange is based on 25 t.

(f) The exchange defines for each listed commodity the acceptable quality required in fulfilling the exchange contract. This may involve chemical analysis for metals or visual inspection for other commodities. For metals, the exchanges limit delivery to material produced from refineries specifically approved and named by the exchange.

(g) Acceptable delivery requires the seller to place the material in a designated warehouse approved by the exchange authorities. The receiving warehouse issues a certificate verifying the weight and quality of the material delivered. Even when a certificate has been issued, there have been instances in which the buyer on taking delivery has claimed quality or quantity deficiencies.

(h) Approved exchange warehouses may be located in cities geographically remote from each other. The seller has the option of making delivery to the warehouse of his choice. For example, delivery against a copper contract on the London Exchange may be made in Stockholm, London, Rotterdam or Genoa, among many locations.

The great majority of exchange transactions, however, do not result in physical delivery of the material. Prior to the specified contract date, the seller has bought back the quantity he has previously sold, or the buyer has resold the quantity he has previously bought. The transactions that result in actual delivery being made represent only a small fraction of the volume of transactions.

Thus, in early 1990, the daily volume of copper transactions on the New York Commodity Exchange regularly exceeded 10 000 contracts, involving 125 000 t of copper. However, the amount of copper physically held in exchange warehouses during the first quarter of the year was never more than 10 000 t and at times was much lower. The number of open contracts at the end of each trading day was in the order of 35 000 to 40 000, in contrast with a monthly trading volume of more than 200 000 contracts. Clearly, a high percentage of all transactions were being closed out within a few days of the transaction date. This reflects the fact that much of the activity on the exchanges is on the part of speculators and investors, as discussed in Sect. 2.

2. *Trading Practice*

For industrial organizations, most sales or purchases of a commodity on an exchange are undertaken to hedge price risks. By making a commitment to deliver a commodity at an agreed price by a specified date, a seller establishes the revenue it will derive from the disposal of its product.

In the same way, by undertaking to take delivery of a needed raw material at an agreed price by a specified date, a buyer fixes the cost of that material which it requires in the production of its manufactured items.

Unless the exchange transaction involves immediate delivery (which is rarely the case) the amount of money transferred is limited to the margins required (see Sect. 2.2). Thus, the transaction is not entered into for revenue purposes, but primarily to establish a price to apply to a future delivery, and is therefore considered a hedging transaction.

In most cases the seller prefers to deal directly with a user of the product when the time for delivery on the exchange approaches. Therefore, shortly before the date of contract maturity, the seller buys back the exchange contract and simultaneously sells directly to a user. These two transactions are based on price conditions prevailing at the time and offset each other. Thus the net effect is that the seller realizes on his product the price of the original exchange transaction.

From the standpoint of the user, there are distinct advantages in obtaining raw materials from a specific commodity producer rather than taking delivery from an exchange warehouse. The producer supplies a known brand and delivers it directly to the point of consumption, saving the user the cost of transportation from the warehouse. Therefore, consumers will usually sell back the contracts they previously bought and simultaneously buy at a corresponding price from a normal supplier.

This description of trading practices involves what are commonly referred to as "trade" interests—producers, consumers and merchants whose operations are tied directly to the specific commodity. However, a substantial share of the activity on commodity exchanges is linked to transactions by speculators and investors. These parties are not involved in the production or consumption of the commodities. They trade in the commodities because they believe they can profit through anticipating price changes—either buying when they expect the price to rise or selling when they expect the price to fall. Still another group of traders are those who derive income through purchase of prompt or nearby positions and simultaneously sell the far forward positions when the spread between the prices equals or exceeds the prevailing interest rates (as described in Sect. 2.1).

2.1 Contangos and Backwardations

Significant price differences are posted in exchange trading, depending on the time when a particular contract matures. In the jargon of the exchange these differences are described as either a contango or a backwardation.

A contango is a premium on the price for future delivery as compared with the price for prompt delivery. A backwardation is a discount on the price for future delivery as compared with the price for prompt delivery.

Usually, commodities trade with premiums for the future months—in other words, with contangos. If substantial quantities of a commodity are held in exchange warehouses, the owners of those inventories are losing interest the longer they hold the materials. The contango therefore reflects this loss of interest plus the actual costs of storage. In periods of recession (such as 1974–1975 or 1982–1983) large surpluses of commodities are likely to develop and to move into the warehouses. The purchasers ("longs") who hold title to these inventories paid cash for them when they took title from the sellers ("shorts"). What happens is that offers are made to sell the materials for forward delivery and that bids for the spot or prompt position are reduced until a normal contango is established.

The precious metals—gold and silver—provide classic illustrations of the workings of the contango. Known stocks of gold and silver are enormous since most of these metals ever mined still exist as bullion bars, coins, jewelry or works of art. There is thus no prospect of a physical shortage of either. As a consequence, quotations for both metals on the exchanges invariably show premiums for future delivery.

Because of the absence of *force majeure* in exchange transactions, the purchase of spot commodities and simultaneous sale for forward delivery at an adequate contango is a relatively risk-free method for financial institutions to obtain a return on liquid funds. This explains how the large warehouse stocks are financed during periods of industrial recession. However, the larger the stocks the greater the pressure on the price of the commodity.

A backwardation develops when the market senses a potential or actual shortage of a given commodity. If stocks of the commodity are limited, consumers become concerned over access to adequate supplies and are prepared to pay premiums for spot material. Strikes, wars or other international disturbances, and unusual weather conditions in the case of agricultural commodities, are the usual causes that lead to development of backwardations. In some cases a sharp increase in demand for a commodity will absorb all available supplies, as was the case with nickel in 1988, when huge backwardations developed.

Whereas in the case of the contango the amount of the spread between prompt and future prices is restricted as a rule to the prevailing interest and storage costs, in the case of the backwardation there are no limitations on the spread that may occur unless exchange authorities intervene in an effort to maintain orderly markets. Several instances of such intervention did occur during the period 1988–1990 when heavy demand absorbed available production of several metals and depleted visible inventories to extremely low levels.

2.2 Margin Requirements and Variation Margins

Any corporation or individual having the required credit standing may buy or sell on commodity exchanges by dealing through a member firm. In the USA, commodity exchanges establish margin requirements. These are deposits that traders must make at the initiation of a transaction, in order to insure that they will perform the contract when it comes due. The amount of the margin to be deposited is not less than a specified amount for each contract traded. The exchange authorities periodically review the margins required for each commodity traded on their exchange. In recent years, the practice has been to set margins at between 5% and 10% of the value of an individual contract. As the price of the commodity fluctuates, the authorities either raise or lower the amount of the initial margin per contract. They may also establish lower margins for the so-called "trade" interests and higher margins for speculators.

Commodity exchange margins are much less than the 50–60% margin required for security transactions. The rationale for the lower margins on futures trades is that it is important to encourage a maximum volume of trading in commodities. The larger the volume, it is argued, the greater the likelihood that the price of a given commodity will represent a true consensus of its value.

In addition to the initial margin required when a transaction is initiated, if prices change significantly so-called variation margins may be required. A seller who has sold at a given price may be required to provide additional funds if the price of the commodity rises sharply. The rationale for this is that the seller may be tempted to ignore his obligation to deliver, forfeiting his initial margin, because by selling the contract quantity through another channel he can net a higher price. Similarly, in the event of a sharp decline in price a buyer will be called on for a variation margin above the initial margin posted at the time of transaction.

In times of violent price movements, such as those that have occurred in many commodities in recent years, some commodity traders have experienced grave difficulties when called on for extensive variation margins. Should the additional funds not be forthcoming, the member firm through whom the deal was initiated will close out the client's position. This usually results in total loss of the initial margin.

For many years the London Metal Exchange did not require margins. Member firms could at their own option require margins from traders whose credit ratings were not well established. The severe losses experienced in 1985 at the time of the collapse of the International Tin Agreement plus pressure from the governmental authorities has led to adoption of margin and clearinghouse arrangements at the London Exchange since then.

3. Unlisted Commodities

Many significant industrial raw materials are not traded on commodity exchanges. Among them are steel, steel scrap, bauxite, iron ore, manganese, molybdenum, tungsten, all of the important non-metallic minerals (sulfur, potash, phosphates, asbestos, graphite, talc, boron, etc.), plastics and wool.

There are various reasons that make commodity exchange trading unsuitable for these commodities. This can be illustrated by considering the case of steel scrap—one of the major commodities in industrial use. Steel scrap has a history of extremely volatile prices as well as large volume, both factors that would normally attract the interest of the exchanges. However, establishing specifications for steel scrap is difficult and inspection and grading of the material on delivery would create enormous problems. The handling of scrap into and out of a warehouse would be difficult and costly in comparison with materials marketed in standard size and uniformly packaged, as is the case with refined metals. Moreover, the bulky nature and relatively low value of steel scrap would mean that a high proportion of the trading value would be absorbed by transportation costs into and out of the warehouse—costs that are not involved in direct delivery of scrap from its source to consuming points.

The extent of these disadvantages is evident when comparison is made with the heavily traded precious metals (gold, silver and platinum). They possess high value, their quality is readily determined, they are cast in uniform shapes, they can be stored easily and transportation costs are an insignificant fraction of their value.

Many of the commodities not traded are, like steel scrap, bulky and have relatively low value. Others are difficult to grade for purposes of quality determination. In short, although many economists believe that trading on an exchange is the best means of determining the price of a commodity, practicalities indicate that for many commodities it is not a viable procedure.

4. Summary

A major feature of commodity exchange trading is that regardless of the state of the business cycle listed commodities can be bought or sold. In times of recession this means that cash can be raised by a seller who might otherwise be accumulating surplus metal. In times of business booms it means that a buyer can obtain badly needed supplies above the quantities available from regular suppliers.

However, these advantages are obtained at the cost of wide price fluctuations. When business is poor the price is set by the offers made by the most desperate seller. When business is good the price is set by the most eager buyer.

In theory, the presence of speculators is supposed to act as a buffer to ameliorate price swings. However, with the simultaneous availability of information worldwide regarding developments affecting supply and demand, in practice in recent years there have been few speculators prepared to take positions contrary to the accepted view of the business outlook. Thus, there are few speculator buyers in times of recession and few speculator sellers in times of economic recovery. Some would argue that speculators have accentuated rather than buffered price swings.

See also: Markets for Materials; Prices of Materials: Theory; Trade in Materials, International

Bibliography

Cordero H G, Tarring L H 1960 *Babylon to Birmingham: An Historical Survey of the World's Non-Ferrous Metal and Iron and Steel Industries and of the Commerce in Metals Since the Earliest Times.* Quin, London

Economist Intelligence Unit 1958. *The London Metal Exchange.* The Economist, London

Gibson-Jarvie R 1976 *The London Metal Exchange: A Commodity Market.* Woodhead-Faulkner, Cambridge

Slade M E 1988 *Pricing of Metals.* Queen's University, Kingston, Ontario, Canada

Thewles R J, Harlow C V, Stone L H 1977 *Commodity Futures Game: Who Wins? Who Loses? Why?* McGraw-Hill, New York

S. D. Strauss
[New Rochelle, New York, USA]

V

Value Added in Processing

Production is ultimately for the purpose of satisfying the wants of consumers—to provide the food, the clothing and housing, the theaters and schools, the transport and communication systems and the many other products required for the satisfactory conduct of personal and social life. The public attaches value to these products and is therefore prepared to pay for the resources and the factors of production required to produce them. Each of the processes whereby raw materials are progressively converted into the final products required by consumers adds value to the original raw materials; the sum total of the values added by the individual processes plus the original raw material value is the value of the final product.

The value attached by consumers to particular products is a matter of subjective judgement and cannot therefore be quantified. However, the term *value added* has a precise quantifiable meaning in economics: the value added by a production process is the sum of the costs incurred in the operation of that process. The value added in the manufacture of a final product is the total cost of processes which contributed to the manufacture of that product. All costs can ultimately be traced back to the payments made to the so-called primary factors of production: capital and manpower. These costs include the wages and salaries of employees, the profits of owners and the rents paid to the owners of land.

National income can be estimated by summing the incomes of all individuals, but the foregoing shows that it can equally be estimated by summing process costs—by summing the values added in all production processes—after making due allowances for international transactions. Value is added by all economic activity, whether this be manufacturing, banking or music-making. This article describes how value is added to raw materials in their progressive processing into physical objects such as automobiles, chairs, houses and bridges.

1. Inputs, Outputs and Value Added

Three essential elements are required for the production of goods:

(a) natural resources—land, ores, mineral fuels and timber, for example;

(b) the human knowledge, skills and energy to convert natural resources into products;

(c) capital—the factories, process plant and machine tools, the tankers and ore ships, the telephones and computers necessary to convert natural resources into products on the scale and with the efficiency required to provide the standard of living expected in advanced economies.

The conversion of natural resources into products is carried out in processes such as mining, steelmaking, extrusion, molding, casting and painting, each of which combines the above three elements so as to produce the desired output. Figure 1 illustrates this point for a single process company. The company purchases the classes of input listed on the left; to these it applies the resources within its own control, its capital and labor, to produce an output of goods and/or services as indicated by the outward-pointing arrow at the base of the diagram. The value added by the process, or rather by the company operating the process, is the difference between the money received from the sale of its outputs—its goods and services—and the money paid to other companies for its inputs—its supply of purchased goods and services.

The manufacture of most final products requires not just one single process stage, but a whole series of connected process stages. The manufacture of a motor car presupposes the manufacture of intermediate products, such as steel sheet, plastic moldings and zinc diecastings; it presupposes the operation of the processes by which the intermediate products are made; and it presupposes operation of all the processes from extraction to refining which provide the materials and energy for making these intermediate products.

A sequence for the manufacture of a product requiring three stages for the conversion of natural resources into a final product is illustrated in Fig. 2. The first stage typically involves the digging or pumping of a raw material from the ground. It is the first stage of adding value in the processing of the raw material. The output from this stage is termed an

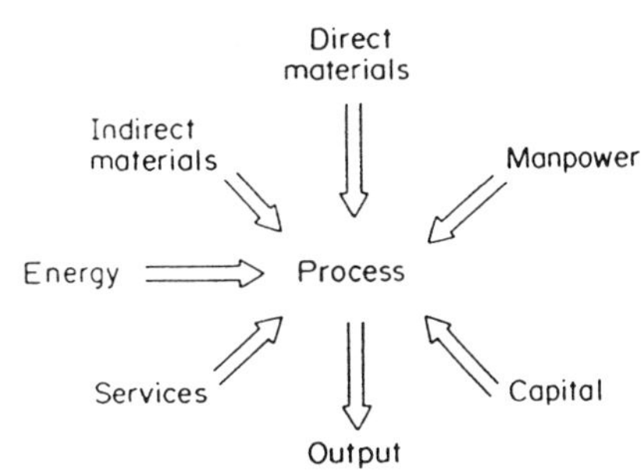

Figure 1
Symbolic representation of inputs to a typical industrial process (after Pick 1972)

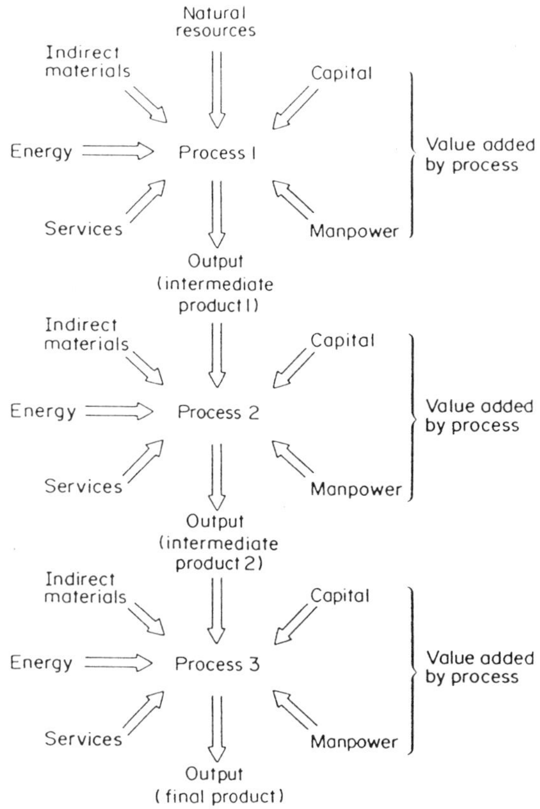

Figure 2
The flow from natural resources to a final product produced in a three-process sequence

intermediate output—an output intended for processing by another company, rather than for sale to final demand. It is in this case sold to the company operating process 2, which combines it with intermediate supplies from other companies and with its own primary resources to produce its output. This is again an intermediate output and is sold to the company operating process 3, in which further intermediate and primary inputs are processed to produce the final product of this particular sequence. This could be a can of beer or an automobile, a building, a bridge or a machine tool—any finished product for use by consumers, either for current consumption or for capital accumulation; it could also be any finished product used by industry for capital accumulation or, in a trading economy, any product, finished or intermediate, sold for export, or any good "consumed" by a branch of government.

In the foregoing description, process inputs were classified either as purchases or as primary inputs. It is not difficult to see that the cost of each of the purchased inputs, indirect materials, energy or services can also be traced back to the primary inputs which had, directly or indirectly, to be used in their production; the cost of the coke used in ironmaking is ultimately attributable to the labor and capital required for coal mining and coal transport and for the operation of the coke ovens. The cost of producing the output from any process stage is equal to the sum of all the primary input costs incurred either directly within that stage or indirectly in the chain of supply processes to that stage. The cost of a final product is the sum of the costs of all the primary inputs directly or indirectly used in producing that final product.

Domestic product is the sum of the value of all the final products produced in an economy; from the foregoing it follows that it must also be equal to all the incomes paid to these primary inputs. (The profit element appears in these calculations as an income to the owners of capital.)

2. *Buildup of Value in Materials Processing*

The basic relations between inputs, outputs and value added may be illustrated by reference to what is still one of the most important process sequences for the conversion of raw materials to final products: the conversion of iron ore first into iron and steel products and then into the ferrous components of final products, such as automobiles, bridges and tin cans. The sequence is illustrated in Table 1, which shows some of its major products and processes. Of particular interest is the low unit price of ore compared with the price of downstream semifinished products and components and of the finished products. This increase in price per unit weight results partly from yields less than unity in most of the process stages in the sequence, but is mainly due to the previously mentioned fact that value, costs and prices simply reflect the cost of primary inputs in processing: that the bulk of processing costs in the manufacture of automobiles and other final products is incurred downstream of the primary extraction process.

The way in which products, prices, weights, output values and value added relate to each other for another particular materials processing sequence, that of aluminum, is illustrated in Table 2. This again shows the small value of the raw material compared with that of the highly processed downstream products. In 1963 (the date of a comprehensive UN report on the aluminum industry) the value of world trade in bauxite costing US$8 per tonne was estimated to be US$240 million, compared with a trade of US$6000 million in wrought semifinished products and castings costing US$1000 per tonne. The origin of the relationship between raw materials and other costs may be seen from Table 3, which illustrates the range of the major inputs to the processes of aluminum conversion. Mining and exploration costs accounted for a smaller fraction of the cost of bauxite than did shipping;

Table 1
Sequence of processes and intermediate products involved in the manufacture of final products from steel. UK 1978 prices per tonne for some of the products in the sequence are also shown. The £5000 per tonne quoted for a final product is the approximate figure for an automobile in a UK showroom

Process	Value (£ t^{-1})
Mining and beneficiation	16
↓	
Iron ore (plus coke, limestone and sinter)	
↓	
Blast furnace	
↓	
Pig iron (plus ferroalloys, scrap and fluxes)	80
↓	
Steelmaking processes	
↓	
Molten steel	
↓	
Teeming	
↓	
Ingot	
↓	
Primary cogging mill	
↓	
Bloom	
↓	
Rerolling	
↓	
Billet	175
↓	
Rerolling	
↓	
Hot rolled products (black bar, hot rolled strip)	225
↓	
Pickling	
↓	
Cold roll or cold draw	
↓	
Cold finished products	285
↓	
Machining, pressing	
↓	
Engineering components (car bodies, machine parts)	1250
↓	
Assembly and finishing	
↓	
Final products	5000

Source: Pick (1976)

bauxite accounted for less than a quarter of the cost of alumina, which in turn accounted for only a quarter of the cost of aluminum ingot. The remaining costs were spread over a variety of items including indirect materials, energy, labor and capital charges.

There are important implications from this for the response of the economic system to changes in the price of a commodity, as occurred during the first oil crisis in the early 1970s, and this may be illustrated by the example of plastics. The processes involved in the progressive conversion of crude oil into plastics products, some of which are illustrated in Fig. 3, also involve a whole range of inputs in addition to that of the primary raw material. The result is that a sudden surge in oil prices does not have a proportionate effect on plastics product prices, even though they will be unsettling for international trade and finance. Figure 4 shows that the 300% increase in the price of crude oil led, for example, to an increase of only 30% in the price of polyethylene bags.

Table 2
Processes, products, prices and values in the progressive conversion of bauxite to semifinished aluminum products in 1963

Process	Product	Price (US$ t^{-1})	Weight (Mt)	Value of output (10^6 US$)	Increases in value from previous process stage (10^6 US$)
Mining	bauxite	8	30	240	240
Ore refining	alumina	75	12	900	660
Aluminum smelting and refining	primary aluminum ingot	450	6	2700	1800
Fabricating and casting	wrought semis and castings	1000	6	6000	3300

Source: Pick (1972)

Table 3
The 1963 cost of inputs in aluminum manufacture

Product	Price (US$ t^{-1})	Typical inputs per ton output	1963 (US$)
Bauxite	8	mining, including exploration	1.0–3.0
		beneficiation	0.0–1.0
		drying	0.8–1.5
		shipping	3.0–4.5
		local taxes, mining and shipping company profits	balance
			8.0
Alumina	75	bauxite	16.8
		caustic soda	5.1
		steam	3.3
		electric power	1.0
		fuel for calcination	2.6
		labor, operating, maintenance and indirect	12.5
		maintenance, materials and equipment	3.0
		capital cost:	
		depreciation on fixed capital	11.8
		interest on fixed capital	10.3
		miscellaneous supplies and general expenses	7.0
			73.4
Primary aluminum ingot	450	alumina	150
		fluorides	25
		carbon	25
		operating and maintenance supplies	18
		power	61
		labor	42
		miscellaneous and general expenses	38
		capital charges:	
		depreciation	53
		interest on fixed capital	38
			450

Source: Pick (1972)

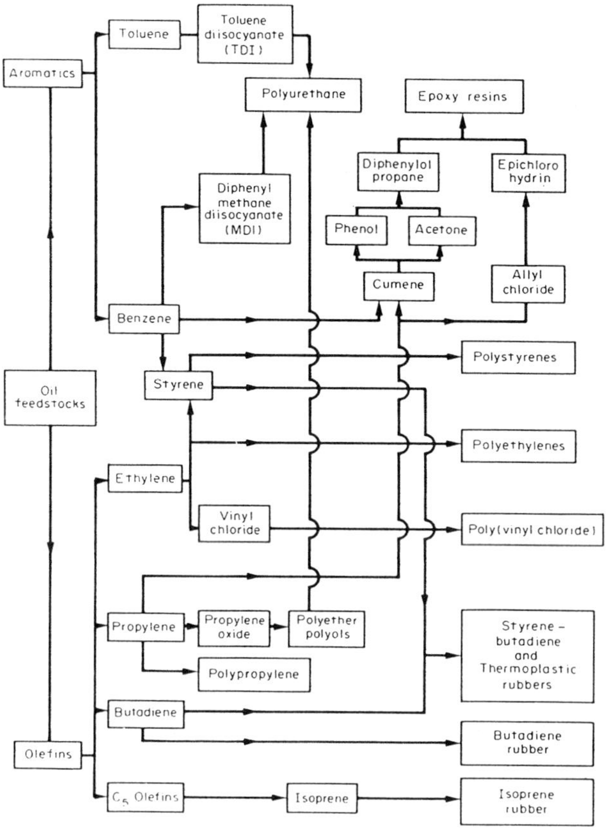

Figure 3
Production of polymers from oil feedstocks (courtesy of Shell International)

Energy costs in manufacturing are on average modest compared with other manufacturing costs. In 1978, for example, total expenditure by the manufacturing industries of the USA on fuel and electrical energy was US\$37.7 billion (US\$20.7 billion on fuels, US\$17.0 billion on electricity), compared with a value added by these industries of US\$518.7 billion. However, some industries are much more energy-intensive than others, particularly some primary industries such as the materials-producing industries. The average energy consumption by manufacturing industry as a whole was 44 MJ per 1972 dollar of value added, but the figure for the steel industry was 299 MJ, for the cement industry 434 MJ and for the aluminum industry 401 MJ. Energy consumption in the final product industries, such as the automobile industry, tends to be relatively small. Energy costs in process sequences such as that illustrated in Table 1 tend to be concentrated in the stages which take place in the materials industries.

3. Raw Materials, Processing Costs and National Product

The whole of manufacturing and construction may be considered as a series of interrelated process sequences as illustrated in Fig. 5. The present article is concerned with the flows indicated by arrows. The basic classes of natural resources are listed on the left, the basic classes of final users on the right. The sum of the sales of the listed industries to final users constitutes their contribution to national product.

It is of interest to show how value builds up in the totality of sequences which constitute the extraction, conversion, manufacturing and construction industries of any economy. This is done in Fig. 6, based on 1978 census data for the US economy. The figure shows that the enormous ratios of final product values to primary product values illustrated above for steel and aluminum also apply to the US economy as a whole. Domestic and imported raw materials together had a value of US\$23 billion but the processing of these gave rise to manufacturing and construction output of US\$609 billion—a ratio of 26.5:1. Gross national product was US\$2156 billion, of which domestic plus imported raw materials accounted for 1.1%.

4. Value Added in Processing and the Structure of Industry

A company sufficiently integrated to produce all the raw materials directly or indirectly required in the

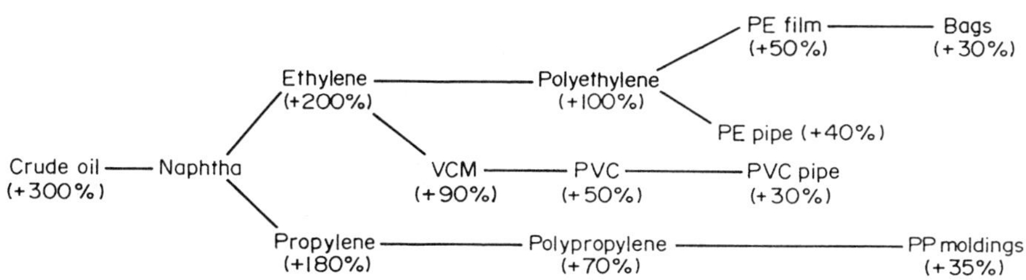

Figure 4
Conversion sequence for the manufacture of some plastics products, showing the percentage increase in price of downstream products resulting from a 300% increase in the price of crude oil (courtesy of ICI Ltd.)

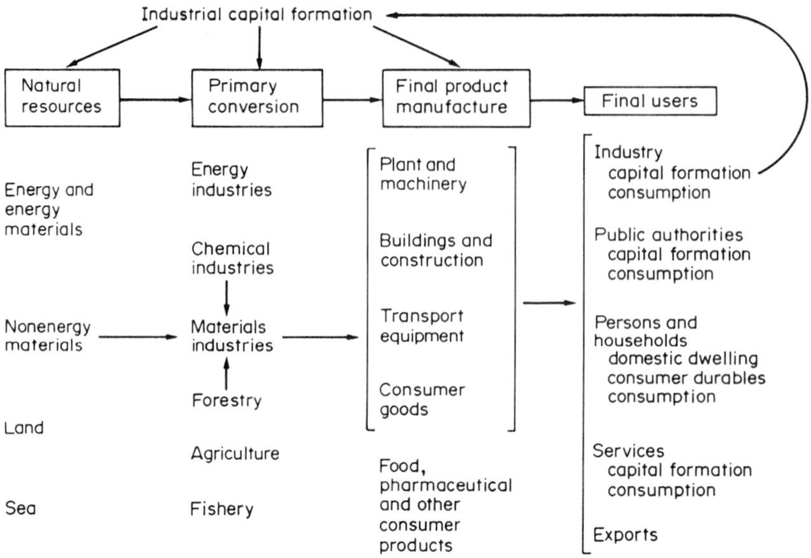

Figure 5
The flow from natural resources to final products through the manufacturing and construction industries

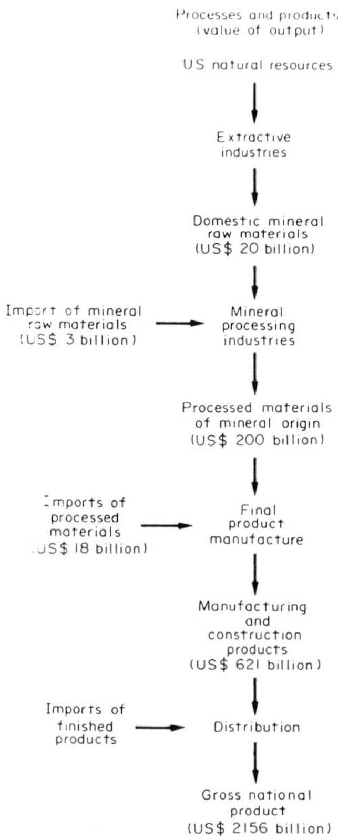

Figure 6
The buildup of value in the US economy in 1978 (after Morgan J D 1978 *Resour. Policy* 4)

manufacture of its products and to build and equip all its factories would make no purchases. Its only inputs would be primary inputs. Its only costs would be the payments to its employees and owners. In practice no firms, not even the most integrated such as an automobile manufacturer who makes his own steel, are so self-contained. In a completely self-contained firm, the whole value of the output would be accounted for by the value added within the company. In a manufacturing company confining its activities to the design and assembly of bought components, the bulk of costs would be accounted for by purchases, the bulk of the value of the products of that company would be attributable to value added by its suppliers.

The percentage of the value of a company's output added within the company itself varies from industry to industry. The metal container industry, for example, operates on such a large scale and with such efficiency that only relatively small costs are incurred in converting its purchased materials into products. In the UK in 1974, 75% of the costs of this industry were attributable to purchases; only around 25% of the value of the output of this industry was accounted for by value added within the industry itself (22% as payment to employees, 3% as profit and other trading income). In the manufacture of small engineering tools, on the other hand, only about 43% is accounted for by purchases; some 57% of this industry's output was, in 1974, attributable to value added within the industry itself, over four-fifths of which was payment to employees. In the UK construction industry in 1974,

43% of the output costs of the industry was accounted for by purchases, the remaining 57% by value added within the industry itself.

There are two points to note about these ratios. First, there is nothing fundamental about them. They simply reflect the structure of industry—the extent to which it has been found economically expedient for firms or industries either to become self-sufficient or to rely on other firms or industries to supply them with materials, components or services. A decision by a firm to operate its own fleet of trucks rather than to subcontract its transport, or a decision to buy its forgings or moldings rather than to make them, will alter these ratios. What is found to be economically expedient will vary from country to country, from one period to another and with changes in technology and with the scale of operations (see *Materials in the Industrial System*).

The second point arises from the fact that many firms find it expedient to purchase rather than to make a considerable fraction of their inputs. This implies that they are in effect also traders in the intermediate products that they purchase; it implies that the effective specification and purchasing of intermediate products is a matter of major importance for their cost-effective operation. In the US automobile industry, for example, the cost of wages decreased from 28% of total costs in 1898 to 11% in 1967 as changes in manufacturing technology led to other, greater improvements in labor productivity; the cost of purchased materials as a proportion of total costs increased from 38% to 73% over the same period.

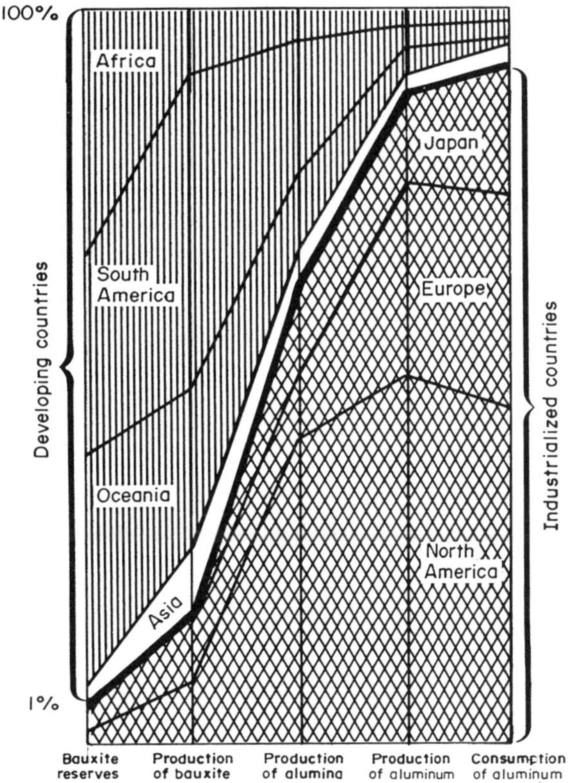

Figure 7
Structure of the aluminum industry (Western world) (after Altenpohl D 1983 *Radex Rundsch.* 1/2)

5. *Some International Implications*

A fraction of the company purchases referred to in Sect. 4 is composed of imports. These may be direct imports—the purchase by the automobile industry of components and subassemblies manufactured in other countries, for example; or they may be indirect imports—for instance, the iron ore or crude oil embodied in the steel and in the plastics purchased from domestic suppliers. Such imports are in effect a purchase of "value added," of primary inputs, from other countries; they represent a payment of incomes to employees and owners of industrial facilities located in other countries. Conversely, exports are a payment of income, of purchasing power, to employees and owners of domestic facilities by residents of other countries.

There are important implications here for international patterns of value creation. Until not very long ago it was assumed that the underdeveloped parts of the world acted primarily as suppliers of primary raw materials, with industrialized countries engaging mainly in downstream activities. The effect in terms of income generation can be seen in Fig. 6 and Tables 1 and 2. The situation for the case of aluminum is summarized in Fig. 7.

In an internationally competitive environment, the major concern of a firm making and marketing final products is to ensure the competitiveness of these products by minimizing its costs. Materials and components will therefore be purchased internationally, from any firm, wherever situated, that offers competitive prices. Many automobiles nominally of UK origin, for example, contain a proportion of components made in other countries: a significant proportion of the finished automobile has value added (generated income) not in the UK but in other countries.

One of the characteristics of industrial development in the past few decades has been the fact that industry in the raw-materials-producing countries has progressively been extended downstream so as to create an increasing amount of value domestically—they set up steelworks and petrochemical plants, later to be followed by final product manufacturing plants, often with capacities in excess of short-term domestic requirements.

Table 4
Relative importance of US industries by shares of output and employment in 1979

Industry title	1972 SIC code	Constant dollar value added (% of total) 1979
All industries[a]		100.00
Agriculture, forestry and fishing	01–09	2.66
Mining		1.42
Coal	11–12	0.27
Oil and gas extraction	13	0.89
Other	10, 14	0.26
Construction	15–17	3.92
Manufacturing		24.80
Nondurable goods		9.74
food and kindred products	20	2.19
tobacco manufactures	21	0.31
textile mill products	22	0.76
apparel and other textiles	23	0.84
paper and allied products	26	0.92
printing and publishing	27	1.13
chemicals and allied products	28	2.02
petroleum refining	29	0.60
rubber and miscellaneous plastic products	30	0.81
leather and leather products	31	0.15
Durable goods		15.06
lumber and wood products	24	0.76
furniture and fixtures	25	0.38
stone, clay and glass products	32	0.76
primary metal industries	33	0.54
fabricated metal products	34	1.74
machinery, except electrical	35	2.96
electric and electronic equipment	36	2.55
motor vehicles and equipment	371	2.07
other transportation equipment	372–9	1.12
instruments and related products	38	0.79
miscellaneous manufacturing industries	39	0.40
Transportation		3.77
Railroad transportation	40	0.74
Trucking and warehousing	42	1.59
Transportation by air	45	0.82
Other transportation	41, 44, 46, 47	0.62
Communications	48	3.39
Electric, gas and sanitary services	49	2.35
Wholesale and retail trade		16.72
Wholesale trade	50–51	6.97
Retail trade	52–59	9.75
Finance, insurance and real estate	60–67	15.33
Services		12.03
Business services	73	2.17
Health services	80	3.39
Other services	70–89 exc. 73, 80	6.47
Government and government enterprises		11.78

Source: *US Economic Outlook*, 1982 a The all industries figures are not weighted sums of the industry detail because they include data on "rest of the world," a balancing item

There are major problems for world capacity, for prices and for the potential for income generation in different parts of the world. Many of the problems experienced by the world economy in the mid-1980s occurred as a consequence of these developments.

Table 5
Relative importance of the values added by the manufacturing and by the construction industries in the USA in 1979 in descending order of industrial output

	Percent of GNP	Percent of manufac-turing
Construction	3.92	
Manufacturing		
machinery, excepting electrical	2.96	11.93
electrical and electronic equipment	2.55	10.27
motor vehicles and equipment	2.07	8.34
chemicals and allied products	2.02	8.14
fabricated metal products	1.74	7.01
primary metal industries	1.54	6.20
other transportation equipment	1.12	4.54
paper and products	0.92	3.70
rubber and plastics	0.81	3.26
instruments and allied products	0.79	3.18
stone, clay and glass products	0.76	3.06
lumber and wood products	0.76	3.06
furniture and fixtures	0.38	1.53
leather and leather products	0.15	0.60
others	6.25	25.18
Manufacturing total	24.82	
Manufacturing and construction total	28.72	

Source: Kennedy and Pick (1984)

6. The Contribution of the Materials-Producing and Materials-Using Industries to the Value of US Industrial Output

The opening paragraphs of this article drew attention to the fact that all production is ultimately for the purpose of satisfying the wants of consumers. In conclusion, it is intended briefly to demonstrate this point by reference to Table 4, which gives the percentage contribution to GNP of the values added by the range of US industries. GNP is seen indeed to originate in a wide range of industries satisfying needs for food and tobacco, for consumer goods and housing, for transportation and health care and so on.

The energy-related industries—coal, oil and gas extraction, petroleum refining, electric and gas services—together contributed a little over 4% of GNP. The manufacturing and construction industries together accounted for 28.72%. Table 5 highlights the specific contribution of the materials industries: the primary metal industries headed the list, contributing 6.2% of manufacturing value added; rubber and plastics, stone, clay and glass, and lumber and wood each contributed a little over 3%.

The output of the materials industries becomes, as was shown earlier, an input to industries producing final products, who in turn sell them to final markets; to industries such as that making nonelectrical machinery, which contributed 11.93% to manufacturing output; to the motor vehicle industries, which contributed 8.34%; and last but not least to the construction industry, which is a major consumer of a wide range of "physical structure" materials.

The final destination of the output of the materials industries in the USA was analyzed by Becker, using 1967 input–output tables, with the result shown in Fig.

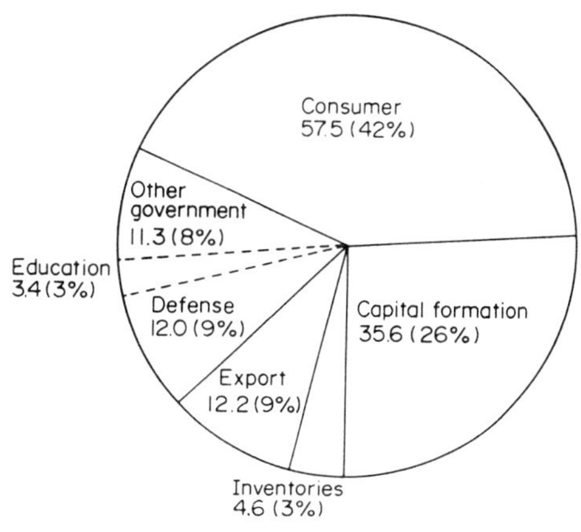

Figure 8
Destination of the output of the US materials industries by categories of final demand (figures are billions of 1967 dollars) (after Becker 1976)

Table 6
Materials content of the purchases by various categories of final demand

	Total expenditure (10^6 US$)	Material content ($\$^{-1}$)
Consumer	490.660	12
Capital formation	110.443	32
Inventories	10.034	46
Export	45.923	27
Defense	71.333	17
Education	39.512	9
Other government	68.274	17
Total	795.388	17

Source: Becker (1976)

8. Forty-two percent of the materials in that year had their destination in consumer goods; a further 26% went to capital formation, which included housing as well as private and public construction projects and the plant and machinery required by all industries from agriculture and mining to manufacturing, retailing and health care.

The "value added" contribution of the materials industries to the products purchased by the various sectors of final demand (sometimes described as the "materials content") was also analyzed, with the results reproduced in Table 6. The materials content of consumer products, at 12 cents per dollar, was very much less than the 32 cents per dollar in capital formation. However, the very much larger value of consumer expenditure led to the position of this class of final expenditure as the largest single market for the output of the materials industries.

See also: Input–Output Analysis; Input–Output Analysis: Applications; Intensity of Use of Materials

Bibliography

Becker P E 1976 The economic role of engineering materials in the UK. *Resour. Policy* 2(3): 193–205

Kennedy A J, Pick H J 1984 *Materials, Industry and Technical Change.* The Technical Change Centre, London

Pick H J 1972 Some economic consequences of technical decisions. *Mater. Sci. Eng.* 10: 301–23

Pick H J 1976 Decision-making in industry and its implications for energy and other resources. *Proc. 4th Natl. Conf. Materials Policy.* US Congress, Office of Technology Assessment, Washington, DC

Pick H J 1980 R & D for materials and energy conservation. *Mater. Sci. Eng.* 45(1): 147–64

Pick H J 1989 The build-up of value in manufacturing and construction and in the economy. *Eng. Costs Prod. Econ.* 16: 49–68

Stone R, Stone G 1983 *National Income and Expenditure*, 10th edn. Bowes and Bowes, London

US Department of Commerce 1969 *Raw Materials in the United States Economy 1900–1966*, Working Paper No. 30. US Department of Commerce, Washington, DC

Additional information can be found in the Reports of the Censuses of Production for the USA, UK and indeed most countries. The *US Industrial Outlook* and the *Statistical Abstract of the USA*, both published annually by the US Department of Commerce, also provide relevant information

H. J. Pick
[Sutton Coldfield, UK]

Value Analysis

The term "value analysis" is often thought limited in use to a rigorously formal procedure in which options are evaluated on their direct impact on cost. Only hard numbers are used to generate a hard-number answer. However, it is often true that many "value" elements cannot be quantified in hard numbers, particularly when complex situations are encountered. It follows that the concepts of value analysis are most often applied without the discipline of a rigorous mathematical approach. This article deals principally with the less structured approach.

1. Concepts

A common basis for choosing between inputs to a final product of competing materials, components or products is their relative cost. When it is recognized that the varying characteristics of competing inputs impact on other costs and/or on the value of the final product, consideration of these other factors as relative "values" becomes the basis for choice between inputs. This is value analysis. In the simplest case, one may limit consideration of the impact of the varying values of competing inputs on the in-house cost of manufacture of the final product. Example 1 deals with such a comparison. However, the competing characteristics of inputs often impact also on the value of the final product. There may be effects on product quality, durability or reliability or more generally on saleability. In such cases added "value" may be defined in terms of increased sales potential, higher prices realized or by decreased warranty costs. The selection of a competing input then involves a much more complex value analysis. Examples 2 and 3 include analysis of such factors.

2. Example 1

Figure 1 illustrates a semiconductor lead frame as used in very large quantities by producers of integrated circuits. It is an intricate flat stamping and, because of the demands of subsequent automated assembly operations, precision is required. It is manufactured from copper alloy strip, 0.25 mm thick and 25 mm wide.

Some years ago one manufacturer of copper alloy strip increased his maximum coil length capability for this product from 800 m to 2000 m. As will be seen, this had direct impact on the lead frame stampers costs. Typically, about 1.5 m of material is scrapped by the stamper at each end of the coil, representing the length of the feed mechanism on the press and of the progressive tools. Since one long coil replaced two and one half of the original shorter coils, the gross amount of material consumed per unit of output was reduced. The 3 m of material scrapped weighed 0.17 kg. The net cost of this loss to the stamper was US$0.42. Thus, using a single long coil instead of $2\frac{1}{2}$ shorter coils saved US$0.63. Dividing this saving by the 113 kg weight of the long coil, it was found that the added value of the long coil to the stamper equated to US$0.0055 per kilogram used. This was only 0.14% of the purchase price of US$3.84 per kilogram which appeared to be of small significance although, because of the large annual production involved, annual totals were of significance.

A second factor to consider was the cost of press downtime associated with each coil change. As the end of each coil is reached, the press must be slowed down and stopped, the scrap piece of materials removed and the new coil positioned and fed into the press and jogged through the tools. The press must then be restarted and accelerated to its normal operating speed. How long does this take? Most stampers do not have "standards" which define the "normal" time interval for a coil change. Instead, coil change time is generally lumped together with a host of other delay factors in establishing an overall standard for machine efficiency—the relation between theoretical continuous operation and acceptable results.

Machine efficiencies of 75% are generally considered very good. Although it was not known how much downtime could be attributed to coil changes, the total is large enough that if coil change represents a significant fraction of all downtime, potential savings which could be attributed to reducing the number of coil changes would be significant. It was estimated that each coil change required five minutes. A fair charging rate for a high-speed precision press is US$40 per hour. Thus, each coil change costs US$3.33. If this estimate is accepted, the savings attributable to running one long coil instead of two and one half of the shorter coils is US$5.00 per long coil or US$0.044 per kilogram. This saving is eight times as great as the saving in scrap generated. One must point out that because of the general absence of standards this value is based on an estimate rather than on rigorous application of "hard" numbers.

Another saving associated with the use of long coils is reduction in tool maintenance cost. The tools used to produce parts such as those illustrated in Fig. 1 are quite complex. Typical costs are US$60 000 or more and the tool set contains over 100 individual punches and an equal number of dies, all of which are fabricated from tungsten carbide and most of which are small and of odd shape. There is no doubt that tool maintenance is a major cost element. Qualitatively, it is well known that most tool breakage occurs at start-up—not during steady-state running. No hard numbers were available and generally it was not even possible to obtain an educated guess as to the value of reduced tool maintenance cost attributable to the use of long coils. However, it was freely conceded that tool maintenance savings were of substantial magnitude—and were probably the most important of the three elements.

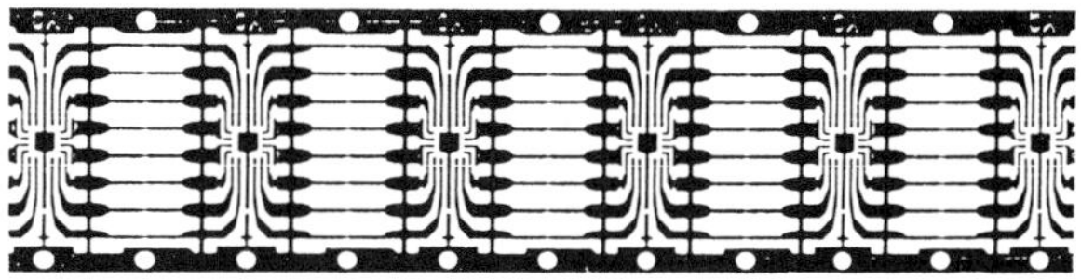

Figure 1
A typical semiconductor lead frame

The easy arithmetic quantifying the savings in scrap generated was thought to represent only 5–10% of the total savings; the remainder was poorly quantified or not quantified at all. The reality of major savings was documented by the fact that in many cases it was necessary for the stampers to make a substantial investment in upgrading press feed mechanisms to accommodate the longer coils, an added investment which they were quite willing to make. This example illustrates the point that rigorous application of a formal procedure recognizing only those benefits that could be fully documented with available data may capture only a small fraction of the total savings to the user. The larger savings had to be estimated and even "negotiated" on a largely judgmental basis.

3. *Example 2*

The second example also comes from the semiconductor industry. It may be surprising that, even though integrated circuits have no moving parts to wear out, they have a finite lifetime. The circuitry produced on the surface of a silicon chip consists of multiple very thin layers of other materials which gradually interdiffuse with each other or with the silicon chip itself. The result of such interdiffusion is modification of the characteristics of the device, the end point being functional failure. Ultimately, all integrated circuits will fail for such reasons. Diffusion rates are affected by the temperature of the diffusing couple. The usual rule of thumb is a doubling of diffusion rate for each 10 °C increase in temperature. When operating, integrated circuits generate heat in the circuit on the face of the chip. The problem of dissipating heat from the chip has become increasingly serious as the complexity of the circuitry has become greater and as more and more functions are crowded into a given chip area. Producers of integrated circuits have become increasingly concerned with the temperature on the face of the chip (usually called the junction temperature) and with finding means of conducting heat away to the outside to keep junction temperatures at acceptable levels.

The great majority of integrated circuits are "packaged" in plastic nonhermetic packages. The chip is mounted on the central pad in the lead frame shown in Fig. 1. After eventual removal of the supporting webs, the pad is electrically and thermally isolated from the external leads. Electrical connection to the outside is provided by fine (25 μm diameter) wire leads bonded to the circuitry on the chip and to the individual legs or leads on the lead frame. The central portion of the assembled device is encapsulated in transfer molded plastic. In this configuration, the lead frame plays a major role in heat dissipation, both as a direct heat conductor to the outside and as a means of more evenly distributing heat within the package, thereby eliminating hot spots. Initially, to minimize thermal stresses on the chip, the lead frames were manufactured from a nickel–iron alloy having a low coefficient of thermal expansion. The thermal conductivity of the nickel–iron alloys is low compared with that of high-conductivity copper alloys. Moreover, the cost of the nickel–iron alloys is over twice that of high-conductivity copper alloys. In the case of a relatively simple (but very high volume) integrated circuit using a lead frame such as that illustrated in Fig. 1, the cost saving per device to be realized from the use of copper alloys in place of nickel–iron is approximately US$0.0027, or about 1% of the sales price of the high-volume simple and inexpensive integrated circuits. Since the semiconductor industry is very competitive and since major manufacturers produce billions of integrated circuits per year, the total savings are of consequence. However, changing lead frame materials entails some costs, since molds built for plastic encapsulation of devices using nickel–iron alloy lead frames cannot be used with copper alloys because of the difference in coefficient of thermal expansion of the two materials. Furthermore, it was necessary to develop more compliant means of mounting the chip on the central pad to accommodate the high coefficient of thermal expansion of the copper alloys and to make some changes in the electroplating systems applied to the lead frame.

At the time at which copper alloy lead frames were first introduced to integrated circuit manufacturers, the devices were of such simplicity and limited function that heat dissipation from the package was not generally a significant problem. It was also felt that, were it assumed that the production level of integrated circuits was kept constant, the cost of tooling and other changes and development costs outweighed the savings to be derived from making a change. However, if new tools are expected to be required to support increased production and changed designs, the cost of building a set of tools to accommodate copper alloy lead frames is the same as one for nickel–iron lead frames. In effect, it may be concluded that if the production of circuits increases, the change of materials will ultimately pay off.

Thus, the semiconductor manufacturer was called on to make a value judgment, the major factor in which was his perception of future integrated circuit production growth. At that time two of the five major merchant integrated circuit manufacturers in the USA elected to convert to copper alloy in order to capture cost savings and ultimately better their competitive position. Eventually one of the three remaining producers sought to replace nickel–iron with stainless steel and thereby capture material cost savings without retooling cost.

As time passed, integrated circuits became more complex and functional density increased causing thermal performance to become more important. Another two of the major producers converted to copper alloy in response to growing market pressure for reduced junction temperatures—and ultimately the fifth began to gradually convert.

This is an example of the same decision being made at two different times for two different reasons—the earlier decision for internal economic reasons which could be quantified but which still depended on growth forecasts and the later decisions for technical performance reasons dictated in the market place.

4. *Example 3*

The third example comes from the automotive industry. Figure 2 shows cross sections of a typical connector system as used in automotive wiring harnesses as well as elsewhere. As the number of electrical and electronic devices used in a typical automobile has increased, terminations used have multiplied. The automotive environment is demanding since the terminations are exposed to extremes of temperature, corrosive environments and to substantial dynamic stresses. In a connector system, such as that illustrated, contact is established by inserting the blade member into the female termination. The interface between two mating terminals is pressurized by the elastic deformation (in bending) of the female termination. Substantial contact pressure must be maintained in order to assure the continuing integrity of the electric circuit. Practically all automotive connector systems use copper alloys in their components since both electrical conductivity and corrosion resistance are required.

Figure 2
Cross section of typical box connector components

When stressed, metals are subject to gradual permanent deformation, the extent of which depends on the stress level imposed, the temperature to which they are exposed and the metallurgical condition of the metal. In the case of automotive connector systems, in which contact pressure is supplied by elastic deformation of the female terminal, gradual permanent deformation of the material will result in a reduction in the stress level and hence in a reduction in contact pressure. The term used to describe this phenomenon is stress relaxation in bending. Although the stress level in the material will not decay to zero and some contact pressure will be maintained, the automotive system designer must be concerned with the potential for excessive stress relaxation and reduction in contact pressure.

If contact pressure is reduced by too great an extent as the result of stress relaxation, the connection may become unreliable and circuit failure may result. In order to obtain adequate contact pressure and to accommodate the effect of manufacturing tolerances, maximum initial design stress typically is close to the yield stress of the copper alloys used; indeed, in some systems, designs are such that the initial stress may exceed the yield stress, the connector being expected to quickly yield to the point that the remaining stress is close to the yield stress. This strategy tends to provide uniform final contact pressure over the range of component manufacturing tolerances. Design operating temperature for "under the hood" applications are now of the order of 105 °C and an increase to the level of 125 °C in the future is expected by some.

Test methods have been developed to measure stress relaxation in bending. It has been found that the stress relaxation behavior of copper alloys at high stress levels varies substantially from one alloy to another as temperature increases and the differences are quite substantial as 105 °C is approached. Historically, 70Cu–30Zn brass, C 260, has been widely used in automotive connector systems. The alloy is relatively low cost, provides good strength and formability, and electrical conductivity is adequate for most applications. Unfortunately, at temperatures much above 75 °C, stress relaxation of C 260 increases rapidly and stress remaining after prolonged exposure at 105 °C is quite low. Because of this problem (which also affects a broad range of other applications), the phenomenon of stress relaxation in bending has been extensively studied by major US producers of connector systems and by their US materials suppliers. Copper alloys providing adequate resistance to stress relaxation at temperatures in this range have been identified and their performance characterized. New alloys have been developed specifically for elevated temperature service. However, all such "superior" alloys are considerably more costly than C 260.

In this manner, the automobile industry was confronted with the need to make a value analysis judgement in which the value of greater reliability is weighed against the cost of using more expensive materials. This value is certainly not such that it can be precisely quantified since its effect is manifested in improved product quality and ultimately in greater saleability of the final product. To a substantial extent, US automobile manufacturers have elected to replace C 260 in critical applications with more costly stress-relaxation-resistant alloys.

See also: Cost Estimating; Life-Cycle Costing

Bibliography

Breedis J F, Caron R N, Khan A A 1980 *Comparison of Alloys for Use in Leadframes*, presented at NEPCON East, New York City, June 17–19, 1980

Butt S H 1970 *Designing for Use of Copper Alloys in Plastic Packaged Semiconductor Devices*, presented at NEREM Conference, 1970

Mendenhall J H 1977 *Understanding Copper Alloys.* Wiley, New York

Parikh P, Shapiro E 1976 *Stress Relaxation in Bending*, Special Technical Publication 608. American Society for Testing and Materials, Philadelphia, PA

Schuder D 1982 *Leaded Chip Carriers: Selection Considerations for Leadframe Materials*, presented at Second Annual International Packaging Conference, San Diego, California, November 15–17, 1982

Shapiro E, Hummel H 1984 *High Temperature Stress Relaxation: Mechanisms and Data*, presented at 17th Annual Connector Symposium, Cherry Hill, NJ

Shapiro E, Hummel H 1984 Stress relaxation in spring materials. *Residual Stress and Stress Relaxation.* Plenum, New York

Zarlingo S P, Scott J R 1981 *Leadframe Materials for Packaging Semiconductors*, presented at First Annual International Packaging Conference, Cleveland, Ohio, November 9 and 10, 1981

S. H. Butt
[Godfrey, Illinois, USA]

W

Wood-Based Industries in the USA: Economic Structure in a Worldwide Context

The US wood-based industry grows and processes timber, a renewable forest resource. The effectiveness of the industry is in large measure a function of how it is organized to undertake these activities. Important considerations are the character of major groups and subindustries which form the wood-based industry, the number of establishments which are available to manufacture timber into usable products and the degree to which assets, value added and employees are concentrated within a few companies. Identification of the role and importance of specific corporations in the industry also enhances comprehension of the complex structural character of the industry. Finally, since timber is such an important material to industrial and final consumer groups, the extent of land devoted by industry to timber production and the volume of the material produced become major considerations to a sound understanding of the role of the wood-based industry as a processor of material.

1. US Wood-Based Industry in General

The wood-based industry is a major sector of the US economy and a processor of 5.83×10^8 m^3 (1986) of industrial and consumer material, namely, wood fiber. In 1987, the industry's activities added US\$$8.65 \times 10^{10}$ in value to shipments that totalled US\$$1.916 \times 10^{11}$. These figures represent approximately 7.6% of the value added and shipment values generated by all manufacturers in 1987. In order to process wood fiber as a material, new capital expenditures by the industry approached US\$$8 \times 10^9$ in 1987, or 1.2 times the amount expended in 1982 (see Table 1). The 45 805 establishments engaged in the manufacture of products from wood fiber represented approximately 12% of all manufacturing establishments in the USA, and a modest gain of 600 over 1982 numbers. Employment in the wood-based industry was 1.5×10^6 people in 1987 (8% of all US manufacturing employment), an increase of 6% over the 1982 level.

2. Structural Characteristics of the Industry

2.1 Major Groups and Industries

The US wood-based industry is composed of three major groups, which in turn are composed of nearly 40 industries. The major groups and selected industries within them are as follows.

(a) Lumber and wood products:
  (i) logging camps and contractors;
  (ii) sawmills and planing mills;
  (iii) millwork, plywood and structural wood members;
  (iv) wood containers;
  (v) wood buildings and mobile homes; and
  (vi) miscellaneous wood products.

(b) Paper and allied products:
  (i) pulpmills;
  (ii) paper mills;
  (iii) paperboard mills;
  (iv) paperboard containers and boxes; and
  (v) miscellaneous converted paper products.

(c) Wood furniture and fixtures:
  (i) wood household furniture;
  (ii) wood office furniture; and
  (iii) wood partitions and fixtures.

2.2 Establishments

An establishment is an economic unit (generally at a single physical location) where business is conducted or where services or industrial operations are performed. The number of establishments processing wood fiber is a major structural characteristic of the industry. The lumber and wood products group accounted for more establishments (33 987) in 1987 than existed in the remainder of the wood-based industry (Table 1). Although enduring a 12% decline in establishments between 1977 and 1982, the group experienced a gain of over 1000 establishments between 1982 and 1987. Within this group, the logging camps and contractors industry is the largest subgroup (35% of the establishments in the group). The fastest growing industries in the group are the pallets and skids industry and the millwork industry, each of which experienced a gain of approximately 430 establishments between 1982 and 1977, on 1977 bases of 1290 and 2333 establishments, respectively.

The number of establishments in the paper and allied products group increased by nearly 1% annually between 1954 and 1987 (Table 1). In 1987, 6292 manufacturing establishments operated in the paper and allied products field. The largest number of establishments in this group is in the corrugated paper products industry (2425 in 1987). This latter industry has been steadily adding establishments since 1950, when they numbered 805.

The wood furniture and fixtures group accounted for 12% of the establishments in the wood-based industry in 1987, namely 5526 (Table 1). Since 1954, the number of establishments in this group has increased by a factor of 2.1. The wood household furniture industry

Table 1
Trends in general economic characteristics of the US wood-based industry, by selected years 1954–1987

Items and year	Industry: Lumber and wood products	Paper and allied products	Wood furniture and fixtures
Number of establishments			
1987	33987	6292	5526
1982	32984	6381	5844
1977	37302	6545	6475
1972	33948	6038	5464
1967	36795	5890	3292
1963	36150	5713	3517
1958	37789	5271	4727
1954	41484	5004	2677
Value added by manufacture (US$$10^6$)			
1987	28664.6	50488.8	7305.9
1982	15376.6	33375.7	6333.9
1977	16222.8	22170.9	4825.2
1972	10309.4	13064.1	3443.1
1967	4973.4	9756.3	1043.7
1963	4020.6	7395.7	782.9
1958	3176.6	5707.5	1134.2
1954	3241.6	4630.2	946.0
New capital expenditures (US$$10^6$)			
1987	1825.2	5752.5	317.9
1982	1342.4	5098.3	255.1
1977	1562.6	3295.0	182.0
1972	930.9	1335.3	179.9
1967	426.0	1585.3	30.9
1963	394.7	708.5	22.0
1958	277.0	618.5	38.1
1954	217.4	532.7	24.7
Number of employees in operating establishments (thousands)			
1987	698.4	611.3	213.4
1982	576.4	605.6	252.2
1977	692.4	628.7	286.0
1972	691.0	633.4	263.2
1967	554.0	638.9	108.7
1963	563.1	588.0	98.6
1958	585.4	551.3	192.6
1954	645.9	527.7	191.9

accounted for the largest number of establishments (2771) in the group in 1987.

2.3 Concentration

Concentration is a key indicator of economic structure and an important factor in judging materials processing efficiency. Concentration is typically measured by "concentration ratios" (the percentage of an industry's shipment values, total assets or total employment which is accounted for by the industry's largest firms). With some exceptions, the wood-based industry is not highly concentrated. Consider the value of industry shipments (see Table 2). The 1982 four-company concentration ratios for industries in the lumber and wood products group ranged from 5% in the wood pallets and skids industry to 43% in the particleboard industry. Nearly half the industries in this group have four company ratios of less than 20%. Industries in the wood furniture and fixture group are also "unconcentrated." The major industry in this group, wood household furniture, increased to 16% at the four company level in 1982, after experiencing no change between 1972 and 1977 (when it was at a level of 14%). The paper and allied products group is characterized by substantially higher concentration ratios. At the four company level, the ratios range from 15% in the set-up paperboard box industry to 93% in the molded pulp goods industry. The pulpmill industry exhibits substantial concentration at the 20 company level (99%). This is in part reflective of the large capital investments

Table 2
Share of shipment values accounted for by largest companies, by selected US wood-based industries

Industry and year	Percentage of industry shipment values accounted for by: 4 largest companies	8 largest companies	20 largest companies	50 largest companies	Total shipment values (US$10^6)
Logging camps and logging contractors					
1982	30	37	47	52	8274.0
1977	29	36	45	50	6230.1
1972	18	25	35	41	2529.5
1967	14	22	30	35	1476.3
1963	11	19	25	31	1154.7
1958	13	19	25	32	833.8
Sawmills and planing mills					
1982	17	23	34	46	10065.2
1977	17	23	36	49	10866.7
1972	18	23	33	45	6420.8
1967	11	15	22	31	3506.4
1963	11	14	20	29	3156.3
Millwork					
1982	15	20	30	44	4248.3
1977	14	20	31	46	3928.1
1972	10	15	25	41	2426.8
Hardwood veneer and plywood					
1982	25	39	57	74	1304.4
1977	27	40	60	78	1250.4
1972	34	43	61	77	911.8
Softwood veneer and plywood					
1982	41	56	74	92	3221.5
1977	36	53	72	91	3804.8
1972	37	50	67	87	2011.5
Particleboard					
1982	43	69	94	100	547.4
1977	48	70	95	100	452.9
1972	48	69	92	100	295.4
Pulpmills					
1982	45	70	99	100	3110.4
1977	48	76	99	100	2091.1
1972	59	83	99	100	709.9
1967	45	70	97	100	730.5
1963	48	72	97	100	609.1
1958	46	68	94	100	452.0
Paper mills (except building paper)					
1982	22	40	71	93	20994.6
1977	23	42	70	92	12613.3
1972	24	40	66	88	6395.2
1967	26	43	65	86	4844.0
1963	26	42	63	85	3824.9
Paperboard mills					
1982	28	45	75	96	9531.1
1977	27	42	70	95	7124.3
1972	29	44	69	93	4153.5
1967	27	42	67	92	2907.0
1963	27	43	67	90	2315.2
Corrugated and solid fiber boxes					
1982	19	33	58	73	10558.1
1977	19	33	58	75	7351.1
1972	18	32	58	76	4308.8
1967	18	33	61	78	2959.6
1963	20	36	63	79	2166.1
Wood household furniture					
1982	16	23	37	54	5056.6
1977	14	23	39	56	4148.8
1972	14	22	40	60	2870.0

required to achieve economies of scale necessary to process pulp material.

Assets can be another focus of concentration. In the wood-based industries, asset concentration is substantial. In the lumber and wood products group, just 12 corporations, and in the paper and allied products group, just 23 corporations, accounted for approximately two-thirds of the assets and taxable income in their respective groups in the early 1980s. Over three-quarters of the corporations in the lumber and wood products group had total assets less than US\$500 000 and accounted for only 5.6% of the total assets for the group. At the other extreme, only 0.1% of the corporations had assets at or above US\$2.50 × 10^8 yet they (12 in number) accounted for 61.6% of the total assets of the group. In the paper and allied products group, the picture was much the same in 1981–1982. Over 90% of the corporations had under US\$5.0 × 10^6 in assets and accounted for only 7.3% of groupwide assets. In contrast, 0.7% (12 corporations) accounted for 68.1% of total assets in the group.

Concentration by employee size class is yet another measure of structure. It provides insight to the economic scale at which firms must operate to efficiently process materials. In the lumber and wood products group, 57% of all establishments in 1977 had fewer than 5 employees and accounted for only 4% of employment, value added and materials purchased by the group. Large firms with 1000 or more employees accounted for less than 0.1% of the establishments in the group and less than 4% of groupwide value added, materials purchased and employees in the same year. Between these extremes were the majority of the companies in the group (60% of the employees in the group were in medium sized establishments with 20–249 employees). A similar situation existed in the paper and allied product group. The very small and very large firms accounted for relatively few employees groupwide and accounted for a small percentage of the key economic variables just discussed. In 1977, over three-quarters of the employees in the group were employed in just over one-quarter of the establishments, with 100–2499 employees.

3. Major Corporations in the US Industry

The wood-based industries are composed of public and private companies. In the aggregate, little is known about the latter. Of the 9831 public companies that filed annual reports with the Securities and Exchange Commission in 1980, 193 were in the lumber and wood products or paper and allied products groups. The 1989 leading ten US lumber producers accounted for 31% of the approximately 1.10 × 10^8 m^3 of lumber produced in the USA in 1985 (see Table 3). Paper and allied product companies in 1988 were led in sales by the International Paper Company (see Table 4). The twentieth ranked company (Federal Paper Board) had paper and allied product sales that were only 9% that

Table 3
Ten leading lumber producing companies in the USA and Canada by company, 1989

Company	Lumber production (10^3 m^3)
USA	
Weyerhaeuser Company	7011.0
Louisiana–Pacific Corporation	5760.3
Georgia–Pacific Corporation	5724.9
WTD Industries Incorporated	2932.9
Sierra Pacific Industries	2241.8
Boise Cascade Corporation	1894.9
International Paper Company	1887.8
RSG Forest Products Incorporated	1855.0
Champion International Corporation	1583.4
Simpson Timber Company	1451.3
Canada	
Canfor Corporation	3757.6
Noranda Forest Incorporated	3451.2
Fletcher Challenge Canada	3057.8
West Fraser Timber Company	2763.3
Weldwood of Canada	2721.1
MacMillan Bloedel	1869.0
Slocan Forest Products	1746.2
Donohue	1463.1
International Forest Products	1418.2
Westar Timber	1416.6

Table 4
Twenty sales-leading paper and allied product companies in the USA by company, 1988

Company	Sales (US\$$10^6$)	Percentage of sales from paper operations
International Paper Company	8197.0	86
James River Corporation	5971.8	100
Kimberly–Clark Corporation	5286.2	98
Scott Paper Company	4726.4	100
Champion International Corporation	3983.7	78
Stone Container Corporation	3641.0	97
Great Northern Nekoosa Corporation	3588.1	100
Weyerhaeuser Company	3525.0	35
Georgia–Pacific Corporation	3436.0	36
Boise–Cascade Corporation	2406.1	59
Mead Corporation	2090.8	47
Union Camp Corporation	2952.3	77
Westvaco Corporation	1948.9	91
Temple-Inland Incorporated	1429.1	81
Bowater Incorporated	1410.4	100
Sonoco Products Company	1325.2	75
Willamette Industries Incorporated	1174.7	69
Consolidated Papers Incorporated	891.9	100
Potlatch Corporation	852.3	79
Federal Paper Board Incorporated	777.8	64

of the sales leader. Wood-based companies are frequently involved in the processing of wood fiber for both paper and lumber products. In 1988, paper sales made up an average of almost 79% of the total wood-based sales of the largest 20 paper companies in the industry.

4. *Major Corporations Worldwide*

4.1 Canada

Many wood-based firms operating in Canada have corporate linkages via subdivisions or subsidiaries with wood-based firms in the USA. In some respects, the US and Canadian wood-based industries are best viewed from a North American perspective. The ten largest Canadian producers of lumber were led in 1989 by the Canfor Corporation (Table 3). These firms accounted for approximately 40% of Canada's 1989 lumber production.

The sales-leading Canadian producer of paper and allied products in 1988 was Noranda Forest Incorporated (US\$$2.3645 \times 10^9$). Other leading 1988 Canadian producers in sales were: Canadian–Pacific Forest Products (US\$$2.1542 \times 10^9$), Consolidated–Bathurst (US\$$1.4907 \times 10^9$), MacMillan Bloedel (US\$$1.4728 \times 10^9$), Abitibi–Price (US\$$1.4626 \times 10^9$), Domtar (US\$$1.3740 \times 10^9$), E. B. Eddy Forest Products (US\$$5.856 \times 10^8$), Donohue (US\$$5.268 \times 10^8$), Quebec and Ontario Paper Company (US\$$4.479 \times 10^8$) and Cascades Incorporated (US\$$3.973 \times 10^8$).

4.2 Worldwide

Manufacture and sale of wood products is a significant economic activity of some of the world's leading corporations. Many of these firms are active participants in Canadian and US wood-based markets. In 1990, the Stora Kopparbergs Berg Group of Sweden and the Noranda Corporation of Canada were sales leaders among wood-based firms headquartered outside the USA. Of 20 firms identified (see Table 5), all are large fully integrated operations involving the manufacture of a wide variety of products from wood fiber.

5. *Industrial Control of Timber as a Raw Material*

5.1 Industrywide

The raw material, timber, is an essential ingredient to the manufacture of products produced by the wood-based industries. An important portion of the timber used by the industry comes from forestland owned by wood-based manufacturing companies. In 1987, industrial ownership of timberland in the USA accounted for 15% of the nation's total, or 2.857×10^5 km^2. Timberland is forestland capable of producing at least 141.5 m^3 of industrial wood per square

Table 5
Large wood-based firms headquartered and operating outside the USA, 1990

Company	Country	Sales (US$$10^6$)	Assets (US$$10^6$)	Employees
Stora Kopppararbergs Berg	Sweden	11070	14973	69700
Noranda	Canada	8121	12854	56000
Fletcher Challenge	New Zealand	7134	10354	40000
Repola	Finland	6040	9737	33987
Svenska Cellulosa	Sweden	5334	8597	30139
Oji Paper	Japan	4514	4728	NA
Jujo Paper	Japan	3901	4142	7151
Kymmene	Finland	3540	6931	17567
Sanyo–Kokusakn Pulp	Japan	3333	3202	7748
Mo Och Domsjo	Sweden	3189	4818	12961
Honshu Paper	Japan	3158	3593	6261
Buhrmann–Tetterode	Netherlands	3052	2251	12954
Daishowa Paper Manufacturing	Japan	2810	5087	7690
Enso–Gutzect	Finland	2778	5449	15180
Abitibi–Price	Canada	2658	2133	14300
Reed International	Britain	2633	4384	18700

kilometer and not reserved for uses that are incompatible with timber production. Regionally, the south has the largest concentration of industrially owned timber. The 1.547×10^5 km^2 of industrial timberland in that region is 2.37×10^4 km^2 more than industry owns in all other regions combined. Since the 1950s, there has been an upward trend in industrial timberland ownership: the 1987 level was 19.6% larger than that of 1952. Recent ownership trends have, however, been slight and are not expected to change dramatically in the years ahead. This attests to the growing difficulties involved in locating large tracts of timberland and securing financial resources for their purchase and subsequent management.

Timberland in the USA contains over 2.1×10^{10} m^3 of growing stock—raw wood-fiber material (see Table

Table 6
Hardwood and softwood growing stock on forest industry land by species group and region, 1952, 1962, 1970, 1977 and 1987

Species group and region	Growing Stock (10^6 m^3) 1952	1962	1970	1977	1987
Softwood					
North	0.17	0.23	0.32	0.36	0.31
South	0.46	0.58	0.62	0.66	0.67
Pacific Coast	1.36	1.16	1.02	0.94	0.92
Rocky Mountains	0.19	0.18	0.17	0.14	0.15
Total softwood[a]	2.18	2.15	2.13	2.10	2.05
Hardwood					
North	0.19	0.24	0.28	0.31	0.35
South	0.31	0.38	0.42	0.47	0.49
Pacific Coast	0.06	0.09	0.11	0.11	0.15
Rocky Mountains	b	b	b	b	b
Total hardwood[c]	0.56	0.71	0.81	0.89	0.99
Total industry	2.74	2.86	2.94	2.99	3.04
Total US stock	17.27	18.69	18.11	20.52	21.36

a Projected total softwood growing stock: 2000, 1.57; 2010, 1.72; 2020, 1.81; 2030, 1.94; and 2040, 2.01 b Less than 0.01×10^6 m^3 c Projected total hardwood growing stock: 2000, 0.71; 2010, 1.02; 2020, 0.97; 2030, 0.92; and 2040, 0.89

Table 7
Ratio of annual timber removals to net annual timber growth on forest industry land by species group, 1952, 1962, 1970, 1976, 1986, with projections to 2040

Year	Softwood	Hardwood	Total industry	All US ownerships
1952	1.5	0.8	1.3	0.7
1962	1.0	0.8	0.9	0.6
1970	1.2	0.6	1.0	0.6
1976	1.2	0.5	1.0	0.5
1986	1.4	0.8	1.2	0.6
2000	1.1	1.0	1.1	0.8
2010	0.9	1.2	0.9	0.8
2020	0.9	1.2	1.0	0.8
2030	0.9	1.2	0.9	0.8
2040	1.0	1.1	1.0	0.8

6). Forest industry accounts for 3×10^9 m^3 or 14% of this national total. In 1987, it was regionally distributed as follows: 22% in the North, 38% in the South, 35% on the Pacific Coast and 5% in the Rocky Mountains. By 2040, growing stock volumes on industrial timberland are expected to decline by 4.6% from 1987. However, growing stock inventories for all forest ownerships are expected to increase by 5.5% by 2040.

Approximately 20% of the 6.31×10^8 m^3 of timber growth occurring annually has been on industry-owned forestland. In 1986, softwood growth—especially in the South—exceeded hardwood growth by 5.48×10^7 m^3.

Timber removals from growing stock on industrial timberland are substantial. Removals are primarily softwoods (82%) which originate primarily in the South and the Pacific Coast. Nationally, 38% of timber removed from forestland originates from that owned by forest industry (1.47×10^8 m^3). Considering industrial timberlands nationwide, timber removal in 1986 was 1.2 times net annual growth (see Table 7). Species and regional disparities are, however, quite dramatic. Hardwood removals on industry lands were only 80% of net annual growth in the same year, while softwood removals were 140% of net annual growth. For industry-owned timberland, removals are expected to nearly equal net annual growth over the period from 2000 to 2040.

Nationwide, annual timber growth averaged 3.15 m^3 ha^{-1} in 1987 (see Table 8). Industrial timberland productivity exceeded this national average

Table 8
Average net annual timber growth and potential timber growth in the USA by region and forest ownership, 1987

	Timber growth (m^3 ha^{-1})				
Region	All owners	National forest	Other public	Forest industry	Farmer and other private
North					
Current	2.45	3.07	2.65	2.38	2.38
Potential	4.61	4.40	4.12	5.17	4.61
South					
Current	3.71	3.78	3.50	4.06	3.64
Potential	5.39	4.97	4.97	5.81	5.39
Rocky Mountains					
Current	2.45	2.66	2.10	3.01	2.10
Potential	4.20	4.48	3.85	5.18	3.57
Pacific Coast					
Current	4.20	2.24	4.69	8.05	4.27
Potential	6.79	6.37	6.16	8.33	6.93
Total					
Current	3.22	2.73	3.22	4.34	3.15
Potential	5.18	5.18	4.76	6.09	5.04

by nearly 1.2 $m^3\ ha^{-1}$. Regionally, industrial forestland growth rates range from a low of 2.38 $m^3\ ha^{-1}$ in the North to a high of 4.06 $m^3\ ha^{-1}$ in the South. In all but the North region, growth rates of industrial timberland exceed that occurring on all other ownerships. Growth rates for all ownerships and regions are below potentials characterized by fully stocked natural stands of timber. Nationally, 62% of potential growth was being realized in 1987—industrial lands were growing at over 71% of their potential. Pacific Coast industrial forests were being managed so as to capture 97% of their potential (fully stocked natural stands of timber).

5.2 Corporate Patterns

Wood-based companies may control access to timber by a variety of means, including fee ownership, timber purchase agreements, public timber sales contracts and

Table 9
The 40 largest industrial forestland ownerships in the USA in 1979

	Area owned (km^2)					
Company	Northwest[a]	Intermountain[a]	South	Central	New England	Total
International Paper	1655	417	19850	0	6851	28773
Weyerhaeuser	9069	2428	12472	0	0	23969
Georgia–Pacific	3213	0	9907	1340	2254	16714
St Regis Paper	619	1255	6022	1930	3039	12865
Champion International	1129	3517	5767	1756	0	12169
Great Northern Nekoosa	0	0	1194	1044	8737	10975
Boise–Cascade	1153	4650	2064	1323	1493	10683
Scott Paper	979	0	2974	0	3484	7437
Crown Zellerbach	2845	737	3456	0	0	7038
Union Camp	0	0	6969	0	0	6969
Time Incorporated	0	0	6192	0	0	6192
Burlington Northern	506	5532	0	0	0	6038
Continental Group	0	0	5957	0	0	5957
Diamond International	0	1655	0	631	3594	5880
Mead	0	0	2946	2456	0	5402
Potlatch	0	2096	2210	1000	0	5306
Westvaco	0	0	2590	2351	0	4941
St Joe Paper	0	0	4452	0	0	4452
ITT (Rayonier)	1416	0	2922	0	0	4338
Bowater	0	0	2631	1619	0	4250
Owens-Illinois	0	0	2958	1093	0	4051
Procter and Gamble	0	0	3913	0	0	3913
Louisiana–Pacific	639	2072	559	194	0	3464
Mobil (Container Corporation)	0	0	3233	0	0	3233
Santa Fe Industries (Kirby)	0	0	2647	0	0	2647
Gulf and Western	0	0	0	0	2448	2448
Johns-Manville (Olinkraft)	0	0	2363	0	0	2363
Kimberly–Clark	0	0	2270	0	0	2270
Willamette	874	0	1028	316	0	2218
American Can	0	0	947	1077	0	2024
Masonite	433	0	1421	158	0	2012
Southwest Forest Industries	0	49	1720	85	0	1854
Southern Pacific	0	1821	0	0	0	1821
Tenneco (Packaging Corporation of America)	0	0	1206	490	0	1696
Longview Fibre	1651	0	0	0	0	1651
Federal Paperboard	0	0	1473	0	0	1473
Chesapeake Corporation of Virginia	0	0	1433	0	0	1433
Hammermill Paper	0	0	708	724	0	1432
Southern Natural Resources	0	0	1327	0	0	1327
Cleveland Cliffs	0	0	0	1255	0	1255

a Northwest region is limited to Washington and Oregon (west of the Cascades) and California Redwood Region; Intermountain region includes eastern Washington and Oregon and remaining portion of California

indirect influence over other private timberland. Determining the nature and extent of timberland control for specific companies is very difficult. In 1979, the total area of forestland owned by the 40 largest industrial timberland owners was estimated to be over 2.347×10^5 km^2 or 84% of the nation's 1979 industrial timberland (see Table 9). The top 90 companies in that year accounted for approximately 91% of the timberland held by industry. Since 1979, there have been numerous shifts in ownership of industrial timberland among companies (e.g., the mid-1980s merger of Champion International and St Regis Paper).

Concentrations of fee-owned timberland can be significant. In 1979, the four largest landowning companies accounted for nearly three out of every ten square kilometers of timberland owned by wood-based firms, while the largest 40 landowning companies retained nearly 85% of such land. Considering all timberland ownership in the USA, the largest wood-based firms accounted for a relatively small portion in 1979—the largest 40 owners owning only 12% of the nation's commercial timberland.

Although land is the base for the wood-based industry's raw material (timber), it is not a particularly good reflector of industrial control over wood fiber. Timber inventory on the land is much more relevant. For example, even though the International Paper Company owned the largest acreage of timberland in 1979, the volume and value of timber owned by other companies was substantially greater. The US\$$9.5 \times 10^9$ worth of Weyerhaeuser timber in 1979 was nearly equal to the combined timber values of International Paper, Georgia–Pacific and Crown Zellerbach, yet the latter three had nearly twice the fee-owned land of Weyerhaeuser. Weyerhaeuser's highly valuable old-growth timber explained this discrepancy.

In 1979, the 40 largest corporate timber holdings had a combined volume of nearly 2.3×10^9 m^3 or 75% of industrywide growing stock. The largest fee owner of timber was Weyerhaeuser—3.26×10^8 m^3 valued at US\$$9.5 \times 10^9$. Relatively few Northwest regional holdings exceeded 2.8×10^7 m^3 while 13 of the top 40 had Southern timber inventories that ranged from 3.7×10^7 m^3 to 1.56×10^8 m^3.

Of the 3×10^9 m^3 of industrial fee-owned timber in the USA, nearly 30% was accounted for by the four largest timber inventory holders in 1979—Weyerhaeuser, International Paper, Georgia–Pacific and Crown Zellerbach. The eight largest accounted for 42%, while 75% of the industrially owned timber volumes could be attributed to the 40 largest companies. Of the timber inventory on all ownerships nationwide in 1979, only 11% was owned in fee by the 40 largest timber owning companies.

See also: Structure of Materials Industries; Timber: Trade, Prices and Consumption in the USA; Wood: Future Availability

Bibliography

Anon 1991 Five hundred largest U. S. industrial corporations: 1990. *Fortune* 123(4): 279–330

Anon 1991 Largest worldwide industrial corporations: 1990. *Fortune* 124(3): 237–280

Bilek E M, Ellefson P V 1987 *Organizational Arrangements Used by U. S. Wood-Based Companies Involved in Direct Foreign Investment: An Evaluation*, Agricultural Experiment Station Bulletin 526–1987 (AD-SD-3207). University of Minnesota, St Paul, MN

Bureau of the Census 1986 *1982 Census of Manufacturers (Concentration Ratios in Manufacturing)*. US Department of Commerce, Washington, DC

Bureau of the Census 1991a *1987 Census of Manufacturers (General Summary: Industry, Product Class, and Geographic Area Statistics)*. US Department of Commerce, Washington, DC

Bureau of the Census 1991b *1987 Census of Manufacturers (Subject Series: Type of Organization)*. US Department of Commerce, Washington, DC

Clephane T P, Carroll J 1980 *Timber Ownership, Valuation, and Consumption Analysis for 87 Forest Products, Paper, and Diversified Companies*. Morgan Stanley, New York

De Braal J P, Majchrowicz T A 1981 *Foreign Ownership of U. S. Agricultural Land (1979–1980)*, Agricultural Information Bulletin 448. US Department of Agriculture Economics and Statistics Service, Washington, DC

Dyer H 1990 *Lockwood-Post's Directory of Pulp, Paper and Allied Trades: 1990*. Miller Freeman, San Francisco, CA

Ellefson P V, Stone R N 1984 *U. S. Wood-Based Industry: Industrial Organization and Performance*. Praeger, New York

Enk G A 1975 *A Description and Analysis of Strategic and Land-Use Decision Making by Large Corporations in the Forest Products Industry*. Dissertation Abstracts, Ann Arbor, MI

Espe C 1986 *International Factbook of Forest Products Industry*. Miller Freeman, San Francisco, CA

Irland L C 1976 Foreign ownership and control of U. S. timberland and forest industry. In: *Foreign Investment in US Real Estate*. US Department of Agriculture Economic Research Service, Washington, DC

LeMaster D C 1977 *Mergers Among the Largest Forest Products Firms, 1950–1970*, Research Center Bulletin 854. Washington State University, Pullman, WA

McKeever D B, Meyer G W 1984 *The Softwood Plywood Industry in the United States, 1965–82*, Resource Bulletin FPL-RB-18. US Department of Agriculture, Madison, WI

McKeever D B 1987 *The United States Woodpulp Industry*, Resource Bulletin FPL-RB-18. US Department of Agriculture, Madison, WI

Mammen M 1990 *International Pulp and Paper Directory: 1991*. Miller Freeman, San Francisco, CA

Mead W J 1966 *Competition and Oligopsony in the Douglas-fir Lumber Industry*. University of California Press, Berkeley, CA

Meyer P 1979 Land rush: a survey of America's land: who controls it; how much is left? *Harpers* 258(1544): 45–60

Mikulenka J *Pulp and Paper North American Fact Book: 1989*. Miller Freeman, San Francisco, CA

Office of Management and Budget 1987 *Standard Industrial Classification Manual*. US Government Printing Office, Washington, DC

O'Laughlin J, Ellefson P V 1981 U. S. wood-based industry structure: top 40 companies. *For. Prod. J.* 31(10): 55–62

O'Laughlin J, Ellefson P V 1982a *New Diversified Entrants Among U. S. Wood-Based Companies: A Study of Economic Structure and Corporate Strategy*, Minnesota Agricultural Experiment Station Bulletin 541. University of Minnesota, St Paul, MN

O'Laughlin J, Ellefson P V 1982b Strategies for corporate timberland ownership and management. *J. For.* 80(12): 784–8

P. V. Ellefson
[University of Minnesota, St Paul, Minnesota, USA]

Wood: Future Availability

Projections of the future availability of wood necessarily depend on assumptions. These assumptions are conditioned by developments that have led to current conditions for availability of wood and on judgments as to what developments there may be in the future. In this article, key developments that have affected the availability of wood during the 1970s and 1980s are reviewed as a prelude to presentation of judgments about future prospects. As will become apparent, prospects for future availability of wood vary markedly, depending on the country or region and product being considered, especially the contrast between industrialized and developing countries.

1. Evolution of the Current Situation

1.1 Industrialized Countries

Worldwide consumption of industrial roundwood (wood used for lumber, plywood, pulp and other end products) increased by more than 31% between 1970 and 1988 (Food and Agriculture Organization 1990). Some 39% of this increase was in the industrialized countries of North America. There were also significant increases in Western Europe, China and Brazil. The world forest industries are cyclical, however, and there were pronounced downturns in 1975 and 1982.

The pattern of consumption in the 1970s and 1980s was attributable in part to rising demands brought about through expansionary economic policies in the industrialized countries and China, a lowering of barriers to trade in finished products, and the emergence of Brazil as a major producer of wood pulp. The increased consumption was also attributable in part to the availability of previously untapped timber resources in Canada, the western USA, Siberia and Southeast Asia. Production of industrial roundwood also increased in the southern USA and Western Europe. Most of the increased output in the industrialized countries consisted of softwood species used in the manufacture of lumber, plywood and pulp. In the USA between 1970 and 1987, however, production of industrial roundwood of softwood species increased by 31% and of hardwood species by 32%. The increased use of hardwood species was attributable in part to the use of hardwoods in pulp manufacture and for manufacture of structural panels with new technologies that can use both softwood and hardwood species.

Japan was the major exception to increased output of industrial roundwood in the industrialized countries. Harvest in Japan declined from 45.3×10^6 m^3 in 1970 to 27.8×10^6 m^3 in 1988. While domestic harvest declined, demand increased rapidly. Imported industrial roundwood accounted for about 66% of Japanese consumption in 1988, up from 46% in 1970.

With the major exception of tropical hardwood logs, trade in timber products during the 1970s and 1980s increased primarily among the industrialized countries. For example, in 1970, North America, Japan and Europe (including the Soviet Union) accounted for 86% of the value of world exports of forest products and 87% of the value of world imports of these products. In 1988, the industrialized countries accounted for 85% of exports and 83% of imports. The value of world trade expanded fivefold during this period.

In the industrialized countries, rising demands for timber products resulted in substantial increases in prices for timber products, especially softwoods in the early 1970s. These price increases were responsible in part for developments in technology and industry practices that had the effect of stretching timber supplies. Examples include the substitution of plywood for lumber in the 1960s and 1970s, substitution of particleboard and other fiberboards for plywood, and pulping of hardwoods in the 1970s and 1980s. These market developments greatly affected prices for timber products in the 1980s. For example, despite record consumption in the USA in the 1980s, prices for timber products were relatively stable.

In the importing countries, relatively few barriers affected trade in chips, logs and other raw materials, but various types of restrictions affected trade in lumber, plywood and other processed products. In the 1970s and 1980s, various types of pressures developed in exporting countries concerning the propriety of exporting raw materials, especially logs. First the Philippines, then Indonesia and finally Malaysia banned exports of tropical hardwood logs, for example. Indonesia and Malaysia have the intention of exporting processed products rather than logs. These bans have caused structural shifts in the tropical hardwood trade in Southeast Asia. For example, Japanese imports of logs from these three countries declined from a peak of 26.8×10^6 m^3 in 1973 to 13.1×10^6 m^3 in 1988.

In the industrialized countries, forest management generally became more intensive and utilization improved in response to rising prices. General awareness of the environmental effects of forestry practices and noncommodity forest outputs affected land use, especially that of public land. For example, concern for threatened and endangered species led to curtailment of timber harvest and management on public lands in the USA.

1.2 Developing Countries

For many of the developing countries, the period of the 1960s and 1970s was one of profound social and economic change. Living standards in some countries increased, but remained low in others. In general, food and the other necessities took priority over timber products in patterns of consumption and plans for development. There was a wide range of attitudes toward the role of forests and forestry among the developing countries. These included the exploitation of timber reserves, the shifting of agriculture with the effect of destroying both forest and the land, and relatively intensive afforestation and reforestation with plantations of introduced species, as in Brazil and Chile.

Individual countries participated in the growth of world trade in timber products, primarily through the export of tropical hardwood logs to industrialized countries. This pattern was especially evident in shipments from the Philippines, Malaysia and Indonesia to Japan. With the ban on log exports in place for Indonesia, plywood has replaced logs as the primary export to Japan.

Especially in the late 1970s and extending through the 1980s, concern developed over the exploitation of reserves of tropical hardwoods. Although few data are available to document either the extent of remaining reserves or the rate of depletion, a general consensus exists that current rates of harvest cannot be maintained indefinitely, especially for preferred species (Schmidt 1987). Tropical deforestation has been linked to concerns about global warming and the world environment.

The end of relatively inexpensive energy in 1974 precipitated what had been a slowly developing crisis in wood for fuel in the developing countries. The jump in prices for petroleum-based fuels in the 1970s has all but placed them out of reach of people in many countries. The result has been increased demands for wood as fuel with many implications for the daily lives of people in the affected areas (Anon 1990).

Sustainable development has been defined as not a fixed state of harmony, but rather a process of change in which the exploitation of resources, the direction of investments, the orientation of technological development and institutional change are made consistent with future as well as present needs (World Commission on Environment and Development 1987). The use of wood for fuel weighs heavily in the consideration of the sustainability of development in some countries.

One of the purposes of the newly formed International Tropical Timber Organization is to facilitate cooperation between consuming and producing countries in regard to reforestation and other management of tropical forests after harvest (Anon 1987). Much remains to be learned, however, about silvicultural and other management practices for tropical hardwoods. Also, tropical hardwood timber grown under management with relatively short rotation will probably result in commercial timber with strength and other properties that are different from the properties of the old-growth timber currently being harvested.

2. The Future

2.1 Industrialized Countries

The economies of the industrialized countries are mature by the historical standards of the 1960s and 1970s. Growing populations, expansionary fiscal and monetary policies, and reductions in barriers to international trade all contributed to growth in demands for timber products. In the 1990s, rates of growth in population will be lower. Alternative scenarios could be drawn for future rates of economic growth. If a consensus exists, it is probably that continued gains can be expected in real economic growth in the industrialized countries. Also, general expectations are for continued growth in demand for major timber products—higher for pulp, paper and other fiber-based products than for solid lumber and plywood.

The sweeping political developments in Eastern Europe and the countries of the former Soviet Union may lead to increased demands for timber products in that part of the world. Much uncertainty surrounds the market implications of these developments, however.

Further unification of the economies of member nations of the European Community in 1992 may lead to stimulation of these economies with resulting increases in demands for timber products. Many questions remain to be answered, however, and they involve mainly building codes and grades and standards for timber products.

The situation for supplies to meet growth in demand will probably differ markedly over the coming two decades compared with the 1970s and 1980s. In the future, any expansion of output in Canada is likely to be at slower rates than in the past (Prins 1987), production of softwood lumber, plywood or both may decline on the West Coast of the USA (Haynes and Adams 1990), expansion of output in Scandinavia will probably be limited (Food and Agriculture Organization and Economic Commission for Europe 1986) and the ability of the countries of the former Soviet Union to expand or even maintain shipments of softwood lumber and logs is open to question (Prins 1987). This contrasts with the general expansion of supplies from these areas in the 1960s and 1970s.

Further expansion of supplies in the industrialized countries will depend in part on the intensification of timber management. Because of the economic importance of forestry to the Canadian economy, intensification of forest management in Canada has the support of government, industry and labor. The European forest is expected to expand in area, growing stock, increment and removals, at least until the year 2020 (Prins 1987).

The government of Japan expects that the demand for lumber and plywood will grow slowly, if at all, and that the demand for pulp and paper products will increase over the coming decades (Japanese Ministry of Agriculture, Forestry and Fisheries 1980). Although the potential exists for significant expansion of Japanese production after the year 2000, the consensus seems to be that a significant share of Japanese needs for timber products will continue to be met by imports.

A report by the US Department of Agriculture Forest Service (1988) suggests the potential for significant expansion of timber supplies if programs could be developed that would encourage owners of small woodlands to take advantage of economic opportunities to intensify management of forestlands.

Projections of supplies and demands for timber products in the industrialized countries generally assume that historical linkages with the developing countries will continue. This generally means the expectation that exports of tropical hardwood logs, veneer or both and of lumber and plywood will continue to be available from the developing countries. Exports of timber products from the developed to the developing countries are generally not considered a significant prospect for growth of markets. There may be some shift to processed products rather than logs, but consumption of imported tropical hardwood products is becoming increasingly controversial in some countries.

Plantations of exotic species in New Zealand, Australia and other areas of the world will reach maturity in the year 2000 and beyond (Prins 1987). These plantations have the potential to affect traditional patterns of world trade in timber products. Realization of the potential, however, requires resolution of problems related to costs other than wood, market acceptance of end products and distance from final markets.

2.2 Developing Countries

Several issues bear on the prospects for future availability of wood in the developing countries—wood for fuel, the conflict between forestry and agriculture for land use, forest management and depletion of timber inventories, and development of processing facilities to meet domestic demands as well as to meet goals related to employment and economic development.

In the past, nearly every developing country with a timber resource has proposed to construct the capacity to process its timber. In general, these proposals have two intentions: first to stimulate economic development through employment and construction of the transportation infrastructure necessary to manage the timber resource; and second, to substitute domestically produced timber products for similar items that otherwise would have to be imported. Among the problems to be faced in these ventures are availability of capital, a trained labor force and equipment. In addition, problems in the manufacture of pulp and paper are associated with procuring chemicals and other additives necessary to the manufacturing process.

Developing countries vary in their ability to implement development plans centered on forestry and timber processing. Joint ventures have been a common mode of development of timber resources. For example, joint ventures have been one of the responses of Japanese investors to the tightening of log-export restrictions by countries in Southeast Asia.

Increased construction of processing capacity in the developing countries with the intention of substituting domestically produced items for imported ones will limit markets for the products of developed countries. This capacity may also lead to increased trade among the developing countries.

2.3 Final Outlook

Up to the year 2000, little basis exists for expecting significant change in the availability of wood, especially in the industrialized countries. Expected price increases for roundwood will continue to lead to the substitution of one wood-based product for another and act as incentives for improved utilization and intensified timber management. Major world trade flows in timber products will continue to be among the industrialized countries and from Southeast Asia to Japan, with some increase in wood pulp exports from Brazil to Europe and Japan and from Chile to the Pacific Rim.

In the developing countries to the year 2000, the problems of availability of wood for fuel and shifting patterns of agriculture may be insurmountable in some areas.

Beyond the year 2000, almost any scenario for the future availability of wood can sound plausible. The price of saw-timber-sized trees will probably continue to increase. Innovations in fiber-processing technology, however, may tend to blur the distinctions between solid and fiber-based products in the industrialized countries. Probably more processing of timber will be in developing countries with an indigenous resource base and in countries such as New Zealand with a resource based on exotic plantations. Mounting population pressures will exacerbate the problems of wood for fuel and shifting agriculture, however, and these problems may exist well into the twenty-first century. If there are significant changes in the global climate, it could force restructuring of world trade in timber products.

See also: Timber: Trade, Prices and Consumption in the USA; Wood-Based Industries in the USA: Economic Structure in a Worldwide Context

Bibliography

Anon 1987 The international tropical timber organization. *Unasylva* 39(2): 61–4

Anon 1990 The impact of fuelwood scarcity on dietary patterns: hypotheses for research. *Unasylva* 41(1): 29–34

Food and Agriculture Organization 1990 *Yearbook of Forest Products, 1988.* Food and Agriculture Organization, Rome

Food and Agriculture Organization and Economic Commission for Europe 1986 *European Timber Trends and Prospects to the Year 2000 and Beyond*, ECE/TIM 30, Vol. 1. United Nations, New York

Haynes R W, Adams D M 1990 Future development of the Douglas-fir region forest economy under changing public policies and private resources. *West. J. Appl. For.* 5(3): 64–9

Japanese Ministry of Agriculture, Forestry and Fisheries 1980 *Basic Plan for Japan's Forest Resources and Long-Range Demand and Supply Projections for Important Forest Products.* Japanese Ministry of Agriculture, Forestry and Fisheries, Tokyo

Prins K 1987 European timber: 2000 and beyond. *Unasylva* 39(2): 51–60

Schmidt R 1987 Tropical rain forest management. *Unasylva* 39(2): 2–16

US Department of Agriculture Forest Service 1988 *The South's Fourth Forest: Alternatives for the Future.* Forest Resource Report No. 24. US Department of Agriculture, Washington, DC

World Commission on Environment and Development 1987 *Our Common Future.* Oxford University Press, New York

Zivnuska J A 1989 Resources of timber worldwide. In: Schniewind A P (ed.) 1989 *Concise Encyclopedia of Wood and Wood-Based Materials.* Pergamon, Oxford, pp. 238–42

D. R. Darr
[US Department of Agriculture, Forest Service, Washington, DC, USA]

LIST OF CONTRIBUTORS

Contributors are listed in alphabetical order together with their addresses. Titles of articles that they have authored follow in alphabetical order. Where articles are coauthored, this has been indicated by an asterisk preceding the title.

Abert, J. G.
School of Business Administration
Georgetown University
Washington, DC 20057
USA
Resource Recovery: Economic and Institutional Aspects

Anderson, D. W.
Manager Materials Engineering
Melting Technology
Corning Glass Works
Decker Building-ME-2
Corning, NY 14831
USA
**Risk Management*

Ayres, R. U.
Professor of Economics
INSEAD
Boulevard de Constance
F-77305 Fontainebleau Cédex
France
Materials and the Environment

Bambrick, S. C.
University of New England
43 Gordon Street
PO Box 1570
Coff's Harbour
2450 New South Wales
Australia
Policy: Australia

Behrman, J. R.
Department of Economics
University of Pennsylvania
McNeil Building, 160/CR
3718 Locust Walk
Philadelphia, PA 19104-6297
USA
Prices of Materials: Stabilization

Bever, M. B.
[deceased; late of Massachusetts Institute of Technology, USA]
**Productivity and Materials*

Bode, R. H.
Management Counseling Section
Arthur D. Little, Inc.
20 Acorn Park
Cambridge, MA 02140-2390
USA
Materials Handling: Unit Materials

Burke, J. E.
33 Forest Road
Burnt Hills, NY 12027
USA
Materials and Innovation

Busch, J. V.
IBIS Associates, Inc.
Suite 220
55 Williams Street
Wellesley, MA 02181
USA
**Markets for Materials: Modelling*

Butt, S. H.
1904 Arrowhead Lane
Godfrey, IL 62035
USA
Value Analysis

Carter, A. P.
Department of Economics
Brandeis University
Waltham, MA 02254-9110
USA
Materials in the Industrial System

Church, A. M.
Department of Economics
University of New Mexico
1915 Roma North East
Albuquerque, NM 87131
USA
Taxation: Effect on Materials Policy

Clark, J. P.
Room 8-407
Massachusetts Institute of Technology
Cambridge, MA 02139-4307
USA
**Critical and Strategic Materials*
**Markets for Materials: Modelling*
**Multiattribute Utility Analysis*

Cochran, C. N.
22 Oakglen Drive
Oakmont, PA 15139
USA
Life-Cycle Analysis of Energy Consumption and Carbon Dioxide Emissions

Crandall, R. W.
Senior Fellow
The Brookings Institution
1775 Massachusetts Avenue, North West
Washington, DC 20036-2188
USA
Recycling of Materials: Economics
Structure of Metals Industries

Curlee, T. R.
Energy Division
Oak Ridge National Laboratory
Building 4500-N
PO Box 2008
Oak Ridge, TN 37831
USA
Recycling of Plastics

Darr, D. R.
US Department of Agriculture
Forest Service
PO Box 96090
12th and Independence, South West
Washington, DC 20090-6090
USA
Wood: Future Availability

Ellefson, P. V.
Department of Forest Resources
University of Minnesota
110 Green Hall
1530 North Cleveland Avenue
St Paul, MN 55108-1030
USA
Timber: Trade, Prices and Consumption in the USA
Wood-Based Industries in the USA: Economic Structure in a Worldwide Context

Field, F. R. III
Room E40-227
Massachusetts Institute of Technology
Cambridge, MA 02139-4307
USA
**Multiattribute Utility Analysis*

Finan, W. F.
Law and Economics Consulting Group, Inc.
Suite 1200
1775 Pennsylvania Avenue, North West
Washington, DC 20006
USA
Economics of the Semiconductor Industry

Gellman, I.
Executive Vice President
National Council of the Paper Industry for Air and Stream Improvement, Inc.
260 Madison Avenue
New York, NY 10016
USA
Paper and Paperboard: Environmental Considerations

Glade Jr, E. H.
Economic Research Service
US Department of Agriculture
Room 1034
1301 New York Avenue, North West
Washington, DC 20005
USA
**Natural Fibers in The World Economy*

Gold, B.
Graduate Management Center
Claremont Graduate School
925 North Dartmouth Avenue
Claremont, CA 91711-6184
USA
Factor Dominance in Industry
Productivity and Technological Improvements: Managerial Evaluations

Gordon, R. L.
Department of Mineral Economics
221 Eric A. Walker Building
Pennsylvania State University
University Park, PA 16802
USA
Markets for Materials: Principles and Public Policy Implications of Competition
Structure of Materials Industries

Greene, A. M.
Operation Management Section
Author D. Little, Inc.
20 Acorn Park
Cambridge, MA 02140-2390
USA
Materials Handling: Automation, Unit Materials

Herz, N.
Department of Geology
University of Georgia
Athens, GA 30602
USA
**Mineral Resources: Definitions, Uses, Classifications and Future Availability*

Hoffman, R. B.
Robbert Associates
Suite B-3
340 McLaren Street
Ottawa
K2P 0M6
Canada
**Process Analysis*

Kaplan, R. S.
US Bureau of Mines
810 7th Street, North West
Washington, DC 20241
USA
**Recycling of Metals*

Kellogg, H. H.
Henry Krumb School of Mines
Seeley W. Mudd Building
Columbia University
New York, NY 10027
USA
Energy Considerations in Metals Production

Kelly, W. S.
Cyprus Minerals Company
9100 East Mineral Circle
Englewood, CO 80112
USA
**Materials Handling: Bulk Materials*

Labys, W. C.
Agricultural and Resource Economics
2034 Agricultural Science Building
West Virginia University
PO Box 6108
Morgantown, WV 26506-6108
USA
Stockpiling: Economics
Trade in Materials, International

Lagrone, B. D.
President
US Rubber Reclaiming Company, Inc.
PO Box 820165
Vicksburg, MS 39182
USA
**Reclaiming of Elastomers*

Laurila, M. J.
Process Technology, Inc.
1122 Calumet Avenue
Calumet, MI 49913-1951
USA
**Materials Handling: Bulk Materials*

Leontief, W. W.
Institute for Economic Analysis
New York University
2nd Floor
269 Mercer Street
New York, NY 10003
USA
Input–Output Analysis

Lorey, F. W.
Senior Vice President
Garden State Paper Company
Park 80 Plaza East
Saddle Brook, NJ 07662
USA
**Recycling of Paper Products*

McCarl, H. N.
Department of Economics
University of Alabama at Birmingham
University Station
Birmingham, AL 35294-4460
USA
Prices of Industrial Minerals: History

McInnis, B. C.
Robbert Associates
Suite B-3
340 McLaren Street
Ottawa
K2P 0M6
Canada
**Process Analysis*

McKelvey, V. E.
[deceased; late of St Cloud, Florida, USA]
**Mineral Resources: Definitions, Uses, Classifications and Future Availability*

Malenbaum, W.
8592 Roswell Road
Apartment 615
Atlanta, GA 30350
USA
Intensity of Use of Materials

Martin, W. C.
President
Wilmart Associates, Inc.
514 Mark Road
Allendale, NJ 07401
USA
**Recycling of Paper Products*

Miller, C. G.
President
The Mining Association of Canada
1105-350 Sparks Street

Ottawa
Ontario
K1R 7S8
Canada
**Policy: Canada*

Morgan, J. D.
Chief Staff Officer
US Bureau of Mines
MS 1003
810 Seventh Street, North West
Washington, DC 20241-0002
USA
Stockpiling in the USA

Nasar, S.
81 Paulding Avenue
Tarrytown, NY 10591
USA
**Productivity and Materials*

Newcomb, R.
Department of Geological Engineering
University of Arizona
Tucson, AZ 85721
USA
Economic Theory of Mineral Resources and Markets
Economics of Materials, Energy and Growth

Page, W. S.
Private Consultant
Kemptville
Ontario
Canada
**Process Analysis*

Passaglia, E.
3601 Albemarle Street, North West
Washington, DC 20008
USA
Economic Effects of Corrosion and Other Degradative Processes
Economic Effects of Fracture

Petrick, A.
Petrick Associates
5544 South Hatch Drive
Evergreen, CO 80439
USA
By-Products and Coproducts

Pick, H. J.
72 Sherifoot Lane
Sutton Coldfield
West Midlands
B75 5DU
UK
Value Added in Processing

Piehler, H. R.
Department of Metallurgical Engineering and Materials Science
Carnegie Mellon University
Pittsburgh, PA 15213
USA
Product Liability and Materials: General and USA

Pindyck, R. S.
Sloan School of Management
E52-454
Massachusetts Institute of Technology
Cambridge, MA 02139-4307
USA
Cartels and Producer Associations

Reddy, B.
National Economic Research Associates
1 Main Street
Cambridge, MA 02142
USA
**Critical and Strategic Materials*

Reynard, K. W.
Wilkinson Consultancy Service
Stable Cottage
Broad Lane
Newdigate
Surrey
RH5 5AZ
UK
**Data Sources for Materials Economics, Policy and Management*

Rose, A.
Department of Mineral Economics
221 Eric A. Walker Building
Pennsylavania State University
University Park, PA 16802
USA
**Input–Output Analysis: Applications*

Schlabach, T. D.
4 Adams Drive
Whippany, NJ 07981
USA
Conservation of Materials: Technology
Substitution of Materials: Technology

Schottman, F. J.
US Bureau of Mines
810 7th Street, North West
Washington, DC 20241

USA
**Recycling of Metals*

Seldon, M. R.
[deceased; late of General Dynamics Corporation, Pomona, CA, USA]
Cost Estimating
Life-Cycle Costing

Slade, M. E.
Department of Economics
Room 997
1833 East Mall
University of British Columbia
Vancouver
British Columbia
V6T 1Z1
Canada
Prices of Metals: History and Current Problems

Smith, D. J.
ITP Systems, Inc.
8 Streeter Road
Paxton, MA 01612
USA
Materials Requirements Planning

Smith, F. G.
Environmental Technologies Alternatives, Inc.
2511 Glen Arbor Drive
Lima, OH 45805
USA
**Reclaiming of Elastomers*

Smith, V. K.
Department of Economics
North Carolina State University
PO Box 8109
Raleigh, NC 27695-8109
USA
Conservation of Materials: Economics

Spofford Jr, W. O.
Resources for the Future
1616 P Street, North West
Washington, DC 20036
USA
Residuals Management

Stanley, T. W.
International Economic Studies Institute
1064 Papermill Court, North West
Washington, DC 20007
USA
**Risk Management*
Stockpiling

Starbird, I. R.
10152 Strawflower Lane
Manassas, VA 22110
USA
**Natural Fibers in The World Economy*

Strauss, S. D.
597 Pinebrook Boulevard
New Rochelle, NY 10804
USA
Markets for Materials
Trading of Materials on Commodity Exchanges

Tilton, J. E.
Department of Mineral Economics
Colorado School of Mines
Golden, CO 80401
USA
Availability of Materials
Substitution of Materials: Economics

Tinkham, E. I.
Chief, Life Cycle Costing
General Dynamics Corporation
Pomona Division
PO Box 2507 (MZ 4-99)
Pomona, CA 91769
USA
**Cost Estimating*
**Life-Cycle Costing*

Vogely, W. A.
RR1
235 Meadowood Lane
Warrior's Mark, PA 16877
USA
Policy: Europe
Policy: General
Policy: Japan
Policy: USA
Prices of Materials: Theory
**Risk Management*

Wang, K.-L.
US–China Council for International Exchange, Inc.
Suite 9
11228 Georgia Avenue
Wheaton, MD 20902
USA
**Input–Output Analysis: Applications*

Westbrook, J. H.
Brookline Technologies
5 Brookline Road
Ballston Spa, NY 12020
USA
**Data Sources for Materials Economics, Policy and Management*

Wilson, D. G.
Department of Mechanical Engineering
Room 3-455D
Massachusetts Institute of Technology
Cambridge, MA 02139-4307
USA
Recycling of Demolition Wastes

Wise, M. J.
Department of Geography
London School of Economics and Political Science
Houghton Street
London
WC2A 2AE
UK
Economic Geography of Materials

Youngs, R. L.
Department of Wood Science and Forest Products
Virginia Polytechnic Institute and State University
210 Cheatham Hall
Blacksburg, VA 24061-0323
USA
Timber Use: History

Yudelman, D.
Ontario Hydro
700 University Avenue
Toronto
M5G 1X6
Canada
**Policy: Canada*

Zekoll, J.
Tulane Law School
Joseph Merrick Jones Hall
Tulane University
New Orleans, LA 70118-5670
USA
Product Liability and Materials: Europe

Zwartendyk, J.
119 Dorothea Drive
Ottawa
Ontario
K1V 7C6
Canada
Resource Appraisal

SUBJECT INDEX

The Subject Index has been compiled to assist the reader in locating all references to a particular topic in the Encyclopedia. Entries may have up to three levels of heading. Where there is a substantive discussion of the topic, the page numbers appear in ***bold italic*** type. As a further aid to the reader, cross-references have also been given to terms of related interest. These can be found at the bottom of the entry for the first-level term to which they apply. Every effort has been made to make the index as comprehensive as possible and to standardize the terms used.